Intelligent Systems and Smart Infrastructure

About the Conference

The International Conference on Intelligent Systems and Smart Infrastructure (ICISSI-2022) was jointly organised by Shambhunath Institute of Engineering and Technology, Prayagraj UP India, Institute of Engineering and Technology (IET) Lucknow, U.P India, and Manipal University Jaipur, Rajasthan India during May 21-22, 2022 with the aim of providing a platform for researchers, scientists, technocrats, academicians and engineers to exchange their innovative ideas and new challenges being faced in the field of emerging technologies. The conference (ICISSI-2022) was held in the campus of Shambhunath Institute of Engineering and Technology Prayagraj, Uttar Pradesh in virtual mode during May 21-22, 2022.

The conference provided an opportunity to exchange ideas among global leaders and experts from academia and industry in developing domains such as machine learning, intelligence systems, smart infrastructure, advanced power technology, and so on.

The conference covered all broad disciplines of electronics, electrical and computer science engineering. Papers were invited in the following domains:

Track-1: Intelligent Wireless Systems

Track-2: High-Speed Electronic Devices

Track-3: Artificial Intelligent and Machine Learning

Track-4: Advanced Power and Energy Systems

A total of 249 research papers were received through the easychair system by researchers from India and abroad. Out of 249 research articles, 14 papers were received from Canada, Indonesia, Mexico, Pakistan, Singapore, and the United Kingdom. The similarity scores of the submitted papers were checked through Turnitin software and papers having similarity scores within an acceptable limit were sent to the potential reviewers (at least two reviewers) for review process. After recommendations by the reviewers, only 82 research articles were accepted for the publication in the book volume of the Talyor & Francis Group.

Intelligent Systems and Smart Infrastructure

Proceedings of ICISSI 2022

Dr. Brijesh Mishra
Shambhunath Institute of Engineering and Technology,
Prayagraj (UP), India
https://orcid.org/0000-0002-2535-7159

Dr. Rakesh Kumar Singh
Shambhunath Institute of Engineering and Technology,
Prayagraj (UP), India

Dr. Subodh Wairya
Institute of Engineering & Technology,
Lucknow (UP), India
https://orcid.org/0000-0003-2072-5781

Dr. Manish Tiwari
Manipal University Jaipur,
Rajasthan, India
https://orcid.org/0000-0002-6758-3939

CRC Press
Taylor & Francis Group
Boca Raton London New York

CRC Press is an imprint of the
Taylor & Francis Group, an **informa** business

First edition published 2023
by CRC Press
4 Park Square, Milton Park, Abingdon, Oxon, OX14 4RN

and by CRC Press
6000 Broken Sound Parkway NW, Suite 300, Boca Raton, FL 33487-2742

© 2023 Shambhunath Institute of Engineering and Technology

CRC Press is an imprint of Taylor & Francis Group, an Informa business

ISBN: 978-1-032-41287-0 (pbk)
ISBN: 978-1-003-35734-6 (ebk)

DOI: 10.1201/9781003357346

Typeset in: Times LT Std
Typeset by: Aditiinfosystems

Table of Contents

Intelligent Systems and Smart Infrastructure – Brijesh Mishra et al. (eds)
© 2023 Taylor & Francis Group, London, ISBN 978-1-032-41287-0

List of Figures

Intelligent Systems and Smart Infrastructure – Brijesh Mishra et al. (eds)
© 2023 Taylor & Francis Group, London, ISBN 978-1-032-41287-0

List of Tables

Intelligent Systems and Smart Infrastructure – Brijesh Mishra et al. (eds)
© 2023 Taylor & Francis Group, London, ISBN 978-1-032-41287-0

Foreword

This book covers the proceedings of ICISSI 2022 (International Conference on Intelligent Systems and Smart Infrastructure), held at one of the most ancient and holy cities in India, Prayagraj, Uttar Pradesh, on May 21–22, 2022. The conference was jointly organised by Shambhunath Institute of Engineering and Technology, Prayagraj, UP India, Institute of Engineering and Technology (IET) Lucknow, U.P India, and Manipal University Jaipur, Rajasthan, India with the aim of providing a platform for researchers, scientists, technocrats, academicians, and engineers to exchange their innovative ideas and new challenges being faced in the field of emerging technologies. The papers presented at the conference have been compiled in the form of chapters to focus on the core technological developments in the emerging fields like machine learning, intelligence systems, smart infrastructure, advanced power technology, etc.

Intelligent Systems and Smart Infrastructure – Brijesh Mishra et al. (eds)
© 2023 Taylor & Francis Group, London, ISBN 978-1-032-41287-0

Preface

Artificial Intelligence and Internet of Things have made a significant position in the industry. Intelligent Systems & Smart Infrastructure make the work of the human beings a piece of cake. They are developed and trained in such ways, so that they can mimic a human, make decisions like a human brain does, and study data in enormous amounts to come up with decisions. Thus, making things cost and time effective. This progress we have made today, would not have been even imaginable without the right concepts, discovered by the right people, with the right implementation. But since we have reached here, we can now envision even higher technologies, algorithms, and procedures paving our ways to effortless living. Being a part of the technical community, it is our responsibility to accomplish our goals of making human life easier with this boon that has been bestowed on us, the technologies. Therefore, to put this all-in action, the details of the International Conference on Intelligent Systems and Smart Infrastructure (ICISSI-2022) is presented hereby.

The conference (ICISSI- 2022) was held in the seminar hall of Shambhunath Institute of Engineering & Technology during the month of May, 2022. This conference was conducted for two days, collaboratively with Institute of Engineering & Technology, Lucknow & Manipal University, Jaipur.

More than 200 individuals participated in the conference that includes the invited speakers, contributing authors, researchers, and the attendees. The participants were enlightened with a broad range of tech stacks, and issues critical to our society and industry in the related areas. The conference provided a platform and a chance to exchange ideas among reputed leaders and global experts from academia and industry in topics like Artificial Intelligent, Deep Learning, Data Science, Human-machine interaction, Smart Cities and Smart Mobility, Wearable Computing, RF & microwaves, software-defined and cognitive radio, signal processing for wireless communications, antenna systems, Fuzzy Logic Controller and their Applications, Power Electronics for Smart Grids, IoT Based Smart Energy Meter, Smart Grid Technologies, vehicular communications, wireless sensor networks, machine-to-machine communications, cellular Wi-Fi integration, etc.

Apart from high-quality contributed papers presented by the authors from all over the country and abroad, the conference participants also witnessed the informative demonstrations and technical sessions form the industry as well as invited talks from renowned experts aimed at advances in these areas. The overall response to the conference was quite encouraging. A large number (249) of research papers were received for consideration for publication in conference proceeding. After a rigorous editorial and review process and oral presentation, 83 papers were selected for inclusion in the conference proceedings. We are confident that the papers presented in these proceedings shall provide a platform for young as well as experienced professionals to generate new ideas and networking opportunities.

The editorial team members would like to extend gratitude and sincere thanks to all contributed authors, reviewers, panellists, organizing committee members, volunteers, and the session chair for paying attention to the quality of the publication. We are thankful to our sponsors (ISTE and IETE) for generously supporting this event and institutional partner (Institute of Engineering & Technology, Lucknow (UP), India, and Manipal University Jaipur (Rajasthan), India) for providing motivation, support, and encouragement on the different stages.

We look forward to seeing all contributors in next edition of ICISSI conference proceedings.

Editors

Dr. Brijesh Mishra

Shambhunath Institute of Engineering and Technology, Prayagraj (UP), India

Dr. Rakesh Kumar Singh

Shambhunath Institute of Engineering and Technology, Prayagraj (UP), India

Dr. Subodh Wairya

Institute of Engineering & Technology, Lucknow (UP), India

Dr. Manish Tiwari

Manipal University Jaipur, Rajasthan, India

Intelligent Systems and Smart Infrastructure – Brijesh Mishra et al. (eds)
© 2023 Taylor & Francis Group, London, ISBN 978-1-032-41287-0

Editors Profile

Dr. Brijesh Mishra

Dr. Brijesh Mishra received his M. Tech. (Electronics Engineering) and PhD (RF & Microwaves) degrees from the University of Allahabad in 2012 and 2018, respectively. Dr. Brijesh Mishra worked as an Assistant Professor in the Department of Electronics and Communication Engineering at Shambhunath Institute of Engineering and Technology (SIET) during 2012-2013 and again from 2017-2018, and as an Assistant Professor (NPIU-MHRD) in the Department of Electronics and Communication Engineering at Madan Mohan Malaviya University of Technology during 2018-2021. He is currently working as an Associate Professor in the Department of Electronics and Communication Engineering at Shambhunath Institute of Engineering and Technology (SIET). He has published 1 patent and more than 45 research papers in journals of international repute, international conferences and book chapters. Dr. Brijesh Mishra has successfully completed two projects funded by the NPIU & the World Bank. He is the recipient of awards like Excellence in Performance and Outstanding Contributions. He has served as potential reviewer of more than 20 SCI & Scopus indexed journals, Organising Track Chair and Organising Secretary for the IEEE Conference (ICE3-2020) and Springer Conference (ICVMWT-2021) respectively. He is a member of IEEE, ISTE, IE(I), IETE, IAENG and IFERP. His research interests include modelling, simulation and fabrication of RF and microwave devices, as well as their applications.

Dr. Rakesh Kumar Singh

Rakesh Kumar Singh Profile: Dr. Rakesh Kumar Singh received his M.Tech. (Electronics Engineering) and PhD (Electronics and Communication) degree from Sam Higginbottom University of Agriculture, Technology and sciences during 2013 and 2018 respectively. He has completedhis B.Sc. and M.Sc. degree from University of Allahabad in 1999 and 2003 respectively. Currently, he is working as Director at SIET, Prayagraj Uttar Pradesh, affiliated from AKTU. He has teaching and research experience of about 16 years. Dr. Rakesh Kumar Singh has published more than 14 papers in International Journals and Conferences and supervised more than three M.Tech. thesis. His research interest includes modelling and simulation of VLSI devices, RF and microwave devices and its applications.

Dr. Subodh Wairya

Dr. Subodh Wairya received his PhD degree Electronics Engineering from MNNIT, Allahabad in 2012 and M.TECH. degree in Tele-Communication Engineering from Jadavpur University, Kolkata, India in 2002. Dr. Wairya has been worked at Defence Research & Development Organization (DRDO), Lucknow (One Year), Hindustan Aeronautical Limited, Lucknow (One Year) and Govt. Polytechnics, Lucknow (Six Months). Currently he is working as Professor in the department of electronics and communication at IET Lucknow and Dean UGSE at AKTU, U.P. India. He has published more than 115 research papers in journals of international repute, international conferences and books chapters. Dr. Wairya has been produced 39 M.TECH. and 1 PhD degree in his supervision and 8 PhD students are enrolled under his name.

His research interest includes design and modelling of low power VLSI system and Wireless communication system.

Dr. Manish Tiwari

2020 OSA Ambassador and SPIE Community Champion, Dr. Manish Tiwari received Ph.D. in ECE in the field of Photonics from MNIT Jaipur. Presently, he is Professor and Head of Department of ECE at Manipal University Jaipur. He has more than 22 years of professional experience in various colleges and universities of repute. He has published more than 100+ research papers in reputed journals and conferences, and 8 book chapters. He has also authored and edited more than half a dozen of books. He has also supervised more than 29 projects sponsored by DST Rajasthan. He is principal investigator in a BRICS project on Quantum Satellite and Fiber Communication (QuSaF) alongwith research partners in UKZN Durban, PSUTI Samara and USTC Shanghai. He has also served on panel of experts and editor in various workshops by CSTT, MHRD, Govt. of India. He has been visiting researcher to City University, London under UKIERI project in Microstructured Optical Fibers during 2010 and 2011 and Tsinghua University, Beijing during 2016. Dr. Tiwari has presented talks in PIERS at NTU Singapore, APMP at PolyU-Hong Kong, TJMW at KMUT-Bangkok and Kasetsart University-Bangkok, City University- London and several UKIERI workshops. He is on the panel of Traveling Lecturer Program of reputed societies like SPIE (USA) and OSA (USA) under which he has delivered invited talks in USTC Shanghai, UESTC Chengdu, Tsinghua University Beijing, ITS Indonesia, University of KwaZulu-Natal, Durban (SA) on photonics technologies and career development. He has been Technical Program Committee member of many IEEE conferences and reviewer of IEEE/Elsevier/OSA/Springer/T&F journals. He is 2020 OSA Ambassador and former Chairman of IETE Rajasthan Centre. He is a Senior Member of IEEE and IEEE Photonics Society, Life Fellow of the Optical Society of India (OSI-India), Senior Member of OSA, Member of SPIE and Fellow of Institution of Electronics and Telecommunication Engineers (IETE) India. His current research interest includes Micro/Nano-structure photonic devices, Photonic ICs, fiber optics, numerical modelling, nonlinear optics and photonic crystal fibers.

Intelligent Systems and Smart Infrastructure – Brijesh Mishra et al. (eds)
© 2023 Taylor & Francis Group, London, ISBN 978-1-032-41287-0

Details of Programme Committee

- Dr. Niraj Shukla, SIET, Prayagraj (UP), India
- Dr. Rajeev Tripathi, MNNIT, Prayagraj (UP), India
- Dr. Arun Verma, MNIT, Jaipur, India
- Dr. Sudhan Majhi, Indian Institute of Science, Bangalore, India
- Dr. M V Reddy, New graphite world (NMG), Montreal, Canada.
- Dr. Ashok Kumar, GOEC, Ajmer
- Dr. Cher Ming Tan, Engineering Department, Chang Gung University, Taiwan
- Dr. Neelam Srivastava, IET, Lucknow, India
- Dr. Radha Raman Chandan, SIET, Prayagraj (UP), India
- Dr. Ram Racksha Tripathi, EC Dept. SIET, India
- Dr. Sudhanshu Kumar Jha, JKIAPT, Prayagraj (UP), India
- Dr. Kamal Prakash Pandey, SIET, Prayagraj (UP), India
- Dr. Somak Bhattacharyya, IIT BHU, India
- Dr. Ghanshyam Singh, MNIT, Jaipur, India
- Dr. V. S. Tripathi, EC Dept. MNNIT Allahabad, UP, India
- Dr. Manish Goswami, IIIT Allahabad, UP, India
- Dr. Manoj Shukla, HBTU Kanpur, UP, India
- Dr. M. K. Meshram, EC Dept. IIT, BHU, India
- Dr. Omjee Pandey, IIT BHU, India
- Dr. K. V. Srivastava, EE Dept, IIT Kanpur
- Dr. Rajneesh Srivastava, EC Dept. University of Allahabad, UP,
- Dr. D. K. Srivastava, EC Dept. BIET, Jhansi

- Dr. Anand Sharma, MNNIT Allahabad
- Dr. Pinku Ranjan, IIITM Gwalior
- Dr. Satyendra Kumar Mourya, BITS Pilani
- Dr. Shishir Maheshwari, BITS Pilani
- Dr. Brijesh Kumar, MMMUT Gorakhpur
- Dr. Akash Gupta, ECE Department, The LNM Institute of Information Technology, Jaipur
- Dr. Nikhil Sharma, ECE department, The LNM Institute of Information Technology, Jaipur
- Dr. Rajesh Kumar, Electrical Engineering Department, DTU, Delhi
- Dr. Tawfik Ismail, Nile University, Egypt
- Dr. Nagendra Yadav, School of Electronics and optical Engineering; China
- Shri Gaurish Kumar Tripathi, Joint Director, DRDO, India
- Shri Mirza Mohammad Zaheer, Scientist SG, ISRO, India
- Dr. Brij N. Singh, IEEE PELS DL Fargo, United States
- Dr. Santosh Kumar, Liaocheng University, China
- Dr. Tatiana Martins, Federal University of Goiã S, Brazil
- Dr. Dmitriy Titov, A. A. Baikov Institute of Metallurgy and Materials Science, Russian Academy of Sciences, Russian Federation
- Dr. Gilberto Medeiros Ribeiro UFMG, Brazil
- Dr. Alexander Burkov Russia Russian Academy of Sciences, Ioffe Institute
- Dr. Krishna Bisetty, Durban University of Technology, South Africa
- Dr. A. D. Darji, SVNIT, Surat, India
- Dr. Poornima Mittal, DTU, India
- Dr. Ramesh Chandra, IIT Roorkee, India

Intelligent Systems and Smart Infrastructure – Brijesh Mishra et al. (eds)
© 2023 Taylor & Francis Group, London, ISBN 978-1-032-41287-0

CHAPTER

1

Free-Space Optical Communication Under Varying Atmospheric Conditions

Priyanshu Singh[1], Deepa Singh[2], Gaurav Agarwal[3] and Nitin Garg[4]

Department of Electronics and Communication Engineering,
Galgotias College of Engineering and Technology,
Greater Noida, Uttar Pradesh 201310, India

Abstract

A line-of-sight (LOS) method that encapsulates lasers offering optical bandwidth in communication link is known as free-space optical communication. Free-space optics (FSO) correspondence has antagonistically filled prevalently as of late on the grounds that to its one-of-a-kind characteristics, for example, high information rate, basic and fast wretchedness, low power, and low mass. One of the most powerful elements influencing the performance of bi-directional FSO is potential disturbance caused by weather conditions because the attenuation factor is high and the factors such as haze, rainy & misty and foggy atmospheric conditions directly affect the properties of FSO, resulting in a rise in the bit error rate (BER) and degrading Quality (Q)-factor to unsatisfactory levels. In this study, the authors examine the various channel ranges that have been proposed previously and come up with a two-channel enhanced FSO system model with mux, demux, and erbium-doped fiber amplifier amplifiers. This design has shown that it is capable of overcoming the attenuation caused by weather conditions. By measuring the Q-factor, performance has improved, as the results of the findings show. Furthermore, by using 1550 nm with many channels, BER could significantly be improved and further communication distance could be reached. The simulation result has demonstrated that the work done in this paper has improved the efficiency.

Keywords: Free-Space Optics, Line-of-Sight (LOS), Bit Error Rate (BER), Erbium-Doped Fiber Amplifier, Atmospheric Turbulence.

Corresponding author: [1]singhpriyan92@gmail.com, [2]deepasingh060609@gmail.com,
[3]agrawal.ga2811@gmail.com, [4]nitin.garg@galgotiacollege.edu

DOI: 10.1201/9781003357346-1

1. Introduction

The demand for bandwidth has increased adversely in today's scenario for wireless communication. The most difficult task is to build a network that is flexible and scalable of providing lot of information that meet the ever-increasing bandwidth demand To address the high information-carrying capacity, several wireless technologies have been developed [1]. FSOC stands for free-space optical communication, which uses LASER to enable optical bandwidth communications. Free-space optics (FSO) correspondence has filled in prominence as of late on account of its novel qualities; for example, high information rate, simple and speedy disgracefulness, and less power utilization [2]. Weather fluctuations like dust storms, fog, rain, dust, and haze are common causes of power outages that can weaken atmospheric system [2]. Various aerosols and gas molecules can disrupt systems by absorbing and scattering laser photons in the atmosphere. In fog, optical attenuation can exceed 340 dB/km, according to studies with a field of view of 50 meters [3]. Comparison has been made between all these different atmospheric conditions. In this enhanced design, two FSO channels with two erbium-doped fiber amplifier optical amplifiers at receiver and transmitter region are used [2]. Also, low bit error rate (BER), enhanced Q factor, and better opening eye diagram have been obtained.

2. System Model

In FSO, the transmitter, atmospheric channel, and receiver are all part of the wireless optical link. The optical modulator that is a component of transmitter side converts an electrical signal into an optical signal, which is then transmitted across the wireless medium [13]. The receiver, on the other side, is made by an optical amplifier, a photodetector, a low-pass filter, a power meter, and a BER analyzer. It then improves the signal strength that is received by the receiver in the exact way that the optical amplifier used in the transmitter does. Out of all the bits transmitted, the BER averages correctly received bits probability as [7]

$$\text{BER} = \frac{\text{Number of error in bits}}{\text{Total number of bits send}}$$

In the suggested approach, the optical amplifier transforms the electrical signal to an optical signal without the requirement for conversion. The optical amplifier enhances the average power of the laser output and amplifies weak signals before the photodetector detects the optical signal [10]. Increasing the optical power level before information is mixed with noise is extremely important in optical communication with longer visibility distances. This is significant in wireless optics link communication as atmospheric conditions have a significant impact on the signal. For diagnostic purposes, the quality (Q) factor is a measure of how loud a pulse is. Typically, the eye pattern oscilloscope will provide a report indicating the Q factor number. The Q factor is a critical metric that defines how well a communication channel performs [6]. There are two modes to calculate the Q factor: The Q factor from BER is calculated numerically:

$$P_e = \frac{1}{2} erfc\left(\frac{Q}{\sqrt{2}}\right)$$

where the Q factor is calculated:

$$Q = \frac{|\mu_1 - \mu_0|}{\sigma_1 - \sigma_0}$$

Here, erfc is the error function while μ 0,1 are the mean and σ 0,1 are the standard deviation on the zeros and the ones. In this study, we examined the performance of a WDM free-space optical transmission system employing NRZ and pin photodiode on the receiver side that is linked to the BER generator under various atmospheric conditions. In the transmitter and receiver parts, WDM is employed to extend the link range. To attain the desired performance, the design includes additional components such as forks, power combiners, and optical adders. Opti-System 18, a software program, is used here. The system is also put through its paces in the presence of various weather conditions. The goal is to create and simulate an FSO link as well as investigate its parameters. In the transmitter and receiver parts, WDM is employed to extend the communication range. The transmitter part consists of pseudo random bit generator, NRZ pulse generator, CW laser, and MZ modulator. CW laser used here is a continuous wave that acts as a carrier wave. In the receiver section, there is WDM DEMUX, followed by PIN photodiode, Gaussian filter, and a BER analyzer.

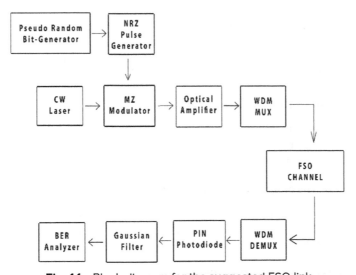

Fig. 1.1 Block diagram for the suggested FSO link

WDM is made up of mux and demux, and as the name suggests, it is a multiplexer itself. The mux essentially combines the datastream received from the trans-receiver into a single beam of light comprising several optical wavelengths that will be sent simultaneously through a single fiber. The demultiplexer (demux) is located at the other end of the fiber link, at the distant site, and it takes the multiple wavelengths that it receives and divides them back into individual data channels [9].

3. Simulation and Result

Fog is defined as a dense cloud that constitutes small water droplets floating in the sky, resulting in degraded visibility. Another atmospheric phenomenon is haze in which the sky is filled by extremely small particles. Dust is basically a thin powder which is composed of extremely minute bits of soil or sand. Few observations have been made over different weather conditions and their corresponding attenuations are mentioned in Table 1.1 [5].

Table 1.1   Different weather and their attenuation

Atmospheric conditions	Attenuation (dB/km)
Haze	11–21
Rain	6–31
Mist	29.1–31.4
Snow	40
Fog	70

The suggested FSO is used to build the end-to-end optical design, as illustrated in Fig. 1.1. The NRZ pulse generator is then used to encode it. The CW laser is used as an optical source, at the transmitter with a power of 60 dBm and a different wavelength of 1550 nm, 785 nm, and 450 nm [4]. By keeping this in mind, the visibility distance has been increased and Q-factor mechanisms were tested under various atmospheric turbulences, which include haze, rain, mist, snow, and fog. The simulation design is shown in Fig. 1.2.

A. Under Haze Atmospheric Conditions

Because the optical signal is significantly attenuated by the haze in the wireless environment, without the need of repeaters, the suggested adaptive and iterative technique has enabled 1800-m distance propagation [3], as shown.

B. Under Rainy & Misty Atmospheric Conditions

Despite the adverse conditions (rainy and cloudy), we were able to communicate over 2 kilometers in an unlicensed band, thanks to the innovative optical communication concept. As illustrated in Fig. 1.5,

C. Under Fogy Atmospheric Conditions

Despite the fact that the FSO signal is significantly attenuated (70 dB/km) due to the fog, with more than 2 km of distance [11], FSO communication can be possible with no use of addition of repeaters using the newly proposed design, as shown in Fig. 1.7.

The atmospheric conditions such as misty, rainy, and foggy have a similar conclusion. To summarize, the proposed trans-receiver design allows for improved communication performance in the FSO domain. The FSO could use the design in any atmospheric environment because of the new optimization and adaptable characteristics of the trans-receiver.

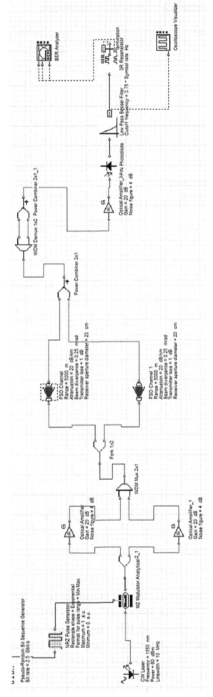

Fig. 1.2 A schematic diagram of the adaptive FSO architecture

Table 1.2 Simulation Parameter

Parameter	Value
Modulator type	Mach-Zehnder
Laser type	CW
Power	60 dBm
Linewidth	10 MHz
Pulse generator	NRZ
Divergence	0.25 mrad.
Amplifier	Optical
Gain (optical amplifier)	20 dB
Noise figure (optical amplifier)	4 dB
Distance between Tx and Rx	1100–2000 m
Transmitter loss	1 dB
Receiver aperture diameter	20 cm

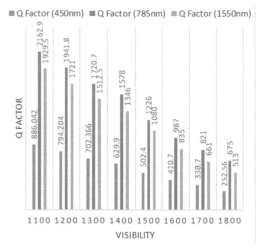

Fig. 1.3 Summary result for haze under different wavelength conditions

Fig. 1.4 Eye diagram for haze at 1550-nm wavelength

4. Conclusion

The optical signal is readily deteriorated by the wireless link, despite the FSO's attractive communication characteristics [15]. An improved connection architecture and an iterative optimization design have been proposed in this study with the help of WDM, where mux is used at transmitter part and demux is used at receiver section to enhance the bandwidth; therefore, the range is increased by the attenuation caused by atmospheric turbulences such as hazy, misty & rainy, and foggy atmospheric conditions. The proposed design has shown better results in terms of bandwidth, enhanced Q-factor, and better opening eye diagram. WDM is

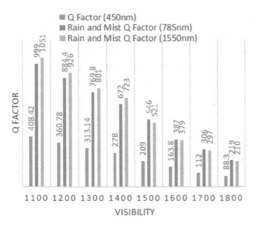

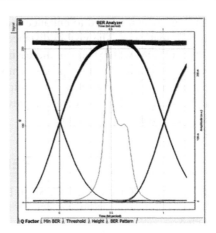

Fig. 1.5 Summary result for rain & misty under different wavelength.

Fig. 1.6 Eye diagram for rain & misty at 1550-nm wavelengths.

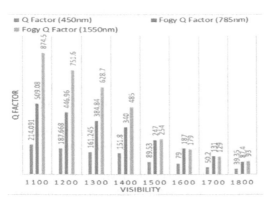

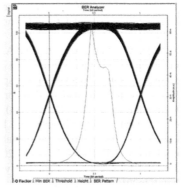

Fig. 1.7 Summary result for fog under different wavelengths

Fig. 1.8 Eye diagram for fog at 1550-nm wavelengths

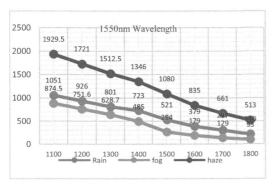

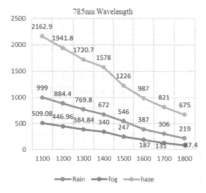

Fig. 1.9 Q-factor for rain, haze, and fog at 1550-nm wavelength

Fig. 1.10 Q-factor for rain, haze, and fog at 785-nm wavelength

a technology which is used to enhance the bandwidth by allowing various datastreams to be sent at different rates. The visible distance is increased by optimizing the optical link design by adding a WDM which helps in increasing the bandwidth. The atmospheric influence on the FSO was demonstrated using Opti-System version 18. An innovative suggested design allows for adaptation of the atmospheric condition. In summary, the simulation results demonstrate that at the sacrifice of low system complexity, higher bandwidth, Q-factor, better opening eye diagram, and BER may be attained.

REFERENCE

1. Srivastava D, Kaur G, Singh G, Singh P. Evaluation of atmospheric detrimental effects on free space optical communication system for Delhi weather. Journal of Optical Communications. 2020 Feb 8.
2. Lema GG. Free space optics communication system design using iterative optimization. Journal of Optical Communications. 2020 Jul 13.
3. Shumani MM, Abdullah MF, Suriza AZ. The effect of haze attenuation on free space optics communication (FSO) at two wavelengths under Malaysia Weather. In 2016 International Conference on Computer and Communication Engineering (ICCCE) 2016 Jul 26 (pp. 459–464). IEEE.
4. Alnajjar SH, Noori AA, Moosa AA. Enhancement of FSO communications links under complex environment. Photonic sensors. 2017 Jun; 7(2): 113–22.
5. Kaur S, Kakati A. Analysis of free space optics link performance considering the effect of different weather conditions and modulation formats for terrestrial communication. Journal of Optical Communications. 2020 Oct 1; 41(4): 463–8.
6. Das S, Chakraborty M. ASK and PPM modulation based FSO system under varying weather conditions. In2016 IEEE 7th Annual ubiquitous computing, electronics & mobile communication conference (UEMCON) 2016 Oct 20 (pp. 1–7). IEEE.
7. Jain D, Mehra R. Performance analysis of free space optical communication system for S, C and L band. In2017 International conference on computer, communications and electronics (Comptelix) 2017 Jul 1 (pp. 183–189). IEEE.
8. Ashraf M, Baranwal G, Prasad D, Idris S, Beg MT. Performance analysis of ASK and PSK modulation based FSO system using coupler-based delay line filter under various weather conditions. Optics and Photonics Journal. 2018 Aug 27; 8(8): 277–87.
9. Huang XH, Li CY, Lu HH, Su CW, Wu YR, Wang ZH, Chen YN. WDM free-space optical communication system of high-speed hybrid signals. IEEE Photonics Journal. 2018 Nov 16; 10(6): 1–7.
10. Chaudhary S, Amphawan A. The role and challenges of free-space optical systems. Journal of Optical Communications. 2014 Dec 1; 35(4): 327–34.
11. Ali, Mazin Ali A. "FSO communication characteristics under fog weather condition." *International Journal of Scientific and Engineering Research* 6, no. 1 (2015).
12. Ahmed, Ashad, Soni Gupta, Yatin Luthra, Konark Gupta, and Sanmukh Kaur. "Analysing the Effect of Scintillation on Free Space Optics Using Different Scintillation Models." In *2019 6th International Conference on Signal Processing and Integrated Networks (SPIN)*, pp. 799–804. IEEE, 2019.

13. Eid MM, Sorathiya V, Lavadiya S, Shehata E, Rashed AN. Free space and wired optics communication systems performance improvement for short-range applications with the signal power optimization. Journal of Optical Communications. 2021 Mar 15.
14. Hassan, M. Mubasher, and G. M. Rather. "Free space optics (FSO): a promising solution to first and last mile connectivity (FLMC) in the communication networks." *IJ Wireless and Microwave Technologies* 4 (2020): 1–15.
15. Ivanov H, Leitgeb E, Kraus D, Marzano F, Jurado-Navas A, Dorenbos S, Perez-Jimenez R, Freiberger G. Free space optics system reliability in the presence of weather-induced disruptions. InGuide to Disaster-Resilient Communication Networks 2020 (pp. 327–351). Springer, Cham.

Intelligent Systems and Smart Infrastructure – Brijesh Mishra et al. (eds)
© 2023 Taylor & Francis Group, London, ISBN 978-1-032-41287-0

CHAPTER

2

Encryption Algorithm for High-Speed Key Transmission Technique

R. S. Ramya[1], S. J. K. Jagadeesh Kumar[2], D. Pavai[3], K. N. Jaya Priya[4]

Kathir College of Engineering, Coimbatore, Tamilnadu, India, 641062

Abstract

Security plays a highly crucial role in business transaction. Nowadays, online transaction is used as the inevitable payment method in transferring funds or money. It is super easy for the transactions to be done online as the whole process is much speedier than one can ever imagine. As a potential outcome, users of the Internet are constantly exchanging data. Cryptography is the elementary category of computer security, which changes data from its usual structure into a jumbled structure. DES, AES, and Blowfish are all commonly used symmetric key encryption algorithms that are utilized in online transactions. This study claims to give a unique comparison of these algorithms. In this study, measures including block size, key size, average latency, throughput, energy consumption, and encryption and decryption durations are utilized as analogies. Millions of people throughout the globe use the Internet to communicate, and a similar number of people shop online on a regular basis. For this reason, the security level has become the burning question and the issues concerned with online transaction also increase day by day. This paper tries to bring out a solution for which it has taken cryptography, the method of concealing information or data, as its principal task.

Keywords: e-commerce security methodology, cryptography, DES, AES, Blowfish, encryption, decryption

Corresponding author: [1]rsramyarenu04@gmail.com, [2]jagadeesh66.sk@gmail.com, [3]pavaidhanasekaran@gmail.com, [4]jayapriya@kathir.ac.in

DOI: 10.1201/9781003357346-2

1. Introduction

Cryptography is a process that purports to transform the information so that it can be accessed to provide divergent security-associated ideas that include privacy, data reliability, validation, endorsement, and non-reputation. It may be deduced by analysing two fundamental parts. The former is known as the algorithm and the latter is called a key. The entire process is a mathematics-oriented methodology in which the key is accessed for data transformation. This process affirms cryptographic security by making use of encryption and decryption. Cryptography is reasonably associated with encryption; for instance, the converting of data from the effortlessly comprehensible phase to the junk. This helps to maintain a strategic distance from unnecessary users who are capable of understanding the data with ease and the senders who hold the capability to encode the data. Its fundamental and central objective is to keep the information secured protecting them from illegal hacking.

2. The Baseline Study

A. Algorithms for Analogy

Data Encryption Standard

Data encryption scheme is the principal encryption standard which has input and output of 64 bits along with 56 bit key. Keys are really 64-bit; however, just one bit in each octet is utilized to guarantee that the octets are equally dispersed. Numerous assaults and techniques have been reported so far, which are marked as the shortcomings of DES.

Advanced Encryption Standard

The Rijndael algorithm is another name for the Advanced Encryption Standard. Using 128-bit information blocks and symmetric keys of 128, 192, or 256, it is able to move quickly and efficiently. Replaced by the DES, the AES is acknowledged. This method has been the target of a brute-force attack, which has been widely successful.

Blowfish

Asymmetric block cypher Blowfish is the next, and it has been successfully accessed for the purpose of obtaining encryption and data protection. It takes an uneven length key, which is approximately computed between 32 bits and 448 bits, transforming it to be fit for authenticating data. As an open source, in 1993, Bruce Schneier proposed Blowfish algorithm as a quick alternative to the current industry standard for protecting secret information. Blowfish is an open-source alternative to that standard. All users of Blowfish are allowed to use it for free since it is not trademarked. Although it has a problem with weak keys, no attack has been proved to work against it (Bruce, 1996; Nadeem, 2005).

Objectives

The objectives of this paper are cited as follows:

(a) Upgrade a secure and effective payment transaction in order to avoid fraudulent activity.

(b) In e-commerce, expand the unbeaten and effective information retrieval.

(c) Reduce the amount of time it takes to encrypt and decrypt a transaction, as well as the amount of time it takes to complete the transaction, in order to make online transactions more secure.

3. Key Size and Encryption System

In the realm of cryptography, a "Key" refers to numeric or alphanumeric data that contain an inquiry character. "Key size" or "Key length" corresponds to the quantity of bits in a key that is utilized in a cryptographic operation. The sole constraint on an algorithm's security is the size of its key. In the face of adversity, it has stood the test of time. This is because brute-force attacks might jeopardize the safety of all algorithms. Of primary importance is that the key length is in agreement with the algorithm's security lower limit. Encryption happens on plain text using a key, whereas decoding uses a key to decode cypher text. Because the security of an encryption technique rests on the choice of key in cryptography, it is of paramount importance. The unique aspects of the encryption technique are the secret key, key length, and initialization vector. Increasing the key's size improves security at the expense of complicating the search for the key.

4. Key Length Comparison

This study used dynamic panel models to curb endogeneity issue due to the unobserved heterogeneity and simultaneity (Wooldridge, 2013). Under dynamic panel models, two-step system-generalized method of moments is considered. This econometric tool eliminates the endogeneity problem through internally generated instrumental variables. Subsequently, certain model specification tests like Arellano–Bond test, Sargan test. and Wald Chi-square (χ^2) test are applied to check the serial correlation and over-identification issues. The insignificant autoregressive terms of Arellano–Bond test indicate the absence of serial correlations. The insignificant p-values of Sargan test indicate no over-identifications issues. Wald test with a significant p-value implies the overall robustness of the model results.

DES

Data protection is the most significant disadvantage of DES. The DES does not give strong protection due to its key length of 56 bits. DES might fully break by brute-force assault. At first, DES was acknowledged as the standard algorithm with strong protection; however, after a period of time, Brute power assault cracked DES. Consequently, DES is certifiably not a secured encryptions algorithm. Plaintext blocks of 64 bits and 56-bit keys are used in the DES algorithm. The 8-bit equality bit in DES is used to detect errors. DES encrypts data by dividing each block into two halves and then applying 16 rounds of processing to each half. Development, key blending, substitution, and permutation are all included in function f's four steps. The 56-bit key length of DES is the primary problem for security.

AES

AES is very secure because it uses a variable-length key, such as 128 bits. Feistel network-based block cypher AES also employs 128-bit blocks with an unstable key length of 128, 192, or 256 bits, depending on the block size. The number of cryptographic rounds varies depending on the length of the key. Every AES round performs key development, substitution bytes, shifting rows, mixing columns, and adding a new round of key development. AES gives a high data protection. Different kinds of assault attempt to split AES like square assault, key assault, differential assault, and enhanced square assault. However, none of them is possible to break this technique. Thus, AES is an extremely protected encryption method. AES can also protect information against future assault.

Blowfish

A 64-bit block and a key size ranging from 32 to 2448 bits are used in Blowfish block encryption. Blowfish provides a high degree of security since it uses a 448-bit key. Because each bit of the master key comprises a distinct round key that is independent, Blowfish is safe against differential key attacks. [1], [7] Blowfish goes through 16 stages of processing. Data encryption and key development are the two primary functions of this kind of encryption. The keys have no effect on substitution boxes. Blowfish keys' varying lengths necessitated more processing time. The time-consuming sub-key creation procedure increases the difficulty of a brute-force attack. There are no known backdoor vulnerabilities; hence, it provides long-term information security for sensitive data. Blowfish's dependability has been compromised as a result of the widespread usage of weak keys. On the basis of a second request for differentiated attacks, the first four rounds of the legal procedure are laid out. Consider the symmetric algorithms DES, AES, and Blowfish, and their block and key sizes. In comparison to trading tactics, the Blowfish approach is superior.

Table 2.1 Key Length Comparison

Algorithm	Key size (in bits)	Block size (in bits)	Round	Structure	Feature
DES	64	64	16	Feistel	Not structure, Enough
AES	128,192,256	128	10,12,14	Substitution, Permutation	Replacement for DES, Excellent security
Blowfish	32–448	64	16	Feistel	Excellent security

Table 2.1 interprets the Blowfish technique that offers implausible insurance plans to examine symmetric algorithms; for instance, DES and AES. Figure 2.1 displays the analogy between key size and block size.

DES

It has a key size of 64 bits and a block size of 64 bits. In some specific cases, there are a very few assaults, which are acknowledged to be the negative aspects of DES.

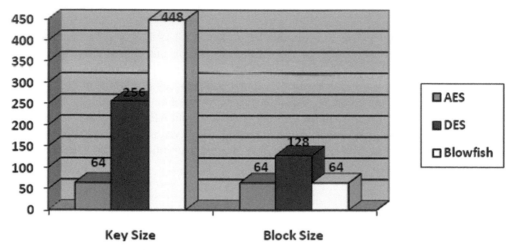

Fig. 2.1   Key size and block size comparison

Algorithm:

Function DES_Encrypts (D, E) where M = (L, R)

D ←IP(D)

For round ←1 to 16 do

E_i ←SI (E, round)

L ←L xor F(R, Ei)

swap(L, R)

end

swap(L, R)←IP^{-1}(D)

return D End

AES

In general, NIST recommends AES as a replacement for DES. The most profitable attack against it is the brute-force attack, in which the attacker attempts to crack the encryption by checking and mixing all characters. It is a block cypher like AES or DES that is used for secure communications. AES has a 256-bit key length by default, although it may be as long as 128, 192, or even more bits. Depending on the key size, this encrypts 128-bit blocks in rounds of 10, 12, and 14. AES is a lightning-fast and nimble cypher. It may be particularly well done in small devices with extraordinary architecture. The use of AES for security reasons has been strangely noticed.

Algorithm

Ciphers(byte[] input, byte[] output)

```
{
byte[4,4] State;
copy inputs[] into states[]
AddRoundKeys for (rounds = 1; rounds < Nr-1; ++rounds)
{
SubBytesShiftRowsMixColumnsAddRoundKeys
}
SubBytesShiftRowsAddRoundKeys
Copy states[] to outputs[]
}
```

Blowfish

Divergent tests and research have established the outcome of Blowfish algorithm as beneficial and also as thoroughly worthy in terms of time consumption. Blowfish is, on the whole, a largely fulfilling and an exceptional algorithm in throughput and strength utilization. Number of evaluation techniques for an intense scrutiny demonstrates the effectiveness of Blowfish approach over a variety of techniques to the extent the managing time. Blowfish is larger than a number of strategies in throughput.

Algorithm

Divide x1 into two 32-bit halve: x_{LL}, x_{RR} for j= 1 to 16:

x_{LL} = x_{LL} XOR P1$_j$

x_{RR} = F1(x_{LL}) XOR x_{RR}

swap x_{LL} and x_{RR}

next j

swap x_{LL} and x_{RR}(undo the last swap step)

x_{RR} = x_{RR} XOR P1$_{17}$

x_{LL} = x_{LL} XOR P1$_{18}$

Recombine into x_{LL} and x_{RR}

From Table 2.1, it sums up that the Blowfish technique is safer to analyze other symmetric key algorithms. It brings about valuable results within very less processing time and rounds. While boosting the key size of Blowfish method from 128 to 448, it becomes more protective to the messaging. It ensures high-end information protection at the time of communications over some risky medium.

Secure payment is ensured by the use of DES, AES, and Blowfish. As demonstrated in Table 2.2, it also features methods for reducing the average delay as well as encryption and decryption times, energy usage, and throughput times. It is wholly understood that Blowfish is doing its best on every specific restraint and functions veraciously within the parameters that are pertinent.

Table 2.2 Average Delay, Throughput, Energy Consumption, Decryption, and Encryption Time for E-Commerce

Algorithm	Average Delay (s)	Throughput (Kbps)	Energy Consumption (Joules)	Encryption Time (s)	Decryption Time (s)
DES	48.7766	28.13	83.087	0.434	0.451
AES	49.1367	5.27	72.087	0.314	0.321
BF	48.7349	35.2	85.544	0.262	0.253

Figure 2.2 represents the average hold-up in milliseconds and the future algorithm Blowfish is likened to the extant algorithms – DES and AES.

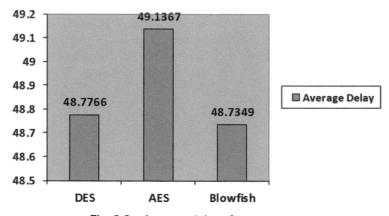

Fig. 2.2 Average delay of e-commerce

Figure 2.3 pinpoints the throughput in kbps, and the BF algorithm is compared to subsisting algorithms such as DES and AES.

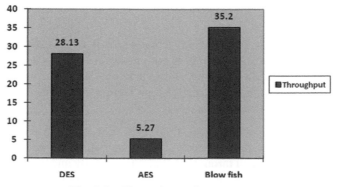

Fig. 2.3 Throughput of e-commerce

Figure 2.4 refers to the energy consumption in joules, and BF algorithm is computed to existing algorithms such as DES and AES.

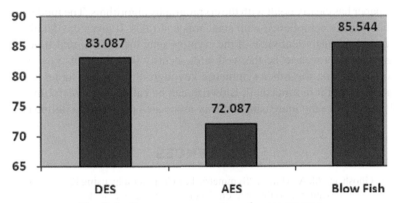

Fig. 2.4   Energy consumption of e-commerce

Figure 2.5 illustrates the encryption and decryption time in milliseconds, and the BF algorithm is estimated. Figure 2.5 notes the dissimilarities between BF and the subsisting algorithms such as DES and AES.

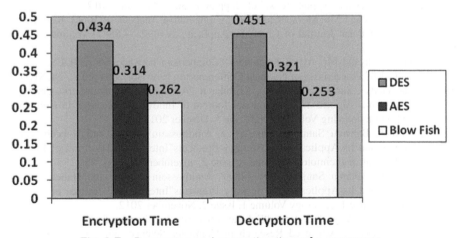

Fig. 2.5   Decryption and encryption time of e-commerce

With DES and AES, Blowfish is calculated based on criteria such as average latency, throughput, and encryption and decryption time. AES is a challenger that provides data secrecy and integrity in close proximity. Its primary goal is to reduce complexity; however, the security it provides is often unreliable. BF augments the protection of the safe and successful payment transactions, and by the way, it prevents fraudulent happenings. Blow BF 0.40 AD, 2.45 EC, 0.52 ET, and 0.68 DT qualify 7.02% throughput. To sum up, this paper standardizes the fact that the Blowfish methodology is the best algorithm.

5. Conclusion

This research paper has clearly dealt with the cryptography algorithms. The model is interpreted through the developer's compliance with standards and rules. It gives the detailed analogy of the key length and the analytical view of the cryptography methods in real-time applications. Blowfish algorithm is proven to be the well-built security because of its key size. Likewise, when it is compared with the other symmetric key algorithms, it consumes less processing time and rounds. To put it in a nutshell, Blowfish can be called the numero uno method as it tremendously helps solve the increasing security-associated problems in online transaction.

REFERENCES

1. A. Nath, S. Ghosh and M. A. Mallik, "Symmetric key Cryptography using Random Key Generator", Proceedings of International conference on Security and Management, , Las Vegas ,USA, Vol. 2, pp.239-244, 12-15 July, 2010.
2. "Advanced Encryption Standard", http://en.wikipedia.org/wiki/Advanced_Encryption_ Standard". (accessed on November 8, 2015).
3. AL. Jeeva, V. Palanisamy and K. Kanagaram, "Comparative Analysis of Performance Efficiency and Security Measures of Some Encryption Algorithms", Proceedings of International Journal of Engineering Research and Applications, Vol. 2, pp.3033-3037, May-Jun 2012.
4. Mr. Mukta Sharma and Mr. Moradabad R. B. "Comparative Analysis of Block Key Encryption Algorithms"International Journal of Computer Applications (0975 – 8887) Volume 145 – No.7, July 2016
5. Mr. Milind Mathur and Mr. Ayush Kesarwani "Comparison between DES, 3DES, RC2, RC6, Blowfish and AES" Proceedings of National Conference on New Horizons in IT - NCNHIT 2013
6. Annapoorna Shetty , Shravya Shetty K , Krithika K "A Review on Asymmetric Cryptography – RSA and ElGamal Algorithm" International Journal of Innovative Research in Computer and Communication Engineering Vol.2, Special Issue 5, October 2014
7. Maryam Ahmed, Baharan Sanjabi, Difo Aldiaz, Amirhossein Rezaei, and Habeeb Omotunde "Diffie-Hellman and Its Application in Security Protocols"International Journal of Engineering Science and Innovative Technology Volume 1, Issue 2,November 2012.
8. Maryam Ahmed, Baharan Sanjabi, Difo Aldiaz, Amirhossein Rezaei, and Habeeb Omotunde "Diffie-Hellman and Its Application in Security Protocols"International Journal of Engineering Science and Innovative Technology Volume 1, Issue 2, November 2012.
9. Dr. Chander Kant and Yogesh Sharma, "Enhanced Security Architecture for Cloud Data security" International Journal of Advanced Research in Computer Science and Software Engineering, Volume 3, Issue 5, May2013, pp. 571–575
10. Method Applied to Cloud Computing" International Journal of Information & Computation Technology. ISSN 0974-2239 Volume 4, Number 15 (2014), pp. 1519–1530.
11. Sivakumar, N., Balasubramanian, D. R. "Credit Card Fraud Detection: Incidents, Challenges And Solutions." International Journal of Advanced Research in Computer Science and Applications, 2015.
12. Sun, J., Zhu, Q., Liu, Z., Liu, X., Lee, J., Su, Z., Xu, W. 2018. "FraudVis: Understanding Unsupervised Fraud Detection Algorithms." In Pacific Visualization Symposium (Paci Thakur, J., Kumar, N. 2011. "DES, AES and Blowfish: Symmetric key cryptography algorithms simulation based performance analysis." International journal of emerging technology and advanced engineering, Vol. 1, No. 2, pp. 6–12.

Intelligent Systems and Smart Infrastructure – Brijesh Mishra et al. (eds)
© 2023 Taylor & Francis Group, London, ISBN 978-1-032-41287-0

CHAPTER

3

Gender and Age Detection System for Customer Movement Analysis

Madhav Goel[1], Mansi Jain[2], Nainsi Jain[3], Mukesh Rawat[4]

Department of Computer Science and Engineering, Meerut Institute of Engineering & Technology, Meerut, Uttar Pradesh., India

Abstract

In today's scenario, classification on the basis of age and gender plays a vital role for helping many organizations and malls. These types of tools are used by various companies for various purposes, which will reduce the overhead for the companies so that they can collaborate with customers and fulfill their needs in a finer way, which will create a great accomplishment for customer relation and business point of view. The needs and demands of the people can easily be recognized on the basis of their gender and age. This research will work on the same path but at the same time, it will extend functionality by not only detecting gender and age but also creating a database and present it in such a way that the stakeholders would be able to understand the current status without any stress of knowledge on technicalities and make better decisions towards growth.

Keywords: Classification, Age and gender detection, Recognition, Creating database, Technicalities

1. Introduction

Facial analysis from images gathered from streaming data attains a lot of interest because it provides solutions to different problems like in providing better ad which customers wants to target, or in security surveillance purposes or in better content recommendation system, etc.

Corresponding author: [1]madhav.deepak.cs.2018@miet.ac.in, [2]mansi.jain.cs.2018@miet.ac.in, [3]nainsi.jain.cs.2018@miet.ac.in, [4]mukesh.rawat@miet.ac.in

DOI: 10.1201/9781003357346-3

Age and gender play a vital role in determining the facial attributes, and identifying these attributes is the first step in facial analysis. This research deals with deep learning techniques, specifically object detection and sense motion, detect face from frame and classify them on the basis of age and gender, and this information is used by mall's supervisor to do better customer's data analysis [2]. After considering all the aspects, a research on GENDER AND AGE DETECTION FOR CUSTOMER MOVEMENT ANALYSIS would prove to be of great help in analyzing customer. AI and deep learning are growing technologies and have a lot of use nowadays. Everything is going for automation so that machine can work as much tasks as human can do in no time. This research will automate the task of keeping the track of customers along with their age and gender details so that insights can be made on data. This information will definitely help to make better decisions. We have used deep learning instead of machine learning (ML). Deep learning extract features manually from an image and it always tries to acquire knowledge for high-level features from the data. This abolishes the need of domain expertise and hardcore feature extraction and also it always uses problem-solving approach. Techniques like deep learning solve the problem end-to-end, whereas ML breaks down the problem statement in parts and then solves these problems one by one, and after, that results are merged. The best thing is that the CNN learns the filters automatically without mentioning it explicitly [3].

2. Literature Review

1. Facial identification is used in biometric, ordinarily as a part of facial identification. Face detection is utilized by some current digital devices for automatic focus too. It is also helpful for selecting areas of interest in frames. A security camera is plugged into a monitor, which detects any facial-object that walks by. Then the system will evaluate the gender and age range of facial image. After the data is being collected, a stream of declarations can be exposed to view, which is distinct with respect to the detected gender/age. Conservation of energy can also use the concept of face identification. Series of steps for the facial identification is based on the method of coding and de-coding the image as discussed in Searala A. Dabadey & Murunal Bewor (2012) [4]. The proposed methodology is composed of two phases – Haar-Based Cascade recognition and classifier using Principal of Component Analysis. Execution of facial detection using principal of partial analysis using 4 distance classifiers is suggested in Hussain Raddy (2010). Different face recognition and detection methods have been assessed in Fazai Ahmed et al. (2012).

2. Lanites [6] suggested the initial technique putting in Convolutional neural network (CNN) to gender recognition, which puts out cranio-facial growth and aging during adulthood and childhood.

   (a) Gender-specific evaluation, which is based on the concept that the aging activity is similar for everyone.

   (b) Specific-appearance evaluation, which is based on the belief that the human faces which look identical tend to have the same process of aging. Zhange [6] put together the conclusion of every person's gender and age to a warped Gaussian process (WGP) evaluation issue, and evolved a multi-task update of WGP to resolve the issue. Since each individual has distinct aging processes, personalization is advantageous for age detection. Earlier researches also convey that the personalization can improve the implementation of gender detection.

3. Methodology and Model Specifications CNN

Usually there are various types of methods which can be used to solve this type of problem. Algorithms like "Fisher faces" and "Eigenface" are developed for facial recognition methods, but they don't provide the satisfactory result which is required. Using CNN, we can design finer solutions, which have come out to be the most favorable model for computer vision tasks. These models have proved to be the most fruitful solution when dealing with image datasets and found to be the most important ML computer vision models [4].

Convolutional layer – This is a first layer of CNN and it is used to take out various features from the input images. It helps to transform the input image in order to extract features from it [3].

Pooling layer – Pooling layer helps to decrease the number of the features maps or we can say it helps to reduce the number of parameters to learn and the amount of computation performed.

Fully connected layer – These are the layers in which the inputs from one layer are connected to next layer's unit of activation [6].

Workflow

1. Capture frames from the video source.
2. Identification of the face from the frame.
3. Feed that identified image to the classification model.
4. Make an object after classification and store that into a database.
5. Create visualizations on that data.

4. Empirical Results

Three terms are used to assess performance: map, -recision, and recall [3]. Map is a combined measure of recall and accuracy used to identify object accuracy. It is derived using a mean precision value for recall greater than 0 to 1 and an IOU that is the convergence of the two values between 0.5 and 0.95.

$$AP = \Sigma Pr = 1/R$$

$$map = 1/N(\Sigma)$$

Preciseness is used to determine how precisely the results are expected. Precision can define how accurate the program is at evaluating the class.

$$Precision = TP/TP + FP$$

$$IoU = Area\ of\ overlap/Area\ of\ union$$

Accuracy can be found using mAP, which defines the AP used in the computation. AP correlates to average AP for IoU starting from 0.10 to 0.90 in 0.05 step.

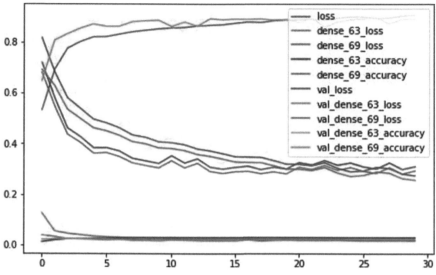

Fig. 3.1 Accuracy graph, x-axis: epoch, y-axis: accuracy

5. Conclusion

In this paper, we have proposed gender and age detection system for customer movement analysis system using CNN that can detect face, classify them on the basis of gender and age, stores this information and create visualizations on a dashboard. This system has various real-time applications such as customer count and time-based analysis. This system is very effective in detecting gender and age on CPU GPUs. This system needs local system with GPU, which may be inappropriate for all systems, but with cloud sources like Google Colab, we can use GPU and also can create a working system that can create custom benefactors. Hence, using the CNN algorithm, this system can keep track of different subjects that have been trained [7].

REFERENCES

1. Jeff skoll, gender and age detection, data flare, 20 april 2015. [accessed 09 january 2022].
2. Yamashita, R., Nishio, M., Do, R.K.G. et al. Convolutional neural networks: an overview and application in radiology. Insights Imaging 9, 611–629 (2018). [accessed 12 January 2022]
3. Sonia Singla , Age and gender detection using deep learning, analytics vidhya, July 27, 2021. [accessed 04 january 2022]

4. Nick Johnson, computer vision, tech vidvan, january 21, 2017[accessed 06 january 2022]
5. vincent tatan, Understanding CNN (Convolutional Neural Network), july 20 2020.[accessed 10 jaunuary 2022].
6. Alzubaidi, L., Zhang, J., Humaidi, A.J. et al. Review of deep learning: concepts, CNN architectures, challenges, applications, future directions. J Big Data 8, 53 (2021). [accessed 05 January 2022]
7. Deep Learning from Scratch: Building with Python from First Principle, Book by Seth Weidman, 2019, [Accessed: December 29, 2021]
8. National Key Lab for Novel Software Technology Nanjing University, China wujx2001@gmail.com May 1, 2017,[accessed 10 january 2022]
9. Ian Goodfellow and Yoshua Bengio and Aaron Courville,MIT Press, august 2016. [accessed on 29 december 2022]

Intelligent Systems and Smart Infrastructure – Brijesh Mishra et al. (eds)
© 2023 Taylor & Francis Group, London, ISBN 978-1-032-41287-0

CHAPTER

4

LTRACN: A Method for Single Human Activity Recognition

Prashant Sharma[1], Amartya Mishra[2], Nilutpol Kashyap[3],
Muheed Muzamil[4], Rahul Singh Rawat, Ali Imam Abidi*

Department of Computer Science and Engineering,
School of Engineering and Technology, Sharda University,
Greater Noida, India

Lokendra Singh Umrao[5]

Department of Computer Science & Engineering,
Institute of Engineering and Technology,
Dr. Rammanohar Lohia Avadh University,
Ayodhya, India

Abstract

In recent years, we have seen that two-stream models (dealing with both the spatial and temporal domain) have performed well and also have achieved state-of-the-art performance with time-series data specifically when working with videos for activity recognition, activity prediction, video captioning, etc. In this paper, we propose a single human activity recognition architecture, Long-Term Recurrent Attention Convolutional Network (LTRACN), based on Convolutional Neural Network (CNN), Attention Model, and Long Short-Term Memory (LSTM). The CNN is used to extract the spatial features from the RGB video frames. An Attention model is added to learn which parts in a particular frame are important for the prediction of classes and later on the information is being processed by the LSTM and prediction is done.

Keywords: Human Activity Recognition, Long-Term Recurrent Attention Convolutional Network (LTRACN), Convolutional Neural Network, Attention, LSTM

*Corresponding Author: ali.abidi@sharda.ac.in

[1]2018013038.prashant@ug.sharda.ac.in, [2]2018007592.amartya@ug.sharda.ac.in, [3]2018002860.nilutpol@ug.sharda.ac.in,
[4]2018000355.muheed@ug.sharda.ac.in, [5]lokendra.manit@gmail.com

DOI: 10.1201/9781003357346-4

1. Introduction

The field of human activity recognition has been an interesting topic for many researchers since the last two decades. Human activity recognition finds vast applications ranging from smart homes to human assistance robotics [1]. In the field of surveillance, it has become essential to record the behaviour of the people to monitor and analyse their activities for abnormal behaviour. With the recent developments in Internet of Things and Machine Learning, researchers and developers are intrigued by the vast potential applications in the real world [2]. The main aim of a Human Activity Recognition (HAR) system is to be able to recognize various types of activities such as jumping, running, sitting, and many more. For achieving this, there are mainly two approaches – non-visual sensor-based approach and visual sensor-based approach. The primary difference between them is the sensors used to collect the human behaviour data for interpretation. In visual sensor-based approach, data is provided in the form of images or videos using 2D or 3D stereo cameras, whereas in non-visual sensor-based approach, data is provided in the form of 1D signals collected using gyroscopes and accelerometer. Non-visual sensors have to be worn continuously by the user whose activity is being tracked, which poses the problem of battery life. On the other hand, visual sensor-based systems face difficulties like variations in body shape and size, and changes in illuminations and angles. In case of complex activities such as drinking and eating, sensor-based approaches find it difficult to differentiate between them [3]. In this case, visual sensor-based approach has achieved favourable results [4].

2. Materials and Methods

A. Related Works

The field of Human Activity Recognition had an extensive research work done on it. Most of the up-to-the-minute models are based on two stream models, that is, models that deal with both spatial and temporal domain of the video clips. One of them is PERF-NET where the author Yinxiao Li et al. [5] combined flow-based input stream through distillation technique and the pose stream with the standard RGB. In the year 2016, Jeff Donahue et al. [6] proposed Long-term Recurrent Convolutional Networks (LRCNs), where Convolutional Neural Network (CNN) and Long Short-Term Memory (LSTM) were used to deal with the visual features and sequence learning of a video clip to make the prediction. Xingjian Shi et al. [7] proposed the ConvLSTM architecture, where it excelled in dealing with spatio-temporal data due to its underlying convolutional structure. In 2018, Lei Wang et al. [8] proposed an activity recognition architecture based on CNN, LSTM, and temporal-wise Attention model. In [8], the author used a CNN and passed the knowledge extracted from frames of the video clips to two separate LSTM models. One was Fully Connected LSTM model and other was Convolutional LSTM model. Later, both separate modules are joined and prediction is made. The work of [6] and [7] laid out a foundation for a lot of the models that were developed after using CNNs and LSTM. Kwang Eun Ko et al. [9] used YOLOv2 in real time to extract images of the subjects and they are labelled the same. The extracted images are then fed into a CNN-Kalman filter-integrated network to distinguish the subjects. Kamal Kant Verma et al. [10] made use of both CNN and LSTM by combining them for gaining both spatial and temporal learning. They used a CNN which learns the correlations between limbs by taking an informative 3D skeleton representation of the input from the RGB-D camera. The extracted features are then input for the LSTM to figure out temporal dependencies in the posture of a person during a sequence. Recurrent Convolution Neural Networks (RCNNs), which is a combination of convolution operations and recurrent links by Zhenqi Xu et al. [11], is able to learn local and dense features from the image frames and also learn temporal features between consecutive frames for video classification. Lin Sun et al. [12] proposed a lattice LSTM architecture. In this, two CNNs are followed by two LSTMs. The CNNs are used for extracting features of both spatial and temporal domains. The double neural network architecture helps to preserve the time-domain features as well as the spatial domain features.

Dataset

For training and evaluation of our model, we have used HMDB-51: A large video database for human motion recognition by Hildegard Kuehne et al. [15]. The dataset consists of a massive collection of realistic videos gathered from various sources like YouTube, movies, etc. It consists of 51 categories/classes like jump, laugh, punch, hug, etc. where each class has at least 101 video clips. It contains a total of 6,849 video clips.

Methodology

Data Pre-processing

Initially, all the video clips in the dataset were resized to a height × width of 64 × 64. With the help of OpenCV, the entire video clip is broken down into a sequence length of 30. The features of all the frames processed in a particular video clip is then stored against its corresponding label. This entire processed data is then split into 70% training data and 20% testing data. The training data is split into 80% training data and 20% validation data.

Long-Term Recurrent Attention Convolutional Network

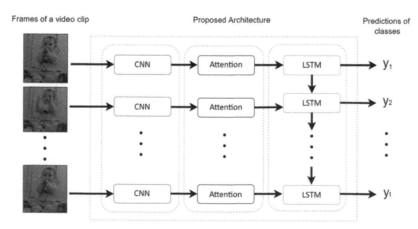

Fig. 4.1 Proposed architecture of LTRACN method

The proposed architecture comprises mainly three models that are CNN, Attention, and LSTM. CNN is used to extract the spatial information from an individual frame of a RGB video clip. The size of frame input in the model is $64 \times 64 \times 3$, where frame's height and width are 64, and 3 is the number of channels. After passing the frame from various time-distributed layers of CNN, the data is then flattened into a 1-dimensional array. This 1-D array is then passed into an Attention model, which teaches the network to emphasize meaningful features along the channels in a frame. The main aim of this network is to selectively concentrate on few relevant things in the network while ignoring the other less relevant things. Finally, an LSTM model is used to learn about the temporal features of the video clips using the result received from the previous Attention Model and a prediction is made based on the output of the LSTM model.

3. Results

The model is trained solely on HMDB51 dataset and no extra training data was used. It reached an overall accuracy of 81.36%. When the same dataset was trained with LRCN model [6], the highest accuracy achieved by the model was 77.12%, adding an Attention model between the two, which boosted the overall accuracy of the model by 4.24%.

Table 4.1 Comparison with state-of-the-art methods on HMDB-51*

Model Name	Accuracy	Year
PERF-NET [5]	83.2	2021
LTRACN (proposed)	**81.36**	**2022**
REPFlow-50 [15]	81.1	2018
Multi-stream I3D [22]	80.92	2019
LGD-3D Two-Stream [23]	80.5	2019
D3D [24]	79.3	2018
HATNet [25]	76.5	2019
LGD-3D RGB [26]	75.7	2019
ADL+ResNet+IDT [27]	74.3	2018

4. Conclusions

In this paper, we have devised a novel single human activity recognition architecture called Long-Term Recurrent Attention Convolutional Network (LTRACN) based on a combination of existing reliable Neural Network architectures, namely CNNs and the LSTMs. The spatial features of an RGB video frame are extracted using CNN while temporal features over the entirety of a video clip are processed by LSTM. In this model, the addition of the Attention model ensures that important parts of the video are given more weight, while feeding them to the LSTM, these can be frames where major activity-defining movements are taking place, etc. This addition of the Attention model boosted the combined efficiency by 4.24%. The best accuracy achieved by our model was 81.36 without using any extra training data, which is comparable with the state-of-the-art models.

REFERENCES

1. R. Poppe, "A survey on vision-based human action recognition", Image Vis. Comput., vol. 28, no. 6, pp. 976–990, 2010.
2. Shih, C.S.; Chou, J.J.; Lin, K.J. WuKong: Secure Run-Time environment and data-driven IoT applications for Smart Cities and Smart Buildings. J. Internet Serv. Inf. Secur. 2018, 8, 1–17.
3. Agac, S.; Shoaib, M.; Incel, O.D. Context-aware and dynamically adaptable activity recognition with smart watches: A case study on smoking. Comput. Electr. Eng. 2021, 90, 106949.
4. Eisa Jafari Amirbandi and Ghazal Shamsipour. Exploring methods and systems for vision based human activity recognition. 1st Conference on Swarm Intelligence and Evolutionary Computation, CSIEC 2016 - Proceedings, pages 160–164, 2016.
5. Yinxiao Li, Zhichao Lu, Xuehan Xiong, Jonathan Huang "PERF-Net: Pose Empowered RGB-Flow Net" arXiv:2009.13087, 2021.
6. Jeff Donahue, Lisa Anne Hendricks, Marcus Rohrbach, Subhashini Venugopalan, Sergio Guadarrama, Trevor Darrell, Kate "Long-term Recurrent Convolutional Networks for Visual Recognition and Description" arXiv:1411.4389, 2016.
7. Xingjian Shi, Zhourong Chen, Hao Wang, Dit-Yan Yeung, Wai-kin Wong, Wang-chun Woo "Convolutional LSTM Network: A Machine Learning Approach for Precipitation Nowcasting" arXiv:1506.04214.
8. Lei Wang, Piotr Koniusz (2021). Self-supervising Action Recognition by Statistical Moment and Subspace Descriptors. arXiv:2001.04627
9. Kwang Eun Ko and Kwee Bo Sim (2018). Deep convolutional framework for abnormal behavior detection in a smart surveillance system. Engineering Applications of Artificial Intelligence 67:226–234, 2018
10. Kamal Kant Verma, Brij Mohan Singh, H.L Mandoria, and Prachi Chauhan, "Two-Stage Human Activity Recognition using 2D-ConvNet" Vol.6, N°2, 2020, doi: 10.9781/ijimai.2020.04.002
11. Zhenqi Xu, Jiani Hu, Weihong Deng, Recurrent Convolutional Neural Network for Video Classification.
12. Lin Sun, Kui Jia, Kevin Chen, Dit Yan Yeung, Bertram E. Shi, and Silvio Savarese. Lattice Long Short-Term Memory for Human Action Recognition. Proceedings of the IEEE International Conference on Computer Vision, 2017-Octob:2166–2175, 2017.
13. I Sutskever, J. Martens, and G. E. Hinton, "Generating text with recurrent neural networks," in ICML, 2011.
14. O. Vinyals, S. V. Ravuri, and D. Povey, "Revisiting recurrent neural networks for robust ASR," in ICASSP, 2012.
15. H. Kuehne, H. Jhuang, E. Garrote, T. Poggio and T. Serre, "HMDB: A large video database for human motion recognition," 2011 International Conference on Computer Vision, 2011, pp. 2556-2563, doi: 10.1109/ICCV.2011.6126543.
16. AJ Piergiovanni and Michael S. Ryoo "Representation Flow for Action Recognition" arXiv: 1810.01455, 2018.
17. S. Hochreiter and J. Schmidhuber, "Long short-term memory," in Neural Computation. MIT Press, 1997.

18. Lei Wang, Piotr Koniusz "Self-supervising Action Recognition by Statistical Moment and Subspace Descriptors" arXiv:2001.04627, 2021.
19. Piotr Koniusz, Lei Wang, Ke Sun "High-order Tensor Pooling with Attention for Action Recognition" arXiv: 2110.05216, 2021.
20. B. Igor L. O., M. Victor H. C. and W. R. Schwartz, "Bubblenet: A Disperse Recurrent Structure To Recognize Activities," 2020 IEEE International Conference on Image Processing (ICIP), 2020, pp. 2216-2220, doi: 10.1109/ICIP40778.2020.9190769.
21. Jingran Zhang, Fumin Shen, Xing Xu, Heng Tao Shen "Cooperative Cross-Stream Network for Discriminative Action Representation" arXiv:1908.10136v1,2019.
22. Joao Carreira, Andrew Zisserman "Quo Vadis, Action Recognition? A New Model and the Kinetics Dataset" arXiv: 1705.07750.
23. Hong J, Cho B, Hong YW, Byun H. Contextual Action Cues from Camera Sensor for Multi-Stream Action Recognition. Sensors. 2019; 19(6):1382. https://doi.org/10.3390/s19061382.
24. Zhaofan Qiu, Ting Yao, Chong-Wah Ngo, Xinmei Tian and Tao Mei "Learning Spatio-Temporal Representation with Local and Global Diffusion" arXiv: 1906.05571, 2019.
25. Jonathan C. Stroud, David A. Ross, Chen Sun, Jia Deng, Rahul Sukthankar "D3D: Distilled 3D Networks for Video Action Recognition" arXiv: 1812.08249, 2019.
26. Ali Dibal, Mohsen Fayyaz, Vivek Sharma, Manohar Paluri, Jurgen Gall, Rainer Stiefelhagen, Luc Van Gool "Large Scale Holistic Video Understanding" arXiv: 1904.11451, 2019.
27. Zhaofan Qiu, Ting Yao, Chong-Wah Ngo, Xinmei Tian, and Tao Mei "Learning Spatio-Temporal Representation with Local and Global Diffusion" arXiv: 1906.0557, 2019.
28. Jue Wang and Anoop Cherian "Contrastive Video Representation Learning via Adversarial Perturbations" arXiv: 1807.09380, 2018.

Intelligent Systems and Smart Infrastructure – Brijesh Mishra et al. (eds)
© 2023 Taylor & Francis Group, London, ISBN 978-1-032-41287-0

CHAPTER

5

A Survey on Artificial Intelligence and its Impact on Education System

Anshika Trisal[1], Paninee Bharti[2], Pooja Dehraj[3]

Noida Institute of Engineering and Technology, India

Abstract

Artificial intelligence's (AI) imminent advent and advancement was not a surprise. The deeper the impact of AI on individuals, the more urgent it is for us to comprehend it. This paper begins by providing a complete overview of AI's past and present, taking into account local, regional, and global views. This research examines AI research to determine if any new optimization methodologies or strategies can be used to progress AI research, education, policy, and practise in order to improve human wellbeing. This survey contains two parts. The first part contains the history of AI, and the second part contains impact of AI on education system.

Keywords: Artificial Intelligence, Intelligent Machine, Smart Education System

1. Introduction

Technology is rapidly evolving, and everyone is becoming more familiar with new technology on a daily basis. AI is a burgeoning technology that is poised to develop intelligent machines. Artificial intelligence (AI) is one of the most exciting and general areas of computer science, with a bright future ahead of it. AI has the tendency to make a machine work like a person.

Figure 5.1 explains that Artificial refers to something created or manufactured by humans rather than something that occurs naturally, especially when it is a duplicate of something natural, and the ability to acquire and use knowledge and skills is referred to as intelligence. AI

Corresponding author: [1]anshutrisal@gmail.com, [2]panineebharti26@gmail.com, [3]drpooja.cse@niet.co.in

DOI: 10.1201/9781003357346-5

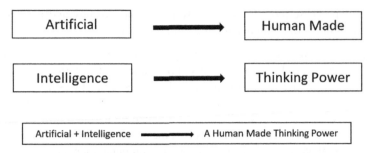

Fig. 5.1 Artificial Intelligence

arises when a machine learns, has a reason, and solves problems in the same way that humans do. According to Greek myth, mechanical men that could work and behave like humans existed in ancient times, hence AI is generally thought to be an old technology.

2. Literature Review

In several ways, the world we live in today resembles the Wonderland envisioned in British mathematician Dodgson's classic works, sometimes referred to as Carroll's works. Artificial intelligence has allowed for the development of robotics, smart devices, and self-driving cars. AI has been a science mired in scientific ambiguity and of limited practical utility since its inception as an educational subject in the 1950s. Along with this, autonomic computing is the AI branch which came into existence in 2001 by IBM. In this approach, the computing system is designed and developed in such a way that the system will automatically handle its internal activities like time-to-time software updates, recovery of system from failure condition, optimization of system resources during processing, and protecting system from unwanted corrupted files. All these activities will be done by the system using self-adaptive features which is a property of AI systems [1-5].

From the 1950s, scientists have predicted that we will achieve Artificial General Intelligence within a few years—systems that display actions that cannot be distinguished from human actions in all domains and contain cognitive, emotional intellect. Only time will tell if this is correct or not. To obtain a better understanding of what is conceivable, we study AI from two perspectives: the path already taken and the road ahead of us. In this editorial, we want to do just that. With the help of the 4 seasons, i.e., AI spring, AI summer and winter, AI fall and The present, we begin by looking at AI's past to see how far this field has progressed, then in the present to accept the confrontation that businesses are facing today, and at last in the future to assist one and all to be ready for the confrontation that lies ahead.

AI Spring: In 1940, a robot was developed that followed the three laws of robotics by Powell and Donavan. The first law states that the humans must not be harmed by the robots, the second law states that the orders given by humans must be followed by the robots in contradiction to the first law, the third law states that the existence of robots must be protected unless first and second laws fail. These three laws were given by Isaac Asimov and it motivated many researchers in the field of AI, computer science, and robotics.

Around the same time, Alan Turing was working on fanciful concerns and designed "The Bombe," a code-breaking machine for the British government to break the Enigma code and is considered as the first working automatic machine. Before Bombe, it was not possible to break the Enigma code even for the greatest researchers. During 1950, Turing published the seminal paper "Computing Machinery and Intelligence," in which he narrated how to develop smart computers and how to access them. If a person interacts with another person and a computer and it is able to differentiate between the person and the computer, then the computer is deemed to be smart; by this test, one has access to check the smartness of an artificial computer.

In 1956, AI first originated when Minsky and McCarthy financed the 6-to-8-week Dartmouth Summer Research Project on AI. This research, which started off the AI Spring, brought together individuals who are now known as "Fathers of Artificial Intelligence". Contestants included Nathaniel Rochester, who then created the 1st commercial research machine "IBM 701", and Claude Shannon (a mathematician) who initiated information theory. The reason for the research was to bring together academics from many disciplines to establish a new research field focused on developing robots that could replicate the intelligence of a human being.

AI Summer and Winter: The famous ELIZA computer programme, developed by Weizenbaum at Massachusetts Institute of Technology between 1964 and 1966, is a classic illustration of The Dartmouth Conference. One of the first computers to pass the Turing Test was ELIZA, an NLP programme that could mimic human speech. The General Problem Solver (GPS) programme, created by Shaw and Newell in the early days of AI, could solve some fundamental issues spontaneously.

But unfortunately, this was not the case. The United States Congress began to express strong resistance to the large investment on research in AI in the year 1973. Again in 1973, Lighthill (The British mathematician) questioned the positive attitude of AI researchers in a report for the British Science Research Council. Robots can only reach the height of a "seasoned amateur" at games like chess, according to Lighthill, and normally utilized reasoning would always be beyond them. As a result, the Britain government cut all but three institutions' funding for AI research. The AI Winter started. In the 1980s, the Japanese government began substantially sponsoring the research in AI and the DARPA responded by increasing money, but little progress was made after that.

AI Fall: One clarification for less growth in the field of AI and the real life lags substantially behind expectations is the exact way like ELIZA and the GPS sought to emulate human intellect in early systems. They all were Expert Systems, which consist of rules that conclude human intellect can be codified and rebuilt from the top to down as a sequence of "if–then" rules. To let Expert Systems do well in fields that adds itself to formalization, let us take an example and consider Deep Blue chess-playing AI of IBM that defeated world champion Gary Kasparov in 1997, disproving Lighthill's assertions. Deep Blue could think two hundred million different moves and identify the finest next move 20 steps ahead utilizing a technique known as tree search.

Expert Systems have difficulty in domains where formalization is not viable. For example, it is not an easy task for Expert System to train for recognizing faces or to differentiate between two images. For such tasks, a system must be able to comprehend every information and to learn from that information and use it whenever required to achieve specified goals and through flexible adaptation, all of which are characteristics of AI. Expert Systems require these characteristics and are hence not considered as true Artificial Intelligence Systems. Since 1940, methods to achieve true AI have been researched, when Donald Hebb (Canadian psychologist) developed Hebb's Rule, a theory of learning that replicates the human brain. As an outcome, the science of Artificial Neural Networks was established. In 1969, Minsky and Paper proved that machines require the processing power to perform the job required by ANN, and the Endeavour was put on hold.

In 2015, ANN made a reappearance in the shape of deep neural networks when Google's AlphaGo defeated the world champion in the board game (Go). Go is far more difficult than chess, with only 20 possible movements to begin with compared to 361 in chess, but computers have never been able to overcome humans in this game. AlphaGo uses Deep Learning, a sort of artificial neural network, to achieve its outstanding results. The majority of AI applications that we are familiar with currently use DL and ANN. They are used in picture identification algorithms on Facebook, as well as speech recognition algorithms in smart speakers and self-driving automobiles. The current AI Fall season is the fruit of prior statistical discoveries.

The Present: These articles begin by examining the interaction between employers and employees, as well as the impact of AI on the labor market as a whole. HR management is characterized by an upper degree of difficulty and relatively uncommon occurrences, both of which have severe repercussions for both personnel and the company. These features provide difficulties in the data production, ML, and decision-making stages of AI systems. The researchers examine these issues, provide suggestions about when humans should take the command and when the AI and explore how workers are likely to respond to various tactics.

The relative significance of mechanical activities, cognitive tasks, and feeling tasks for different job categories is examined in this article. Because thoughtful jobs will be taken over by computers, these authors show through actual research that human employees will be more engaged with sentiment activities in the future.

The papers in this special issue examine the relationship within a company and its consumers, with a focus on the use of AI in trade. V Kumar, Bharath Rajan, Rajkumar Venkatesan, and Jim Lecinski explain how AI can help with the automated robot choice of items, costs, web pages, and marketing campaigns that are tailored to the preferences of each individual customer in their paper "Recognizing Artificial Intelligence's Place in Customized Interaction Advertising." They have to get into great detail about how personalization and data duration affect advertising and customer-oriented schemes for companies in both growth and emerging markets.

Finally, in "Allowing Computers to Take Over: Using Artificial Intelligence to Solve Key Challenges," Gijs Overgoor, Manuel Chica, William Rand, and Anthony Weishampel present a model for how AI may help us with our marketing. Three case studies are used to apply this concept to problems that many businesses confront today. It is focused on the gathering of business and data intelligence, data preparation and modeling, as well as the analysis and implementation of products.[6].

3. Impact on Education System

Impact of Artificial Intelligence on Education System: AI now offers a huge impact on the educational system. AI in the education system seems to have become easier to apply due to the rapid advancement of computer technologies. AI in the education system explains how AI technology can be applied in educational contexts to restructure activities, training, and evaluation easier. In education, AI is used to support a smart learning system.

AI and robotics are also expected to change the world the way education is delivered, affecting how students learn, the roles of instructors and researchers, and the way universities operate as institutions. AI and robotics will very certainly have a tremendous influence on higher education [11]. Cobots in collaboration with teachers or colleague robots (cobots), according to Timms, are being used to teach children basic skills while also adapting to their nature [12]. Natural language translation, face unlocking, and virtual reality are just a few of the new applications that AI and machine learning have recently been researching with the purpose of enhancing computing quality and allowing new applications [13].

The development and adoption of new technologies in learning and teaching has accelerated in the previous 30 years. AI) is also enhancing daily equipment and instruments in cities and universities around the world. Self-driving technology has advanced to the point where certain large firms, like Tesla, Volvo, Mercedes-Benz, and Google, have made it a top priority for development. Trials on public roads in Australia began in 2015 [14].

Vision: While AI in education has been recognized as a main research subject, the varied environment represents a major challenge for academics and in areas of computers and education. This paper explains the purpose of AI in education as a Smart Learning System from the viewpoint of educational demands. We present a framework to demonstrate the discussion of AI deployment in Education to assist students in learning, faculty members in teaching, and administrators in making decisions. In education, AI has opened up new possibilities in order to create effective designs and teaching methods, especially generating improved technology-assisted education apps or contexts. Most academicians and researchers in the disciplines of technologies and education find it difficult to put relevant activities or systems in place.

Challenges: A primary goal of AI in education is to provide individualized learning assistance or help to specific pupils based on their current academic standing, desire, or individual components. The most significant difficulty of AI in education is preparing instructors for AI-powered education and preparing AI to understand education. In the field of education, one of the most significant purposes of AI is to deliver tailored learning ideas or aids to individual pupils based on their educational achievement and beliefs. Computer programming skills are not the only requirement; strategies to replicate the brains of particular experts are also obstacles in the development of smart tutoring and adaptive learning systems. An AI programmer, for example, may act as a tutor, observing students' learning processes, measuring their efficacy, and providing timely support based on their needs. A multidisciplinary team (for example, composed of technology and cognitive researchers) can develop a smart learning system that allows students to study and perform, especially engage with classmates or instructors while providing hints, assistance, and support to individuals based on their needs and conditions. Knowing the capabilities and features of AI technologies, it also aids instructors in incorporating appropriate AI technologies into their lectures to improve students' learning outcomes, enthusiasm, or involvement, while educationists may investigate the effects of AI technologies. In the next portion, we provide a context for understanding the functions of AI in education as a smart learning system, as well as the explanation and qualities of AI technologies [7].

Roles and Framework of Artificial Intelligence in Education: In terms of educational applications, AI can serve as a smart tutor, smart tutee, smart assessment, smart learning tools, or policy-maker, as indicated in Figure 5.2. Many AI education investigations have been reported by researchers in recent decades. These studies can be classified into five categories:

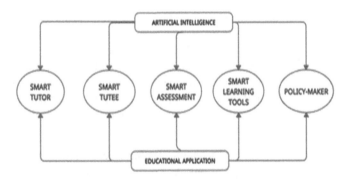

Fig. 5.2   Framework of Smart Learning System

1. **Smart Tutor:** This could be the most prominent AI in Education application category. This area includes smart learning systems, adaptive/personalized learning systems, and recommendation systems. Teachers can use the Smart Learning System to constantly modify teaching material so that it can be personalized to students' evolving understandings, as well as analyze student performance so that teachers can improve their instruction. Students can take charge of their learning, see how they compare to their peers, and take tests to keep up with their colleagues' development, while teachers can use the smart learning technology to prevent differences in learners' prerequisite expertise and crucial study abilities. Smart learning technology may encourage learners to gain experience and competencies in a more individualized and imaginative manner, giving them a clearer picture of their progress and what they need to do to achieve their educational objectives.

Assessments, a more modern example, combines smart learning and assessment capabilities to deliver instantaneous feedback to students while they work on assignment, as well as information summaries for teachers on each assignment [8].

2. **Smart Tutee:** Most AI-based educational platforms focus on supporting learners rather than giving possibilities for learners to serve as tutors or advisors, hence studies in this category are rare. Fascinating students in the context of assisting everyone (i.e. AI tutees or smart tutees) in understanding complicated concepts, on the other hand, could be a great way to improve their relatively high reasoning skills and knowledge levels [9].

3. **Smart Assessment:** Enhanced assessment of students' learning process, such as pre-class previews, observation, oral presentations, assignments, specific courses, and programme writing, as well as the inclusion of an examination procedure in the smart evaluation system, can aid in smart assessment [10].

4. **Smart Learning Tools:** The deployment of smart learning tools is a critical effect from the constructivism and student-centered learning perspectives. Students can use technology to more efficiently and effectively collect and analyze data, allowing them to concentrate upon more important issues or high reasoning skills (for example, observation, and estimation) instead of just reduced operations (e.g., correction and computation). Certain technologies even has the ability to evaluate and present data in a "smarter" manner, allowing learners to go deeper into the data and uncover significant insights. Traditional problem-solving technologies, for example, let learners organize their knowledge by automatically modeling the relationships between concepts, which helps them achieve organizational goals. A smart concept mapping tool, on the other hand, may suggest as well as provide advice to participants as well as analyze the visualizations produced, however, during the process.

5. **Policy maker:** In recent years, AI technologies have been used to inform and influence the development of policy or regulations. As a result, developing a policy-making advisor for educational policy development is both viable and realistic. Policymakers may benefit from understanding educational opportunities and issues from both simple and complex perspectives with the use of AI technology, which can aid them in developing and evaluating effective educational policies.

4. Conclusion

The aim of this survey was to examine the influence of AI on education and explain roles and framework of AI in education. As a result of the use of various platforms and technologies, teachers' efficacy and efficiency have increased or improved, resulting in a higher instructional quality. The preceding explanation demonstrates that AI would develop into a much a part of daily experiences as the World Wide Web or social networking sites did previously. As a result, AI will have a significant impact not only on our personal lives, but also on education system. Similarly, AI has improved learning experiences for students by allowing them to customize and personalize learning materials depending on their needs and abilities. AI has had a considerable impact on education, notably in the learning sectors of the education system and inside individual learning institutions.

REFERENCES

1. Dehraj, Pooja, and Arun Sharma. "Complexity assessment for autonomic systems by using a neuro-fuzzy approach." Software Engineering. Springer, Singapore, 2019. 541–549.
2. Dehraj, Pooja, and Arun Sharma. "Autonomic computing based trustworthiness attributes weight estimation using Electre-Tri method." Journal of Interdisciplinary Mathematics 23.1 (2020): 21–29.
3. Dehraj, Pooja, and Arun Sharma. "A new software development paradigm for intelligent information systems." International Journal of Intelligent Information and Database Systems 13.2-4 (2020): 356–375.
4. Dehraj, Pooja, and Arun Sharma. "Autonomic Provisioning in Software Development Life Cycle Process." Proceedings of International Conference on Sustainable Computing in Science, Technology and Management (SUSCOM), Amity University Rajasthan, Jaipur-India. 2019.
5. Dehraj, Pooja, and Arun Sharma. "An empirical assessment of autonomicity for autonomic query optimizers using fuzzy-AHP technique." Applied Soft Computing 90 (2020): 106137.
6. Haenlein, Michael, and Andreas Kaplan. "A brief history of artificial intelligence: On the past, present, and future of artificial intelligence." *California management review* 61.4 (2019): 5–14.
7. Hwang, Gwo-Jen, et al. "Vision, challenges, roles and research issues of Artificial Intelligence in Education." (2020): 100001.
8. Yang, Stephen JH, et al. "Human-centered artificial intelligence in education: Seeing the invisible through the visible." Computers and Education: Artificial Intelligence 2 (2021): 100008.
9. Wolf, Marty J., Keith W. Miller, and Frances S. Grodzinsky. "Why we should have seen that coming: comments on microsoft's tay "experiment," and wider implications." *The ORBIT Journal* 1.2 (2017): 1–12.
10. Hsiao, Chia-Chang, ct al. "Exploring the effects of online learning behaviors on short-term and long-term learning outcomes in flipped classrooms." *Interactive Learning Environments* 27.8 (2019): 1160–1177.
11. Cox, Andrew M. "Exploring the impact of Artificial Intelligence and robots on higher education through literature-based design fictions." International Journal of Educational Technology in Higher Education 18.1 (2021): 1–19.
12. Michael J. "Letting artificial intelligence in education out of the box: educational cobots and smart classrooms." *International Journal of Artificial Intelligence in Education* 26.2 (2016): 701–712.
13. Chen, Lijia, Pingping Chen, and Zhijian Lin. "Artificial intelligence in education: A review." *Ieee Access* 8 (2020): 75264–75278.
14. Popenici, Stefan AD, and Sharon Kerr. "Exploring the impact of artificial intelligence on teaching and learning in higher education." *Research and Practice in Technology Enhanced Learning* 12.1 (2017): 1–13.

Intelligent Systems and Smart Infrastructure – Brijesh Mishra et al. (eds)
© 2023 Taylor & Francis Group, London, ISBN 978-1-032-41287-0

CHAPTER

6

Role of Sensor-Based Insole as a Rehabilitation Tool in Improving Walking among the Patients with Lower Limb Arthroplasty: A Systematic Review

Sumit Raghav*

Department of Physiotherapy,
Lovely professional University,
Punjab, India
Subharti College of Physiotherapy,
Swami Vivekanand Subharti University,
Uttar Pradesh, India

Anshika Singh

Subharti College of Physiotherapy,
Swami Vivekanand Subharti University,
Uttar Pradesh, India

Suresh Mani*, Amber Anand

Department of Physiotherapy,
Lovely Professional University, Punjab, India

Shashwat Pathak

Department of Electronics and Communication
Engineering, MIET, Uttar Pradesh, India

Gokulakannan Kandasamy

Department of Psychology,
Sports and Exercise
Teesside University, Middlesbrough, UK

Mukul Kumar

Central Research and Incubation Centre,
Swami Vivekanand Subharti University, Uttar Pradesh, India

Abstract

The primary issue that a patient has lower limb arthroplasty is difficulty in walking. Recent advancements in sensor-based wearable technology enable the monitoring and quantification of gait in patients undergoing lower limb arthroplasty in both the clinical and home environment.

*Corresponding author: suresh.22315@lpu.co.in

DOI: 10.1201/9781003357346-6

The goal of this systematic review would have been to offer an overview of sensor-based insole technology, sensor types, and their usefulness as a rehabilitation tool for enhancing and monitoring walking. Two reviewers, "SR" and "AS", both physiotherapists, did a thorough search utilizing several electronic databases, such as PubMed, Science Direct, Scopus, Google Scholar, and Web of Science, and reviewed titles and abstracts for eligibility. The present systematic review includes fifteen studies (Eligible – 33, Excluded – 18, and Included – 15). All included studies covered the various types of sensor-based insole technologies, outcome measures, data-processing algorithms, and study population. All included studies were categorized and put in a tabular form on the basis of types of sensor-based insole technologies used, type of sensor, and outcome measure that were taken to monitor and quantify the data of gait and activities of lower limb in the patients with mobility impairments. This review summarizes the uses of sensor-based insole technology as a rehabilitation tool for monitoring and quantifying lower extremity activity in individuals who have had lower limb arthroplasty or have another form of mobility limitation.

Keywords: Sensor-based insole, Lower limb arthroplasty, Rehabilitation

1. Introduction

The protocol for this systematic review was previously published, and the following section is an updated and expanded version of that methodology (1). Rehabilitation is a core service in healthcare system after knee arthroplasty and hip arthroplasty as means of enhancing the functional status (2).

Arthroplasty is a very common surgical procedure performed of large peripheral joints such as hip and knee joints of lower extremity in elderly population to reduce the symptoms and improve the function of hip and knee joint (3). After arthroplasty, the common complaints are faced by the patient such as difficulty in symmetrical loading on bilateral lower extremity, difficulty in walking, balancing during walking, and abnormal gait parameters (4). Such complaints are managed through rehabilitation with various therapies and training (5). There are various rehabilitation programs like gait training, balance training, and weight transfer activities delivered to the patients who underwent hip and knee arthroplasty (6).

Due to lack of appropriate and advanced resources, the recovery rate after surgery shows negative impact in terms of independency and functional status of the patients (7). Evidence-based practice is critical and well-established in the discipline of physiotherapy (8). Gait labs, sensor-based mat, and force plates are already available in the reputed institutions and hospitals to provide the facilities for patients in order to improve their walking pattern and functions on real-time basis (9). Fully equipment-assisted gait labs are very costly from the point of view utilization of these technologies at clinical setting for the patients (10). Nowadays, sensor-based shoe and insole are in trend to obtain the real-time data of gait during static and dynamic condition of lower extremities in order to speed up the recovery rate and improve the quality of rehabilitation (11). Recently, innovative technology-assisted rehabilitation services have been introduced and used by healthcare workers and patients such as e-health, wearable sensor technology, telemedicine, tele-rehabilitation services, and online educational tools (12). Recent artificial intelligence (AI) research suggests that it can also help in exercise tracking, and the development of non-invasive sensors for monitoring the walking pattern (13). Many literatures have published on design, development, and testing of wearable and non-wearable sensor insole to monitor health outcomes such as gait parameters, limb loading, kinetics, and kinematics of spine and joints of lower limb and plantar pressure mapping (14). Thus, the goal of this review is to summarize the available information and to assess the significance of sensor-based insole therapy in addressing mobility deficits in patients who have had lower limb arthroplasty, such as knee or hip arthroplasty.

2. Methodology

A. Participants

Human participants involved only adults between 45 and 75 years of age undergone hip, knee, and ankle arthroplasty in this systematic review. Both genders, i.e. males and females, were included. Studies involving patients suffering from other serious acute/chronic illness were excluded from this systematic review.

B. Included Study Design

All types of study that described the role of sensor-based insole or sensor insole technology such as wearable sensor device or non-wearable sensor device in improving walking were included in this systematic review. Studies that involved technology-assisted intervention for rehabilitation were included in this review. The gait parameters and function were included as first-line measures in included studies of this review, whereas secondary outcome parameters were pain, quality of life, user's satisfaction, and safety.

C. Interventions and Comparators

Some studies were conducted on monitoring the gait parameters and functional status in patients with mobility impairments due to musculoskeletal disorders using wearable sensor technology in gait labs. Few studies were done on application and uses of sensor-based device to evaluate the gait and posture among the patients with hip and knee arthroplasty. Accelerometer, gyroscope, and magnetometer are the most common devices used for monitoring the different phases of gait cycle in patients with walking impairments due to various musculoskeletal or neuromuscular disorders

D. Outcome Measures

From all included investigations, the outcome measures for measuring and improving gait characteristics such as step length, stride length, step duration, step time and cadence, the kind and location of sensor technology, as well as the participants and research design were collected. The best suggested methodology or method covered in this evaluation examined many wearable sensor technologies in conjunction with a data-driven algorithm.

E. Search Strategy

All required databases were checked from years 2001 to 2021. We performed a comprehensive search of the following electronic databases: Google Scholar, PubMed/Medline, Science Direct, Scopus, and Web of Science. A thorough search was conducted at the library of Lovely Professional University (LPU), Punjab and the Library of Swami Vivekanand Subharti University (SVSU), Meerut, India during a span of approx. 2 years. Readily accessible published research articles were assessed from the mentioned databases.

F. Study Records

All the searched results were combined using "Mendeley" information management software. The results of online searches were also saved to the PUBMED account of the researcher. A shared folder on "Google drive" was developed by the principal investigator/researcher to promote and facilitate collaboration among reviewers and make it accessible by all researchers.

G. Selection and Data Collection of Studies

A thorough search was conducted using various electronic databases and screened the eligibility of titles and abstracts by two reviewers "SR" and "AS", both physiotherapists. The full text of selected potentially relevant articles was obtained to minimize duplication, and multiple articles from the same study were linked on different databases. Both reviewers must extensively analyze full text articles to verify their conformity according to the requirements for inclusion and exclusion.

H. Best Evidence Synthesis

Due to the possible heterogeneity of the outcome measures, a narrative synthesis of the selected studies from the search results was given. Population and outcome measures for patients were described in a summary narrative. Information on protocol adherence, resources used, compliance monitoring, and expenditure where available were extracted from the selected studies.

3. Results

The systematic review identified 247 papers, 11 of which were included using citation scanning of previous published publications. A total of 33 papers were retained for full-text screening, and ultimately, 15 studies in Table 6.1 met the systematic review's stated inclusion criteria. Fig. 6.1 depicts the screening method in its entirety. The study population was the primary cause for exclusion, accounting for 33% of all excluded publications. Comprehensive and systematic overview of using sensor-based wearable technology, parts of body sensors applied, target population and the sensors used to evaluate the outcome measure related to mobility of lower limb (16).

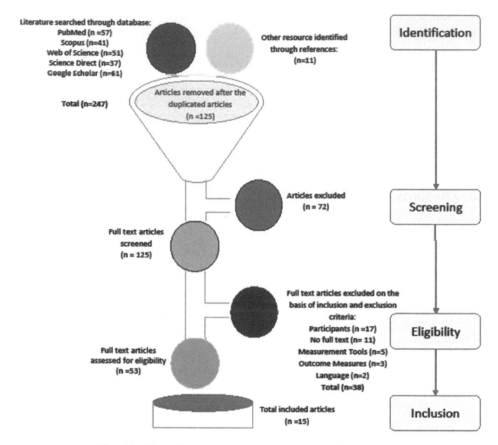

Fig. 6.1   Flow diagram of process adopted to filter articles

They provide good opportunities to overcome the economic burden in order to establish gait labs for assessment and rehabilitation purpose. Accelerometer, magnetometers, gyroscope, biomarkers, and wearable cameras measure the physical responses in addition to walking (17). Fig. 6.2 provides an outline of the sensor technologies employed, the sensor location on the body areas, and sample population used to analyze the performance metrics.

A. Sensor Technologies

Pressure sensors and motion sensors (18) were often used in previous published literatures. Sensor-based devices such as accelerometer, gyroscope, or IMUs were used in the previous studies. Three studies were conducted on the use of temperature sensor adjunct with pressure sensor for monitoring the body's segment temperature along with mobility and pressure mapping of foot (19) simultaneously. Five studies were done on the working and testing of the accelerometer along with gyroscope for motion analysis and detecting orientation of foot in the patients with mobility impairment (20) after surgical procedure performed in lower limb.

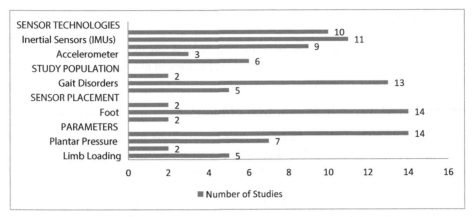

Fig. 6.2 Frequency distribution of the used sensor technologies, sensor placement on the body parts of the study population, and the parameters used

B. Sensor placement

The sensors such as FSRs, accelerometers, gyroscope, and IMUs were most frequently placed in heel, mid-foot, and great toe region of the foot (21). The location of the sensor on the body was determined by the resultant measurements. In investigations that analyzed gait metrics, researchers chose an insole sensor or a combo of ankle sensors. In comparison, studies that examined outcomes linked to hip, knee, and ankle joint mobility in the lower extremities put sensors on the hip, with a particular fondness for knee as well as ankle sensors (22).

C. Study Population

Wearable sensors, FSRs, accelerometers, gyroscope, and IMUs were most frequently applied in underwent hip, knee, and ankle joint surgeries as well as after hip, knee, and ankle arthroplasty (23). Large range of application of wearable sensor-based technology to quantify the gait parameters in people with hip, knee, and ankle arthroplasty were identified inn six different previously published articles (24).

D. Parameters

All outcome measures related to gait parameters such as spatiotemporal parameters such as step length, stride length, double-limb support, single-limb support, cadence, step width, and detecting the range of motion of the joints (25) of lower extremity were often described in the previous studies. The various parameters related to kinetics and kinematics of lower limbs and gait of the patients with mobility impairments were detected in different positions. One study included the single measurement in terms of assessing the angle of knee. Besides obtaining the period of these different positions of body, it was also to assess the transition of body from sitting to standing and walking (26).

4. Discussion

This review provides the comprehensive data that reflects on the role and importance of existing wearable and non-wearable sensor-based technology in monitoring and rehabilitating the patients with mobility problems.

A. Sensor-Based Insole in Prevention of Injuries

Sensor-based insole helps in monitoring the gait impairments and thus helps in changing the walking patterns. These have shown the changes in angle of joints, loading response of lower limb, range of motion, and kinetics of hip and knee joint. Usage of insoles affects the gait parameters in such patients and has effect on preventing gait-related abnormalities (27). Acceleration-based assessment accurately recognized acute changes in gait parameters (e.g., speed, step length, step duration, periodicity, vertical distance, asymmetries, and irregularities) amid simulated operational knee restrictions with high repeatability (28).

B. Sensor Insole Technology for Gait Analysis

Few studies advocated that gait analysis being used as the primary evaluation method for managing the patients with osteoarthritis of knee and hip joint (29). Fully instrumented traditional gait lab is very expensive and time-consuming and requires specialist to operate while attempting the patients. Body markers, cameras, body-worn electrodes, and sensor mat are used to assess the gait analysis and data recorded by the computer for post-evaluation analysis. Utilizing ubiquitous internet connectivity, the smart insole system may transmit data from the patient's shoe to centralized servers inside the medical sector, allowing caretakers to monitor the user's stride in real time and over extended periods of time (30).

C. Sensor-Based Insole as a Rehabilitation Tool

Recently, sensor-based rehabilitation in form of motion device, smart insole, smart shoe, smart socks and other wearable and non-wearable technology has been developed and used in healthcare setup by clinicians and at home by the patients for self-monitoring of their health status. Pressure sensors of the FSR type are placed into the shoe insole to collect and analyze data on gait characteristics, such as step length, stride length, and velocity before and after rehabilitation. Some challenges were also faced, such as low-quality connection of internet at patient's home and delaying in the process of technology installation. Additionally, old-age patient may experience problems while using the technology due to lack of feasibility with such type of technology. These challenging factors highlighted the need for cost-effective and user-friendly technology, especially for elderly population and sufficient features (31).

5. Conclusion

The present review summarizes the uses of sensor-based insole technology as a rehabilitation tool for monitoring and quantifying lower extremity activity in persons who have had lower limb arthroplasty. This review discussed pertinent literature and the essential sensor-based insole methodologies. Thus, this systematic review summarizes established sensor insole-based applications, and offers instant access to pertinent literary work for other researchers or readers intrigued in monitoring and quantifying gait on real time.

Conflict of Interest: The authors have no conflict of interest.

REFERENCES

1. Raghav S, Singh A, Mani S, Kandasamy G, Anand A. The Role of Sensor Based Insole as a Rehabilitation Tool in Improving Walking among the Patients with Lower Limb Arthroplasty: A Protocol for Systematic Review. Asian J Med Heal. 2020;

2. Fink J, Iorio R, Clair AJ, Inneh IA, Slover JD, Bosco JA, et al. CARE REDESIGN: PLANNING FOR REHAB IN A BUNDLED MODEL OF CARE WORLD. GeriNotes. 2013;

3. Galitzine S. Refresher course: Regional anaesthesia and postoperative pain management for total hip and knee replacement. Reg Anesth Pain Med. 2015;

4. Than P, Kránicz J, Bellyei A. Surgical complications and their treatment options in total knee replacement. Orv Hetil. 2002;

5. Leijendekkers RA, van Hinte G, Nijhuis-van der Sanden MW, Staal JB. Gait rehabilitation for a patient with an osseointegrated prosthesis following transfemoral amputation. Physiother Theory Pract. 2017;

6. Koseki K, Mutsuzaki H, Yoshikawa K, Endo Y, Kanazawa A, Nakazawa R, et al. Gait Training Using a Hip-Wearable Robotic Exoskeleton After Total Knee Arthroplasty: A Case Report. Geriatr Orthop Surg Rehabil. 2020;

7. Dunbar M, Fuentes A, Macdonald H, Whynot S, Richardson G. Gait analysis - objective differentiator of surgical and non-surgical candidates for arthroplasty. J Orthop Res. 2016;

8. Saha P, Yangchen T, Sharma S, Kaur J, Norboo T, Suhail A. How Do Physiotherapists Treat People with Knee Osteoarthritis and their evidence awareness: A cross-sectional survey among Indian Physiotherapists. Int J Physiother Res. 2021;

9. Psaltos D, Chappie K, Karahanoglu FI, Chasse R, Demanuele C, Kelekar A, et al. Multimodal Wearable Sensors to Measure Gait and Voice. Digit Biomarkers. 2019;

10. Novel Tools for the Delivery and Assessment of Exercise Programs Adapted to Individuals With Parkinson's Disease. Case Med Res. 2019;

11. Nagano H, Begg RK. Shoe-insole technology for injury prevention in walking. Sensors (Switzerland). 2018.

12. De Geest S, Sabaté E, Walsh JC, Corbett TK, Hogan M, Duggan J, et al. Interventions for enhancing medication adherence (Review) Interventions for enhancing medication adherence. JMIR mHealth uHealth. 2017;

13. Sharma KK, Pawar SD, Bali B. Proactive Preventive and Evidence-Based Artificial Intelligene Models: Future Healthcare. In 2020.

14. Small SR, Bullock GS, Khalid S, Barker K, Trivella M, Price AJ. Current clinical utilisation of wearable motion sensors for the assessment of outcome following knee arthroplasty: a scoping review. BMJ Open. 2019.

15. Ostadabbas S, Nourani M, Saeed A, Yousefi R, Pompeo M. A knowledge-based modeling for plantar pressure image reconstruction. IEEE Trans Biomed Eng. 2014;

16. Kataoka Y, Homan K, Takeda R, Tadano S, Chiba T, Tohyama H. The effects of unweighting by a lower body positive pressure treadmill on 3-D gait kinematics. J Orthop Res Conf. 2017;

17. Sim I. Mobile Devices and Health. N Engl J Med. 2019;

18. Jung Y. Hybrid-aware model for senior wellness service in smart home. Sensors (Switzerland). 2017;

19. Lutjeboer T, Van Netten JJ, Postema K, Hijmans JM. Validity and feasibility of a temperature sensor for measuring use and non-use of orthopaedic footwear. J Rehabil Med. 2018;

20. Staab W, Hottowitz R, Sohns C, Sohns JM, Gilbert F, Menke J, et al. Accelerometer and gyroscope based gait analysis using spectral analysis of patients with osteoarthritis of the knee. J Phys Ther Sci. 2014;

21. Giovanelli D, Farella E. Force Sensing Resistor and Evaluation of Technology for Wearable Body Pressure Sensing. J Sensors. 2016;

22. Totaro M, Poliero T, Mondini A, Lucarotti C, Cairoli G, Ortiz J, et al. Soft smart garments for lower limb joint position analysis. Sensors (Switzerland). 2017;

23. ZHOU Z, WANG J. Shoes and Leg Braces Integrated System for Abnormal Gait Detection. DEStech Trans Comput Sci Eng. 2017;

24. Rapp W, Brauner T, Weber L, Grau S, Mündermann A, Horstmann T. Improvement of walking speed and gait symmetry in older patients after hip arthroplasty: A prospective cohort study. BMC Musculoskeletal Disorders. 2015.

25. Mentiplay BF, Perraton LG, Bower KJ, Pua YH, McGaw R, Heywood S, et al. Gait assessment using the Microsoft Xbox One Kinect: Concurrent validity and inter-day reliability of spatiotemporal and kinematic variables. J Biomech. 2015;

26. Beauchet O, Launay CP, Sekhon H, Gautier J, Chabot J, Levinoff EJ, et al. Body position and motor imagery strategy effects on imagining gait in healthy adults: Results from a cross-sectional study. PLoS One. 2018;

27. Abdul Razak AH, Zayegh A, Begg RK, Wahab Y. Foot plantar pressure measurement system: A review. Vol. 12, Sensors (Switzerland). 2012. p. 9884–912.

28. Zhao H, Wang Z, Qiu S, Shen Y, Wang J. IMU-based gait analysis for rehabilitation assessment of patients with gait disorders. In: 2017 4th International Conference on Systems and Informatics, ICSAI 2017. 2017.

29. Wesseling M, Meyer C, Corten K, Desloovere K, Jonkers I. Longitudinal joint loading in patients before and up to one year after unilateral total hip arthroplasty. Gait Posture. 2018;

30. Xu W, Huang MC, Amini N, Liu JJ, Electrical LH, Sarrafzadeh M. Smart insole: A wearable system for gait analysis. In: ACM International Conference Proceeding Series. 2012.

31. Lie DYC, Nukala BT, Jacob J, Tsay J, Shibuya N, Lie PE, et al. The design of robust real-time wearable fall detection systems aiming for fall prevention. In: Activities of Daily Living (ADL): Cultural Differences, Impacts of Disease and Long-Term Health Effects. 2015.

Intelligent Systems and Smart Infrastructure – Brijesh Mishra et al. (eds)
© 2023 Taylor & Francis Group, London, ISBN 978-1-032-41287-0

CHAPTER

Junctionless Field Effect Transistor: A Technical Review

Mohd. Shadan[1], Hritik Goel[2], Aryan Singh Rana[3], M. M. Singh[4], V. Varshney[5]

Department of ECE, Meerut Institute of Engineering and Technology, Meerut-250005, India

D. K. Singh[6]

Shambhunath Institute of Engineering and Technology, Jhalwa, Prayagraj 211015, India

Abstract

In this article, we have studied the different junctionless field effect transistors (JL-FETs). A list of papers has been studied and concluded on the basis of their working and applications. Some of them are analysed using SILVACO TCAD. Also, these are derived and reviewed using TCAD, which confirms the technical review on a junctionless transistor. It is derived from all the papers on how it has become a more special device in that particular genre.

Keywords: Junctionless transistors (JLTs), high mobility, silicon FET, FinFET, TCAD

1. Introduction

Limitations of scaling on the previously used field-effect transistors (FETs) are increasing in this era (M. Maiti et al., 2021; A. Rassekh et al., 2022; M. F. M. Rasol et al., 2020; A.

Corresponding author: [1]mohd.shadan.ec.2018@miet.ac.in; [2]hritik.goel.ec.2018@miet.ac.in;
[3]aryan.singh.ec.2018@miet.ac.in; [4]macky4589@gmail.com, manmohan.singh@miet.ac.in;
[5]vikrant.varshney@miet.ac.in; [6]dsdineshsingh012@gmail.com

DOI: 10.1201/9781003357346-7

Baro and K. C. Deva Sarma, 2020, M.M. Singh and M.J. Siddiqui, 2017, M.M. Singh et al., 2017). But the applications of the transistors include the researcher's interest in junctionless transistors (JLT). Junctionless field effect transistor (JL-FET) is an excellent device for this application as it has both the properties of being junctionless to reduce the junction's problems and field effect for introducing the properties of FET. As the name mentions "junctionless", it has no junction and no requirement of doping concentration gradient. JLTs are also known as gated resistor and don't have junction (C. W. Lee et al., 2009). Although it is complex with fundamentals, but due to being junctionless, it overcomes the fabrication limitation with exceptional controllability of gate with providing larger current drive counter to short channel effects.

Previously, junctions are formed between substrate and drain, & substrate and source, that make extremely difficult to fabricate below 90 nm with sharp doping concentration (Colinge J.P et al., 2009). A lot of advantages have been considered like short channel effect, fewer drain current variation on varying doping concentration, and degradation of subthreshold slop (SS) by junctionless transistor over MOSFET's. Carrier concentration typically happens at the centre of its channel of the junctionless-FET device (Asif Islam Khan et al., 2007; Anisul Haque and Mohammad Zahed Kauser, 2002).

Transistor is a device which transfers resister from input to output of the device and its carrier density controlled by the gate (Jacob Millman, 1985). Diffusion is not taken place in the JLT structure as same doping concentration throughout the device from source to drain. Junctionless-FET has been studied and investigated with different geometries and with single, double, gate all around and multigate architecture, and it is found that, it gives suitable switching characteristics. The size of any device plays a vital role in the investigation of its electrical and structural properties, so its small size up to 30–40 nm also makes it very useful but nowadays but nowadays technology beyond 32 nm also exist and on comparing, it is bigger in size and complex in nature that make it doubtful.

Remainder of this article is systematized as follows. Section II discusses the basic junctionless structure and its simulation. Section III discusses the review of different types of JLTs and its characteristics at varying device geometry. Lastly, Section IV summarizes the paper.

2. Basic Junctionless Structure

In the JL-FETs, we introduce the junctionless bar of silicon (Si), which creates the source and drain over buried oxide layer of undoped substrate; its structure is shown in Fig. 7.1.

On the basis of the above-designed JLT, we have derived the characteristics with drain current versus gate voltages, as shown in Fig. 7.2.

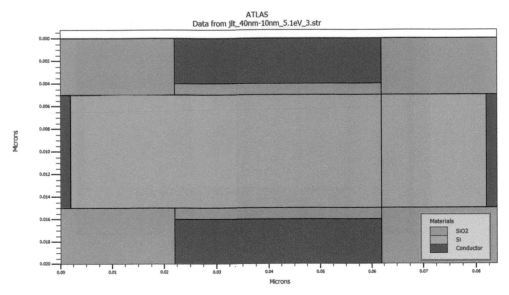

Fig. 7.1   Basic junctionless transistor (JLT) with Si diffused over undoped oxide silicon (SiO$_2$)

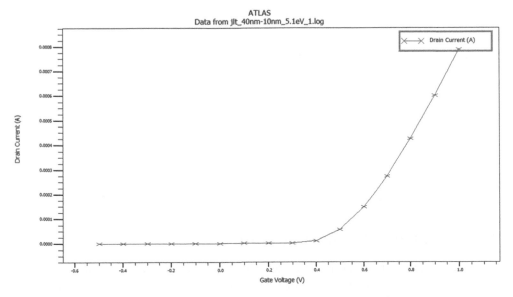

Fig. 7.2   Drain Current versus Gate Voltage Graph

The doping concentration profile and conduction current density profile [Fig. 7.3(a)] provide the overall current conducted along the x-axis of the device, and the electric field throughout the device [Fig. 7.3(b)] shows how the electric field spreads throughout the device on applying voltage; valance and conduction band energy are also shown and studied in the basic structure of device.

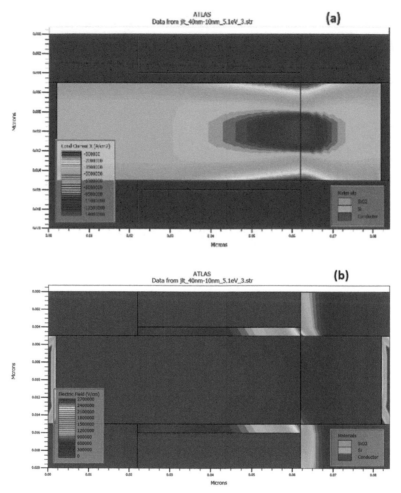

Fig. 7.3   (a) Conduction Current Density versus Device Geometry. (b) Electric Field Throughout the Device at Gate Voltage (VGS = 0 volts)

3. Review of Junctionless Structure

The properties and feasibility of the proposed JLT have been evaluated for both silicon and germanium channels. The performance of silicon junctionless transistors with 22 nm gate length has been characterized at elevated temperature and stressed conditions. It is observed that the floating body in JLT is relatively dynamic compared with that in IM (Inversion Mode) device structure, which may further reduce the voltage drop for a sub-60 mV/dec subthreshold slope. For providing delay in analog circuits, non-planer junctionless FETs have been used with double gate with bulk current (M. Sen et al., 2019; M.M. Singh and M.J. Siddiqui, 2016; Saxena, A. et al., 2014). To get the high current density at very low voltage, JLTs has been used over conventional field effect transistor in different modes of operations.

Yan et al. found that the characteristics of a diode with junctionless transistors are identical to the normal PN junction diode. For junctionless transistors, the device is fully depleted below threshold. The device performed is determined by its short channel characteristics and analog performance. For every short channel device, junctionless transistors exhibit improved short channel characteristics. A very small bias of gate gives a flat band in accumulation that hastily rises the drain current. Similarly, gate controls the resistance of JL-FETs and it acts as gated resistor for any junctionless transistor.

Table 7.1 Summary of Various Articles

S. No.	Abstraction of article	References
1	It is originating that EDD-JLT is more stable and also suggested better inherent gain and saturation physical characteristics that make it capable for application in ultra-low power circuits	Raushan M.A et al. (2019)
2	Junctionless transistors are suitable for future endeavours the same as normal FETs. JLTs can also be used in communication systems	Nowbahari, A etal. (2020); P. Agnihotri and N. Tripathi (2019)
3	With increase in thickness of gate dielectric and channel, reduction in electric field takes place	A. Talukdar and K. C. Deva Sarma (2020)
4	Compatible with different VLSI-based devices and circuits and power electronics as its electrical characteristics and geometry vary accordingly	N. Das and K. C. Deva Sarma (2020)
5	10–20 nm gate length with GaAs-based JGAA JLT gives better performance over existing transistor	M. F. M. Rasol et al. (2020)
6	Because of DP penetrating in the channel region, the reduction in leakage current takes place in off state	M. Maiti et al. (2020)
7	The OBG (Oversized Black Gate) technique shows notable ambipolar behaviour as associated to other technique used to overwhelm ambipolarity like DMG, electrode with higher work function, and DMG DLFETs	M. A. Raushan et al. (2018)
8	A few significant constraints of a parallel gated JL-FET had been studied at geometry/device level	A. K. Raibaruah and K. C. Deva Sarma (2020)
9	For low I_{OFF}, this article uses nanowire of silicon-based junctionless-FET, along with this ION should be high for low-power circuits. The proposed device is an alternative to FinFET devices	P. Cadareanu and P. E. Gaillardon (2021)
10	Architecture and working of junctionless transistor are very simple due to its junction elimination fabricated on non-SOI	S. I. Amin and R. K. Sarin (2013); A. K. Raibaruah and K. C. Deva Sarma (2020)
11	In this article, to enhance the performance of device, we can vary the thickness and work function to achieve higher gain	V. Narula and M. Agarwal (2019)
12	We can calculate/evaluate electron and hole concentrations, electrostatic potential under varying physical parameters	A. Goel et al. (2021)
13	Junctionless can fashion ultra-low leakage currents and remarkable short channel characteristics at petite gate spans	S. I. Amin and R. K. Sarin (2013)

4. Conclusion

The transistor without junction has been used to increase the speed of the device at low power dissipation and low gate applied voltages. Study of different types of JL-FETs is successfully completed and concluded that this device has tremendous advantages and highly appreciable over the conventional MOSFETs. A very low leakage current and less short channel effect has been derived with varying doping concentrations. Hence, junctionless field effect transistor is a promising device for simple architecture, low power consumption, and highly controlled device, which make this device a future of its particular genre.

5. Acknowledgement

Authors are grateful to the Meerut Institute of Engineering and Technology and the Department of Electronics and Communication Engineering for providing us the basic requirements for pursuing research for this research article.

REFERENCES

1. M. Maiti, M. Jain and C. K. Pandey. (2021). Enhanced DC Performance of Junctionless Field-effect Transistor Using Dielectric Engineering. Devices for Integrated Circuit (DevIC), 6: 107–111.
2. A. Rassekh, F. Jazaeri and J. -M. Sallese. (2022). Nonhysteretic Condition in Negative Capacitance Junctionless FETs. Transactions on Electron Devices 69(2): 820–826. IEEE.
3. M. F. M. Rasol, F. K. A. Hamid, Z. Johari, R. Arsat and M. F. M. Yusoff. (2020). Performance Analysis of Silicon and III-V Channel Material for Junctionless-Gate-All-Around Field Effect Transistor. Student Conference on R & D (SCOReD), 1–5.
4. A. Baro and K. C. Deva Sarma (2020). Study on Electrical Characteristics of Double gate Junctionless Field Effect Transistor with Triangle Shaped Spacer. International Conference on Computational Performance Evaluation (ComPE), 601–603.
5. Singh, M.M. and Siddiqui, M.J. (2017). Electrical characterization of triple barrier GaAs/AlGaAs RTD with dependence of operating temperature and barrier lengths. Materials Science in Semiconductor Processing, 58: 89–95. Elsevier.
6. Singh, M.M., Siddiqui, M.J. and Alvi, P.A. (2017). Theoretical investigation of delta-doped double barriers GaAs/AlGaAs RTDs at varying device geometry and temperature dependence. Journal of New Technology and Materials, 7(2): 47–55.
7. C. W. Lee, A. Afzalian, N. D. Akhavan, R. Yan, I. Ferain, and J. P. Colinge. (2009) Junctionless multigate field effect transistor. Applied Physics Letters, 94: 053511.
8. Colinge J.P., Lee, C.W., Afzalian, A., Dehdashti, N., Yan, R., Ferain, I., Razavi, P., O'Neill, B., Blake, A., White, M., Kelleher, A.M., McCarthy B., and Murphy R. (2009). SOI Gated Resistor: CMOS without Junctions. Int. SOI Conf., California, USA.1-2. IEEE.
9. Asif Islam Khan, Mohammad Khalid Ashraf and Quazi Deen Mohammad Khosru. (2007). Effects of wave function penetration on gate capacitance modeling of nanoscale double gate MOSFETs. International Conference on Electron Devices and Solid-State Circuits, 137–140. IEEE.
10. Anisul Haque and Mohammad Zahed Kauser. (2002). A Comparison of Wave Function Penetration effects on Gate Capacitance in Deep submicron n and p-MOSFET. Transaction on Electron Devices, 49(9): 1580–1587. IEEE

11. Jacob Millman (1985). Electronic devices and circuits. Singapore: McGraw-Hill. 384–385. ISBN 978-0-07-085505-2.

12. M. Sen, A. Gatait, S. Ghosh, M. Chanda, S. Roy and P. Debnath. (2019). Verilog-A Modeling of Junction-less MOSFET in Sub- Threshold Regime for Ultra Low-Power Application. Women Institute of Technology Conference on Electrical and Computer Engineering (WITCON ECE), 172–176.

13. Singh, M.M. and Siddiqui, M.J. (2016). Effect of Si-delta doping and barrier lengths on the performance of triple barrier GaAs/AlGaAs resonant tunneling diode. Int. Conf. on Elect. Devices and Solid-State Circuits (EDSSC), 12: 30–34. IEEE.

14. Saxena, A., Singh, M.M. and Singh, I.V. (2014). The state of art of MEMS in automation industries. Proc. of the Inno. Trends in App. Phy., Chem., Math. Sci., and Emerging Energy Tech. for Sust. Develop. (APCMET '14), 1–5.

15. Raushan, M.A., Alam, N. & Siddiqui, M.J. (2019). Electrostatically doped drain junctionless transistor for low-power applications. J Comput Electron 18: 864–871.

16. Nowbahari, A.; Roy, A.; Marchetti. L. (2020). Junctionless Transistors: State-of-the-Art. *Electronics*, 9: 1174.

17. P. Agnihotri and N. Tripathi. (2019). Junctionless Transistors for Future Applications. ITEE Journal, Inf. Tech. & Electrical Engg, 8(4): 50–53.

18. A. Talukdar and K. C. Deva Sarma. (2020). An Analytical Potential Model for Normally on Double Gate Junctionless Field Effect Transistor. International Conference on Computational Performance Evaluation (ComPE), 464–468.

19. N. Das and K. C. Deva Sarma. (2020). Surrounded Channel Junctionless Field Effect Transistor. International Conference on Computational Performance Evaluation (ComPE), 608–610.

20. M. F. M. Rasol, F. K. A. Hamid, Z. Johari, R. Arsat and M. F. M. Yusoff. (2020). Performance Analysis of Silicon and III-V Channel Material for Junctionless-Gate-All-Around Field Effect Transistor. Student Conference on R & D (SCOReD), 1–5. IEEE.

21. M. A. Raushan, N. Alam and M. J. Siddiqui. (2018). Dopingless Tunnel Field-Effect Transistor with Oversized Back Gate: Proposal and Investigation. Transactions on Electron Devices, 65(10): 4701–4708. IEEE.

22. A. K. Raibaruah and K. C. Deva Sarma. (2020). Parallel Gated Junctionless Field Effect Transistor. International Conference on Computational Performance Evaluation (ComPE), 178–181.

23. P. Cadareanu and P. -E. Gaillardon. (2021). A TCAD Simulation Study of Three-Independent-Gate Field-Effect Transistors at the 10-nm Node. Transactions on Electron Devices, 68(8): 4129-4135. IEEE.

24. S. I. Amin and R. K. Sarin. (2013). Junctionless transistor: A review. Third International Conference on Computational Intelligence and Information Technology (CIIT 2013), 432–439.

25. V. Narula and M. Agarwal. (2019). Correlation between Work Function and Silicon Thickness of Double Gate Junctionless Field Effect Transistor. Women Institute of Technology Conference on Electrical and Comp. Engg. (WITCON ECE), 227–230.

26. A. Goel, S. Rewari, S. Verma, S. S. Deswal and R. S. Gupta. (2021). Dielectric Modulated Junctionless Biotube FET (DM-JL-BT-FET) Bio-Sensor. Sensors Journal, 21(15): 16731–16743. IEEE.

Intelligent Systems and Smart Infrastructure – Brijesh Mishra et al. (eds)
© 2023 Taylor & Francis Group, London, ISBN 978-1-032-41287-0

CHAPTER

8

Dual-Port MIMO Antenna with Gain and Isolation Enhancement for 5G Millimeter Wave Applications

Amrees Pandey*, Navendu Nitin, Piyush Kumar Mishra, Aditya Kumar Singh, J. A. Ansari

Department of Electronics and Communication, University of Allahabad, Prayagraj, Uttar Pradesh 211002, India

Abstract

A low-profile ($30 \times 25 \times 1.575$ mm^3), SK dual-port MIMO (multiple-input-multiple-output), chalice glass-shaped microstrip antenna is inspected for bandwidth, gain, and isolation enhancement. Whole antenna models (A_1–A_4) are methodically calculated, where A_4 is the proposed model and preferred antenna for desired applications. The proposed band characteristics at (23.69–25.23) & (27.46–29.57) GHz with impedance bandwidth of 6.29 & 7.39%, respectively, for port-1 and (23.53-25.26) & (27.43–30.46) GHz with impedance bandwidth of 7.09 & 10.46%, respectively, at port-2 are observed. Parallel slits are introduced on the radiating patch to achieve dual-resonances of 24.30 & 28.30 GHz with peak gain of 9.79 & 9.08 dBi, respectively, at both ports-1 & 2 being the same and perfectly matched. Isolation less than −27 dB, Diversity Gain (DG) between 9.988 and 9.999, and envelope correlation coefficient less than 0.003 are observed. Simulation and optimization are approved through HFSS electromagnetic tool.

Keywords: Isolation, TARC, Diversity gain, MIMO, and ECC

*Correspondence Author: amrishpandey19@gmail.com

DOI: 10.1201/9781003357346-8

1. Introduction

Major functions of communication system are developing the superiority in data transmission, which is possible in MIMO systems [1]. The conservative way to enhance isolation or reduce mutual coupling is to locate antennas at an appropriate distance, often half a wavelength from each other [2].

In this interval, a MIMO antenna is designed for improving stability and capability of the method. Isolation is usually a loss as it characterizes the instructions of the MIMO antenna and instructions of the antenna array [3]. Communication between array components/MIMO features reduces system presentation by cumulative mutual coupling; changes in channel capacity loss, envelope correlation coefficient (ECC), radiation patterns, and falling DG [4]. Placing antennas too close can exacerbate this problem. The MIMO antenna consists of at least two dispersive radiation cores for the purpose of achieving a respectable separation between them. In contrast, the attainable planet is unsuitable for useful transportable equipment. The proliferation methods have acknowledged enhancing the isolation among the radiating elements [5-6]. Through the goal of advanced dependability and functionality, several methods have been described anew to reduce the mutual coupling [7].

On the other hand, in predictable broadcast systems, the transmitter and receiver aspect dilapidation is problematic in the transmission standard that is based on multipath broadcast. Therefore, the fourth most practicable choice for unraveling this multi-path problem is MIMO expertise [8]. The possibility of multiple approaches has discussed for improved reliability of communication by reducing far-field degradation [9]. The Some MIMO antennae have developed a variety of variants for enhanced separation between ports, [10], scrounging foundations [11], wandered line shreds [12], etc.

In MIMO, there is a trade-off between coupling and size. For MIMO communication, a small antenna with excellent isolation is recommended due to the limited amount of area available [13-14]. Several approaches have been discussed for advanced separation between mechanisms, as well as the feasting of DGS (defected ground structures), diversity techniques, neutralization lines, partially reflecting surface, and orthogonal orientation-based designs [15]. MIMO technology has certain advantages, including high data removal, less economical rate, and comfort of manufacture. The leading amount of expertise suffers as of decline at physical operations [16]; several strategies are adopted for broad-band uses and mostly for satellite, radiolocation, mobile, etc. [17].

In this communication, a parasitic chalice glass-shaped MIMO antenna with dual-band, enhanced gain and high isolation is proposed and discussed.

2. Methodologies and Evolvement of Proposed Structure

The proposed prototype geometry, with dimensions of antennas (A$_1$–A$_4$) on a millimeter scale, is revealed in Fig. 8.1. Top (patch) and bottom (ground) views are shown in green and yellow

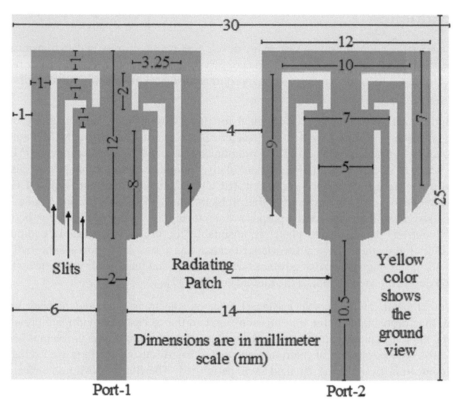

Fig. 8.1 Arrangement of the proposed model (A$_4$); immersion with top (green) and bottom (yellow) views

color, respectively. Desired strategy has the dual-patches with 2-mm broader microstrip lines through 50 Ω characteristic impedance.

Evolution strategies of the proposed structure are mentioned in Fig. 8.2 (a)and Table 8.1. A$_1$ obtained through manipulation of two symmetrical structures by a glass-shaped chalice. A$_2$ succeeded through introducing dual analogous equal inverted L-shaped rectangles slit scratch in patches on A$_1$. A$_3$ acquired from A$_2$ through dual inverted L-shaped rectangular slits are etched in the radiating elements at A$_2$. A$_4$ is obtained as of A$_3$ through dual analogous and equal rectangular slits on A$_3$. Whole antennas (A$_1$–A$_4$) determination has been described in the inset figure geometry of Fig. 8.2(a) as well as discussed in Table 8.1. Desired structures are detached through an arrangement (4 mm) of the patches that construct a MIMO configuration, as well as whole antennas (A$_1$–A$_4$) are designed with Rogers RT/duroid 5880 (tm) substrate (ε = 2.2, h = 1.575 mm) with loss tangent 0.009 and simulated using ANSOFT HFSS version 13.

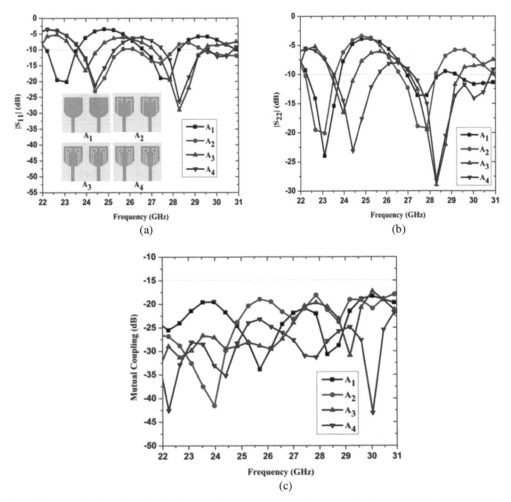

Fig. 8.2 Analysis of (a) S_{11}, (b) S_{22}, and (c) mutual coupling of antenna versus frequency curve (A_1–A_4)

3. Results and Discussion

The entire prototype remains considered in rapports of DG (variety gain), mutual coupling, current distribution, TARC (total active reflection coefficient), radiation pattern, ECC, and return loss. Table 8.1 reports the isolation in dB, operative-band in GHz, IBW (impedance bandwidth) in %, peak reflection coefficient in dB, resonant frequency in GHz and peak gain (A1-A4) in dBi of antennas at port 1 and 2 respectively.

Table 8.1 Ports physiognomies of the Antennas (A1-A4)

Antenna	Port No.	Number of bands	Operating band (GHz) / IBW (%)	Isolation (cB)	f_r (GHz)	Peak reflection coefficients (dB)	Peak gain (dBi)
A_1	1	1	22.15-23.52 / 5.99	< -19	23.01	-19.72	5.09
		2	26.57-28.27 / 6.1	< -20	27.82	-19.71	7.5
	2	1	22.23-23.85 / 7.03	< -19	23.04	-23.49	5.1
		2	26.93-28.64 / 6.15	< -20	27.54	-13.42	6.73
A_2	1	1	23.65-25.67 / 8.19	< -18	24.38	-22.47	8.36
		2	26.16-27.99 / 6.75	< -18	27.38	-13.98	9.87
	2	1	22.15-23.52 / 5.99	< -18	23.01	-19.72	8.5
		2	26.57-28.27 / 6.19	< -18	27.82	-19.71	9.68
A_3	1	1	23.32-24.50 / 4.93	< -26	23.89	-16.11	8.32
		2	27.01-29.37 / 8.37	< -19	28.23	-28.60	3.13
	2	1	23.36-24.25 / 3.73	< -26	23.89	-13.04	8.32
		2	27.54-30.75 / 11.01	< -19	28.23	-21.06	3.13
					30.42	-14.06	1.62
A_4 (Proposed)	1	1	23.69-25.23 / 6.29	< -28	24.30	-21.22	9.79
		2	27.46-29.57 / 7.39	< -27	28.30	-26.28	9.08
	2	1	23.53-25.26 / 7.09	< -28	24.30	-23.12	9.79
		2	27.43-30.46 / 10.46	< -27	28.30	-28.18	9.08

A. Isolation and Return Loss of (A_1-A_4)

Figure 8.2(a, b & c) shows the reflection coefficient ($|S_{11}|$ & $|S_{22}|$) as well as mutual coupling ($|S_{12}|$ & $|S_{21}|$) among the antenna mechanisms at ports 1 & 2. The band characteristics of A_4 at (23.69–25.23) & (27.46–29.57) GHz having IBW of 6.29 & 7.39%, respectively, for port-1 and (23.53–25.26) & (27.43–30.46) GHz with IBW of 7.09 & 10.46%, respectively, for port-2 are observed. For the proposed entire band, $|S_{11}|$ & $|S_{22}|$ are < -10 dB and $|S_{12}|$ & $|S_{21}|$ are < -27 dB.

For 5G mm-wave applications have provided the operating band in Korea 26.5-29.5 GHz, USA 24.25-28.35 GHz, China 24.25-27.25 GHz, Japan 27.5-28.28 GHz, Europe 24.25-27.5 GHz, and India (24.5–29.5) GHz and hence, required return loss (<-10 dB), gain (> 0 dBi) and element isolation (< -15 dB) [8]. Therefore, the proposed MIMO antenna occupies the mentioned 5G millimeter wave operating band frequency range. Later, antenna A_4 is appropriate for 5G millimeter wave purposes.

B. Scattering Parameter & Gain

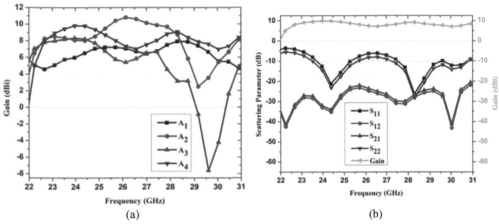

(a)   (b)

Fig. 8.3   Analysis of (a) gain of antennas (A_1–A_4) and (b) S factor & proposed gain vs. frequency characteristics of A_4

Regarding antennas (A_1–A_4), presentation of antenna A_4 is improved (cf. Fig. 8.3(a) & (b) and consuming dual-resonances of 24.30 & 28.30 GHz with peak gain of 9.79 & 9.08 dBi, respectively, at both ports 1 & 2 are the same and perfectly matched. From the survey of Fig. 8.3(b) and Table 8.1, has a slight variation in ports 1 and 2, which is perceived by the small gap (4 mm) between the radiation elements and the low-profile ($30 \times 25 \times 1.575$ mm^3) of the antenna.

Table 8.1 and an examination of Figs. 8.2(a, b and c) and 8.3 (a and b) clearly show that gain and band physiognomies of A_4 remain equal with minimal difference on all ports, although for A_4, the band and gain improvements are better as related to A_1, A_2, and A_3.

C. MIMO Performances and Radiation Efficiency

The evaluation of a MIMO antenna design on the basis of three crucial physiognomies has been investigated: TARC, DG, and ECC features. The allowable criteria of TARC < 0, DG is near 10 dB, and ECC < 0.5.

ECC, DG, and TARC by way of a utility are presented in Fig. 8.4(a-b). ECC & DG calculations with antennas agree with (1) and (2). The TARC of model reflects on the basis of return loss at the entire MIMO classification, for instance, now agreed with calculation (3).

$$\text{ECC} = \frac{|S_{11} * S_{12} + S_{21} * S_{22}|}{\left(1 - |S_{11}|^2 - |S_{21}|^2\right)\left(1 - |S_{22}|^2 - |S_{12}|^2\right)} \qquad (1)$$

$$\text{Diversity Gain} = 10 \times \sqrt{1 - (\text{ECC})^2} \qquad (2)$$

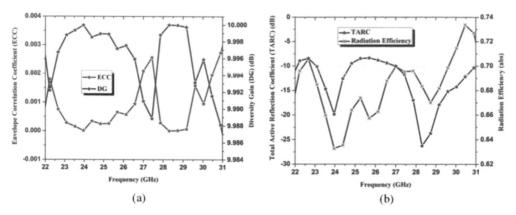

Fig. 8.4 Analysis of (a) ECC & DG (b) TARC & radiation efficiency vs. frequency characteristics of A_4

$$\text{TARC} = \frac{\sqrt{(S_{11} + S_{12})^2 + (S_{22} + S_{21})^2}}{\sqrt{2}} \qquad (3)$$

The ECC differs by the amount that the ports are connected through each other, and this can be calculated by the scattering of the MIMO method, the range (0–0.05) satisfying the ECC anomaly of A_4. Enhanced isolation of A_4 leads towards an enhanced data amount and well DG (9.979–9.999 dB). For MIMO outline, the satisfactory significance of TARC is below 0 dB. Fig. 8.4(b) describes an improved TARC significance, appropriate for satellite applications. TARC is pre-processed to give a respectable assortment of MIMO antenna.

D. Far-Field Analysis and Current Distribution of A_4

The distribution of current at A_4 has resonated on 24.30 GHz & 28.30 GHz with 54.75 A/m & 40.03 A/m, respectively, and is presented in Fig. 8.5(a & b). The maximum density of surface current of 54.75 A/m is observed at 24.30 GHz.

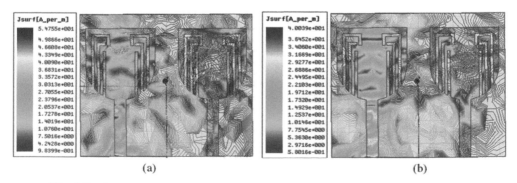

Fig. 8.5 Distribution of surface current by (a) 24.30 GHz & (b) 28.30 GHz of A_4

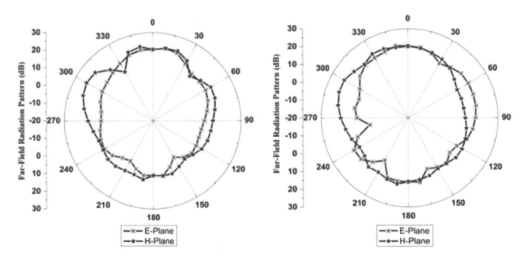

Fig. 8.6 Analysis of far-field pattern for E and H-plane at (a) 24.30 GHz and (b) 28.30 GHz of A_4

Antenna (A_4) for the E and H-plane patterns in the primary flat surface: azimuthal ($\Phi = 0$) and altitude ($\Phi = 90$) has shown in Fig. 8.6 (a–d). In Fig. 8.6, the radiation patterns (a) and (b) are at 24.30 GHz and 28.30 GHz, respectively. The proposed antennas (A_4) are intended for all resonant frequencies, suitable for omnidirectional wide radiation pattern characteristics as illustrated.

4. Examination with Existing Antennas

A relative outline of chalice glass-shaped expressions, compatibility of antenna, elements used, material, ECC, peak gain, mutual coupling/isolation, as well as TARC is available in Table 8.2. However, the compact-ability of A_4 has a smaller than the reported literature [2, 3, 7, 8, 9 and 10].

However, the proposed antenna exhibits superior peak gain, improved ECC, high isolation, and better TARC, in agreement with antennas described in references [2, 3, 7, 8, 9, 10 & 15]. In the proposed work, the material used is Rogers RT/duroid 5880, while in the reported references [2, 3, 7, 8, 9, 10 & 15], FR4 has been used. The use of Rogers RT/duroid 5880 substrate has led to better and more consistent properties such as high gain, low loss, and better isolation.

5. Conclusion

This paper presents a new and compact ($30 \times 25 \times 1.575$ mm^3) MIMO antenna design that is suitable for dual-band operation and is inspected for bandwidth, gain, and isolation improvements. It is appropriate with respect to K and Ka band uses and especially for 5G

Table 8.2 Relative analysis of A_4 with existing antennas

Ref. No.	Size of Antenna (mm³)	Elements used	Material used	Peak gain (dBi)	ECC	Isolation (dB)	TARC (dB)
[2]	37 × 44 × 1.6	2	FR4 (ε_r = 4.4, tan δ = 0.02)	4	< 0.10	< -15	NA
[3]	50 × 28 × 1.6	2		NA	< 0.12	< -18	NA
[7]	50 × 82 × 1.6	2		4.1	< 0.04	< -15	NA
[8]	35 × 68 × 1	2		3.4	< 0.03	<-20	NA
[9]	48 × 48 × 0.8	2		2.8	< 0.005	< -15	NA
[10]	22 × 36 × 1.6	2		2.7	< 0.06	< -15	NA
[15]	25 × 25 × 1.6	4		1.84	< 0.14	< -15	< -4
Proposed work	30 × 25 × 1.575	2	Rogers 5880 (ε_r = 2.2, tan δ = 0.009)	9.79	< 0.003	<-27	< -8

millimeter wave uses. The proposed antenna consumes maximum and minimum peak gain of 9.79 dBi & 9.08 dBi at ports 1 & 2, respectively; DG having the usual range from 9.988–9.999 dB, and TARC less than −8 dB is observed. The proposed gain and band physiognomies are identical at ports 1 & 2 with minimal deviancy, which obeys noble mutual coupling and admirable isolation (< −27dB) between antenna fundamentals; ECC is consistently improved (in the range between 0 and 0.003). The proposed antenna has an omni-directional radiation pattern, which is useful even for vehicular applications.

REFERENCES

1. Jaglan, N., S. D. Gupta, B. K. Kanaujia, S. Srivastava, and E. Thakur, "Triple band notched DGCEBG structure based UWB MIMO/diversity antenna," Progress In Electromagnetics Research C, Vol. 80, 21–37, 2018.
2. Iqbal A, Saraereh O. A, Bouazizi A, Basir A.: Meta-material based highly isolated MIMO antenna for portable wireless applications. Electronics 7(10), 267, 1–8, 2018.
3. Ibrahim A. A, Machac, J, Shubair, M. Raed.: Compact UWB MIMO antenna with pattern diversity and band rejection characteristics. Microwave and Optical Technology Letters, 59(6), 1460-1464, 2017.
4. Kamal, S. and A. A. Chaudhari, "Printed meander line MIMO antenna integrated with air gap, DGS and RIS: A low mutual coupling design for LTE applications," Progress in Electromagnetics Research C, Vol. 71, 149–159, 2017.
5. Toktas, A. G-shaped band-notched ultra-wide-band MIMO antenna system for mobile terminals. IET Microw. Antennas Propag., 11, 718–725, 2017.
6. Wang, F.; Duan, Z.; Wang, X.; Zhou, Q.; Gong, Y. High isolation millimeter-wave wideband MIMO antenna for 5G communication. Int. J. Antennas Propag., 2019, 4283010, 2019.
7. Gao, P.; He, S.; Wei, X.; Xu, Z.; Wang, N.; Zheng, Y. Compact printed UWB diversity slot antenna with 5.5-GHz band-notched characteristics. IEEE Antennas Wirel. Propag. Lett., 13, 376–379, 2014.

8. Kayabasi, A., A. Toktas, E. Yigit, and K. Sabanci, "Triangular quad-port multi-polarized UWB MIMO antenna with enhanced isolation using neutralization ring," AEU-International Journal of Electronics and Communications, 2017.

9. Zhang, S. and G. F. Pedersen, "Mutual coupling reduction for UWB MIMO antennas with a wideband neutralization line," IEEE Antennas and Wireless Propagation Letters, Vol. 15, 166–169, 2016.

10. K. A. I. D. A. Xu and J. Zhu, "Wideband patch antenna using multiple parasitic patches and its array application with mutual coupling reduction," IEEE Access, Vol. 6, 42497–42506, 2018.

11. Nirdosh, C. M. Tan, and M. R. Tripathy, "A miniaturized T-shaped MIMO antenna for X-band and Ku-band applications with enhanced radiation efficiency," 27th Wirel. Opt. Commun. Conf. WOCC, 2018, Vol. 1, 1–5, 2018.

12. Pouyanfar, N., C. Ghobadi, J. Nourinia, K. Pedram, and M. Majidzadeh, "A compact multiband MIMO antenna with high isolation for C and X bands using defected ground structure," Radio engineering, Vol. 27, No. 3, 686–693, 2018.

13. Dwivedi, A. K., Sharma, A., Tripathi, P. N., & Singh, A. K.: Dual Placed CDRA- Based MIMO Antenna for Wi-f/WLAN Applications. Singapore: Springer. 2020.

14. Wang,F., Duan, Z., Wang, X., Zhou Q., Gong,Y.: High Isolation Millimeter-wave Wideband MIMO Antenna for 5G Communication," International Journal of Antennas and Propagation 10(1), 1–12, 2019.

15. Tathababu, A., Vaddinuri, R.:Compact Two-Port MIMO Antenna with High Isola- tion Using Parasitic Reflectors for UWB, X and Ku Band Applications Progress In Electromagnetics Research C 102(8), 63–77, 2020.

16. Mishra B.: An ultra-compact triple band antenna for X/Ku/K band applications. Microw Opt Technol Lett, 61(9), 1857–1862, 2019.

17. Biswas K., Chakraborty U.:Textile Multiple Input Multiple Output Antenna for X- Band and Ku-Band Uplink-downlink Applications. 2020 National Conference on Emerging Trends on Sustainable Technology and Engineering Applications 13(9), pp. 1-4, 2020.

Intelligent Systems and Smart Infrastructure – Brijesh Mishra et al. (eds)
© 2023 Taylor & Francis Group, London, ISBN 978-1-032-41287-0

CHAPTER

9

Performance Analysis of Various Fast and Low-Power Dynamic Comparators

Nashra Khalid[1], Anurag Yadav[2], Subodh Wairya[3]

Department of Electronics and Communication Engineering,
Institute of Engineering and Technology, Lucknow, A.K.T.U., Lucknow, India

Abstract

With digitization in every field, there is need for efficient analog-to-digital converters (ADCs). Comparator is an important block of ADCs. Dynamic comparators have replaced static comparators as these are fast and also overcome the problem of huge static power dissipation in static comparators. Therefore, ADCs, which require high speed and less power consumption, used dynamic regenerative comparators. These comparators are clock based and has two stages: a preamplifier stage followed by latch stage. Latch provides positive feedback to the output of preamplifier stage that ensures higher speed to the circuit. In recent years, many of dynamic comparators have been introduced. Fast operation, less power consumption, high accuracy, and less offset are some desirable features of a comparator. Any one topology of comparator cannot provide all these features. Thus, an effort has been made in this paper to analyze performance of some topologies of dynamic comparators based on parameters like power, delay, offset voltage, and number of transistors. So, it is easy to select suitable comparator topology based on required application. This paper comprises comparative analysis of these four topologies of dynamic comparators, namely, conventional dual tail comparator, modified dual tail and fully differential dual tail and shared charge dual-tail comparator. For applications requiring low-power and high-speed shared-charge, dual-tail comparator is best, and for less offset requirement, fully differential configuration is best suited. These comparators are simulated on Cadence Virtuoso simulation software using 90-nm technology.

Keywords: ADC, Clocked regenerative comparator, Conventional double-tail dynamic comparator, Delay, Power, Offset

Corresponding author: [1]khalidnashra@gmail.com, [2]anuraglko2014@gmail.com, [3]swairya@gmail.com

DOI: 10.1201/9781003357346-9

1. Introduction

Since the world is moving towards digitalization, there is need to store and process data digitally. In the health industry, engineering field, and electronic communication field and other areas, there is a growing need for transportable, and minimal power-consuming devices as the semiconductor industry evolves (Moyal and Tripathi, 2016)(Gupta, Singh and Agarwal, 2021). These low-power devices employ ADCs as the main component. Comparator is often also referred as 1-bit ADC.

Comparators find extensive application in analog circuits, which can operate at high speed at lower power consumption. For this purpose, speedy dynamic regenerative comparators are used. By dynamic, it means these comparators are clocked comparators and, by regenerative, it means it uses positive feedback in the form of cross-coupled latch. Dynamic latch is a typical analog circuit architecture. It is commonly used to meet the requirement for fast speed while maintaining the adequate level of precision. The important parameters that define the performance of a comparator are its power consumption, delay offered by it, offset, and area (Figueiredo *et al.*, 2011). Over the time, various topologies of dynamic comparator have been developed, with each having their own tradeoffs. Dual-tail comparator (DoTDC)(Schinkel *et al.*, 2007) overcomes problem of delay and power. Modified dual-tail comparator (M-DoTDC) (Babayan-Mashhadi and Sarvaghad-Moghaddam, 2014) further reduces the delay. A new shared-charge idea is developed in shared-charge dual-tail comparator (SC-DoTDC) (Varshney and Nagaria, 2020), which reduces the delay as well as the power consumption. For low-offset consideration, fully differential dual-tail comparator (Gandhi and Devashrayee, 2018) is available.

Besides these configurations, some other configurations that are based on DoTDC are clock gating, new gate-biased cross-coupled latch is reported in Wang *et al.* (2019), and bulk-driven MOS technique is used in Joseph and Shahul Hameed (2021). The further flow of the paper is that the section provides the comparative analysis, which encompasses working, advantage and disadvantages of four different dynamic comparator topologies SC-DTC, DoTDC, M-DoTDC, and FD-DoTDC. Section III consists of performance analysis of the above-mentioned comparator topologies based on four parameters that include delay, power consumption, number of transistors, and offset voltage. Comparison has been presented in tabulated as well as graphical form. Section IV provides the conclusion to the above comparative analysis presented in this paper.

2. Clocked Regenerative Comparators

Comparators find extensive application in analog circuits. Dynamic comparators are used to increase the speed and accuracy. Due to the fact that regenerative latch has robust positive feedback, clocked and regenerative comparators have found widespread use in a variety of high-speed ADCs. In this section, we have studied four different topologies of clocked regenerative comparators that are conventional DoTDC, modified DTDC, SC-DoTDC, and fully differential DTDC.

A. Conventional Dual-Tail Comparator (DoTDC)

Figure 9.1(a) shows a schematic of a traditional DoTDC (Schinkel *et al.*, 2007). In comparison to the traditional dynamic comparator, this design can function at lesser supply voltages. This circuit in the latching stage allows larger current. At the same time, moderate current at input stage, that is small Mtail, results in low offset. The circuit's operation can be broken down into two parts. To begin, when clock = 0, the tail transistors of the circuit Mtail1,2 are turned off, while transistors M3,4 are turned on, causing nodes Outn and Outp to be pulled to Vdd shown in Fig. 9.1(b). When clock =1 during the decision-taking phase, transistors Mt1 and Mtail are turned on, whereas M3,4 are turned off. As a result, voltage at the fn and fp nodes begins to decline. The in-between stage created by Mr1,2 passes Vfn(p) to the cross-coupled inverters while also insulating the input and output, decreasing kickback noise (Schinkel *et al.*, 2007).

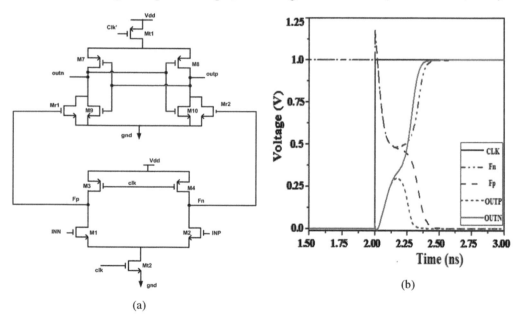

(a)   (b)

Fig. 9.1   (a) Schematic diagram of standard DoTDC. (b) Transient result of the standard DoTDC

This configuration has the advantage that it can work at lower voltages since it has less stacking than a single-tail transistor. It uses two MOS transistors as the tail transistors, and thus provides two current channels for preamplifier and the latch stage. The latching is faster and independent of the common-mode input due to larger current. Due to these advantages, this topology is used as the base topology for many new comparator configurations. The disadvantage of this circuit is that in the middle stage, transistors in this comparator will get cut-off in the end, and thus they will have no effect on the latch's effective transconductance. Furthermore, at the time of reset phase, output nodes must charge from zero voltage to VDD, which consumes power (Babayan-Mashhadi and Sarvaghad-Moghaddam, 2014).

B. Modified Dual-Tail Comparator (MDoTDC)

The circuit diagram of M-DoTDC is represented in Fig. 9.2(a) (Babayan-Mashhadi and Sarvaghad-Moghaddam, 2014). The design of this comparator is centered on dual-tail topology as it has better performance at lower supply voltage. It aims at speeding of latching is ex. So, transistors MC1 and MC2 are introduced in parallel to transistors M3/M4 to the first stage in a cross-coupled way. The working of comparator can be seen in Fig. 9.2(b) and is stated as follows. When the clock is zero, this is the reset stage, where tail transistor Mt1,2 are off and the PMOS transistors M5 and M6 are on. So, the nodes Fn and Fp charge to Vdd. Due to this, Mr1,2 gets "on". So, the outputs Outn,p are pulled down to zero voltage.

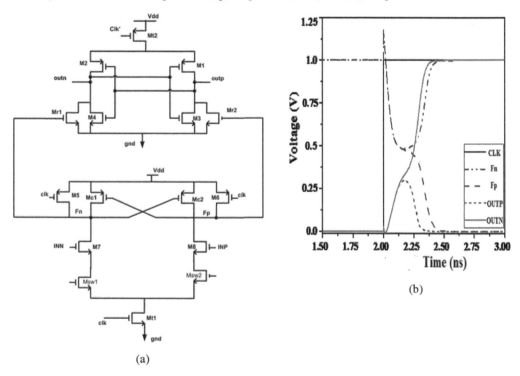

Fig. 9.2 (a) Schematic diagram of final structure of modified DoTDC. (b) Transient response of the modified DoTDC

Now, when the comparator enters in the second stage in which the clock is "1" high, the tail transistor gets on and M5,6 MOS transistors are off. So, now nodes Fn, Fp start to discharge with uneven rates that is decided by the current in transistors M7,8. Further, this decides which transistor among Mc1, Mc2 will get on. Suppose if INP>INN, Fp discharges more quickly than Fn and transistor MC1 is switched ON. The node Fn is again charged to supply voltage . In this way, Fn remains at Vdd and Fp discharges fully; this in turn will switch on Mr1, and Mr2 will be switched off. This will initiate the latching operation and Outn will be at zero level while Outp will be at Vdd level.

This comparator has the advantage of using inside positive feedback, which makes the entire latch regeneration firmer. The increase in speed is much more noticeable at lower supply voltages (Babayan-Mashhadi and Sarvaghad-Moghaddam, 2014). It provides two channels for the current in the preamplifier and the latch stage since it uses two MOS transistors as tail transistors. The energy required per conversion is also lowered. One of the drawbacks of the circuit is that when any of the control transistor turns on (e.g., Mc2), there is static power dissipation through input transistors M7, M8 and tail transistor Mt. To eliminate this problem, NMOS switches are added [Fig. 9.3(b)].

C. Fully Differential Dual-Tail Comparator (FDDoTDC)

The circuit diagram of FDDoTDC is depicted in Fig. 9.3(a) (Gandhi and Devashrayee, 2018). The operation of this comparator can be realized by its two stages. First one is the reset stage and the second stage is the decision-taking stage.

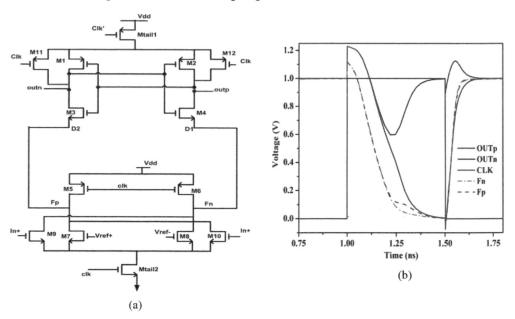

(a)

(b)

Fig. 9.3 (a) Schematic diagram of fully differential DoTDC. (b) Transient response of the FDDoTDC

When clock is at "0" level, this is known as the reset stage. In this stage, M5, M6, M11, and M12 are the transistors which are turned on. Outp and Outn then charge to supply voltage VDD.

The transistors on which reference voltages are applied i.e. M7 and M8 are switched ON which causes the intermediate nodes D1(Fn) and D2(Fp) to be charged to Vdd. Comparison Mode: When the clock is at "1", i.e., voltage is high, MA, MB, M9, and M12 are the transistors which are switched off. Also, Fn and Fp begin to discharge to ground level. Output nodes Outp and

Outn are discharged to low voltage level ground through M1–M4 at different rates. If Vinp > Vinn, Outn is discharged earlier than Outp. This can be seen in Fig. 9.3(b).

Fully differential dual-tail comparator decreases the power dissipation and reduces input offset, which are basically the advantages of FDDoTDC. The input offset is also minimized due to the double tail. However, reduction in power consumption is not that much remarkable. Also, the delay is magnified extensively.

D. Shared Charge Dual-Tail Comparator (SCDoTDC)

The schematic of charge-shared DoTDC (Varshney and Nagaria, 2020) is shown in Fig. 9.4. This comparator topology is based on charge sharing idea in the reset phase. In this technique, a pass transistor (PT) is connected between the output nodes Xn and Xp. When clock is low in the reset phase, this transistor (PT) gets ON and shares the charge between output nodes. Therefore, when clock transitions from 0 to 1, comparator enters the comparison phase. The output nodes do not have to begin from Vdd or ground voltage.

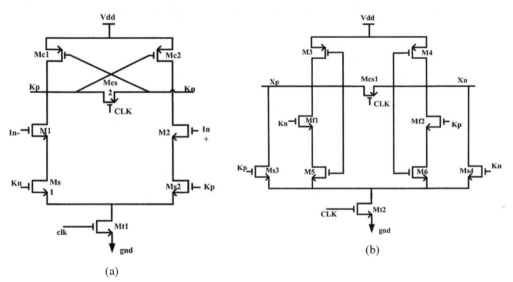

Fig. 9.4 Schematic diagram of shared-charge DoTDC: (a) Reset stage and (b) latching stage

Working of this comparator can be seen in Fig. 9.5. When the clock is "0" low, both the tail transistors are in cutoff mode to prevent the static power dissipation. During this period, clock mcs2 is ON and the charge is shared between the (Kn, Kp) and (Xn, Xp) charges to vdd. When the clock is high, i.e., "1", the tail comparator gets on and switch transistors were off in this. Dn, Dp and outn, outp start falling and rate of falling depends on the input signal. If In ® In+, then dp falls at start and then get charged again. This topology provides the advantage of fast operation, i.e. reduced delay and less power consumption.

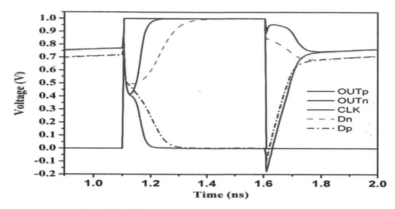

Fig. 9.5 Transient response of the SCDoTDC

3. Simulation Results and Performance Analysis

The above-discussed four configurations of dynamic are compared below. The performance comparison parameters are power, delay, PDP, and number of transistors. The summary of the performance comparison of these four comparators can be seen in Table 9.1. From Fig. 9.6(a), we can conclude that fully differential comparator offers the highest delay, whereas shared charge dual tail provides least delay with a reduction of 84.26%. Power consumption of shared-charge dual tail topology is least with the reduction of 46.66% with respect to modified dual tail.

Table 9.1 Comparison of Performance Results of Different Topologies

Parameter	DoTDC	MDoTDC	FDDoTDC	SCDoTDC
CMOS technology (nm)	90	90	90	90
Supply voltage (V)	1	1	1	1
Power (μW)	91.38	132.40	98.12	48.74
Total delay (ps)	102.9	69.28	181.45	28.56
PDP	0.94	0.91	1.78	0.13
1-σ Offset voltage (mV)	3.78	4.46	1.68	2.44
No. of transistors	12	16	13	18

Figure 9.6(b) shows that fully differential dual-tail comparators provide least offset and modified dual-tail configuration offers the most offset.

4. Conclusion

From the comparative analysis of the various dynamic comparators, the following conclusions are drawn. To overcome the drawback of single tail comparator, standard dual-tail comparators

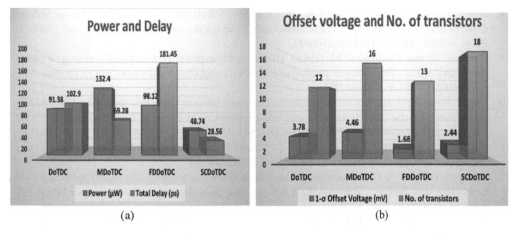

Fig. 9.6 Comparison of comparator topologies on the basis of (a) power and delay. (b) Offset and number of transistors.

are used, which provides fast response and reduced power dissipation. SC-DoTDC offer less delay. It provides 58.77% and 84.26% reduction in delay when compared to M-DoDTC and FDDoTDC, respectively. So, when delay is the main consideration, SC-DoTDC is the best choice. FD-DoTDC offers the least offset with 31.14% and 55.55% reduction compared to SC-DoDTC and DoTDC. But this reduction in offset is at the cost of increased delay.

REFERENCES

1. Babayan-Mashhadi, S. and Sarvaghad-Moghaddam, M. (2014) 'Analysis and design of dynamic comparators in ultra-low supply voltages', *22nd Iranian Conference on Electrical Engineering, ICEE 2014*, pp. 255–258.
2. Bindra, H.S. *et al.* (2018) 'A 1.2-V Dynamic bias latch-type comparator in 65-nm CMOS with 0.4-mV input noise', *IEEE journal of solid-state circuits*, 53(7), pp. 1902–1912.
3. Canal, B. *et al.* (2021) 'Low-voltage dynamic comparator using positive feedback bulk effect on a floating inverter amplifier', *Analog Integrated Circuits and Signal Processing 2021 108:3*, 108(3), pp. 511–524.
4. Dubey, A.K. and Nagaria, R.K. (2019) 'Low-power high-speed CMOS double tail dynamic comparator using self-biased amplification stage and novel latch stage', *Analog Integrated Circuits and Signal Processing*, 101(2), pp. 307–317.
5. Figueiredo, M. *et al.* (2011) 'A two-stage fully differential inverter-based self-biased CMOS amplifier with high efficiency', IEEE Transactions on Circuits and *Systems I: Regular Papers*, 58(7), pp. 1591–1603. doi:10.1109/TCSI.2011.2150910.
6. Gandhi, P.P. and Devashrayee, N.M. (2018) 'A novel low offset low power CMOS dynamic comparator', *Analog Integrated Circuits and Signal Processing*, 96(1), pp. 147–158. doi:10.1007/S10470-018-1166-9.
7. Gupta, A., Singh, A. and Agarwal, A. (2021) 'A low-power high-resolution dynamic voltage comparator with input signal dependent power down technique', *AEU - International Journal of Electronics and Communications*, 134. doi:10.1016/J.AEUE.2021.153682.

8. Joseph, G.M. and Shahul Hameed, T.A. (2021) 'A Sub-1 V Bulk-Driven Rail to Rail Dynamic Voltage Comparator with Enhanced Transconductance', *https://doi.org/10.1142/S0218126622500645*, 31(4). doi:10.1142/S0218126622500645.

9. Khorami, A. and Sharifkhani, M. (2016) 'High-speed low-power comparator for analog to digital converters', *AEU - International Journal of Electronics and Communications*, 70(7), pp. 886–894. doi:10.1016/J.AEUE.2016.04.002.

10. Moyal, V. and Tripathi, N. (2016) 'Adiabatic Threshold Inverter Quantizer for a 3-bit Flash ADC', *Proceedings of the 2016 IEEE International Conference on Wireless Communications, Signal Processing and Networking, WiSPNET 2016*, pp. 1543–1546. doi:10.1109/WISPNET.2016.7566395.

11. Razavi, B. (2015) 'The StrongARM latch [A Circuit for All Seasons]', *IEEE Solid-State Circuits Magazine*, 7(2), pp. 12–17. doi:10.1109/MSSC.2015.2418155.

12. Schinkel, D. *et al.* (2007) 'A double-tail latch-type voltage sense amplifier with 18ps setup+hold time', *Digest of Technical Papers - IEEE International Solid-State Circuits Conference* [Preprint]. doi:10.1109/ISSCC.2007.373420.

13. Varshney, V. and Nagaria, R.K. (2020) 'Design and analysis of ultra high-speed low-power double tail dynamic comparator using charge sharing scheme', *AEU - International Journal of Electronics and Communications*, 116, p. 153068. doi:10.1016/J.AEUE.2020.153068.

14. Wang, Y. *et al.* (2019) 'A Low-Power High-Speed Dynamic Comparator With a Transconductance-Enhanced Latching Stage', *IEEE Access*, 7, pp. 93396–93403.

Intelligent Systems and Smart Infrastructure – Brijesh Mishra et al. (eds)
© *2023 Taylor & Francis Group, London, ISBN 978-1-032-41287-0*

CHAPTER

10

Pose Data Improvement for Skeleton-Based Action Recognition Using Graph Convolutional Networks

Gunagya Singh Mamak*, Anshita Jain, Anuj Verma, and Anil Singh Parihar

Department of Computer Science and Engineering,
Delhi Technological University, Delhi, India

Abstract

Within computer vision, action recognition using joint position points has recently attracted a lot of interest. In previous works, a prevalent way of recognising actions from human pose data is graph convolutional networks, which use both action and structural data to generate features along both spatial and temporal dimensions for classifying human activity. We explore the popular dataset, kinetics-skeleton, extracted from video sequences in Kinetics 400, and analyse the pose data. The analysis brought forth that the pose data in the Kinetics skeleton dataset obtained via applying OpenPose is noisy and consists of large number of missing joints. Hence, the proposed method is also applied on a subset of the Kinetics dataset where we extract pose data using detectron2 to draw a comparison between the two pose detection algorithms. The model achieves significantly better performance with the newly extracted pose data, which we use to bolster the need to improve pose data in kinetics-skeleton.

Keywords: Human activity recognition, graph convolutional networks, spatio-temporal relations, pose detection, kinetics-skeleton

1. Introduction

Human activity recognition (HAR) can be referred to as a study of retrieving characteristics of actions performed by humans when engaged in activities. These actions or activities can

*Correspondence Author: gunagyamamak2@gmail.com

DOI: 10.1201/9781003357346-10

be classified as simple or complex but are always in a real-world scenario. Human activities can be characterised inherently in a hierarchical structure that has 3 different levels, namely, primitive actions level (constitute complex human activities), action or activity level, and complex interactions level (involves more than two persons and objects). The research in HAR has led to easier lives for several walks of people such as those in senior care, rehabilitation, people with cognitive disabilities, etc. It also assists people on a daily basis like life-logging and personal fitness [1].

Some challenges that are faced in HAR are the unreal assumptions that the scene is figure-centric with uncluttered background and free of issues such as image being partially obstructed, and the images will have differences in lighting, scale, frame resolution, etc. Another challenge is distinguishing between intraclass and interclass similarities. This basically points to the fact that different people perform the same actions in a different manner, hence creating intraclass disparities in one human action, and in some cases like falling or stumbling, there can be multiple actions that can have similar aspects of sorts, hence making it difficult to distinguish interclass disparities [2].

We plan on working on the above-mentioned problem statement using graph convolutional networks. In recent years, CNNs have become popular in the literature of a lot of applications related to the problem of computer vision. This is largely due to their ability to capture semantic information with a minute set of parameters, using the convolution operation. These networks have been able to show results comparable to industry standards on several image recognition tasks.

CNNs are useful for well-ordered data present in the form of a matrix, such as the pixels of an image. However, many non-regular datasets have given rise to a more general network, known as the Graph Convolutional Network (GCN). Just as a regular convolution, the graph convolution operation utilises the values adjacent to each node to calculate its value for the next layer; however, it does not require each node to be uniform. GCNs have a lot of potential power because they are capable of learning and using the relation between arbitrary pairs of nodes. There are many different types of GCNs that have been used across different fields.

A large number of previous works rely on the kinetics-dataset, which consists of poses extracted from video sequences in the Kinetics 400 dataset. Though it acts as a convenient, lightweight option as compared to reprocessing the entire Kinetics 400 dataset, the pose data in kinetics-skeleton is very noisy and consists of several missing joints. We aim to improve this data by extracting fresh poses using a more recent pose extraction model, namely detectron2.

Thus, our main contributions can be summarized as follows:

- We improve the pose data in kinetics-skeleton by extracting new, better quality data using detectron2.
- Exploration, visualization, and evaluation of the kinetics-skeleton dataset using carefully chosen metrics.

The paper has several sections where we talk about our work. In Chapter 2, we explore work done in recent years related to HAR using GCNs. In Chapter 3, the proposed methodology related to our work via GCNs is elaborated in detail. The same chapter tabularizes and explains the results and, finally, gives a conclusion. In Chapter 4, the paper elaborates on experimentation to draw the difference between the pose data extracted by openpose and detectron2 as well as use metrics to compare them. In Chapter 5, the paper is concluded and has ideas for future work.

2. Related Works

In this section, we discuss the different methodologies that have been applied in recent years in the field of HAR via GCNs.

Several papers that were referred to extracted spatio-temporal features and elaborated on structuring methodologies on the basis of these features, due to the importance of the joints in space as well as of the ordering with respect to time. In one such paper, the authors designed graph convolution kernels useful for human skeletons. This research was the first to combine human body skeletons with GCNs for the purpose of HAR. In fact, this introduced GCN operations not only on spatial domains (in a single frame) but also on the same node across different frames [6]. There is research that uses practical temporal reasoning graph (TRG), where they create trainable temporal relation graphs and a multi-head temporal adjacent matrix where each investigates temporal relationships across multiple scales and represents multiple sorts of temporal relations to aid multi-scale temporal relation extraction, respectively. This TRG is used for capturing the visual features and temporal relationships between surveillance videos at many temporal frames at the same time [8].

Methodologies like manifold regularization aim to set up information in the best way possible, which is used to add new data while simultaneously deleting outdated data. The basis for the model is the manifold regularization component of the currently used cost function, which might cause the function to alter as the data given as a sample for the distribution manifold changes. MRDGCN tries to iteratively improve its manifold structure information as well as the weight matrix of the neural net independently till a proper fit of the model is obtained [9]. Research has also been conducted specific to actions of people driving for driver activity recognition. The spatial features are modelled using a GCN and the temporal features are extracted using LSTMs. Being the first instance of applying GCN to driver activity recognition, the paper aims to extract activity information even from low-quality monocular cameras, noisy videos in real time [3].

GCNs when paired with high order features also have implications, which suggest an activity recognition methodology using the skeleton data features such as spatial and temporal ones. Some of these features include relative distance between 3D joints, velocity features, and acceleration features. The features that are extracted in the earlier stage are also converged using multi-stream feature fusion [14]. Pseudo-GCNs are used in some research which proposes the use of a learnable adjacency matrix rather than a fixed one where only edges along bones are present, to allow the network to capture the relationship between distant joints as well. Additionally, the model uses an attention mechanism to focus on key frames and input channels for action recognition [15].

3. Methodology

In this paper, we make use of concepts like GCNs and spatio-temporal feature extraction for the purpose of HAR.

A. Dataset

The Kinetics 400 dataset [7] is a popular dataset for human action recognition. It contains around 300,000 separate scenes obtained from YouTube clips. The videos are divided into 400 activities, which include day-to-day tasks, outdoor activities, as well as more complicated ones which depend on objects. Every clip is 10 seconds long and no skeleton data is included in the dataset. Being a video-based dataset, it is very heavy (>300 GB). In many cases, it is not practical to train models on the entire dataset. Fortunately, skeleton-based action recognition has gained popularity, allowing lightweight models such as GCNs to be trained only on human pose data and still obtain good performance.

Previously, this dataset has been used in [6], where the authors extracted pose data from the Kinetics dataset and have made the processed pose sequences available through a new dataset, which has come to be known as "kinetics-skeleton". This kinetics skeleton is extracted via OpenPose frame-by-frame at a frame rate of 30 fps. These sequences of 300 frames are each labelled with 18 joints of the 2 most confident people. It is a much more practical dataset of 7.5 GB.

A quick sampling-based qualitative analysis of kinetics-skeleton reveals that in most sequences, the pose data is very noisy and consists of several missing joints. It would be hard for even a human to recognise the action solely from the information provided by the extracted poses. The conclusion drawn from this points to the fact that this is a significant contributor in the large discrepancy between the SOTA models on Kinetics and Kinetics-skeleton dataset. This disparity is quite clear since state-of-the-art models can achieve an accuracy of 89.1% [10] on the Kinetics dataset, whereas the highest performance for Kinetics-skeleton is only 47.7% [11].

Our work proposes to improve the kinetics-skeleton dataset by extracting fresh pose data from the videos of the dataset using a more modern model, namely detectron2 [12]. Detectron2 is a product of the research team at Facebook released in 2019. It is a modification of the Detectron object detection platform. Some of the features that make detectron2 our model of choice include flexibility, the fact that it includes high-quality implementations of state-of-the-art object detection algorithms, open-source availability, and existence of models that can work on smoothening detectron2 pose data.

The first step of our dataset preparation is pre-processing of videos in Kinetics-400. The resolution of all videos are made to be 340 ´ 256 and the frame rate is adjusted to 30 fps prior to pose extraction. This is similar to the pre-processing done by Yan et al. [6]. The pose extraction module of detectron2 yields 17 joints for every human figure detected in an image. The data of a joint consists of its 2-dimensional pixel location (X, Y) along with its labelling confidence C, forming a single 3-dimensional vector (X, Y, C). This method is run across all 300 frames of a video. To achieve uniformity in the number of human figures detected in different frames, we label only those figures which have the 2 highest average confidence across their joints. Thus, a single data tuple consists of a channels' dimension (C = 3), the temporal dimension (T = 300), a joints' dimension (V = 17), and the number of figures detected (M = 2). The entire training and validation datasets consist of many such training tuples stacked together. A structural comparison between OpenPose and detectron2 poses is depicted in Fig. 10.1(a) and (b).

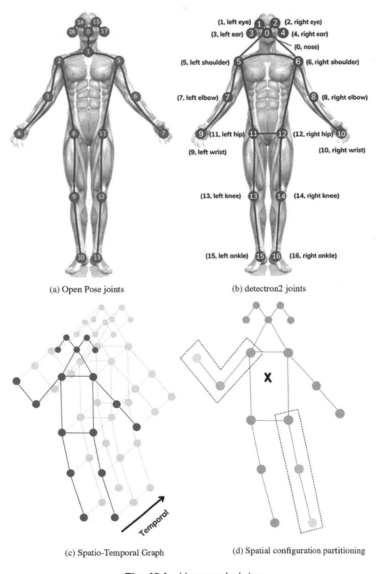

(a) Open Pose joints

(b) detectron2 joints

(c) Spatio-Temporal Graph

(d) Spatial configuration partitioning

Fig. 10.1   Human skeleton

Since the original dataset does not contain pose labels itself, we cannot assure better performance of our poses than the original kinetics-skeleton dataset. However, we provide the following metrics as evidence in support of our claim for pose improvement:

- Smoothness: It is a physics based metric that uses average magnitude of joint accelerations to measure the smoothness of the generated pose sequence [13]. Since our dataset consists of missing joints, we calculate smoothness only over those frames where the joints are detected.
- Missing joints: This metric is measured by the number of unidentified joints. All joints which are positioned at the origin (0,0) of the frame are considered missing joints.

B. Graph Convolutional Network

Our baseline GCN is largely based on the work of Yan et al. [6]. We summarise the various components in this section.

Graph Construction

Within a single frame, the human-skeleton graph consists of V joints as its nodes and the edges defined by body links and bones, along with a self-link on each node. The entire structure is represented as an adjacency matrix which is passed through the network. Across multiple frames, nodes labelling the same joint in adjacent frames are connected. The entire spatio-temporal graph is depicted in Fig. 10.1(c).

Partitioning Strategy

Partitioning brings about the uniformity required in a graph for employing a convolution operation. Each partition set is associated with the learnable weight of a filter. We use spatial configuration partitioning. In this partitioning scheme, the neighbours of each node are divided into 3 sets (refer Fig. 10.1(d):

- The root node, which contains only the root.
- The centripetal set, which contains nodes closer than the root node to the gravity centre of the skeleton.
- The centrifugal set, which contains nodes further than the root node to the gravity centre of the skeleton.

Graph Convolution Operation

The spatial graph convolution can be summarised by the following equation [16]:

$$f_{out} = \Lambda - 1/2(A + I)\Lambda - 1/2 f_{in} W \tag{1}$$

where f_{in} and f_{out} represent the input and output to the graph convolutional layer, respectively, A represents the adjacency matrix of the graph, I is the identity matrix which introduces self-links, Λ is the diagonal degree matrix which is used for normalization, and W is the learnable weight matrix.

Spatio-Temporal Graph Convolution

Along the temporal axis, we define a parameter called the temporal kernel size. The neighbour set in the spatial domain is interpolated to include the corresponding nodes of the frames which are at most distance away from the current frame. Each frame is assigned an independent group of partition sets, and associated with a separate learnable weight. Since the temporal axis is uniform, the implementation of a convolution operation is straightforward and is done using a simple 2D convolution.

Model Architecture

The model starts with batch normalisation of inputs. A total of 9 layers of ST-convolution operations are applied, where the first 3 layers have 64 output channels, the next 3 have 128 output channels, and the last 3 have 256 channels for output. Residual connections as well as learnable masks for edge importance weighting are added in each layer. The temporal kernel size is 9, i.e. 9 consecutive frames are taken for the temporal convolution. Dropout is employed on every layer with probability 0.5. The 4th and 7th layers are pooling layers that have a stride of 2. Finally, a global pooling layer is added to produce the final output feature vector. Softmax function is used for classification.

Training and Testing

For model training, we use categorical cross-entropy as our loss function and stochastic gradient descent is used as the optimizer with a base learning rate of 0.1 and weight decay of 0.0001. We perform data augmentation on the input data sequences in the form of small, stochastic affine transformations during training to prevent overfitting. The model was written and trained using the PyTorch framework and a batch size of 64. Model training is halted after 25 epochs when the accuracies stabilise. We use top-1 and top-5 accuracies while training and testing, as suggested by the authors of [7]. Our model performances are reported in the following section.

4. Experimental Results and Discussion

A. Qualitative Experiments

We perform a random-sampling-based qualitative analysis on sequences in which poses have been extracted using both OpenPose and detectron2. The visualisations of these poses are shown in Fig. 10.2. In general, we observed that poses extracted from detectron2 were less noisy and more recognisable than those extracted from OpenPose.

In Fig. 10.2(a), detectron2 recognizes both humans in the frame both with distinctive, recognizable positions as compared to OpenPose where the individuals are neither recognizable nor segregated. This supports the hypothesis that the new poses are better not only for the most confidently labelled person, but also for the second most confidently labelled person. In Fig. 10.2(b), the action to be recognised is golf driving, where the movement of arms is the most important. OpenPose is only able to identify one arm while detectron2 is able to identify

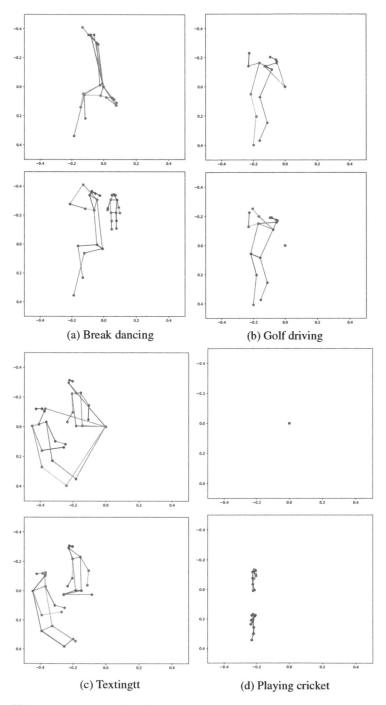

Fig. 10.2 Pose data by OpenPose (top) and detectron2 (bottom) for 4 actions

both, making the action more clear. In Fig. 10.2(c), the many missing joints connect the skeleton to the origin, hence making the pose data incoherent to the human eye. Detectron2 has achieved half as many missing joints, and this improves the coherency of the pose data, which can be seen clearly by the bottom visualization. Finally, in Fig. 10.2(d), we see an instance of the human pose being very small with respect to the frame size, a scenario in which OpenPose struggles while detectron2 is able to successfully detect the poses of two people playing cricket. With fewer number of missing joints, the model has more information to learn and perform better.

B. Quantitative Experiments

The bottleneck in the pose extraction pipeline was running the video sequences through the detectron2 model. In the interest of time and computational resources, we extract a subset of 20 classes from the kinetics dataset in order to run and train our models on the two sources of pose data. The names of the 20 classes are as follows: drawing, folding clothes, high jump, brushing hair, throwing ball, making a sandwich, cartwheeling, breakdancing, playing cricket, drinking, clapping, texting, cleaning floor, climbing tree, jogging, golf driving, taking a shower, catching fish, slapping, and wrestling, totalling 10,348 training samples and 994 validation samples.

The classes were chosen so as to act as a representative set and provide a sufficient coverage over the general dataset with 400 classes, including single person actions, person–person interactions as well as person–object interactions. Any performance variations observed in the model for the new pose data on this subset should be reflected similarly when observed for the entire dataset. Under this assumption, we provide evidence in the form of our chosen metrics to validate our hypothesis that the pose data extracted from detectron2 is of higher quality and greater use than the existing poses.

As can be seen from Table 10.1, detectron2 is able to detect significantly more joints than the existing OpenPose data; around 1.5 times more in the single person case and around 1.9 times more overall, with comparable smoothness values. The similarity between the metrics of the small and full dataset in case of OpenPose provides evidence that the chosen subset of classes is well representative of the entire dataset.

Further, the smoothness of poses extracted from detectron2 can be easily improved using models such as VideoPose 3D [17]. We leave this as an exercise for future works.

Table 10.1 Pose Estimation Comparison

Metric	OpenPose		Detectron2
	Full Dataset	Small Dataset	Small Dataset
Missing Joints % [Single person] (↓)	54.08	53.78	31.09
Missing Joints % [Two persons] (↓)	68.13	68.01	40.75
Smoothness [Two persons] (↓)	114.23	116.61	133.01

C. Model Performance

During our experiments, we first train a baseline model consisting of ST-GCN units on the full OpenPose kinetics skeleton data. We then take the OpenPose data for the subset of classes and train the model on this small dataset. Finally, we train the model on detectron2 data that has been extracted for the same subset of classes and compare it with the baseline using top-1 and top-5 accuracy measures. We report our results in the table that follows.

We find that the use of detectron2 has led to better pose extraction than the use of OpenPose, which is the pose extractor used in current research involving the Kinetics dataset. Despite downsampling the dataset, our research takes a large enough pool for it to prove that spatio-temporal GCNs perform better with kinetics skeletons extracted by detectron2 (see Table 10.2), where the paper has discussed the accuracy in both cases.

The top-1 accuracy improves by 5.5% and the top-5 accuracy improves by 5.2%.

Table 10.2 Model Performance

Dataset Used		Top-1 Accuracy (%)	Top-5 Accuracy (%)
OpenPose	Full	25.44	47.43
	Small	36.87	77.40
Detectron2	Small	41.35	82.60

5. Conclusion and Future Work

In this paper, we explored various ways to achieve HAR via GCNs through methods like studying related works and learning the current phase of research in building and practically implementing an initial model where spatial and temporal relations are created. We also projected how the use of detectron2 greatly enhanced the use of GCNs on the kinetics skeletons and showed a comparison between the skeletons themselves via parameters such as smoothness and number of identified joints, as well as via the results of the same GCN on both skeletons. We show that detectron2 could be used to improve current work in the field of human action recognition instead of the use of OpenPose for pose extraction.

Some of the future scopes of work being planned in terms of research include facilitating the experimentation on accuracy fluctuations with an increase of classes as well as similar activities with similar gestures or objects. The GCN could be integrated with pre-made models like ResNets and MobileNets to further improve performance and facilitate ease of use. Work could also be done on improving the smoothness of the extracted detectron2 data using other models. The applications of our work can be used in any construction site to detect if somebody has fallen and to get them help immediately. They could also be used to ensure proper exercise formations and ensure well-being, as well as with security systems to recognize illicit activities and alert authorities.

REFERENCES

1. Golestani, Negar, and Mahta Moghaddam. "Human activity recognition using magnetic induction-based motion signals and deep recurrent neural networks." Nature communications 11, no. 1 (2020): 1–11.
2. Zhang, Shugang, Zhiqiang Wei, Jie Nie, Lei Huang, Shuang Wang, and Zhen Li. "A review on human activity recognition using vision-based method." Journal of healthcare engineering 2017 (2017).
3. Pan, Chaopeng, Haotian Cao, Weiwei Zhang, Xiaolin Song, and Mingjun Li. "Driver activity recognition using spatial-temporal graph convolutional LSTM networks with attention mechanism." IET Intelligent Transport Systems 15, no. 2 (2021): 297–307.
4. Vrigkas, Michalis, Christophoros Nikou, and Ioannis A. Kakadiaris. "A review of human activity recognition methods." Frontiers in Robotics and AI 2 (2015): 28.
5. Wang, Xiaolong, and Abhinav Gupta. "Videos as space-time region graphs." In Proceedings of the European conference on computer vision (ECCV), pp. 399–417. 2018.
6. Yan, Sijie, Yuanjun Xiong, and Dahua Lin. "Spatial temporal graph convolutional networks for skeleton-based action recognition." In Thirty-second AAAI conference on artificial intelligence. 2018.
7. Smaira, Lucas, Jo~ao Carreira, Eric Noland, Ellen Clancy, Amy Wu, and Andrew Zisserman. "A short note on the kinetics-7002020 human action dataset." arXiv preprint arXiv:2010.10864 (2020).
8. Zhang, Jingran, Fumin Shen, Xing Xu, and Heng Tao Shen. "Temporal reasoning graph for activity recognition." IEEE Transactions on Image Processing 29 (2020): 5491–5506.
9. Liu, Weifeng, Sichao Fu, Yicong Zhou, Zheng-Jun Zha, and Liqiang Nie. "Human activity recognition by manifold regularization based dynamic graph convolutional networks." Neurocomputing 444 (2021): 217–225.
10. Yan, Shen, Xuehan Xiong, Anurag Arnab, Zhichao Lu, Mi Zhang, Chen Sun, and Cordelia Schmid. "Multiview Transformers for Video Recognition." arXiv preprint arXiv:2201.04288 (2022).
11. Duan, Haodong, Yue Zhao, Kai Chen, Dian Shao, Dahua Lin, and Bo Dai. "Revisiting skeleton-based action recognition." arXiv preprint arXiv:2104.13586 (2021).
12. Chen, Xinlei, Ross Girshick, Kaiming He, and Piotr Doll´ar. "Tensormask: A foundation for dense object segmentation." In Proceedings of the IEEE/CVF International Conference on Computer Vision, pp. 2061–2069. 2019.
13. Yuan, Ye, and Kris Kitani. "3d ego-pose estimation via imitation learning." In Proceedings of the European Conference on Computer Vision (ECCV), pp. 735–750. 2018.
14. Dong, Jiuqing, Yongbin Gao, Hyo Jong Lee, Heng Zhou, Yifan Yao, Zhijun Fang, and Bo Huang. "Action recognition based on the fusion of graph convolutional networks with high order features." Applied Sciences 10, no. 4 (2020): 1482.
15. Yang, Hongye, Yuzhang Gu, Jianchao Zhu, Keli Hu, and Xiaolin Zhang. "PGCN-TCA: Pseudo graph convolutional network with temporal and channel-wise attention for skeleton-based action recognition." IEEE Access 8 (2020): 10040–10047.
16. Kipf, Thomas N., and Max Welling. "Semi-supervised classification with graph convolutional networks." arXiv preprint arXiv:1609.02907 (2016).
17. Pavllo, Dario, Christoph Feichtenhofer, David Grangier, and Michael Auli. "3d human pose estimation in video with temporal convolutions and semi-supervised training." In Proceedings of the IEEE/CVF Conference on Computer Vision and Pattern Recognition, pp. 7753–7762. 2019.
18. Parihar, Anil Singh, Kavinder Singh, Hrithik Rohilla, and Gul Asnani. "Fusion-based simultaneous estimation of reflectance and illumination for low-light image enhancement." IET Image Processing (2020).

19. Singh, Kavinder, and Anil Singh Parihar. "Variational optimization based single image dehazing." Journal of Visual Communication and Image Representation 79 (2021): 103241.

20. Parihar, Anil Singh, Gaurav Jain, Shivang Chopra, and Suransh Chopra. "SketchFormer: transformer-based approach for sketch recognition using vector images." Multimedia Tools and Applications 80, no. 6 (2021): 9075–9091.

21. Jain, Gaurav, Shivang Chopra, Suransh Chopra, and Anil Singh Parihar. "Attention-Net: An Ensemble Sketch Recognition Approach using Vector Images." IEEE Transactions on Cognitive and Developmental Systems (2020).

Intelligent Systems and Smart Infrastructure – Brijesh Mishra et al. (eds)
© 2023 Taylor & Francis Group, London, ISBN 978-1-032-41287-0

CHAPTER

Driver Drowsiness Using Image Processing

Rajat Garg[1], Sunil Kumar[2], Anshul Kumar[3], Vijay[4], and Aparna Goel[5]

Department of Computer Science & Engineering
Meerut Institute of Engineering and Technology, Meerut, India

Abstract

Road safety is a major concern around the world. Most of the accidents occur due to driver drowsiness and fatigue. Driver drowsiness or short sleep can cause many road accidents, such as inability to pause when needed, loss of control of the car on highways when speeding, stopping car during overtake, etc. Almost all drivers will experience this condition for reasons such as long driving, weather effects, etc. Ensuring current safety is very important as one driver's mistake can put many lives at risk. Although human error cannot be completely erased, it can be reduced to a significant amount. This research paper is widely used to find out if the driver is tired or drowsy or not.

In this case, the driver will be under the computer to test him by reading his face and seeing with the help of Open CV image processing method. The alarm will sound if the driver is found to be in any of the above conditions or traveling in that condition. This will alert the driver and thus prevent further road accidents.

Keywords: Drowsiness, Fatigueness, Computer-vision, Open CV, Image Processing, Safety, Accidents, Driving, Facial landmark detection, EAR (Eye Aspect Ratio), ECR (Eye Closure Ratio).

Corresponding author: [1]rajat.garg.cs.2018@miet.ac.in, [2]sunilymca2k5@gmail.com,
[3]anshul.kumar.cs.2018@miet.ac.in, [4]vijay.kumar@miet.ac.in, [5]aparna.goel.cs.2018@miet.ac.in

DOI: 10.1201/9781003357346-11

1. Introduction

Safety and security are very important while traveling. This includes not only the strength of the vehicle and its safety features but also the presence of road accidents and other driving skills. The increase in the number of deaths due to road accidents is one of the biggest problems in the world. According to a recent study by the World Health Organization, about 3700 people lose their lives every day in road accidents. In this case, about 20% of the causes of death are the driver's unconsciousness on the road such as convulsions, drowsiness, fatigue, etc. [1]

Drowsiness is a state of "need for sleep" and fatigue is defined as "fatigue due to overwork". In any case, one is not aware of local events, such as drunkenness. Drowsiness of the driver can lead to the risk of many lives on the road. This short sleep can cause car crashes, unnecessary road cuts, and collisions. To eliminate this problem and provide safety for drivers, our project detects driver drowsiness using image processing and ML. The aim of our project is to get the driver drowsy fast and to alert the driver [1,7,9].

The purpose of this paper is to increase the prototype of the sleep alert warning system. Our complete recognition and awareness can be based on device design to accurately reflect the open and closed world of the driver in real time. Through continuous eye tracking, it may be seen that symptoms of inflammatory fatigue may appear early to distance themselves from spontaneous. This discovery can be made using a series of eye pictures and movements of the face and head. Observing eye movements and edges for detection may be used. Hitting tools when drivers fall asleep and giving them warning about threats, or even controlling vehicle movements, have been a challenge to good education and development.

The cause of fatigue is a major complication that leads to many road injuries every 12 months. It is not always possible to calculate a wide variety of sleep-related risks due to problems finding out whether fatigue is a problem or not and to assess the level of fatigue. Yet, research shows that up to 25% of road accidents in India are related to fatigue. Studies in other countries additionally suggest that driver fatigue is a major problem. To experience this international suffering, a sequential photographic response transmits real-time driving information to a server, and determines drowsiness. The use of EAR (Eye Aspect Ratio) and ECR (Eye Closure Ratio) is proposed and implemented using Android software. The price calculated by the machine makes the dynamic force do damage or rest for a while. The techniques used have no natural power; therefore, no additional costs will be incurred during the drowsiness route. This activity directly detects a person's drowsiness by continuously scanning the eyes and face and recording the blink of attention. The alarm can be set if the person's eyes close for a long time. Similarly, the cause further increases the alarm if the other person's eye is willing to be distracted by using a few objects, away from the road. This increase in alarm will alert the driver, and for that reason, important steps can be taken to avoid inconsistencies. A few other methods can be used to assess drowsiness and impaired motivation. The paper is divided into six sections; the first part involves the presentation of paper content. An overview of the research papers is discussed in the section outlining the various photography techniques. The third section provides the triggers for drowsiness. Section IV provides paper comparisons based on result accuracy, frequently used classification algorithms, and data sets. Finally, the discussion, conclusion, and consideration of future research work on the drowsy diagnosis program are discussed in Sections V and VI, respectively. [1,3,7]

2. Literature Review

To tackle the above difficulties, in this study, we are thinking about convolutional neural networks (CNNs). CNN has made fast growth in the area of desktop vision, in particular facial recognition. Viola and Jones and Yang et al. have developed the use of the AdaBoost algorithm with Haar elements to teach a number of vulnerable-type dividers, entering the stable levels of discovering non-human faces.

Motamedi-Fakhr et al., in their audit paper, gave a structure of in excess of 15 generally utilized elements and strategies for human rest examination. Highlights are partitioned into brief, visual, time, and backhanded highlights. Notwithstanding these elements, they additionally audit the exploration papers on the characterization of rest classes. Rashid et al. evaluated the present status, difficulties, and potential arrangements of the EEG-based cerebrum PC interface. Inside their work, they additionally momentarily talked about the most broadly utilized parts of cerebrum and PC associations that are coordinated by time region, recurrence, time recurrence, and area.

Sahayadhas et al. assessed vehicle-based measures, social based measures, and body estimations for driver languor. The physiological measures area comprises 12 papers with components of recurrence space as it were. Sikander and Anwar assessed rest designs and isolated them into five classes — subject report, driver organic highlights, driver's actual elements, vehicle highlights while driving, and half breed highlights. With regards to lack of sleep utilizing EEG flags, the writers center more around characterizing the recurrence attributes utilized for rest apnea as opposed to introducing research that has proactively been done in this field.

Balandong et al. isolate the driver's rest frameworks into six classes in light of the technique utilized — (1) self-gauges, (2) vehicle-based estimations, (3) driver-based conduct frameworks, and (4) rest and wake-up factual models, (5) frameworks in view of the qualities of the human body, and (6) crossover frameworks. The creators underscored frameworks in view of the attributes of the human body, yet just frameworks that depend on a set number of EEG anodes, as these sorts of frameworks are exceptionally practical in true applications. The creators infer that the best outcomes were gotten when the issue was recognized as a parallel division issue and that the incorporation of EEG highlights with other actual pointers ought to prompt better precision.

Different papers auditing driver's rest frameworks are accomplished in a specific field, e.g., Hu and Lodewijsk center around recognizing non-recognizable weakness, practical weariness, and rest-based signals, free testing, driving way of behaving, and visual measurements. Soares et al. inspect the good example for tiredness, Bier et al. put an emphasis on repetitiveness-related exhaustion, and Philips et al.explored execution measures (e.g., ranking staff, plan of proper timetable) that diminish the gamble of lack of sleep.

3. Factors Causing Drowsiness Driver

Driving power fatigue is often the result of 4 important things: sleep, work, time of day, and exercise. People are always trying to do more in the day and that they are losing precious sleep because of this. Usually by drinking caffeine or various stimulants, people continue to fall asleep. Insomnia lasts for a few days and after that happens the body eventually collapses and the person falls asleep. The elements of the time of day can always have an effect on the body. The human mind is prepared to think in times when the body ought to be snoozing. Expanding the length of rest will prompt the breakdown of the edges. The last part of an individual's body condition consumes medications that cause laziness or have sensitivities that cause these issues. Absence of actual wellness, being both underweight or stout, will cause weariness. Moreover, passionate injury will advance edge exhaustion rapidly. Individuals frequently disregard weariness as a deterrent to their lackluster showing. This can prompt numerous perilous circumstances particularly in the event that drivers dread their lifestyle more than the existences of various individuals out and about. Lazy driving is a hazardous mix of driving while worn out and sleeping. [2,6] The quantity of reasons for weakness are as follows:

1. Driver exhaustion frequently happens while the persuading factor does not rest soundly inside the most recent 24 hours. The typical individual ought to rest no less than 7 hours every day to remain sound.

2. Various variables including rest issues, for example, sleep deprivation, sleep apnea, and shift work rest jumble (SWSD) might be because of uncommon hours in the drawings.

3. Excessive number of drawings can cause difficulty and anxiety that often lead to insomnia during normal hours.

4. In line with observation, those who suppress industrial vehicles, especially heavy vehicles, are generally treated in a sleepy manner more often than uninvited drivers [5]. This is mainly due to improved staff telephony and higher pay for extra hours in drawings at regular times of the day.

5. The human mind can join lay down with the hours of the evening. This prompts an expansion in rest driving episodes as drivers rest all the more effectively even around evening time contrasted with daytime.[8]

4. Proposed Methodology

Various methods are advised by the special authors of many research papers to detect fatigue in drivers properly. Open CV library from Python can be used for facial and facial scratches as it should be for recognizing fatigue. This makes the machine exceptionally smooth to work; at the same time, it makes the face recognition process more slow. Different strategies, for example, inspecting changes in progressive edges for looks, can make this interaction no less than multiple times quicker [6]. The natural eye framework that utilizes round Hough modify (CHT) to decide the right iris can make total rest apnea more dependable [7]. CHT is utilized to work out the region between the sweep of the iris, which is significant for ascertaining the distance between the eyelids. Any remaining frameworks use video contribution to filter each eye and mouth to screen the eyes and mouth to more readily anticipate rest sluggishness. To see that different shaded faces are recognized by utilizing light just, YCbCr which is a result of added substances Luminance (EyeMapL) and Chrominance (EyeMapC) can assist with observing an extraordinary hued surface higher than after evacuation of light. Conceal space, for example, the HSV chart is utilized to identify eye position for example opening or shutting which can be utilized to work out the PERCLOS boundary to pass judgment on sleepiness. Level of structural similarity (SSIM) can be utilized for eye location since it has better execution than any typical advances. Consolidating the caution discovery with this impact gives knowledge that permits you to decide whether the alert should be set off by really taking a look at the languid classes.[8,9]

A. Proposed System

The main methods that meet all of these requirements are Open CV, Tensor Flow, HaarCascades acquisition model, and face-to-face visualization. The whole system works in this process only, as explained in Fig. 11.1.

The working of the program is mainly categorized into the following parts:

1. Capturing the image
2. Pre-process the picture
3. Face extraction and 2D image conversion
4. Find facial landmarks on the image
5. Calculate EAR from the image
6. Predict drowsiness from calculations.
7. Raising an alert.

1. Capturing the Image:

Primarily, the image is automatically captured as input by a web camera or specific hardware through Open CV function.

2. Pre-process the picture:

Converts an Red-Green-Blue (RGB) image to a black-and-white image and converts the image into a binary form for data processing. After a photo conversion is successful, we use that data to predict the shape of that image.

3. Face extraction and 2D image conversion:

In this case, the Eigen-based face algorithm is used to convert 3D image to 2D image taken from a webcam. The human face and other objects are then extracted using a Gaussian Skin model and a rectangle box technique to create the Eigen faces. These Eigen faces are then divided into frames for further steps.

4. Find facial landmarks on the image:

For local facial features – eyes, mouth, and ears on the face, we have supported the Scpy python module. This module helps us to calculate the Euclidean distance from one side of the face to the other in various directions so that we can accurately identify the face. The main algorithm used here is the Viola Jones algorithm. Global face signals are generated with the help of the DLIB python module. These landmarks are created as a reference list. Therefore, in order to produce optical circuits from manufactured sets of landmarks, we need to know the indicators of the exact fragment that will enable us to produce optical circuits from the same component parts. Figure 11.1 shows facial identification.

Fig. 11.1   Facial Landmark

5. Calculate EAR from the image:

Next, we will calculate the EAR (Eye Aspect Ratio). EAR is an eye-opening measure. It is a fixed value when the eye valve is opened and descends to the egg when the eyelids are closed. Calculating a computer blink in the EAR involves the following combination: Imaging of the eye area, the limit of finding the white spot of the eye, which reflects the blink of an eye by detecting the disappearance of white eyes. [4,7]

6. Predict drowsiness from calculations:

The calculated EAR is tested and compared with a predefined value of 0.3 for this system. If the calculated EAR is greater than 0.3, the system predicts a drowsy state to be true, some false. For the analysis of yawning, if the oral dose is found to be greater than 0.5 and reaches

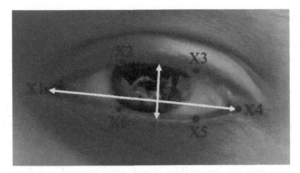

Fig. 11.2   Principal Component (Eye Opening)

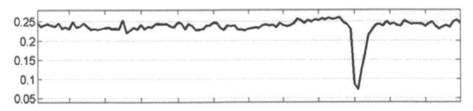

Fig. 11.3   The peak indicates closure and blinking of eye

6 frames per second, the driver is thought to be entering a drowsy state. Figure 11.4 shows the graph designed for the EAR over time. If the maximum value in the graph stays longer than the maximum, the system predicts a sleepy state.

7. Raising an alarm:

The beep sound of frequency 3500 is finally struck with the help of a sys module whenever the trigger is triggered after a drowsiness is detected in the system. The system will be in alarm mode until the driver does not return to normal.

Table 11.1   Comparison between proposed and existing system

MEASURES	OUR SYSTEM	EXISTING SYSTEM
ACCURACY	EAR ratio and Object detection: (Dim light – 92%, 88% and Bright light – 98%, 93%)	EAR ratio: (Dim light – 93%, and Bright light –97%)
BEHAVIORAL MEASURES	Yawning, Object detection, Eye blinking, Closure of eyes, Position of head, Working in dim light	Yawning, Eye blinking, Closure of eyes, PERCLOS
HARDWARE	2 GHz Processor, Memory 20GB/4GB, Display – 800 × 600	Arduino UNO SMD. HDMI Cable, Adapter, GSM Module. Buzzer
SOFTWARE	Python (OpenCV, TensorFlow)	Cross Compiler Arduino-1.5.5v. Android Programming

B. Techniques Used

So one can understand the effects of drowsiness, with high-tech technology that does not require the use of non-disruptive and cost-effective solutions. There is a lot of research to enforce such strategies. There are a number of software such as Tensor Flow and Open CV that can help you know the face and its unusual features to make fatigue recovery easier to use. There are techniques such as tracking the distance between the eyelids by the length to judge drowsiness. Others include features such as blink rate and view acquisition. The following is a mouth-watering experience in people when there is more than one yawn. Various techniques include tracking car information to strike over sleep within the dynamic force. It contains abnormal directional movements, reversible mood, unexpected acceleration or unexpected descent, lateral position within the track, and so on.

Image-Based Processing:

(i) Eyelid movement-based technique: There are four primary sorts of eye activities: sacades, smooth pursuit activities, retaliatory activities, and vestibule-visual activities. The elements of each kind of eye development are introduced here; in ensuing stages, brain hardware liable for 3 of these sorts of activities is given an extra component. Saccades are exchanged on an immediate point very quickly. They differ long from little developments even to examination; for instance, numerous enormous developments made while checking a room out. Saccades can be stimulated consequently; however, the Saccades moves very rapidly as soon as eyes are opened at any time. Rapid eye movement in all important parts of sleep is also in saccades. After the beginning of the saccade objective, it takes around 2 ms for the eye development to start. During this suspension, the objective region with the appreciate in fovea is determined (i.e., how far the eye ought to go), and the distinction between the first and planned position, or "engine deserts", is changed over into an engine order that opens extraocular muscles to move the eyes ideal distance on the correct way. Saccadic eye developments are supposed to be ballistic in light of the fact that the saccade creation framework can't answer ensuing changes inside the objective capacity during eye bearing. Assuming that the objective development is rehashed right now (request 15–100 ms), the saccade will be left unattended, and saccade 2 ought to be made to address the mistake [2,4,6,9]

(ii) Eye Blinking-Based Technique: At this rate, the blink of an eye and the duration of blindness are measured to cause drowsiness because when the driving force feels sleepy then his blinking eye and the look between the eyelids are different from normal conditions so without any problems they get drowsy. Figure 11.4 shows the blink based on getting completely drowsy in this gadget; the location of the irises and eye conditions are monitored over time to measure the frequency of the blink of an eye and its immediate length. This type of machine is also used as a remote camera to collect video and portable visual acuity techniques and then use it to line face, eyes, and eyelids according to the degree of closure. Using these eyes near the blink of an eye is easy to detect. Such a system, installed in a smart car environment, can show any signs of tilt of the head, loose eyes, or simultaneous yawning. The following separations show the blinking sight. [1,3,4,5,8,9]

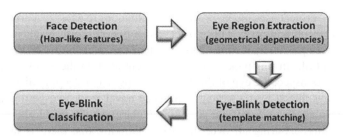

Fig. 11.4 Proposed algorithm for eye-blink detection

(iii) Yawning-Based Technique: Drowsiness is primarily based on the level of yawn. This includes various advances that incorporate continuous recognition and checking of the driver's face, location, oral following, and identification of yawning signs in light of estimating each worth and measure of rectification inside the objective region. The car's APEX™ smart camera platform has greatly improved through Connive Corp. In our approach, the face of the driving force is always captured using a front-facing video camera that shows inside the car finding the next drowsiness involves two important steps to get to the right level changes in facial expressions that indicate drowsiness.

V. Experimental Results

We have successfully implemented our proposed method on a Corei5 8th generation portable computer with a clock speed of 2.6 GHz and 8GB of RAM under the TENSORFLOW environment using a built-in web camera of 0.9MP (1280 × 720). In this test, we tested our system for different people in different situations to confirm the output. The system works with high accuracy even with major obstacles such as mirrors, dim light, and sound.

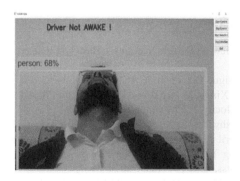

Fig. 11.5 Drowsiness Snapshots

6. Conclusion & Future Scope

Driver sleep systems are being analyzed and a driver alert system is being developed. In this paper, the main focus is on avoiding accidents due to drowsiness and alert drivers when the situation arises. This project uses image processing techniques to analyze a driver's drowsiness by continuing to detect blinking eyes. This project has been successfully tested in all strict testing conditions. It has been proven by testing that the system works at almost 100% accuracy when the lights are on and off. In this paper, we have concentrated on the extraordinary strategies that can be utilized to inspect the eye of the driver to recognize a sleeping disorder. This paper momentarily portrays the bit-by-bit strategies for each novel-made eye-getting method. Also, it portrays the advantages and disadvantages of particular procedures in light of the exactness of results and genuine enduring all over the planet thinking that there is no informational index right now with eccentric techniques that are remarkably difficult to assess the impacts of genuine elite methodologies completely. This paper has made sense of the significance of the outcomes revealed by different creators and how they might connect with the objective of this paper on tracking down the most effective way to get languor.

Since that is a genuine issue and is presently unbelievable, from here on out, it needs a reasonable arrangement. As made sense of in the remainder of the paper, there are a couple of innovations accessible to coincidentally find driving weariness; however, they have their shortcomings. As the best execution of the main impetus locator with the most dependable illustrations should have been accomplished. When these types of parameters are used, it can improve accuracy in many ways. We plan to further the work-related drawings with the help of a sensor to sing the heartbeat along the way to prevent accidents caused by sudden heart attack in drivers. The same version and techniques that can be used for a variety of applications such as Netflix and other live streaming services which can be logged in while the user is asleep and block the video, as a result, can also be used for sleep deprivation. For the future, we are planning a similar study on this topic with a view to having a positive response to help limit and further eliminate this problem, and, as a result, designing a low-cost device that can detect drowsiness to protect the highway.

REFERENCES

1. Bagus G. Pratama, IgiArdiyanto, Teguh B. Adji, A Review on Driver Drowsiness Based on Image, Bio Signal, and Driver Behavior, IEEE, July 2017
2. SacidFazli, Parisa Esfehani, Tracking Eye State for Fatigue Detection, ICACEE, November 2012. Gao Zhenhai, Le DinhDat, Hu Hongyu, Yu Ziwen, Wu Xinyu, Driver Drowsiness Detection Based on Time Series Analysis of Steering Wheel Angular Velocity, IEEE, January 2017.
3. Fouzia, Roopalakshmi R, Jayantkumar A Rathod, Ashwitha S, Supriya K, Driver Drowsiness Detection System Based on Visual Features., IEEE, April 2018.
4. Kyong Hee Lee, Whui Kim, Hyun Kyun Choi, Byung Tae Jan. A Study on Feature Extraction Methods Used to Estimate a Drivers Level of Drowsiness, IEEE, February 2019
5. M. Ali, S. Abdullah, C. S. Raizal, K. F. Rohith and V. G. Menon, "A Novel and Efficient Real Time Driver Fatigue and Yawn Detection-Alert System," 2019 3rd International Conference on Trends in Electronics and Informatics (ICOEI), 2019, pp. 687–691, doi: 10.1109/ICOEI.2019.8862632.
6. K. Satish, A. Lalitesh, K. Bhargavi, M. S. Prem and T. Anjali., "Driver Drowsiness Detection," 2020 International Conference on Communication and Signal Processing (ICCSP), 2020, pp. 0380-0384, doi: 10.1109/ICCSP48568.2020.9182237.

7. V B Navya Kiran, Raksha R, Anisoor Rahman, Varsha K N, Dr. Nagamani N P, 2020, Driver Drowsiness Detection, INTERNATIONAL JOURNAL OF ENGINEERING RESEARCH & TECHNOLOGY (IJERT) NCAIT – 2020 (Volume 8 – Issue 15)

8. Diwankshi Sharma, Sachin Kr Gupta*, Aabid Rashid, Sumeet Gupta, Mamoon Rashid, Ashutosh Srivastava "A novel approach for securing data against intrusion attacks in unmanned aerial vehicles integrated heterogeneous network using functional encryption technique", Transactions on Emerging Telecommunications Technologies, Wiley, 32(7) pp: 1–32, 2020.

9. Akshita Gupta, Sachin Kr Gupta*, Mamoon Rashid, Amina Khan, Manisha Manjul "Unmanned aerial vehicles integrated HetNet for smart dense urban area", Transactions on Emerging Telecommunications Technologies, Wiley, pp: 1–22, 2020.

Intelligent Systems and Smart Infrastructure – Brijesh Mishra et al. (eds)
© 2023 Taylor & Francis Group, London, ISBN 978-1-032-41287-0

CHAPTER

12

5G in Healthcare: Revolutionary Use-cases and QoS Provisioning Powered by Network Slicing

Monika Dubey[1] and Richa Mishra[2]

Department of Electronics and Communication,
University of Allahabad, Prayagraj, India

Abstract

The healthcare industry radically shifted from conventional hospitals to AI-based medical assistance that requires more intelligent equipment and fast internet connectivity. This paper briefly discusses the four potential use-cases of healthcare scenarios, including telesurgery, telemonitoring, teleconsultation, and connected ambulance. It is found these services have diverse service requirements and KPIs (bandwidth, latency, reliability, and throughput). For example, the connected ambulance has strict mobility requirements but mobility is not the priority in the case of remote surgery. Achieving this diverse requirement is not possible with the legacy network's "one-fit-all" model. 5G-enabled network slicing (NS) plays a pivotal role in QoS provisioning. To deal with the above-mentioned issue, this paper proposed a QoS-aware NS-based model that could help to provide network flexibility. The performance of the proposed model is evaluated on the Mininet emulator.

Keywords: Network Slicing, eHealth, QoS Provisioning, SDN

1. Introduction

The healthcare sector undergoes radical change due to the sudden increase in emergency healthcare services due to COVID-19 and its variants. With the technical advancement, the

Corresponding author: [1]mdubey.452@gmail.com and [2]richa_mishra@allduniv.ac.in

DOI: 10.1201/9781003357346-12

e-healthcare system has proven as a promising vehicle for the healthcare sector. It privileges not only patients to reach the doctor remotely but also helps in minimizing medical error, life risk, and treatment costs. It has been forecasted that by 2025, the healthcare and fitness sectors could have an economic impact of $ 170 billion to $ 1.6 trillion per year. (James et al., 2015). Before moving forward, it is important to discuss major restrictions of the e-healthcare system. Firstly, the current healthcare system is not personalized. Usually, doctors are prescribing medications based on general symptoms based on the medical history stated by a patient. Secondly, advanced healthcare services like connected ambulances and teleconsultation services demand communication between high mobility and static setup (Emergency ambulance). Thirdly, telesurgery and connected ambulances need ultra-low latency (near 1ms) and reliability (99.999%). Communication latency (delay more than 20 ms) and reliability is the major issue with the existing network. The legacy network "on-fit-all" paradigm is not suitable to deal with these diverse and continuous requirements. The emergence of the fifth-generation wireless network (5G) and IoTs has motivated a major paradigm shift in the healthcare industry. 5G-enabled NS plays a pivotal role in QoS provisioning. Software-Defined Network (SDN) and Network Function Virtualization (NFV) allow the NS paradigm to manage complex scenarios in the healthcare industry.

The major contributions of this paper are mentioned below:

- Review the role of 5G and IoT in the advanced e-healthcare system.
- Highlights of four potential use-cases (telemonitoring, teleconsultation, telesurgery, and connected ambulance), their working principles, and QoS requirements in a healthcare scenario.
- Proposed NS-based QoS provisioning for the future e-healthcare system. This approach could help to provide network flexibility.
- The performance of the proposed model is evaluated on the Mininet emulator.

The remainder of this paper is organized as follows: Section 2 discusses the role of 5G and IoT in e-healthcare systems. Section 3 includes the working principle of advanced healthcare use-cases and associated QoS requirements. Section 4 includes the contribution of NS to achieving QoS requirements. Section 5 proposed a QoS-aware model for QoS provisioning. Section 6 evaluates the performance of the proposed model and the concluding remarks are given in Section 7.

2. Role of 5G and IoT in e-heathcare System

The e-healthcare system is an example of human-to-machine (H2M) and machine-to-machine (M2M)-type communication (Mohanta et al., 2019). It is heterogeneous in terms of hardware, software, and network resource requirements. The emergence of the 5G and IoTs has motivated a major paradigm shift in the healthcare industry. 5G is aiming to enhance the network performance in terms of bandwidth, reliability, efficiency, etc. (Monserrat et al., 2015). Massive-MIMO, beamforming, and millimeter wave (mmWave) are the advanced underpinning technologies associated with the 5G network. It breaks the traditional paradigms and enables networks to serve diversified vertical industries based on the QoS (bandwidth, reliability, latency, etc.) (Akyildiz et al., 2016). ITU has categorized use-cases into three broad categories: (i) enhanced Mobile Broadband (eMBB), (ii) massive Machine-Type Communication (mMTC), and (iii) Ultra-Reliable and Low Latency Communication (URLLC) (ITU-R M.2083, 2015). The complex e-healthcare scenario utilizes all three service categories. eMBB provides a stable connection and high data rate required for patient real-time data gathering during emergency ambulance services and remote surgery. mMTC supports massive IoT device requirements for patient healthcare monitoring . URLLC focuses on ultra low latency and reliable services required for services like remote surgery. IHS Markit analyzed that by 2035, 5G will bring the opportunity of $13.2 trillion in global output in the healthcare sector (Campbell et al., 2019). 5G-enabled e-health services with the combination of other advanced technologies (AI, big data, ML, and IMoT) have the potential to revamp the challenges of the existing healthcare system (Ahad et al., 2020).

3. Healthcare Scenarios

The next-generation wireless network leverages the healthcare industry to ensure specific QoS provisioning. Tele-surgery, telemonitoring, teleconsultation, and connected ambulance are the potential use-cases discussed in this section. The working principle and QoS requirements associated with these advanced use-cases are discussed in Table 12.1. Tele-monitoring refers to the continuous monitoring of the change in the critical profile of the patient remotely. It helps the patients to receive medical services without visiting the hospital physically. It mainly includes the population suffering from chronic illness (cardiopulmonary disease, heart disease, asthma, etc.) or fragile people (Dzogovic et al., 2020). Fig. 12.1 shows a model telemonitoring system. The system demands a high connection density, reliability, and security as it is related to the private medical data of individuals. The current 4G network is not suitable to deal with the massive demand for connecting terminals (Lloret et al., 2017).

Teleconsultation refers to the use of ICT to support remote clinical services to patients. Fig. 12.2 shows the framework that involves patient–doctor communication. It allows patients to take virtual appointments via two-ways HD conference.

Telesurgery privileges a surgeon to remotely perform surgery. It allows the surgeon to guide and support in-person or robot through experience over the internet. Fig. 12.3 describes the architecture of the telesurgery system. It consists of three domains: master domain, slave domain, and network domain (Gupta et al., 2019).

Medically equipped 5G-enabled "connected ambulance" acts as a connecting hub between patient and hospital. According to the instruction, paramedic staff can perform urgent treatment before they reach the hospital. This use-case has strict QoS (reliability, bandwidth, latency, and mobility) demands (Zhai et al., 2021).

Like other industries, the healthcare industry also has diverse requirements. Table 21.1 broadly describes the components and requirements of advanced healthcare services. It can be observed from the table that the healthcare use-case has very diverse sub-use cases. For example, the connected ambulance has strict mobility requirements but mobility is not the priority in the

Table 12.1 Elements of e-healthcare sub-use cases (Indicator – Low (↓), Medium (↔), High (↑))

Parameter		Tele-monitoring	Tele-consultation	Tele-surgery	Connected ambulance
QoS Requirements		Capacity, Energy efficiency, Reliability and security, Mobility	Bandwidth Mobility Security	Extremely low latency Reduced jitter High bandwidth Reliability	Sensing vital signs (High availability, security, reliability, and energy efficiency) Remote multi- media services (Low latency ,high bandwidth and mobility)
Resources		Sensors for vital signs for collecting patient's health attribute. Wearable (on body) Sensors (onbed)	Multimedia for video conferencing. Communication network for two-way multimedia communication.	Multimedia service (real time video communication, AR/VR headset)	On body sensors and wearable's for sensing vital signs (heart monitoring, infusion pumps, ECG etc.) Multimedia equipment's (video camera and audio devices) 5G infrastructure for communication between ambulance and hospital Generic hardware and video conferencing components (HD cameras and AR/VR headset)
Users		Patient getting remote services. Healthcare professional.	Patient receiving healthcare services. Healthcare professional.	It consists of three domains: master domain: the remote surgeon slave domain: the robot network domain: the virtual reality interface between the master domain and the slave domain	Patient receiving remote care. Paramedic staff for assistance Hospital staff located at hospital performing diagnosis on the basis of patient's medical condition.
KPI	Reduced latency	↔	↔	↑	↑
	High bandwidth	↔	↑	↑	↑
	Mobility	↑	↑	↓	↑
	Capacity	↑	↔	↔	↔
	Reliability	↑	↑	↑	↑

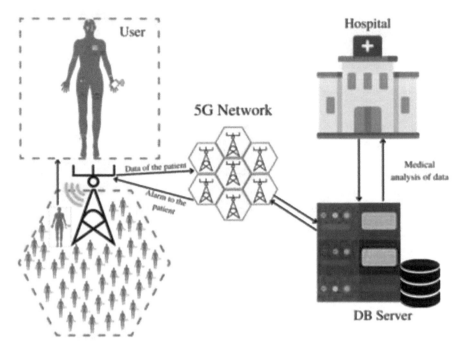

Fig. 12.1 Tele-monitoring system.

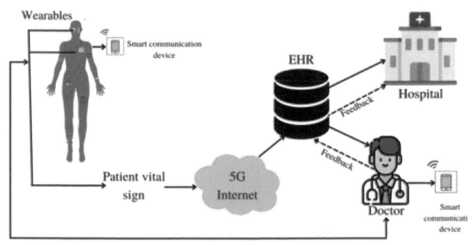

- Audio conferencing
- Video conferencing
- Feedback

Fig. 12.2 Tele-consultation system

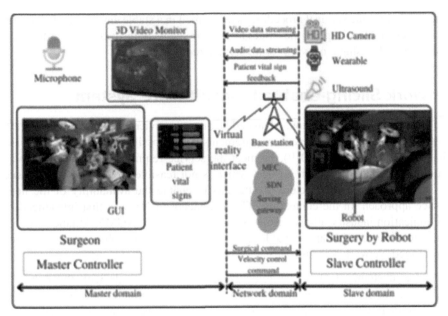

Fig. 12.3 Tele-surgery system

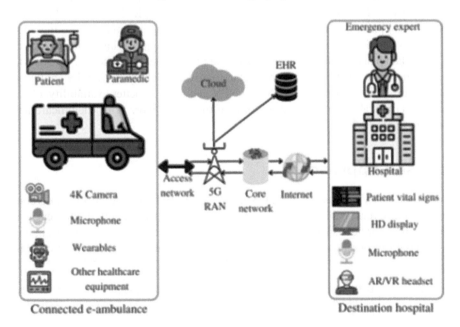

Fig. 12.4 Connected ambulance

remote surgery since the surgical procedure is performed in a stable environment. A traditional network is not suitable to support the strict QoS requirements of the healthcare industry as it is designed to provide a single network for all service requirements.

4. Network Slicing-enabled e-healthcare System

As discussed in the previous section, advanced healthcare services have diverse and strict QoS requirements. Achieving these service requirements is not possible with the legacy network's "one-fit-all" model (Su et al., 2019). To meet the service requirements, a new and separate infrastructure is required for each service, which is not feasible. To deal with this issue, the concept of NS came into the picture. It attempts to enable the availability of the network-as-a-service approach. In the context of 5G, the concept of NS was first introduced by the telecommunication industry alliance NGMN in February 2015 (NGMN Alliance, 2015). It is an architectural concept that allows the logically cutting of a single physical layer in the form of slices, termed Network Slice Instance (NSI).

Each NSI has customized QoS requirements of different verticals with the help of QoS flow based on service level agreement (SLA) between the mobile operator and business customers. To deal with heterogeneous and dynamic requirements, NS is strengthening with new innovative technologies such as SDN and NFV. It allows the NS paradigm to manage and provide network flexibility. SDN is the approach to decouple the control plane and data plane. The control plane is responsible for network monitoring and making routing decisions on how packets should flow. Subsequently, based on the control plane routing policy, the forward plane/data plane forwards packets from one place to another. On the other hand, NFV decouples the network functions from proprietary hardware. It helps to run the application on generic hardware. A combination of SDN and NFV can provide better availability, efficiency, and scalability (Barakabitze et al., 2020). Some of the recent work done (healthcare and 5G) is discussed in Table 12.2.

5. Proposed Approach

This paper proposes an NS-based novel QoS-aware approach in order to meet the diverse QoS requirements of different healthcare services. It mainly includes VoIP, multimedia services, AR, and delay-sensitive services such as remote surgery. Although the proposed approach, physical networks and their resources are partitioned into three pre-configured slices mentioned below:

Slice Type	Description
Emergency Slice (NSres)	Dedicated resources (resource isolation) are required for emergency services. To ensure the uninterrupted services, selection of admission control technique also plays an important role.
Dynamic Slice (NSdyn)	This type of slice scheduling plays an important role for eMBB-type services. It aims to share resources to achieve the guaranteed throughput.
On-demand Slice (NSod)	It is preferred for URLLC services. It is mainly focused on delay-sensitive resource utilization.

Table 12.2 Research work done in the field of healthcare

Reference	Aim	Description	Approach and Outcomes
(Liu et al., 2020)	Heterogeneous network slicing model for tele-medicine scenarios	• The model categorizes telemedicine service requirements into four categories: Medical Data Transmission, Audio and Video Transmission (large bandwidth and low latency), Real-Time Medical Guidance (high bandwidth and ultralow latency), and telemedicine operations. • Applied resource allocation-based classification for inter-slice and intra-slice resource allocation.	• Radial-based function • 98% resource utilization
(Vergütz et al., 2020)	NS-based framework for reliable smart healthcare applications.	• ML-based approach to classifying traffic type for smart healthcare applications.	• ML-based fingerprinting approach. • 90% accuracy
(Celdrán et al., 2019)	SDN and NFV based approach for dynamic NS management	• Highlighted challenges of resource allocation. • Proposed SDN-based architecture to manage the NS life cycle in a dynamic environment.	• Applied policy-based approach defining intra-slice and inter-slice policies.
(Vergütz et al., 2020)	Proposed PRIMS architecture to meet the low latency and high reliability of smart healthcare applications.	• A PRIM consists of three modules: Decision-making, NS management, and Resource allocation. • Emulation result shows that NS and SDN provide flexibility to meet requirements of healthcare applications.	• A PRIM consists of three modules: Decision-making, NS management, and Resource allocation. • Emulation result shows that NS and SDN provide flexibility to meet requirements of healthcare applications.
(Kammoun et al., 2018)	Presented architecture for two-way communication for in-ambulance patient treatment.	• Presented architecture for two-way communication between in-ambulance patient and healthcare expert in hospital. • Two-way communication includes video streaming for patient's medical condition monitoring and vital sign monitoring.	• Described how NS leverages the eMBB and URLLC services for healthcare emergency services.

The first is the emergency slice (NS_{res}). Resources of the reserved slice will not be shared with any other slice. It is reserved for emergency services in massive communication breakdowns like massive earthquakes and flood situations. Rest two slices are shared NS. One is for delay-sensitive services (NS_{od}) and another is for services that seek high data rates (NS_{dyn}) (Schmidt, et al., 2019). In this scenario, the common slice is considered for eMBB and mMTC services. When a user request is made, admission control mechanism based on SLA is applied to select the appropriate slice to meet QoS requirements. However, SLA bases services are out of consideration in this work.

Constraints for slice (i) creation for network resource (NR_j) (Banchs et al., 2020):

$$\text{Reliability: } R_j \geq R_{ij}$$

$$\text{Delay: } D_j \leq D_{ij} \tag{1}$$

$$\text{Availability: } A_j \geq A_{ij}$$

Slice-type resources represented in equation (1) are allocated to user requests only when its availability A_j is equal to or higher than the required user availability. In this work, QoS-aware priority queuing and admission control strategy is applied for resource allocation. In this work, the NS_{res} has the highest priority ($w < -\infty$) for critical services. The second and third priority is given to NS_{od} and NS_{dyn}.

6. Experimental Setup and Result Analysis

The performance of the proposed work is emulated on the Mininet emulator under the SDN paradigm. It supports NS and virtualization technique. This emulator runs on Linux Ubuntu 18.04 operating system. Formally, let $N = (1,2,3,\ldots,N)$ be a set of requests generated from a different application. Specific requirements (RNS) = $[B_{min}, L_{max}]$ of different applications are defined through QoS-module on the controller in terms of delay (L) and bandwidth (B). Other QoS parameters are mentioned below.

Parameter	Value
Number of slices	3
Slice type	{Emergency Slice(NSres),URLLC(NSod) and eMBB(NSdyn)}
Total no. of user requests (N)	100
Emergency Slice (NSres)	B = [10 kbps, 1 Gbps], L= [1 ms, 50 ms]
URLLC (NSod)	B = [50 Mbps, 1 Gbps], L= [1 ms, 10 ms]
eMBB (NSdyn)	B = [200 Mbps, 1 Gbps], L= [1 ms, 50 ms]

In this setting, all three slices are managed by SDN-based controller. SDN controller manages the QoS-flow through OpenvSwitches using MininetQoS-module, using priority queuing to guarantee the QoS in the network. We applied three different admission control algorithms to optimize network resources. NS_{res} has the highest priority, so no restriction is applied to this request. NS_{od} aims to serve delay-sensitive services, so delay threshold is applied for admission on this slice. NS_{dyn} aims to provide data-driven services. In this paper, the performance of the proposed model is based on two parameters delay and packet loss rate for three different slices. NS_{res} initially performs well and achieved a latency of 5 ms. Later, with increased user request (100_h request), it decreases to 87 ms. As resources are reserved, it works efficiently and packet loss of only 0.83%. With the increased number of requests, loss percent also increased up to 37%. On the other hand, for NS_{od}, latency and the packet loss rate are observed 9 ms and 2.8%, respectively, initially. NS_{dyn} has the highest latency in comparison with the other two slices initially, i.e. 25 ms, and packet loss increases up to 80%.

7. Conclusion

This paper investigates the importance of the 5G in the healthcare vertical. It concentrates on QoS provisioning and how it can be achieved. For this purpose, this paper highlights four potential use-cases: tele-surgery, tele-monitoring, tele-consultation, and connected ambulance. The key capability of e-health services is ultra-low latency, bandwidth, capacity, reliability, mobility, and energy efficiency. It is observed that different use-cases in healthcare verticals have diverse and strict QoS requirements. For example, the connected ambulance has strict mobility requirements but mobility is not the priority in the case of remote surgery. Currently, 4G technologies lack in providing reliable connection and not enough to meet strict QoS requirements. To deal with this issue, this paper highlights that the network slice-based healthcare system with QoS-aware scheduling algorithm could help to provide network flexibility. The performance of the proposed model is evaluated on Mininet emulator. The result is evaluated based on two parameters delay and packet loss rate for all three slices Emergency Slice (NSres), Dynamic Slice (NSdyn), and On-demand Slice (NSod). It is observed that this proposed model works well initially for all three different slices but its performance decreases with the number of requests increases. In the future, we will try to improve its performance by modifying admission control, routing, and scheduling strategies. Deep reinforcement-based learning will be taken into account to enhance QoS provisioning.

REFERENCES

1. Manyika, J., Chui, M., Bisson, P., Woetzel, J., Dobbs, R., Bughin, J., & Aharon, D. (2015). Unlocking the Potential of the Internet of Things. McKinsey Global Institute, 1.
2. Mohanta, B., Das, P., & Patnaik, S. (2019, May). Healthcare 5.0: A paradigm shift in digital healthcare system using Artificial Intelligence, IOT and 5G Communication. In 2019 International Conference on Applied Machine Learning (ICAML) (pp. 191–196). IEEE.
3. Monserrat, J. F., Mange, G., Braun, V., Tullberg, H., Zimmermann, G., & Bulakci, Ö. (2015). METIS research advances towards the 5G mobile and wireless system definition. EURASIP Journal on Wireless Communications and Networking, 2015(1), 1–16.
4. Akyildiz, I. F., Nie, S., Lin, S. C., & Chandrasekaran, M. (2016). 5G roadmap: 10 key enabling technologies. Computer Networks, 106, 17-48. Series, M. (2015). IMT Vision–Framework and overall objectives of the future development of IMT for 2020 and beyond. Recommendation ITU, 2083, 21.

5. Campbell, K., Diffley, J., Flanagan, B., Morelli, B., O'Neil, B., & Sideco, F. (2019). The 5G economy: How 5G technology will contribute to the global economy. IHS economics and IHS technology, 4, 16.

6. Ahad, A., Tahir, M., Aman Sheikh, M., Ahmed, K. I., Mughees, A., & Numani, A. (2020). Technologies trend towards 5G network for smart health-care using IoT: A review. Sensors, 20(14), 4047.

7. Dzogovic, B., Santos, B., Jacot, N., Feng, B., & Van Do, T. (2020, May). Secure healthcare: 5G-enabled network slicing for elderly care. In 2020 5th International conference on computer and communication systems (ICCCS) (pp. 864-868). IEEE.

8. Lloret, J., Parra, L., Taha, M., & Tomás, J. (2017). An architecture and protocol for smart continuous eHealth monitoring using 5G. Computer Networks, 129, 340-351.

9. Gupta, R., Tanwar, S., Tyagi, S., & Kumar, N. (2019). Tactile-internet-based telesurgery system for healthcare 4.0: An architecture, research challenges, and future directions. IEEE Network, 33(6), 22–29.

10. Zhai, Y., Xu, X., Chen, B., Lu, H., Wang, Y., Li, S., ... & Zhao, J. (2021). 5G-network-enabled smart ambulance: architecture, application, and evaluation. IEEE Network, 35(1), 190–196.

11. Su, R., Zhang, D., Venkatesan, R., Gong, Z., Li, C., Ding, F., ... & Zhu, Z. (2019). Resource allocation for network slicing in 5G telecommunication networks: A survey of principles and models. IEEE Network, 33(6), 172–179.

12. Alliance, N. G. M. N. (2015). 5G white paper. Next generation mobile networks, white paper, 1(2015).

13. Barakabitze, A. A., Ahmad, A., Mijumbi, R., & Hines, A. (2020). 5G network slicing using SDN and NFV: A survey of taxonomy, architectures and future challenges. Computer Networks, 167, 106984.

14. Liu, Y., Wang, L., Wen, X., Lu, Z., & Liu, L. (2020, August). A Network Slicing Strategy for Telemedicine based on Classification. In 2020 IEEE/CIC International Conference on Communications in China (ICCC Workshops) (pp. 191–196). IEEE.

15. Vergutz, A., Noubir, G., & Nogueira, M. (2020). Reliability for smart healthcare: A network slicing perspective. IEEE Network, 34(4), 91–97.

16. Celdrán, A. H., Pérez, M. G., Clemente, F. J. G., Ippoliti, F., & Pérez, G. M. (2019). Dynamic network slicing management of multimedia scenarios for future remote healthcare. Multimedia Tools and Applications, 78(17), 24707–24737.

17. Vergütz, A., G Prates, N., Henrique Schwengber, B., Santos, A., & Nogueira, M. (2020). An Architecture for the Performance Management of Smart Healthcare Applications. Sensors, 20(19), 5566.

18. Kammoun, A., Tabbane, N., Diaz, G., Dandoush, A., & Achir, N. (2018, May). End-to-end efficient heuristic algorithm for 5G network slicing. In 2018 IEEE 32nd International Conference on Advanced Information Networking and Applications (AINA) (pp. 386–392). IEEE.

19. Schmidt, R., Chang, C. Y., & Nikaein, N. (2019, December). Slice scheduling with QoS-guarantee towards 5G. In 2019 IEEE Global Communications Conference (GLOBECOM) (pp. 1–7). IEEE.

20. Banchs, A., de Veciana, G., Sciancalepore, V., & Costa-Perez, X. (2020). Resource allocation for network slicing in mobile networks. IEEE Access, 8, 214696-214706.

Intelligent Systems and Smart Infrastructure – Brijesh Mishra et al. (eds)
© *2023 Taylor & Francis Group, London, ISBN 978-1-032-41287-0*

CHAPTER

13

Design and Analysis of High-Speed Low-Power Dynamic Comparator

Kunal[1], Anurag Yadav[2], Subodh Wairya[3]

Department of Electronics and Communication Engineering,
Institute of Engineering and Technology, Lucknow, A.K.T.U., Lucknow, India

Abstract

Now a days less power consumption and high-speed devices are need; therefore, there is a need to grow in the domain in which there is less trade-off between power and delay. A new topology of standard single-tail dynamic comparator (STDC) is presented in the paper. The comparison is done on the basis of delay, power, and PDP. The proposed comparator runs on clock frequency of 500 MHz. When compared with the previously defined STDC and standard double-tail dynamic comparator (DoTDC), this circuit gets less power consumption and delay. This is due to addition of a p-type metal oxide semiconductor (PMOS) transistor in the latch as well as in the amplification stage of the dynamic comparator. An 85% reduction in power and a 52% reduction in delay are observed after simulation. The proposed comparator works on the 90-nm technology. All the simulation results are carried out using CADENCE software at 90-nm technology. The total power dissipation is 54.89 µW and the delay is 107.10 ps.

Keywords: Standard single-tail dynamic comparator (STDC),Standard double-tail dynamic comparator (DoTDC), pass transistors, low power dissipation, less delay

1. Introduction

Comparator is the heart of analog-to-digital converter (ADC). These days ADC is used in many electronics devices, including medical kits. So the lowest power consumption is one of

Corresponding author: [1]kunal243001@gmail.com,[2]anuraglko2014@gmail.com, [3]swairya@gmail.com

DOI: 10.1201/9781003357346-13

the basic parameters, because these devices operate on battery. Another parameter is that the device should be less bulky (means less die area). Last but not, least speed of the comparator should be very high, which means less delay. It is an electronic device that compares two signals (voltage or current) applied to its input and generates an output specifying which signal is larger. Comparator is the elementary block of an ADC and is often referred to as 1-bit ADC. Comparators are devices that compare two signals at its input and produce an output accordingly. Operation of a comparator can be stated as follows: if the non-inverting terminal has higher input applied to it than the inverting terminal, then it produces logic high at the output. Similarly, if the input at the inverting terminal is greater, then it produces a logic zero as the output. Nowadays we are continuously scaling down the CMOS technology but the change in threshold voltage is not possible at the same matching pace, which restricts the scaling of the supply voltage. So high-speed comparator has low power supply problem. This problem creates more challenges and difficulties for the designer those who deal with mixed signal. There are basically two types of comparators: (a) Dynamic comparator and (b) Static comparator. Power consumption in the static comparator is more as compared to dynamic comparator as these are always ON. The speed is also limited due to the absence of negative feedback and clock signal. So there will be no static power dissipation. The symbol of the basic comparator is shown Fig. 13.1. High-speed ADC requires high-speed comparators and the speed of comparators basically depends on several parameters. The performance of ADC is totally depending on the comparators so that high performance comparator is need. For good performance of the comparator, the combination of high speed and accuracy is need.

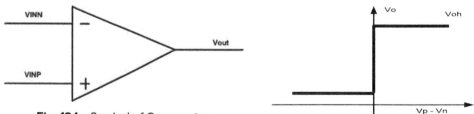

Fig. 13.1 Symbol of Comparator

Fig. 13.2 Ideal voltage transfer characteristic (Razavi 2015)

The accuracy is increased when the peak input error or the kickback noise is reduced. This article is bifurcated into four different segments. Figure 13.2 shows the voltage transfer characteristic (ideal) of opamp (see equations 1 and 2) (Razavi 2015). The graphical curve of the characteristic states that the output of the opamp is directly related to the difference between applied input voltages.

$$\text{If } V_p > V_n, \text{ then } V_o = V_{oh} \tag{1}$$

$$\text{If } V_p < V_n, \text{ then } V_o = V_{ol} \tag{2}$$

According to the above-mentioned equation, when the voltage at positive terminal Vp is larger then the Vn negative terminal, the output voltage Voh reaches on high voltage, i.e Voh and vice versa. As a result, the voltage transfer curve is a straight line until the output voltage hits saturation. Following that, output remains constant, as shown in Figure 13.2. Section (I) contains the basic introduction of the comparator and its application in modern world. Section (II) accommodates the previously defined architecture of the different comparators. Here is the discussion on standard single-tail and standard double-tail design of the dynamic comparators with their transient analysis. Section (III) carries the discussion on the proposed work of the dynamic type of comparator. This is based on single-tail technology. The main advantage of the proposed design can be noticed in the power dissipation and delay. Section (IV) is result and discussion in which the comparison of the parameters used for the dynamic comparator is summarized in the tabular form. Section (V) holds the conclusion.

2. Different Topologies

In this section, we have simulated and analyzed two topologies of clocked dynamic comparator, that are standard single-tail dynamic comparator (STDC) and standard double-tail dynamic comparator (DoTDC). Based on this study, we have proposed a new circuit which uses single tail topology.

A. Standard Dynamic Comparator

Figure 13.3 contains the circuit diagram and Fig. 13.4 shows the waveform of the standard STDC. This is known as the STDC comparator because there is only single discharging path for the current. In single-tail comparator, both the operations latch as well as amplification is not separate, because of this property, the kickback noise of the comparator increases and also the delay is high due to this structure. The working of the standard STDC when the clock is zero in this stage is known as the reset stage and the transistor Mtail is off, so there is no static

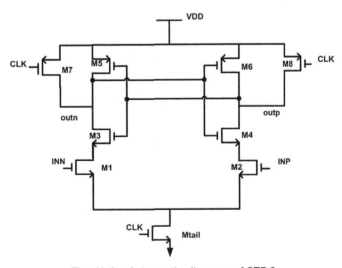

Fig. 13.3 Schematic diagram of STDC

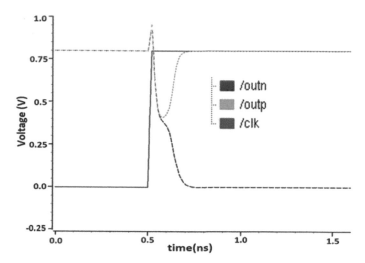

Fig. 13.4 Transient waveform of standard STDC

power dissipation when the clock is low. Transistors (M7–M8) are known as reset transistor and they are on during this period. These transistors pull up the outp and outn to vdd to have a valid logic level. Now when the clock is vdd or high, the reset transistors (M7–M8) are off. Now the output nodes outp and outn start to discharging in accordance to the voltage applied on the input of the comparators. So, when the Vinp > Vinn, the outp discharges faster as compared to outn. When the voltage at outp node is low, then the transistor M5 will get switched ON which helps outn node to be charged upto vdd while the outp node continues to discharge fully.

Kickback noise is defined as the fluctuation in the input signal of the comparator. These small spikes in the input signal are known as the kickback noise. The parameter which affects the kickback noise are capacitors and the supply voltage of the comparator.

Delay of this comparator comprises two parts which are listed below. t_0 represents the time taking in discharging of the load capacitor, represented by equation 3 from Babayan-Mashhadi and Sarvaghad-Moghaddam (2014), and equation 4 represents t_{latch}; this delay is defined as the time taken by latched transistor to make decision in the second stage of the operation. Equation 5 represents the total delay.

$$t_0 = \frac{C_L |V_{thp}|}{I_2} \cong 2 \frac{C_L |V_{thp}|}{I_{tail}} \tag{3}$$

$$t_{latch} = \frac{C_L}{g_{m.eff}} \cdot \ln\left(\frac{\Delta V_{out}}{\Delta V_0}\right) = \frac{C_L}{g_{m.eff}} \cdot \ln\left(\frac{V_{dd}/2}{\Delta V_0}\right) \tag{4}$$

$$t_{delay} = t_0 + t_{latch} = \frac{C_L |V_{thp}|}{I_2} + 2\frac{C_L |V_{thp}|}{g_{m.eff}} \ln\left(\frac{V_{dd}}{4 |V_{thp}| \Delta V_{in}} \sqrt{\frac{I_{tail}}{\beta_{1,2}}}\right) \tag{5}$$

B. Double-tail Dynamic Comparator

The diagrammatic portrayal and waveform is exhibited in Fig. 13.5 and 13.6. This topology is different from the standard dynamic comparator.

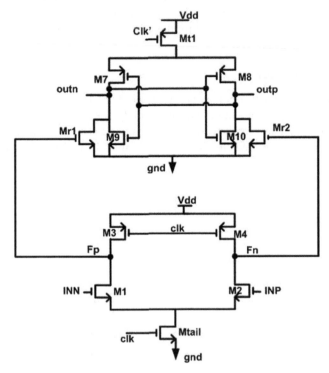

Fig. 13.5   Schematic diagram of DoTDC

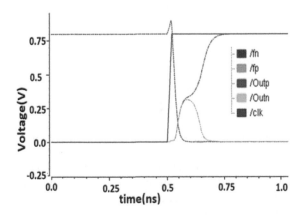

Fig. 13.6   Transient waveform of the standard DoTDC (Babayan-Mashhadi and Sarvagh-ad-Moghaddam 2014)

When the clock is low or at zero voltage, the intermediate output nodes Fn, Fp and the outn, outp charge to vdd. Since when the clock is '0' tail transistors Mtail1,2 are off; in this condition, transistors Mr1,2 get on and the output nodes start to discharge ground. Now when the clock is high, the tail transistor gets on and the MOS Mr1,2 off. The output nodes start to discharge with a defined rate, which is depending on the input voltage applied on the circuit. When the Vinp > Vinn, then fn discharges faster than fp; therefore, Mr2 gets off and the outp charges to vdd and the outn discharges to gnd.

C. Proposed Comparator

In Fig. 13.7, we describe the new topology based on the standard STDC. The delay is less than the standard dynamic comparator and fully differential dynamic comparator. The power of the novel circuit is also less. This circuit works on low voltage. The schematic diagram and the transient waveform of the proposed circuit is given in Figs 13.7 and 13.8, respectively.

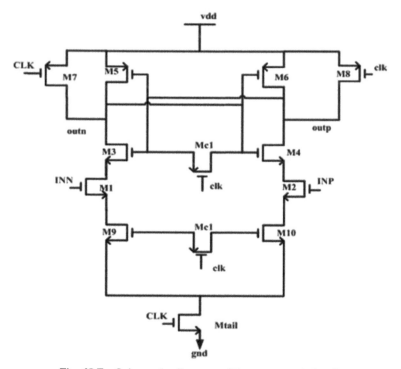

Fig. 13.7 Schematic diagram of the proposed circuit

The operation of the proposed circuit is basically alike to the standard STDC. When we provide clock = '0', transistors M7, M8 and Mc1,2 get on and the tail comparator is off. So, there is very less static power consumption in the circuit. So, when the clock is '0', outn and outp charge to vdd. In the conclusive phase, when the clock is high, '1' nodes of output outn and outp start to discharge with distinct pace totally depending on the applied voltage on M1 and

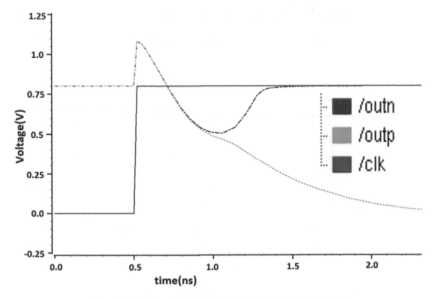

Fig. 13.8 Transient waveform of the comparator

M2 transistors. If the INP > INN, then the outp discharges faster than outn. Therefore, the gate voltage of transistor M5 is low and it gets on and because of that outn charges to vdd again.

If the voltage INN>INP, the process is vice versa; from our calculation, it is inferred that both delay and power are reduced compared to the above-discussed standard dynamic circuit.

D. Results and Analysis

This section contains the discussion of the results of our proposed circuit on the basis of different parameters that are delay, power, PDP, number of transistors, EPC, and the operating clock frequency. This is shown in Table 13.1. Further, Table 13.2 contains collation of our novel circuit with existing configurations, namely standard STDC and standard DoTDC.

Table 13.1 Summary of Performance Analysis

Parameter	Value
CMOS Technology(nm)	90
No. of transistor	13
Clock frequency (GHz)	500
Total delay (ps)	107.1
Power dissipation (μw)	54.87
PDP (fJ/s)	5.85
EPC (fJ)	0.109
Efficiency (fJ)	1.467
No. of transistor	13

Table 13.2   Sizing of the Proposed Comparator

Transistor	Aspect ratio (W/L) nm
M1,2	240/90
M3,4	240/90
M5,6	360/90
M7,8	360/90
M9,10	240/90
Mc1,2	500/90

Table 13.1 shows the performance summary of the proposed comparator and Table 13.2 showcases the aspect ratio of different transistor used in the proposed circuit.

From Table 13.3, comparison can be done and conclude that the proposed comparator performs better as the parameters such as power dissipation, delay and energy per conversion are reduced when compared to the configurations like STDC and DoTDC.

Table 13.3   Simulation results and Comparison of Different Topologies of the Comparators

Topology	Delay (ps)	Power (μW)	PDP (ps)	EPC (fJ)	Efficiency (fJ)
STDC (Babayan-Mashhadi and Sarvaghad-Moghaddam 2014)	224.23	65.88	14.77	0.131	3.692
DoTDC (Babayan-Mashhadi and Sarvaghad-Moghaddam 2014)	142.81	171	24.42	0.342	6.105
Proposed circuit	107.108	54.85	5.874	0.109	1.468

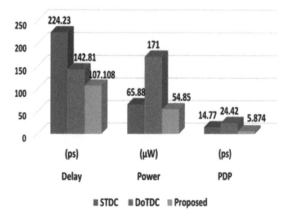

Fig. 13.9   Delay, power and PDP comparison

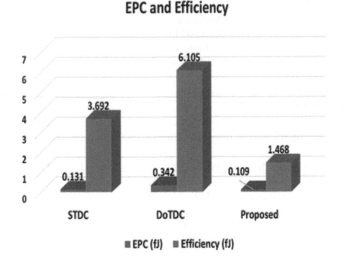

Fig. 13.10   Comparison of EPC and efficiency

Figure 13.10 shows the comparison of the different comparator on the basis of EPC and efficiency. The efficiency of our proposed circuit is increased by 76% compared to the double-tail comparator. From Fig. 13.9, conclusion can be made that our proposed circuit offers 52% less delay than single and provides 85% improvement in power when compared to DoTDC (Babayan-Mashhadi and Sarvaghad-Moghaddam 2014). These show that the novel comparator is working better in terms of delay, power, PDP, EPC, and efficiency.

3. Conclusion

The proposed dynamic comparator is which has PMOS transistor in the latch and amplification stage; this modification shows the significant improvement in the parameters, i.e. power dissipation, time delay, energy per conversion and power delay product. This design is based on single-tail topology. These parameters are simulated on the 90-nm technology of the CADENCE software and comparison of these parameters of proposed design is done with standard STDC, standard DoTDC, and fully differential DoTDC. This circuit works on differential input voltage of 5 mV, common mode voltage of 0.7 V, and supply voltage of 0.8 V. After the comparison, it is noticed that this novel circuit is faster than the discussed standard circuits. This circuit has the delay of 107.10 ps and total power dissipation is 54.87 µw. The EPC and the efficiency of the proposed circuit are also less than the discussed standard circuits.

REFERENCE

1. A. K. Dubey and R. K. Nagaria, "Low-power high-speed CMOS double tail dynamic comparator using self-biased amplification stage and novel latch stage," *Analog Integr. Circuits Signal Process.*, vol. 101, no. 2, pp. 307–317, Nov. 2019, doi: 10.1007/S10470-019-01518-7.

2. A. M. Maghraby, I. T. Abougindia, and H. N. Ahmed, "A Low-Noise, Low-Power, Dynamic Latched Comparator Using Cascoded Structure," *2020 12th Int. Conf. Electr. Eng. ICEENG 2020*, pp. 335–338, Jul. 2020, doi: 10.1109/ICEENG45378.2020.9171746

3. Babayan-Mashhadi, S. and Sarvaghad-Moghaddam, M. (2014) 'Analysis and design of dynamic comparators in ultra-low supply voltages', *22nd Iranian Conference on Electrical Engineering, ICEE 2014*, pp. 255–258.

4. Bindra, H.S. *et al.* (2018) 'A 1.2-V Dynamic bias latch-type comparator in 65-nm CMOS with 0.4-mV input noise', *IEEE journal of solid-state circuits*, 53(7), pp. 1902–1912.

5. Canal, B. *et al.* (2021) 'Low-voltage dynamic comparator using positive feedback bulk effect on a floating inverter amplifier', *Analog Integrated Circuits and Signal Processing 2021 108:3*, 108(3), pp. 511–524.

6. Dubey, A.K. and Nagaria, R.K. (2019) 'Low-power high-speed CMOS double tail dynamic comparator using self-biased amplification stage and novel latch stage', *Analog Integrated Circuits and Signal Processing*, 101(2), pp. 307–317.

7. Figueiredo, M. *et al.* (2011) 'A two-stage fully differential inverter-based self-biased CMOS amplifier with high efficiency', IEEE Transactions on Circuits and *Systems I: Regular Papers*, 58(7), pp. 1591–1603. doi:10.1109/TCSI.2011.2150910.

8. Gandhi, P.P. and Devashrayee, N.M. (2018) 'A novel low offset low power CMOS dynamic comparator', *Analog Integrated Circuits and Signal Processing*, 96(1), pp. 147–158. doi:10.1007/S10470-018-1166-9.

9. Gupta, A., Singh, A. and Agarwal, A. (2021) 'A low-power high-resolution dynamic voltage comparator with input signal dependent power down technique', *AEU - International Journal of Electronics and Communications*, 134. doi:10.1016/J.AEUE.2021.153682.

10. H. Yousefi and S. M. Mirsanei, "Design of a novel high speed and low kick back noise dynamic latch comparator," *2020 28th Iran. Conf. Electr. Eng. ICEE 2020*, Aug. 2020, doi: 10.1109/ICEE50131.2020.9260821.

11. Joseph, G.M. and Shahul Hameed, T.A. (2021) 'A Sub-1 V Bulk-Driven Rail to Rail Dynamic Voltage Comparator with Enhanced Transconductance', *https://doi.org/10.1142/S0218126622500645*, 31(4).

12. Khorami, A. and Sharifkhani, M. (2016) 'High-speed low-power comparator for analog to digital converters', *AEU - International Journal of Electronics and Communications*, 70(7), pp. 886–894. doi:10.1016/J.AEUE.2016.04.002.

13. K. Yoshioka, "VCO-Based Comparator: A Fully Adaptive Noise Scaling Comparator for High-Precision and Low-Power SAR ADCs," *IEEE Trans. Very Large Scale Integr. Syst.*, vol. 29, no. 12, pp. 2143–2152, Dec. 2021, doi: 10.1109/TVLSI.2021.3119691.

14. M. Figueiredo, R. Santos-Tavares, E. Santin, J. Ferreira, G. Evans, and J. Goes, "A two-stage fully differential inverter-based self-biased CMOS amplifier with high efficiency," *IEEE Trans. Circuits Syst. I Regul. Pap.*, vol. 58, no. 7, pp. 1591–1603, 2011, doi: 10.1109/TCSI.2011.2150910.

15. Razavi, B. (2015) 'The StrongARM latch [A Circuit for All Seasons]', *IEEE Solid-State Circuits Magazine*, 7(2), pp. 12–17. doi:10.1109/MSSC.2015.2418155.

16. S. Chevella, D. OrHare, and I. OrConnell, "A Low Power 1V Supply Dynamic Comparator," *IEEE Solid-State Circuits Lett.*, 2020.

17. Schinkel, D. *et al.* (2007) 'A double-tail latch-type voltage sense amplifier with 18ps setup+hold time', *Digest of Technical Papers - IEEE International Solid-State Circuits Conference* [Preprint]. doi:10.1109/ISSCC.2007.373420.

18. S. K. Sharma, R. S. Gamad, and P. P. Bansod, "Design and Analysis of Low Power, High Speed, Leakage Control Comparator using 180nm Technology For A/D Converters," *Proc. IEEE Madras Sect. Int. Conf. 2021, MASCON 2021*, 2021

19. V. Jain, S. Tayal, P. Singla, V. Mittal, S. Gupta, and J. Ajayan, "An Intensive Study of Thermal Effects in High Speed Low Power CMOS Dynamic Comparators," *Proc. 6th Int. Conf. Commun. Electron. Syst. ICCES 2021*, pp. 250–254, Jul. 2021,.

20. V. Moyal and N. Tripathi, "Adiabatic Threshold Inverter Quantizer for a 3-bit Flash ADC," *Proc. 2016 IEEE Int. Conf. Wirel. Commun. Signal Process. Networking, WiSPNET 2016*, pp. 1543–1546, Sep. 2016, doi: 10.1109/WISPNET.2016.7566395.

21. Varshney, V. and Nagaria, R.K. (2020) 'Design and analysis of ultra high-speed low-power double tail dynamic comparator using charge sharing scheme', *AEU - International Journal of Electronics and Communications*, 116, p. 153068.

22. Y. Wang, M. Yao, B. Guo, Z. Wu, W. Fan, and J. J. Liou, "A Low-Power High-Speed Dynamic Comparator With a Transconductance-Enhanced Latching Stage," *IEEE Access*, vol. 7, pp. 93396–93403, 2019, doi: 10.1109/ACCESS.2019.2927514.

Intelligent Systems and Smart Infrastructure – Brijesh Mishra et al. (eds)
© 2023 Taylor & Francis Group, London, ISBN 978-1-032-41287-0

CHAPTER

14

Tomato Leaf Diseases Detection Using Deep Learning—A Review

Vishal Seth[1], Rajeev Paulus[2] and Anil Kumar[3]

Vaugh Institute of Agricultural Engineering and Technology (VIAET), Sam Higginbottom
University of Agriculture Technology and Sciences Prayagraj, U.P, India

Abstract

Tomatoes are India's popular vegetable crop. Its growth is best suited to a tropical environment. Climate change and plant diseases, on the other hand, have wreaked havoc on agricultural output, resulting in financial losses. Disease diagnosis in the traditional sense is sluggish and occurs late in the ripening life cycle. Early detection has the potential to produce better results than the standard technique. Thus, use of deep learning (DL) with computer vision can be used for early detection. DL is a cutting-edge image processing technology that has demonstrated its worth and yielded promising outcomes. This study examines forty studies that employed DL approaches to solve issues in tomato plants. The examination included data processing technique, applied transfer learning, augmentation technique, and various DL models used DL approaches that outperform all other methods in image processing, according to the findings. However, it is largely dependent on the dataset utilized.

Keywords: Tomato Leaf Disease, Deep Learning, CNN, Feature Extraction, Data Augmentation

1. Introduction

Agriculture is crucial to both the world and Indian economies. The growth in population along with urbanization has rapidly lowered the value of cultivated land, which has increased the concern over the agriculture techniques. The affair of plant diseases has an adverse effect on

Corresponding author: [1]sethvishu.gp@gmail.com, [2]rajeev.paulus@shiats.edu.in, [3]anil.kumar@shiats.edu.in

DOI: 10.1201/9781003357346-14

productivity of agriculture lands [1]. Timely plant diseases detection is the foundation for control and elimination of diseases. This plays an important step in management and resolution of agriculture production. In most cases, the leaves of plants are the primary source of information on plant health, including disease and insect infestation, as most of the indication of disease may start to appear on the leaves [2]. The recognition of plant disease is mainly done by on-site framers based on their knowledge. This process is not reliable, time-consuming, and also inefficient for classification of diseases in plants or trees.

In order to combat this process and fulfil the ever-increasing demand for food, machine learning methods and computer vision are currently being used to lower the amount of worry and improve the quality of food management [3]. These intelligence-based systems are able to show optimistic results in various applications; recently, it has also been introduced in the agriculture domain [4]. There are various applications of machine learning and deep learning (DL) techniques in real-world scenario such as health care [5], social network [6], visual data processing [7], and audio processing [8]. The efficiency of DL models like the convolution neural network (CNN) [9], as well as long short term memory (LSTM) [10] and auto encoders [11], has been demonstrated to solve the daily life problems.

In the realm of biomedicine, DL automation has shown to be an excellent technique for picture categorization and illness diagnosis. For example, detection of tumors (benign, malignant), skin diseases, blood cancers, anomalies of heart, eye, and chest, etc. [12]-[13] The image classification approach uses multiple properties of plant leaves, pests, and diseases to design an imaging model which can be used for further classifications. The working of CNN model can be classified into two phases: firstly, to capture the image of the plant or crop accordingly to the area of interest and, secondly, to provide these capture image into deployed model to get the output and future analysis.

The implication of this paper is to gather work related to tomato plant like classification, tomato detection, and disease identification, which can help the further research work in generating a more accurate and clear DL model which can help to fill the gap in research and agriculture domain [14].

Section 2 of the article will provide a quick overview of DL, shared database, transfer learning, and data augmentation. Section 3 consists of a table containing detail analysis of all the existing work. Section 4 consists of a summary of what is achieved from the existing work and Section 5 is used to pen down the conclusion along with suggestion for future models.

2. Breif Description of Deep Learning

A. Deep Learning's History

The term "deep learning" refers to a subset of machine learning; it came into picture around the year 1943 and developed with the help of three stages. The initial stage consisted of a linear model that could only be used to solve a linear classification issue. In 1986, the second stage of neural networks introduced the notion of backward propagation, which was suited for multilayer perceptron (MLP), which provided an efficient solution for non-linear classification. The third stage of neural network started from the year 2006 continues till now and since then CNN has fascinated the researchers community.

B. Evaluation Matrices

These are the statistical data used for analytic performance measure of the classification and regression problem with DL. The measurement grouped images into 4 categories as follows:

1. TP (True Positive) – The number of genuine positive picture samples that are perfectly identified as tainted.
2. FP (False Positive) – The number of false positive input images that are incorrectly recognised as polluted.
3. FN (False Negative) – The number of picture samples incorrectly labelled as non-contaminated.
4. TN (True Negative) – The number of picture samples successfully categorised as healthy/non-contaminated.

C. Common Tomato Database

Noticing the database used by most of the authors for their models is the publicly available Plant Village [17]-[22] database which can be seen in Table 14.1. It has 39 classes of diverse crop pictures such as apple, maize, tomato, potato, grapes, and a few more, with 10 classes dedicated to tomato plants.

Table 14.1 Plant Village Dataset Summary of Tomato Classes

Classes (Tomato Plant)	Images
Tomato Bacterial spot disease	2127
Tomato Septoria leaf spot disease	1723
Mosaic virus disease	325
Leaf mold disease	904
Target spot disease	1356
Early blight disease	952
Yellow leaf curl virus disease	4032
Tomato Late blight disease	1781
Two spotted spider mite	1628
Healthy	1591

3. Examining Previous Work

Table 14.2 gives the detailed analysis of all the existing work.

Table 14.2 Analysis of all the existing work related to tomato plant

Reference No	Data Source			Data Augmentation	Transfer Learning	Architecture	Result	Analysis	Comparison
	Class	Image	Ratio						
31	2	43,843	50/25/25	No	Yes	ResNet50	Precision – 94.6 %	Defects in tomatoes were identified but all images were with black background in lab setup condition. Imbalance data set results in high precision, whereas the rottenness or bruises were not identified	ResNet101, ResNet104
17	4	1,20,000	80/16/4	Yes	No	Residual Deep CNN	Accuracy – 98 %	Used plant village data and only three diseases were detected using residual learning plus attention mechanism. High detection rate as compared to existing methods	CNN – 84% accuracy
18	7	13,262	N/A	No	Yes	AlexNet	Accuracy – 97.4 %	Used AlexNet and VGG16 model, altered batch size to decrease accuracy. AlexNet shows high accuracy with less execution time. Low data set with only 373 images per class. Transfer learning using ImageNet weight	VGG16
19	3	2343	80/20/0	No	Yes	AlexNet	Accuracy – Feature Extraction 93.4 %	Using pre trained weights on a very small data sets and lacks with real images of field. Among the three models, AlexNet proves to be better suited for fine-grained image classification.	SqueezeNet, Inception V3

Reference No	Data Source			Data Aug-menta-tion	Transfer Learn-ing	Architecture	Result	Analysis	Comparison
	Class	Image	Ratio						
24	5	4,178	60/10/80	No	No	Improve FRCNN	Accuracy – 2.71% higher than FRCNN	An accuracy of 98.54% is achieved. Fast detection speed as compared to original faster RCNN. Used ResNet for feature extraction. Lack real images, including fruits and stem and combination of characteristics for further development	FRCNN
35	11	286	60/20/20	No	No	FRCNN, Mask RCNN	ResNet 101,Shortest detection time.	The mask RCNN along with ResNet 101 shows higher identification rate and accuracy of 99.64% mAB but mask RCNN training time is longer than faster RCNN. Low image resolution cause detection failure	MobileNet, Resnet 101
36	3	5624	80/20/0	No	No	Customized FRCNN, K-means clustering, Soft NMS	Accuracy-Tomato flowers 90.5 %, green tomato 90.8 %, red tomato90.9 %	The training model can be converted into tomato picking robot device in future. Average pulling source higher accuracy than max pulling. Data set only contains healthy images of tomatoes, which makes this model enable to detect any type of disease or physical damage which occur generally in real scenario	With various architecture and NMS
37	12	15000	N/A	No	No	Improved Yolo V3	Accuracy – 92.39%	The datasets is self-created with different weather and light conditions, resulting in high accuracy but increased computational complexity	FRCN, YOLO V3

Refer-ence No	Data Source			Data Aug-menta-tion	Transfer Learn-ing	Architecture	Result	Analysis	Comparison
	Class	Image	Ratio						
20	10	16419	N/A	No	No	PVA, WOA	Accuracy – 94%	This paper does not consider various illumination conditions required for classification. The whale optimization algorithm provides most valuable features used in CNN resulting	N/C
29	11	2000	N/A	Yes	Yes	Super Resolution network	Accuracy – 81.11%	The model suggested in this paper does not use any pre-processing resulting in low accuracy, whereas the predicted time was less than 0.01 sec which is good for practical implementation.	Varying augmented Datasets
38	2	1656	N/A	No	No	SVM and Focal Loss function	F1 score – 92.15%	The small datasets result in high accuracy as using machine learning algorithm and low space of model result in mobile deployment but the datasets only work on red tomatoes not discolored or green tomatoes	DenseNet and ResNet
27	10	1800	77/6/16	Yes	No	CNN Model Customized	Accuracy – 98.7%	The model results in a good accuracy but the dataset used is very small and accuracy also decreases significantly when tested in real-field conditions	SVM,KNN
33	10	16419	80/20/0	No	Yes	GoogleNet	Accuracy – 99.72%	Proposed solution is a 22-layer model which provides a good accuracy but the only drawback is the datasets lacks the life cycle of diseases	ResNet, Inception V3

Refer-ence No	Data Source			Data Aug-menta-tion	Transfer Learn-ing	Architecture	Result	Analysis	Comparison
	Class	Image	Ratio						
21	10	14828	N/A	No	Yes	GoogleNet	Accuracy – 99.18%	This leads to good accuracy but requires high computation and deep neural network which can be reduced	AlexNet
39	5	1000	N/A	Yes	No	CNN	Accuracy – 91.9%	The proposed model received good accuracy and implementation time but the dataset is very small for any type of implementation	Various augmentation
22	5	1500	80/20/0	Yes	No	GAN with GoogleNet	Accuracy – 94.33%	The paper proposed use of DCGAN for data augmentation to achieve higher accuracy, whereas the dataset is very small, which can be used for implementation	GAN with AlexNet, Vgg16, ResNet
40	NAP	640	N/A	No	Yes	FRCNN with ResNet-101	Precision– 87.83	The overlapping size estimation of tomatoes is the main feature of this dataset, so as to develop a model for the robot picking tomatoes	FRCN, ResNeetV2
41	NAP	966	N/A	Yes	No	YOLO – Customized	Accuracy – 94.58	The model decreases the effect of illumination variation and overlapping with the help of proposed C-B box, thus maintaining a high accuracy which can be used for deployment	YOLOv3, FRCNN

Reference No	Data Source			Data Augmentation	Transfer Learning	Architecture	Result	Analysis	Comparison
	Class	Image	Ratio						
42	3	60	70/30/0	No	No	CNN	Accuracy – 100%	The dataset is very small and lacks variation of images as it has very low number of ripe, overripe and rotten photos	ANN,LVQ,SVM
43	2	2365	N/A	No	Yes	MobileNet-v2+YOLOv3	Average F1 Score – 93.07	This proposed model detects only one disease, attaining good result, and is a lightweight model that can be used for mobile deployment but lacks the different variation of diseases	RCNN , SSD
44	38	54306	60/40/0 (Varying)	No	Yes	GoogleNet	Accuracy – 99.35%	The proposed GoogleNet model provides a high accuracy with very fast classification, which allows it to deploy model in the mobile but requires a larger time from training	AlexNet
45	15	4438	92/8/0	Yes	Yes	CNN	Accuracy – 96.30%	The proposed method gives an accurate classification of plant leave disease but require fine-tuning and argumentation to improve accuracy	N/A
46	6	384	90/10/0	No	No	Decision Tree	Accuracy – 97.30%	The proposed model provides good accuracy with supervised learning algorithm but dataset too small	N/A
47	10	N/A	N/A	Yes	Yes	AlexNet	Accuracy – 95.65%	The two model was proposed for mobile computation and computer computation separately but AlexNet requires long training time and small batch size	AlexNet, GoogleNet

Refer-ence No	Data Source			Data Aug-menta-tion	Transfer Learn-ing	Architecture	Result	Analysis	Comparison
	Class	Image	Ratio						
48	11	7040	62.5/25/12.5	Yes	Yes	Vgg-16	Accuracy – 89%	The proposed model is effective but the training includes only very high-quality image, which is not possible during the implementation in the field	SVM
49	3	N/A	N/A	Yes	Yes	Modified CNN	Accuracy – 89%,	In the proposed model, data augmentation and transfer learning is used for the detection	N/A
50	30`	87848	70/30/0	No	Yes	VGG	Accuracy – 99.53%	All the dataset is in laboratory condition, resulting in poor performance	AlexNet
51	6	9000	N/A	No	N/A	CNN	Accuracy – 99.84%	The proposed CNN model tested on both grayscale and colored photo and it receives best accuracy on color dataset but lacks in recognition of diseases	N/A
52	9	5550	80/20/0	Yes	Yes	ResNet	Accuracy – 97.28%	The proposed model has a depth of 37 layers leading to higher accuracy and robustness but this also increases the response time	AlexNet, GoogleNet
53	10	18160	74/26/0	No	Yes	LeNet	Accuracy – 94%	The model has 94.8% accuracy, which can occur due to the use of SGD, whereas various other optimizers such as Adam can be used to increase accuracy	AlexNet, GoogleNet

Reference No	Data Source			Data Augmentation	Transfer Learning	Architecture	Result	Analysis	Comparison
	Class	Image	Ratio						
54	4	4923	80/20/0	No	Yes	F-RCNN	Accuracy – 91.67%	The model uses transfer learning to achieve high accuracy and the best part is it uses field images in real condition	AlexNet
55	5	500	80/20/0	No	No	CNN + LVQ	Accuracy – 88%	High accuracy with the help of CNN and LVQ algorithm. But demerit is the classification rate which is slow	N/A
56	10	14903	80/10/10	No	Yes	VGGNet	Accuracy – 99.11%	The VGGNet outperformed all the other architecture but the demerit is that it requires high-end hardware setup	LeNet, Xception Net
57	10	4585	80/20/0	Yes	Yes	ResNet-50	Accuracy - 97%	The proposed model achieves a high accuracy but it also requires high-end hardware for training	ResNet,GoogleNet

4. Discussion

In this section, an analysis is depicted on all the above-mentioned papers. From the survey, it can be concluded that CNN methods are commonly used for plant leaf disease detection. The detection rate increases significantly in CNN by using object detection models, hence the classification process was notably increased.

Some improvements were noticed based on the observations collected, such as the fact that most researchers focused on identifying diseases and pests, whereas other areas such as macronutrient insufficiency and weed detection in tomato plants can be addressed in future research.

5. Conclusion

The goal of this survey is to inspire new researchers to use DL techniques to agricultural issues, particularly those involving tomato plant categorization, prediction, and data analysis. The use of real-field dataset, augmentation method, and different models has also been examined. Despite the achievements in this field, there still exist some challenges such as

1. mild symptom of plant diseases in their early life-cycle;
2. nutrition imbalance due to multiple factors;
3. plant health consideration by monitoring growth and ripeness life-cycle of fruits;
4. some lesion spots have no determined shape.

The future development of DL technology will also help in solving diverse challenges faced by ongoing research and implementing it in real-world scenario.

REFERENCES

1. Hasan, Reem I., Suhaila M. Yusuf, and Laith Alzubaidi. 2020. "Review of the State of the Art of Deep Learning for Plant Diseases: A Broad Analysis and Discussion" *Plants* 9, no. 10: 1302.
2. Ali Masri, 2020, "MarkTechPost". Available at: https://www.marktechpost.com/2019/06/14/data-pre-processing-for-deep-learning-modelsdeep-learning-with-keras-part-2/.
3. Liakos, K.G.; Busato, P.; Moshou, D.; Pearson, S.; Bochtis, D, Sensors 2018, "Machine Learning in Agriculture: A Review".18, 2674.
4. Kamilaris A and Prenafeta-Boldu F,2018, "Deep learning in agriculture: A survey Computers and Electronics in Agriculture" 147, pp. 70–90.
5. Esteva, A., Robicquet, A., Ramsundar, B., Kuleshov, V., DePristo, M., Chou, K., Cui, C., Corrado, G., Thrun, S., & Dean, J, 2019, "A guide to deep learning in healthcare. *Nature Medicine*", 25, pp. 24–29.
6. Tan, Q.; Liu, N.; Hu, X, 2019, "Deep Representation Learning for Social Network Analysis". Frontiers in Big Data.
7. Meiguins A, Santos Y, Santos D, Meiguins B, Morais J. 2019, "Visual Analysis Scenarios for Understanding Evolutionary Computational Techniques" Behavior Information, pp. 929–931.
8. Purwins, H.; Li, B.; Virtanen, T.; Schlüter, J.; Chang, S.-Y.; Sainath, T.N, 2019, "Deep Learning for Audio Signal Processing". IEEE J. Sel. Top. Signal Process. 13, pp 206–219.

9. Siddharth Das, "Medium" . Available at: https://medium.com/analytics-vidhya/cnns architectures-lenet-alexnet-vgg-googlenet-resnetand-more-666091488df5. Accessed on 21 September 2020.

10. Wang, Y, 2017, "A new concept using LSTM Neural Networks for dynamic system identification". In Proceedings of the 2017 American Control Conference (ACC), Seattle, WA, USA, pp. 5324–5329.

11. Debnath, T.; Biswas, T.; Ashik, M.H.; Dash, S, "Auto-Encoder Based Nonlinear Dimensionality Reduction of ECG data and Classification of Cardiac Arrhythmia Groups Using Deep Neural Network". In Proceedings of the 2018 4th International Conference on Electrical Engineering and Information & Communication Technology (iCEEiCT), Dhaka, Bangladesh, pp. 27–31.

12. Litjens, G.; Kooi, T.; Bejnordi, B.E.; Setio, A.A.A.; Ciompi, F.; Ghafoorian, M.; Van Der Laak, J.A.; Van Ginneken, B.; Sánchez, C.I, 2017, "A survey on deep learning in medical image analysis". Med. Image Anal, pp. 60–88.

13. Alzubaidi, L.; Fadhel, M.A.; Al-Shamma, O.; Zhang, J.; Duan, Y, 2020, "Deep Learning Models for Classification of Red Blood Cells in Microscopy Images to Aid in Sickle Cell Anemia Diagnosis". Electronics, pp 427.

14. Boulent, J.; Foucher, S.; Théau, J.; St-Charles, P.-L, 2019, "Convolutional Neural Networks for the Automatic Identification of Plant Diseases". Front. Plant Sci. pp 941.

15. Elsalamony, H.A, 2016, "Healthy and unhealthy red blood cell detection in human blood smears using neural networks". Micron, pp 32–41.

16. Das, P.K.; Meher, S.; Panda, R.; Abraham, A, 2020, "A Review of Automated Methods for the Detection of Sickle Cell Disease". IEEE Rev. Biomed. Eng, pp 309–324.

17. Karthik R, Hariharan M, Anand S, Mathikshara P, Johnson A, and Menaka R, 2020, "Attention embedded residual CNN for disease detection in tomato leaves Applied Soft Computing J". pp 86.

18. Rangarajan A K, Purushothaman R, and Ramesh A, 2018, "Tomato crop disease classification using pre-trained deep learning algorithm Procedia Computer Science", pp 1040–1047.

19. Verma S, Chug A, and Singh A P, 2020, "Application of convolutional neural networks for evaluation of disease severity in tomato plant" Journal of Discrete Mathematical Sciences and Cryptography 23, pp. 273–82.

20. Gadekallu T R, Rajput D S and Praveen kumar M, 2021, "A novel PCA–whale optimization-based deep neural network model for classification of tomato plant diseases using GPU" Journal of Real-Time Image Processing. pp. 1383–1396.

21. Brahimi M, Boukhalfa K, and Moussaoui A, 2017, "Deep Learning for Tomato Diseases: Classification and Symptoms Visualization Engineering Applications of Artificial Intelligence" 31, pp. 299–315.

22. Wu Q, Chen Y, and Meng J, 2020, "Dcgan-based data augmentation for tomato leaf disease Identification" IEEE Access 8. pp. 98716–98728.

23. Zhao, Jiayue and Jianhua Qu, 2019, "A Detection Method for Tomato Fruit Common Physiological Diseases Based on YOLOv2." in Proceeding of 10th International Conference on Information Technology in Medicine and Education (ITME). pp. 559–563.

24. Zhang Y, Song C, and Zhang D, 2020, "Deep Learning-Based Object Detection Improvement for Tomato Disease" IEEE Access 8. pp. 56607–56614.

25. Gutierrez A, Ansuategi A, Susperregi L, Tubio C, Rankic I, and Lenza L, 2019, "A Benchmarking of Learning Strategies for Pest Detection and Identification on Tomato Plants for Autonomous Scouting Robots Using Internal Databases Journal of Sensors".

26. [26] Fuentes A, Yoon S, Kim S C and Park D S, 2017, "A robust deep-learning-based detector for realtime tomato plant diseases and pests recognition Sensors".

27. Agarwal M, Gupta S K and Biswas K K, 2020, "Development of Efficient CNN model for Tomato crop disease identification Sustainable Computing: Informatics and System" 28/100407.
28. Nithish E, Kaushik M, Prakash P, Ajay R, and Veni S, 2020, "Tomato Leaf Disease Detection using Convolutional Neural Network with Data Augmentation" Proceeding of 5th International Conference on Communication and Electronics Systems, pp. 1125–1132.
29. Zhang L, Jia J, Li Y, Gao W and Wang M, 2019, "Deep learning based rapid diagnosis system for identifying tomato nutrition disorders KSII Trans. Internet Inf. Syst." 13, pp. 2012–2027.
30. Sharpe S M, Schumann A W and Boyd N S, 2018, "Goosegrass Detection in Strawberry and Tomato Using a Convolutional Neural Network", Scientific Reports 10, pp. 1–8.
31. da Costa A Z, Figueroa H E H and Fracarolli J A, 2020, "Computer vision based detection of external defects on tomatoes using deep learning Biosystems Engineering", pp. 131–44.
32. Xu Z F, Jia R S, Liu Y B and Zhao C Y, 2020, "Fast method of detecting tomatoes in a complex scene for picking robots", IEEE Access 8, pp. 106890–106898.
33. Maeda-Gutierrez V, Galvan Tejada E and Zanella-Calzada A, 2020, "Comparison of convolutional neural network architectures for classification of tomato plant diseases", Applied Sciences 10.
34. R. G. de Luna, E. P. Dadios and A. A. Bandala, 2018, "Automated Image Capturing System for Deep Learning-based Tomato Plant Leaf Disease Detection and Recognition," in Proceeding of TENCON 2018 - 2018 IEEE Region 10 Conference, pp. 1414–1419.
35. Wang Q, Qi F, Sun M, Qu J, and Xue J, 2019, "Identification of Tomato Disease Types and Detection of Infected Areas Based on Deep Convolutional Neural Networks and Object Detection Techniques Computational Intelligence and Neuroscience".
36. Sun J, He X, Ge X, Wu X, Shen J and Song Y, 2018, "Detection of key organs in tomato based on deep migration learning in a complex background Agriculture", pp. 8–12.
37. Liu J and Wang X, 2020, "Tomato Diseases and Pests Detection Based on Improved Yolo V3 Convolutional Neural Network Frontiers in Plant Science", 11, pp. 1–12.
38. Liu J, Pi J, and Xia L, 2020, "A novel and high precision tomato maturity recognition algorithm based on multi-level deep residual network Multimedia Tools and Applications", 79, pp. 9403–9417.
39. Zhang L, Jia J, Gui G, Hao X, Gao W and Wang M, 2018, "Deep Learning Based Improved Classification System for Designing Tomato Harvesting Robot", IEEE Access 6, pp. 67940–67950.
40. Mu Y, Chen T S, Ninomiya S, and Guo W, 2020, "Intact detection of highly occluded immature tomatoes on plants using deep learning techniques Sensors", (Switzerland) 20, pp. 1–16.
41. Liu G, Nouaze J C, Mbouembe P L T, and Kim J H, 2020, "YOLO-tomato: A robust algorithm for tomato detection based on YOLOv3 Sensors", (Switzerland) 20, pp. 1–20.
42. M. Haggag, S. Abdelhay, A. Mecheter, S. Gowid, F. Musharavati and S. Ghani, "An Intelligent Hybrid Experimental-Based Deep Learning Algorithm for Tomato-Sorting Controllers," in IEEE Access, vol. 7, pp. 106890-106898.
43. Liu J and Wang X, 2020, "Early recognition of tomato gray leaf spot disease based on MobileNetv2", YOLOv3 model Plant Methods, pp. 1–16.
44. Mohanty, Sharada P., David P. Hughes, and Marcel Salathe. 2016, "Using deep learning for image-based plant disease detection", Frontiers in plant science 7, pp. 1419.
45. Sladojevic, S., Arsenovic, M., Anderla, A., Culibrk, D., and Stefanovic, D. 2016, "Deep Neural Networks Based Recognition of Plant Diseases by Leaf Image Classification". Computational Intelligence and Neuroscience.
46. H. Sabrol and K. Satish, 2016, "Tomato plant disease classification in digital images using classification tree," 2016 International Conference on Communication and Signal Processing (ICCSP), pp. 1242–1246.

47. Durmus, Halil, Ece Olcay Gune§, and Murvet Kirci. 2017, "Disease detection on the leaves of the tomato plants by using deep learning." in Proceeding of 6th International Conference on Agro-Geoinformatics. IEEE.

48. Shijie, Jia, Jia Peiyi, and Hu Siping. 2017, "Automatic detection of tomato diseases and pests based on leaf images." Chinese Automation Congress (CAC). IEEE.

49. Adhikari, Santosh, ct al. 2018, "Tomato plant diseases detection system using image processing." in Proceeding of 1st KEC Conference on Engineering and Technology, Lalitpur. Vol. 1.

50. Ferentinos, Konstantinos P. 2018, "Deep learning models for plant disease detection and diagnosis." Computers and Electronics in Agriculture 145, pp. 311–318.

51. A. Kumar and M. Vani, "Image Based Tomato Leaf Disease Detection," 2019 in Proceeding of 10th International Conference on Computing, Communication and Networking Technologies (ICCCNT), pp. 1–6.

52. Zhang, Keke, et al. 2018, "Can deep learning identify tomato leaf disease?." Advances in Multimedia.

53. TP. Tm, A. Pranathi, K. SaiAshritha, N. B. Chittaragi and S. G. Koolagudi, "Tomato Leaf Disease Detection Using Convolutional Neural Networks," in Proceeding of 2018 Eleventh International Conference on Contemporary Computing (IC3), pp. 1–5.

54. R. G. de Luna, E. P. Dadios and A. A. Bandala, 2018, "Automated Image Capturing System for Deep Learning-based Tomato Plant Leaf Disease Detection and Recognition," in Proceeding of TENCON 2018 - 2018 IEEE Region 10 Conference, pp. 1414–1419.

55. M. Sardogan, A. Tuncer and Y. Ozen, 2018, "Plant Leaf Disease Detection and Classification Based on CNN with LVQ Algorithm," 2018 3rd International Conference on Computer Science and Engineering (UBMK), pp. 382–385.

56. A. Kumar and M. Vani, "Image Based Tomato Leaf Disease Detection," 2019 in Proceeding of 10th International Conference on Computing, Communication and Networking Technologies (ICCCNT), pp. 1-6.

57. N. K. E., K. M., P. P., A. R. and V. S., "Tomato Leaf Disease Detection using Convolutional Neural Network with Data Augmentation," 2020 5th International Conference on Communication and Electronics Systems (ICCES), pp. 1125–1132.

Intelligent Systems and Smart Infrastructure – Brijesh Mishra et al. (eds)
© 2023 Taylor & Francis Group, London, ISBN 978-1-032-41287-0

CHAPTER

Identifying the Relation Between Pollution Before and During Lockdown Using ANOVA

Shubham Singhal*, Chiranjeev Garg, Ujjwal Parashar, Tanveer Ikram, Anjali Sharma and Rajesh Singh

Department of Computer Science & Engineering (CSE),
Meerut Institute of Engineering & Technology (MIET),
Meerut, India

Abstract

The main focus of this study is to detect the variation of pollution before lockdown and during lockdown 1 and lockdown 2 in New Delhi and various cities around New Delhi using ANOVA test. During the COVID-19 pandemic, the whole nation is shut down initially for 21 days started from 24th March 2020 to 14th April 2020 and it has been continued up to 3rd May 2020. Due to the mandatory and strict restrictions on whole nation, there has been a steep fall in the level of pollution across the country in just few days, which pulls of the debate regarding shutdown to be an efficient alternate measure to be taken for reducing pollutant in the air. The existing research eventually operates in this way to focus upon the standard of air pollution before and during the shutdown period scientifically to the New Delhi and cities around New Delhi, with the assistance of data on air standard of pollutant parameter particulate matter (PM2.5) for 13 monitoring stations spread in Delhi and surrounding cities.

Keywords: Pollution, Lockdown, ANOVA, Air Quality, Particulate Matter (PM2.5)

*Corresponding author: shubham.singhal.cs.2018@miet.ac.in

DOI: 10.1201/9781003357346-15

1. Introduction

The pandemic outbreak has been announced by World Health Organization (WHO) as a worldwide pandemic on 11th March 2020. The virus was initially discovered from Wuhan city, China on December 31st, 2019 (World Health Organization, 2020); by January 30th, 2020, the WHO announced the outbreak as a global pandemic, which is dangerous for public health. In our country, the initial case of virus was first discovered on January 30th, 2020 in south state (Kerala), when a college-studying student from China returned back to the nation. Even though the cases had been limited in the initial days, looking the threat appearing from this Covid virus and supplemented with the aid of the studies and classes learnt via other nations, the administration of India has declared complete shut down throughout the nation from March 24th, 2020 within the first period accompanied by a second period of extension. The authorities of India made a formidable response on imposition of complete shutdown despite the huge financial losses, but on the other hand, there is a formidable improvement in air quality due to lockdown. For the reason that covid virus particularly affects the respiration organ, different people from different cities facing excessive air pollutants are greater susceptible for viral infection, which could result in higher chance and mortality. In view of that, a residents of India has extended contact to the pollutants like PM10 and NO_2, they own excessive risk of large number of deaths to COVID-19 virus. Consequently, it is anticipated that the locations with negative or poisonous air quality index may additionally revel in huge losses of lives. Keeping in mind the above discussions, the current research has been formulated to look into the result of shutdown on extensive level of air pollution in few parts of India before lockdown and during lockdown.

2. Literature Review

The novel Coronavirus, first discovered in China, causes light to medium respiratory or lungs infections; due to its extensive growth, the WHO announced it a global disease on January 30th, 2020. Since then, almost every country in the world has affected with this virus. The Covid virus can be spread as a result of close contact with a Covid-infected person or by contact with any infected person but the main problem with the covid virus is its undiagnosed symptoms such as colds, flu, sores, sore throat, and dry cough [5]. The research found that COVID-19 patients in major cities were linked with overcrowding, history of travel, and air pollutant levels. As the covid virus affects the respiratory organs, people in country with huge levels of air pollutant are at greater risk of infection and higher mortality [7]. Specifically, the people above the age of sixty years and who already have some medical issues are at higher possibility of exposure of this virus. According to WHO, the major symptoms of this virus are high fever, fatigue, and dry cough [13]. The condition is out of control and there is no proven cure for the virus. Housing confinement and social isolation are the only security measures taken across the country [3]. On the one side, this deadly virus threatens our lives and, on the other side, the improvement of environment is also going on.[1] Moreover, 8 of the 10 worst cities in the world in terms of pollution are found in India, with capital state Delhi topping the particulate matter (PM10) pollution list among global major cities (WHO, 2018) [8]. Therefore, worldwide concern for air pollution has led to major attention to the analysis of air pollution during the pandemic [1].

Following the implementation of strong measures to close the door on the Coronavirus disease, the decrease of non-road and roadway traffic, less important industrial activities, and businesses have led to a major decline in PM concentration [2]. Among the pollutants, PM10 and PM2.5 saw a significant fall followed by CO, NO_2, and NH [1] The unprecedented decline in economic activity resulting from the closure of COVID-19 represents unique opportunities to study the contribution of major sources of pollution and to understand changes in air chemistry under reducing greenhouse gas emissions in cities [2]. Emission of carbon has reduced, and the standard of air has spotted unusual progress. It is surprising to watch 85.1% decrease in concentration of PM2.5 in the capital city of India (Delhi), compared to the concentration 3 months ago [3]. In this regard, performance variance analysis has been shown to be useful for monitoring air quality emergencies before and during job closures and for evaluating the performance equity of each group, which is geographically classified into measurement areas [6]. Air pollution is the world's greatest environmental hazard. Approximately seventy lakh people loss their life prematurely every year worldwide due to contact to air pollutants, with more than ninety one percent of the earth's population living in areas where the quality of air crosses the guidelines of the WHO. PM2.5 is a major reason for bad health effects such as pulmonary disease and respiratory tract infections, resulting in the deaths of nearly thirty lakh people worldwide. Nitrogen dioxide (NO_2) has major contribution in childhood asthma in urban areas worldwide [9]. It is being investigated by this observation that even in the case of clear skies, membership formation may increase a significant amount of PM2.5. Small drops of mist (or fog) evaporate rapidly with the rising of the sun's radiation within 2–3 hours [10]. This also highlighted water quality parameters that are available beyond the permissible distance during and after the most advanced closure during closure [12]. Changes in pollution are measured by comparing concentration during the tight closure (March 25 to 3 May 2020) at the same time in 2017 and 2019. Significant and important reductions were seen in both PM2.5 and PM10 in all areas of country [4]. Implementing the closure of the function at this interim and regional rate is not possible, but circumstances have proven that environmental recovery is much faster when the human intervention is very limited [11].

3. Methodology

The independent variable considered in the experiment was air pollution in three different conditions used to identify level of PM2.5 (three values considered).

The dependent variable was a level of PM2.5 in air.

Three conditions of air pollution were investigated using secondary data.

BLD – Air pollution before lockdown;

LD1 – Air pollution during lockdown 1;

LD2 – Air pollution after lockdown 2;

Data: Air pollution data was taken from government website before and during lockdown 1 & 2.

Thirteen areas in and around Delhi were chosen to collect data. This data was organized into three columns in .csv file, representing PM2.5 before lockdown in column 1, during lockdown 1 in column 2, and during lockdown 2 in column 3. A python program was developed to implement ANOVA and identify relations between three different air pollution conditions.

4. Results

Table 15.1   PM$_{2.5}$ values at different locations in and around Delhi

Area	BLD (Before Lockdown)	LD1 (Lockdown 1)	LD2 (Lockdown 2)
Anand Vihar	78.1	30.09	50.6
Narela	71.9	42.52	60.2
IGIA T3	55.3	34.60	38.7
Najafgarh	69.3	50.60	42.0
Okhla	61.5	38.00	43.2
Jahangirpuri	104.5	122.29	96.6
Sonipat	33.2	22.64	34.9
Panipat	38.7	38.81	42.7
Karnal	51.2	32.25	51.9
Bahadurgarh	61.6	36.7	47.1
Gurugram	71.3	38.4	48.4
Faridabad	71.9	31.47	27.2
Noida	57.4	33.6	49

Relation between BLD and LD1

Table 15.2   Relation between BLD and LDI using one-way ANOVA test

	df	sum_sq	Mean_sq	F	PR(>F)
LD1	1.0	2121.612642	2121.612642	12.932253	0.004198
Residual	11.0	1804.615051	164.055914	NaN	NaN

Relation Between BLD and LD2

Table 15.3   Relation between BLD and LD2 using one-way ANOVA test

	df	sum_sq	Mean_sq	F	PR(>F)
LD1	1.0	1872.174925	1872.174925	10.025996	0.008976
Residual	11.0	2054.052767	186.732070	NaN	NaN

If value of PR is greater than F, then there is a considerable difference between the two values – two values are not the same.

If value of PR is less than F, then there is no much difference between the two values – two values are the same.

In this case, PR < F, then it means there is a relation between two values. It means there is an effect of one value on another. In all cases, value of PR is less than F, which means lockdown has direct relation with pollution.

5. Graph

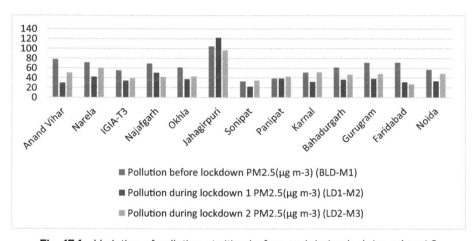

Fig. 15.1 Variation of pollution at cities before and during lockdown 1 and 2

Graph in Fig. 15.1 clearly shows variation in pollution due to lockdown, especially lockdown 1.

6. Conclusion

The outcome illustrates that during strict lockdown, the quality of air is drastically enhanced. The concentrations of air pollutants have observed maximum reduction in comparison to before lockdown. In comparison to before strict lockdown, the reduction was high as about 33% during first lockdown. About 23% improvement in air quality is identified in second lockdown. Overall, the study showed that there is an effect of lockdown on pollution and lockdown is a control source for pollution and hence improves the quality of air. Such temporary steps should be taken regularly to restore the environment but other aspects must also be considered before using it.

REFERENCES

1. Susanta Mahato, Swades Pal, Krishna Gopal Ghosh.: Effect of lockdown amid COVID-19 pandemic on air quality of the megacity Delhi, India. Department of Geography, University of Gour Banga, West Bengal, India, 2020.

2. Pierre Sicard a , Alessandra De Marco b, *, Evgenios Agathokleous c, Zhaozhong Feng c, *, Xiaobin Xu d, Elena Paoletti e, José Jaime Diéguez Rodriguez f, Vicent Calatayud.: Amplified ozone pollution in cities during the COVID-19 lockdown : ARGANS, 260 route du Pin Montard, Biot, France, 2020

3. Albrecht, A. J.: Indirect impact of COVID-19 on environment: A brief study in Indian context. Department of Environmental Science & Technology, Shroff S.R. Rotary Institute of Chemical Technology, Ankleshwar, Gujarat, 393135, India, 2020

4. Vikas Singh a, Shweta Singh a, Akash Biswal a, b, Amit P. Kesarkar a, Suman Mor b, Khaiwal Ravindra c.: Diurnal and temporal changes in air pollution during COVID-19 strict lockdown over different regions of India, National Atmospheric Research Laboratory, Gadanki, AP, India, 2020

5. Chimurkar Navinya1, Girish Patidar1, Harish C. Phuleria: Examining Effects of the COVID-19 National Lockdown on Ambient Air Quality Across Urban India, Interdisciplinary Program in Climate Studies, Indian Institute of Technology Bombay, Mumbai – 400076, Maharashtra, India, 2020

6. Christian Acal1, Ana M. Aguilera1, Annalina Sarra2, Adelia Evangelista2, Tonio Di Battista2 Sergio Palermi.: Functional ANOVA approaches for detecting changes in air pollution during the COVID-19 pandemic, Department of Philosophical, Pedagogical and Economic Quantitative Sciences, University G. d'Annunzio, V.lePindaro, 42, 65127 Pescara, Italy, 2020

7. Anchal Garg, Arvind Kumar & N. C. Gupta.: Impact of Lockdown on Ambient Air Quality in COVID-19 Affected Hotspot Cities of India: Need to Readdress Air Pollution Mitigation Policies, University School of Environment Management, Guru Gobind Singh Indraprastha University, New Delhi, India, 2020

8. Arindam Datta Md. Hafizur Rahman R. Suresh.: Did the COVID-19 lockdown in Delhi and Kolkata improve the ambient air quality of the two cities?, Centre for Environmental Studies, EarthSciences and Climate Change Division, The Energy and Resources Institute, New Delhi, India, 2020

9. Asheshwor Man Shrestha1, Uttam Babu Shrestha1, Roshan Sharma2, Suraj Bhattarai1 , HanhNgoc Thi Tran3, Maheswar Rupakheti4.: Lockdown caused by COVID-19 pandemic reduces air pollution in cities worldwide, Interdisciplinary Program in Climate Studies, 1 Global Institute for Interdisciplinary Studies, Kathmandu, Nepal 2 Centre for Urban Research ,RMIT University, Melbourne, VIC, Australia 3 School of Marketing and Management, The University of Adelaide, Adelaide, SA, Australia 4 Institute for Advanced Sustainability Studies, Potsdam, Germany

10. Surendra K. Dhaka1*, Chetna2, Vinay Kumar1, Vivek Panwar1,A. P. Dimri3, Narendra Singh4, Prabir K. Patra5, Yutaka Matsumi6, Masayuki Takigawa5, Tomoki Nakayama7,KazuyoYamaji 8, Mizuo Kajino9, Prakhar Misra10 & Sachiko Hayashida1.: PM2.5 diminution and haze events over Delhi during the COVID-19 lockdown period: an interplay between the baseline pollution and meteorology, 1 Radio and Atmospheric Physics Lab, Rajdhani College, University of Delhi,, 2020

11. Mohan Sarkar1, Anupam Das2, Sutapa Mukhopadhyay1.: Assessing the immediate impact of COVID-19 lockdown on the air quality of Kolkata and Howrah, West Bengal, India, Department of Geography, Visva-Bharati, Santiniketan, West Bengal 731235, India, 2020

12. National Library of Medicine National Center of Biotechnology Information https://www.ncbi.nlm.nih.gov/pmc/articles/PMC8276728/

13. Vaishali Deshwal, Vimal Kumar.: Study of Coronavirus Disease (COVID-19) Outbreak in India, UttarPradesh , India, Department of Computer Science and Engineering, Meerut Institute of Engineering and Technology, 250005 Meerut, India

Intelligent Systems and Smart Infrastructure – Brijesh Mishra et al. (eds)
© 2023 Taylor & Francis Group, London, ISBN 978-1-032-41287-0

CHAPTER

16

Dual-Square Slot-Shaped Monopole Antenna Using DGS with Protruding Stub in the Ground Plane for Multiband Applications

Parimal Tiwari[1], K. K. Verma[2]

Department of Physics and Electronics,
Dr. Rammanohar Lohia Avadh University,
Ayodhya, U.P., India

Chandan[3]

Department of Electronics and Communication Engineering,
Institute of Engineering & Technology,
Dr. Rammanohar Lohia Avadh University,
Ayodhya, U.P., India

Abstract

In this research article, a dual-square-shaped slots microstrip-fed multiband antenna with protruding stub and defect structure in ground plane is presented. The prototype antenna is drafted using Rogers TMM3 dielectric substrate with relative permittivity 3.27 and dissipation factor 0.002. This antenna is fed by 50Ω microstrip feed lines. The dimensions of the prototype antenna are $34 \times 30 \times 0.8$ mm^3. The three different frequency ranges as (2.3–3 GHz)/2.6 GHz (6–6.2 GHz)/6.1 GHz and (6.8–7.3 GHz)/7 GHz having the return loss of −39 dB, −18 dB, and −38 dB, respectively, are covered by the antenna, which are much suitable for the applications like GSM, Bluetooth, WLAN, WiMAX, and some long-distance radio telecommunication applications. The gains achieved are 2.63 dBi, 3.023 dBi, and 3.23 dBi, respectively. The

Corresponding author: [1]parimal.tiwari1@gmail.com, [2]kkverma23@gmail.com, [3]chandanhcst@gmail.com

DOI: 10.1201/9781003357346-16

various parametric analyses, Return Loss, Current Distribution, Pattern of radiation, and Gain, simulated on ANSYS HFSS, are discussed in this paper.

Keywords: Microstrip feed lines, Square-shaped slots, WLAN, WiMAX, Rogers TMM3.

1. Introduction

For the past few years, the popularity of multiband microstrip patch antenna of different designs for accessing the wireless communications is increasing day by day and has become an important issue. The compact and low-cost antennas with radiation patterns of omnidirectional type with large bandwidth are the requirement of commercial wireless systems nowadays. And it is already known that the printed slot antennas are simple, compact in size, and are of low cost for fabrication. The multiband microstrip antenna is a device which covers different frequencies in a single design for accession of different wireless applications at a time (Du et al., 2001). There is a need of high-speed transmission and greater capacities to make all the services available to the multiple users, and to have this thing done, the design of higher performance systems is required like the multi- and wide-band antennas in Chen and Lin (2003) and Li et al. (2003).

In this paper, dual-square slots cut with modified microstrip feed lines with defected ground structure and protruding stub in the ground plane are presented. This low-profile antenna is simulated on ANSYS HFSS with microstrip feed of 50 Ω. The frequencies are ranging from (2.3–3 GHz)/2.6 GHz, (6–6.2 GHz)/6.1 GHz, and (6.8–7.3 GHz)/7 GHz, having the return loss of −39 dB, −18 dB, and −38 dB, respectively, and suitable for applications like GSM, Bluetooth, WLAN, WiMAX, and some long-distance radio telecommunication applications.

2. Literature Survey

In Eldek (2005), an antenna having square slot with two feedlines, which are orthogonal to each other for dual-band applications, is discussed, which is having large size than the proposed antenna. A square slot with radiating patch antenna is presented in Hadizafar and Azarmanesh (2011), which also has size greater than the proposed antenna. A multiband antenna with square shape slots and microstrip feedline with defected ground structure are presented with larger size when compared to the proposed antenna by Kumawat et al. (2017). In Ali et al. (2016), design of etched slot antenna fed with microstip line for wireless applications producing two bands is discussed. The antenna size is much larger than that of the proposed antenna. When some more reviews are carried out [8-10], it is found that the proposed antenna is having the compact and smaller size. Some more literature survey is done to get the knowledge of the design implementation of the proposed antenna given in [11-18]. Hence, it is found that the proposed antenna is of compact size when compared to different reference antennas.

3. Antenna Representation

The dual-square slot-shaped monopole microstrip antenna using defected ground structure with extended stub in the ground is engraved on Rogers TMM3 substrate constant of dielectric 3.27, thickness 0.8 mm, and loss tangent of 0.002 is demonstrated in Fig. 16.1. A microstrip line of 50 Ω is used for the excitation. The dielectric substrate of length and width denoted as Lsub and Wsub, respectively, is taken. After that, a rectangular patch of length and width Lp and Wp is designed on to the substrate. This antenna is fed with modified microstrip feedlines; the first feed length and width are given by LF1 and WF1 and second feed just after the first one has the dimensions as LF2 and WF2 as the length and width, respectively. A step is added between the feed line and the patch of dimensions P1L and P1W. Onto the patch, two symmetric square slots are cut with sides of the same measure, i.e., S1 = S1 = 10 mm and the slots are 1-mm thick. Defected ground structure technique is used in designing the proposed antenna to provide multi-resonant frequencies with the length and width as Lg and Wg, respectively. A square slot is made with the sides G1 = G2 = 4 mm. A protruding stub is also added in the middle of the ground edge with dimensions LST and WST. The dimension of the proposed antenna is $34 \times 30 \times 0.8$ mm^3.

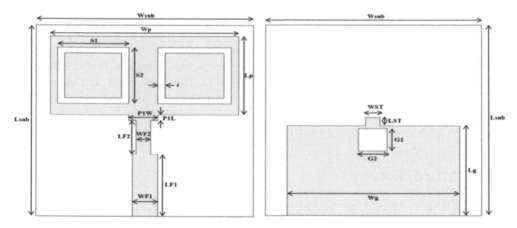

Fig. 16.1 The Antenna Prototype with Dimensions

Table 16.1 gives the optimum geometrical values that are taken to design the proposed antenna. All the parameters are in mm and are fixed.

Table 16.1 Optimum Values of Antenna Geometry

Lsub	Wsub	Lp	Wp	h	Lg	Wg	WF1	LF1	WF2
34 mm	30 mm	14 mm	26 mm	0.8 mm	16 mm	24 mm	3.5 mm	11 mm	2 mm
LF2	**P1$_W$**	**P1$_L$**	**L$_{ST}$**	**W$_{ST}$**	**G1**	**G2**	**S1**	**S2**	**t**
6 mm	4 mm	1 mm	1.5 mm	2 mm	4 mm	4 mm	10 mm	10 mm	1 mm

Source: Author's compilation.

Figure 16.2 illustrates reflection coefficient of the antenna designed with dual-square shaped slot with defect in ground structure and protruding stub in ground plane. The graph shows that the antenna generates three bands at 2.6 GHz, 6.1 GHz, and 7 GHz. The return loss (S11 $\leq$ −10 dB) and the optimised parametric values are provided in Table 16.1.

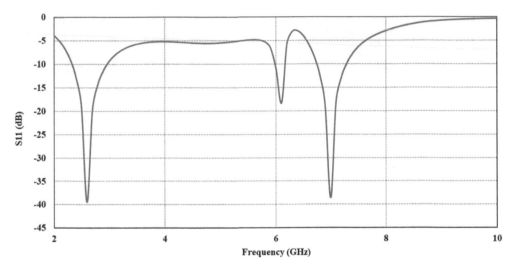

Fig. 16.2 S11 Parameters of Multiband Proposed Antenna

The bands are ranging from (2.3–3 GHz)/2.6 GHz, (6–6.2 GHz)/6.1 GHz, and (6.8–7.3 GHz)/7 GHz with the bandwidth of 27%, 3.3%, and 7.14%, respectively.

4. Antenna Design Evolution

Figure 16.3 demonstrates the different possible configurations of the proposed antenna through which good results can be obtained. The final proposed antenna is shown in Fig. 16.3(d). Antenna 1 in Fig. 16.3(a) is a simple rectangular patch with the microstrip feed lines. This antenna generates only a single resonant frequency as shown in Fig. 16.4, which illustrates the S_{11} parameters of all the configured antennas. Now, a single square-shaped slot is cut from the patch and a protruding stub from the ground is positioned as shown in Fig. 16.3(b).This antenna now generates dual-band with less return loss and impedance bandwidth as shown in the figure as Antenna 2. After that, again a square-shaped slot is cut on the patch and ground is defected by a square slot cut on the ground structure, as shown in Fig. 16.3(c).

This antenna produces three bands but with less return losses at all the frequencies, denoted by Antenna 3 in the figure. Finally, after the parametric analysis, the width of the first feed WF1 is increased by a factor with the increase in the length of the ground plane Lg from the previous design. This is the final proposed antenna shown in Fig. 16.3(d) and its S_{11} parameter is shown in Fig. 16.4 as Antenna 4. It is evident from the figure that the three bands generated

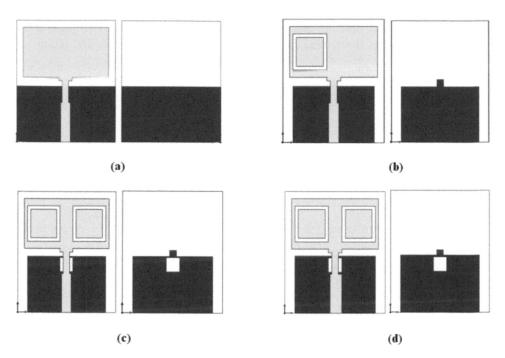

Fig. 16.3 Different Configurations of the Prototype Antenna: (a) Antenna 1, (b) Antenna 2, (c) Antenna 3, and (d) the Final Antenna Prototype

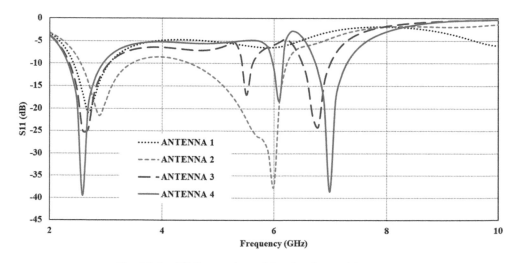

Fig. 16.4 S11 Parameters of the Configured Antennas

have high return losses and good impedance matching with better bandwidth than that of the previous ones.

5. Results and Discussions

The total dimensions of the antenna is $34 \times 30 \times 0.8$ mm^3 and resulted in good return loss and better impedance matching. There is an omnidirectional type of radiation pattern obtained from this antenna, which is just the same as that of the monopole antennas and covers the frequencies from (2.3–3 GHz)/2.6 GHz, (6–6.2 GHz)/6.1 GHz, and (6.8–7.3 GHz)/7 GHz with −39 dB, −18 dB, and −38 dB as return loss, respectively, and suitable for applications like GSM, Bluetooth, WLAN, WiMAX, and some long-distance radio telecommunication applications with the impedance bandwidth of 27%, 3.3%, and 7.14%, respectively.

Figure 16.5 illustrates the reflection coefficient curves for variation in the Lg from 15.75 mm to 16.25 mm. It is evident from the graph that when the Lg is taken as 15.75 mm, the three bands are generated but with less return losses at the obtained frequencies. After that, when the value has become 16.25 mm, antenna resonates dual-band that too with low return losses. But, when the value 16 mm is given to Lg, the antenna resonates three bands with much better return losses than the previous taken values. Hence, the Lg with the value 16 mm gives out the optimum results.

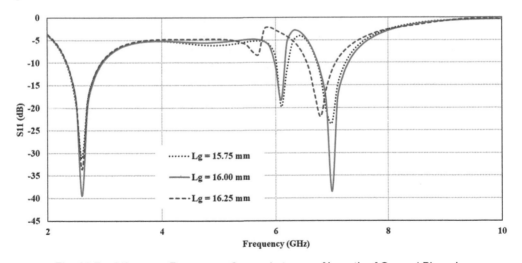

Fig. 16.5 S11 versus Frequency Curves in terms of Length of Ground Plane Lg

In Fig. 16.6, the successive values for the width of the feed 1 WF1 and its effect on the S_{11} graph are illustrated when the values are changed from 3 mm to 4 mm. The graph clearly demonstrates that at all the three values taken during the parametric analysis, antenna generates three bands but it is much more evident from the figure that at the value WF1 = 3.5 mm, the proposed antenna resonates with much high return losses at the obtained frequencies when

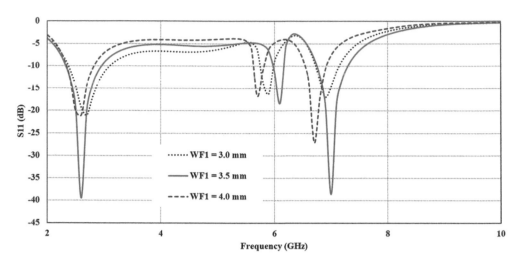

Fig. 16.6 S11 versus Frequency Curves in terms of Width of Feed WF1

compared to the values taken WF1 as 3 mm and 4 mm, in which the return losses are very low when compared. Hence, the value WF1 = 3.5 mm is the optimum value to get the desired results.

Figure 16.7 shows the reflection coefficient curves for the variation in the values of the thickness of the substrate which is taken as h = 0.8 mm and 1.6 mm. The antenna resonates only single band of frequencies when the value of h is taken as 1.6 mm. When h = 0.8 mm, antenna generates three bands with better return losses. Hence, the value of h = 0.8 mm gives the optimum results.

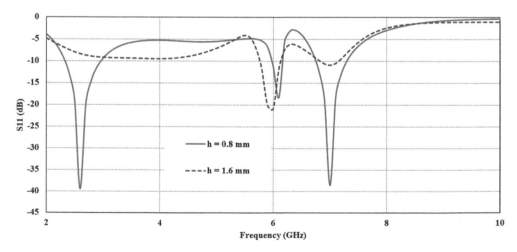

Fig. 16.7 S11 versus frequency Curves in terms of Thickness of the Substrateh

Figure 16.8 shows the surface current distributions at 2.6 GHz, 6.1 GHz, and 7 GHz of the prototype antenna. The indication in red shows the maximum current distribution at that point. At 2.6 GHz, 6.1 GHz, and 7 GHz, the red indication is mainly on the strip line, which gives high current distribution.

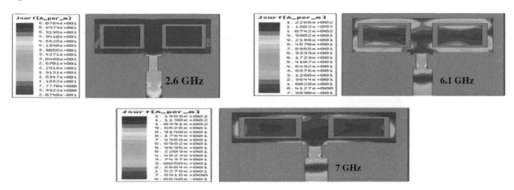

Fig. 16.8 Current Distribution of Antenna at 2.6 GHz, 6.1 GHz, and 7 GHz

Table 16.2 describes the comparison in terms of the size of the antenna proposed and that of the reference antennas. It is found and evident from the table and the literature survey done that the planned antenna is smaller in size than the reference antennas.

Table 16.2 Comparitive Study of Prototype Antenna and Reference Antennas Size

Reference	No. of Bands	Size of Antenna
[4]	Two	50×50 mm^2
[5]	One	35×35 mm^2
[6]	Three	50×50 mm^2
[7]	Two	50×50 mm^2
[8]	Three	30×40 mm^2
Prototype Antenna	Three	34×30 mm^2

Source: Author's compilation.

Figure 16.9 demonstrates the pattern of radiation of the proposed antenna at 2.6 GHz, 6.1 GHz, and 7 GHz. The patterns are taken at phi = 0 degree and phi = 90 degrees. The radiation patterns obtained at three different frequencies are the same as that of the omnidirectional antennas.

The gain of the antenna against the frequencies is shown in Fig. 16.10 at 2.6 GHz, 6.1 GHz, and 7 GHz. The antenna resonates triple bands with coverage of the frequencies simultaneously with the gain achieved as 2.63 dBi, 3.023 dBi, and 3.23 dBi, respectively, at 2.6 GHz, 6.1 GHz, and 7 GHz.

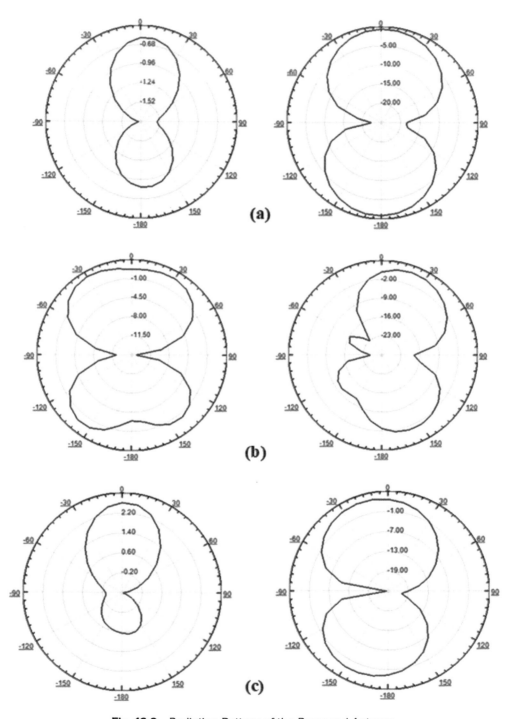

Fig. 16.9 Radiation Pattern of the Proposed Antenna

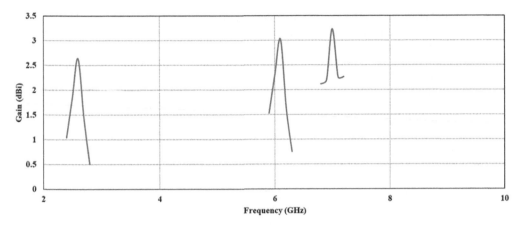

Fig. 16.10 Gain Plot of the Antenna

6. Conclusion

The paper presents all the outcomes of this multiband proposed antenna, which covers the parametric analysis, radiation patterns, surface current distribution, and gain. The three different frequency ranges as (2.3–3 GHz)/2.6 GHz, (6–6.2 GHz)/6.1 GHz, and (6.8–7.3 GHz)/7 GHz having the return loss of −39 dB, −18 dB, and −38 dB, respectively, are covered by the antenna, which are much suitable for the applications like GSM, Bluetooth, WLAN, WiMAX, and some long-distance radio telecommunication applications. The gain achieved are 2.63 dBi, 3.023 dBi, and 3.23 dBi, respectively. The size of the proposed compact antenna is **34 × 30 × 0.8 mm³**. ANSYS HFSS simulation tool is handled for the simulation.

REFERENCES

1. Du, Z., Gong, K., Fu, J. S. and Gao, B. (2001). Analysis of Microstrip Fractal Patch Antenna for Multi-Band Communication. Electronics Letters, 37: 805–806.
2. Chen, H. M. and Lin, Y. F. (2003). Printed Monopole Antenna for 2.4/5.2 GHz Dual-Band Operation. IEEE Antennas and Propagation Society International Symposium, 3: 60–63.
3. Li, R. L., Dejean, G.,Tendtzeris, M. M. and Laskar, J. (2003). Novel Multi-Band Broadband Planar Wire Antenna for Wireless Communication Handheld Terminals. IEEE Antennas and Propagation Society International Symposium, 3: 44–47.
4. Eldek, A. A., Elsherbeni, A. Z. andSmith, C. E. (2005). Square Slot Antenna for Dual Wideband Wireless Communication Systems. Journal of Electromagnetic Waves and Applications, 19(12): 1571–1581.
5. Hadizafar, L. and Azarmanesh, M. N. (2011). Enhanced Bandwidth Double-Fed Microstrip Slot Antenna with a pair of L-Shaped Slots. Progress in Electromagnetics Research C, 18: 47–57.
6. Kumawat, B. P.,Meena, S. and Yadav, S. (2017). Square Shape Slotted Multiband Microstrip Patch Antenna using Defect Ground Structure. International Conference on Information, Communication, Instrumentation and Control, 1–4.

7. Ali, J., Abdulkareem, S., Hammoodi, A., Salim, A., Yassen, M., Hussan, M. and Al-Rizzo, H. (2016). Cantor fractal-based printed slot antenna for dual-band wireless applications. International Journal of Microwave and Wireless Technologies, 8(2): 263–270.

8. Verma, S. and Kumar, P. (2014). Compact triple-band antenna for WiMAX and WLAN applications. Electronics Letters, 50(7): 484–486.

9. Pushkar, P. and Gupta, V. R. (2016). A metamaterial based tri-band antenna for WiMAX/WLAN application. Microwave and Optical Technology Letters, 58(3): 558–561.

10. Chandan, Srivastava, T. and Rai, B. S. (2016). Multiband monopole U-slot patch antenna with truncated ground plane. Microwave and Optical Technology Letters, 58(8): 1949–1952.

11. Chandan, Srivastava, T. and Rai, B.S. (2017). L-Slotted Microstrip Fed Monopole Antenna for Triple Band WLAN and WiMAX Applications. Springer in Proceedings Theory and Applications, 516: 351–359.

12. Chandan, Ratnesh, R. K. and Kumar, A. (2021). A Compact Dual Rectangular Slot Monopole Antenna for WLAN/WiMAX Applications. Springer Cyber Physical Systems, Lecture Notes in Electrical Engineering book series (LNEE), 788: 699–705.

13. Singhal, S., Sharma, P. and Chandan. (2021). A Low-Profile Three-Stub Multiband Antenna for 5.2/6/8.2 GHz Applications. Springer Cyber Physical Systems, Lecture Notes in Electrical Engineering book series (LNEE), 788.

14. Vashisth, S., Singhal, S. and Chandan. (2021). Low-Profile H Slot Multiband Antenna for WLAN/Wi-MAX Application. Springer Cyber Physical Systems, Lecture Notes in Electrical Engineering book series (LNEE), 788.

15. Chandan, Bharti, G.D., Srivastava, T. and Rai, B.S. (2018). Miniaturized Printed K shaped Monopole Antenna with Truncated Ground Plane for 2.4/5.2/5.5/5.8 Wireless LAN Applications. AIP Conference Proceedings, American Institute of Physics, 200371–200377.

16. Chandan, Bharti, G.D., Srivastava, T. and Rai, B.S. (2018). Dual Band Monopole Antenna for WLAN 2.4/5.2/5.8 with Truncated Ground. AIP Conference Proceedings, American Institute of Physics, 200361–200366.

17. Chandan, Bharti,G. D., Bharti, P. K. and Rai, B.S. (2018). Miniaturized Pi (π) - Slit Monopole Antenna for 2.4/5.2.8 Applications. AIP Conference Proceedings, American Institute of Physics, 200351–200356.

18. Chandan.(2020). Truncated Ground Plane Multiband Monopole Antenna for WLAN and WiMAX Applications. IETE Journal of Research, 1–6.

Intelligent Systems and Smart Infrastructure – Brijesh Mishra et al. (eds)
© 2023 Taylor & Francis Group, London, ISBN 978-1-032-41287-0

CHAPTER

17

Evaluation of the WDM-FSO System's Effectiveness Under Chronic Weather Conditions

Nitin Garg[1], Adarsh Singh Tomar[2],
Abhyuday Dubey[3] and Anurag Kumar[4]

Electronics and Communication Department
Galgotias College of Engineering and Technology
Greater Noida, India

Abstract

Free space optics based on wavelength division multiplexing (WDM OWC) has become a possible candidate for a communication network for long-distance connections and other uses. Due to the spectrum conundrum and other concerns, OWC has gotten a lot of attention in recent years as a complement and alternative to radio frequency-based communication. WDM, on the other hand, has been suggested as one of the solutions for enhancing bandwidth efficiency and data throughput in next-generation optical access networks. As a result, a new research path that integrates these two technologies in a hybrid network has evolved. In this study, the history of WDM OWC, its progression, and its current state will be discussed. The acronym MIMO stands for "Multiple Input Multiple Output," and it refers to a method that improves the performance and consistency of free-space fiber optics under a variety of environmental circumstances. The goal of this research is to create a MIMO-based WDM-based OWC communication system that receives various independent records of the data and analyzes them for different data streams. The BER and Quality Factor for different MIMO data streams under diverse atmospheric conditions are investigated using the OptiSystem for WDM-based OWC communication with MIMO. This model is being tested for different windows for effective transmission.

Corresponding author: [1]nitingarg@galgotiacollege.edu, [2]adarshsinghtomar.19gcebec036@galgotiacollege.edu, [3]abhyudaydubey.19gcebec031@galgotiacollege.edu, [4]anuragkumar.19gcebec180@galgotiacollege.edu

DOI: 10.1201/9781003357346-17

Keywords: Optical wave communication (OWC), Wavelength division multiplexing (WDM), Bit error rate, Free space optics, Q factor, Atmospheric attenuation

1. Introduction

OWC stands for "Optical Wave Communication," which is a wireless communication technology that employs light to transmit data. Due to the ever-increasing needs for bandwidth, researchers are increasingly turning their attention to this type of wireless communication. Like any other communication system, OWC has a transmitter and receiver. However, the channel itself is considered "free space" or air. In the transmitter, the information is modulated over a light signal that is produced by an optical source such as a LASER or an LED. In the receiver, a photo detector, which may be an Avalanche Photo Detector (APD) or a PIN, transforms light impulses into electrical signals so that the data may be interpreted further. It is analogous to the conventional optical fibre communication technology in that both make use of optical signals as the information carrier. On the other hand, OWC links do not need the deployment of optical fibre cable or the costly construction of rooftop optical fibre installations [1-3]. High data transfer rates, a large amount of licence-free spectrum, no security upgrade requirements, a large modulation bandwidth, immunity to electromagnetic interference, low power consumption, low deployment cost, and an easy and quick installation process are just a few of the benefits of OWC links [4-5]. Due to the multiple advantages of OWC connections, they may be utilized for a number of reasons, including inter-aircraft communications, inter-satellite communications, ground communications, and military applications. Although OWC looks to be the most viable technology owing to its significant advantages over wireless alternatives, it does have several limitations that must be addressed in order to assure uninterrupted connection. Atmospheric instabilities such as scintillation, absorption, and dispersion, as well as meteorological phenomena such as rain, fog, snow, and mist are examples of these restrictions. To lessen the negative impacts of attenuation in the surroundings and obstacles in the course of the laser beam, the Multiple Input Multiple Output (MIMO) technique was used. More than a beam of light of the multiplexed signal is conveyed across the transmission medium and eventually reaches the receiving end through the free space optics channel. Each beam follows a distinct route, resulting in differences in the degree of attenuation. The source generates the data that must be transferred across the wireless optical system [6-8]. The data is modulated by the modulator, which makes it appropriate for transmission. MZ modulation, amplitude modulation, frequency modulation, and other modulation techniques can all be employed. Because of changes in weather, buildings move. This causes misalignment or pointing errors. This causes even more misalignment or pointing mistakes, because the changes in weather make it even more likely that buildings will move. Owing to factors such as fog, beam dispersion, air absorption, precipitation, snowfall, shadowing, and pollution, the field of view of the OWC is restricted to a very narrow region. One of the major disadvantages of OWC, particularly in tropical regions, is nonselective scattering. Some of the techniques offered to overcome these challenges include careful selection of modulation schemes, multiple input and multiple output (multibeam notion), aperture averaging, and wavelength division multiplexing (WDM). A considerable fading margin is also present in certain cutting-edge transceivers.

2. System Design

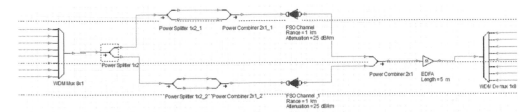

Fig. 17.1 Circuit diagram of simulation

In Optisystem v7.0, a single-beam WDM-OWC system was established. The technology is also evaluated based on its performance in adverse weather conditions such as hazy, rainfall, snowfall, and foggy. In the section devoted to analysis, the transmitter has sixteen channels, each of which is equipped with a continuous wave (CW) laser that generates a wave light beam that is unbroken. WDM (Mach Zehnder) free-space optics transmission system modulator that controls the intensity of the optical wave and changes the binary information into an optical signal, an optical amplifier to boost signal strength, PRBS (Pseudo Random Bit Sequence generator) that chooses the flow of data bits at random, and NRZ pulse generator that makes pulses. A WDM demultiplexer is used in the receiver sector, and it has the same parameters as the WDM multiplexer. A demultiplexer is a device that divides a signal into wavelengths. The optical signal is then transformed to an electrical signal using a detector, which is one of the essential components of the receiver portion. A 3R regenerator then processes the data pulse, carrying out tasks such as resizing, reassessing, and intensification. A low-pass Gaussian filter allows only signals in a certain frequency band to pass through. Finally, the variables lowest BER, maximum quality factor, eye peak, and acceptability value are tested using an eye diagram analyzer.

In spite of the fact that OWC and WDM systems have each been the subject of a significant amount of study on their own, the ever-present need for bandwidth has led to the convergence of the two technologies becoming a new area of investigation. [10-12]. In hybridization strategies, using WDM to transport data across free space helps increase the amount of bandwidth that can be used by broadband applications. In recent years, the data-carrying capacity of WDM has increased to levels that have reached orders of terabits per second, and it can be easily paired with OWC to greatly boost bit rates [13]. As a consequence of this, WDM-OWC offers a workable alternative for satisfying the exponentially expanding global demand for broadband services [15]. OWC may be made to be totally tightly aligned with optical fiber communication by making use of a couple of OWC terminals that are capable of being transparently attached to single-mode fiber [16]. Figure 17.2 depicts the main concept of WDM OWC.

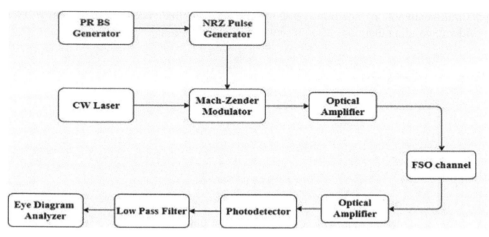

Fig. 17.2   Schematic diagram of OWC communication system

Table 17.1   BER at different optical windows

Attenuation (dB/km)	850 nm	1310 nm	1550 nm
3	0.0	0.0	0.0
22	0.0	0.0	0.0
45	0.0	0.0	10.16421 × 10-22
65	3.43586 × 10-15	7.66465 × 10-15	0.00164508

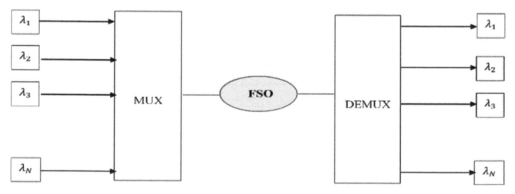

Fig. 17.3   Schematic of a WDM OWC system

3. Results and Discussion

The BER analyzer is used to check the performance of the FSO system in this part. A BER value of 10^{-6} means that on average, one mistake is made for every million bits. A reliable, long-distance communication system should have a maximum error rate of 10^{-9} while receiving bits.

BER is calculated by dividing the number of errors by the total number of bits that were sent. The signal-to-noise ratio of a received information transmission may also be used to calculate BER.[18]

$$BER = \frac{2}{\pi} \times SNR \times e^{\frac{-SNR}{8}}$$

A. 850 nm

At 850 nm, the transmitter is operational, which was the initial light waveguide window available for testing the performance of the OWC connection. Figures 17.1–17.4 show the result of the OWC system with an active signal luminosity of 850 nm. As the signal is sent via the atmosphere, it is subjected to attenuation due to the atmosphere. The outcomes are unmistakably evident that even at 65 dB attenuation, the required value is 5.8, which is smaller than the value of Q factor. The eye's peak at 65 dB is calculated to be 1.31245×10^{-5} and the collected signal strength is determined to be 9.39843×10^{-6}.

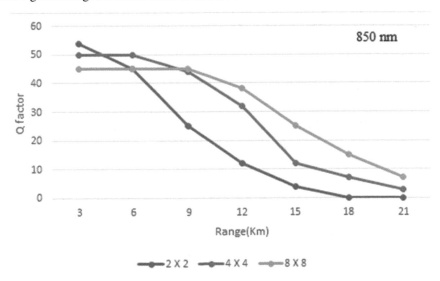

Graph 17.1   Q factor at different channels at 850 nm

B. 1310 nm

It was decided to use 1310 nm as the second light waveguide window for the assessment of the OWC connection. When operating at this wavelength, the output of the OWC system is seen in Fig. 17.4. When the signal attenuation reaches 65 dB, and the best performance is provided by the broadcaster performing activities at 1310 nm, however it is not as excellent as

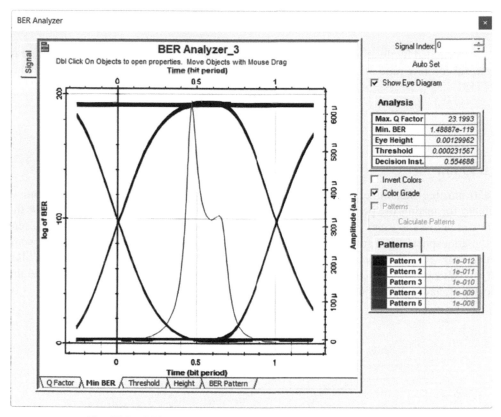

Fig. 17.4 BER analyzer's output in rainy conditions (22 dB km)

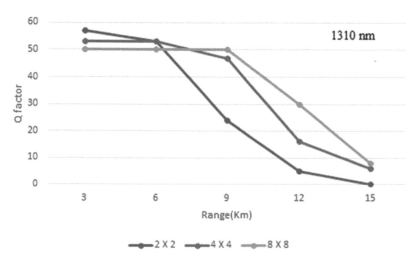

Graph 17.2 Q factor at different channels in 1310 nm

the transmitter device operating at 850 nm. The Q factor dropped to 7.45 as a result, and the lowest BER is 8.41912×10^{-5}. The predicted elevation of the eye diagram at this wavelength is 1.22742×10^{-5}, and the incoming signal power is 9.23167×10^{-6}.

C. 1550 nm

The output of the OWC system is seen at 1550 nm in Figs 17.2–17.4. The Q factor is now at 2.79, and the BER is at its lowest ever recorded value of 0.00146908. The transmitter device operates at 1550 nm when the attenuation is set to 65 dB; however, the results are not ideal since the Q factor falls to 2.89 and the minimum BER is 0.00194608 in this configuration. The apex of the eye diagram is not very noteworthy. When there is a significant amount of attenuation, the performance of the first two light transmission windows is superior to that of 1550 nm. This window's received signal power is the same as 1310 nm, or 9.39843×10^{-6}.

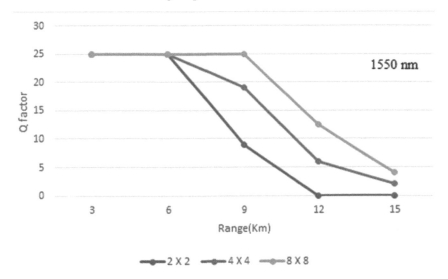

Graph 17.3   Q Factor at different channels in 1550 nm

Table 17.1 compares the BER analysis of a connection with dissimilar light windows at random attenuation levels. In comparison to both high optical transmission windows, the first optical window has the lowest BER.

In all three optical transmission windows, BER rises as the value of attenuation increases, as seen in Figs 17.2–17.4. In comparison to the 1550-nm window, the 1310-nm window has a reduced bit error rate. Because the BER for all three windows is insignificant at low attenuation, 1550 nm should be favoured for user safety. However, under high attenuation situations, 1310 nm is advantageous since it has a lower BER, a higher Q factor, and is safer than the 850 nm transmission window.

The performance of the OWC connection is compared for various optical windows using the pictorial depiction of the quality factor given. When compared to the other two optical windows, it is evident that the window at 1310 nm produces superior results. In a similar vein, the distinction of eye peak for different optical windows with various degrees of weakness or attenuation lends to the idea that this window is best used in environments with significant levels of attenuation.

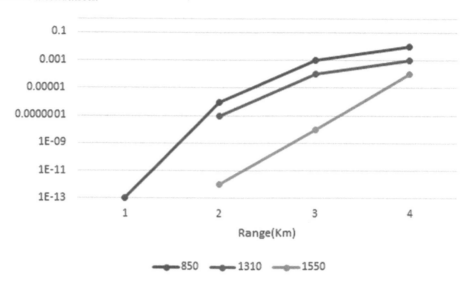

Graph 17.4 BER Comparison at different Optical Windows

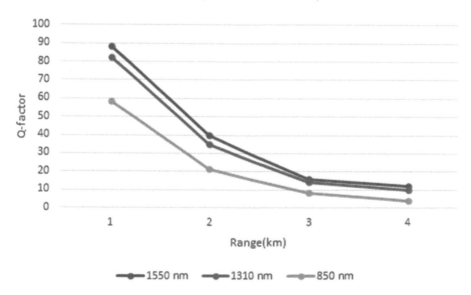

Graph 17.5 Quality Factor 's comparison of system at different Optical Windows

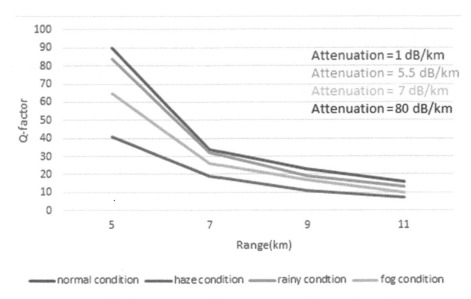

Graph 17.6 Quality factor's comparison at different attenuation

4. Conclusions

The OWC communication connection has been evaluated utilizing multiple optical communication windows located at 850 nm, 1310 nm, and 1550 nm. These windows were examined under a variety of environmental conditions. It has been shown that at a wavelength of 1550 nm, transmission distance is maximized while BER is reduced to the bare minimum under low attenuation circumstances. As a consequence, 1550 nm should be used in low attenuation situations. Attenuation circumstances with strong attenuation were shown to have reduced bit error rates if the 850 nm functioning optical window was used, based on a modelling investigation.

However, the second optical transmission window, 1310 nm, provides superior outcomes under high attenuation circumstances and is safer for consumers when compared to the first optical transmission window, 850 nm. As a consequence of this, the 1310-nm window performs the best in terms of transmission efficiency even when subjected to significant levels of attenuation.

REFERENCES

1. Ciaramella, E., Arimoto, Y., Contestabile, G., Presi, M., D'errico, A., Guarino, V. and Matsumoto, M., 2009. 1.28-Tb/s (32$\times$ $40 Gb/s) Free-Space Optical WDM Transmission System. IEEE Photonics Technology Letters, 21(16), pp. 1121–1123.
2. Khalighi, M. A. and Uysal, M., 2014. Survey on free space optical communication: A communication theory perspective. IEEE communications surveys & tutorials, 16(4), pp. 2231–2258.

3. Epple, B., 2010. Simplified channel model for simulation of free-space optical communications. Journal of Optical Communications and networking, 2(5), pp. 293–304.

4. García-Cela, E., Ramos, A.J., Sanchis, V. and Marin, S., 2012. Emerging risk management metrics in food safety: FSO, PO. How do they apply to the mycotoxin hazard?. *Food Control*, *25*(2), pp. 797–808.

5. Sandalidis, Harilaos G., Theodoros A. Tsiftsis, George K. Karagiannidis, and Murat Uysal. "BER performance of FSO links over strong atmospheric turbulence channels with pointing errors." *IEEE Communications Letters* 12, no. 1 (2008): 44–46.

6. Henniger, H. and Wilfert, O., 2010. An Introduction to Free-space Optical Communications. Radioengineering, 19(2).

7. Sandalidis, H.G., Tsiftsis, T.A., Karagiannidis, G.K. and Uysal, M., 2008. BER performance of FSO links over strong atmospheric turbulence channels with pointing errors. *IEEE Communications Letters*, *12*(1), pp. 44–46.

8. Popoola, W.O. and Ghassemlooy, Z., 2009. BPSK subcarrier intensity modulated free-space optical communications in atmospheric turbulence. Journal of Lightwave technology Ciaramella, E., Arimoto, Y., Contestabile, G., Presi, M., D'errico, A., Guarino, V. and Matsumoto, M., 2009. 1.28-Tb/s (32\$\times \$40 Gb/s) Free-Space Optical WDM Transmission System. IEEE Photonics Technology Letters, 21(16), pp. 1121–1123.

9. Khalighi, M.A. and Uysal, M., 2014. Survey on free space optical communication: A communication theory perspective. IEEE communications surveys & tutorials, 16(4), pp. 2231–2258.

10. Epple, B., 2010. Simplified channel model for simulation of free-space optical communications. Journal of Optical Communications and networking, 2(5), pp. 293–304.

11. Majumdar, Arun K. *Advanced free space optics (FSO): a systems approach*. Vol. 186. Springer, 2014.

12. Nadeem, Farukh, Vaclav Kvicera, Muhammad Saleem Awan, Erich Leitgeb, Sajid Sheikh Muhammad, and Gorazd Kandus. "Weather effects on hybrid FSO/RF communication link." *IEEE journal on selected areas in communications* 27, no. 9(2009).

13. Henniger, H. and Wilfert, O., 2010. An Introduction to Free-space Optical Communications. Radioengineering, 19(2).

14. Nadeem, F., Kvicera, V., Awan, M.S., Leitgeb, E., Muhammad, S.S. and Kandus, G., 2009. Weather effects on hybrid FSO/RF communication link. *IEEE journal on selected areas in communications*.

15. Popoola, W.O. and Ghassemlooy, Z., 2009. BPSK subcarrier intensity modulated free-space optical communications in atmospheric turbulence. Journal of Lightwave technology, 27(8), pp. 967–973.

16. Alzenad, Mohamed, Muhammad Z. Shakir, Halim Yanikomeroglu, and Mohamed-Slim Alouini. "FSO-based vertical backhaul/fronthaul framework for 5G+ wireless networks." *IEEE Communications Magazine* 56, no. 1 (2018): 218–224.

17. Samimi, H. and Uysal, M., 2013. End-to-end performance of mixed RF/FSO transmission systems. *Journal of Optical Communications and Networking*, *5*(11), pp. 1139–1144.

18. Wang, J., Yang, J.Y., Fazal, I.M., Ahmed, N., Yan, Y., Huang, H., Ren, Y., Yue, Y., Dolinar, S., Tur, M. and Willner, A.E., 2012. Terabit free-space data transmission employing orbital angular momentum multiplexing. Nature photonics, 6(7), pp. 488–496.

19. Ansari, I.S., Yilmaz, F. and Alouini, M.S., 2013. Impact of pointing errors on the performance of mixed RF/FSO dual-hop transmission systems. *IEEE Wireless Communications Letters*, *2*(3), pp. 351–354.

20. Li, G., Bai, N., Zhao, N. and Xia, C., 2014. Space-division multiplexing: the next frontier in optics.

Intelligent Systems and Smart Infrastructure – Brijesh Mishra et al. (eds)
© 2023 Taylor & Francis Group, London, ISBN 978-1-032-41287-0

CHAPTER

18

Facial Expression Recognition Using Simple Attention Mechanism

Harsh Tomar*, Neha Mittal

Department of Electronics and Communications Engineering
Meerut Institute of Engineering and Technology, Meerut, UP, India

Anshuman Bansal, Wakar Ahmad

Department of Information and Technology,
Meerut Institute of Engineering and Technology, Meerut, UP, India

Abstract

Recognising emotions has always been an arduous task for machines. Alternately the human brain can easily discriminate among facial expressions because of its vast network of biological neurons and subtlety. Researchers from around the world are working on methods and algorithm's to train machines to recognise human facial expressions and the most promising method available is a CNN (convolutional neural network). CNN-based models are very much accurate and capable of learning complex expressions and local binary patterns. A CNN-based model can efficiently be trained to recognise facial expressions and their efficiency can be boosted by adding more to CNN and improving the input fed data. There are datasets like CK+ RAF (real-world affective) having labelled images with emotions categorised angry, sad, fear, neutral, disgust, happy, and surprised. The several existing models use different pre-processing techniques to reduce the error in recognising every emotion. Here, VGG-inspired architecture with simple attention module is used to classify facial expression images. We used two datasets CK+ and RAF to calculate the accuracy. Test accuracy is found to be 93.9% on CK+ dataset and 74% on RAF dataset.

Keywords: facial expressions, Convolution Neural Network, Attention module

*Corresponding author: Harsh.tomar.ec.2018@miet.ac.in

DOI: 10.1201/9781003357346-18

1. Introduction

Facial expressions are the capability of humans to express emotions, or, in other words, express inner feelings to another living being without the use of words. These emotions are very helpful when dealing with human–machine interaction, computer vision, and automating several processes like auto drive to learn the behaviour of humans while driving in a different scenario. Interactive AI assistance which uses these expressions along with audio inputs to enhance the quality of interaction between humans and machines similarly, autopilots nowadays use facial expression monitoring to take over from the pilot in case the pilot is in a medical condition, also inspired from these systems body-language recognition systems(in real-time) are also being developed. Facial expression recognition is seeking more attention in the past few years as it is now finding more applications due to largely grown internet. Recognising human emotions artificially involves several challenges like the collection of proper data with more useful information and less futile data. The images in the dataset must be of satisfying resolution and a wide variety. A good quantity of data on specific emotions will contribute to greater accuracy on the same. Developing algorithms to squeeze out useful information from available data is the most dominant task as it involves more experimentation and logical thinking. To recognise emotions, computer has to first detect the geometry of the face followed by detecting changes. Darwin [1] and Suwa [2] tried to scrutinise facial expressions from image sequences in 1872 and 1978, respectively. Researchers have developed complex but logical methods and algorithms, like Caifeng Shan [3] in his paper using local binary pattern introduced by Ojala [4], which detects binary patterns using the kernel of size 3×3. Detecting edges is the most promising method as the variation in the expressions from angry to happy, the geometry of eyes, lips, and wrinkles on the forehead are the most discriminative.

2. Literature Survey

The facial expressions are classified into seven categories: angry, happy, sad, disgust, neutral, surprised, and fear. Deep learning, a part of machine learning, is the most biased method to extract features from image data. It uses convolution as the hidden layer with kernels of different sizes (3×3, 5×5, 7×7) to obtain feature maps and fed them to a classifier with a dense network of neurons [5,6].

The use of pre-processing filters has also been the choice of researchers to remove noise from the images and enhance features like edges and curves. [7] use a Gaussian process to enhance invariant features. Pre-processing data before feeding into the convolutional neural network (CNN) model leads to greater accuracy. Ritanshi Agarwal [8] uses a series of Gaussian kernels followed by Laplacian of Gaussian (LoG) kernel and then feeding pre-processed data to CNN, resulting in the gradual increase in accuracy on datasets CK+. The majority of methods available uses some form of illumination correction.

ILSVRC (ImageNet Large Scale Visual Recognition Challenge) [9] used a deep convolutional neural network (AlexNet) with five convolution layers and three fully connected dense layers. This model laid the foundation for other succeeding models like googleNet [10] which used kernels of different sizes and obtained a common feature-map and lastly performed $1 \times 1 \times$ dept convolution; this model reduced the computation as it used 1×1 convolution. ResNet further reduced the error to 3.57% as it used the inception layer to preserve the existing features with the new extracted features.

The convolution operation extracts features from the entire image irrespective of regions in the image where no useful data is present [11,12,13,14]. In 2015, Vaswani et al. [15] proposed the first attention mechanism in a text converter, and later, Li et al. [16] proposed an occlusion-aware model where the model was integrated with an attention mechanism and it pays attention to several patches or areas in an image. The attention mechanism [17] is applied after a convolution layer and it increases the weights of the most significant features in the image and either it highlights the area (soft attention) or crops the area (hard attention). Fernandez et al. [18] used attention mechanism to create an attention feature map and feature extraction map and combining them and using reconstruction module to create a focused attention map. This network pays attention mostly to the face and crops the other part and this produces less error and also prevents computation on non-required areas.

Li Jing [19] used two feature extraction modules, one of which is dimensionality based and the sum of these two feature maps is fed into the attention branch and reconstruction module, which uses inception layers followed by again convolution and finally to fully connected dense neuron network as a classifier.

3. Problem Statement

Facial expression is a very significant feature to recognise human emotion. Humans have adapted this feature over a long period of evolutions and training a machine to do so is a task many people have attempted and their results were very good. Human emotions are divided into seven classes that are happy, angry, sad, disgust, fear, surprise, and neutral. The facial data is divided according to the different shapes of eyes, nose, lips, etc. Now the task of the computer is to detect these edges and combine these edges to form useful patterns and remember the changing patterns according to different classes of emotions.

4. Proposed Method

The proposed model uses a VGG 16-inspired architecture with a simple attention module to improve its efficiency. The model has two modules: a conventional CNN architecture for feature extraction with convolution layers of kernel size 3×3 followed by ReLu (rectified linear unit) followed by a max-pooling layer with kernel size 2×2 and an attention module for increasing the weights detected in the previous layers.

A. Convolutional Layers

The architecture aims to detect edges of different shapes, or, technically, we refer to them as features. The convolution layer has parameters such as the number of filters and kernel size;

the kernel size defines the small square array whose size can vary according to the necessity like 3×3, 5×5, or 7×7. This matrix has numbers arranged differently according to the type of edge to be detected; for example, horizontal and vertical edges, etc. The number of filters defines the number of feature maps at the output. These feature maps contain those edges detected by the kernel in form of numbers.

In our case, we have two convolutional layers with 16 and 32 filters, respectively, and kernel size of 3×3. This configuration uses less computational power and extracts the required features.

$$z = \sum_{i=1}^{q} W_i * x^i + b$$

(1)

B. Activation Function

The activation function is used to remove non-linearity in the output feature map. Non-linearity refers to values less than zero; to remove these non-linearities, ReLu function is used. ReLu replaces all the values less than zero with zero.

$$\text{ReLU}(x) = \max(x, 0)$$

(2)

C. Pooling

The output feature map contains lots of data and thus if the size of the feature map is not reduced, it will take lots of computation in the later process. To lower the size of the feature map, pooling layer is introduced. Max-pooling is a famous pooling method used; it uses a matrix of the size of 2×2 and keeps the maximum value among the four.

$$\left[\frac{\text{input shape} - \text{pool kernal size}}{\text{stride}} \right] + 1$$

(3)

D. Attention Module

The attention module is an additional architecture introduced after the convolution layer to increase the weights of previously detected edges. In convolutional neural network, when we start, the feature map of the first layer consists of the most basic edges, and as we go deeper, the model starts detecting micro edges and it is the point where the model starts collecting unwanted edges or noise which resist in the accuracy of model.

In our case, the most significant feature to detect emotion is in the shape of eyes, nose, lips, or simply the features of the central face; to achieve this, we just have to increase the weights of these regions of the picture.

The attention module takes to input the output feature maps of current and previous convolution layers now since the shapes of these feature maps are different ($48 \times 48 \times 16$ and $24 \times 24 \times 32$). 1×1 convolutions are used to change the shape of the feature map without affecting the values.

Previous Layer: In the previous layer, the size is $48 \times 48 \times 16$, so we apply 1×1 convolution with 32 filters and stride of (2, 2) so the resulting output of feature map is of size $24 \times 24 \times 32$.

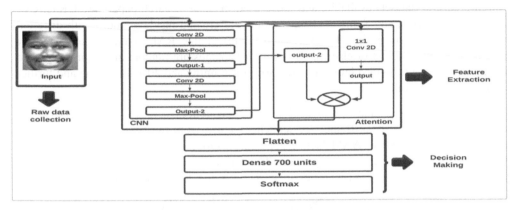

Fig. 18.1 Flowchart of our Model

Now since the dimensions of both the feature maps are the same, we multiply the corresponding weights of both the feature maps and more significant weights become higher and less significant become lower in value.

Table 18.1 Network Architecture

Input	Type	K-size/stride	Activation	Output
$48 \times 48 \times 1$	Conv.2D	$3 \times 3/(1,1)$	ReLU	$48 \times 48 \times 16$
$48 \times 48 \times 16$	Ma $\times$ -Pool	$2 \times 2/(2,2)$		$24 \times 24 \times 16$
$24 \times 24 \times 16$	Conv.2D	$3 \times 3/(1,1)$	ReLU	$24 \times 24 \times 32$
$24 \times 24 \times 12$	Ma $\times$ -Pool	$2 \times 2/(2,2)$		$12 \times 12 \times 32$
Attention Module				
$24 \times 24 \times 16$	Conv.2D	$1 \times 1/(2,2)$		$12 \times 12 \times 32$
$12 \times 12 \times 32$	Multiply			$12 \times 12 \times 32$
$12 \times 12 \times 1$				
Flatten				
700 units	Dense		ReLU	
7 units	Dense		Soft-max	One of seven emotions

E. Fully Connected Layer

Now the feature map is flattened and passed to a fully connected or dense layer with 700 neurons and at last a soft-max classifier with 7 units or classes of emotions.

5. Datasets

There are several datasets available, and those used in most research papers are CK+, RAF (real-world affective), etc. Some of these datasets are created inside a studio with different volunteers and models posing the expression. Datasets like RAF contain images mostly collected from the internet or else we can say most images are random and expressions are natural. The dataset is divided into two parts: 80% for training and 20% for testing and further 20% of training for validation.

The extended Cohn–Kanade (CK+) [20] has 593 sequences across 123 subjects, which are FACS coded at the peak frame. All sequences reach from the neutral face to the max expression. All sequence of AAM (active appearance model) [21] were tracked along with 68-point landmarks for each and every image. The dataset contains 981 grey images with only one channel divided into seven categories anger, sad, disgust, fear, surprise, happy, and neutral.

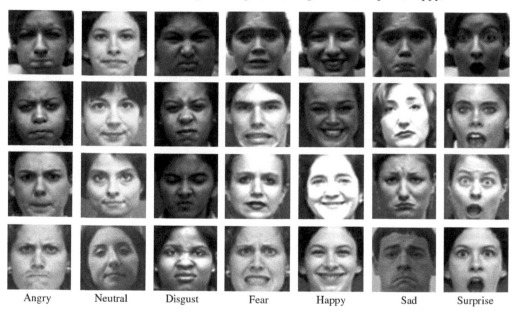

| Angry | Neutral | Disgust | Fear | Happy | Sad | Surprise |

Fig. 18.2 Sample Images of CK+ Dataset.

The RAF [22] is based on crowdsourcing annotation; each image has been independently labelled by 40 annotators. The dataset has 15339 3-channel RGB images. The images require some modifications in the architecture since the images are RGB images (see Fig. 18.3).

6. Hardware

We have verified the model by using two datasets CK+ and RAF .We have used Tensor-flow framework on system with 512Gb SSD, 8GB RAM, Windows 10 Home and with GTX 1650 graphics.

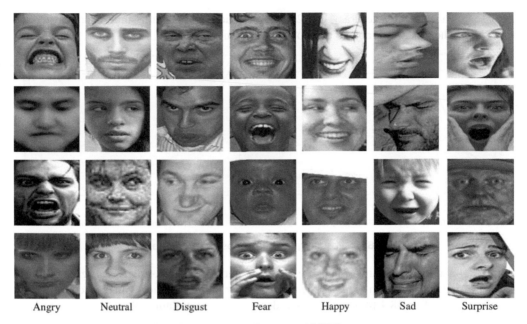

| Angry | Neutral | Disgust | Fear | Happy | Sad | Surprise |

Fig. 18.3 Sample Images of RAF Dataset

7. Experimental Results

The model proposed uses an attention branch along with the conventional CNN architecture to increase the accuracy. The conventional model detects edges on entire images, that is, areas of an image where there is no useful data or edges of no use; when a model learns these edges and when we pass test images to the model, the model tries to search those edges or noise in those images and thus it makes less accurate results. The attention module plays a vital role in removing this vulnerability. When closely monitored, the edges detected in the first convolutional layer are mostly large edges; in our case, the edges around the eyes, lips, and nose (mostly around the central face); when the next convolution layer is applied, the model starts detecting micro edges but it detects along the entire area of the feature map. Now we pass the output feature map of the first convolutional layer and second convolutional layer to the attention module; attention module applies 1×1 convolution to bring both features maps to the same dimensions and multiplies the weights of both the feature maps &&parallelly and, as a result, weights of edges detected in previous layer increase and subsequently edges with lesser weights become more less. Now we have a feature-map increased weight around the central face and now when it is flattened and passed to the dense layer, and then to the soft-max classifier, we observe more accurate results. The model was able to achieve 93.9% accuracy on the CK+ dataset and, with some modifications, 74% accuracy on the RAF dataset.

After testing the model on CK+ test images, a validation accuracy of 93.9% and a test accuracy of 93.9% were achieved, and when testing the model on RAF dataset test images, an accuracy of 74% and a validation accuracy of 72% were achieved, and with some changes in the architecture, the model was able to overcome over-fitting.

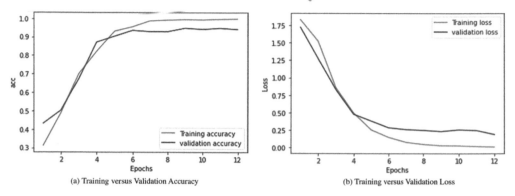

(a) Training versus Validation Accuracy   (b) Training versus Validation Loss

Fig. 18.4   Resulting Graphs of Loss and Accuracy

Figure 18.4 shows the training graphs for accuracy and loss on CK+ dataset. Figure 18.4(a) shows the model's validation accuracy increases along training accuracy with respect to epochs and, similarly, in Fig. 18.4(b), validation loss decreases along training loss.

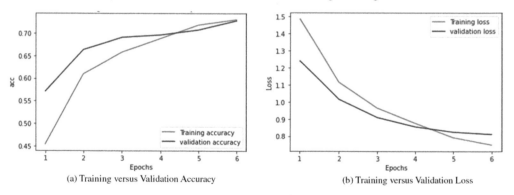

(a) Training versus Validation Accuracy   (b) Training versus Validation Loss

Fig. 18.5   Resulting Graphs of Loss and Accuracy.

Figure 18.5 shows the training graphs for RAF dataset. From Fig. 18.5(a), it can be seen that in case of RAF validation accuracy increases along with training accuracy. Figure 18.5(b) shows the reduction in validation loss along with training loss.

Figure 18.6 shows the individual accuracies attained on datasets simultaneously. In Fig. 18.6(a), it can be seen that the model performs better on CK+ dataset and degrades only to recognise surprise and angry face since a surprised and angry face share a bit similar features. In case of RAF, Fig. 18.6(b) recognition is good since the faces in images are more natural and images are coloured.

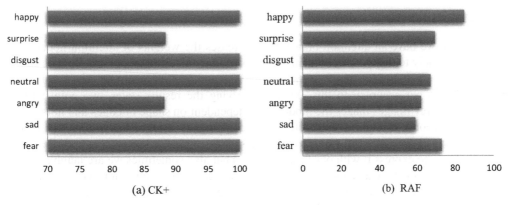

Fig. 18.6   Individual Accuracy on Datasets

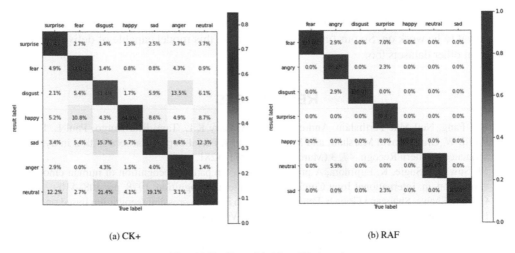

Fig. 18.7   Error Matrix of Datasets

Figure 18.7 shows the error matrix (also known as confusion matrix) for both the datasets. Confusion matrix is important because it actually shows how much the model went wrong. In case of CK+, Fig. 18.7(a) in the model was bit confused in case of angry and surprise. Figure 18.7(b) model seems to be more confused but still the results are satisfying.

Table 18.2   Accuracies on test data

Dataset	Train accuracy (%)	Test accuracy (%)
CK+	93.4	93.4
RAF	74.1	74.0

8. Conclusion

In this paper, we have an attention branch that increases the weights of attention nodes that paid attention only to the most significant areas. Experimental results of our architecture show impressive accuracy in CK+ and RAF, which are 93.4 and 74.0, respectively. Here, attention model helps to get accuracy to this extent. Observing the results states two points: first, working with grey images makes the model more dependent on shape of edges and not on the texture, and second, it requires mode efforts for computer to recognise emotions that come out naturally and not forced in studio with modes; for example, when we get surprised, we are more likely to express it with our face and hands but also sometimes we are laugh and get surprised at the same time and also sometimes we cry with neutral expression. The accuracy is high on a dataset which contains emotions that are forced and different emotions are expressed by a same person, and it subsequently drops on datasets which have images clicked randomly where people are actually laughing or sad or surprised, etc.

In this paper, we have used CK+ and RAF datasets but other datasets like FER2013, JAFFE, etc. are yet to be work on. So our future work is to make this model capable of delivering best results on other datasets too.

References

1. Hui Fang, Neil Mac Parthaláin, Andrew J. Aubrey, Gary K.L. Tam, Rita Borgo, Paul L. Rosin, Philip W. Grant, David Marshall, and Min Chen. 2014. Facial expression recognition in dynamic sequences. Pattern Recogn. 47, 3 (March, 2014), 1271–1281.
2. M. Suwa, N. Sugie, K. Fujimora, A preliminary note on pattern recognition of human emotional expression, in: International Joint Conference on Pattern Recognition, 1978, pp. 408–410.
3. Caifeng Shan a, Shaogang Gong b, Peter W. McOwan Facial expression recognition based on Local Binary Patterns: A comprehensive study, Image and Vision Computing 27 (2009) 803–816
4. T. Ojala, M. Pietikäinen, D. Harwood, A comparative study of texture measures with classification based on featured distribution, Pattern Recognition 29 (1) (1996) 51–59.
5. M. Verma, S. K. Vipparthi, and G. Singh, "HiNet: Hybrid inherited feature learning network for facial expression recognition", IEEE Letters of the Computer Society, Vol. 2, No. 4, pp. 36-39, 2019.
6. Aurelien Geron, Hands-On Machine Learning with Scikit-Learn, Keras, and TensorFlow, 2nd Edition, O'Reilly Media, Inc., September 2019.
7. S. Eleftheriadis, O. Rudovic, and M. Pantic, "Discriminative shared Gaussian processes for multiview and view-invariant facial expression recognition", IEEE Transactions on Image Processing, Vol. 24, No. 1, pp. 189–204, 2015.
8. Ritanshi Agarwal, Dr Neha Mittal, Hanmandlu Madasu, Convolutional Neural Network Based Facial Expression Recognition Using Image Filtering Techniques, International Journal of Intelligent Engineering and Systems, Vol. 14, No. 5, 2021
9. Krizhevsky, Alex, Ilya Sutskever, and Geoffrey E. Hinton. "Imagenet classification with deep convolutional neural networks." Advances in neural information processing systems 25 (2012): 1097–1105.

10. Szegedy, Christian, Wei Liu, Yangqing Jia, Pierre Sermanet, Scott Reed, Dragomir Anguelov, Dumitru Erhan, Vincent Vanhoucke, and Andrew Rabinovich. "Going deeper with convolutions." In Proceedings of the IEEE conference on computer vision and pattern recognition, pp. 1–9. 2015.

11. W. Wang, Q. Sun, T. Chen, et al., A Fine-Grained Facial Expression Database for End-to-End Multi-Pose Facial Expression Recognition, arXiv preprint arXiv:1907.10838, 2019.

12. C.M. Kuo, S.H. Lai, M. Sarkis, A compact deep learning model for robust facial expression recognition, Proceedings of the IEEE Conference on Computer Vision and Pattern Recognition Workshops, 2018, pp. 2121–2129

13. C. Turan, K.M. Lam, X. He, Soft Locality Preserving Map (SLPM) for Facial Expression Recognition. arXiv preprint arXiv:1801.03754, 2018.

14. Z. Zhang, M. Lyons, M. Schuster, S. Akamatsu, Comparison between geometrybased and Gabor-wavelets-based facial expression recognition using multilayer perceptron, in Third IEEE International Conference on Automatic Face and Gesture Recognition, 1998. Proceedings, Nara 1998, pp. 454_9.

15. Ashish Vaswani, Noam Shazeer,Niki Parmar, Jakob Uszkoreit,Llion Jones,Aidan N. Gomez,Łukasz Kaiser Illia Polosukhin, Attention Is All You Need, 31st Conference on Neural Information Processing Systems (NIPS 2017), Long Beach, CA, USA.

16. Y. Li, J. Zeng, S. Shan and X. Chen, "Occlusion Aware Facial Expression Recognition Using CNN With Attention Mechanism," in IEEE Transactions on Image Processing, vol. 28, no. 5, pp. 2439–2450, May 2019, doi: 10.1109/TIP.2018.2886767.

17. Yinghui Kong, Zhaohan Ren, Ke Zhang, Shuaitong Zhang, Qiang Ni, and Jungong Han "Lightweight facial expression recognition method based on attention mechanism and key region fusion," Journal of Electronic Imaging 30(6), 063002 (2 November 2021).

18. Fernandez, Pedro D. Marrero, Fidel A. Guerrero Pena, Tsang Ing Ren, and Alexandre Cunha. "Feratt: Facial expression recognition with attention net." arXiv preprint arXiv:1902.03284 3 (2019).

19. Li Jing, Kan Jin, Dalin Zhou, Naoyuki Kubota, and Zhaojie Ju. "Attention mechanism-based CNN for facial expression recognition." Neurocomputing 411 (2020): 340–350.

20. https://www.kaggle.com/datasets/shawon10/ckplus

21. T.F. Cootes, G.J. Edwards, C.J. Taylor, Active appearance models, IEEE Trans. Pattern Anal. Mach. Intell. 23 (6) (2011) 6815

22. Li, Shan and Deng, Weihong and Du, JunPing. Reliable crowdsourcing and deep locality-preserving learning for expression recognition in the wild. Computer Vision and Pattern Recognition (CVPR), 2017 IEEE.

Intelligent Systems and Smart Infrastructure – Brijesh Mishra et al. (eds)
© 2023 Taylor & Francis Group, London, ISBN 978-1-032-41287-0

CHAPTER

19

Performance Evaluation and Analysis of Round Robin Load Balancing in SDN

Abhishek Kumar Pandey[1], Niraj Kumar Tiwari[2], Shivam Bhardwaj[3]

Computer Science & Engg.,
Shambhunath Institute of Engineering & Technology, Prayagraj, India

Kamal Prakash Pandey[4]

Electronics & Communication Engg.
Shambhunath Institute of Engineering & Technology, Prayagraj, India

Abstract

Software-defined networking (SDN) is a novel networking method that aims to improve the performance and reducing the cost of numerous networking systems over traditional networks that have less resource utilization. The load balancer in SDN is implemented in this work to optimize throughput utilization. Network is created by Mininet software. "This was done using a floodlight controller and the Python programming language". For TCP and UDP protocols, two scenarios were analyzed before and after the load balancer. Using a load balancer with SDN improves network performance dramatically.

Keywords: SDN, Load balance, Mininet, Open stack

1. Introduction

New network architectures are required as cloud services and virtualization (Deeban Chakravarthy and Amutha 2019) grow in number. To fulfil the demands of today's networks,

Corresponding author: [1]abhi_007cs@yahoo.co.in, [2]nirajt131@gmail.com, [3]shibambhardwaj@gmail.com, [4]pandeykamal.1976@gmail.com

DOI: 10.1201/9781003357346-19

a new paradigm termed software-defined networking (SDN) was introduced (Mandal et al. 2021), which assists in maximizing resources consumption, maximizing performance, reducing response time, and preventing overloading of any particular resource. Many strategies on the servers are employed to regulate excessive network traffic and load balancing and eventually form a one point of failure and congestion, such as a switch of multilayer or a domain name server (Gusenbauer and Haddaway 2020). In a traditional network, typically, devices are configured individually. Server load balancing is used to manage extremely large data volumes (Kang and Choo 2018). P to P networks and server user are two common models. SDN separates the control and data planes and introduces a new device called a controller to regulate traffic and provide a broader perspective on the network (Bronzino et al. 2019). domain name network server is a database that stores names and is used to locate and translate domain names on the internet into Internet Protocol (IP) addresses (Xie et al. 2019). In a traditional network, large data demands are generally met through server load balancing (Seng and Ang 2022).

2. Research Problem

The most complex problem is that, if certain servers have greater processing power, memory, or other characteristics than others, the algorithm is unable to submit any additional requests to those servers. A traditional network also has hardware provided by a single vendor, protocols that can only be used by that vendor and devices that are usually set up one at a time.

3. Methodology

This study investigates and compares the various load-balancing techniques offered. Comparing SDN-based server load balancing to traditional techniques could enhance load-balancing efficiency and save installation time (Cui et al. 2018). Simple open flow tables can reliably monitor and the experiment was conducted under the Ubantu environment (CPU: 32 Cores, RAM: 8 GB) and required installation of Mininet and floodlight. The study includes the following stages: Create an SDN-based network topology in Mininet by writing a Python. Measure these parameters (transfer, throughput, delay, jitter and packet loss). Many studies are conducted to assess efficiency, so the network is divided into two pools, one for TCP requests and the other for UDP requests, and measuring the equal parameters with installing the load balancer (Li and Cheng 2020). Efficiency is measured in a relationship of client/server. Finally, compare the results with and without the load balancer.

4. Experimental Setting

In the experimental setting, the model of network was designed to include 1 controller coupled to 4 switches, with Mininet serving as the network emulation; the SDN simulation enables the development of SDN solutions, in addition to the creation and management of a system prototype. Mininet will be in responsible of constructing the SDN network using conventional

Linux hosts, switches, connections, and controllers. It lets you set up a network on a single PC. Using a command-line tool, a network is developed. Mininet includes a programming interface that enables a Python to design a customizable network architecture. In return, a Python builds a network topology. The experiment's network topology is given below.

5. Result and Discussion

The Virtual Environment's Performance Test Results

This section deals with virtual finds. The parameters obtained here are used to assess SDN-based load balancing. During the test, the transfer and bandwidth numbers were found. The test had an interval of 0.9 sec. So, they took out the middle. Due to the difficulty of using decimals in such tests, there were a lot of figures that were rounded to the nearest number. Because recommendations examine for clear tendencies by executing all situations with and without the load balancing, a 0.9-second difference between two solutions is not decisive or part of the solution.

Results of TCP: This section compares with and without TCP load balance results.

Discussion of TCP with and without load balancer

IPERF was used to perform the first test, which is depicted in Table 19.1, which shows how it worked with no load balancing and with the load-balancing section on 2 servers. In the next set of figures, you can see the results. It should be noted that because Floodlight uses default forwarding, there is no further check for both modules being activated. This is due to the fact that load balancing is accomplished through the use of a curl script rather than the traditional forwarding. The tests should be done this way because this is the best way to use the SDN controller.

Table 19.1 TCP output from IPERF

Condition of the test	Interval time in second	Data transfer rate in Mbyte	Avg. throughput in Mbit/sec	Minimum throughput in Mbit/sec	Maximum throughput in Mbit/sec
Without load balancer	0–16	76.0	9.87	0.12	15.83
With load balancer	0–16	139.38	18.75	2.12	18.63

The throughput performance is depicted in Fig. 19.1, where the vertical line represents the throughput in Mbits per second and the horizontal line represents the number of tests. Benchmarks at throughput, IPERF is designed to run 16 tests in a 1-second interval, and the controller is never idle since the TCP buffer is constantly full. Figure 19.1 shows the IPERF test results for hosts 1 and 2 to server 5. To get the minimum throughput when only one server is being used is 0.12 Mbit/sec because all of the loads are controlled with one server. Because all loads are handled by two servers, the minimum throughput is now 2.12 Mbit/Sec when the competitor parameter is doubled. In comparison to the single-component testing, in Fig. 19.1,

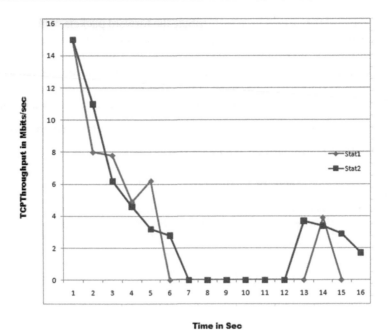

Fig. 19.1 TCP throughput without load balancer

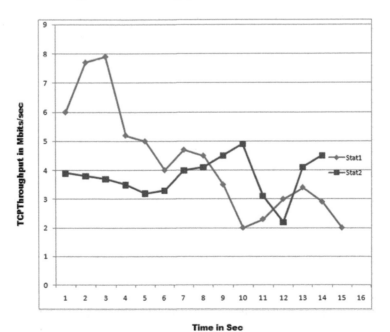

Fig. 19.2 TCP throughput with load balancer

in this scenario, the load balancer is used to distribute the forward packets, which includes both components loaded. In addition, as seen in Table 19.1, having a load balancer allows you to send more data than if you did not use one. As seen in Fig. 19.2, there is a significant difference without and with the load balancer, indicating that the load balancer technique reduces the cost and improves stability. Similarly, the average of the other hosts can be determined. According to Table 19.1, the transmission of packets is more exact and efficient after load balancer.

Results of UDP: This section compares with and without UDP load balance results.

Discussion of UDP with and without load balancer

Table 19.2 UDP output from INPERF

Condition of the test	Interval time in second	Data transfer rate in Mbyte	Avg. throughput in Mbit/sec	Minimum through put in Mbit/sec	Maximum throughput in Mbit/sec
Without load balancer	0–16	86.9	12.11	1.4	21.34
With load balancer	0–16	130.7	18.13	1.05	22.13

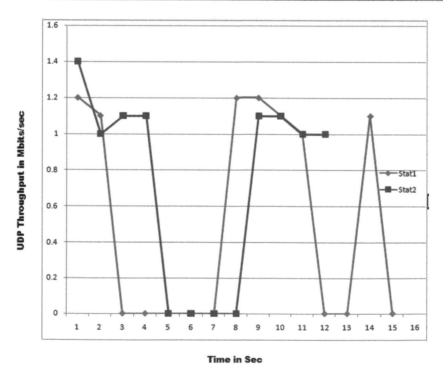

Fig. 19.3 UDP throughput without load balancer

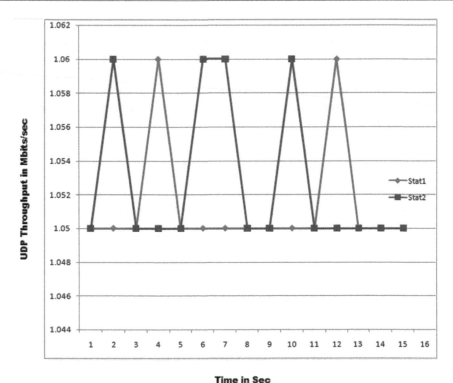

Fig. 19.4   UDP throughput with load balancer

The results of the trials for UDP traffic are shown in the second series of tests in Table 19.2. Resultant value in same topology are assessing better significance including mininet created simulation evironment. The graphs showed the results of running a single server without a load balancer and a second server with a load balancer running. The numbers for transfer and bandwidth were obtained with a 1-second interval in this test. Figures 19.3 and 19.4 illustrate the throughput before the load balancer and after the load balancer, respectively. The y-axis shows the throughput in Mbit/sec. Figure 19.3 compares the performance of hosts 1 and 2 to server 7. Because only one server is loaded at a time, the throughput is inconsistent, ranging from 1.4 Mbit/sec to 21.34 Mbit/sec. With two servers, maximum throughput is 21.34 Mbps, with a minimum of 1.4 Mbit/sec.

6. Conclusion

SDN is a modern network model that can be used today. SDN is a modern network model that can be used in architectural decouples, control plan and network control plan. Dynamic load balancing increases web server response time and enables for more efficient resource distribution and load balancing. A single fully programmable network is therefore given. It is possible to use the controller as the primary open source if the administrator wants to build a secure system. According to the published scientific article, using a load balancer improved network performance by increasing throughput, reducing latency, packet loss, and jitter. The load balancer with an open virtual switch and a Floodlight controller is more flexible and cost-effective. Assuring greater stability, protocol-based load balancing was created to efficiently disperse server load. The testing focused on Quality of service characteristics like throughput, latency, packet loss, and jitter between servers and hosts in the network. The load balancer enhanced network efficiency.

REFERENCES

1. Bronzino, Francesco, Ivan Seskar, Sumit Maheshwari, and Dipankar Raychaudhuri. 2019. "NoVN: Named-Object Based Virtual Network Architecture." ACM International Conference Proceeding Series, 90–99. https://doi.org/10.1145/3288599.3288637.
2. Cui, Jie, Qinghe Lu, Hong Zhong, Miaomiao Tian, and Lu Liu. 2018. "A Load-Balancing Mechanism for Distributed SDN Control Plane Using Response Time." IEEE Transactions on Network and Service Management 15 (4): 1197–1206. https://doi.org/10.1109/TNSM.2018.2876369.
3. Deeban Chakravarthy, V., and B. Amutha. 2019. "Path Based Load Balancing for Data Center Networks Using SDN." International Journal of Electrical and Computer Engineering 9 (4): 3279–85.https://doi.org/10.11591/ijece.v9i4.pp 3279–3285.
4. Gusenbauer, Michael, and Neal R. Haddaway. 2020. "Which Academic Search Systems Are Suitable for Systematic Reviews or Meta-Analyses? Evaluating Retrieval Qualities of Google Scholar, PubMed, and 26 Other Resources." Research Synthesis Methods 11 (2): 181–217. https://doi.org/10.1002/jrsm.1378.
5. Kang, Byung Seok, and Hyun Seung Choo. 2018. "An SDN-Enhanced Load-Balancing Technique in the Cloud System." Journal of Supercomputing 74 (11): 5706–29. https://doi.org/10.1007/s11227-016-1936-z.
6. Li, Jiawei, and Yuanming Cheng. 2020. "Design and Implementation of Voice-Controlled Intelligent Fan System Based on Machine Learning." Proceedings of 2020 IEEE International Conference on Advances in Electrical Engineering and Computer Applications, AEECA 2020, 548–52. https://doi.org/10.1109/AEECA49918.2020.9213552.
7. Mandal, Sumit K., Jie Tong, Raid Ayoub, Michael Kishinev sky, Ahmed Abou Samra, and Umit Y. Ogras. 2021. "Theoretical Analysis and Evaluation of NoCs with Weighted Round-Robin Arbitration." IEEE/ACM International Conference on Computer-Aided Design, Digest of Technical Papers, ICCAD 2021-Novem. https://doi.org/10.1109/ICCAD51958.2021.9643448.
8. Seng, Kah Phooi, and Li Minn Ang. 2022. "Artificial Intelligence Internet of Things: A New Paradigm of DistributedSensorNetworks" 18(3). https://doi.org/10.1177/15501477211062835.
9. Xie, Junfeng, F. Richard Yu, Tao Huang, Renchao Xie, Jiang Liu, Chenmeng Wang, and Yunjie Liu. 2019. "A Survey of Machine Learning Techniques Applied to Software Defined Networking (SDN): Research Issues and Challenges." IEEE Communications Surveys and Tutorials 21(1): 393–430. https://doi.org/10.1109/COMST.2018.2866942.

Intelligent Systems and Smart Infrastructure – Brijesh Mishra et al. (eds)
© 2023 Taylor & Francis Group, London, ISBN 978-1-032-41287-0

CHAPTER

20

Energy Scheduling of Residential and Commercial Appliances with Demand Response Technique Using Grid, PV, and EV as a Source

Ajay Kumar Prajapat[1] and Sandeep Kakran[2]

NIT Kurukshetra, Haryana 136119, India

Abstract

Demand-side management programs encourage the users to revamp their energy consumption according to the real-time electricity prices so that peak hour demand can be reduced. So, to maintain the equilibrium between the demand and power supply, peak load management is imperative. Peak load reduction is a process that allows the utility to reshape the load curve, i.e., flattens the load curve, increases the system reliability, efficiency, stability, and reduces the grid fluctuations, overall operation cost, and environmental pollution. To achieve these goals such as cost-saving and reduction of peak demand, an incentive-based demand response (IB-DR) program and energy scheduling of residential and commercial appliances are used. This paper presents an algorithm for the reduction of peak demand and electricity bills by scheduling the appliances of residential and commercial areas with the use of electric vehicles (EV), photovoltaic units (PV), and the grid as a source. EVs considered in this paper are working in the vehicle-to-grid (V2G) mode, and their batteries are used as a storage system at the time of parking. The batteries are discharging during peak hours. The two scenarios considered in this paper are random scheduling with incentive-based demand response (IB-DR) and without IB-DR and user-preferred duration scheduling with and without IB-DR. For the optimization, mixed-integer linear programming is used, and a CPLEX solver is selected for modeling and solving the equations in GAMS software.

Corresponding author: [1]ajaykumarprajapat777@gmail.com, [2]skakran@gmail.com

DOI: 10.1201/9781003357346-20

Keywords: Demand-Side Management, Energy Scheduling, Electric Vehicle, Peak Load Management, Vehicle to Grid, PV, Incentive-Based Demand Response, GAMS, CPLEX

Nomenclature

T_{ari}	Entering time of EV		SOC_n^{mx}	EV maximum SOC [kWh].
T_{dpp}	Exit time of EV		SOC_n^{mn}	EV minimum SOC [kW].
N	Set of EV, $n \in N$		α	Efficiency at which EV discharges [%].
Z_n	Interval when EV is plugged ()		β	Efficiency at which EV charges [%].
ΔT	Duration of time intervals		$R^{ch,EV}$	Charging rate of EV [kW].
X	Maximum discharging limit of EV [%]		$R^{ds,EV}$	Discharging rate of EV [kW].
B_n	Battery capacity [KWh].		$P_{t,n}^{g'EV}$	Power is given by the EV to load during time t
$P_n^{mx, EV}$	Maximum instantaneous EV power [KW].			
$SOC_{Ini, n}$	Initial value of SOC of the EV during [kWh].		$SOC_{t,n}^{EV}$	EV state of charge at time t.

1. Introduction

At present, the population is increasing very fast, so the energy demand is also going high. To fulfill these high demands, utilities cannot rely on conventional resources of energy to fulfill the demand of the consumer because conventional energy sources are limited and very harmful to the environment. So, utilities have to contemplate non-conventional energy resources like renewable energy resources (RES). Various researches have been done in the field of RES, which is discussed in Chandra and Chanana (2018). By participating the users in the smart grid, the energy and the cost of the energy can be saved. To reduce the cost, the demand response technique can also be used. Demand response is mainly classified as incentive-based and price-based (Rehmam.et.al, 2021). Non-schedulable and schedulable appliances are the two classification of home and commercial appliances. Schedulable appliances can be categorized further into two parts, i.e., interruptible appliances and non-interruptible appliances. Peak demand in the residential and commercial sectors has been decreasing with the use of EVs, which is shown in Mahmud, Hossain, and Ravishankar (2018). The reason behind the use of EVs is their exponential growth in the modern world. Various researches have been done to reduce the peak hour demand and cost by using EV (Achour, Oammi, Zejli, 2021) and scheduling the appliances. In Catalin and Marinescu (2019), the researcher looks into the best way to schedule EV charging at home based on RES, with a focus on both renewable energy and microgrid battery usage. In Kakran and Chanana (2018), the microgrid generation has been scheduled considering incentive-based demand response program to fulfill the demand of a residential sector. Modeling of renewable energy sources and electric vehicles (EVs) integration into micro-grid is discussed in Hassan et al. (2014). In Alavi et al. (2016), the study describes an energy management system using fuzzy-based intelligent and EV batteries for energy storage services in a microgrid. While cars are only utilized for transportation for a short period, the stored energy in the onboard hydrogen tanks of fuel cell vehicles may be used to supply electricity when the vehicles are parked, as discussed (Sengor. et al., 2019). In Kakran and Chanana (2019), for EV parking lots, an optimum energy management strategy considers the reduction in peak load with DR and also considers the erratic behavior of EVs, such as the departure and arrival time of the EV. In Mahmud. et al. (2017), a controller based on an artificial neural network (ANN) has been employed to regulate the peak load demand on the domestic side in the proposed system. Also, in Kim, Son, and Yong (2013), EV has been considered in both modes like grid-to-vehicle (G2V) and vehicle-to-grid (V2G) mode, charging the EV during the low price time and discharging the EV during high price time. In Zhou et al. (2016), a grid-tied residential microgrid and a battery management method is created utilizing a non-linear optimization algorithm to decrease grid power. Energy scheduling has been done for some residential appliances in Balakrishna and Kakran (2019). The aforementioned research shows that most of the work has been done only for residential load via EV, where EV is used in V2G and G2V modes. In this work, only discharging power of the EV is considered. To decrease the cost, the DR technique is used, but no work has been done for residential and commercial consumers.

In this paper, fixed load and the variable load of the residential and commercial consumer are considered to reduce the peak with the use of EV, and IB-DR is used to reduce the cost. Energy scheduling of commercial and residential appliances is done. The appliances are scheduled considering the two scenarios. First, the appliances are scheduled randomly, and second, the appliances are scheduled considering user-preferred duration. In this paper, GAMS software is used, and CPLEX solver is used to optimize the load and to optimize energy cost during peak load demand.

2. Problem Formulation

A. EVs Integration into the Load

The following equations represent the EV utilized as a V2L

$$P_{t,n}^{g\,EV} = \alpha \cdot P_{t,n}^{EV,ds} \qquad\qquad t \in [T_{ari}, T_{dpp}], n \in N \qquad\qquad (1)$$

$$0 \le P_{t,n}^{EV,ds} \le R^{ds,EV}(1 - a_{t,n}^{EV}) \qquad\qquad t \in [T_{ari}, T_{dpp}], n \in N \qquad\qquad (2)$$

$$0 \le P_{t,n}^{EV,ch} \le R^{ch,EV} a_{t,n}^{EV} \qquad\qquad t \in [T_{ari}, T_{dpp}], n \in N \qquad\qquad (3)$$

$$SOC_{t,n}^{EV} = SOC_{Ini,n} - \alpha \cdot P_{t,n}^{EV,ds} \qquad\qquad t \in [T_{ari}, T_{dpp}], n \in N \qquad\qquad (4)$$

$$SOC_n^{mn} \le SOC_{t,n}^{EV} \le SOC_n^{mx} \qquad\qquad t \in [T_{ari}, T_{dpp}], n \in N \qquad\qquad (5)$$

$$SOC_n^{mn} \leq SOC_{t,n}^{EV} \leq SOC_n^{mx} \qquad\qquad t \in [T_{ari}, T_{dpp}], n \in N \qquad (6)$$

$$P_{t,n}^{EV,ds} \geq (x.SOC_{Ini,n})/t \qquad\qquad t = T_{ari} \qquad (7)$$

$$SOC_{t,n}^{EV} = SOC_{Ini,n} - (x.SOC_{Ini,n}) \qquad\qquad t = T_{dpp} \qquad (8)$$

where

$a_{t,n}^{EV}$ decides the charging and discharging status of the EV.

If $a_{t,n}^{EV} = 1$, then EV will charge.

$a_{t,n}^{EV} = 0$, then EV will discharge.

Equation (1) demonstrates that the EV's effective energy may be utilized to meet the needs of a section of the commercial and residential loads. Equations (2) and (3) represent the EV's limits of discharging and charging power. EV's SOC in the present hour t is equivalent to earlier hours SOC minus the transferred of effective energy from the EV in the discharging interval, as shown in equation (4). Equation (5) depicts the lowest discharging and highest charging level of EV. Equation (6) represents the discharging SOC limit of the EV according to the EV's owner preferences. Equation (7) shows the SOC's value at the instance of vehicle's arrival, and equation (8) shows the minimum SOC of the EV at the instance of departure.

B. Objective Function

The primary purpose of this study is to create load appliance scheduling in a way that the consumer makes a massive profit on their energy bill with minimizing the effect on the user's preferences. For this, the objective function is given by

$$C_{min} = \sum_{t=1}^{24} \sum_{apl=1}^{4} N_{apl}^{t} \times Pr_g \qquad (9)$$

$$N_{apl} = \left[N_1^1, N_2^2, \ldots N_9^{24} \right] \qquad (10)$$

$$\sum_{t=1}^{24} N_{apl}^{t} = N_{apl}^{total} \qquad (11)$$

C_{min}: Minimized energy cost of all appliances

N_{apl}^{t}: Consumption of the energy for each appliance at a particular time slot

$$N_{apl}^{t} = d_{t,apl} \times W_{apl}^{mx} + (1 - d_{t,apl}) \times W_{apl}^{mn} \qquad (12)$$

$$\sum_{t=1}^{24} d_{t,apl} \times g_{apl,t} = u_{apl} \qquad (13)$$

$d_{t,apl}$: A binary variable that is responsible for the ON/OFF time of an appliance

$g_{apl,t}$: A binary variable that is responsible for the ON/OFF duration of an appliance

u_{apl}: Total number of slots when the appliances are ON

$$\sum_{n \in N} k_{n,apl} = 1 \tag{14}$$

k is a binary variable, and 'N' denotes the number of possible appliances 'apl' schedule in the duration preferred by the user, and it is calculated as N = Total duration preferred by the user – Appliance's total ON time + 2

$$N_{apl}^t = \sum_{n \in N} k_{n,apl} \times N_{n,t} \tag{15}$$

$N_{n,t}$: t^{th} hour of schedule n energy consumption

$$W_{apl}^{mn} \le N_{apl}^t \times g_{apl,t} \le W_{apl}^{mx} \tag{16}$$

$$\sum_{a=1}^{4} N_{apl}^t \le H^{mx} \tag{17}$$

H^{mx}: the limit put by the utility to consume energy at each hour

C. Solar Power Equation

$$P_{pv}(t) = P_r \times P_d \times \frac{Rd(t)}{R_{ref}} \tag{18}$$

where P_{pv}: rated power of PV;

P_d: Derating factor;

Rd: Solar radiation;

R_{ref}: Reference solar radiation.

D. Load Balance Equation

Without IB-DR

$$P_s(t) + P_v(t) + P_{gri}(t) = P_{resi}(t) + P_{come}(t) \tag{21}$$

where $P_s(t)$: PV's active power at time t;

$P_v(t)$: EV's active power at time t;

$P_{gri}(t)$: Grid's active power at time t;

$P_{resi}(t)$: total residential demand at time t;

$P_{come}(t)$: total commercial demand at time t.

With IB-DR

$$P_s(t) + P_v(t) + P_{gri}(t) = P_{resi}(t) + P_{come}(t) - X_t^{dr} \tag{22}$$

Equation (22) demonstrates that the requirement of energy of the residential and commercial appliances following the IB-DR program can be fulfilled by the grid, EV, and PV alone or by the combination of all these.

3. Results

In this paper, for the implementation of incentive-based response, the assumption is taken that 40% of customers are participating, and of that, 33% of the load is available for participation in demand response. The incentive rate assumed is 1.5 cents. Fig. 20.1 shows the fixed residential and commercial load. Depending on the number of PV panels placed, each user's rooftop PV panel capacity may change. Fig. 20.2 shows the solar irradiation. The tariff rates change according to the time of day, with Fig. 20.3 depicting the hourly rate. (https:// hourlypricing.comed.com/live-prices/,2021). Fig. 20.4 shows the power supply by the grid, PV, and EV to fulfill the load demand without IB-DR. With the participation of some users in demand response, the total load is reduced, which is shown in Fig. 20.5. Table 20.1 shows the controllable residential and commercial appliances. Fig. 20.6 and Fig. 20.7 show the variation in cost for different scenarios considering random scheduling and user-preferred duration. Fig. 20.8 shows the scheduling of the constant load combined with the variable appliances of residential and commercial sectors for user-preferred duration. Fig. 20.9 shows the fixed load of the residential and commercial sectors and the random scheduling of variable appliances.

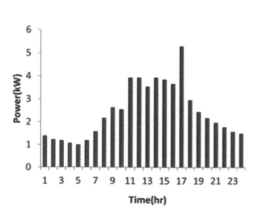

Fig. 20.1 Fixed load of residential and commercial

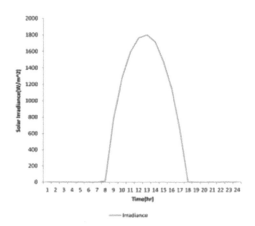

Fig. 20.2 Solar irradiation data

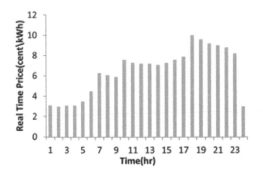

Fig. 20.3 Tariff rate variations

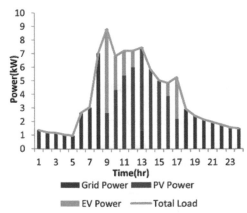

Fig. 20.4 Load supplied by the grid, PV, and EV without IB-DR

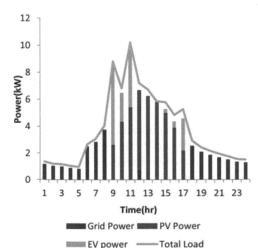

Fig. 20.5 Load supplied by the grid, PV, and EV with IB-DR

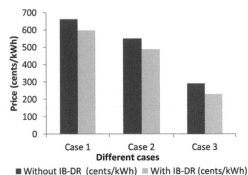

Fig. 20.6 Variation in cost with the different source combinations in user preferred duration

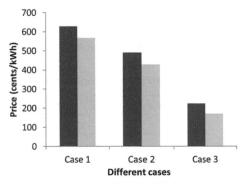

Fig. 20.7 Variation in cost with the different source combinations in random scheduling

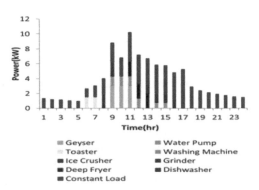

Fig. 20.8 Fixed load and scheduling of variable appliances considering the user-preferred duration

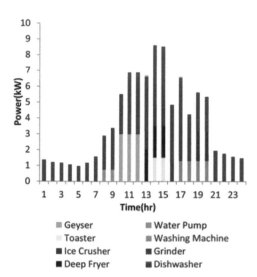

Fig. 20.9 Fixed load and Random scheduling of variable appliances

Table 20.1 Controllable residential and commercial appliance

Appliance	Energy consumed (kWh)	User-preferred duration
Controllable residential appliances		
Geyser	3	9 am–1 pm
Water Pump	0.75	4 pm–8 pm
Toaster	1.5	6 am–9 am
Washing machine	1.3	9 am–3 pm
Controllable commercial appliances		
Ice crusher	0.4	10 am–1 pm
Grinder	1.5	12 pm–3 pm
Deep fryer	2	11 am–4 pm
Dishwasher	1.2	9 am–2 pm

Every consumer in this work has three supply sources at a particular instant. One is supplied by the grid, and the other two are renewable sources which are EV and PV. In Fig. 20.6, scenario 1 represents the cost when the only grid is used as a source in user-preferred duration with and without IB-DR, scenario 2 represents the cost when both grid and EV are used, and scenario 3 represents the cost when the grid, EV, and PV are used as a source. In Fig. 20.7, scenario 1 represents the cost when the only grid is used as a source in random scheduling of appliances with and without IB-DR, scenario 2 represents the cost when both grid and EV are used, and scenario 3 represents the cost when the grid, EV, and PV are used as a source.

Fig. 20.6 and Fig. 20.7 show that in scenario 1, the cost of electricity is highest, i.e., when the only grid is used as a source, and the cost is lowest in scenario 3, i.e., when all the three sources, grid, PV, and EV are connected. The reduction in cost with the use of EV and PV shows that with the use of renewable sources, cost decreases and dependency on the grid decreases as there are more sources available to fulfill the load demand. Similarly, the difference in cost with and without IB-DR is highest when all the three sources are connected and lowest when the only grid is connected. Therefore, utility dependency decreases, and utility saves some energy which is a benefit for it.

4. Conclusion

In this paper, EV, PV, and grid are used as a source to supply power to the residential and commercial load. With the results, it can be concluded that with the use of IB-DR, the consumption of grid power and cost of electricity consumption reduce. At the same time, the different residential and controllable commercial appliances are scheduled considering the user-preferred duration. The reduction in total cost can be stated when appliances are scheduled at a minimum price rate. Total cost during random scheduling of appliances with the sources of the grid, PV, and EV is reduced by 21.60%, and when appliances are scheduled considering user-preferred duration, the cost is reduced by 24.62%. In upcoming research, extension of this work can be done by taking more controllable commercial and residential appliances into consideration, along with the inclusion of industrial appliances.

REFERENCES

1. Lokesh Chandra and Saurabh Chanana Energy Management of Smart Homes with Energy Storage, Rooftop PV and Electric Vehicle IEEE International Students' Conference on Electrical, Electronics & Computer Science2018.
2. Ateeq Ur Rehman, Ghulam Hafeez, Fahad R. Albogamy, Zahid Wadud, Faheem Ali, Imran Khan, Gul Rukh, and Sheraz Khan An Efficient Energy Management in Smart Grid Considering Demand Response Program and Renewable Energy Sources IEEE 2021.
3. Khizir Mahmud, M. J. Hossain and Jayashri Ravishankar Peak-Load Management in Commercial Systems With Electric Vehicles IEEE Systems Journal 2018.
4. Catalin ION and Corneliu Marinescu Optimal Charging Scheduling of Electrical Vehicles in a Residential Microgrid based on RES International Conference 2019.
5. Sandeep Kakran and Saurabh Chanana Operation management of a renewable microgrid supplying to a residential community under the efect of incentive-based demand response program International Journal of Energy and Environmental Engineering 2018.
6. Yasmine Achour, Ahmed Ouammi and Driss Zejli Model Predictive Control Based Demand Response Scheme for Peak Demand Reduction in a Smart Campus Integrated Microgrid IEEE 2021.
7. A. S. Hassan, A. Firrincieli, C. Marmaras, L.M. Cipcigan and M.A. Pastorelli Integration of Electric Vehicles in a Microgrid with Distributed Generation IEEE 2014.
8. Femina Mohammed Shakeel and Om P. Malik Fuzzy Based Energy Management System for a Micro-grid with a V2G Parking Lot IEEE 2020.

9. Farid Alavi, Esther Park Lee, Nathan van de Wouw, Bart De Schutter and Zofia Lukszo Fuel cell cars in a microgrid for synergies between hydrogen and electricity networks journal 2016.

10. Ibrahim Sengor, Ozan Erdinc, Bans Yener, Akın Tas¸cıkaraoglu and Joao P. S. Catalao Optimal Energy Management of EV Parking Lots Under Peak Load Reduction Based DR Programs Considering Uncertainty IEEE 2019.

11. Sandeep Kakran and Saurabh Chanana Operation Scheduling of Household Load, EV and BESS Using Real Time Pricing, Incentive Based DR and Peak Power Limiting Strategy International Journal 2019.

12. K. Mahmud, S. Morsalin, M. J. Hossain and G. E. Town Domestic Peak-load Management Including Vehicleto-Grid and Battery Storage Unit Using an Artificial Neural Network IEEE 2017.

13. Bin Zhou, Wentao Li, Ka Wing Chan, Yijia Cao, Yonghong Kuang, Xi Liu, Xiong Wang,Smart home energy management systems: Concept, configurations, and scheduling strategies,Renewable and Sustainable Energy Reviews,Volume 61, 2016.

14. Yu, T. & Kim, Dong & Son, Sung-Yong. (2013). Optimization of scheduling for home appliances in conjunction with renewable and energy storage resources. International Journal of Smart Home.

15. B. Balakrishna and S. Kakran, "Energy Scheduling of a Household Considering Different Dynamic Pricing Schemes," 2019 3rd International Conference on Trends in Electronics and Informatics (ICOEI), 2019, pp. 876–879.

16. Real time electricity prices. Available at: https://hourlypricing.comed.com/live-prices/ [last accessed October,2021]

Intelligent Systems and Smart Infrastructure – Brijesh Mishra et al. (eds)
© 2023 Taylor & Francis Group, London, ISBN 978-1-032-41287-0

CHAPTER

21

Chronic Kidney Disease Prediction Using Machine Learning Classifiers

Pooja Rani[1]

MMICTBM, Maharishi Markandeshwar
(Deemed to be University), Mullana, Ambala, Haryana, India

Rohit Lamba[2]

Department of ECE, MMEC, Maharishi Markandeshwar
(Deemed to be University), Mullana, Ambala, Haryana, India

Ravi Kumar Sachdeva[3]

Chitkara University Institute of Engineering and Technology,
Chitkara University, Punjab, India

Rajneesh Kumar[4]

Department of Computer Engineering, MMEC,
Maharishi Markandeshwar (Deemed to be University),
Mullana, Ambala, Haryana, India

Priyanka Bathla[5]

Department of Computer Science & Engineering,
Chandigarh University, Gharuan, Mohali, Punjab, India.

Abstract

Chronic kidney disease (CKD) is one of the most critical health issues. Poor dietary habits and limited water intake contribute to this disease. In this disease, kidneys are gradually

---

Corresponding author: [1]pooja.rani@mmumullana.org, [2]rohitlamba14@mmumullana.org,
[3]ravisachdeva1983@gmail.com, [4]drrajneeshgujral@mmumullana.org, [5]priyankabathla85@gmail.com

DOI: 10.1201/9781003357346-21

damaged by several factors, resulting in irreversible kidney dysfunction. Machine learning has a significant impact on the healthcare industry since it allows for more precise diagnosis and treatment of several of chronic diseases. In this paper, a methodology is proposed to predict CKD using machine learning. Four classifiers, naive bayes (NB), logistic regression (LR), support vector machine (SVM), and decision tree (DT), are used to predict CKD. LR achieved 99.37% accuracy, SVM achieved 88.37% accuracy, DT achieved 98.12% accuracy, and NB achieved 100% accuracy.

Keywords: Logistic Regression, Support Vector Machine, Chronic Kidney Disease, Decision Tree, Naive Bayes.

1. Introduction

The kidney is a vital organ in human body. Osmoregulation and Excretion are two of the most important functions of kidney. In layman's terms, the renal and excretion systems gather and eliminate all hazardous and useless material from the body (Nishanth and Thiruvaran 2017). In India, 1 million instances of CKD are diagnosed each year. Renal failure is another name for CKD. It is a serious disease that causes a gradual loss of renal function. A person's kidneys will fail permanently (Reshma et al. 2020). CKD is a serious condition that can lead to heart disease also and early mortality. CKD is now the 18th biggest cause of death worldwide. Over 500 million individuals worldwide are suffering from this disease (Rajarajeswari, S., T., Tamilarasi 2021). It is gradually escalating into a worldwide medical crisis. Unhealthy lifestyle and a lack of water intake are both major causes of this disease (Pathak et al. 2020). A person survive for only 18 days without their kidneys. The two most prevalent risk factors for CKD are high blood pressure and diabetes. Diabetes damages various parts of the body, along with the kidneys. Uncontrolled high blood pressure is also a significant cause of CKD (Wang et al. 2018). CKD is a life-threatening disease that impacts over 14% of the worldwide people, and precisely predicting it allows people to discover it early and receive low-cost, low-risk treatment. Machine learning's impact on the healthcare industry is growing and it is used to diagnose various diseases (Lamba et al. 2021a). In this paper, a methodology is proposed to predict CKD using four classifiers SVM, LR, DT, and NB. The motivation behind this work is to accurately predict CKD so that timely and effective treatment can be provided to the patient. The structure of this paper is broken as follows. Literature survey is done in Section 2. Section 3 includes materials and methods. Results are discussed in Section 4. In Section 5, conclusion and future scope are presented.

2. Literature Survey

Gunarathne et al. (2017) predicted CKD using 14 attributes and achieved 99.1% accuracy. Aljaaf et al. (2018) predicted CKD using a multilayer perceptron, with an accuracy of 99.5%. Tekale et al. (2018) predicted CKD with various classifiers and achieved the highest accuracy of 96.75% with SVM. Rady and Anwar (2019) used MLP, probabilistic neural network (PNN), and SVM algorithms to predict CKD. PNN provided the highest performance with 98.9% accuracy. Elhoseny et al. (2019) developed a framework using the ant colony optimization algorithms. It achieved accuracy of 95%. Ekanayake and Herath (2020) predicted CKD with various classifiers and achieved 100% accuracy with k nearest neighbor (KNN). De and Chakraborty (2020) used LR, KNN, random forest (RF), DT, NB, and SVM classifiers to predict CKD. The highest accuracy of 100% was achieved using RF. Snega et al. (2020) used RF and neural network to detect CKD. RF achieved 88.7% accuracy and neural network achieved 98.40% accuracy. Ifraz et al. (2021) used various machine learning methods, including KNN, LR, and DT to predict CKD. The highest accuracy of 97% was achieved with LR. Almustafa (2021) predicted CKD with 99.25% accuracy using DT classifier. Ilyas et al. (2021) predicted CKD using RF and J48. J48 performed better with 85.5% accuracy. Senan et al. (2021) used four classifiers SVM, RF, DT, and KNN to predict CKD. Relevant features were selected using recursive feature elimination. Kim et al. (2021) developed neural network for CKD diagnosis. Weights of neural network were optimized using genetic algorithm.

3. Materials and Methods

A. Dataset

The study was conducted with the help of the CKD dataset available on UCI (University of California, Irvine). This dataset contains 400 records and 26 fields. The value of the classification column is either 1 indicating CKD or 0 indicating not CKD. Figure 21.1 shows the distribution of CKD and non-CKD entries. Total amount of data for CKD is 250, while total data for non-CKD is 150. The correlation of features in CKD dataset is illustrated in Fig. 21.2.

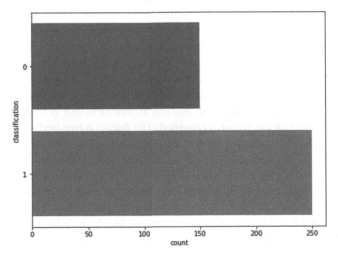

Fig. 21.1 Distribution of CKD and Non-CKD Entries

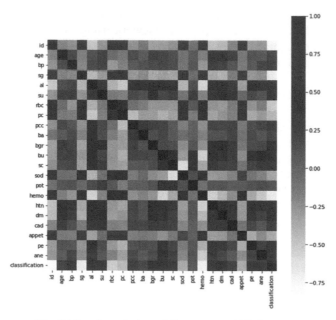

Fig. 21.2   Correlation of Features in CKD Dataset

B. Classifiers

Methodology for predicting CKD has used four classifiers: LR, SVM, DT, and NB. In this section, these four classifiers are described. For the analysis and classification of data sets, researchers use logistic regression (LR) as one of the most important approaches. LR can be used for binary classification as well as multi-classification problems (Rani et al. 2021a). The SVM algorithm draws a line to determine which classes the data belongs to. A hyperplane is a line that serves as a decision boundary. There are two sorts of algorithms: linear and nonlinear (Lamba et al. 2021b). When the dataset has two separable classes, linear SVM is used. If the dataset is inseparable, the approach uses a nonlinear SVM to turn the coordinate region into a separable space. Multiple hyperplanes can exist, and optimum hyperplane is decided based on the maximum margin (Rani et al. 2021b). A tree structure is used in DT algorithm. Decisions are made based on the features provided. The decision tree performs the comparison of features with root node to make a decision and continues to the next node depending on the result of comparison. It compares the attributes of second node with those of subnodes, and so on until leaf node is reached (Lamba et al.). NB classifier is a probabilistic classifier in which each feature has an equal contribution in predicting the target class. It assumes that all features are independent and they don't have any interaction with each other. This classifier has fast computation speed, and it can give good performance on big datasets having a large number of features. The NB classifier is also resistant to noise (Rani et al. 2022).

C. Methodology

Methodology for predicting CKD is shown in Fig. 21.3. Records having blank values in any field are deleted. LR, SVM, DT, and NB classifiers are used to perform classification. The classification was done to classify each record belonging to CKD class or non-CKD class. The performance of each classifier is measured using common performance parameters. NB achieved the highest accuracy of 100%.

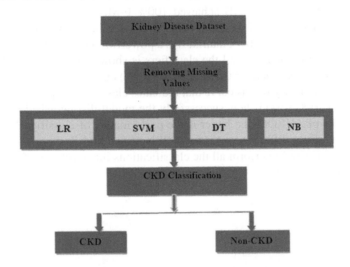

Fig. 21.3   Methodology for Predicting CKD

4. Results and Discussion

Experiments are done using LR, SVM, DT, and NB classifiers on CKD dataset. These four classifiers are used to predict CKD. The ten-fold validation method is utilized to evaluate prediction performance of classifiers. Accuracy, specificity, sensitivity, F-Measure, and precision of classifiers are evaluated. Performance of different classifiers in predicting CKD is shown in Table 21.1.

Table 21.1   Performance of Classifiers in Predicting CKD

Classifier	Accuracy	Sensitivity	Specificity	Precision	F-Measure
LR	99.37	97.67	100	100	98.82
SVM	87.37	53.48	100	100	69.69
DT	98.12	93.02	100	100	96.38
NB	100	100	100	100	100

Accuracy measures correct predictions done by classifier. LR achieved 99.37% accuracy, SVM achieved 87.37% accuracy, DT achieved 98.12% accuracy, and NB achieved 100% accuracy.

Sensitivity measures the percentage of times the person having CKD is correctly classified. LR achieved 97.67% sensitivity, SVM achieved 53.48% sensitivity, DT achieved 93.02% sensitivity, and NB achieved 100% sensitivity. Specificity measures the percentage of times the person not having CKD is correctly classified. All the classifiers achieved 100% sensitivity. Precision measures the percentage of times the system had produced relevant results. All the classifiers achieved 100% precision. F-Measure calculates the harmonic mean of precision and sensitivity. LR achieved 98.82% F-Measure, SVM achieved 69.69% F-Measure, DT achieved 96.38% F-Measure, and NB achieved 100% F-Measure. A confusion matrix is a table that illustrates predictions of model. It displays the number of wrong and right guesses. The matrix is represented as an x-by-x matrix. It is a powerful tool for calculating a classifier's accuracy [10]. Confusion matrix of all the classifiers is shown in Figs 21.4, 21.5, 21.6, and 21.7, respectively. Receiver Operator Characteristic (ROC) curve is a classification evaluation metric. AUROC is a summary of this curve that indicates classifier's capability to differentiate between classes. AUROC reveals how successfully the model distinguished between classes. The larger the value of AUROC, the better is the performance of classifier. If AUC = 1, the classifier can successfully differentiate between all Negative and Positive class points. If the AUC= 0, the classifier would perform all the classifications incorrectly. ROC curve of all the classifiers is shown in Fig. 21.8.

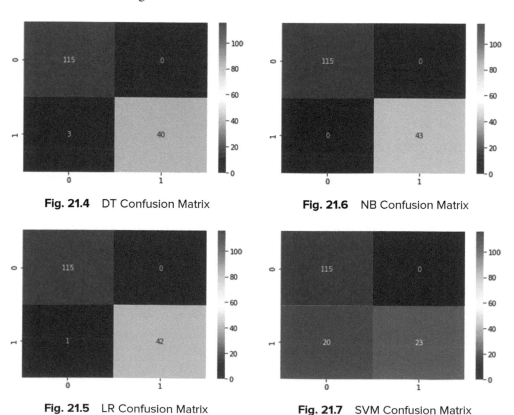

Fig. 21.4 DT Confusion Matrix   **Fig. 21.6** NB Confusion Matrix

Fig. 21.5 LR Confusion Matrix   **Fig. 21.7** SVM Confusion Matrix

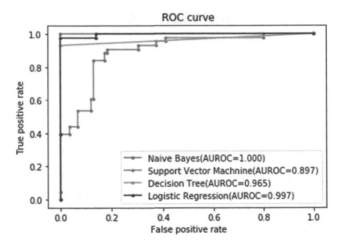

Fig. 21.8   ROC Curve of Classifiers

Comparative analysis of the accuracy of classifiers in predicting CKD is shown in Fig. 21.9. NB has achieved 100% performance in all the parameters.

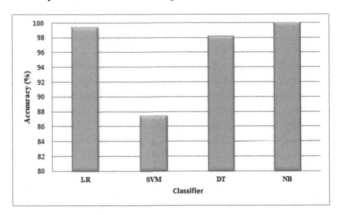

Fig. 21.9   Accuracy of Classifiers

5. Conclusion & Future Work

CKD is a life-threatening disease and its early prediction helps individuals to receive treatment at a low cost and risk. In this paper, four machine learning classifiers LR, SVM, DT, and NB are used to predict CKD. LR achieved 99.37% accuracy, SVM achieved 87.37% accuracy, DT achieved 98.12% accuracy, and NB achieved 100% accuracy. In the future, system can be enhanced by combining it with IoT devices so that it can be used in real time.

REFERENCES

1. Nishanth, A., Thiruvaran, T. (2017). Identifying important attributes for early detection of Chronic Kidney Disease. IEEE reviews in biomedical engineering. 11: 208–216. https://doi.org/10.1109/RBME.2017.2787480.

2. Reshma, S., Shaji, S., Ajina, S.R., Priya, V. and Janisha, A. (2020). Chronic Kidney Disease Prediction using Machine Learning. International Journal of Engineering Research & Technology. 9(7): 137-140. http://dx.doi.org/10.17577/IJERTV9IS070092.

3. Rajarajeswari, S. and Tamilarasi, T. (2021). Chronic Kidney Disease (CKD) Prediction Using Supervised Data Mining Techniques. International Journal of Advanced Networking and Applications. 12(6): 4776-4780.

4. Pathak, A., Gani, A., Tasin, A.H., Sania, S.N., Adil, M. and Akter, S.. (2020). Chronic Kidney Disease (CKD) Prediction Using Data Mining Techniques. In International Conference on Intelligent Computing & Optimization. 2020: 976-988. Springer, Cham. http://dx.doi.org/10.1007/978-3-030-68154-8_82.

5. Wang, Z., Chung, J.W., Jiang, X., Cui, Y., Wang, M. and Zheng, A. (2018). Machine learning-based prediction system for chronic kidney disease using associative classification technique. International Journal of engineering & Technology. 7(4.36): 1161-1167.

6. Almasoud, M. and Ward, T.E.. (2019). Detection of chronic kidney disease using machine learning algorithms with least number of predictors. International Journal of Soft Computing and Its Applications. 10(8): 89-96. https://dx.doi.org/10.14569/IJACSA.2019.0100813.

7. Lamba, R., Gulati, T., Alharbi, H.F and Jain, A. (2021). A hybrid system for Parkinson's disease diagnosis using machine learning techniques. International Journal of Speech Technology. 2021. https://doi.org/10.1007/s10772-021-09837-9.

8. Gunarathne, W., Perera, K. and Kahandawaarachchi, K.. (2017). Performance evaluation on machine learning classification techniques for disease classification and forecasting through data analytics for chronic kidney disease (ckd). In IEEE 17th International Conference on Bioi informatics and Bioengineering (BIBE) (pp. 291–296). IEEE. https://doi.org/10.1109/BIBE.2017.00-39.

9. Aljaaf, A.J., Al-Jumeily, D., Haglan, H.M., Alloghani, Baker, T., Hussain, A.J. and Mustafina, J. (2018). Early prediction of chronic kidney disease using machine learning supported by predictive analytics. In IEEE Congress on Evolutionary Computation (CEC) (pp. 1–9). IEEE. https://doi.org/10.1109/CEC.2018.8477876.

10. Teklae, S., Shingavi, P., Wandhekar, S. and Chatorikar, A. (2018). Prediction of Chronic Kidney Disease Using Machine Learning Algorithm. International Journal of Advanced Research in Computer and Communication Engineering. 7 (10): 92-96. https://doi.org/10.17148/IJARCCE.2018.71021.

11. Rady, E.A. and Anwar, A.S. (2019). Prediction of kidney disease stages using data mining algorithms. Informatics in Medicine Unlocked. 15: 100178. https://doi.org/10.1016/j.imu.2019.100178.

12. Elhoseny, M., Shankar, K. and Uthayakumar, J. (2019). Intelligent Diagnostic Prediction and Classification System for Chronic Kidney Disease. Scientific Reports. 9: 9583. https://doi.org/10.1038/s41598-019-46074-2.

13. Ekanayake, I.U. and Herath, D. (2020). Chronic Kidney Disease Prediction Using Machine Learning Methods. In Moratuwa Engineering Research Conference (MERCon) (pp. 260-265). https://doi.org/10.1109/MERCon50084.2020.9185249.

14. De, S. and Chakraborty, B. (2020). Development of Chronic Kidney Disease Prediction System (CKDPS) Using Machine Learning Technique. In Hemanth D., Shakya S., Baig Z. (eds) Intelligent Data Communication Technologies and Internet of Things. ICICI 2019. Lecture Notes on Data

Engineering and Communications Technologies. 38. Springer, Cham. https://doi.org/10.1007/978-3-030-34080-3_18.

15. Snegha, J., Tharani, V., Preetha, S.D., Charanya, R. and Bhavani, S. (2020). Chronic Kidney Disease Prediction Using Data Mining. In International Conference on Emerging Trends in Information Technology and Engineering (ic-ETITE) (pp. 1–5). https://doi.org/10.1109/ic-ETITE47903.2020.482.

16. Ifraz, G.M., Rashid, M.H., Tazin, T., Bourouis, S. and Khan, M.M.. (2021). Comparative Analysis for Prediction of Kidney Disease Using Intelligent Machine Learning Methods. Computational and Mathematical Methods in Medicine. 2021: 1-10. https://doi.org/10.1155/2021/6141470.

17. Almustafa, K.M. (2021). Prediction of chronic kidney disease using different classification algorithms. Informatics in Medicine Unlocked. 24: 100631. https://doi.org/10.1016/j.imu.2021.100631.

18. Ilyas, H., Ali, S., Ponum, M., Hasan, O., Mahmood, M.T., Iftikhar, M. and Malik, M.H.. (2021). Chronic kidney disease diagnosis using decision tree algorithms. BMC Nephrology. 22: 273. https://doi.org/10.1186/s12882-021-02474-z.

19. Senan, F.M., Al-Adhaileh, M.H., Alsaade, F.W., Aldhyani, T.H.H., Alqarni, A.A., Alsharif, N., Uddin, M.I., Alahmadi, A.H., Jadhav, M.E. and Alzahrani, M.Y. (2021). Diagnosis of Chronic Kidney Disease Using Effective Classification Algorithms and Recursive Feature Elimination Techniques. Journal of Healthcare Engineering. 2021: 1004767. https://doi.org/10.1155/2021/1004767.

20. Kim, D.H. and Ye, S.Y. (2021). Classification of chronic kidney disease in sonography using the GLCM and artificial neural network". Diagnostics 11 No. 5, 864. https://doi.org/10.3390/diagnostics11050864.

21. https://archive.ics.uci.edu/ml/datasets/chronic_kidney_disease.

22. Rani, P., Kumar, R. and Jain, A. (2022). A Hybrid Approach for Feature Selection Based on Correlation Feature Selection and Genetic Algorithm. International Journal of Software Innovation. 10(1): 1–17. http://doi.org/10.4018/IJSI.292028.

23. Lamba, R., Gulati, T., Al-Dhlan, K.A. and Jain, A. (2021). A systematic approach to diagnose Parkinson's disease through kinematic features extracted from handwritten drawings. Journal of Reliable Intelligent Environments. 7: 253–262. https://doi.org/10.1007/s40860-021-00130-9.

24. Rani, P., Kumar, R. and Jain, A. (2021). Coronary artery disease diagnosis using extra tree-support vector machine: ET-SVMRBF". International Journal of Computer Applications in Technology 66 No. 2, 209–218.

25. Lamba, R., Gulati, T. and Jain, A. (2022). A Hybrid Feature Selection Approach for Parkinson's Detection Based on Mutual Information Gain and Recursive Feature Elimination. Arabian Journal for Science and Engineering. https://doi.org/10.1007/s13369-021-06544-0.

26. Rani, P., Kumar, R. and Jain, A. (2022). A Novel Hybrid Imputation Method to Predict Missing Values in Medical Datasets. In: Marriwala N., Tripathi C., Jain S., Kumar D. (eds) Mobile Radio Communications and 5G Networks. Lecture Notes in Networks and Systems. 339. Springer, Singapore. https://doi.org/10.1007/978-981-16-7018-3_16.

Intelligent Systems and Smart Infrastructure – Brijesh Mishra et al. (eds)
© 2023 Taylor & Francis Group, London, ISBN 978-1-032-41287-0

CHAPTER

22

Automatic Head Position Detection in Computer Tomography for Prevention of Wrong-Side Treatment

Vibha Bora Bafna*

G H Raisoni College of Engineering, Nagpur, India

Ashwin Kothari[1], Avinash Keskar[2]

Visvesvaraya National Institute of Technology, Nagpur, India

Abstract

Accurate labeling of Computer Tomography (CT) images is significant to avoid misdiagnoses or wrong-site surgery. In CT scanner, if data about patient's orientation is incorrectly entered into device control software, then medical error can occur. Without a double check in place, the system is vulnerable to errors that could lead to wrong-site surgery. In proposed method, morphological operation, bounding box technique, and pattern-based radon transform are applied to identify the position of the scout CT images automatically. For any mismatch between the supine position in software and actual prone position or vice versa, warning message will be given to CT device operator to check and correct the position. The algorithm was tested on a total of 310 head scout images, which were taken from local hospital, out of which 217 supine and 93 prone were used. In 310 scout views, 256 scout views belonged to persons above 18 years, 45 scout views belonged to persons in the age group of 8 –18 years and 9 scout views were of children below 7–8 years. The accuracy of the algorithm was 97.09% for scout images of all age group patients and 100% for patients above 7–8 years of

*Corresponding author: vibha.bora@raisoni.net,
[2]ashwinkothari@ece.vnit.ac.in, [3]avinashkeskar@ece.vnit.ac.in

DOI: 10.1201/9781003357346-22

age groups. Due to its accuracy and fast response, the method can be used for correcting the patient position in control software of CT to eliminate the risk of wrong-site surgery.

Keywords: Medical Imaging, Computer Tomography, Radon Transform, Pattern Detection, Wrong-Side Treatment

1. Introduction

Over the past several years, technological advances have resulted in numerous innovations in medical imaging. The application of computers to help radiologists in the acquisition (e.g. Mammography, Computer Tomography (CT), Magnetic Resonance Imaging, Ultra Sonography), storage, and reporting of medical images is well recognized. However, in spite of the advances, medical errors are still prevalent leading to mortality. The estimated rate of wrong-site surgical treatment varies broadly ranging from 0.09 to 4.5 per 10,000 surgeries carried out (DeVine, Chutkan, Norvell, & Dettori, 2010).

The WHO "Safe Surgery Checklist" introduced in 2008 (Rothmund & Rothmund, 2008) is compulsory in almost all hospitals around the world, which has been quite effective in reducing the wrong-site surgery (Hanchanale, Rao, Motiwala, and Karim, 2014) but this is not supported by scientific evidence (Lin, Wernick, To-lentino, and Stawicki, 2018). Further, many computer programs have been established and approved for use in medical practice that supports radiologists in noticing potential possible abnormalities on diagnostic radiology exams or to decrease observational oversights. Wrong-site surgery refers to the assemblage of medical errors comprising surgical procedures on the wrong location of the body, wrong person, or wrong procedure (Steinhauer, 2001). The prevalence of wrong-site surgery has been assessed in 1:112,994 procedures; however, the number of unreported cases is estimated to be more (Geraghty, Ferguson, McIlhenny, & Bowie, 2020)

A CT scanner is an X-ray device that builds a three-dimensional image of a subject by assembling and organizing a series of two-dimensional X-ray images as shown in Fig. 22.1. In the present procedure, one technologist input the patient's orientation prior to scan into the computer via console (Fig. 22.2) and this is the only input used to tag the CT image. Next when the patient is positioned in the machine for scan, however, in some cases, it may happen that patient position needs to be changed for the sake of patient's ease to enter the gantry. This change in position if not reflected in the computer may result in wrong-site labelling, thus leading to wrong-site diagnosis or surgery (Castellino, 2005; DeSantis, Ma, Bryan, & Jemal, 2014; Agrawal, & Bajaj, 2021). Wrong-aspect (left as opposed to right) treatments are not frequent; however, when it does, the result may be severe and life-threatening for the patient and may cause a bad reputation for the hospital in which the case takes place (Doi, 2007). In spite of WHO guidelines in the treatment environment, wrong-site treatment still remains a concern (Lee, Yang, & Suh, 2015).

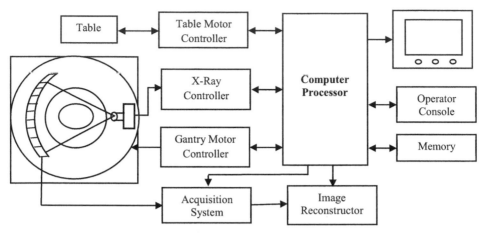

Fig. 22.1 Computed tomography (CT) imaging system

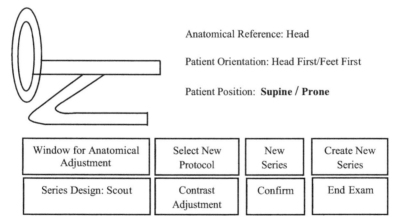

Fig. 22.2 Operator console

2. Literature Review

Wrong-site surgery is common, especially in cases where organs are symmetrical, For, e.g., Head (Christensen, Hutchins, & McDonald, 2006) Otolarynalogy (Liou & Nussenbaum, 2014), Spine (Epstein, 2021), Urology (Dyer, 2004). In Otolaryngology procedures, incorrect-site surgical events were found to be 0.3%–4.5% and wrong-site surgery accounts for 04%–06% of all scientific mistakes (Liou & Nussenbaum, 2014). Around 09% to 21% otolaryngologists have reported experiences with incorrect-site surgical treatment due to inverted imaging and ambiguity in site marking over their career and the procedures most frequently lead to temporary injuries to the patient with few cases of permanent disability or even loss of life.

In Spine, wrong-site spine surgery (WSSS) occurs for 12.8% (Epstein, 2021), whereas in Urology, 1 in 454 cases (Geraghty et al., 2020).

In head CT imaging, Christensen et al. (2006) had proposed algorithm using center of mass and inertia for figuring patient head alignment from the CT image data and detecting oversights in image orientation tagging. Another method was proposed by Wu & Ramaseshan (2014) using correlations between tabletop, axial lasers, Chester field longitudinal shifting direction, and imaging plane to determine the deviations of spatial orientations. One of the root causes of wrong-site surgery is found to be the mismatch in actual orientation and entered CT scan orientation of affected person. This paper proposes a novel algorithm to automatically detect patient's head position with the help of test brain images to figure out any mismatch between the entered and actual position. The algorithm will signal the device operator through a pop-up message regarding the detected inconsistency for correction of the position. This algorithm can be implemented with the help of scout images acquired before CT scan. The scout photo is a mandatory step of performing CT as shown in Fig. 22.3. It is also known as a scanogram, topogram, test projection radiograph, or pilot experiment. They are the projections received to aid planning of the subsequent CT examination. A lateral scout view of the head is always obtained while performing head CT.

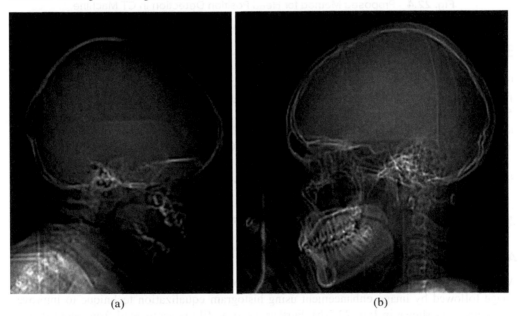

Fig. 22.3 Head Scout images: (a) Supine Position and (b) Prone Position

In the proposed work, feature descriptor was extracted from the biological shape of the head scout image. The detection of head position from scout image is done automatically using bounding box algorithm and pattern-based radon transform. Due to typical hemispherical shape of the brain, radon transform of the scout image gives projection which gives the actual position of the patient in CT machine, as shown in Fig. 22.4.

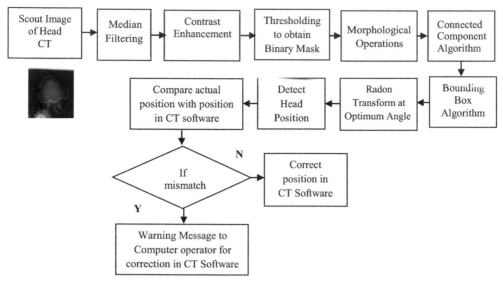

Fig. 22.4 Proposed Method for Head Position Detection in CT Machine

3. Data and Variables

The proposed method for automated orientation detection of head in CT using scout view was implemented on MATLAB ver. R2020b. The computer used was a Dell PC with Intel core i5 processor clocked at 4.10 GHz, with 8GB RAM. A total of 310 head scout images were acquired from 3D Vidharbha Diagnostic Imaging Centre, Nagpur, India, out of which 217 supine and 93 prone were used for experimentation. Out of 310 scout views, 256 scout views belonged to persons above 18 years, 45 scout views belonged to persons in the age group of 8–18 years, and 9 scout views were of children below 7–8 years.

4. Methodology and Model Specifications

A. Pre-Processing

As a first step in pre-processing, median filter was applied to eliminate noise from head scout image followed by image enhancement using histogram equalization technique to improve the contrast, as shown in Fig. 22.5(b). Further, head profile is segmented from scout image with the help of optimum thresholding. The optimum threshold T_{opt} for each scout image is calculated from histogram using the first-order derivative (R & R, 2008; Kapgate, Kalbande, & Shrawankar, 2019) as shown in Fig. 22.5(c). The gray intensity with maximum gradient in the empirical range of 50–150 is selected as T_{opt} for all the images in the database, which gives binary head mask with some unwanted background pixels, as shown in Fig. 22.5(d). The undesirable regions were eliminated using morphological opening operation, as shown in Fig. 22.5(e). This step was followed by connected component algorithm (A, 1989), which

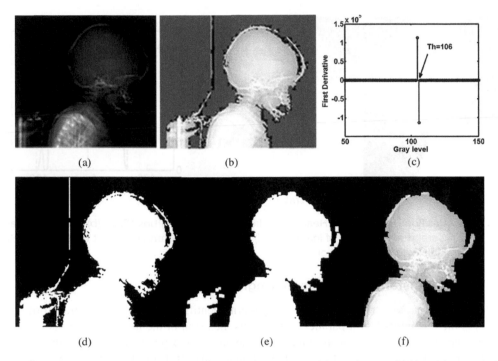

Fig. 22.5 Pre-processing of Supine Scout Image: (a) Original Scout Image; (b) Histogram-equalized Image; (c) Optimum Threshold T_{opt} Selected from First Derivative of Histogram in the Gray Range of 50–150; (d) Binary Image Using T_{opt}; (e) Image After Morphological Opening Operation; (f) Largest Scout Profile Segmented Using Connected Component Algorithm from Background

removes all connected components (objects) that have fewer than P pixels, where P is the number of pixels in the largest profile (lateral view of head), as shown in Fig. 22.5(f)

B. Bounding Box Algorithm

After preprocessing, lateral scout view is segmented from the background. The cerebral head mask region was extracted with the help of bounding box technique. A bounding box is a rectangular region of interest whose edges are parallel to the image boundary, and is thus defined by its minimum and maximum extents, i.e. a 2D box is given by all (x, y) coordinates specified by the extreme points (x_{min}, y_{min}) and (x_{max}, y_{max}), which were its top-left and bottom-right corners, respectively. Hence, the *bounding box* is the box with minimal area that contains complete lateral scout views. Mathematically, the bounding box *(Bob)* is given by

$$Bob = \{(x, y) \mid x_{min} \leq x \leq x_{max} \ \& \ y_{min} \leq y \leq y_{max}\} \tag{1}$$

From *Bob* image, as shown in Fig. 22.6(a), the hemispherical cerebrum part of the brain is extracted as a region of interest (ROI), as shown in Fig. 22.6(b) The ROI extracted from *Bob* is (2)

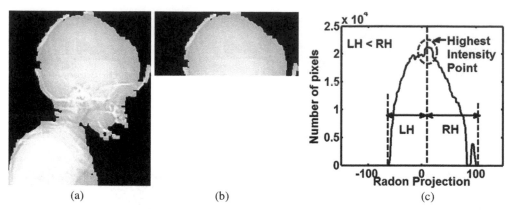

(a)   (b)   (c)

Fig. 22.6   Position Detection: (a) Head Scout Image in Bounding box (*Bob*); (b) Hemispherical Cerebrum as Region of Interest (ROI); (c) Radon Transform of ROI at $\theta = 0°$; LH < RH, indicating supine position

C. Radon Transform and Optimum Angle

Radon transform is a transform used for pattern identification. The radon transform of $f(x, y)$ is the line integral of it parallel to the y-axis. i.e., sum of the image along a radial line in the spatial domain (x, y) of the image corresponds to the point in the radon projection $f(\rho, \theta)$, oriented at a specific angle θ (A, 1989; Rangayyan & Rolston, 1998; Bafna, Kothari, & Keskar, 2015; Khekare & Verma, 2020). Radon transform was applied on preprocessed scout image. Mathematically,

$$f(\rho, \theta) = \int\limits_{-\infty}^{\infty}\int\limits_{-\infty}^{\infty} f(x,y)\, \delta(x\cos\theta + y\sin\theta - \rho)\,dx\,dy$$

$$\text{where } -\infty < \rho < \infty, \; 0 \le \theta \le \pi \tag{3}$$

where $\rho = x\cos\theta + y\sin\theta$ is the equation of straight line in polar form.

Due to anatomical structure of the brain, the pattern generated using radon transform at $\theta = 0°$ is unique and indicates the patient position either as supine or prone. The highest peak corresponds to the posterior side of the lateral scout view, which dictates the position of the patient in CT machine as follows.

At $\theta = 0°$, the radon transform equation becomes

$$ROI(\rho, 0°) = \int\limits_{-\infty}^{\infty}\int\limits_{-\infty}^{\infty} ROI(x,y)\, \delta(x - \rho)\,dx\,dy$$

$$= \int\limits_{-\infty}^{\infty} ROI(\rho, y)\,dy \quad -\infty < \rho < \infty \tag{4}$$

Radon transform at an optimum angle $\theta = 0°$ is applied on ROI to detect the highest intensity point, as in Fig. 22.6(c). Then, the distance between left edge and right edge of *Bob* image with respect to the maximum intensity point was computed and abbreviated as LH and RH, respectively. If LH<RH, then the head or nose was facing towards right, indicating the supine position of the patient, whereas if LH>RH, then the head or nose was facing towards left, indicating the prone position of the patient. Thus, due to biological structure of the brain, distinctive pattern produced with scout view indicates the patient position either as supine or prone.

For any mismatch between prior entry of patient position in CT console and actual position of patient in CT gantry, pop-up message can be given to avoid mislabeling.

5. Empirical Results

The proposed algorithm was applied on 310 head scout CT views. Fig. 22.7(a–d) shows the position detection in prone scout image and Fig. 22.7(e–h) shows the position detection in supine scout image. Table 22.1 shows the quantification rules for position detection from head

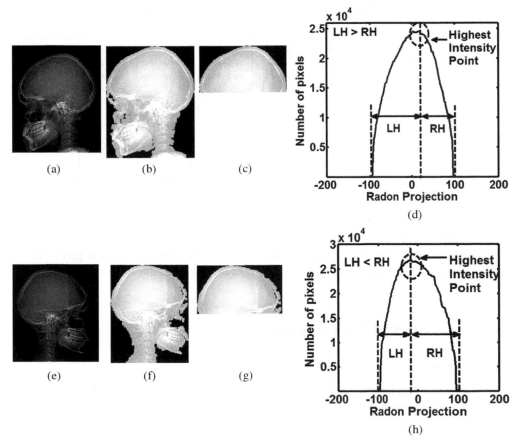

Fig. 22.7 Position Detection from Head Scout Image. (a–d) For Prone Scout Image. (e–h) For Supine Scout Image

Table 22.1 Position Detection Algorithm from Head Scout images.

Actual patient position on CT table	Input of patient position in CT software	Compare LH and RH from head scout view	Position detected	Labeling	Warning message to operator for correction
Supine	Supine	LH<RH	Supine	Left brain as left and right brain as right	N
Prone	Prone	LH>RH	Prone	Left brain as left and right brain as right	N
Prone	Supine	LH>RH	Prone	Left brain as right and right brain as left	Y
Supine	Prone	LH<RH	Supine	Left brain as right and right brain as left	Y

Table 22.2 Results of the Proposed Method on Scout Images for Supine and Prone Position Detection

	No. of images	Above 18 years	8-18 years	Below 7–8 years	Supine detected as Supine	Prone detected as prone	Accuracy over all age groups	Accuracy above 7–8 years
Total scout images	310	256	45	9	301			
Supine images	217	175	34	8	209	—	97.09%	100%
Prone images	93	81	11	1	—	92		

scout images. Table 22.2 shows the results of the proposed algorithm on scout images for supine and prone position detection. For scout views, over all age groups, the proposed method gives 97.09% accuracy and for patients above 7–8 years, the proposed method gives 100% accuracy. The average processing time required for position detection from scout view was 2.15 seconds.

6. Discussion and Conclusion

A. Discussion

With the help of scout images, the position of the head was detected using radon transform, which can help the operator to confirm the patient position in CT scan gantry. For all scout images, cerebrum mask is extracted from bounding box as region of interest to which radon transform was applied to obtain the highest intensity point. Due to asymmetrical hemispherical shape of the brain, radon projection gives unique pattern, which identifies the landmark region to indicate the location of the head in CT machine.

B. Exception

The algorithm may fail on scout images of patients below 7–8 years age group. The distance between left and right edges with respect to the highest intensity point may not vary in scout views of small age group, giving incorrect results.

C. Conclusion

Wrong-site surgery is a cause of concern during surgical procedures. The proposed method is a novel method for automatic position detection from head scout images in CT. A crosswise scout view of the head is always obtained when performing head CT. The current CT scan process relies on a technologist to note the patient's position in the computer console prior to the actual positioning of the patient. Any mismatch in the actual position and the entered position may lead to wrong-site labelling. The proposed algorithm for scout images properly marks the patient position (left and right) on the CT scan. Due to asymmetrical and hemispherical shape of the brain, radon transform at $\theta = 0°$ gives unique projection pattern, which dictates the position of the patient in CT machine. The proposed algorithm is fast, automatic and can be easily integrated into the CT machine. On detection of mismatch, the operator can be signaled regarding any mismatch between entered position and actual position in gantry. This will enable the correction of mistakes at the source where they first occur and before they may be integrated into the picture archiving and communication system (PACS).

REFERENCES

1. A, J. (1989). Fundamentals of digital image processing. Prentice Hall.
2. Bafna, V. B., Kothari, A., & Keskar, A. (2015). A heuristic approach for automated mammogram orientation in computer aided diagnosis. In 2015 7th international conference on emerging trends in engineering & technology (icetet) (pp. 56–61).
3. Castellino, R. A. (2005). Computer aided detection (cad): an overview. Cancer Imaging, 5 (1), 17.
4. Christensen, J. D., Hutchins, G. C., & McDonald, C. J. (2006). Computer automated detection of head orientation for prevention of wrong-side treatment errors. In Amia annual symposium proceedings (Vol. 2006, p. 136).
5. DeSantis, C., Ma, J., Bryan, L., & Jemal, A. (2014). Breast cancer statistics, 2013. CA: a cancer journal for clinicians, 64 (1), 52–62.
6. DeVine, J., Chutkan, N., Norvell, D. C., & Dettori, J. R. (2010). Avoiding wrong site surgery: a systematic review. Spine, 35 (9S), S28–S36.
7. Doi, K. (2007). Computer-aided diagnosis in medical imaging: historical review, current status and future potential. Computerized medical imaging and graphics, 31 (4–5), 198–211.
8. Dyer, O. (2004). Doctors suspended for removing wrong kidney. BMJ, 328 (7434), 246. Epstein, N. (2021). A perspective on wrong level, wrong side, and wrong site spine surgery. Surgical Neurology International, 12.
9. Geraghty, A., Ferguson, L., McIlhenny, C., & Bowie, P. (2020). Incidence of wrong-site surgery list errors for a 2-year period in a single national health service board. Journal of Patient Safety, 16 (1), 79.
10. Hanchanale, V., Rao, A. R., Motiwala, H., & Karim, O. A. (2014). Wrong site surgery! how can we stop it? Urology Annals, 6 (1), 57.

11. Lee, Y. H., Yang, J., & Suh, J.-S. (2015). Detection and correction of laterality errors in radiology reports. Journal of digital imaging, 28 (4), 412–416.

12. Lin, A., Wernick, B., Tolentino, J. C., & Stawicki, S. P. (2018). Wrong-site procedures: Preventable never events that continue to happen. In Vignettes in patient safety-volume 2. IntechOpen.

13. Liou, T.-N., & Nussenbaum, B. (2014). Wrong site surgery in otolaryngology head and neck surgery. The Laryngoscope, 124 (1), 104–109.

14. R, G., & R, W. (2008). Digital image processing. Pearson Prentice Hall.

15. Rangayyan, R. M., & Rolston, W. A. (1998). Directional image analysis with the hough and radon transforms. Journal of the Indian Institute of Science, 78 (1), 3.

16. Rothmund, M., & Rothmund, M. (2008). Safe surgery saves lives 2008. Mitt Dtsch Ges Chir, 37, 363–364.

17. Steinhauer, J. (2001). So, the brain tumor's on the left, right? (seeking ways to reduce mix-ups in the operating room; better communication is one remedy, medical experts say). New York Times, 23, 27.

18. Wu, M. C., & Ramaseshan, R. (2014). An approach for measuring the spatial orientations of a computed-tomography simulation system. Journal of Applied Clinical Medical Physics, 15 (2), 138–150.

Intelligent Systems and Smart Infrastructure – Brijesh Mishra et al. (eds)
© 2023 Taylor & Francis Group, London, ISBN 978-1-032-41287-0

CHAPTER

23

Dealing with Data Imbalance in Speaker Accent Recognition Using Fuzzy kNN and IQR

Sandipan Bhowmick[1], Ashim Saha[2]

Computer Science and Engineering,
National Institute of Technology,
Agartala

Abstract

The classification problems in the real world are often provided with skewed and unequal datasets. There is a significant problem with machine learning regarding unbalanced datasets. Correctly categorizing examples from the minority class is a problem for any machine learning system when learning from an unbalanced dataset. An unbalanced dataset is undesirable in real-world classification applications and decreases algorithm performance because machine learning algorithms mistakenly select cases from the minority class. Speaker accent recognition is a real-world problem. The dataset provided with this problem is an excellent example of a real-world dataset that is skewed and imbalanced with data outliers. Due to these issues, conventional machine learning algorithms provide poor results. The data outlier and class imbalance in this dataset are frequently overlooked in research on speaker accent recognition tasks. The authors investigated the possible enhancement of classification performance using Fuzzy kNN with Interquartile range threshold in this research work. This fuzzified variant of kNN considers the fuzziness inherent in neighbour proximity. The authors' primary objective was to determine the effectiveness of Fuzzy kNN against imbalanced data. It has

Corresponding author: [1]sandipanbhowmick3598@gmail.com1, [2]ashim.cse@nita.ac.in

DOI: 10.1201/9781003357346-23

been discovered that the proposed technique improves classification performance while also providing higher accuracy and reliability.

Keywords: Speaker accent recognition, Fuzzy kNN, Interquartile range, Multiclass classification, Data imbalance

1. Introduction

One of the critical problems in pattern recognition is classification. Using classification, datasets can be classified into a variety of categories. The classification works best when applied to a dataset in which each class is roughly the same size. A balanced dataset was the initial idea for popular classification algorithms like support vector machine, naive Bayes, and nearest neighbour. These classifiers could produce unsatisfactory classification results for the minority class if they are employed to deal with unbalanced databases. Since typical classification algorithms are built to deal with classes that are equally sized, they always produce results that favour the larger of the two classes, which is why they fail when used in datasets with unbalanced classes. This is one of the prime reasons for learning from imbalanced datasets, one of the ten most difficult problems in data mining (Yang, Q et al., 2006). Real-world applications face a higher cost if instances from the minority class are incorrectly classified (Mollineda et al., 2007). kNN (kth nearest neighbour) method, while it is considered a top ten data mining method (Wu etal., 2008) and effective against real-world data (Govindarajan et al., 2010; Choi S et al., 2014), also suffers from imbalanced datasets. Many studies have been undertaken in recent years to address the problem of data imbalance. Class imbalance can be addressed using various methods, including data-level techniques (pre-processing), algorithm-level approaches, and their hybrid versions. Research at the algorithm level has focused on the algorithm of the classification and attempted to omit the algorithm's sensitiveness to the majority class to prevent the classification algorithm from drifting in preference of the majority class. Fuzzy logic has proven better accurate findings and more explainability among the several algorithmic approaches developed in recent years (Zhang, X et al, 2013). The Fuzzy version of kNN performs well in the case of data imbalance (Sun B et al., 2021). Assuming that class boundary lines are perfectly characterized, the kNN algorithm assigns weightage to each neighbor, which causes the low performance of the traditional kNN. In order to address this problem, Fuzzy-kNN has been developed, which uses fuzzy sets to improve kNN. In fuzzy kNN, rather than a crisp class label, the membership of every class is assigned to every instance, which preserves a multitude of classification information and allows for a complete classification. The data for this study came from the Speaker Accent Recognition dataset, which was obtained from the UCI Machine Learning dataset source (Ma et al., 2015). The dataset is the real-valued, skewed, and multi-class dataset. Any real-valued dataset contains outliers, which might have an unintended effect on statistical results when tested using a machine learning algorithm (Gupta et al., 2014). Interquartile range (IQR) is a prominent tool for separating outlier from datasets since it operates well for datasets with skewed distributions (Wilcox et al., 2011). While there is much work done with this dataset, most authors have been concerned with the binary classification that predicts a given accent in two classes (US accent or non-US accent). *Our contribution* is the proposal of improved multi-class classification of the mentioned dataset using IQR thresholding and Fuzzy kNN algorithm to provide good accuracy with better F-Measure compared to the traditional kNN algorithm.

2. Literature Review

Early over the past half-century, studies on voice and speech recognition systems have been increasingly popular. Speech and speaker recognition have been the subject of much research (Van Leeuwen et al., 2006). To achieve the best accuracy in computer recognition of human speech, speaker recognition utilizes parametric sound information derived from the voice. Acoustically, it is not straightforward to tell human voices apart. That is why speech recognition uses the structure of the speaker's vocal cords as a primary distinguishing factor and their method of speaking, accent, gender, and age to identify speakers (Faria et al., 2006). Feature extraction algorithms are used to derive these characteristics from voice data. One of the most efficient algorithms for feature extraction in speech recognition is Mel-Frequency Cepstral Coefficients (MFCC) (Huang et al. 2001). Ma et al. (2015) used the original time-domain soundscape of the maximum 1 s of reading the words for binary classification of accent detection in the United States. MFCC was used to find 12 explanatory variables from it. They have used various machine learning algorithms to compare; notably the accuracy of the binary classification using kNN was 85.48%. The dataset used in this paper is provided by the authors of this paper (Ma et al., 2015). Ayranci et al. (2020) used the same dataset to compare various machine learning for multiclass classification, and their results showed the highest accuracy achieved using kNN is 80.49% with 0.808 F-Measure. At the same time, testing with a 75:25 train test spilled. Authors of the paper (AYRANCI et al., 2021) performed multiclass classification with this dataset with an additional 37 Turkish accents using various machine learning techniques. Their result showed the highest accuracy achieved by kNN with 82.78% with 0.830 F-Measure while testing with 10-fold cross-validation. While classical kNN has been utilized for speaker accent recognition, no work has been done combining fuzzy kNN with IQR. The goal of this work is to fill in a research gap and compare previous studies in the field.

3. Materials and Methods

A. Dataset

As mentioned above, authors have used speaker accent recognition dataset provided by Ma et al. (2006) The dataset consists of 12 explanatory variable attributes X1 to X12 and one class attribute 'language'. The total number of classes is 6. This dataset consists of 45 English accent, 29 Spanish accent, 30 French accent, 30 Germen accent, 30 Italian accent, and 165 US accent. The class imbalance is shown in Fig. 23.1.

The attributes of this dataset are also skewed. The skewness of attribute X6 is shown in Fig. 23.2. The details of skewness in the dataset are given numerically in Table 23.1, where

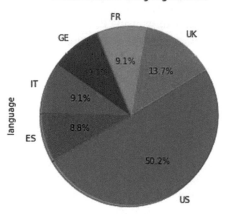

Fig. 23.1 Class imbalance in dataset

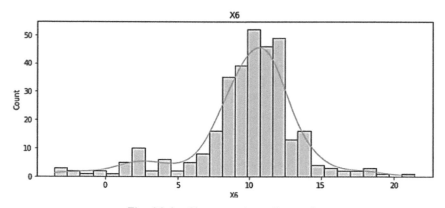

Fig. 23.2 Skewness in attribute x6

Table 23.1 Skewness of every attributes

Attributes	Skewness	Attributes	Skewness
X1	0.1923	X8	0.3942
X2	-0.2266	X9	-1.6376
X3	1.1317	X10	1.3590
X4	-0.7432	X11	-1.6868
X5	1.1800	X12	0.3213
X6	-0.9856	language	Categorical Values
X7	0.4832		

Table 23.2 Result of Experiment

Values of K	Accuracy	Precision	Recall	F-Measure	kappa	RMSE
K=1	79.33%	0.828	0.800	0.814	0.707	0.2625
K=3	82.06%	0.824	0.821	0.821	0.745	0.2445
K=5	83.89%	0.842	0.839	0.839	0.769	0.2317
K=7	80.54%	0.809	0.805	0.805	0.7205	0.2546
K=9	80.24%	0.806	0.802	0.802	0.7147	0.2566
K=11	79.63%	0.799	0.796	0.796	0.706	0.2605
K=13	79.027%	0.793	0.790	0.788	0.6935	0.2699
K=15	79.027%	0.796	0.790	0.789	0.6939	0.2699
K=17	76.89%	0.776	0.766	0.76	0.6524	0.2793
K=19	77.20%	0.778	0.772	0.766	0.6621	0.2757

Table 23.3 Statistics of data

Attribute	Min	Max	Mean	SD
X1	−6.608	17.75	5.645	5.105
X2	−14.973	3.571	−4.271	3.514
X3	−6.187	17.066	2.635	3.635
X4	−8.844	16.179	7.2	4.31
X5	−15.657	7.913	−5.649	4.596
X6	−3.529	21.446	9.81	3.625
X7	−15.366	−0.424	−9.408	2.484
X8	−2.874	13.846	5.117	2.651
X9	-15.511	4.79	-1.229	3.635
X10	-11.429	16.326	-2.362	5.042
X11	-13.664	9.166	2.431	3.478
X12	-13.724	5.259	-3.98	2.986

positive value means distribution tail on right side and negative means on left side. Table 23.3 shows the statistics of the dataset.

B. Interquartile Range (IQR)

The IQR can be used to identify outliers in dataset consist of real values. It is the discrepancy between the first and third quartiles. In a summary, an IQR exists between dataset entries in the top 25% and 75%. The values L_1 and L_3 denote the 25% and 75%, while L_2 denotes 50%. The IQR is often referred to as the mid-spread. For the IQR thresholding approach, the following is a basic formula for determining the threshold (Yang et al., 2019):

$$P_{min} = L_1 - e \times IQR \tag{1}$$

$$P_{max} = L_3 + e \times IQR \tag{2}$$

$$IQR = L_3 - L_1 \tag{3}$$

e is typically set to 1.5 as a default (Simmons et al. 2011). Data's extreme upper and lower boundaries were adjusted using IQR thresholding. Authors have utilized P_{max} and P_{min} to replace the extreme upper and lower bounds of each attribute within dataset in order to remove the outlier from the data. The results are shown in Figs. 23.3 and 23.4 for X6 attribute. The data outlier from each attribute is removed by the IQR thresholding.

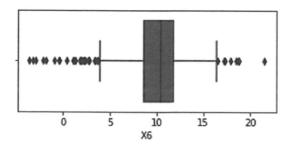

Fig. 23.3 Outlier in X6 attribute before IQR thresholding

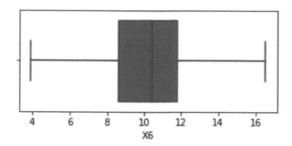

Fig. 23.4 Removal of outlier in X6 attribute after IQR thresholding

C. Fuzzy kNN

Instead of identifying complete connectedness, the fuzzy K-nearest neighbour approach determines memberships of data samples into classes. This enhances the classification performance. For fuzzy memberships of training data instances into classes, Keller et al. (1985) developed this fuzzy version of kNN.

if $a \in P$ and $P = n$, then,

$$\mu_p(a) = \begin{cases} 0.51 + (m_p/l) * 0.49 & \text{if } P = n \\ (m_p/l) * 0.49 & \text{otherwise} \end{cases} \tag{4}$$

Here, m_p is the nearest neighbour of "a" from class P. $\mu_p(a)$ is membership of "a" in class P. And for any membership of test instance x_j,

$$\mu_p(x_j) = \frac{\sum_{b=1}^{l} \mu_{ab} \left(\dfrac{1}{\| x_j - x_b \|^{2/(c-1)}} \right)}{\sum_{b=1}^{l} \left(\dfrac{1}{\| x_j - x_b \|^{2/(c-1)}} \right)} \tag{5}$$

The authors have used this approach created by the Keller et al. and have utilised similarity function as indicated below:

$$\text{Similarity} = X_i - X_j^2 \tag{6}$$

And the Kleene–Dienes T norm is used for this research work

$$T(X, Y) = \text{Min}(X, Y) \tag{7}$$

4. Results and Discussion

Odd values of k, ranging from 1 to 20, were used in this study. We were able to forecast the test data using 10-fold cross-validation. Cross-validation is a technique in which a model is trained using a variety of data points. As a rule of thumb, it is employed in situations where we wish to see how accurately a prediction algorithm performs when put to the test with unseen data. Machine learning prediction systems are commonly evaluated using metrics based on the classification technique. When dealing with an imbalanced dataset, the accuracy paradox (Zhang et al., 2020) is by far the most important consideration. When dealing with imbalanced datasets, accuracy is not a useful criterion for evaluating classification in predictive modelling. It is possible that a simple model is accurate, yet it is also ineffective. The use of multiple performance matrices is required in these cases (Abma, B. J. M., 2009). Precision, recall, F-Measure, kappa score, RMSE (root mean squared error), and accuracy were some of the metrics used to evaluate proposed approach's performance.

The Fuzzy kNN provides a good classification performance. The accuracy achieved at k = 1 is 79.33% with kappa score 0.707, which demonstrates the reliability of the classification algorithm. The accuracy of the performed task increases with increased valued of k. The highest accuracy achieved in this task is 83.89% with F-Measure 0.839 and kappa score 0.769. The F-Measure and kappa score validate the accuracy of this imbalanced multiclass classification. The confusion matrix for k = 5 is given in Fig. 23.6. Here 0 refers the Spanish accent, 1 refers the French accent, 2 refers to the German accent, 3 refers to the Italian accent, 4 refers to the English accent, and 5 refers to the US accent. At k = 5, 27 out of 29 Spanish accent, 26 out of 30 French accent, 22 out of 30 German accent, 20 out of 30 Italian accent, 37 out of 45 English accent, and 144 out of 165 US accent are correctly classified by the algorithm. The chosen test method was 10-fold cross-validation, which shows which experiments work best based on the

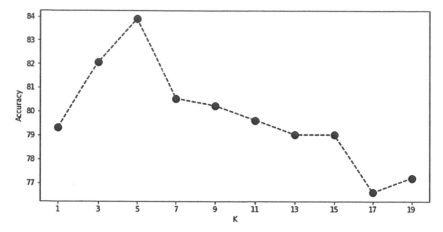

Fig. 23.5 Accuracy vs K plot

```
*** Confusion Matrix ***

    a   b   c   d   e   f   <-- classified as
   27   0   0   0   0   2 |  a = 0
    0  26   0   0   1   3 |  b = 1
    0   0  22   2   1   5 |  c = 2
    0   1   5  20   0   4 |  d = 3
    0   0   1   2  37   5 |  e = 4
    3   3   7   0   8 144 |  f = 5
```

Fig. 23.6 Confusion matrix

correctly classified instances. Fig. 23.5 shows the accuracy vs K value graph. The study finds that the Fuzzy kNN is able to handle the class imbalance present in the dataset.

Removing the data outlier also helps the algorithm to perform better. The achieved accuracy is 83.89% with F-Measure 0.839 for k=5, which is more than the traditional kNN.

5. Conclusion

In this research work, the authors investigate the algorithmic approach of solving data imbalance for multiclass classification. The authors investigated capabilities of Fuzzy kNN for performing with class imbalance. The speaker accent recognition dataset is taken as example. The dataset consisted of 329 instances. The dataset itself consisted of data outliers, and thus the authors performed IQR thresholding to remove the data outliers. The objective of this research was to enhance the performance of classification, which is achieved. The capabilities of Fuzzy kNN against imbalanced dataset is also tested. But there must be a thorough research of performance-enhancing approaches when dealing with a real-world dataset like speaker accent recognition dataset. The authors of this paper hope to work more in this domain to make progress in this area in the future.

REFERENCES

1. Yang, Q., & Wu, X. (2006). 10 challenging problems in data mining research. *International Journal of Information Technology & Decision Making, 5*(04), 597–604.
2. Mollineda, R., Alejo, R., & Sotoca, J. (2007, September). The class imbalance problem in pattern classification and learning. In *II Congreso Espanol de Informática (CEDI 2007). ISBN* (pp. 978–84).
3. Wu, X., Kumar, V., Ross Quinlan, J., Ghosh, J., Yang, Q., Motoda, H., ... & Steinberg, D. (2008). Top 10 algorithms in data mining. *Knowledge and information systems, 14*(1), 1–37.
4. Govindarajan, M., & Chandrasekaran, R. M. (2010). Evaluation of k-nearest neighbor classifier performance for direct marketing. *Expert Systems with Applications, 37*(1), 253–258.
5. Choi S., Ghinita, G., Lim, H. S., & Bertino, E. (2014). Secure knn query processing in untrusted cloud environments. *IEEE Transactions on Knowledge and Data Engineering, 26*(11), 2818–2831.
6. Zhang X., Dong, G., & Bailey, J. (2013). Overview and Analysis of Contrast Pattern Based Classification. *In Contrast Data Mining: Concepts, Algorithms, and Applications; Chapman and Hall/CRC:* Boca Raton, FL, Volume 11, pp. 151–170.
7. Sun B., & Chen, H. (2021). A Survey of Nearest Neighbor Algorithms for Solving the Class Imbalanced Problem. *Wireless Communications and Mobile Computing, 2021.*
8. Ma, Z., & Fokoué, E. (2015). A comparison of classifiers in performing speaker accent recognition using MFCCs. *arXiv preprint arXiv:1501.07866.*
9. Gupta, M., Gao, J., Aggarwal, C., & Han, J. (2014). Outlier detection for temporal data. *Synthesis Lectures on Data Mining and Knowledge Discovery, 5*(1), 1–129.
10. Wilcox, R. R. (2011). *Introduction to robust estimation and hypothesis testing.* Academic press.
11. Van Leeuwen, D. A., Martin, A. F., Przybocki, M. A., & Bouten, J. S. (2006). NIST and NFI-TNO evaluations of automatic speaker recognition. *Computer Speech & Language, 20*(2-3), 128–158.
12. Faria, A. (2006). Accent Classification for Speech Recognition. *In: Renals, S., Bengio, S. (eds) Machine Learning for Multimodal Interaction. MLMI 2005. Lecture Notes in Computer Science,* vol 3869. Springer, Berlin, Heidelberg.
13. Huang X., Acero, A., Hon, H. W., & Reddy, R. (2001). *Spoken language processing: A guide to theory, algorithm, and system development.* Prentice hall PTR.
14. Ayranci, A. A., Atay, S., & Yıldırım, T. (2020, October). Speaker Accent Recognition Using Machine Learning Algorithms. In *2020 Innovations in Intelligent Systems and Applications Conference (ASYU)* (pp. 1–6). IEEE.
15. AYRANCI, A. A., Sergen, A. T. A. Y., & YILDIRIM, T. (2021) Speaker Accent Recognition Using MFCC Feature Extraction and Machine Learning Algorithms. *International Journal of Advances in Engineering and Pure Sciences, 33*, 17–27.
16. Yang J., Rahardja, S., & Fränti, P. (2019, December). Outlier detection: how to threshold outlier scores?. In *Proceedings of the international conference on artificial intelligence, information processing and cloud computing* (pp. 1–6).
17. Simmons, J. P., Nelson, L. D., & Simonsohn, U. (2011). False-positive psychology: Undisclosed flexibility in data collection and analysis allows presenting anything as significant. *Psychological science, 22*(11), 1359–1366.

18. Keller, J. M., Gray, M. R., & Givens, J. A. (1985). A fuzzy k-nearest neighbor algorithm. *IEEE transactions on systems, man, and cybernetics*, (4), 580–585.

19. Zhang, Y., & Sang, J. (2020, October). Towards accuracy-fairness paradox: Adversarial example-based data augmentation for visual debiasing. In *Proceedings of the 28th ACM International Conference on Multimedia* (pp. 4346–4354).

20. Abma, B. J. M. (2009). Evaluation of requirements management tools with support for traceability-based change impact analysis. *Master's thesis, University of Twente, Enschede.*

Intelligent Systems and Smart Infrastructure – Brijesh Mishra et al. (eds)
© 2023 Taylor & Francis Group, London, ISBN 978-1-032-41287-0

CHAPTER

24

IOT-Based Drainage Monitoring System Using Arduino Uno

Ritika Gupta[1], Shruti Jain[2], Muskan Maheshwari[3], Vikrant Varshney[4], Dinesh Kumar Singh[5], Man Mohan Singh[6]

Department of Electronics and Communication,
Meerut Institute of Engineering and Technology,
Meerut, Uttar Pradesh 250005, India

Abstract

India is progressing with the new government which launched "Swatch Bharat Abhiyan". Drainage system of any society is the backbone of cleanliness and sanitation. Many odour and sewage issues lead to poor sanitation, which in turn leads to life-threatening illnesses in humans. You may overcome them in novel and efficient ways. A good city's wastewater management is a representation of its quality. The sewage and drainage system is the most important part of any modern city. There is no such thing as a safe haven. All hatches are compatible. Traffic accidents might occur if the sewage system is damaged. The damaged and opened manholes leads to personal safety danger and thus needed to be prevented by detecting the overflow. Overflow may be detected using the humidity and dryness sensors. The overheating and humidity distance sensors can be used to detect overflow and choked drainage system. To detect dangerous gases such as methane, gas sensors can be used, and temperature can be detected by LM35 sensors. This will help to make clean society and prevent health and sanitation of health workers and citizens.

Keywords: methane, temperature, sensors, sewage line

Corresponding author: [1]rguptaritika@gmail.com, [2]shrutijainmrt2018@gmail.com,
[3]muskanmaheshwari653@gmail.com, [4]varshney.vikrant26@gmail.com, [5]dsdineshsingh@gmail.com,
[6]manmohan.singh@miet.ac.in

DOI: 10.1201/9781003357346-24

1. Introduction

In large cities where millions of people live, drainage systems play a very important role. The drainage system is called the foundation of excess and unused water, rainwater and sewage and drought. Manual drainage monitoring is not possible. Monitoring a million of drainage and sewage system manually is not possible so it leads to overflow in rainy season [2]. This system requires lot of man power to cater manually. This causes flood of sewer water on road and market places. Because the present drainage system is not automated, determining if a blockage has formed at a given area is challenging. Sometimes it is also due of the trash in these systems. When a drainage system or sewage is choked, it starts depositing at a single place and will be a cause for production of methane gas (CH_4) in large quantity, which is hazardous if inhaled in large quantity. The death of sewer workers is caused by this gas alone. To overcome the above problems, we propose a real-time monitoring of sewage system and drainage system, which will monitor the temperature and gas continuously and send data in real time to central server through Wi-fi on cloud [3]. The city's waste management system consists of a main large size underground pipeline, and a sewage tank where waste content collects and flow forward. These tanks are filled in rainy season and choked due to solid and plastic contents. Infrastructure of any modern city can be regulated smoothly by innovative ideas to resolve problems and centralized information centre to monitor all sewage tanks at one place. This is only possible through connecting all sewage tanks to internet or intranet through some monitoring hardware installed at tanks.

2. Existing System

The current waste checking framework is not mechanized. Consequently, when a jam happens, it is hard to decide the specific area of the jam. There is additionally no late-summer cautioning. Hence, it consumes most of the day to distinguish and dispose of the blockage [3,4]. It is quite difficult to manage the pipelines which are totally obstructed. Individuals have a major issue because of the disappointment of the channel pipe.

3. Proposed System

Because most Indian cities have subterranean seepage frameworks, the framework's regular activity is critical to keeping the city immaculate, safe, and sound. If they don't keep up with the seepage framework, pure water may be contaminated by garbage, and infectious diseases may develop. In this approach, several types of work have been done to locate, monitor, and govern these subsurface frameworks. Furthermore, holes and blasts are inescapable components of water circulation framework administrators and may handle a significant amount of water misery in the distribution framework [4]. If the venture is not discovered for a long time, it will address the execution and setup capacity of employing alternative frameworks.

In Fig. 24.1 [5], it is shown how various sensors can be connected to Arduino Uno and also that the results obtained can be taken on Mobile Using Bluetooth Module.

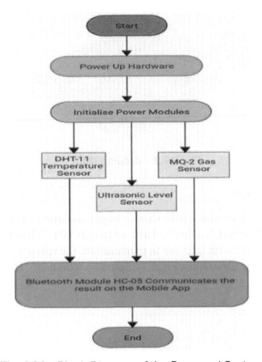

Fig. 24.1 Block Diagram of the Proposed System

We use ultrasonic sensors, gas sensors, and temperature sensors in this architecture. This slew of sensors is linked to the Arduino Uno. A Bluetooth communication device HC-05 is also connected to the Arduino through which we will be able to see the real-time status of a particular manhole whenever required on a Bluetooth app on our mobile phones. This app can be installed on the mobile phones of officials of the Municipal Corporation and they can see the real-time status of a particular manhole, and whenever there is high level or alarming condition, they can immediately take actions to resolve it. And whenever we wish to make that data available on a server, we will be able to easily do that from the mobile app through cloud.

3. Module Description

A. Arduino Uno

The Arduino Uno is a free and open-source microcontroller based on organization's microchip ATmega328P microprocessor. The card includes advanced and easy information/yield (I/O) pin gatherings that may be linked to various extension cards (protection) and circuits [5,6]. The card features 14 advanced I/O pins (six with PWM yield capacity) and 6 basic I/O pins, and it is linked to the Arduino IDE (Integrated Development Environment) by a USB Type B connection Programming. As shown in Fig. 24.2, an Arduino board has multiple pins and a USB port.

Fig. 24.2   Arduino Uno

B. Temperature Sensor

The primary capacity of a thermistor is to show a huge, unsurprising, and precise obstruction change with changes in internal heat level. The thermistors (PTC) have a positive temperature coefficient which means that with increase in temperature, the resistance also increases [6]. As shown in Fig. 24.3, a temperature sensor has 3 terminals, which are VCC, VOUT, and ground.

Fig. 24.3   Temperature Sensor

C. Methane Gas Sensor

The methane gas sensor is a semiconductor-based sensor. The presence of methane gas in the air is detected by this sensor and its output is in the form of an analog voltage. Its sensing ranges from 300 ppm to 1000 ppm. This range is highly suitable for leak detection. It can operate at temperature −10 to 50 degree Celsius. At 5 V, it draws less than 150 mA. As shown in Fig. 24.4, a methane gas sensor has 4 pins, namely, VCC, ground, A0 (Analog output), and D0 (Digital Output).

Fig. 24.4   A methane gas sensor

D. Ultrasonic Sensor

The ultrasonic sensor is a device that, by emitting ultrasonic sound waves, measures the distance of a water level in the manhole by converting the reflected sound into an electrical signal. The ultrasonic signal travels faster than the audible sound. These sensors have 2 components: one is the transmitter and the other is the receiver. As shown in Fig. 24.5, there are 4 terminals in an ultrasonic sensor, which are Vcc, trigger, echo, and ground.

Fig. 24.5   An Ultrasonic Sensor

E. HC-05 Bluetooth Module

HC-05 is a Bluetooth module used for Bluetooth communication, which is a wireless communication technology. This module can be configured as either a master or a slave. As shown in Fig. 24.6, there are 6 pins in a HC-05 module, which are enable, Vcc, ground, transmitter, receiver, and the state.

Fig. 24.6   HC-05 Bluetooth Module

4. Result and Discussion

The data in Table 24.1 is obtained by connecting the system to mobile using an app through the HC-05 Bluetooth module.

In Fig. 24.7, a bar graph representation of recorded data of one week is observed [9].

Table 24.1

Date	ID	LEVEL	GAS LEVEL	Temp	ALERT
14 MARCH	001	1	440	35	HIGH
13 MARCH	001	2	415	33	LOW
12 MARCH	001	1	430	34	HIGH
11 MARCH	001	2	425	32	MID
10 MARCH	001	2	425	33	MID
09 MARCH	001	3	420	33	LOW
08 MARCH	001	3	420	32	LOW
07 MARCH	001	3	420	31	LOW
06 MARCH	001	4	400	30	LOW

Fig. 24.7 Graphical Representation

5. Conclusion

Manually monitoring any underground actions is very difficult. This system includes a completely different way to track and monitor drainage and sewage systems. Our system tracks and monitors underground sewage and drainage [10]. Our system reduces hazardous gas leakage risk and and hence the risk sewage cleaners' life while entering inside the sewage system. Our system is a warning and alarming system for monitoring drainage systems so that we can prevent various health and sanitation problems that can be lead by the damaged Manholes. Officer based on our information can take action to restrain severe pollution levels.

Final Hardware

REFERENCES

1. Jia, Gangyong, Guangjie Han, Huanle Rao, and Lei Shu. "Edge computing-based intelligent manhole cover management system for smart cities."Internet of Things Journal -5, IEEE, 2017.
2. Guo, Xiucai, Bingbing Liu, and Lili Wang. "Design and implementation of intelligent manhole cover monitoring system based on NB-IoT." In 2019 International Conference on Robots & Intelligent System (ICRIS), IEEE, 2019.
3. Sultana, Samiha, Ananya Rahaman, Anita Mahmud Jhara, Akash Chandra Paul, and Jia Uddin. "An IOT Based Smart Drain Monitoring System with Alert Messages." In International Conference on Intelligent Human Computer Interaction. Springer, Cham, 2020.
4. Bhojane, Vishakha, Anjali Bhosale, Ankita Sankpal, Shubhada Bhosale, and Ujjwala Patil. "IOT Based Underground Drainage and Manhole Monitoring System for Cities."
5. Aarthi, M., and A. Bhuvaneshwaran. "Iot Based Drainage and Waste Management Monitoring and Alert System for Smart City." Annals of the Romanian Society for Cell Biology, 2021.
6. Salehin, Saadnoor, Syeda Sabrina Akter, Anika Ibnat, Tasmiah Tamzid Anannya, Nurun Nahar Liya, Manisha Paramita, and Md Mahboob Karim. "An IoT Based Proposed System for Monitoring Manhole in Context of Bangladesh." In 2018 4th International Conference on Electrical Engineering and Information & Communication Technology (iCEEiCT). IEEE, 2018.
7. Patel, Rutvik, Jay Prajapati, Meha Dave, Ishwariy Joshi, and Jagdish M. Rathod. "IoT based wastewater spillage detection system." In Journal of Physics: Conference Series, IOP Publishing, 2021.

8. Drenoyanis, Adam, Raad Raad, Ivan Wady, and Carmel Krogh. "Implementation of an IoT based radar sensor network for wastewater management." sensors 19, 2019.

9. Aly, Hesham H., Abdel Hamid Soliman, and Mansour Mouniri. "Towards a fully automated monitoring system for Manhole Cover: Smart cities and IOT applications." First International Smart Cities Conference (ISC2), IEEE, 2015.

10. Aly, Hesham. "Low Power IoT based Automated Manhole Cover Monitoring System as a Smart City application." PhD diss., Staffordshire University, 2019.

Intelligent Systems and Smart Infrastructure – Brijesh Mishra et al. (eds)
© 2023 Taylor & Francis Group, London, ISBN 978-1-032-41287-0

CHAPTER

25

A Comparative Study on MPPT Techniques for PV System

Yuvan Saroya[1]

Department of Instrumentation and Control Engineering,
Netaji Subhas University of Technology, New Delhi, India

Kundan Anand[2]

Department of Electrical Engineering,
Netaji Subhas University of Technology, New Delhi, India

Dinesh Kumar Singh[3]

Department of Electrical Engineering
Shambhunath Institute of Engineering and Technology, Prayagraj, India

Abstract

The electrical power output of the solar photovoltaic (PV) system varies due to variations in irradiance level, temperature, and various types of shading patterns. To harvest the maximum power from the solar PV system, various maximum power point tracking (MPPT) algorithms are used. In this paper, various MPPT algorithms are discussed to operate the system at MPP under different irradiation levels with respect to time. In this paper, a comparative study is presented between Perturb and observe (P&O), Incremental Conductance (InC), and P&O-Fuzzy logic controller (FLC) MPPT techniques. From the results, it is evident that the P&O-FLC MPPT technique tracks MPP rapidly and effectively compared to P&O and InC. The photovoltaic array is connected with a boost converter in which IGBT is used as a switching device. Furthermore, a constant resistive load is used while simulation.

Keywords: Fuzzy Logic, Incremental Conductance, MPPT, Perturb and Observe, Photovoltaics

Corresponding author: [1]Yuvans.ie20@nsut.ac.in, [2]Kundananand000@gmail.com, [3]dsdineshsingh012@gmail.com

DOI: 10.1201/9781003357346-25

1. Introduction

Renewable energy resources are non-conventional energy resources whose requirement is high in modern times. It contributes to the degradation of greenhouse gas emissions when contrasted with carbon-based fossil fuels. To replace the conventional energy sources, renewable energy sources are making their way into India's energy sector. Also, the demand of renewable energy sources in increasing day by day under 'Make in India' initiative by Government of India. Several renewable energies sources like wind energy, solar energy, hydro-thermal energy, tidal energy, geothermal energy, etc are creating job opportunities and cleaner environment in urban areas as well as in rural areas. This paper highlights how to achieve desired power output from the system under various irradiation levels using various MPPT techniques.

Overall, MPPT circuit works on the maximum power transfer theorem, in which the internal impedance is balanced according to load impedance by a boost converter. Due to a balance in internal impedance and load impedance, maximum power is transferred to the load from PV panels. MPPT is executed in a photovoltaic system to persistently change the impedance of the PV array to keep the PV system working at the maximum power of the PV panel under fluctuating circumstances like changes in irradiation level, load, and temperature (Yadav et al., 2020). There are various algorithms like Cuckoo Search Algorithm (CSA), Particle Swarm Optimization (PSO), Incremental Conductance (INC), Gravitational Search Algorithm (GSA), Flower Pollination Algorithm (FPA), Perturb and Observe (P&O), etc.

Through the literature review, the authors also discussed the limitation of the P&O technique that oscillation occurs all over the MPP and efficiency decreases under a low level of solar irradiance. The advantage of this method is that it can be tested on PV arrays without pre-requisite knowledge of the PV arrays configuration throughout the PV system.

The INC method is implemented in the PV system whose MPP tracking is done in contrast to the PV array's instantaneous and incremental conductance. The problem with INC methodology is the same as P&O in which the size of the step is fixed, that control the response speed and accuracy of the MPPT algorithm. So, the commutation has been created in-between the time required to reach MPP and steady state (Mohamed et al., 2019).

2. Maximum Power Point Tracking

MPPT regulator is compulsory for any sun-oriented power framework that requires withdrawal of maximum power from PV modules (Andrean et al., 2018); it pressurizes PV modules to work at a voltage near the most extreme power highlight to produce the greatest accessible power from the system.

Figure 25.1 shows the system's maximum power point is the operating point where $\dfrac{\Delta P}{\Delta V} = 0$ (Viswambaran et al., 2016).

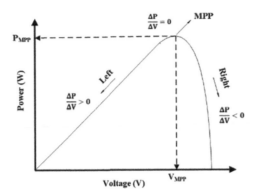

Fig. 25.1 PV panel indicates MPP and operating principle behavior

A. INC MPPT Method

In the incremental conductance method, the regulator computes increment adjustments in the current and voltage of the PV array to determine the impact of a voltage variation. Also, it acquires more determination in the regulator but can track MPP under fluctuating circumstances more rapidly than the P&O algorithm. Similar to the P&O algorithm, it also creates an oscillation in output power (Shang et al., 2018). This approach uses the incremental conductance of the PV

arrays $\left(\dfrac{\Delta I}{\Delta V}\right)$ to determine the alteration in the signal of the power with respect to the voltage

that is $\left(\dfrac{\Delta P}{\Delta V}\right)$. The INC computes the MPP by comparing the incremental conductance $\left(\dfrac{\Delta I}{\Delta V}\right)$

with the conductance of the array $\left(\dfrac{I}{V}\right)$. When these 2 variables are similar $\left(\dfrac{\Delta I}{\Delta V}=\dfrac{I}{V}\right)$, the

MPP voltage (V_{mpp}) is the output voltage. The regulator controls that voltage up till irradiance varies and this procedure is replayed. The INC method is fixed upon the certainty that at the

MPP $\left(\dfrac{\Delta P}{\Delta V}=0\right)$ and $P = V \cdot I$ (Chendi et al., 2016). INC method has only 2 sensors, current

and voltage sensors, needed in an organization to calculate the PV array's output current and output voltage. Fig. 25.2 shows the flowchart of the execution of the INC method.

The corresponding equations mathematically describe the INC algorithm and the output power from the source (Shang et al., 2020), which can be expressed as follows:

$$P = V \cdot I \tag{1}$$

The certainty that $P = V \cdot I$ and the derivative chain rule of product concerning voltage are:

$$\frac{\Delta P}{\Delta V} = \frac{\Delta(V.I)}{\Delta V} \tag{2}$$

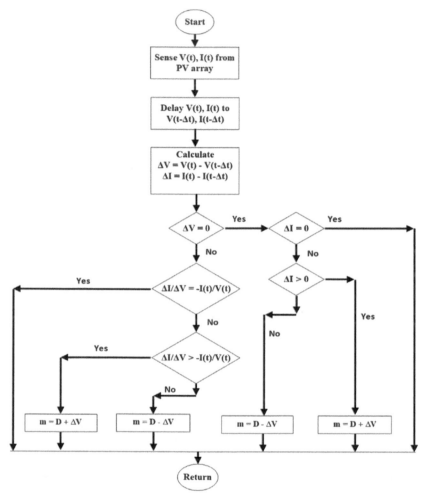

Fig. 25.2 Flowchart of INC method

$$= I \cdot \frac{\Delta I}{\Delta V} + V \cdot \frac{\Delta I}{\Delta V} \tag{3}$$

$$= I + V \cdot \frac{\Delta I}{\Delta V} \left(\frac{1}{V} \right) \frac{\Delta P}{\Delta V} \tag{4}$$

$$= \frac{1}{V} + \frac{\Delta I}{\Delta V} \tag{5}$$

The point of this algorithm is to look through the voltage working place where the conductance is equivalent to steady conductance.

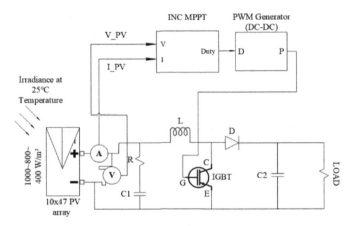

Fig. 25.3   Simulation model of PV array with implemented INC-based MPPT technique

V = Voltage measurement; A = Current measurement; V = Input voltage port of MPPT controller; I = Input current port of MPPT controller; V_PV = Voltage at PV array's side; I_PV = Current at PV array's side; R = Resistor; C1 = Input capacitor; C2 = Output capacitor; L = Inductor; D = Diode; E = Emitter; C = Collector; G = Gate terminal of IGBT.

Figure 25.3 shows that there is a PV array with a 10×47 series-parallel configuration and the whole PV system consists of the boost converter which is linked between the load and photovoltaic array. INC MPPT has two input i.e. PV voltage and current. According to these two input INC MPPT algorithm computes the duty cycle according to which PWM is generated for the IGBT of the boost converter. IGBT is used instead of MOSFET due to its advantages like being operatable on high voltage, narrow load operation, low cost, tolerant electrostatic discharge, and overloads (Raziya et al., 2019). Power, voltage, and current values can be obtained with the help of voltage- and current-measuring meters; hence, power can be calculated by the resultant product of current and voltage across the load.

B. P&O MPPT Method

P&O method is the standard technique adapted for the adjustment of the PV-generating module's MPPT algorithm. Its structure is basic, cheap, implementation is easy, and requires minimum parameters. This algorithm depends on the correlation between the voltage of the photovoltaic module and its output power. Its simulation model is the same as shown in Fig. 25.3; the only difference is that instead of the INC MPPT block diagram, there is a P&O MPPT block diagram. The rest of the components are the same. Power, voltage, and current values can be obtained with the help of voltage- and current-measuring meters; hence, power can be calculated by the resultant product of current and voltage across the load.

Figure 25.4 shows the flowchart for the execution of the P&O method (Azad et al., 2017); initially, the useful current and voltage from the photovoltaic array are determined. After all, the resultant product of current and voltage assigns the existing t-power output of the photovoltaic array. After that, it will verify the statement, whether ΔP is equal or not equal to zero (Azad et al., 2017).

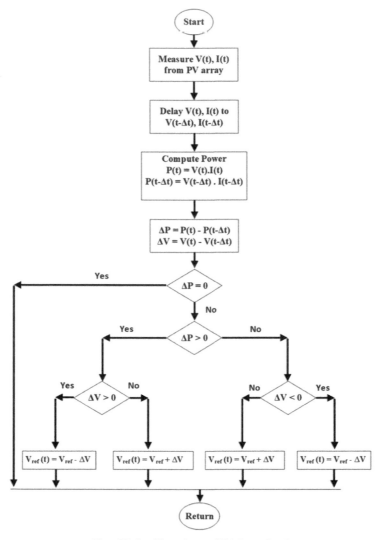

Fig. 25.4   Flowchart of P&O method

Let's consider Figs. 25.1 and 25.4; if this statement is verified, then MPP is the operating point. If it is not verified, then it will verify another statement that is $\Delta P > 0$. If this statement is verified, then it will verify that $\Delta V > 0$. If it is verified, then it demonstrates that the MPP's left side is the operating point. If $\Delta V > 0$ statement is not verified, then it demonstrates that the MPP's right side is the operating point. This process will constantly be repeated up till it arrives at the MPP.

Conventional P&O's disadvantages like lack of knowledge to operate at MPP under insolation can be eliminated by improved P&O MPPT (Ali et al., 2022) in which by changing the value of ΔI (change in current), the drift problem can be eliminated to detect the MPP accurately.

C. P&O-based FLC MPPT

P&O MPPT is already discussed in Section 2.2 and now Fuzzy Logic Controller is used which is based upon the P&O. Fig. 25.5 shows that there is a PV array with a 10×47 series-parallel configuration and the whole PV system consists of a boost converter that is connected between the load and photovoltaic array. FLC-based P&O MPPT controller takes inputs at its input port through which it intakes current and voltage from PV array, and then after computing or run of code, through its output port, the duty cycle is generated and supplied to the gate of IGBT for switching operation of the boost converter. Power, voltage, and current values can be obtained with the help of voltage- and current-measuring meters and hence power can be calculated by the product of current and voltage across the load.

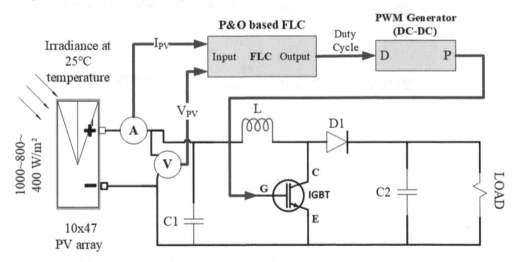

Fig. 25.5   Simulink model of PV array with implemented P&O-based FLC controller technique

A = Current measurement; V = Voltage measurement; IPV = Current at PV array side; VPV = Voltage at PV array side; R= Resistor; C1 = Input capacitor; C2 = Output capacitor; L = Inductor; D1 = Diode; E = Emitter; C = Collector; G = Gate terminal of IGBT.

3. Results and Discussions

This paper gives the comparative result of various MPPT techniques discussed in this paper under different irradiance levels as shown in Table 25.1.

Table 25.1 All MPPT techniques done on different irradiance levels with respect to time on the PV system

Time (seconds)	Irradiance level (W/m²)	Temperature (°C)
0–0.5	800	25
0.5–1.5	400	25
1.5–4	1000	25

A. INC Outputs

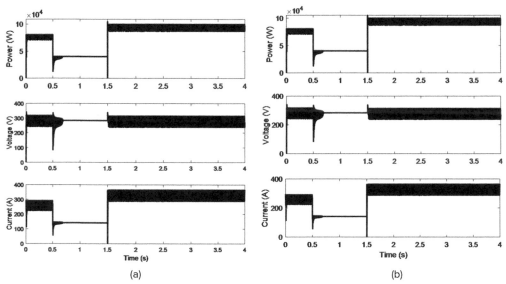

(a) (b)

Fig. 25.6 (a) PV array-side output of power, voltage, and current (b) Load-side output of power, voltage, and current

Both Figs. 25.6(a) and (b) show the output graph of respective variables at the PV array- and load-side of INC MPPT. The load-side graph shows that output power achieved at 800 W/m² irradiances (irr) is 80 kW, at 400 W/m² irr is 40 kW and at 1000 W/m² irr is 100 kW. But steady-state oscillations are observed at every stage. Convergence is fast, which means that the time required to detect the MPP is more rapid than the P&O, which is observable by comparing Fig. 25.6 and Fig. 25.7, which show the result of INC and P&O, respectively.

B. P&O Outputs

Both Fig. 25.7(a) and (b) shows the output graph of respective variables at the PV array- and load-side of P&O MPPT. Both graphs show that achieved output power is the same as INC results at 1000 and 800 W/m² irr, but at 400 W/m², irr is 31 kW with steady-state oscillations. At low irr, P&O is incapable to detect MPP properly due to a lack of knowledge. Convergence

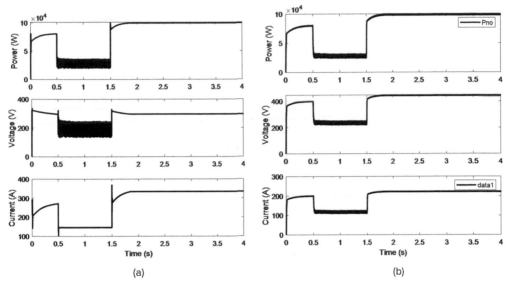

Fig. 25.7 (a) PV array-side output of power, voltage, and current, (b) Load-side output of power, voltage, and current

is slow, which means that more time is required to track MPP as compared to the INC (Choudhury el at., 2015), which is observable when comparing Fig. 25.6 and Fig. 25.7, which shows the result of INC and P&O, respectively.

C. P&O-Based Fuzzy Logic Controller Outputs

Both Fig. 25.8(a) and (b) show the output graph of respective variables at the PV array- and load-side of P&O-based FLC PV system. Both graphs show that the achieved output power at both sides at 800 Watt/m^2 irr is 83 kW, at 400 Watt/m^2 irr is 40 kW and at 1000 Watt/m^2 irr is 104 kW. Here, again the convergence is faster to track the MPP within less time in comparison to conventional P&O.

4. Conclusion

This paper gives the comparative result of INC, P&O, and P&O-based FLC PV systems under different irradiance levels at a constant temperature. The INC MPPT PV system gives the desired output but with steady-state oscillation. The tracking speed of MPP is fast, which means take very less time to detect MPP. The P&O MPPT PV system also gives the desired output but with more oscillations, especially under low irradiance levels. Also, the speed of tracking of MPP is slow and takes more time to detect. P&O-based FLC PV system gives the best results and desired output as compared to INC and conventional P&O MPPT techniques in case of MPPT tracking time and maximum output power.

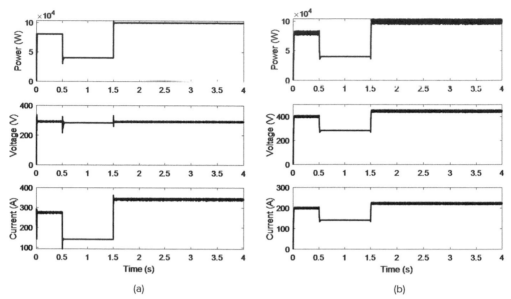

Fig. 25.8 (a) PV array-side output of power, voltage, and current, (b) Load-side output of power, voltage, and current

REFERENCES

1. Yadav, Karan, Bhavnesh Kumar and D. Swaroop. "Mitigation of mismatch power losses of PV array under partial shading condition using novel odd even configuration." Energies Reports 6: pp. 427–437, 2020.
2. Mohamed, Shazly A., and Montaser Abd El Sattar. "A comparative study of P&O and INC maximum power point tracking techniques for grid-connected PV systems." SN Applied Sciences 1, no. 2: pp. 1–13, 2019.
3. Viswambaran, Vidhya K., Arfan Ghani, and Erping Zhou. "Modelling and simulation of maximum power point tracking algorithms & review of MPPT techniques for PV applications." 5th International Conference on Electronic Devices, Systems and Applications (ICEDSA), pp. 1–4, IEEE, 2016.
4. Li, Chendi, Yuanrui Chen, Dongbao Zhou, Junfeng Liu, and Jun Zeng. "A high-performance adaptive incremental conductance MPPT algorithm for photovoltaic systems." Energies 9, no. 4: pp. 288, 2016.
5. Andrean, Victor, Pei Cheng Chang, and Kuo Lung Lian. "A review and new problems discovery of four simple decentralized maximum power point tracking algorithms—Perturb and observe, incremental conductance, golden section search, and Newton's quadratic interpolation." Energies 11, no. 11: pp. 2966, 2018.
6. Shang, Liqun, Hangchen Guo, and Weiwei Zhu. "An improved MPPT control strategy based on incremental conductance algorithm." Protection and Control of Modern Power Systems 5, no. 1: pp. 1–8, 2020.

7. Shang, Liqun, Weiwei Zhu, Pengwei Li, and Hangchen Guo. "Maximum power point tracking of PV system under partial shading conditions through flower pollination algorithm." Protection and Control of Modern Power Systems 3, no. 1: pp. 1–7, 2018.

8. Ali, Ahmed Ismail M., and Hassanien Ramadan A. Mohamed. "Improved P&O MPPT algorithm with efficient open-circuit voltage estimation for two-stage grid-integrated PV system under realistic solar radiation." International Journal of Electrical Power & Energy Systems vol.137: pp. 107805, 2022.

9. Azad, Murari Lal, Soumya Das, Pradip Kumar Sadhu, Biplab Satpati, Anagh Gupta, and P. Arvind. "P&O algorithm based MPPT technique for solar PV system under different weather conditions." International Conference on Circuit, Power and Computing Technologies (ICCPCT), pp. 1–5, IEEE, 2017.

10. Raziya, Fathima, Mohamed Afnaz, Stany Jesudason, Iromi Ranaweera, and Harsha Walpita. "MPPT technique based on perturb and observe method for PV systems under partial shading conditions." Moratuwa Engineering Research Conference (MERCon), pp. 474–479, IEEE, 2019.

11. Choudhury, Subhashree, and Pravat Kumar Rout. "Adaptive Fuzzy Logic Based MPPT Control for PV System under Partial Shading Condition." International Journal of Renewable Energy Research (IJRER) 5, no. 4: pp. 1252–1263, 2015.

12. Ramli, Makbul AM, Ssennoga Twaha, Kashif Ishaque, and Yusuf A. Al-Turki. "A review on maximum power point tracking for photovoltaic systems with and without shading conditions." Renewable and Sustainable Energy Reviews vol. 67: pp. 144–159, 2017.

Intelligent Systems and Smart Infrastructure – Brijesh Mishra et al. (eds)
© 2023 Taylor & Francis Group, London, ISBN 978-1-032-41287-0

CHAPTER

26

Bandwidth and Gain-Enhanced Rectangular Shape Slot and Notch-Loaded Dual-Band Microstrip Patch Antenna for 5G Application

Ravi Kant Prasad[1], Karabi Kalita[2], Manoj Kumar Viswakarma[3]

Department of ECE, Lloyd Institute of Engineering and Technology,
Greater Noida, UP, India

Ramesh Kumar Verma*

Department of ECE, Bundelkhand Institute of Engineering and Technology,
Jhansi, UP, India

Alok Dubey[4]

Department of ECE, GL Bajaj Institute of Technology and Management,
Greater Noida, UP, India

Maninder Singh[5]

Department of ECE, Motilal Nehru National Institute of Technology Allahabad,
Prayagraj, UP, India

Abstract

The technological advancement and development in area of antenna influenced the wireless technology, and with the rapid expansion of 5G application, the demand of microstrip antennas is increasing in the modern communication system. In this work, a rectangular microstrip antenna having dual band consisting enhanced bandwidth and gain has been designed by

*Corresponding author: ramesh85.ec@gmail.com
[1]ravi4prasad@gmail.com1, [2]karabikdas25@gmail.com, [3]manoj.rvsjsr@gmail.com, [4]alok.dubey@gmail.com, [5]manibiet04@gmail.com

DOI: 10.1201/9781003357346-26

loading rectangular shape slots and notches and the different performance of antenna is also analyzed. The proposed antenna has fractional bandwidth 24% (745 MHz) and 16.3% (726 MHz) between frequency bands 2.742 GHz to 3.487 GHz and 4.093 GHz to 4.819 GHz, respectively. This dual-band antenna resonated at a pair of frequencies 3.12 GHz and 4.054 GHz hold reflection co-efficient of −44.48 dB and −26.42 dB, respectively. The proposed antenna has incremented peak gain and directivity of 4.76 dB and 5.04 dB, respectively. This proposed antenna having enhanced bandwidth and gain can be useful for 5G applications.

Keywords: Dual band, Rectangular, Bandwidth, Slots, Notches, IE3D

1. Introduction

Due to excessive utilization and with the advent of multi-function of wireless devices in our everyday life schedule, the progression of modern wireless communications is rising. For the broad development of wireless communication systems for WLAN/5G applications, upgrading of modern wireless electronic devices plays a considerable role. For latest electronic devices, the most significant behavior is multiband operation with broadband and antennas playing an indispensable role in modern communication devices, so a variety of designing techniques of multiband antennas have been invented. For 4th-generation communication Long-Term Evolution band and 5th-generation N78 band, a new low-profile antenna is designed, whose radiation pattern resembles with monopole radiation pattern , which covers frequency band from 1.6 GHz to 3.98 GHz, that can be useful for both 4th- and 5th-generation indoor micro ceiling base station [1]. For 5th-generation application, a dual-band shared surface antenna is designed and the feed is provided by micro strip line. The antenna is fabricated and tested having fractional bandwidth of 23.45% and 9.76% in S and Ka band, respectively, using CMA (characteristics mode analysis) [2]. An L-shaped antenna having dual-band structure is designed to have dual frequencies 2.45 GHz and 5.125 GHz. This can be used as lower and upper resonating frequency containing −10 dB bandwidth of 4.13% and 8.82%, respectively [3]. For obtaining high gain and wide band, a clover-like conductor profile is also very useful [4]. For enhancing the bandwidth of antenna, an inverted T-shaped slot-loaded antenna is presented [5]. Moreover, a pentagonal-shaped CDRA having dual band as well as dual polarized having bandwidth of 5.44% and 24.29% resonating at 3.87 GHz and 6.63 GHz [6], dual-band petal-shaped gap coupled antenna having simulated bandwidth of 3.9% and 4% [7], and dual-band stacked antenna having simulated bandwidth of 9.53% and 6.95% loaded with slot are also proposed [8].

In this manuscript, a low-profile microstrip patch antenna having dual band containing enhanced gain, directivity and bandwidth, which is used for the 5G wireless communication system resonating between frequency bands 2.742 GHz to 3.487 GHz and 4.093 GHz to 4.819 GHz has been designed and discussed. The proposed dual-band antenna is resonated at couple frequencies 3.12 GHz and 4.054 GHz with S_{11} parameters of −44.48 dB and −26.42 dB, respectively. The effect of notch and slot loading has been observed on the antenna characteristics. The proposed antenna is analyzed and simulated with IE3D simulation software [9].

2. Design Aspect

A procedure of design is outlined, whose description is footed on simplified formulation which escorts to realistic rectangular patch antenna design. The designing assumption comprises the resonant frequency (f_r) , dielectric constant (ε_r), and substrate thickness (h). The steps are as follows [10], [11]:

Step 1: The patch antenna which gives high radiation efficiencies for an efficient radiator can be designed and the practical width (W) of that antenna is given by the subsequent formula.

$$W = \frac{\vartheta}{2f_r}\sqrt{\frac{2}{\varepsilon_r + 1}} \tag{1}$$

where ϑ is the velocity of light (3×10^8 m/s).

Step 2: The effective dielectric constant (ε_{reff}) is given as

$$\varepsilon_{reff} = \left\{ 0.5(\varepsilon_r + 1) + 0.5(\varepsilon_r - 1)\left(1 + \frac{12h}{W}\right)^{-0.5}\right\} \tag{2}$$

Step 3: After getting the practical width (*W*), the extension of the length ΔL is computed as

$$\frac{\Delta L}{h} = 0.412 \frac{(\varepsilon_{reff} + 0.3)\left(\dfrac{W}{h} + 0.264\right)}{(\varepsilon_{reff} - 0.258)\left(\dfrac{W}{h} + 0.8\right)} \tag{3}$$

Step 4: Now actual patch length is computed as

$$L = \frac{c}{2f_r\sqrt{\varepsilon_{reff}}} - 2\Delta L \tag{4}$$

3. Design Specifications and Proposed Structure

The proposed structure of designed antenna is shown in Fig. 26.1(a) and (b). For this antenna design, FR4 glass epoxy substrate whose dielectric constant is 4.4, loss tangent of 0.02, and thickness of 1.6 mm [12][13] has been considered. For designing the antenna, the operating frequency is taken as 3.5 GHz, which is used for WLAN and 5G application. The calculated practical width and actual length of the patch are 26 mm and 20 mm, respectively, which are taken for design consideration. In design consideration, width and length of ground plane are taken as 40.4 mm and 32 mm, respectively. For designing the proposed microstrip antenna, the conventional antenna has been loaded with three rectangular slots having dimension (Length × Width) 2×10 mm^2 and two rectangular notch having dimension (Length × Width) 16×2 mm^2.

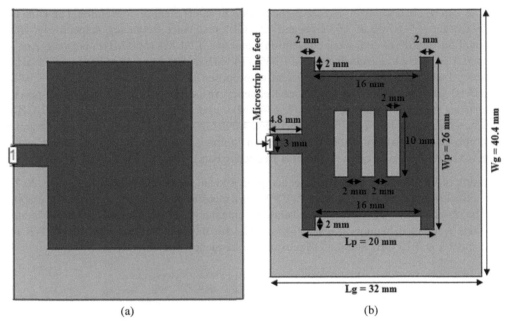

(a)   (b)

Fig. 26.1 Geometry of antenna, (a) Conventional antenna and (b) proposed antenna

The spacing between each slot is 2 mm. The simulation is done by IE3D simulation software in which antenna having impedance of 50Ω is fed through microstrip line feed positioned at left side of the patch having co-ordinate (x = −14.8 mm, y = 0 mm) with dimension of 4.8 × 3 mm^2 to attain maximum bandwidth. The specifications of designed antenna are given in Table 26.1.

Table 26.1 Specifications of antenna design structure

Parameters	Value	Parameters	Value
Patch width Wp	26 mm	Slot width	10 mm
Patch length Lp	20 mm	Notch length	16 mm
Ground plane width Wg	40.4 mm	Notch width	2 mm
Ground plane length Lg	32 mm	Feed length	4.8 mm
Slot length	2 mm	Feed width	3 mm

4. Results and Discussion

The simulation of antenna design structure is performed by means of IE3D simulation software. Fig. 26.2 shows the graph of return loss of the conventional and suggested antenna vs. frequency. The graph of both antennas shows dual resonating band in frequency spam from 1 GHz to 6

GHz. As can be observed from Fig. 26.2, the reflection co-efficient of conventional antenna is −26.79 dB and −29.87 dB at 3.03 GHz and 4.56 GHz resonance frequency, respectively. The fractional bandwidth of dual-band conventional antenna is 21.8% (660 MHz) between 2.687 GHz and 3.347 GHz frequency bands and 16.3% (740 MHz) between 4.163 GHz to 4.904 GHz frequency bands. The reflection co-efficient of proposed antenna is −44.48 dB and −26.42 dB at 3.12 GHz and 4.50 GHz resonance frequency, respectively. This dual-band proposed antenna has fractional bandwidth of 24% (745 MHz) between 2.742 GHz and 3.487 GHz frequency bands and 16.3% (726 MHz) between 4.093 GHz and 4.819 GHz frequency bands. The proposed antenna has enhanced bandwidth in comparison to conventional antenna as well as improved reflection co-efficient in lower frequency band.

The graph of VSWR of both conventional and proposed antenna is also illustrated in Fig. 26.2. To find the level of impedance mismatch, VSWR is a measurement technique. From the graph, it can be observed that the VSWR of conventional antenna is 1.096 and 1.066 at 3.03 GHz and 4.56 GHz resonance frequency, respectively, while for the proposed antenna, the VSWR is 1.012 and 1.1 at 3.12 GHz and 4.5 GHz resonance frequency, respectively.

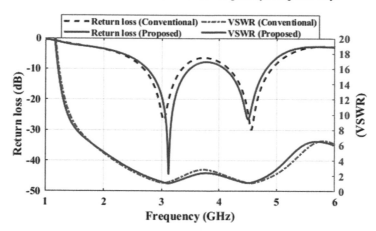

Fig. 26.2 Return loss and VSWR of proposed and conventional antenna

The graph of input impedance of both conventional and designed antenna is shown in Fig. 26.3. It is noticed that the conventional antenna has input impedance (in term of Z-parameter) Z = 54.79-j0.18Ω and Z = 46.96-j0.66Ω at resonance frequency 3.03 GHz and 4.56 GHz, respectively, whereas the proposed antenna has Z = 49.46+j0.25Ω at 3.12 GHz and Z = 45.72-j1.59Ω at 4.5 GHz.

At both resonance frequencies, the conventional and designed antenna gain is illustrated in Fig. 26.4. It has been examined that conventional antenna has gain of 3.27 dB and 4.25 dB at resonance frequencies 3.03 GHz and 4.56 GHz, respectively, whereas the proposed antenna has gain of 3.28 dB and 4.71 dB at 3.12 GHz and 4.5 GHz resonance frequencies, respectively. It is also observed the conventional antenna peak gain is 4.27 dB, but in the proposed antenna, gain has been enhanced and it reaches 4.76 dB. Fig. 26.4 also shows the directivity of the suggested

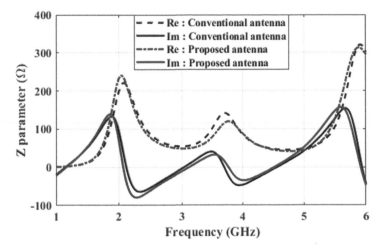

Fig. 26.3   Input impedance of conventional and proposed antenna

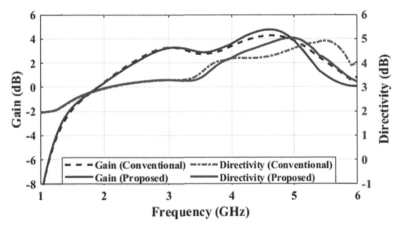

Fig. 26.4   Gain and directivity of conventional and proposed antenna

and conventional antenna at both cut-off frequencies. It is observed that the conventional antenna has directivity of 3.28 dB and 4.25 dB at resonance frequencies 3.03 GHz and 4.56 GHz, respectively, whereas the proposed antenna has directivity of 3.28 dB and 4.72 dB at 3.12 GHz and 4.5 GHz resonance frequencies, respectively. It is also examined the peak directivity of conventional antenna is 4.92 dB, but in the proposed antenna, directivity has been enhanced and it reaches 5.04 dB.

Figure 26.5 illustrates the efficiency of the antenna at both resonance peaks for conventional and proposed antenna. Conventional antenna shows efficiency of 99.72% at 3.03 GHz and 99.89% at 4.55 GHz while the proposed antenna shows efficiency of 99.97% at 3.1 GHz

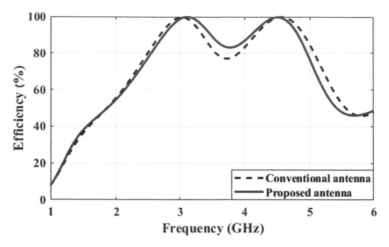

Fig. 26.5 Efficiency of conventional and proposed antenna

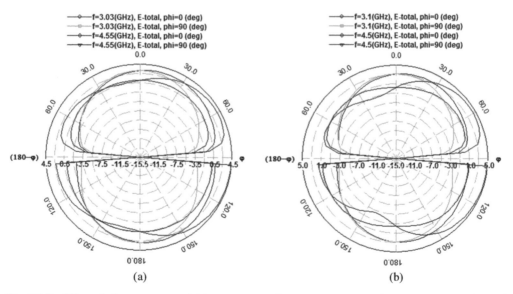

Fig. 26.6 2D radiation pattern of (a) conventional antenna at 3.03 GHz and 4.55 GHz and (b) proposed antenna at 3.1 GHz and 4.5 GHz

and 99.77% at 4.5 GHz. Fig. 26.6 illustrates the 2D pattern of radiation of both the proposed and conventional antennas at both lower and upper resonance frequency for φ = 0° and φ = 90°, respectively. The graph of radiation pattern of both conventional and proposed antenna is bidirectional for φ = 0° and φ = 90°.

5. Conclusion

For design frequency of 3.5 GHz, a rectangular patch antenna is designed successfully by loading rectangular-shaped notches and slots which have dual band of operation in frequency range from 1 GHz to 6 GHz. From the simulated result of reflection co-efficient, it has been concluded that the best possible suggested antenna has improved bandwidth with respect to the conventional antenna. From the simulated result of gain and directivity of the proposed antenna, it has been concluded that it has enhanced gain from 4.27 dB to 4.76 dB in comparison to conventional antenna as well as having directivity of 5.04 dB. For simulated result, simulation is performed by IE3D simulation software tool, and for impedance matching, microstrip line feed is chosen. This proposed antenna can be helpful for application of WLAN wireless communication as well as 5G network applications.

REFERENCES

1. S. Wen, Y. Xu and Y. Dong, " Low-profile wideband omnidirectional antenna for 4G/5G indoor base station application based on multiple resonances", *IEEE Antennas and Wireless Propagation Letters*, vol. 20, no. 4, pp. 488–492, 2021
2. T. Li and Z. N. Chen, "Shared-surface dual-band antenna for 5G applications", *IEEE Transactions on Antennas and Propagation*, vol. 68, no. 2, pp. 1128–1133, 2020
3. A. Mishra, J.A. Ansari, K. Kamakshi, A. Singh, M. Aneesh and B. R. Vishvakarma, "Compact dual band rectangular microstrip patch antenna for 2.4/5.12 GHz wireless applications", *Wireless Networks*, vol. 21, no. 2, pp. 347–355, 2015
4. H. Ozpinar, S. Aksimsek and N. T. Tokan, "A novel compact, broadband, high gain millimeter-wave antenna for 5G beam steering applications", *IEEE Transactions on Vehicular Technology*, vol. 69, no. 3, pp. 2389–2397, 2020
5. R. K. Verma and D. K. Srivastava, "Bandwidth enhancement of a slot loaded T-shape patch antenna," *Journal of Computational electronics,* vol.18, no.1, pp.205–210, 2019
6. C. Rai, A. Singh, S. Singh, A. K. Singh and R. K. Verma, "Dual-band and dual polarized inverted pentagonal shaped hybrid cylindrical dielectric resonator antenna for wireless applications," *Wireless Personal Comm.*, pp.1–19, 2022
7. B. Mishra, V. Singh and R. Singh, "Gap coupled dual-band petal shape patch antenna for WLAN/ WiMAX applications", *Advances in Electrical and Electronic Engineering*, vol.16, pp.185–198, 2018
8. B. Mishra,V. Singh and R. Singh, "Dual and wide-band slot loaded stacked microstrip patch antenna for WLAN/WiMAX applications", *Microsyst Technol* ,vol.23, pp.3467–3475, 2017
9. Zeland Software, Inc., IE3D Simulation Software Version 9.0
10. Antenna Theory, C. A. Balanis, Wiley, 2nd edition (1997)
11. Kumar, G. and Ray, K. P., 2003. *Broad Band Microstrip Antennas.* Artech House, Boston, London
12. R. K. Verma, D. K. Srivastava and B. Mishra, "Circuit theory model-based analysis of triple-band stub and notches loaded epoxy substrate patch antenna for wireless applications," *International Journal of Communication Systems*, vol.35, pp.1–17, 2022
13. R. K. Prasad, D. K. Srivastava and J. P. Saini, "Design and analysis of gain and bandwidth enhanced triangular microstrip patch antenna", *International Journal on Communication Antenna and Propagation (IRECAP)*, vol. 8, no.1, pp. 1–8, 2018.

Intelligent Systems and Smart Infrastructure – Brijesh Mishra et al. (eds)
© 2023 Taylor & Francis Group, London, ISBN 978-1-032-41287-0

CHAPTER

27

IoT-Based Automated Hydroponics System Using CAN Protocol

**Deekshitha M.[1], Chandan G.[2], Sapthagiri H. N.[3],
Anjaneya[4], and ShipraUpadhyay[5]**

Department of Electronics and Communication Engineering,
Atria Institute of Technology, Bangalore, India

Abstract

Agriculture is the backbone of civilization, whereas the human population is rising. Higher crop quality and quantity are essential in current time, and this can only be accomplished through the use of smart farming known as hydroponics. Regular farming uses manual monitoring of a crop to grow, and the crops may die if not properly maintained; however, hydroponics is entirely automated and can be utilized in the agricultural area with introduction of business skills. In our proposed method, the farming technique is further improved by using Internet of Things, which helps in receiving and transferring data. For the automation of growing crops, we have proposed to use Controller Area Network (CAN) along with sensors which are used to monitor and control the physical parameter such as light intensity, pH, temperature, water level, and turbidity. These monitoring parameter values will be sent to the microcontroller using CAN, which in turn sends to the user/connected person through Wi-Fi. The sub-module in the proposed method can be thought of as a mobile application that allows a user to access data sent by sensors to the application over the internet.

Keywords: Hydroponics, Arduino Uno, sensors, CAN transmitter and receiver

Corresponding author: [1]deekshudee99@gmail.com, [2]gchandang72@gmail.com,
[3]sapthagiri.hn.sappu.1998@gmail.com, [4]anji234234@gmail.com, [5]shipra.u@atria.edu

DOI: 10.1201/9781003357346-27

1. Introduction

Rapid improvement in the Internet of Things has helped in many ways, and one such is smart farming. Hydroponics is a smart farming system where the plants are grown in soil-less culture; instead, the plant is grown using nutrient-based solution where the roots are directly placed in the solution which have better nutrients than the soil. Any type of terrestrial plant can be grown using this smart farming. We can consider that hydroponics is ecofriendly, and has the advantage that it uses very less amount of water, almost 90 percent less as compared to traditional farming. By using hydroponics, we can save or conserve water. Hydroponics uses relatively less space; by using this method, farmers can easily detect problems, as we have used smart sensors which help in sensing the physical parameters such as temperature, pH, water level, turbidity, etc., and correct any deficiency found. The growth rate is more in hydroponics system; due to comparatively shorter growing cycle, plants can even grow two times faster than traditional farming. Change in climate can make farming difficult in some region; according to environment protection agency, increase in heat stress, flooding, rainfall, and intensity drought could reduce crop yield and also result in unusable land which leads to food insecurity. Hydroponics remove the need of soil, and hence it can overcome the problem of climate change. Several startups in India are using this smart farming in order to produce organic or healthy food but there is lack of land availability considering the expense of buying a large piece of land. Hydroponics systems helps in providing high-quality, low-cost vegetables in small space for masses promoting healthier life style [1-3].

Presently, in the market, there are few companies which sell home gardening kits, which are useful as they do not require continuous watering or maintenance because it is fully automated. Many startups have already started this business. In future, as there will be increase in population and lack of land availability, hydroponics will have great demand in future.

The contribution of this paper is as follows:

- Review of low-cost hydroponics system along with thorough discussion on advantages and disadvantages of development of such systems.
- Proposal of a new method of smart farming that is simple to set up and manage. Further, this method allows easy monitoring and controlling of physical parameters of farming such as temperature, pH, water level, etc.
- Finally, and most importantly, the controlling of all these parameters using Internet of Things (IoT) is also discussed.

2. Literature Survey

The first paper we referred was written by Henriue Sancher in the year 2019. In this paper, the author analyses that there will be more loss if we grow plants using regular farming, so that there is more demand to find an alternative method. This paper will explain about one of the easy types of hydroponics, that is, Deep water culture. This was achieved by using dried vericompost powder as the nutrient source [4].

The second paper was written by Manv Mehra in the year 2018. Regular farming uses soil, whereas the modern faring does not require any soil. Plants are made to grown using the nutrient water and it is fully automated. To achieve this automation, they have used machine learning algorithm like neutral network and Bayesian network. The system is developed with the intelligence of IoTs [5].

The author of the third paper is Dr. D. K. Sreekanthain in the year 2018. Since there are more disadvantages in traditional farming, he tried to implement IoT in agriculture field, where it helps to gather information about whether, moisture, fertility of soil, etc. Along with wireless sensors that are used to detect or monitor the process of farming, this paper proposed the application protocol for making farming precise [6].

The author Jumras Pitakphongmetha in the year 2019 analyses the effect of the ultraviolet radiation on plants and also feels that growing crops in the uncontrolled environment is very difficult as it has many disturbances. For this reason, planting the crops in greenhouse is more easy to grow and maintain. This was achieved by using IoT and cloud controlling. And the next paper we referred was written by Rajeev Lochan Mishra of the year 2015 [7].

This paper tells about the automation in agriculture field and also about automatic addition of nutrients, which uses electrodes as a benchmark of the amount of nutrients. If there is any deficiency in nutrients, it will be notified, so that the user can correct the problem, which saves time, cost, and labor.

The author Ravi Lakshmanan wrote a paper in the year 2020, which explains about the design of a smart hydroponics system. It was implemented using NodeMcu, MQTT, and sensors. All the data gathered is sent to the cloud so the user can get information using web page or through smartphones [8].

Another paper was written by Srisruthi S in the year 2018. He analyzed that traditional farming requires more usage of natural resources, which includes lots of water, land, and energy. So, he adopted a very easy process of agriculture which uses less amount of land water and energy. This was achieved with the help of green sensors technology and electric control system. The next paper was written by Nisha Sharma in the year 2019. In this paper the commercial use of hydroponics system is discussed which shows that hydroponics is a very efficient process of growing plants [9].

This has been successful throughout the word as it saves up to 80 to 90 percent of water. The author of the next paper is Mamta Deorao Sardare in the year 2013. She found that the fertility of the soil is decreasing day by day due to large application of fertilizers so the plants grown in soil do not have enough nutrients. So she tried to grow plants with no need of soil and was successful [10].

3. Block Diagram

By the definition of hydroponics, it's a smart farming where the plants are grown in soil-less method [11].

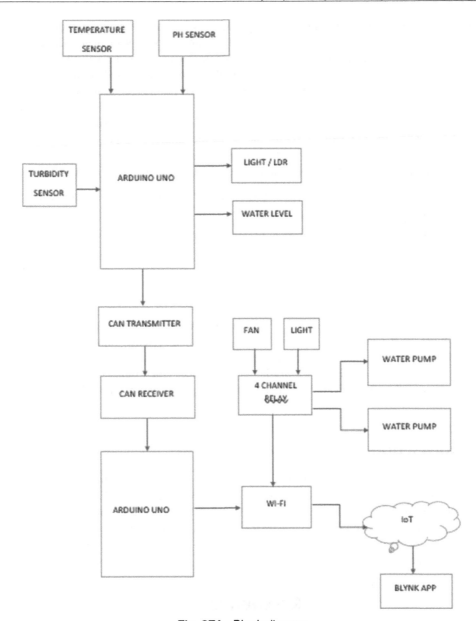

Fig. 27.1 Block diagram

In this paper, a new methodology has been shown for the growth of plants without the aid of soil. Here the plants are grown through the inorganic solvents or nutrients along with monitoring of various parameters essential for the required growth of plants.

Various sensors are employed to monitor the parameters for smart farming without any soil. The sensors which are used in our proposed methods are pH sensor or meter, LDR sensor, water-level sensor, and temperature sensor. A pH meter is a scientific device used to measure hydrogen-ion activity to detect the acidity or alkalinity of water-based solutions. A light-detection sensor (LDR) is a type of sensor that detects light. Photoresistors, also known as light-dependent resistors (LDRs), are light-sensitive devices used to detect the presence or absence of light and to quantify the intensity of light. The water-level sensor measures the water level in the plants, while the temperature sensor measures the ambient temperature. These sensor values are sent to the microcontroller with the help of Controller Area Network (CAN) trans-receiver, which in turn sends the information to the user/concerned person through Wi-Fi.

The subsystem can be considered as the mobile application using which the person can access the data sent by the sensors to the app via internet.

This enables the person to control different parameters such as turn on/off light, fan, or water pump.

4. Conclusion

A smart hydroponics system based on IoT along with CAN protocol is discussed and various literature work on the same topic has been reviewed. In such systems, smart sensors are used to collect the data. This data may be received or transmitted using wireless CAN protocol and also it may be connected to Wi-Fi for better assistance. Due to this, the user can get the notification in the blynk application. Further, user can monitor and control all the parameters using mobile.

5. Futurescope

Hydroponics will be in high demand in the future as the world's population grows and land becomes scarce. As the hydroponics industry grows, more startups are likely to emerge. In India, as well as in the rest of the world, urban farming will gain traction. This strategy does not necessitate high-end technology or a large financial commitment, yet it has both environmental and economic benefits. In future, the proposed system with slight modifications can be used for domestic gardening.

REFERENCES

1. Falmata Modu1, Adam Adam*, 1, Farouq Aliyu2, Audu Mabu1, 2020, A Survey of Smart Hydroponic Systems,Advances in Science, Technology and Engineering Systems Journal, Vol 5, No. 1, pp 233–248, DOI: 10.25046/aj050130.
2. Ravi Lakshmanan, Mohamed Djama, Sathish Kumar Selvaperumal, Raed Abdulla Automated,2020, smart hydroponics system using internet of things, International Journal of Electrical and Computer Engineering (IJECE), Vol. 10, No. 6, pp. 6389~6398, DOI:10.11591/ijece.v10i6. pp 6389–6398.

3. C. J. G. Aliac and E. Maravillas, 2018, IOT Hydroponics Management System, in IEEE 10th International Conference on Humanoid, Nanotechnology, Infromation Technology, Communication and Control, Environment and Management, Philipines, pp. 1–5.

4. P. Sambo, C. Nicoletto, A. Giro, Y. Pii, F. Valentinuzzi, T. Mimmo, P. Lugli,

5. G. Orzes, F. Mazzetto, S. Astolfi et al., 2019, "Hydroponic solutions for soilless pro-duction systems: Issues and opportunities in a smart agriculture perspective," https://doi.org/10.3389/fpls.2019.00923.

6. K. S. Aishwarya, M. Harish, S. Prathibhashree, and K. Panimozhi,2018, "Survey on iot based automated aquaponics gardening approaches," in 2018 Second International Conference on Inventive Communication and Computational Technologies (ICICCT).

7. O. Adrianes and G. M. S. Zaraźua, 2017, "Potassium acrylate: A novelty in hydro-ponic substrates," in 2017 XIII International Engineering Congress (CONIIN).IEEE.

8. C. Peuchpanngarm, P. Srinitiworawong, W Samerjai, and T. Sunetnanta, 2016, Diy sensor-based automatic control mobile application for hydroponics,in 2016 Fifth ICT International Student Project Conference (ICT-ISPC). IEEE.

9. A. N. Harun, R. Ahmad, and N. Mohamed, 2015, "Plant growth optimization using variable intensity and far red led treatment in indoor farming,"International Conference on Smart Sensors and Application (ICSSA) pp 92–97.

10. T. Nishimura, Y. Okuyama, A. Matsushita, H. Ikeda, and A. Satoh, 2017, "A compact hardware design of a sensor module for hydroponics," in 2017 IEEE6th Global Conference on Consumer Electronics (GCCE). IEEE.

11. Mark Griffiths, 2014, The Design and Implementation Of a Hydroponics Control System, thesis, M.S theses, Dept. of Information Technology, Oulu University of Applied Sciences, Finland

12. Vijendra Sahare, Preet Jain, 2015, Automated Hydroponic System using Psoc4 Prototyping Kit to Deliver Nutrients Solution Directly to Roots of Plants on Time Basis, International Journal of Advanced Research in Electrical,Electronics and Instrumentation

Intelligent Systems and Smart Infrastructure – Brijesh Mishra et al. (eds)
© 2023 Taylor & Francis Group, London, ISBN 978-1-032-41287-0

CHAPTER

28

A Comprehensive Review on UWB Multi-notch Antennas

Himani Jain[1], Praveen Chakravarti[2]

Department of Electronics and Communication Engineering,
Meerut Institute of Engineering and Technology, Meerut, India

Ajay Kumar*

IMS Engineering College, Ghaziabad, Uttar Pradesh, India

Abstract

There is rapid growth in the area of ultra-wide band (UWB) wireless communications. The UWB system consists of large bandwidth and low power consumption, which make this system more effective for short-range wireless communications. Basically, for UWB antenna design, monopole antennas are more preferrable due to its ease fabrication, light weight, and low cost. The main drawback of UWB antenna is to avoid the interference with undesired frequency band. The undesired frequency bands are stopped by creating the notch at that frequency in the UWB antenna. This paper covered a comprehensive review on UWB antennas with band-notch characteristic, i.e., an UWB antenna with band rejection characteristic. The paper is not only limited for single-band notches but it covered single-band, dual-band, and tri-band notch in the UWB antenna. This paper also covers different band-notch design technique in the UWB antenna. This work is helpful for the researchers who are working on this field.

Keywords: Antenna, UWB, Notch

1. Introduction

The antenna is an electronic device, which transmits and receives the electromagnetic waves. In the current wireless communication, high-speed data transmission and reception is required.

*Corresponding author: akgangwarr@gmail.com
[1]himanijain1820@gmail.com, [2]praveen.chakravarti@miet.ac.in,

DOI: 10.1201/9781003357346-28

Thus, for high-rate data transmission, large bandwidth is required, which is not only applicable for data transmission but it covers different wireless frequency bands. This is achieved by designing the low-cost and less complex wideband and ultra-wide band (UWB) antennas. For UWB communications, the unlicenced band-width allocated by Federal Communications Commission lies from 3.1 GHz to 10.6 GHz frequency. Due to unlicenced frequency, it created huge interest for researchers to design the UWB antenna [1].

The UWB antennas can be classified in to planner and non-planner categories but researchers are more focused on planner antenna because it is of low cost, small size, and easy to be integrated. The designing of UWB antennas has some challenges related to frequency of interferences. In the design of UWB antennas, there are several narrow frequency bands, which utilize several applications such as 3.3 to 3.6 GHz for WiMAX, and 5.15 to 5.35 GHz, 5.725 to 5.825 GHz for wireless local area network applications [2]-[4]. Thus, there is a great demand to design an UWB antenna that is able to reject the undesired bands, by keeping in mind several UWB antennas with single/multi-notch band have been designed.

2. Various Types of UWB Notch Antennas

A. Single-band-notch UWB Antenna

There are various single-band-notch UWB antennas designed such as in year 2010, S. Barbarino et al. [1] designed a single-notch antenna using an inverted-l notch for 5.0 GHz wireless local area network (WLAN) band. This antenna is matched in 2.883 to 18.604 GHz and rejects 4.844 to 6.190 GHz band. During rejection band antenna consists minimum 4.2 and maximum 19.3 voltage standing wave ratio (VSWR). It has smaller number of freedom and shows higher rejection of frequency band with 5G WLAN communication and, in the year 2012, Shilpa Jangid, Mithlesh Kumar et al. [5] designed a UWB notch antenna for WiMAX and WLAN applications. For the design of the antenna, FR4 substrate is used with 4.4 dielectric constant and thickness of 1.6 mm. The antenna covered the size of 15×14.5 mm^2, fed with 50 ohm feed line than 10 dB and shows band notch to avoid interference caused by WLAN and Wi-MAX. It has good omnidirectional coverage and stable transmission, which indicate well suitability for integration into UWB portable devices; further, Nie Fan, Jin Long et al. [6] proposed a reconfigurable filtering antenna with second-order stop band filter. The antenna is designed and simulated using FR4 substrate. It works on the frequency band from 5.15 to 5.85 GHz and provides filtering capability, and for performing the filter operation, a shape factor is placed. The antenna found good attenuation at notched band and achieved 0.72 shape factor for good filtration. The antenna is designed on multilayer structure, which provides sufficient gain with good shape factor. The antenna consists small radiating patch and ground plane. The radiating patch is connected with two resonator stubs. The designed antenna provides suitable band notch during the operating band. Similarly, in the year 2019, Yaohui Zhang et al. [7] designed multi-band band with dual polarized filtering antenna for WLAN applications. In the antenna, notch band is from 3.4 to 3.6 GHz band. The filtering operation is achieved by altering the dipole arm with split rings and all of this happen with enabled filtering circuit. The antenna operates from 2.39 to 2.69 GHz and 4.98 to 6.36 GHz frequency band with less than two

VSWR and the antenna achieved more than 7.5 dBi and 9.7 dBi gain at lower frequency and higher frequency band. Anees Abbas et al. [8] designed a notch-band UWB rectangular antenna for WLAN applications. The antenna is designed on TLY 5 A substrate and resonates at central frequency and controlled the notched bandwidth. The notch band characteristic was found by tuning the lower end of the radiating patch with EBG structure. In the antenna notch, resonance frequency was tuned by changing the EBG parameter. The antenna covered the size of $16 \times 25 \times 1.52$ mm^3. The antenna provides stable gain and radiation pattern. It has stable rectangular band notch from 5.0 to 6.0 GHz for WLAN and covered operational bandwidth from 3.1 to 12.5 GHz. The band notch antenna and its return loss are shown in Fig. 28.1.

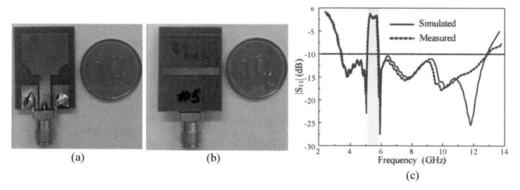

Fig. 28.1 (a) Top view of antenna; (b) back view of antenna; and (c) return loss of the antenna [8]

In the year 2020, V N Koteshwara et al. [9] designed a compact three-notch tapered UWB microstrip-fed antenna for C, X, and Ku-band applications. The antenna is designed by elliptical patch with truncated ground plane. It covered the bandwidth of 3.1 to 18.8 GHz @ −10 dB. During the UWB bandwidth, the antenna has three band notches at 3.7 to 4.2 GHz, 5.2 to 5.85 GHz, and 8.0 to 8.4 GHz for C band, WLAN band, and X band applications. These notches are realized by cutting inverted U-shaped slits on the radiating patch. The small-size antenna is suitable for wireless mobile application. Wahaj Abbas Awan et. al [10] designed a compact size printed UWB notch antenna using a genetic algorithm for compact devices applications. The antenna covered frequency band from 3.75 to 4.85 GHz. The notch in the frequency band is accomplished by cutting two symmetrical slits from radiating patch. The antenna covers the size of $20 \times 15 \times 0.508$ mm^3. The designed antenna is resonant at 5.25 GHz and covered UWB with the fractional bandwidth of 180%. The UWB provides good return loss with omnidirectional radiation pattern. A summary of single-band-notch UWB antennas is given in Table 28.1.

B. Dual-band-notch UWB Antenna

Extensive research has been done to design a dual-band UWB antenna such as in the year 2016, Qi Lui et al. [11] designed a novel planar UWB filtering antenna. It has dual-notch characteristics in (5.5 & 7.5) GHz. For designing and simulation of the antenna, HFSS software is used. The Taconic RF-35 substrate is utilized for antenna design with the permittivity (εr) of

Table 28.1   Summary of single-band-notch UWB antenna

Ref. No	Size (mm^2)	Design and techniques used	Operating frequency (GHz)	Band-notches (GHz)	Applications
1	20.4 × 64	Inverted L-notch filter	5.1–5.8	(2.83,18.604) & (4.844,6.190)	5G WLAN Communication
5	15 × 14.5	Rectangular slot on radiating element	2.97–12.10	5.12-6.10	UWB portable devices
6	17 × 32	Shorted patch & ground jointed with two resonant stubs	3.1–10.6	5.15–5.85	–
7	–	Modifying dipole arm into C shaped split rings	2.39–2.69 4.98–6.36	3.4–3.6	WLAN applications
8	16 × 25	Truncating lower ends of patch & used EBG structure	3.1–12.5	5–6	Wireless positioning systems
9	16 × 26	Three inverted U-shaped slots on patch & truncated ground plane	3.1–18.8	3.7–4.2; 5.2–5.8; 8–8.4;	Mobile applications
10	20 × 15	Two symmetrical slots from pentagonal radiating element	–	3.75–4.85	WiMAX, WLAN, Sub-6 GHz

3.5, thickness 0.79 mm, and loss tangent 0.02. The structure covered the size of 32×20 mm^2. The dual-notch UWB antenna was designed using half-wavelength resonator structure with two U-shaped gaps in coupler feeder. The antenna provides good impedance characteristics at the frequency of 3.5–10.5 GHz. The antenna reduces the interference between UWB antenna and other communication system for good wireless applications; Lin Chusan Tsai et al. [12] designed a heptagonal-shaped UWB antenna, which is designed using FR4 substrate with ε_r of 4.4, loss tangent of 0.0245, and height of 1.6 mm. The proposed antenna covered the size of 26×16.38 mm^2. The antenna is designed on heptagonal-shaped patch with ground plane, in which two L-shaped slots and SRRs are cut from radiating patch and ground plane, respectively. The antenna operates from the frequency of 2.63 to 10.86 GHz @VSWR< 2. The double-notch UWB antenna covers two notches for WiMAX and WLAN. The antenna provides good VSWR, gain, and radiation pattern. Further, Sam Weng Yik et al. [13] designed a compact size reconfigurable twin notch UWB antenna for microwave and RF system applications. The UWB antenna is designed with a couple of reconfigurable L-shaped resonators on the radiating patch and T-shaped notch is attached on the ground plane. It provides the smaller bandwidth 224.8 and 89.90 MHz at 6.0 V input voltage for RF/microwave front-end subsystem. The front view and back view of antenna and their return loss are given in Fig. 28.2.

In the year 2019, A K M Ariful et al. [14] designed a compact planar UWB notch antenna for WLAN and WiMAX applications. It is designed by rectangular patch with slotted partial ground plane and microstrip line is used for feeding the power. The antenna covered the size

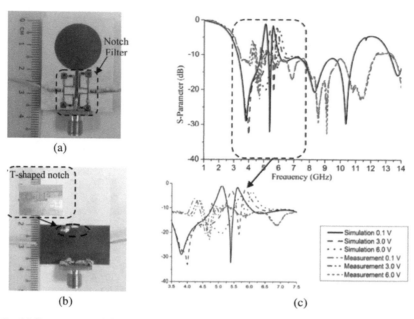

Fig. 28.2 (a) Front view of the antenna; (b) back view of the antenna; (c) and return loss of the antenna [13]

of $30 \times 12 \times 1.6$ mm^3. It used a method of moment-based simulation technology. Its operating bandwidth is from 2.98 to 12 GHz and gain is 3.95 dBi. It has two notch bands at 3.5 & 5.45 GHz and has low cost and profile, which is used as transceiver in UWB communication, and in the following year, a dual-notch UWB coplanar waveguide antenna was designed by Ravichandran Somasundaram et al. [15]. The antenna is designed and simulated using Rogers RT/duroid 5880 dielectric substrate with dimension of $18 \times 21 \times 1.6$ mm^3 and feeding is done using 50 ohm transmission line. The antenna covers dual-notch bands 3.3 to 3.7 GHz and 5.9 to 6.9 GHz. The lower band notch is designed by an inverted pie section slot and higher band notch is designed using EBG structure for WiMAX and satellite uplink applications. The design process is suitable for tunning the frequencies. Further, it can be use for reconfigurable antenna design. Further, in 2021, Mubarak Sani Ellis et al. [16] designed a notch bands antenna with vertical stubs and power is fed by microstrip feedline. The designed UWB notch antenna solves the lack of details and replicable understanding the filtering techniques. It is designed on 25×30 mm^2 dielectric substrate. In this dual-notch antenna, one narrow band notch lies from 3.3 to 3.6 GHz and second band notch from 5.15 to 6.0 GHz. The notch band has good gain with group-delay rejection. The antenna provides stable radiation patterns and has low cross-polarization. A summary of dual-band-notch UWB antennas is given in Table 28.2.

C. Tri-band Notch UWB Antenna

Extensive research is going on for designing a triple-band-notches UWB antenna; in the year of 2013, Yingsong Li et al. [17] designed three notches reconfigurable UWB antenna for

Table 28.2 Summary of dual band notch UWB antenna

S. No	Size (mm^2)	Design and Techniques Used	Operating Frequency (GHz)	Band Notches (GHz)	Applications
11	32 × 20	U-shaped gaps in coupler feeder, resonant structure	3.5–10.5	5.5 & 7.5	–
12	26 × 16.38	Two slots on patch & two split resonator slots on ground plane	2.63–10.86	3.4–3.69 5.15–5.85	UWB applications
13	37.6 × 28	Two pair of reconfigurable L resonator & T-shaped notch on ground plane	3.048–10.561	5.2 & 5.8	RF/ microwave front end subsystem
14	30 × 12	Rectangular patch & slotted partial ground	2.98–12	3.5 & 5.45	UWB transceiver
15	18 × 21	Inverted π section slot & EBG structure		3.3–3.7 5.9–6.9	Satellite uplink application
16	25 × 30	Vertical stubs protruded in feedline	3–10.6	3.3–3.6 5.15–6	–

cognitive radio. The tri-band notch UWB antenna was designed using a defective microstrip structure with stop filter implanted in feedline and inverted pie-shaped slot in radiation patch. The antenna operates on eight modes controlling by ON & OFF switches. The UWB antenna covers frequency band from 3.1–14 GHZ with 3 notched bands from 4.2–6.2 GHZ, 6.6–7 GHz, and 12.2–14 GHz. I provides stable gain during bandwidth, and omni-directional radiation patterns. The antenna is suitable for UWB cognitive radio, and due to reconfigurability, the modes can be changeable, and further in the year 2018, Qurratul Ain et al. [18] had studied and analyzed the band notches characteristics for triple-band notched UWB antenna. A small square UWB antenna of size 24 × 31 mm^2 with triband rejection is designed and simulated using HFSS software. It reduced mutual coupling to the minimum among different slots placed on radiator without affecting the performance. The antenna provides an omnidirectional radiation for WPAN and other UWB applications. All these specifications justify that antenna is of low cost, compact, and ease to integrate in wireless devices. The proposed antenna and its return loss are shown in Fig. 28.3.

In 2019, Abhishek Patel et al. [19] designed an UWB monopole with multiple band notches filtering antenna. It achieves better matching by creating a rectangular slot in ground plane and covers a range of 3.1 GHz–10.6 GHz frequencies. Rejection levels at 5.8 and 8.4 GHz are below 3dB. Gain & efficiency are good for these frequencies. It is designed on CST microwave studio. Its application is in short range communication, and further, Shun Li et al. [20] designed a triple-notch antenna with high selectivity characteristics for UWB. It is designed and simulated using HSFF software and covered the bandwidth up to 9.6 GHz. The antenna filters the three bands of 2.4 GHz, 5.0 GHz, and X-BAND uplink frequency without

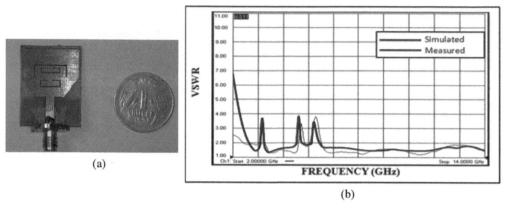

Fig. 28.3 (a) Antenna prototype; (b) VSWR of the antenna [18]

consuming more frequency bands. These notches are achieved by etching different resonators at diverse position of the antenna, and 2021, Warsha Balani et al. [21] designed tri-band notch monopole antenna for wide-band applications. The antenna is designed using C-shaped resonator, E-shaped stub with split elliptical-shaped slot. The antenna covers the bandwidth from 1.6 to 47.5 GHz with three notch bands at 1.8–2.2 GHz, 4–7.2 GHz, and 9.8–10.4 GHz for AWS, C, and X bands, respectively. The antenna achieved acceptable radiation, persistent group delay, and acceptable gain of the passbands. In the antenna, during the notch bands, low gain and high variation in group delay are achieved. So, it is suitable for pulse-based super wide band (SWB) communication. A summary of tri-band-notch UWB antennas is given in Table 28.3.

Table 28.3 Summary of tri-band-notch UWB antenna

S. No	Size (mm²)	Design and Techniques Used	Operating Frequency (GHz)	Band Notches (GHz)	Applications
17	–	Defective microstrip feedline & inverted π-shaped slot in radiation patch	3.1–14	4.2–4.6 6.6–7 12.2–14	UWB cognitive radio communication
18	24 × 31	Different C and inverted C slots placed on radiator	3.1–14	3.3–3.7; 5.1–5.4; 5.7–6	UWB applications
19	32 × 40	Creating a rectangular slot in ground plane	3.1–10.6	5.8 & 8.4	Short-range communication
20	55 × 50	C-shaped slit with CIDCLIR and patch	1.2–10.8	2.4–2.6; 5.1–5.8; 7.8–8.5	–
21	45 × 40	E-shaped stub, slot, & resonator	1.6–47.5	1.8–2.2; 4–7.2; 9.8–10.4	Pulse-based communication

3. Conclusion

In this paper, different types of UWB antennas with notch band are studied. These antennas are designed for reducing the interference with the narrow bandwidth channel. In these papers, single-band, dual-band, and tri-band notches in the UWB antenna are studied, and identifying the design techniques will be helpful for creation of notches in UWB antennas. From the studies of the antennas, it is found that monopole antenna design technique is suitable for wideband/ultrawideband antenna design and for generation of the notches different resonators are incorporated in the radiating patch and/or ground plane.

REFERENCES

1. S. Barbarino, and F. Consoli: UWB circular slot antenna provided with an inverted-l notch filter for the 5 GHz WLAN band. Progress In Electromagnetics Research, PIER 104, 1–13 (2010).
2. A. Gangwar and M. S. Alam: A high FoM monopole antenna with asymmetrical L-slots for WiMAX and WLAN applications. Microwave Optical Technology Letter, 60, 196–202 (2017).
3. Ajay Kumar Gangwar and Muhmmad Shah Alam: A miniaturized quad-band antenna with slotted patch for WiMAX/WLAN/GSM applications. International Journal of Electrons and Communication (AEÜ), 112, 152911, (2019).
4. A.K. Gangwar, M. S. Alam, V. Rajpoot, and A. K. Ojha: Filtering antennas: A technical review. International journal of RF and Microwave computer aided engineering, e22797, (2021).
5. Shilpa Jangid, and Mithilesh Kumar: A Novel UWB Band Notched Rectangular Patch Antenna with Square slot. Fourth International Conference on Computational Intelligence and Communication Networks, pp. 5–9, IEEE, Mathura, India.
6. K. V. Ajetrao, and A. P. Dhande: Phi Shape UWB Antenna with Band Notch Characteristics. Engineering, Technology & Applied Science Research, 8 (4), 3121–3125 (2018).
7. Yaohui Zhang, Yonghong Zhang, Yong Fan and Daotong Li: Band-Notched Filtering Crossed Dipole Antenna Without Extra Circuit. IEEE International Symposium on Antennas and Propagation and USNC-URSI Radio Science Meeting, pp. 1135–1136, IEEE, Atlanta, GA, USA (2019).
8. Anees Abbas, Niamat Hussain, Min-Joo Jeong , Jiwoong Park, Kook Sun Shin, Taejoon Kim and Nam Kim: A Rectangular Notch-Band UWB Antenna with Controllable Notched Bandwidth and Centre Frequency. Sensors, 20 (777), 1–11 (2020).
9. Nam-I Jo, Dang-Oh Kim, and Che-Young Kim: A Compact Band Notched UWB Antenna for Mobile Applications. PIERS, 6 (2), 177–180 (2010).
10. Wahaj Abbas Awan, Abir Zaidi, Musa Hussain, Niamat Hussain, and Ikram Syed: The Design of a Wideband Antenna with Notching Characteristics for Small Devices Using a Genetic Algorithm, Mathematics, 9 (2113), 1–13 (2021).
11. Qi Liu, Chuansong Yu, Xianliang Wu, and Zhongxiang Zhang: A Novel Design of Dual-Band Microstrip Filter Antenna, IEEE Advanced Information Management, Communicates, Electronic and Automation Control Conference (IMCEC), pp. 1573–1576, IEEE, Xi'an, China (2017).
12. Lin-Chuan Tsai: A ultrawideband antenna with dual-band band-notch filters. Microwave Optical Technology Letter, 59, 1861–1866 (2017).
13. Sam Weng Yik, Zahriladha Zakaria, Noor Azwan Shairi: A Compact Design of Reconfigurable Dual Band Notched UWB antenna. IEEE International Workshop on Electromagnetics: Applications and Student Innovation Competition (iWEM), pp. 1–2, IEEE, Nagoya, Japan (2018).

14. A. K. M. Ariful H. Siddique, Rezaul Azim and Mohammad T. Islam: Compact planar ultra-wideband antenna with dual notched band for WiMAX and WLAN. International Journal of Microwave and Wireless Technologies, 11(7), 1–8 (2019).
15. Ravichandran Sanmugasundaram, Somasundaram Natarajan, and Rengasamy Rajkumar: Ultrawideband Notch Antenna with EBG Structures for WiMAX and Satellite Application. Progress In Electromagnetics Research Letters, 91, 25 32 (2020).
16. Mubarak Sani Ellis, Philip Arthur, Abdul Rahman Ahmed, Jerry John Kponyo, Benedicta Andoh-Mensah, Bob John: Design and circuit analysis of a single and dual band-notched UWB antenna using vertical stubs embedded in feedline, Heliyon, 7 (12), e08554 (2021).
17. Yingsong Li, Wenxing Li, and Qiubo Ye: A Reconfigurable Triple-Notch-Band Antenna Integrated with Defected Microstrip Structure Band-Stop Filter for Ultra-Wideband Cognitive Radio Applications. International Journal of Antennas and Propagation (2013).
18. Qurratul Ain, Neela Chattoraj: Parametric study and Analysis of Band Stop Characteristics for a Compact UWB Antenna with Tri-band notches. Journal of Microwaves, Optoelectronics and Electromagnetic Applications, 7 (4), 509–527 (2018).
19. Abhishek Patel and Manoj Singh Parihar: UWB Monopole Antenna with Triple-Band Notch Rejection. IEEE Conference on Information and Communication Technology, pp. 1-4, IEEE, Allahabad, India (2019).
20. Shun Li, Shuxiang Song, Mingcan Cen, and Lihua Yu: Design of a high selectivity fractal antenna with triple band-notched characteristics. International Conference on Microwave and Millimeter Wave Technology (ICMMT), pp. 1–3, IEEE, Shanghai, China (2020).
21. Warsha Balani, Mrinal Sarvagya, Tanweer Ali, Ajit Samasgikar, Saumya Das, Pradeep Kumar, and Jaume Anguera: Design of SWB Antenna with Triple Band Notch Characteristics for Multipurpose Wireless Applications. Applied sciences, 11 (711), 1–21 (2021).

Intelligent Systems and Smart Infrastructure – Brijesh Mishra et al. (eds)
© 2023 Taylor & Francis Group, London, ISBN 978-1-032-41287-0

CHAPTER

29

DigiCure: A Smart Android-Based M-Health Application Using Machine Learning and Cloud Computing

Akshita Agarwal[1], Sunil Kumar[2]

Department of Computer Science & Engineering,
Meerut Institute of Engineering & Technology,
Meerut, India

Ishita Kaushik[3], Harsh Raghav[4], Wakar Ahmad[5]

Department of Information Technology,
Meerut Institute of Engineering & Technology,
Meerut, India

Abstract

Nowadays, healthcare system uses latest technologies to provide best medical services to the patients. Emerging Mobile Health (M-Health)-based technologies use advanced data connectivity, artificial intelligence, cloud computing and machine learning methodologies to provide health-related solution. Here, application collects the huge amount of data that is further stored in cloud storage system. With the help of cloud services, the collected data are further analyzed. Furthermore, various machine learning concepts are being employed for accurate illness predictions. In this paper, an android-based application called DigiCure has been introduced that uses a reliable cloud-based and machine learning model to provide an opportunity to develop an improved custom healthcare solution. The application uses predictive model for classifying diseases based on their symptoms. Moreover, it also provides health

Corresponding author: [1]akshitaagarwal634@gmail.com, [2]sunilymca2k5@gmail.com,
[3]ishitakaushik2908@gmail.com, [4]raghav.harsh09@gmail.com, [5]wakar.ahmad@miet.ac.in

DOI: 10.1201/9781003357346-29

tracking and online appointments booking for qualified health professionals to the patients. Finally, it provides a platform to resolve COVID-19-related queries and serves as a helping hand for the people who require pharmaceutical help.

Keywords: android, cloud computing, healthcare application, machine learning, prediction analysis

1. Introduction

People avail the healthcare facilities and consultation from doctors by visiting hospitals. It is acted as a barrier for people living in small towns and villages where medical facilities are not advanced. Thus, healthcare sector is encountering various issues to provide better and cost-effective medical facilities to the patients [1]. Healthcare professionals have to embrace the fact that technology is widely being used nowadays and required evolution of the traditional ways in which healthcare facilities were provided [2]. The future of healthcare lies in working hand-in-hand with technology. As the COVID-19 virus is having a devastating impact on the healthcare sector, online medication is stepping up and helping the caregivers such as doctors, nurses, and healthcare-providing organizations to respond better and at a much faster pace to the various needs of the people. Online medical facilities are making a very positive contribution to healthcare, especially during the pandemic, and are now being used in a variety of ways.

Currently, many M-Health applications are available but these applications do not contain all healthcare facilities in one spot. For example, if users want to track their health, they must use a different application than if they want to sell or acquire used pharmaceuticals, and so on. Apart from this, these applications have complex functionalities with low performance. As a result, users face unpleasant and laborious experience to access healthcare facilities since they are unable to handle massive volume of patient's queries. Another limitation of the M-Health applications is that they consume a lot of storage and more processing power that is not possible on smartphone-limited resources.

To tackle these limitations, we have proposed an android based M-Health application called "DigiCure" that is built to cope with the issues in Medicare and provide Medicaid. The main purpose of this work is to cope with the concerns associated with the current pandemic due to which the emphasis on healthcare is far more important than ever. It is more effective to avail the various medical and healthcare services in online mode and that is the main objective behind the idea.

The main objectives of this proposed work are as follows:

- To design a secure, attractive, and user-friendly Android-based application.

- To ensure that user experience is smooth, user's private data is safe and unauthorized access is prevented.
- To provide a healthcare model where all the health-related issues may be addressed and resolved in a single place.
- To provide a platform that is user-friendly, easy to use, and interactive.
- To create a model to access a variety of medical and healthcare services in an online mode.

The rest of the proposed paper is structured as follows: Section 2 discusses about existing M-Health systems in detail. Section 3 presents detailed explanation of our proposed model. In Section 4, results and discussion are provided in detail. Finally, Section 5 concludes the paper.

2. Literature Review and Related Works

Digvijay H. Gadhar et al. [25] worked for digitalizing the front-office handling at the hospital and developed a software that is fast and effective. It is associated with the collection of patients as well as diagnosis data, and other similar information. It was done manually before; this new system's major task is to record and store doctors and patients' information.

Muhammad Nazrul Islam et al. [14] introduced a web portal for those who wish to donate their spare drugs to those who are poor or with low income. Using this gateway, authorized medical experts can recommend drugs to persons who are impoverished or have a low income. They have employed an online poll and a focus group interview to analyze the needs of the people.

Sherwin Fernandes et al. [3] created a system that uses Dialog flow and the SVM algorithm at the frontend and backend, respectively, to categorize the data and forecast whether or not the user has heart disease. The major goal of this system's goal is to discover and forecast the occurrence of cardiac disease as accurately and quickly as possible which are both significant aspects of this project.

Godphrey and Khamisi [17] proposed a mobile application for patient appointment scheduling to improve healthcare facilities provided in Tanzanian hospitals. The goal of this study is to make appointment scheduling easy and to prioritize high-priority patients.

Daniel et al. [22] introduced COVID-19 chat bot which is based on combination of learning-based algorithms. It is, however, mostly a rule-based algorithm. This encompasses both the front-end and the backend of the application. The conversational interface is the part of the application that converts data from the users into specified actions and the other way around. The bot's computation and integrations with other web services are referred to as the back-end.

Seema et al. [23] proposed a chat bot to predict whether a person is suffering from heart disease. The data is entered into the system by the user. A number of attributes from the center data set are represented in the data. Dialog flow includes a less complex response to predict whether the user has a cardiac condition or not.

Based on this study, we have proposed an android application called "DigiCure" that provides many healthcare facilities under one umbrella. The detailed explanation of DigiCure application is discussed in the next section.

3. The Proposed Model

Since most of the populations are located far from hospitals, in these hospitals, there is the unavailability of proper and advanced medical facilities. In this situation, M-Health services are extremely helpful and valuable. We, therefore, have proposed a secure Android M-Health application platform that provides user interface for collecting data and provide best healthcare facilities to the users. The application supports a variety of android-supported devices such as mobile phones, smart TVs (Android-based), Desktops (Windows 11 and above), tablets, and smart watches [7]. Fig. 29.1 represents a proposed M-Health model that contains five modules: (a) Online Appointment Booking System, (b) Medicine Donation Box, (c) Illness Prediction Model, (d) COVID-19 Frequently Asked Questions (FAQs) System, and (d) Health Tracking System.

(a) Online Appointment Booking System

The proposed appointment booking system provides online appointment booking facilities to the patients based on symptoms in an easy way. The following steps are performed to get appointment from doctors:

- First, user finds the specialized doctor from list of doctors based on his symptoms.
- In the next steps, users select online payment mode. The application will support a wide variety of different online payment methods available such as Credit cards, Debit cards, UPI, Net banking, etc. using secure Razor Pay Gateway.
- After payment, user can interact with doctor through call or email services.
- Finally, user can write review for doctors based on its satisfaction. He can also give rating to the doctor.

(b) Medicine Donation Box

Because of poverty, poor people cannot buy expensive medicines, whereas many people waste a large number of medicines [16]. Our application acts as a helping hand to the patients who cannot afford to buy expensive medicines with the help of Medical Donation Box service. The following steps are performed to access this service:

- People can post the details of the unused and extra medicines along with their necessary contact details.
- Android framework contains Jetpack library Recycler View, which will be used to display posts of all the medicines. Recycler View is used as it minimizes memory usage and can thus show the enormous amount of data sets without effect on the performance of the application.

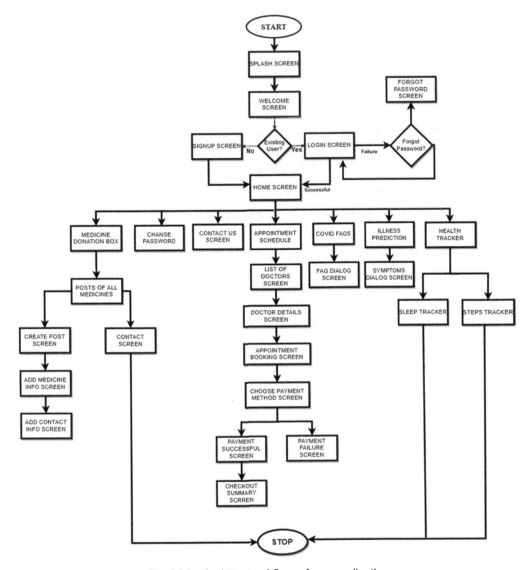

Fig. 29.1 Architectural flow of our application

- NoSQL database Cloud Fire Store is used to store data of posts of medicines by the users.
- Users can communicate with each other using calls and email services.

(c) Illness Prediction System

The implementation of the Illness Prediction System framework was done using the Python language due to the following functionalities: cross-platform and high availability of third-party libraries for tasks relating to machine learning and NLP [5]. The Following steps are taken to implement this service:

- Users can use this application to analyze the symptoms and predict the possible illness u sing Machine Learning with the help of Natural Language Processing.
- Data will be processed for accuracy using a decision tree supervised machine learning algorithm.
- Machine learning agents will be created with the help of Google Dialog Flow and trained with the help of different intents and corresponding responses.
- Google Dialog Flow API will be used for integration with the android application.

(d) COVID-19 FAQ System

The COVID-19 FAQ functionality of our M-Health application has been implemented using the following steps:

- The COVID-19 FAQ system is also based on the machine learning model and will interact with the user using natural language processing.
- It responds to the real-time worldwide COVID-19-related queries to the user using various sources.
- The interface is created using Google Cloud Dialog Flow.
- This will be integrated with the main android application using Google Dialog Flow API.

(e) Health Tracker

The health tracker functionality is responsible for tracking the health record of steps count and sleep time. It has following features:

- Google FIT service is used to track the health of the users.
- Visual reports are provided to the user using graphs and charts.

4. Results and Discussion

Figure 29.2 represents the authentication/login system of our DigiCure android application. Firebase is used to construct authentication system. When a user enters the required mandatory information and clicks the login/signup button, the information is sent to Firebase. In case of new user, it checks the validity of password requirements. It also ensures that the email address has never been used before.

The user's email address and password are validated during login. The user gets redirected to the application's home screen if authentication is successful. If a user forgets their password, they can press the Forgot Password button to reset it. The system then confirms the existence of the user. The user will receive an email with directions on how to update their password on successful verification. Users can either log in through Google sign in or by their email id and password. The Navigation Drawer provides the ability to the user to toggle between various components of the application. It provides user-friendly and easy-to-use interface.

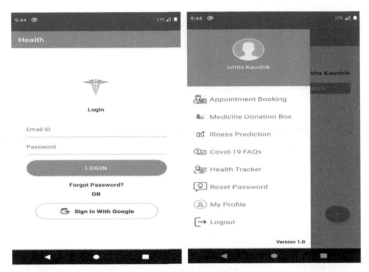

Fig. 29.2 The User Interface of DigiCure M-Health Application

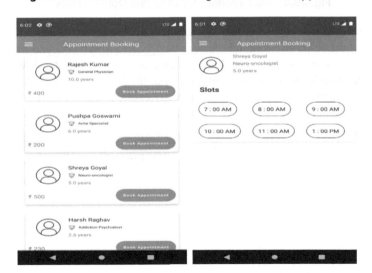

Fig. 29.3 User Interface for Appointment Booking System

Figure 29.3 represents the appointment booking system where users can choose a doctor from a list of doctors along with their necessary details, as shown in the above figure. The user will then choose the slot for their appointment, which will direct them to the various online payment methods provided by RazorPay gateway. The slot that is booked will be automatically removed from the list of slots of the respective doctor on successful completion of the payment.

Figure 29.4 represents the medicine donation system in which a user can post unused medicines which they no longer need by clicking on the "+" button shown in the above figure

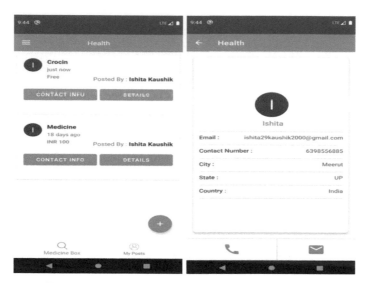

Fig. 29.4   User Interface for Medicine Donation Box

and providing necessary details about the medicine and their contact details. Other interested users can contact the seller by using this feature using contact numbers and email. We have also provided a check where the expired medicines can automatically be deleted. Thus, the expired medicines will be automatically removed from the list of medicines displayed above. Also, users cannot submit medicines that have already expired. Users can also delete and edit their posts by navigating to the "My Posts" option provided.

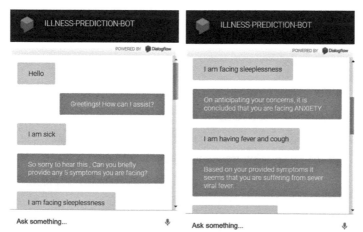

Fig. 29.5   Example of User Interaction with Illness Prediction Bot

Figure 29.5 represents the illness prediction model where users can self-diagnose and predict illness that is possible based on the symptoms.

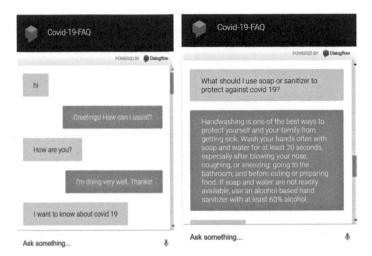

Fig. 29.6   Example of User Interaction with COVID-19 FAQ Bot

Figure 29.6 represents the COVID-19 FAQ chat bot where a user can interact with a virtual assistant to resolve corona virus-related queries.

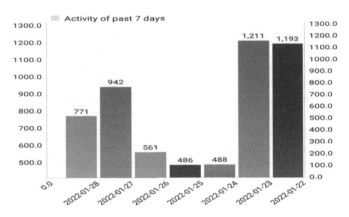

Fig. 29.7(a)   Virtual Report Embedded to Show Step Counter

Figure 29.7(a) represents the virtual chart which is a graphical representation between days of the week and the number of steps counted by the Fitness Client. In this bar chart, the day of the week is taken on the X-axis of the XY plane and the number of steps counted for that respective day is taken on the Y-axis.

Figure 29.7(b) represents the virtual chart which is a graphical representation between days of the week and the number of hours slept on that day counted by the Fitness Client. In this line chart, the day of the week is taken on the X-axis of the XY plane and the number of hours slept

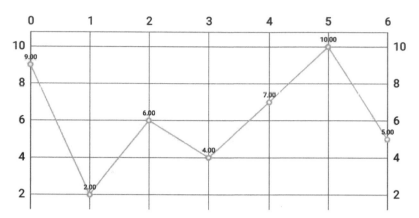

Fig. 29.7(b)   Virtual Report Embedded to Keep Track of Sleep

for that respective day is taken on the Y-axis. This visual approach helps us keep better track of our health and emphasize the trends of sleep and physical activities of the user.

5. Conclusion

Advanced techniques for healthcare services can be provided by new and smart technologies. To monitor, process, evaluate, and store patient data, M-Health systems are being combined with such technology. In this paper, we developed an M-Health android application that provides a secure cloud platform and artificial intelligence-based anticipating framework. This is an advanced and smart application that offers modern emulsions in the healthcare sector. Throughout the collection of user's data, the security of the data is the primitive aspect. We plan to evaluate the suggested predictive model to existing predictive models and do a 10-fold validation in the future. We will also perform the research with other ailments, such as Alzheimer's condition, which is another significant disease that has a global influence on human life.

REFERENCES

1. Kashif Naseer Qureshi, Sadia Din Gwanggil Jeon, Francesco Piccialli (2020) "An Accurate and Dynamic Predictive Model for a Smart M - Health System Using Machine Learning", S0020-0255(20)30 611-3, 2020.
2. S. Doan, C. K. Maehara, J. D. Chaparro et al., "A natural language processing tool to identify patients with high clinical suspicion for Kawasaki disease from emergency department notes," Academic Emergency Medicine, vol. 23, no. 5, pp. 628–636.
3. Sherwin Fernandes, Rutvij Gawas, Preston Alvares, Macklon Fernandes, Deepmala Kale, and Shailendra Aswale, "Doctor Chatbot: Heart Disease Prediction System", ISSN – 2306-708X.

4. B. C. Helsel, J. E. Williams, K. Lawson, J. Liang, and J. Markowitz, "Telemedicine and mobile health technology are effective in the management of digestive diseases: A systematic review" Digestive diseases sciences, vol. 63, no. 6, pp. 1392-1408, 2018.

5. Nicholas A. I. Omoregbe, Israel O. Ndaman, Sanjay Misra, Olusola O. Abayomi-Alli, and Robertas Damasˇevicˇius,(2020) "Text Messaging Based Medical Diagnosis Using Natural Language Processing and Fuzzy Logic", Volume 2020, Article ID 8839524.

6. S. Hema Kumar, J. UdayKiran, V. D. Ambeth Kumar, G. Saranya, Ramalakshmi V (2019), "Effective Online Medical Appointment System", ISSN 2277-8616.

7. Shelar Pooja, Hande Nilima, Dhamak Prajakta, Hingane Nisha, Jadhav Vinayak (2018), "Smart Appointment Generation for Patient", Volume 5, special issued 4.

8. Nitesh A. Godhichor, Saurabhi Nagdeote, Krutakshi Gokhale, Vaishnavi Jaiswal, Aman Asati, Shuvam Kumar (2021), "Online Medicine Donation System", Volume 9, ISSN – 2320-2882.

9. Rohan Bhardwaj, Ankita R. Nambiar, Debojyoti Dutta, "A Study of Machine Learning in Healthcare", 2017 IEEE 41st Annual Computer Software and Applications Conference.

10. Alison Callahan and Nigam H. Shah, "Machine Learning in Healthcare", Key Advances in Clinical Informatics.

11. Min Chen, Yixue Hao, Kai Hwang, Fellow, IEEE, Lu Wang, and Lin Wang, "Disease Prediction by Machine Learning over Big Data from Healthcare Communities" 10.1109/ACCESS.2017.2694446.

12. K. Shailaja, B. Seetharamulu, M. A. Jabbar, "Machine Learning in Healthcare: A Review", IEEE Conference Record # 42487; IEEE Xplore ISBN:978-1-5386- 0965-1.

13. Po-Hsuan Cameron Chen, Yun Liu, and Lily Peng, "How to develop machine learning models for healthcare", VOL 1 8 I MAY 2019 I 410–427.

14. Muhammad Nazrul Islam, Ashratuz Zavin, Sanjana Srabanti, Chowdhury Nawrin Ferdous, SaymaAlamSuha, Lameya Afroze, Nafin Shawon, Naznin Sultana Refath (2017), "GiveMed: A Web portal for Medicine Distribution among Poverty-stricken People", DOI:10.1109/R10-HTC.2017.8288960.

15. Malik, Shafaq, Nargis Khan, Sehrish Sultana, Razia Rauf, Sadaf. (2016). Mr. Doc: A Doctor Appointment Application Sy stem. International Journal of Computer Science and Information Security, 14. 452–460.

16. Ahmed Abdelaziz, Mohamed Elhoseny, Ahmed S. Salama, A.M. Riad (2018), "A Machine Learning Model for Improving Healthcare services on Cloud Computing Environment", S0263 -2241(18)30022-8.

17. Kyambille, Godphrey Kalegele, Khamisi. (2015). Enhancing Patient Appointments Scheduling that Uses Mobile Technology. International Journal of Computer Science and Information Security. 13. 21.

18. V. Akshay, A. Kumar S., R. M. Alagappan and S. Gnanavel, "BOOKAZOR - an Online Appointment Booking System" 2019 International Conference on Vision Towards Emerging Trend s in Communication and Networking (ViTECoN), 2019, pp. 1–6, DOI: 10.1109/ViTECoN.2019.8899460.

19. Irin Sherly. S, Mahalakshmi. A, Menaka. D, Sujatha. R "Online Appointment Reservation and Scheduling for Healthcare-A Detailed Study", International Journal of Innovative Research in Computer and Communication Engineering, 2016, DOI: 10.15680/IJIRCCE.2016. 0402056.

20. Qaffas, Alaa Barker, Trevor. (2012). Online Appointment Management System.

21. Palarimath, Suresh. (2021). Medicine Donation System: An Effective Distributed Tool During Covid -19 Pandemic.

22. Daniel, Emmanuel Eneojo & Daniel, Enoch Ojonimi. (2021). Covid19 I-Sabi Chat-Bot Application Using the Natural Language Processing with Dialog-Flow. 55–72.

23. Jagadeesh, Seema R, Chirag G, Vinay D, Balakrishna. (2021). Doctor Chatbot– Smart Health Prediction. International Journal of Scientific Research in Science and Technology. 751–756. 10.32628/IJSRST2183172.
24. Saimanoj, Kotapati Poojitha, Grandhi Dixit, Khushbu Jayannavar, Laxmi. (2020). "Hospital Management System using Web Technology".
25. Digvijay II. Gadhari, Yadnyesh P. Kadam, Prof.Parineeta Suman (2016). Hospital Management System", Vol-01, Issue 11, ISSN: 2494–9150.

Intelligent Systems and Smart Infrastructure – Brijesh Mishra et al. (eds)
© 2023 Taylor & Francis Group, London, ISBN 978-1-032-41287-0

CHAPTER **30**

A Literature Review: LoRa Technology and Packet Loss Analysis in LoRaWAN Line-up in Our College Campus

Shweta Singh[1], Vipin Kumar Upaddhyay[2], S. K. Soni[3]

Electronic and Communication Engineering,
Madan Mohan Malaviya University of Technology,
Gorakhpur, India

Abstract

Nowadays, monitoring and controlling plays an important role in our day-to-day life. And this led the advanced technology such as IOT to attain its utmost importance and gained high fame in today's world. Embedded systems are becoming key part of our lives. People are controlling, monitoring, and doing a lot many things from remote areas. This is only possible by connecting various objects through network connectivity, which reduces the physical distance. Many a times, this system requires battery-operated system and needs high battery backup. This led the researchers to think about the technology which covers long distance and also utilizes less energy. There are lots of technologies such as Wi-Fi, Zig-Bee, Bluetooth, etc. which are popular and presently being used but they utilize large amount of power, which is not appropriate for battery-functioned system. Therefore, a new technology which fits, into our requirement and promoted as to be the most useful technology for the application in IoT devices is called as LoRa technology. This technology is a long-range low-power technology.

Keywords: LoRa, LoRaWAN, IoT, line-up, PLR (packet loss rate), embedded system

Corresponding author: [1]2020043214@mmmut.ac.in, [2]vipin08120@gmail.com, [3]sksoniec@mmmut.ac.in

DOI: 10.1201/9781003357346-30

1. Introduction

Recent researches found the IoTs (Internet of Things) have been used in various applications which have some specific necessities such as long communication range and very low power consumption within our cost range. There are lots of technologies, such as BLE and ZigBee, are widely used for short-range communication, so this is not good or adopted for long-range transmission. Another technologies solution based on cellular mobile communication, i.e. 2G, 3G, 4G, and 5G, ensures longer transmission distance but it consumes lots of power.

Therefore, IoT applications requirement led to the rise of the new technology of LPWAN (Low-Power Wide Area Network). In the recent years, attention has been put on the LPWAN because IOT scenario requires its extensive radio range, free frequency bands, and low energy consumption.

LPWAN aims to provide long-range communications, connecting devices which are distributed over a large geographical area. It is an alternative option to M2M (Machine-to-Machine) communication using cellular mobile technologies. Most of the LPWAN operates in the ISM (Industrial, Scientific, and Medical), unlicensed frequency band. The different frequencies are 169 MHz, 433 MHz, 868/915 MHz, and 2.4 GHz, which depend on the operating region. Its latency is high; therefore, it is not suitable for the application in which delays are not tolerated or high rate of data transfer is required.

Several LPWAN technologies are already present in the market: *SigFox, NB-IoT, LoRa*, etc.

This paper discusses all about LoRa technology, i.e. its advantages, disadvantages, protocols, its alternative option, etc. We will also discuss small work on packet loss analysis in our college campus. Rest of the paper is organized as follows – Section I briefs about LoRa and LoRaWAN Technology. Section II explains about how the end-devices and gateways are implemented and created. Section III discusses about its application. Section IV explains our work in packet loss analysis in our college campus. Section V is followed by conclusion.

2. LoRa and LoRaWAN Technology Overview

LoRa stands for "**Long Range**", which is a long-range wireless communication system which features low-power operation, i.e. around 10 year of battery lifetime, low data rate, i.e. 27 kb/s, and long communication range (2–5 km in urban areas and 15 km in rural areas). Its spreading factor is 7. It is promoted by LoRa Alliance.

LoRa features includes long range, robustness, multipath resistance, Doppler resistance, low power consumption, and forward error correction (FEC). LoRa can have two distinct layers: *Physical layer* and *MAC layer protocol (LoRaWAN)*.

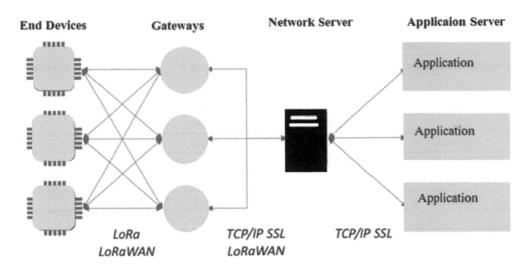

Fig. 30.1 LoRaWAN Network Architecture

There are mainly four basic components of LoRaWAN network.

1. End Devices, 2. Gateways, 3. Network Server, 4. Application Server

3. How the End Devices and Gateways are Created and Implemented?

Even after having complete theoretical knowledge about LoRa, if we do not understand how to use it in practical life and from where to start, we are going to discuss this in this section. Before getting into software packages, hardware needs to be completed. The hardware connections for end devices and gateways are given in Table 30.1. After completing the hardware connection, we have to more onto software packages. First of all, we configure the Raspberry Pi for Lora Modules by using different python packages such as pyLoRa package, NumPy, Pandas, etc. After configuration, we move on the programming section. For more detail in programming, refer to [1]. This was the basic step for creating LoRa communication. To proceed further for distance communication and to make complete IoT package, we need to add sensors on Arduino UNO side and cloud platform on Raspberry Pi side.

4. Applications of LoRa Technology

Some of the works on LoRa Technology's Application is shown in Table 30.2.

Table 30.1 Pin Connection of LoRa Module With Arduino UNO and Raspberry Pi

End Devices		Gateways	
LoRa SX1278 Module	**Arduino UNO Board**	**Raspberry Pi**	**LoRa SX1278 Module**
3.3 V	3.3 V	3.3 V	3.3 V
Gnd	Gnd	Ground	Ground
E_s/N_{ss}	D10	GPIO10	MOSI
GO/DIO0	D2	GPIO9	MISO
SCK	D13	GPIO11	SCK
MISO	D12	GPIO8	Nss
MOSI	D11	GPIO4	DIO0
RST	D9	GPIO17	DIO1
		GPIO18	DIO2
		GPIO27	DIO3
		GPIO22	RST

Table 30.2 Different Applications of LoRaWAN Technology

Ref. No.	Application Field	Outcomes
[2]	Healthcare	Checking the blood temperature at provincial centre
[3]	Smart building	Check and control the room's temperature and humidity in order to reduce the cost of heating ventilation and air conditioning
[4]	Environment	Identify and prevent a destructive landslide
[5]	Safety regards	Vehicle diagnostic system
[6]	Agriculture	Visual monitoring technique specialized

5. Our Work on Packet Loss Analysis in LoRaWAN Line-up in our College Campus

This was a small research that was done in the early stages when we started reading about LoRa technology. This research was done only so that we can see for future projects which, how and how many factors affect the deployment of LoRa network so that we can minimize packet collision be balancing load of adjacent gateways.

This packet loss analysis is needed because there is still gap between the actual networking performance and the specification and there is a lack of analysis on how large-scale networks are deployment.

For this research, we made a gateway and a node with the help of method discussed in Section 2 of this paper, for analysing the different factors affecting the packet loss. This gateway and node are placed at different location in the college campus. Fig. 30.2 shows the network setting in the college campus. For the comparative study, we divided the location into two parts:

1. Around boys' hostel,
2. Around ECE department

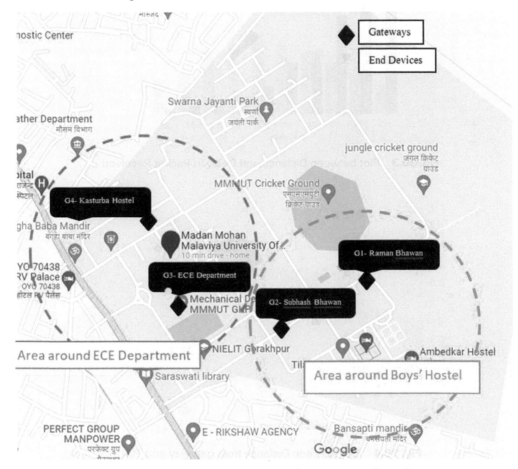

Fig. 30.2 Gateways and End Devices placement in college Campus

To find the result of this research, plotting between different parameters has been done, which is shown in Figs. 30.3–30.7.

From the above result and observations, we can conclude that:

1. The root cause for the packet loss event to occur are

    (a) *Electromagnetic environmental factors*
    (b) *Transmission distance*

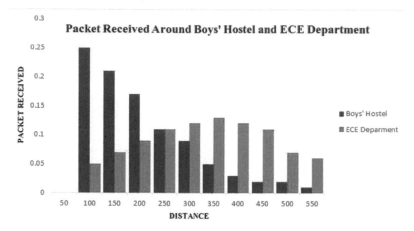

Fig. 30.3 Plot between Distance and Delay in Packet Received

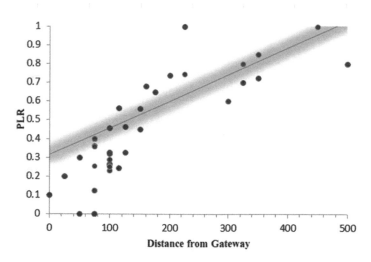

Fig. 30.4 Plot between Distance from gateway and PLR

(c) *Packet collision*

(d) *Type of end devices*

2. On increasing the distance of end devices from gateway, PLR also increases as packet received at gateway decreases.

3. On increasing the height of the gateway, its coverage area increases because ground reflection decreases.

4. End devices which don't have to send packet continuously are more effective because packet collision occurs less in that case.

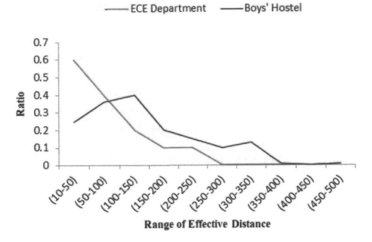

Fig. 30.5   Comparison of PLR around both areas

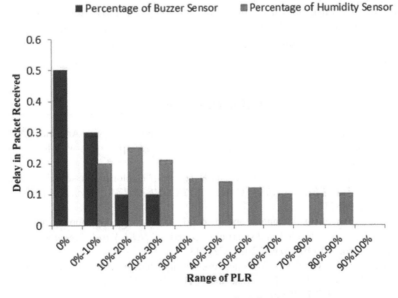

Fig. 30.6   Effects in End Devices

With increase in the number of days, nodes functionality decreases as packet received at gateway decreases due to environmental factors.

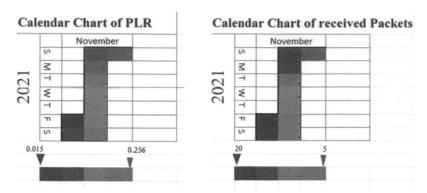

Fig. 30.7 Calendar Chart of PLR

6. Conclusion

In this paper, we presented the literature review along with small analysis on the PLR in LoRaWAN line-up in our campus. It includes all relevant research paper regarding LoRa and LoRaWAN technology. Its aspects such as implementation, limitation, future scope, etc. are mentioned in a sequential order. After reading this complete article, all of us will get complete clarity and satisfactory information about LoRa technology. Firstly, detailed discussion on LoRa technology and its protocol. Then, it is followed by its implementation. Further, we talked about its advantages, limitations, applications, recent works, and its future scope. Finally, this paper ends with the packet loss analysis in LoRaWAN line-up.

REFERENCES

1. A. Raj, "LoRa with Raspberry Pi – Peer to Peer Communication with Arduino RASPBERRY PI," 2019. https://circuitdigest.com/microcontroller-projects/raspberry-pi-with-lora-peer-to-peer-communication-with-arduino

2. E. Aras, G. S. Ramachandran, P. Lawrence, and D. Hughes, "Exploring the security vulnerabilities of LoRa," *2017 3rd IEEE Int. Conf. Cybern. CYBCONF 2017 - Proc.*, 2017, doi: 10.1109/CYBConf.2017.7985777.

3. M. Centenaro, L. Vangelista, A. Zanella, and M. Zorzi, "Long-range communications in unlicensed bands: The rising stars in the IoT and smart city scenarios," *IEEE Wirel. Commun.*, vol. 23, no. 5, pp. 60–67, 2016, doi: 10.1109/MWC.2016.7721743.

4. R. F. Romdhane *et al.*, "Wireless sensors network for landslides prevention," *2017 IEEE Int. Conf. Comput. Intell. Virtual Environ. Meas. Syst. Appl. CIVEMSA 2017 - Proc.*, pp. 222–227, 2017, doi: 10.1109/CIVEMSA.2017.7995330.

5. Y. S. Chou *et al.*, "I-Car system: A LoRa-based low power wide area networks vehicle diagnostic system for driving safety," *Proc. 2017 IEEE Int. Conf. Appl. Syst. Innov. Appl. Syst. Innov. Mod. Technol. ICASI 2017*, pp. 789–791, 2017, doi: 10.1109/ICASI.2017.7988549.

6. M. Ji, J. Yoon, J. Choo, M. Jang, and A. Smith, "LoRa-based Visual Monitoring Scheme for Agriculture IoT," *SAS 2019 - 2019 IEEE Sensors Appl. Symp. Conf. Proc.*, pp. 1–6, 2019, doi: 10.1109/SAS.2019.8706100.

Intelligent Systems and Smart Infrastructure – Brijesh Mishra et al. (eds)
© 2023 Taylor & Francis Group, London, ISBN 978-1-032-41287-0

CHAPTER

31

Unattended E-Vehicle Charging Solution for High-Rise Residential Complex

Sumit Kumar Singh[1], Gulam E. Gous[2],
Puskar Bharti[3], Raju Ranjan[4]

School of Computing Science & Engineering,
Galgotias University, Greater Noida, India

Abstract

We live in a new era where technology is so advanced that human-made robots have reached mars and are very close to the center of the solar system. Still, to date, we have not found any other planet similar to earth where a human can survive. So, humans' responsibility is to keep the earth clean and safe from global warming. When we are talking about global warming, fossil fuel in vehicles plays a vital role in increasing global warming because it produces some toxic gases that are harmful to our ecosystem; for example, carbon dioxide, nitrous oxide, and methane. The percentage of air pollution through the vehicles is 27%, as shown in Fig. 31.1, in India [1]. It is pretty significant in number. The most important thing is that fossil fuel is limited so that humans can search for its alternative. The best option is that humans can use an electric vehicle (EV). It is safe and does not produce any gases. It also does not create any loud sound. Government of India has taken some steps to fight pollution. Promoting non polluting Electrical-Vehicle (e-vehicle) is one of them. In some metropolitan cities, e-vehicle has started plying, and day by day, the number will keep increasing. Nevertheless, the main problem is that we do not have any unattended charging stations to charge our e-vehicle. This article aims to describe our unattended e-vehicle intelligent charging system.

Keywords: e-vehicle, ecosystem, battery

Corresponding author: [1]sumit_singh.scsebtech@galgotiasuniversity.edu.in,
[2]gulam_gous.scsebtech@galgotiasuniversity.edu.in, [3]puskar_bharti.scsebtech@galgotiasuniversity.edu.in,
[4]drraju.ranjan@galgotiasuniversity.edu.in

DOI: 10.1201/9781003357346-31

1. Introduction

E-vehicle is a vehicle running with the help of electricity, either partially or fully. It consists of a chargeable battery that uses electricity for charging. Since e-vehicle does not use fossil fuel, it does not emit any toxic gases like SO_2 and NO_2, which is the primary pollutant of air. So we want to make an e-vehicle charging system, which does not require any attendant. The EV charging station is the equipment used to connect electric vehicles to the source of electricity to recharge the battery of the e-vehicle. These stations provide a special universal connector to the various electric charging connectors described in the middle of the article. An advanced EV charging station is the need of the hour. This system has the facility to select the required time or amount of charging. The user has to pay money only for the consumed energy. The motive is to develop an affordable and easy-to-use charging station with efficient technology.

2. Literature Review

A. Remove Fossil Fuel Vehicle and Use e-vehicle

Hengsong Wang et al. (2010) proposed a new approach for the layout of electric vehicle charging. They proposed a system in which energy crisis and environmental pollution have become a global problem these days. The number of vehicles running on fossil fuel is increasing day by day, which play a major role in the above problem. To overcome these problems, adopting e-vehicle in place of the traditional vehicle is one of the best options. It is energy-conserving and environment-friendly, but the main reason which stops proper functioning of e-vehicle is absent of proper layout method of charging station. An algorithm is designed that updates existing gas stations and makes them fit for installing a charging station.

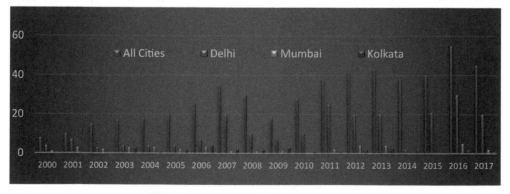

Fig. 31.1 Pollution effect in major cities

B. Introduce e-vehicle Charging Station

S. Ilayaraja and T. V. Narmadha (2016) proposed a model of an e-vehicle charging station using a DC–DC self-lift converter The author presents the design of maximum power tracking

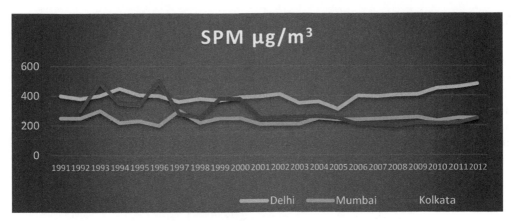

Fig. 31.2 SPM level in major cities

by the use of a self-lift converter. PI controller is used to tracking the maximum power from the solar panel. It will decrease the voltage and current ripple.

C. Introduce e-vehicle Charging Station from Solar Power

G. R. Chandra et al. (2016) proposed a system design for a solar-powered electric vehicle charging station for workplaces. According to the author, this model is for charging the battery of an e-vehicle at the workplace which will be done by solar energy.

3. Data and Variables

A. Charging Station for Electric Vehicle

E-vehicle is dependent on the battery, and the backup system of the battery is limited (in time), which means the battery is running for a specific period. After discharge, we have to charge it again. So if we want to expedite the use of e-vehicle, then there should be an easy and feasible solution for its charging.

In our proposed system, there will be different types of charging stations:

1. Home charging facility
2. Public charging facility

   (a) AC charging facility
   (b) AC charging facility (standard)
   (c) AC charging facility (fast)
   (d) DC charging facility (fast)

B. Home Charging Facility

It is one of the safest and most straightforward charging facilities. It can be installed in one's house or campus, where one can charge an e-vehicle. For example, when the vehicle is plugged, it gets completely charged after a specific period, and one can use the vehicle for the period of its running capacity. However, the biggest problem is that if your battery is low during the driving time (means if you are going somewhere but in the middle, your battery is low), it is not possible to go back home and then charge and use it. So it is necessary to install a public charging station for e-vehicle as per the specification of respective manufacturer.

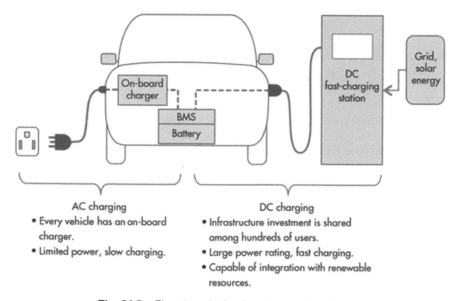

Fig. 31.3 Electric vehicle charging station diagram

C. Public Charging Facility

It is an e-vehicle charging station that is for public use. This station is installed everywhere, just like an ATM. The user will come and connect is charging cable with the station and set it.

AC charging facility:

AC charging means a charging system connected with an alternate current. The AC power grid is connected with our charging station with electrical wire of proper thickness just like our home supplies connects with the nearby grid as shown in Fig. 31.4.

(a) AC charging facility (normal): This charging facility is used for small vehicles like electric bicycles, bikes, cars, etc., whose charging capacity is approx.. 2.5 kW to 3 kW. The average full charging time of these vehicles is approximately two to three hours.

(b) AC charging facility (fast): This charging facility is for medium-sized vehicles whose charging capacity is approx. 7 kW to 22 kW. The average full charging time is approx. one to two hours.

(c) DC charging facility (fast): In an e-vehicle charging station, solar panels play a vital role because they can be used as a power backup. If anyhow alternating current is interrupted, then the solar panel can be used as a backup for the continuous working of the charging station. Simply DC charging system means a charging system connected with a power backup system, as shown in Fig. 31.4. In this system, the power station is connected with a battery which is charged by solar energy.

D. E-vehicle Charging Station Power Backup System

Solar energy is one of the cleanest and non-limited energies. It converts light and heat energy into electrical energy. It is green energy, which means it does not produce any type of harmful gases that pollute our environment. So we use solar panels for power back. We use solar panels, which will convert light energy into electric energy and charge our battery.

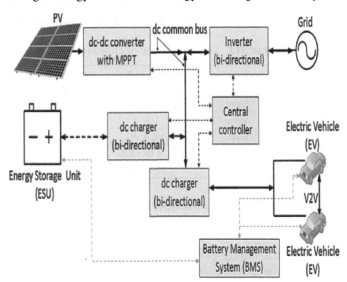

Fig. 31.4 Charging with direct current (power backup)

There are two types of charging methods:

Level 1

Level 1 is a primary charging method in which the supply is 120 V. It uses alternating current having input limit of 15 A to 20 A, and comes under the standard set by Charging Foundation.

Advantages:

(a) Cost of installation is meager.

(b) The impact of electricity is low during high demand.

Disadvantage:

The rate of charging mode is slow (approx. 3 or 5 Ampere/hour).

Level 2:

Level 2 is a basic charging method in which it supplies 208 V–240 V, and alternating current associated with input limit is 50 to 70 A, which is the standard fixed by charging foundation standard. It is a fast-charging technique.

Advantages:

(a) *It is fast charging in comparison of level 1 (approx. 10–20 Ampere/hour)*
(b) *Its energy efficiency level is high.*

E. Electric Vehicle Charger

An electric vehicle charger is a long electric wire cable attached with some socket used to recharge the electric vehicle battery. One side of these charger sockets is connected with the electric board, and another side is connected with the e-vehicle that charges the battery.

Table 31.1   Different types of charger

Charger type	Charging type	Vehicle to be charged	Power output (KWH)	Cost in Rs.
Bharat AC – 001	Slow	2 W	3×3.9	5500
Bharat DC – 001	Slow	4 W	15	7500
Type 2-AC	Fast	2 W, 4 W	50	20,100
CHAdeMO	Fast	2 W, 4 W, HDV	50	30,600
CSS	Fast	4 W, HDV	50	50,500

F. Public Place e-vehicle Charging Station Installation Issues

There are so many issues behind e-vehicle charging installation in a public place that follows:

(a) In public places, there is mostly government land, so before installation, we have to take permission from the government agencies to install the charging station. Without permission, the charging station can be sealed and heavy penalties can be imposed.
(b) It requires a large area for installation because space is not only for installation, it should have a space for parking because approximately 1 to 2 hours is required to charge the battery.
(c) It should be installed in an open area.
(d) Availability of electricity and backup should be ensured.

G. Best Location for Charging Station Installation

It is most important to choose the best location for e-vehicle charging station installation. It should be installed in the region having high footfall so that more people can take benefit from it. The location of e-vehicle charging station installation is divided into two parts:

Primary location:

(a) *Traffic area:* Railways Station, Bus Station, Airport, Highways, Toll Booth.

(b) *Attractive Areas:* Shopping Mall, Multiplex, Parks, Government, and Private Offices.

Secondary location:

(a) Near the power distribution station

(b) Avoiding narrow roads.

H. E-vehicle Charging Station Installation Condition

It is the most important thing to check the condition where the system is installed because it requires huge amount of electricity and 24×7 availability; for power backup, it requires proper sunlight.

These are the factors that should be kept in mind during installation:

(a) It should have proper availability of power.

(b) It should be in a large area.

(c) It should have constructability.

(d) Mounting.

4. Conclusion

This article describes the unmanned electrical charging station and fulfills the requirement of the modern e-vehicle charging problem. It also helps to make a green and pollution-free environment. Basically, this paper provides a solution to charge e-vehicle at public parking places, especially for the residents of high-rise apartments whose parking place is not fixed. It varies according to the first-come-first-serve system. The authors studied several papers related to this problem but none of them provides a fully automated unmanned system and power backup system. The paper proposes an unmanned charging system, which means there is no need for any human being for protecting or taking care of these systems. It is fully automated and has a power backup system. In case of power supply failure, this proposed system provides electricity through solar panels.

REFERENCES

1. Wang, H., Huang, Q., Zhang, C., & Xia, A. (2010, December). A novel approach for the layout of electric vehicle charging station. In The 2010 International Conference on Apperceiving Computing and Intelligence Analysis Proceeding (pp. 64-70). IEEE.
2. Ilayaraja, S., & Narmadha, T. V. (2016, March). Modeling of an e-vehicle charging station using DC-DC self-lift SEPIC converter. In 2016 Second International Conference on Science Technology Engineering and Management (ICONSTEM) (pp. 526-531). IEEE.
3. Mouli, G. C., Bauer, P., & Zeman, M. (2016). System design for a solar powered electric vehicle charging station for workplaces. *Applied Energy*, *168*, 434-443.

4. Kurien, C., Srivastava, A. K., & Molere, E. (2020). Emission control strategies for automotive engines with scope for deployment of solar based e-vehicle charging infrastructure. *Environmental Progress & Sustainable Energy, 39*(1), 13267.

5. Lam, A. Y., Leung, Y. W., & Chu, X. (2014). Electric vehicle charging station placement: Formulation, complexity, and solutions. *IEEE Transactions on Smart Grid, 5*(6), 2846-2856.

6. Csiszár, C. (2019). Demand calculation method for electric vehicle charging station locating and deployment. *Periodica Polytechnica Civil Engineering, 63*(1), 255-265.

7. Bayram, I. S., & Bayhan, S. (2020, July). Location analysis of electric vehicle charging stations for maximum capacity and coverage. In *2020 IEEE 14th International Conference on Compatibility, Power Electronics and Power Engineering (CPE-POWERENG)* (Vol. 1, pp. 409-414). IEEE.

8. Moghaddam, Z., Ahmad, I., Habibi, D., & Phung, Q. V. (2017). Smart charging strategy for electric vehicle charging stations. *IEEE Transactions on transportation electrification, 4*(1), 76-88.

9. Balacco, G., Binetti, M., Caggiani, L., & Ottomanelli, M. (2021). A Novel Distributed System of e-Vehicle Charging Stations Based on Pumps as Turbine to Support Sustainable Micromobility. *Sustainability, 13*(4), 1847.

10. Akila, A., Akila, E., Akila, S., Anu, K., & Elzalet, J. (2019, March). Charging station for e-vehicle using solar with IOT. In *2019 5th International Conference on Advanced Computing & Communication Systems (ICACCS)* (pp. 785-791). IEEE.

Intelligent Systems and Smart Infrastructure – Brijesh Mishra et al. (eds)
© 2023 Taylor & Francis Group, London, ISBN 978-1-032-41287-0

CHAPTER

32

Integrating Robotic Process Automation and Machine Learning for Web Scraping

Aarti Chugh*, Charu Jain and Yojna Arora

Amity School of Engineering and Technology,
Amity University, Gurugram, Haryana, India

Abstract

A web browser can only view data present on most of the websites. Websites do not provide a feature for saving any important or desired data for offline use. Web scraping method helps to pull out large amounts of data from websites, which is further saved in tabular format. Web scraping is commonly used to facilitate online price comparisons, aggregate contact information, extract online product catalog data, extract economic/demographic/statistical data, and create web mashups, among other uses. This paper discusses development of smart web scraping bot for commercial websites using Robotic Process Automation (RPA) and Machine Learning. Different automation activities are used to fully automate the bot such as automatic email access and email generation, automatic CSV (comma separated values) file generation, etc. Machine learning is applied to output produced by RPA as RPA cannot automate judgement procedure. Results showed that the integrated model of bot works effectively and accurately while answering all queries.

Keywords: Web Scraping, RPA, Machine Learning, UiPath

1. Introduction

The World Wide Web (WWW) is an interconnected network of information which users can access through different websites. The data present on these websites is in huge amount, variety, and different forms and not all data is useful to the users.

*Corresponding Author: achugh@ggn.amity.edu

DOI: 10.1201/9781003357346-32

Web scraping is a process of extracting large amount of data, i.e., big data from websites (Vargiu. et. al. 2012). Such data is heterogeneous in nature and web scraping transforms it into useful and structured form of data that can be stored in a database or any type of file. The main goal of web scraping is to extract targeted information from a website and which is stored into table formats such as comma separated values (CSV) file or spreadsheet, etc. Information like product details on shopping sites, stocks details, business contact details such as address, mobile number, etc. and reviews or ratings on different websites can be easily gathered by web scraping. Fig. 32.1 describes the basic idea of web scraping.

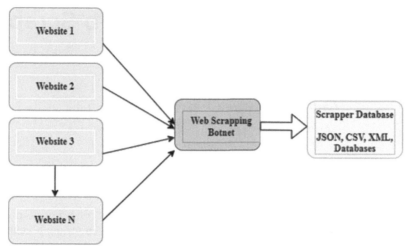

Fig. 32.1 Web Scraping Process Applied to Websites

There are various techniques for extracting data from websites like using libraries Scrapy and BeautifulSoup, and tools rvest and Blueprismt, etc. Web scraping techniques can be classified into three groups, namely programming language libraries, desktop-based environments, and tools or frameworks (Singrodia et. al. 2019). The old technique of web scraping can be efficiently extended in Robotic Process Automation or RPA to automate processes that are repeatedly carried out by human. RPA is a software-based system, which has proved benefits in terms of productivity, costs, speed, and error reduction (Aguirre et al. 2017). Although RPA can automate repeated activities but it is not an intelligent system and cannot lead to complete automation. For instance, a process which involves making judgment, RPA will not be able to automate the process. This is where Machine Learning approach comes in and the solution to this problem can be obtained. This paper discusses design of an integrated automated model for web scraping using UiPath, a RPA tool, and a machine learning algorithm. The proposed automated bot not only performs web scraping but also compares the extracted data with the previously scraped data and gives the changes in the data-enabling user to monitor changes in the website. Section 2 gives literature review in the field of web scraping, methodology is given in Section 3, results are discussed in Section 4, and finally, Section 5 provides conclusion drawn.

2. Literature Review

Eloisa Vargiu et al. (2012) discussed concepts related to web advertising and web scraping. They employed web scraping technique to collect relevant advertisements from web pages. The process begins with a collaborative filtering, which selects related web pages and passes them to web scraper to analyse page content and provide advertisements only.

Katharina Kaiser et al. (2005) published a survey describing the requirements, components, and design approaches of information extraction systems. According to authors, the task of information extraction system is information retrieval and then its structural representation. Since NLP-based systems do not have rich grammar constructs authors suggested use of wrappers. A wrapper is a set of instructions that can extract information through multiple repositories over WWW, merge them, and finally put them in a self-described representation. The authors discussed various manual, semi-automatic, and automatic wrapper generation methods. Out of these, automatic methods are most desirable and research-oriented.

David Mathew Thomas et al. (2019) have designed a web scrapper using a web crawler Scrapy and Python, which extracts the unstructured data from multiple sources over Internet. This data is further cleaned, organized, and analysed through different models and algorithms to achieve desired results. The model was tested over e-commerce sites like Flipkart, Amazon, etc. and the results proved that the proposed architecture works effectively and perform query processing efficiently.

Charu Jain et al. (2020) propounded an email reply automation software using RPA and machine learning algorithm. Here, RPA extracts data from emails and machine learning technique analyses this data to create relevant responses for any input queries. The authors have done comparison in terms of accuracy and positive results rate for various existing machine learning algorithms and the proposed algorithm BERT. The designed hybrid achieved 82% accuracy.

Moaiad Ahmad Khder (2021) talked about web scraping types, technologies, advantages, legal and ethical issues, and most important how it is related to other well-known technologies like AI, Big Data, Business Intelligence, Data Science, and Cyber Security.

3. Methodology

RPA is one of the smartest technologies that came in recent past. RPA is basically the automation of repetitive tasks. The whole concept of RPA is simplified as a process where a bot is trained to do the repetitive task carried out by a human user and the bot does it automatically (Jain et al. 2020; Florentina et al. 2020). Thus, the process would no longer need a human user to execute it. Through RPA, business processes are automated, which results in low cost, lesser errors, and improved human productivity as the workers can focus on other needs within their departments. There are a number of software available in the market such as Automation Anywhere, UiPath RPA, Blue Prism, Pega Platform, etc.

UiPath is chosen here to build an intelligent bot for web scraping as it is deemed as the fastest RPA solution and enables rapid application development through its inbuilt facilities (Fabiana Corredor 2021). The designed bot is capable of automatically logging in to a web site, extract data spanning multiple web pages, and filter and transform it into the format of user's choice, before integrating it into another application or web service.

As the execution of the bot starts, it takes input from the user's mail id using Get IMAP Mail Messages activity. Bot then opens the browser using Open Browser activity and goes to the provided URL. Bot scraps the data. The new scraped data in CSV format for the user search "iPhone" is compared with the old data to help in decision-making process.

The broad steps of methodology are as follows:

(a) To perform web scraping using RPA, a bot is created in UiPath software using its data scraping tool, which selects the desired elements, i.e. data or fields from selected webpage or application or document. Fig. 32.2 shows selection of web elements to prepare bot. The bot runs to take the input from the selected browser window. For example, the experimental bot is prepared to select mobiles available for sales on Flipkart website.

Fig. 32.2   Selection of Web Elements for Preparing Bot

(b) All scrapped data is stored in a DataTable variable and then will be saved into a final list as a CSV file (Fig. 32.3). In experimental setup, the selected mobile product details are saved for further processing.

Fig. 32.3 Creating CSV File

(c) The bot is further programmed to email the extracted list using Get IMAP Mail Message activity, which accesses the mail of the desired user and SMTP mail message activity to send email to that user. Fig. 32.4 shows various important attributes of IMAP activity

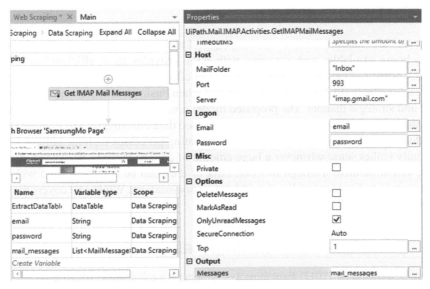

Fig. 32.4 Get IMAP Mail Messages

like MailFolder, Port, Email, and Password, selecting the number of the mails to be accessed and read only unread messages, etc. The Password attribute is important and Uipath provides the application-generated password security also. Port number 993 should be used for working with email automation. The extracted CSV file is attached as attachment in the mail.

(d) The final step is to combine the power of machine learning with RPA to create intelligent web scraping bot. Machine learning is employed to compare and analyze sales through the different lists extracted through bot. For integrating machine learning into UiPath bot, Python Scope and Run Python Script activity are used. Coding is done to compare the new and the old data sets which are present in the form of CSV files. The result is sent to the user's id by using again Send SMTP mail message activity.

4. Results and Discussion

Hence, RPA can efficiently extract data from websites and convert it into a structured form but these bots do not provide any predictions or conclusions from the extracted data. Machine Learning with RPA makes an "intelligent" RPA bot that can analyse, draw conclusion, and make predictions from the data. For example, the proposed ML model can be extended to detect the change in prices of a product where the bot should extract the current prices from the website. This CSV file is sent to user's email. Then, bot runs the Python script which opens the new scraped file and the old scraped file. Both new and old scraped data which are in the CSV file formats are then compared using the Python script. The final result consists of a new CSV file containing the changes in the data stored in the computer.

5. Conclusion

This paper provides an overview on web scraping and its role in data analysis. Despite various techniques being available, web scraping with RPA provides most efficient and effective results. RPA bots have certainly reduced the human effort by performing all repetitive tasks. Infusing machine learning with RPA is meaningful when business automation is pursued in an integrated and strategic manner. The proposed model not only performs web scraping but also compares the extracted data with the previously scraped data enabling user to monitor changes in the products present on website. Therefore, automating processes with the help of RPA and ML especially makes sense whenever a huge amount of data needs to be processed, analyzed, compared, and structured. Through this bot, bulk of data can be extracted from websites in few seconds and can be stored and analyzed for decision-making. While ML covers the task of thinking and learning, RPA automates and executes.

REFERENCES

1. Vargiu, E. and Urru, M. (2012). Exploiting web scraping in a collaborative filtering- based approach to web advertising. Artificial Intelligence Res. 2(1):44.

2. Aguirre, S. and Rodriguez. A. (2017). Automation of a business process using robotic process automation (RPA): A Case Study. Communication Computational Inf. Sci. 742: 65–71.
3. Thomas, D. M. and Mathur, S. (2019). Data Analysis by Web Scraping using Python Enhanced Reader. IEEE Xplore: 450–454.
4. Jain, C., Chugh, A., Chutani, J., Sundar, S. and Thayil, R. (2020). Integrating Robotic Process Automation With Machine Learning For Incoming. J. of Critical Reviews. 7(18): 414–420.
5. Khder, M. (2021). Web Scraping or Web Crawling: State of Art, Techniques, Approaches and Application. International Journal of Advances in Soft Computing and its Applications 13(3): 145–168.

Conferences
6. Singrodia, V., Mitra, A. and Paul, S. (2019) A Review on Web Scrapping and its Applications. 2019 International Conference on Computer Communication and Informatics, ICCCI 2019.

Online Documents/Resources
7. Florentina, M., (2020). Web Data Extraction with Robot Process Automation. Study On Linkedin Web Scraping Using Uipath Studio. Ann. Constantin Brancusi Univ. Targu Jiu, Econ. Ser. Accessed: Nov. 13, 2021. [Online]. 20(1): 14–19
8. Corredor, F. 2021 Guide: Best RPA Tools and Why UiPath is #1 https://www.auxis.com/blog/top-rpa-tools (accessed Nov. 20, 2021).
9. Kaiser, K. and Miksch, S. (2005). Information Extraction. A Survey. Interactive Media Systems. https://www.ims.tuwien.ac.at/publications/tuw-139463 (accessed Nov. 13, 2021).

Intelligent Systems and Smart Infrastructure – Brijesh Mishra et al. (eds)
© 2023 Taylor & Francis Group, London, ISBN 978-1-032-41287-0

CHAPTER

33

Study and Implementation of Efficient Pseudorandom Number Generator

Priya Katiyar[1], Sandeep Kumar[2],
Upendra Kumar Acharya[3], Prabhakar Agarwal[4]

Department of Electronics and Communication Engineering,
National Institute of Technology, Delhi, India

Abstract

Pseudorandom numbers in the present world find a wide number of applications like computers, information security, secure communication, digital signatures, gambling, and many more. It acts as inputs in many protocols, and thus becomes the prime focus. With growing digitalization in present scenario, the need for security is increasing and so is the need for better methods to generate pseudorandom numbers. In this paper, we lay our focus on pseudorandom number generator (PRNG). We have focused on discussing various methods used to generate PRNG in practical systems followed by implementation of an efficient method. The Logistic map methods, chaotic methods such as Bernoulli shift map method, tent map method, and zigzag map method which rely on the theory of chaos, along with the Henon map and Four Wing memristor-based methods, are discussed, following which the Bernoulli shift map method is used to generate random sequence.

Keywords: pseudorandom number generator, random number, seed element, Bernoulli map, Henon map, tent map, chaos.

1. Introduction

In the earlier times before twentieth century, the methodology for generating random numbers included dealing out well-shuffled cards, throwing dice, etc. As we move further towards

Corresponding author: [1]202221010@nitdelhi.ac.in, [2]sandeep@nitdelhi.ac.in, [3]upendraacharya1989@gmail.com, [4]prabhakar@nitdelhi.ac.in

DOI: 10.1201/9781003357346-33

twentieth century, mechanized devices came into being which increased the speed of generating numbers. Later with time, the electronic devices became more prominent generation. However, the methods used till now enabled us to generate small numbers with ease. But in present times, we need larger numbers, which range to thousands or even millions which require much more memory, time, and efficiency.

Random number generation (RNG) has two basic types based on the process used for generation: true random number generation (TRNG) and pseudorandom number generator (PRNG). TRNG uses physical processes listed as photon noise, thermal noise, frequency jitter generated by oscillators, quantum random processes, and chaotic oscillator [1]. The random process values are sampled digitally and followed by postprocessing techniques to produce random output. TRNG is a slower method compared to the other method and hence the sequence generated is non-reproducible and unpredictable. This is because of the fact that the noise cannot have identical value for any two slots of time period which can be predetermined. TRNG finds application wherein slow generation and non-reproducibility are the main demands.

In this paper, PRNG methods based on logistic map, chaotic maps, henon map, and lastly the Four Wing memristive hyperchaotic system have been discussed followed by implementation of an efficient PRNG method. The purpose served by this paper is to introduce the reader to the various generation methods and their key points that define their usability.

This paper is organized as follows: Methods for PRNG followed by its properties and challenges are discussed in Section II. Implementation of Bernoulli method for PRNG is presented in Section III and concluded in Section IV.

2. Methods for Random Number Generation

In this paper, the various methods to generate the random numbers have been discussed such as chaotic maps, logistic map, Henon map, and the Four Wing memristive hyperchaotic system. This paper introduces various generation methods and their key points that define their usability.

A. The Logistic Map Method

It is a one-dimensional map which can be used to study complex behavior, population dynamics, cryptography, PRNG, and complex behavior [2]. It has a simpler implementation as it employs basic equations for generation in recursive manner such as

$$x_{i+1} = ax_i (1 - 3x_i) \tag{1}$$

whereas x_i is the discrete state, and a is the control parameter of chaotic sequence. Here i depicts the number of iterations. The logistic map has chaotic orbit which is verified with the help of Lyapunov exponent. It is a quantitative measure which depicts chaos and orbital divergence. Exponent having a positive value depicts chaos and orbital divergence.

B. Chaotic Map Method

The chaotic maps are easily implementable and in hardware form and have easier mathematical descriptions. The basic chaotic maps that are analyzed in this paper are Bernoulli map, shift map, tent map, and zigzag map [15].

1. Bernoulli shift map

A Bernoulli map is composed of uniform probability function, which comprises dual piecewise linear parts separated by a discontinuity point. It can be presented using two linear functions as shown below:

$$x_{n+1} = \begin{cases} Cx_n - d & \text{if } x_n \geq 0 \\ Cx_n + d & \text{if } x_n < 0 \end{cases} \tag{2}$$

Here, the variable c is the slope parameter which is responsible for controlling the stochastic properties and a is the scaling factor that increases or decreases the product value preceding it and hence binds the output value in the range $[-d, d]$. Since the variables can assume n number of values, its properties can be further improved. This can be done by selecting appropriate number by hit-and-trial method or by performing prior calculations.

2. Tent map

The tent map has a simple shape and is generally studied in dynamical systems in mathematics. This is one of the maps used for studying chaotic maps for discrete nonlinear dynamical systems [12]. The equation for the map is shown below:

$$x_{n+1} = \begin{cases} vx_n, & \text{for } x_n \in \left[0, \dfrac{1}{v}\right] \\ \dfrac{v}{v+1}(1-x_n), & \text{for } x_n \in \left(\dfrac{1}{v}, 1\right] \end{cases} \tag{3}$$

The x_{n+1} amounts to "0" when controlling parameter $v \in [0,1]$. The LE value obtained is positive and hence it has sufficient chaotic properties.

3. Zigzag map

The zigzag map is zigzag in shape as the name suggests and has an invariant fractal set on the vertical line. It poses to be faster as compared to the other maps to generate random number, which also consumes low power and is highly resilient to process variations [4]. It can be realized using the equations as

$$x_{n+1} = \begin{cases} -p\left(x_n + \dfrac{2}{|p|}\right), & \text{for } x_n \in \left(-1, -\dfrac{1}{|p|}\right], \\ px_n, & \text{for } x_n \in \left(-\dfrac{1}{|p|}, \dfrac{1}{|p|}\right), \\ -p\left(x_n - \dfrac{2}{|p|}\right), & \text{for } x_n \in \left[\dfrac{1}{|p|}, 1\right]. \end{cases} \tag{4}$$

Here *LE* is represented by $ln(|m|)$. Hence, for $|p| < 1$, the behavior differs from the chaotic one; subsequently, for intervals p (2, 1), (1, 2), [3, 2), and (2, 3], its behavior becomes chaotic.

C. The Henon Map Method

The Henon method as compared to other has better statistical properties or rather say uniformin nature. It shows the sensitivity with respect to initial conditions and also the control parameters when analysed. It can be realized by using the equations as listed below.

$$x_{n+1} = 1 - ax_n^2 + y_n \tag{5}$$

$$y_{n+1} = bx_n, \tag{6}$$

whereas $(x_n, y_n) \in R^2$ are the values of the map that occur discretely, the control parameters are a and b, n is the number of iterations for which the equation is implemented, and x_0 and y_0 are the initial conditions [7].

D. The Four-wing Memrsitive Hyperchaotic Method

Memrister is a non-volatile, non-linear element. It does the functioning of regulating the flow of electrical current in the circuit and remembers the amount of charge that had flown previously through the conductor. It retains memory without power, i.e., its resistance can be programmed and this remains stored in it. The memristive chaotic system shows variation with initial value of memrister and also shows sensitivity in circuit parameters under study. This system uses the memory characteristics and the non-linearity property of the memrister as feedback. It generated complex dynamic non-linear numbers. Hence, the number generated by this method becomes difficult to predict.

The Four-wing Memrsitive Hyperchaotic Method (FWMHS) can be modelled by using various methods since it just uses differential equations; nonetheless, one of the prominent methods used for this is the Runge–Kutta method. To be specific, a modified version of Runge–Kutta method can be used for this, i.e., fourth-order RK algorithm. The equations are as mentioned below.

$$k_1 = \Delta h f(p_k\, v_k)$$

$$k_2 = \Delta h f(p_k + \Delta h/2,\, v_k) + 1/2 \cdot k_1 \tag{9}$$

$$k_3 = \Delta h f(p_k + \Delta h/2,\, v_k + 1/2 \cdot k_2)$$

$$k_4 = \Delta h f(p_k + \Delta h,\, v_k + k_3)$$

$$u_{k+1} = u_k + 1/6\,(k_1 + 2k_2 + 2k_3 + k_4) \tag{10}$$

where $\Delta h = 0.001$, k_1, k_2, k_3, and k_4 represent the values of slope on $[x_k, x_{k+1}]$. Here, v_k and p_k represent the values at $t = t_k$. Alternatively, v_{k+1} and p_{k+1} are the values of time $t = t_{k+1}$. The initial four equations mentioned in Eq. (9) are substituted in Eq. (10), then state variables are analyzed to solve, as depicted in Eq. (2), where k_{i1}, k_{i2}, k_{i3}, k_{i4} ($i = p, q, r, s$) parameters in Eq. (12) depict the slope of the method for system [1]. Also when $t = 0$, initial values of variables p_k, r_k, q_k, and s_k are chosen as $p_0 = 0.1$, $r_0 = 0.1$, $q_0 = 0.1$, $s_0 = 0.1$.

3. Implementation of Pseudorandom Number Generator Using Bernoulli Method

As discussed previously, the Bernoulli method appears to be more efficient and simpler for generating random numbers. And hence we proceed further to generate random number using the same method. The equations for the generation are as further discussed.

$$x_{n+1} = \begin{cases} Bx_n + 0.5 & x < 0 \\ Bx_n - 0.5 & x \geq 0 \end{cases} \tag{13}$$

The above equation is used to generate the next iteration of the random sequence; here "B" is the parameter that can assume different values to enhance the properties of the generated number. Also "x_n" is the present iteration value. So the present value is operated upon using constant values to generate the next value, which gets chosen based upon the present value of "x".

$$b(n) = B(x_n, T_n) = \begin{cases} 0, & x_n \leq T_b \\ 1, & x_n > T_b \end{cases} \tag{14}$$

The equation mentioned above explains how "0" and "1" binary numbers are generated using the values generated using the previous equation. Here "T_b" is the threshold value that the further decides whether the output should be either "1" or "0". The value further can be used to increase or decrease the uncertainty of the generated sequence. The following flowchart shows the same process for better understanding. The above equations have been implemented using Verilog where the value of B is 177. Here the initial value of "x" is taken as the seed element, which can be varied. The following figure shows the output random sequence using different seed elements and threshold values. The same has been depicted using flowchart shown in Fig. 33.1.

Above the values, certain efficiency or increased uncertainty can be attained by changing the parameters by few numbers. This can be depicted in Fig. 33.2 and Fig. 33.3. This comes in handy, when a simpler random number generation method is needed, which is easy to implement and hence be incorporated in and is rather a complex procedure.

4. Implementation of Pseudorandom Number Generator Using FWMHS

The Four Wing Memristor method involves much more complex equations for the generation of pseudorandom number generation. It basically uses 4th-order Runge–Kutta method for implementation. Runge–Kutta method is an explicit or implicit iterative method used for approximate analysis of simultaneous nonlinear equations. The flowchart depicting the process is as shown in Fig. 33.4.

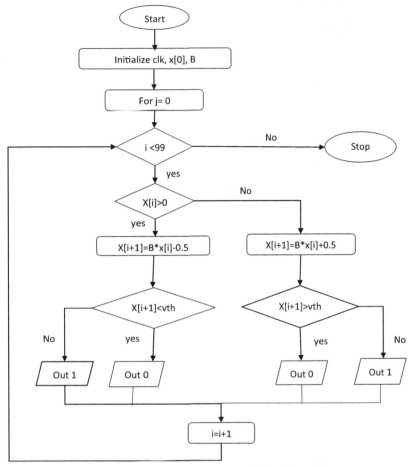

Fig. 33.1 Bernoulli Map based PRNG Flowchart

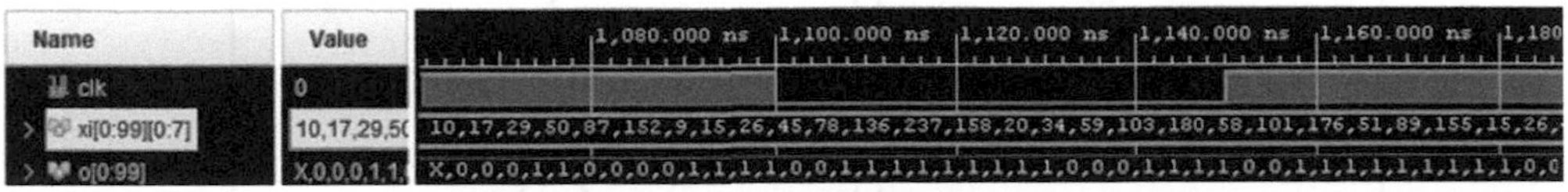

Fig. 33.2 The random sequence generated when V^{th} (threshold) is 80 and $x(i)$ seed element is 5.

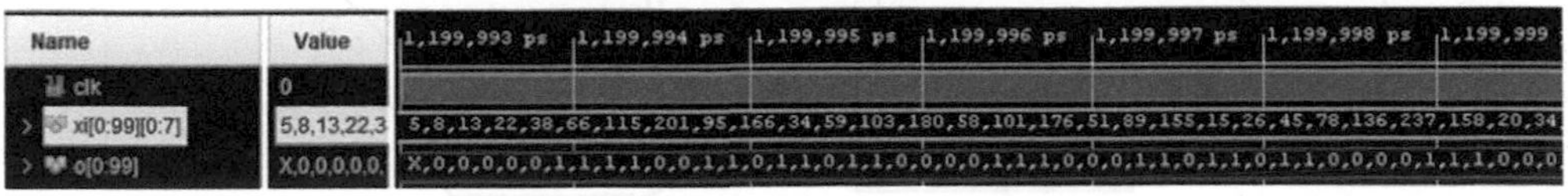

Fig. 33.3 The random sequence generated when V^{th} (threshold) is 50 and $x(i)$ seed element is 10.

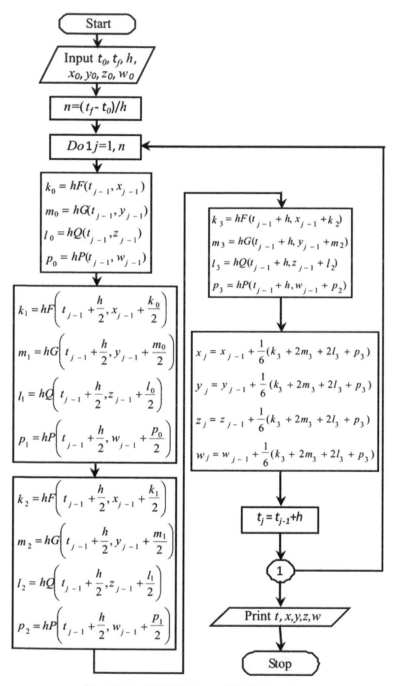

Fig. 33.4   FWMHS-based PRNG flowchart

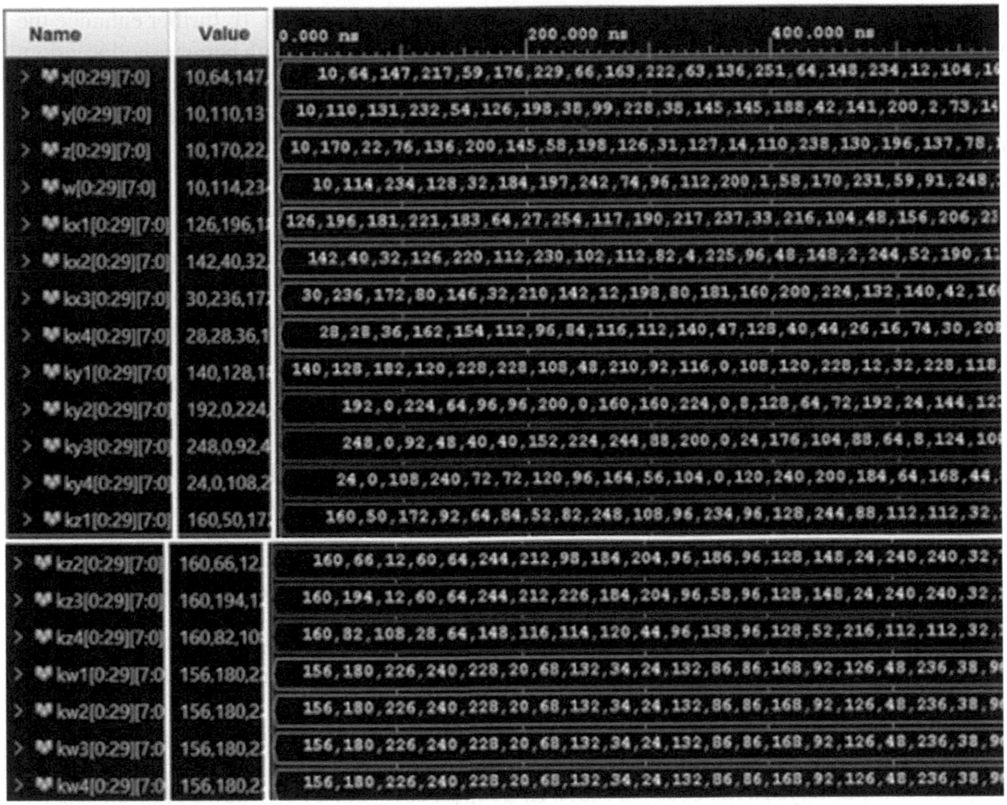

Name	Value	0.000 ns	200.000 ns	400.000 ns
> ❤ x[0:29][7:0]	10,64,147,	10,64,147,217,59,176,229,66,163,222,63,136,251,64,148,234,12,104,1¢		
> ❤ y[0:29][7:0]	10,110,13	10,110,131,232,54,126,198,38,99,228,38,145,145,188,42,141,200,2,73,14		
> ❤ z[0:29][7:0]	10,170,22	10,170,22,76,136,200,145,58,198,126,31,127,14,110,238,130,196,137,78,:		
> ❤ w[0:29][7:0]	10,114,23	10,114,234,128,32,184,197,242,74,96,112,200,1,58,170,231,59,91,248,:		
> ❤ kx1[0:29][7:0]	126,196,1:	126,196,181,221,183,64,27,254,117,190,217,237,33,216,104,48,156,206,2:		
> ❤ kx2[0:29][7:0]	142,40,32	142,40,32,126,220,112,230,102,112,82,4,225,96,48,148,2,244,52,190,1:		
> ❤ kx3[0:29][7:0]	30,236,17:	30,236,172,80,146,32,210,142,12,198,80,181,160,200,224,132,140,42,16¢		
> ❤ kx4[0:29][7:0]	28,28,36,1	28,28,36,162,154,112,96,84,116,112,140,47,128,40,44,26,16,74,30,20¢		
> ❤ ky1[0:29][7:0]	140,128,1	140,128,182,120,228,228,108,48,210,92,116,0,108,120,228,12,32,228,118,		
> ❤ ky2[0:29][7:0]	192,0,224	192,0,224,64,96,96,200,0,160,160,224,0,8,128,64,72,192,24,144,12:		
> ❤ ky3[0:29][7:0]	248,0,92,4	248,0,92,48,40,40,152,224,244,88,200,0,24,176,104,88,64,8,124,10¢		
> ❤ ky4[0:29][7:0]	24,0,108,2	24,0,108,240,72,72,120,96,164,56,104,0,120,240,200,184,64,168,44,		
> ❤ kz1[0:29][7:0]	160,50,17:	160,50,172,92,64,84,52,82,248,108,96,234,96,128,244,88,112,112,32,:		
> ❤ kz2[0:29][7:0]	160,66,12	160,66,12,60,64,244,212,98,184,204,96,186,96,128,148,24,240,240,32,:		
> ❤ kz3[0:29][7:0]	160,194,1:	160,194,12,60,64,244,212,226,184,204,96,58,96,128,148,24,240,240,32,:		
> ❤ kz4[0:29][7:0]	160,82,10:	160,82,108,28,64,148,116,114,120,44,96,138,96,128,52,216,112,112,32,:		
> ❤ kw1[0:29][7:0]	156,180,2:	156,180,226,240,228,20,68,132,34,24,132,86,86,168,92,126,48,236,38,9¢		
> ❤ kw2[0:29][7:0]	156,180,2:	156,180,226,240,228,20,68,132,34,24,132,86,86,168,92,126,48,236,38,9¢		
> ❤ kw3[0:29][7:0]	156,180,2:	156,180,226,240,228,20,68,132,34,24,132,86,86,168,92,126,48,236,38,9¢		
> ❤ kw4[0:29][7:0]	156,180,2:	156,180,226,240,228,20,68,132,34,24,132,86,86,168,92,126,48,236,38,9¢		

Fig. 33.5 The random sequence generated using FWMHS method.

As we can see from the above flowchart, k, m, l, p have been studied for $0,1,2,3$ iterations, and hence qualifies for four orders. Here, K, m, l, p are the slope values used for multiple iterations of x, y, z, and w. The output of the same is shown above. Here, x, y, z, and w shown in the first 4 rows show the random sequence generated. A final sequence which may include a combination of all or selected ones can be used to generate a single random sequence.

5. Conclusion

The methods discussed above have varied properties; some have better statistical properties and some have better power reduction mechanism. These all characteristics are highly dependent upon the equations used for the implementation and also the methods used to analyze them. Moreover, these properties define their application in any industry since all processes do not have similar requirements. The logistic map method requires lesser algorithm, and hence has higher speed in comparison to others. Now the chaotic map used provides better pseudorandom properties, but out of all, the Bernoulli is the one which is used frequently mainly because of the ease of implementation and uniformly distributed random numbers. To further enhance the properties, we had Henon map, which has better key sensitivity along with better pseudorandom properties. The memristive system discussed in the later part was far better than the others; it had all their previously discussed properties along with the ability to be controlled; this comes into picture by the usage of memrister.

REFERENCES

1. F. Yu, L. Li, B. He, L. Liu, S. Qian, Y. Huang, S. Cai, Y. Song, Q. Tan, Q. Wan, and J. Jin. 2019. Design and FPGA Implementation of a Pseudorandom Number Generator Based on a Four-Wing Memristive Hyperchaotic System and Bernoulli Map, IEEE Access.
2. M. A. Murillo-Escobar, C. Cruz-Hernández, L. Cardoza-Avendaño, and R. Méndez-Ramírez, Nonlinear Dyn. 2017. A novel pseudorandom number generator based on pseudorandomly enhanced logistic map, vol. 87, no. 1 (Jan), pp. 407–425.
3. L. G. de la Fraga, E. Torres-Pérez, E. Tlelo-Cuautle, and C. Mancillas-López, Nonlinear Dyn.. 2017. Hardware implementation of pseudo-random number generators based on chaotic maps, vol. 90, no. 3 (Nov), pp. 1661–1670.
4. K. Rajagopal, A. Bayani, and A. J. M. Khalaf, H. Namazi, S. Jafari, and V.-T. Pham. 2018. A no-equilibrium memristive system with four-wing hyperchaotic attractor'', AEU-Int. J. Electron. Commun., vol. 95 (Oct), pp. 207–215.
5. K. Rajagopal, S. Arun, A. Karthikeyan, P. Duraisamy.2018. A hyperchaotic memristor system with exponential and discontinuous memductance function, A. Srinivasan, International Journal of Electronics and Communications.
6. A. Saito, and A. Yamaguchi. 2018. Pseudorandom number generator based on the Bernoulli map on cubic algebraic integer, Chaos 28, 103122.
7. M. Bakiri, C. Guyeux, J.F. Couchot, and A. K. Oudjida. 2018. Survey on hardware implementation of random number generators on FPGA: Theory and experimental analyses, Comput. Sci. Rev., vol. 27 (Feb), pp. 135–153.
8. A. A. Rezk, A. H. Madian, A. G. Radwan, and A. M. Soliman. 2019. Reconfigurable chaotic pseudo random number generator based on FPGA,'' AEUInt. J. Electron. Commun., vol. 98 (Jan), pp. 174–180.
9. Y. Liu and X. Tong. 2016. Hyperchaoticsystem-based pseudorandom number generator, IET Inf. Secur., vol. 10, no. 6, pp. 433–441.
10. W. Z. Wang, X. Q. Wang, J. Wang, N. N. Xiong, S. Cai, and P. Liu. 2020. Ensuring cryptography chips security by preventing scan-based side-channel attacks with improved DFT architecture, IEEE Transactions on Systems, Man and Cybernetics: Systems.

Intelligent Systems and Smart Infrastructure – Brijesh Mishra et al. (eds)
© 2023 Taylor & Francis Group, London, ISBN 978-1-032-41287-0

One Sun, One World, One Grid: Possibilities of Asian Interconnection

Bipasha Basu[1], Gagan Singh[2], Bhanuprakash Saripalli[3]

Department of Electrical and Electronics & Communication Engineering,
DIT University, Dehradun, India

Abstract

The Indian government wants to connect solar power plants all around the world via a shared infrastructure known as "One Sun, One World, One Grid (OSOWOG)". The "Green Grid Initiative: One Sun, One World, One Grid" was launched by the World Bank, the International Solar Alliance, the United Kingdom, and the Government of India to ensure efficient, economical, clean, reliable, and accessible power all over the world by 2030. In this paper, a case study of a few cities in varied climates across the Asian continent was done to demonstrate the applicability of the "OSOWOG" method.

Keywords: One Sun, One World, One Grid (OSOWOG), Solar Generation Forecasting, Solar Interconnection

1. Introduction

At the first assembly of the International Solar Alliance (ISA) in 2018, Honourable Prime Minister Narendra Modi of the Government of India (GOI) proposed the "Green Grid Initiative: One Sun, One World, One Grid (GGI: OSOWOG)" to connect countries through an interconnected solar infrastructure with the help of the World Bank. With 81 countries having already approved "GGI: OSOWOG," OSOWOG's mission is to create efficient, clean, reliable, accessible, and economic energy throughout the world by 2030. Tropical areas received the

---

Corresponding author: [1]bipasha.basu.ofcl@gmail.com, [2]gaganus@gmail.com, [3]saripalli.bhanuprakash@gmail.com

DOI: 10.1201/9781003357346-34

most focus at the start of this unique undertaking. This programme may be a way to reduce reliance on fossil fuels and transition to green energy. A single grid will connect 140 countries around the world to transfer solar energy, according to a draught proposal published by India's Ministry of New and Renewable Energy (MNRE). With India at the centre, the solar spectrum can be separated into two broad zones: Far East and Far West. The Far East will connect Myanmar, Vietnam, Thailand, Laos, and Cambodia. The Far West will connect the Middle East with Africa. The basic plan for the project is separated into three phases: the first, second, and third phases. Keeping the Indian Grid front and centre throughout the first phase, solar and other renewable energy grids in the Middle East, South Asia, and Southeast Asia will be connected. African countries will be connected to the countries that were connected in the first phase and the second phase. Global interconnection will take place in the third phase. "OSOWOG" is entirely powered by renewable energy. As resources become available, solar will be employed first, followed by wind, biomass, tidal, and other sustainable energy sources. Different cities from the Asian continent were used in this work, each with a different climate.

2. Literature Review

It's tough to find a globally efficient location despite geographical constraints. A work [1] briefly discussed how to find an effective location for "OSOWOG." It not only found the site, but it also provided information about the Smart Grid System. On an hourly basis, the Smart Grid System can project power demand and generation, designate regions as Surplus, DeficitR, Balanced, Deficit, and Worst, and transmit energy. To find efficient locations, first get satellite images from a free source, then use the eastern, western, northern, and southern coordinate systems to find them (co-ordinate is the combination of latitude and longitude). It is vital to determine whether the land is barren land or not after collecting satellite pictures. Then the efficiency of the barren land has to be calculated [1], [4]. An Artificial Neural Network approach was utilized for forecasting load and energy consumption on an hourly and daily basis [2] and forecasting of energy applications [10], which showed how daily and hourly energy consumption can be achieved. The Levenberg–Marquardt algorithm has been implemented. In the Levenberg–Marquardt algorithm, the steepest-descent method's robustness is combined with the Gauss–Newton method's quadratic convergence rate. This algorithm is more efficient and faster in networks, has better optimization, and is less difficult. Cluster Analysis was used by the author of a research [3] to classify weather regimes, and this method is meant to be used for forecasting improvement. A Geographic Information System (GIS) can be used to determine the best location and size for a Solar Energy System [5]. The best location for large-scale photovoltaic power generation (LSPPG) will achieve increased utilisation efficiency with the same number of resources (land space, materials, and energy) [6]. Forecasting is another key part of solar energy generating. ARIMA can be used for multi-month forecasting [7] or daily solar radiation forecasting [8]. Other forecasting methodologies, such as the "FoBa", "leapForward", "spikeslab", "Cubist", and "bagEarthGCV" models, can be used to project Daily Global Solar Irradiance [9]. Changes in latitude and longitude have an impact on the Solar Collector's performance [11]. For efficient Solar Radiation, the clearness index should be greater [12].

3. Case Study

A case study has been done to show the possibilities of "OSOWOG" on the Asian Continent. Here nine cities have been chosen throughout Asia [13]. These are: My Tho (Vietnam), Riyadh (Saudi Arabia), Battambang (Cambodia), Jessore (Bangladesh), Churu (India), Chennai (India), Muscat (Oman), Bangkok (Thailand), and Colombo (Sri Lanka). Every city (listed) has a separate time zone, such as GMT+7 in My Tho, GMT+7 in Battambang, GMT+6 in Jessore, GMT+5:30 in Churu and Chennai in India, GMT+4 in Muscat in Oman, and GMT+3 in Riyadh in Saudi Arabia. Bangkok is GMT+7, while Colombo is GMT+5.30. Because "OSOWOG" should be representative of equatorial and tropical regions, those regions were chosen to demonstrate the potential of this unique effort. Yearly Solar Radiation of 9 Cities (Table 34.1) reveals that the Solar Radiation of those cities is nearly identical. If you take the average of those cities, you will get 5.63 km/m^2/day. NSRDB [14] and NREL [15] are the data sources. According to the bar chart of yearly Solar Radiation data presented in Fig. 34.1, the trend line is almost touched every bar. Monthly Solar Radiation of nine cities is reported in Table 34.2, with the monthly average of nine cities.

Table 34.1 Solar radiation data for chosen cities throughout a year (Data Source: NREL)

Location/ Parameters	My Tho	Battam- bang	Jessore	Churu	Chennai	Muscat	Riyadh	Bangkok	Colombo
Latitude	10.37N	13.01N	23.17N	28.25N	13.08N	23.95N	24.7N	13.77N	6.93N
Longitude	106.38E	102.98E	89.22E	74.95E	80.27E	67.05E	46.80E	100.50E	79.86E
Solar Radiation (kWh/m^2/day)	5.24	5.25	5.4	5.98	5.51	6.14	6.56	5.25	5.36

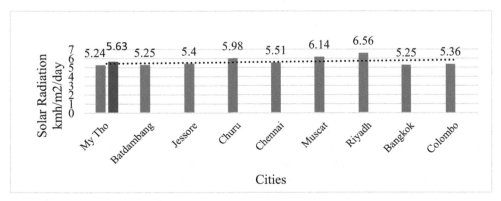

Fig. 34.1 Solar Radiation Comparison Chart of Cities (Yearly)

From Table 34.2, it is clear that despite different climate conditions, by taking solar radiation, solar interconnection has been possible. As per Table 34.2, it is seen that in March and

April, Solar Radiation is increased, then again from May to August, it is decreasing, again in September, it is increased little and it is in the lowest on the November and December.

Table 34.2 Solar radiation data for chosen cities on a monthly basis (Data Source: NREL)

Location/ Parameter	My Tho	Batta-bang	Jessore	Churu	Chennai	Muscat	Riyadh	Bangkok	Colombo	Aver-age
January	5.72	6.08	5.34	4.94	6.82	5.77	5.52	5.64	6.47	5.68
February	6.74	6.51	6.08	5.86	6.86	6.38	6.04	6.26	6.38	5.61
March	6.84	6.08	6.59	6.6	7.26	6.91	6.35	6.41	6.39	6.56
April	6.03	5.93	6.47	6.97	6.61	7.16	6.5	5.79	5.45	6.36
May	4.89	5.03	5.92	6.73	5.67	6.77	7.05	5.31	4.54	5.84
June	4.3	4.73	5.03	6.27	5.04	6.14	7.48	4.5	4.56	5.56
July	4.54	4.48	4.36	5.6	4.6	5.22	7.39	4.6	4.93	5.38
August	4.5	4.72	4.5	6.16	4.57	5.41	7.39	4.68	4.98	5.54
September	4.72	4.74	4.85	6.31	5.07	6.11	7.38	4.77	5.06	5.76
October	5.28	5.19	5.57	6.11	4.89	6.4	7.07	4.99	5.08	5.74
November	5.07	3.85	5.32	5.25	3.73	5.9	5.92	4.78	4.74	5.01
December	4.27	5.63	4.82	4.93	4.99	5.52	4.68	5.27	5.78	5.14

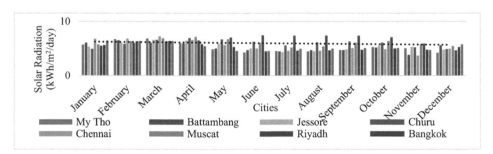

Fig. 34.2 Solar radiation comparison chart of cities (monthly)

Again in January and February, it is increased little. But all that data is not abnormally increasing or decreasing. If 5% range is taken, then it would be 5.35 kWh/m^2/day to 5.92 kWh/m^2/day, and for 10% radiation, it would be from 5.10 kWh/m^2/day to 6.20 kWh/m^2/day. Table 34.2 shows that the monthly solar radiation falls within that range (both 5% and 10%). The monthly solar radiation bar chart with cities in Fig. 34.2 shows that, like the yearly bar chart in Fig. 34.1, the monthly bar chart is in range, and the trend line in Fig. 34.2 is almost linear. Cities in this case study were chosen based on the following criteria: land connected to the same continent, solar radiation matching, different continent of the same hemisphere matching, and if that matched as well, area of different hemisphere matching. My Tho, Battambang, Jessore, Churu, Chennai, Muscat, Riyadh, and Bangkok are among the first Asian cities chosen for this case study. Solar radiation in such places is in the range

($\pm$5%) (range 5.35kWh/m^2/day to 5.92 kWh/m^2/day to Asian countries throughout the year as the yearly average is 5.63 kWh/m^2/day), according to data collected on a yearly and monthly basis. The linear graphs in Figs 34.3 and 34.4 show that "OSOWOG" is a distinct possibility. From the above figures(Fig 34.1 and Fig 34.2), it can be concluded that as the trendline of yearly and monthly solar radiation of selected nine cities is nearer to the trendline, so practical implementation of solar interconnection on Asian basis over geographical diversity is feasible.

4. Challenges and Future Scope

The most difficult part of putting this "OSOWOG" concept into action is gathering solar radiation and choosing a location, preferably lonely terrain that will not harm the environment or humans. Grid and transmission technologies will be exceptionally high voltage since the "OSOWOG" initiative will connect the entire world via a single grid. As a result, drive costs will be immense, and every precaution should be taken to guarantee that this one-of-a-kind OSOWOG effort does not harm humanity. The entire planet will be connected by a single grid as part of the "OSOWOG" initiative, putting the entire world at risk of blackouts if any system in any area fails. This difficulty, however, can be easily solved provided adequate safeguards are followed. To reduce cost and increase the efficiency of this interconnection initiative need to upgrade the design of electrical drives. In later future wireless transmission also can be used.

5. Conclusion

Despite the geographical diversity of the Asian continent, our Sun, according to the case study, serves as a link between cities and countries. Solar radiation in equatorial and tropical locales is almost comparable or within a 5-percentage-point range, according to the example study. It is conceivable to connect the sun all the way around the continent using that solar radiation range. If the process is carried out according to the proposed technique above, any geographical constraints can be handled by utilising the solar radiation of such places, as shown in the case study. The adoption of appropriate methodology and sun radiation enabled the creation of 'OSOWOG.'

REFERENCES

1. Khurpade, J.Mante, V.K Gurap, R.P Chopde, A.P Chandsare, and O.R. Irole. 2021. *"Smart Grid System and Efficient Location Finder for Renewable Power Plant based on One Sun One World One Grid."* Asian Journal for Convergence in Technology (AJCT) ISSN-2350-1146 7, no. 1: 134–136.
2. R. Filipe, C.Cardeira, and J.M.F Calado. 2014. *"The daily and hourly energy consumption and load forecasting using artificial neural network method: a case study using a set of 93 households in Portugal."* Energy Procedia 62: 220–229.
3. Pan, Cheng, and Tan J. 2019. *"Day-ahead hourly forecasting of solar generation based on cluster analysis and ensemble model."* IEEE Access 7: 112921–112930.

4. Khan, Ghazanfar, and Rathi S. 2014. *"Optimal site selection for solar PV power plant in an Indian state using geographical information system (GIS)."* International Journal of Emerging Engineering Research and Technology 2, no. 7 (October): 260–266.

5. Zhang, Ge, Shi Y, Maleki A, and Marc A.R. 2020. *"Optimal location and size of a grid-independent solar/hydrogen system for rural areas using an efficient heuristic approach."* Renewable Energy 156: 1203–1214.

6. Ogbonnaya, Chukwuma, Ali T, and C Abeykoon. 2020 *"Novel thermodynamic efficiency indices for choosing an optimal location for large-scale photovoltaic power generation."* Journal of Cleaner Production 249: 119405.

7. Belmahdi, Brahim, Md L, and A. El Bouardi. 2020. *"One month-ahead forecasting of mean daily global solar radiation using time series models."* Optik 219: 165207.

8. Atique, Sharif, Subrina N, V. Ray, V.Suburaj, S.Bayn, and Joshua M. 2019. *"Forecasting of total daily solar energy generation using ARIMA: A case study."* In 2019 IEEE 9th annual computing and communication workshop and conference (CCWC), pp. 0114–0119. IEEE.

9. Sharma, Amandeep, and Ajay K. 2018. *"Forecasting daily global solar irradiance generation using machine learning."* Renewable and Sustainable Energy Reviews 82: 2254–2269.

10. Noton G, Cyril V, Alexis F, J.L Duchaud, and M.L.Nivet. 2019. *"Some applications of ANN to solar radiation estimation and forecasting for energy applications."* Applied Sciences 9, no. 1 (January): 209.

11. S.Gholamabas, A.L. Pisello, H.Safarzadeh, Miad P, and Mohammad Jowzi. 2020. *"On the effect of storage tank type on the performance of evacuated tube solar collectors: Solar radiation prediction analysis and case study."* Energy 198 (May): 117331.

12. Zhang, Chunxiao, C. Shen, Q. Yang, S. Wei, G.Lv, and C.Sun. *"An investigation on the attenuation effect of air pollution on regional solar radiation."* Renewable Energy 161 (2020): 570–578.

13. FChen, Jintian, YiwenM, and Z.Pang. *"A mathematical model of global solar radiation to select the optimal shape and orientation of the greenhouses in southern China."* Solar Energy 205 (2020): 380–389.

14. NSRBD: National Solar Radiation Database (https://pvwatts.nrel.gov/pvwatts.php)

15. NREL: National Renewable Energy Laboratory (https://www.nrel.gov)

Intelligent Systems and Smart Infrastructure – Brijesh Mishra et al. (eds)
© 2023 Taylor & Francis Group, London, ISBN 978-1-032-41287-0

CHAPTER

35

EMBFL: Ensemble of Mutation-based Techniques for Effective Fault Localization

Jitendra Gora[1], Durga Prasad Mohapatra[2]

Department of Computer Science and Engineering,
NIT Rourkela, Rourkela, India

Arpita Dutta[3]

School of Computing,
National University of Singapore,
Singapore

Abstract

Finding locations of faults in a program is a crucial activity in reliable and effective software development. A large number of fault localization techniques exist; however, none of these techniques outperforms all other techniques in all circumstances for all kinds of faults. Under different circumstances, different fault localization techniques yield different results. In this study, we have proposed Ensemble of Mutation-Based techniques for effective Fault Localization (EMBFL). EMBFL classifies statements of a program into Suspicious and Non-Suspicious sets. The model we have used in our research is straightforward and intuitive because it is based solely on information regarding statement coverage and test case execution results. This helps to reduce the search space significantly. Our proposed EMBFL approach, on average, is 31.34% more effective than the techniques for fault localization that currently exist such as DStar (D^*), Tarantula, Back-Propagation Neural Network, etc.

Keywords: debugging, ensemble classifier, fault localization, mutation analysis

Corresponding author: [1]jitendragora1305@gmail.com, [2]durga@nitrkl.ac.in, [3]arpitad10j@gmail.com

DOI: 10.1201/9781003357346-35

1. Introduction

With the continuously growing usage of software in our daily lives, it has become critical to systems in several industries such as healthcare, teaching, marketing, etc. This has resulted in a substantial scale in the complexity and size of software. With so much of complexity and size, software faults are inevitable and these faults often lead to execution failure of the software. Therefore, testing and debugging the software has become highly crucial part of software development process. Fault localization is a vital step of software testing. It is the activity of finding out the faulty locations in a software and has been an expensive task in terms of manual effort, time, and money, considering the size and complexity of the software. To overcome these limitations, researchers are trying to develop techniques that partially or fully automate this task and assist developers in the debugging. Many fault localization techniques are being used currently but no technique outperforms all available techniques in all circumstances. For instance, few techniques may perform really well for faults that are related to relational and logical operators, whereas few other techniques may perform well for faults that are related to arithmetic operators.

The objective of this study is to develop effective and efficient techniques for fault localization. Our technique is inspired by the two famous domains of fault localization, i.e., Mutation-Based Fault Localization (MBFL) and Spectrum-Based Fault Localization (SBFL). We named this approach as EMBFL since we combined multiple **M**utation-**B**ased **F**ault **L**ocalization techniques using an **E**nsemble classifier.

The rest of the paper is structured in the following way. In the following section, few of the related works are discussed. Our proposed methodology is described in depth in Section 3. The last two sections summarize the experimental studies and conclusion of our work done, respectively.

2. Literature Review

In this section, we have attempted to summarize few of the existing works of authors related to fault localization. Weiser (1984) proposed a technique called Program Slicing technique that reduces the search space for inspection. There are some drawbacks of slicing techniques. For instance, slicing techniques are not suitable for assigning ranks to the statements. SBFL techniques, as mentioned in the works of Jones et al. (2005), Renieres et al. (2003), and other researchers, overcame such limitations by taking different coverage information like statement coverage and branch coverage, along with the test case execution results and producing suspiciousness score for each of the program entities. Then, they used ranking metrics to assign ranks to all of the executable statements in the program. Later, Wong et al. (2013) improved the results and came up with an effective technique named DStar (D^*), which is currently the state-of-the-art.

Problem with SBFL techniques is that they assign the same rank to multiple statements and that increases the inspection domain in order to find the location of faults. To overcome this limitation, MBFL techniques such as Metallaxis and MUSE were introduced by Papadakis et al. (2015) and Moon et al. (2014), respectively. MBFL techniques focus on generating useful mutants to make fault localization more effective. Mutants are copies of the original program under test with the syntax being changed at least once based on mutation operators. Also, MBFL techniques are effective enough to deal with the problem of coincidental correctness (Li et al., 2019). Although these MBFL techniques take impact information into consideration, there may be some cases where they perform poorly, e.g., some entities may not have any mutants to simulate their impacts. To eliminate such limitations, researchers currently use machine learning and deep learning-based techniques (Ascari et al., 2009; Wong et al., 2011).

The novelty of our study lies in that machine learning-based ensemble classifier has been used in our approach to combine SBFL and MBFL techniques. As per research, ensemble classifiers perform much better than the constituent learning techniques alone (Roychowdhury et al., 2012; Dutta et al., 2021). Besides, the existing techniques are not effective enough for large sized programs. Our proposed EMBFL approach for fault localization addresses this problem as well.

3. Proposed Approach: EMBFL

A. Overview

In our approach, the original program is given as input and mutants are obtained for the given program. Then, code coverage information is generated using these mutants along with the test cases. After getting code coverage, we give this information with the test case execution results to statement score generator which uses 40 different MBFL techniques to calculate the sequence of suspicious scores for all executable statements of the program. After getting sequence of score, we perform the normalization to bring the scores in a fixed range. Later, we apply ranking algorithms on all the forty MBFL techniques used, which results in prioritised list of statements that can contain the faults in the program.

B. Detailed Description of EMBFL

In this subsection, we have presented a detailed discussion on Mutator, Program Spectra Generator, Statement Score Generator, and Learning to Rank algorithm. Figure 35.1 depicts the high-level architecture of our proposed methodology.

Mutator: Since we are provided with the original program only, we need to generate mutants for that program in order to get the information about code coverage and results of test case execution. Mutator is a module which we have developed to generate different mutants for the original program. The original program is given as input to the mutator and mutants are obtained as output. Depending upon the available substitutes for a statement, the mutants are

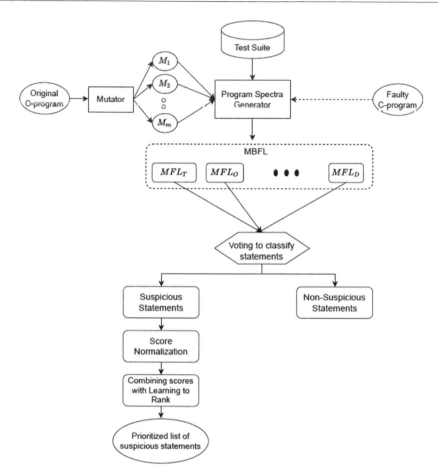

Fig. 35.1 High-level architecture of the methodology proposed in this study

categorized as follows: Category-1 and Category-2. Category-1 contains the mutants which have finite number of possible replacements. These mutants can be generated by replacing an operator by one of its substitutes. Few of the operator-substitutes from Category-1 are given below.

- AOR: Arithmetic Operator Replacement (e.g., multiplication ('*') in place of division ('/'))
- (I/D)OR: Increment/ Decrement Operator Replacement (e.g., decrement ('--') in place of increment ('++'))
- AsOR: Assignment Operator Replacement (e.g., plus equals to ('+=') in place of minus equals to ('-='))

- ROR: Relational Operator Replacement (e.g., less than ('<') in place of greater than ('>'))
- LOR: Logical Operator Replacement (e.g., OR ('||') in place of AND ('&&'))

Category-2 contains the mutants which have numerous numbers of possible replacements. Few of the mutation operations under Category-2 are given below.

- SI: Statement Insertion
- SD: Statement Deletion
- VR: Variable Replacement (e.g., 'x=y+d' in place of 'x=y+z')
- SIE: Swapping of 'if' block with 'else' block statements
- SI: Sign Inversion (e.g., 'x=−y' in place of 'x=y')

Program Spectra Generator: The next module that we developed is the Program Spectra Generator. It produces information regarding code coverage and results of test case execution by taking original program, its mutant, and all accessible test cases as input. If a test case covers a statement, then we use '1' to represent the statement, else we use '0' to represent it. The result of a test case execution indicates whether the particular test case has failed (F) or passed (P). Table 35.1 depicts a sample program spectra along with the results of test case execution. Suppose that the sample program contains eight executable statements (ST1 to ST8) and our test suite contains five test cases (TC1 to TC5). Table 35.1 highlights that three of the five test cases (TC1, TC2, and TC5) passed while the other two failed. Let's pick a test case, say TC3, it can be observed that only the statements ST1 and ST8 are executed by TC3 and the remaining statements are not executed by it.

Table 35.1 Sample program spectra with test case execution results

S. No.	Test Cases	ST1	ST2	ST3	ST4	ST5	ST6	ST7	ST8	Result
1	TC1	1	1	1	1	1	1	1	0	P
2	TC2	0	0	1	0	0	0	1	1	P
3	TC3	1	0	0	0	0	0	0	1	F
4	TC4	1	0	1	1	1	1	1	0	F
5	TC5	1	0	1	0	0	0	0	1	P

Statement Score Generator: Statement Score Generator (SSG) is a module which we have developed to obtain statement score sequences. It takes program spectra and the results of test case execution as input and produces sequence of statement scores depending upon multiple MBFL techniques as output. Since we have used forty MBFL techniques in our approach, SSG produces forty score sequences. Table 35.2 enlists the formulas of few of the SBFL techniques which we have used in our approach. N is the total number of test cases available; N_p is the number of passed test cases among those available and N_f is the number of failed test cases among those available. The number of succeeded test cases that executed a statement is N_{ep}, whereas the number of failed test cases that executed a statement is N_{ef}. Similarly, N_{np} represents the number of succeeded test cases that did not execute a specific statement, while N_{nf} represents the number of failed test cases that did not execute a specific statement.

Table 35.2 Definitions of some SBFL techniques

S. No.	Name of Technique	Formula
1	Tarantula	$(N_{ef}/(N_{ef}+N_{nf})) / (N_{ef}/(N_{ef}+N_{nf}) + N_{ep}/(N_{ep}+N_{np}))$
2	Ample1	$\mid (N_{ef}/N_{ef}*N_{ep}) - (N_{ep}/N_{ep}*N_{np}) \mid$
3	Ample2	$N_{ef}/N_f - N_{ep}/N_p$
4	Lee	$N_{ef} + N_{np}$
5	Goodman	$(2*N_{ef}-N_{nf}-N_{ep})/2*N_{ef}+N_{nf}+N_{ep}$
6	Wong1	N_{ef}
7	Wong2	$N_{ef}-N_{ep}$
8	DStar (D*)	$(N_{ef})^*/(N_{ep}*N_{nf})$
9	Jaccard	$N_{ef}/(N_f+N_{ep})$
10	Euclid	$\sqrt{(N_{ef} + N_{np})}$

Learning to Rank: This algorithm considers a collection of objects and assigns a suitable rank to them. We are using forty MBFL techniques in our approach and each technique assigns a different score to all the statements of the input program for each mutant. Therefore, we have used Learning to Rank algorithm to all of the techniques and this results in assigning higher rank to the right technique in every different scenario. After applying this Learning to Rank algorithm, we get a prioritised list of statements.

4. Experimental Studies

This section presents the data set, evaluation metrics used, and the empirical results obtained in our approach.

A. Data Set Used

We have used Siemens suite as our input data set, which contains 7 different programs named "printtokens", "printtoken2", "replace", "schedule", "schedule2", "tcas", and "totinfo". Table 35.3 shows the characteristics of seven Siemens suite programs. We have generated different category-1 mutants for each program in the Siemens suite.

B. Evaluation Metrics

For evaluating the performance of our approach EMBFL and comparing it with that of the existing fault localization techniques, Exam-Score evaluation metric has been used. Exam-Score is the percentage of statements that must be examined in order to locate faults in a program. It is computed using Equation (1).

$$\text{Exam} - \text{score} = (\mid S_{examined} \mid \div \mid S_{total} \mid) * 100 \qquad (1)$$

where $\mid S_{examined} \mid$ denotes the count of statements required to be examined to find the faulty locations, and $\mid S_{total} \mid$ denotes the total number of statements in the program. The fault localization technique that has Exam-Score less than that of all others is the most effective technique.

Table 35.3 Characteristics of programs available in Siemen suite

S. No.	Name of Program	Count of Faulty Versions	Count of Executable LOC	Count of Test Cases	Count of Mutants
1	Print_Tokens	7	195	4130	285
2	Print_Tokens2	10	200	4115	314
3	Replace	32	244	5542	508
4	Tcas	41	65	1608	216
5	Tot_info	23	122	1052	571
6	Schedule	9	152	2650	406
7	Schedule2	10	128	2710	350

C. Empirical Evaluation

We have compared the effectiveness of EMBFL with four other prominent existing fault localization techniques. Among these four techniques, Tarantula and DStar (D*) are taken from SBFL family. On the other hand, BPNN and RBFNN belong to the neural network family of fault localization techniques. Figures 35.2 to 35.5 depict the effectiveness comparison of EMBFL with other techniques for fault localization using the line graphs. Best effectiveness means the statement containing fault is examined "first" among those with the same score of suspiciousness. Worst effectiveness, on the other hand, means that the statement containing fault gets examined "last" among those with the same score of suspiciousness.

Figure 35.2 compares the effectiveness of Tarantula and EMBFL for Siemens suite. The figure suggests that by inspecting only 3% of program statements, EMBFL localizes faults in 25.77% of faulty versions. Tarantula (Worst) and Tarantula (Best), respectively, localize faults in 10.31% and 20.61% of faulty versions only. On average, Tarantula (Worst) and Tarantula (Best) require 31.71% and 19.98% of program statement examination to localize bugs. On the other hand, EMBFL requires only 16.08% of code examination, on average. Our proposed EMBFL method is 49.29% and 19.54% more effective than Tarantula (Worst) and Tarantula (Best), respectively.

Figure 35.3 presents the effectiveness comparison of DStar and EMBFL for Siemens suite. It is clearly evident from the figure that EMBFL is performing equally well as DStar (Best) for most of the program points. However, there are several points present on which EMBFL is performing much better than DStar (Best). We observe that there is a huge gap between the effectiveness of DStar (Worst) and EMBFL. EMBFL requires 20% and 7.5% less code inspection than DStar (Worst) and DStar (Best), respectively, in the worst scenario. On average, EMBFL outperforms DStar (Worst) and DStar (Best) by 43.53% and 4.38%, respectively.

Figure 35.4 compares the results of BPNN and EMBFL for Siemens suite. According to the graph, EMBFL only requires 8% code inspection to locate defects in 50% of the faulty versions, whereas BPNN requires at least 14.47% code inspection. In the worst-case scenario, EMBFL requires 3% less code inspections than BPNN, and it is 21.15% more effective on average.

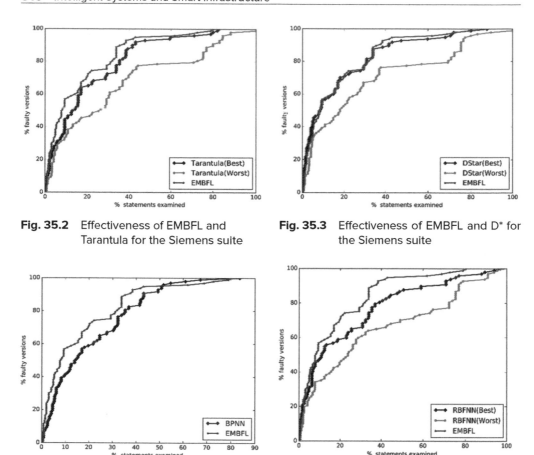

Fig. 35.2 Effectiveness of EMBFL and Tarantula for the Siemens suite

Fig. 35.3 Effectiveness of EMBFL and D* for the Siemens suite

Fig. 35.4 Effectiveness of EMBFL and the Siemens suite

Fig. 35.5 Effectiveness of EMBFL and BPNN for and RBFNN for the Siemens suite

Figure 35.5 compares the effectiveness of RBFNN and EMBFL for Siemens suite. The figure indicates that EMBFL locates errors in 38.14% of faulty programs by evaluating only 5% of statements in the program, but RBFNN (Worst) and RBFNN (Best) locate faults in only 24.74% and 30% of faulty versions, respectively. On average, EMBFL outperforms RBFNN (Worst) and RBFNN (Best) by 52.01% and 29.49%, respectively. EMBFL also has to examine 18.14% and 16.60% less code in the worst scenario than RBFNN (Worst) and RBFNN (Best).

Table 35.4 presents the pairwise comparison of EMBFL with Tarantula, DStar, RBFNN, and BPNN. The table illustrates the percentage of faulty versions on which EMBFL performs more effectively, equally effectively, and less effectively than the respective fault localization techniques in rows 2, 3, and 4, respectively. It can be observed that EMBFL performs at least as effective than all the other techniques in more than 55% of faulty versions. There are substantially a smaller number of faulty versions present on which EMBFL is lesser effective than the existing methods of fault localization.

Table 35.4 Pairwise comparison between EMBFL and existing fault localization techniques with respect to Exam-Score

	EMBFL vs. Tarantula (Best)	EMBFL vs. Tarantula (Worst)	EMBFL vs. DStar (Best)	EMBFL vs. DStar (Worst)	EMBFL vs. RBFNN (Best)	EMBFL vs. RBFNN (Worst)	EMBFL vs. BPNN
More Effective	49.48	81.44	18.56	68.04	43.3	69.07	58.76
Equally Effective	32.99	5.15	39.18	10.31	8.25	7.22	6.19
Less Effective	17.53	13.4	42.27	21.65	48.45	23.71	35.05

5. Conclusion and Future Scope

Fault localization is a crucial part of developing reliable and effective software. To make fault localization easier and more effective, in this study we have proposed an ensemble classifier-based technique. We have combined different mutation-based fault localization techniques. Our method is able to effectively identify locations of common as well as intrinsic faults present in the program. From our empirical evaluation, we have observed that, on average, EMBFL performs 31.34% more effectively in terms of less code examination than the related fault localization techniques such as Tarantula, DStar, BPNN, and RBFNN.

As part of future scope of our work, we intend to apply EMBFL on multiple fault programs and also attempt to improve its performance. We also intend to further improve our approach such that it makes use of the individual fault-exposing capabilities of a test case.

References

1. Weiser, M. (1984). Program slicing. IEEE Transactions on software engineering. 4: 352–357.
2. Jones, James, A., and Harrold, M. J. (2005). Empirical evaluation of the tarantula automatic fault-localization technique, In Proceedings of the 20th IEEE/ACM International Conference on Automated software engineering (pp. 273–282).
3. Renieres, M. and Reiss, S. P. (2003). Fault localization with nearest neighbor queries, In 18th IEEE International Conference on Automated Software Engineering, 2003. Proceedings. (pp. 30–39). IEEE.
4. Wong, W. E., Debroy, V., Gao, R. and Li, Y. (2013). The DStar method for effective software fault localization. IEEE Transactions on Reliability. 63(1): 290–308.
5. Papadakis, M. and Traon, Y. L. (2015). Metallaxis-FL: mutation-based fault localization. Software Testing, Verification and Reliability. 25(5–7): 605–628.
6. Moon, S., Kim, Y., Kim, M. and Yoo, S. (2014). Ask the mutants: Mutating faulty programs for fault localization, In 2014 IEEE Seventh International Conference on Software Testing, Verification and Validation (pp. 153–162). IEEE.
7. Ascari, L. C., Araki, L. Y., Pozo, A. R. and Vergilio, S. R. (2009). Exploring machine learning techniques for fault localization, In 2009 10th Latin American Test Workshop (pp. 1–6). IEEE.

8. Roychowdhury, S. (2012). Ensemble of feature selectors for software fault localization, In 2012 IEEE International Conference on Systems, Man, and Cybernetics (SMC) (pp. 1351–1356). IEEE.

9. Wong, W. E., Debroy, V., Golden, R., Xu, X. and Thuraisingham, B. (2011). Effective software fault localization using an RBF neural network. IEEE Transactions on Reliability. 61(1): 149–169.

10. Li, X., Li, W., Zhang, Y. and Zhang, L. (2019). Deepfl: Integrating multiple fault diagnosis dimensions for deep fault localization, In Proceedings of the 28th ACM SIGSOFT International Symposium on Software Testing and Analysis (pp. 169–180).

11. Dutta, A., Srivastava, S. S., Godboley, S. and Mohapatra, D. P. (2021). Combi-FL: Neural network and SBFL based fault localization using mutation analysis. Journal of Computer Languages. 66: 101064.

Intelligent Systems and Smart Infrastructure – Brijesh Mishra et al. (eds)
© 2023 Taylor & Francis Group, London, ISBN 978-1-032-41287-0

CHAPTER

36

Facial Emotion Recognition Using Deep Learning: A Case Study

Ratna Patil*

AIML Department,
Noida Institute of Engineering & Technology, Greater Noida, India

Ananya Tripathi, Yashasewi Singh, Gaurav Prajapati

Computer Science Department,
Noida Institute of Engineering & Technology, Greater Noida, India

Abstract

Humans feel and express a lot of different emotions. They can not only experience these emotions but also tell other person's emotions just by looking at their face, making it easier for humans to communicate. The concept of facial emotion recognition (FER) using neural networks is used to simplify the interaction between humans and machines. Despite how effortlessly we do it in our daily lives, the same does not go for machines. For a machine to understand these emotions, a network must be trained with various datasets containing diverse images so it can detect the emotion with the highest accuracy. This is done in multiple steps in which we first remove the background and isolate the target face from the image. Then, we study various features of the face to detect the emotion being expressed. In this project, we will be using the Keras library to train our network and classify the image into one of the seven emotions. A lot of work is already done and is also going on in this field which helped us through this project. This paper discusses FER, how it works, and what are the various tools available for you to work with FER. Apart from this, it also focuses on the prospects of this technology so that machines can understand human emotions and respond accordingly just as humans do.

Keywords: FER (Facial Emotion Recognition), Convolution Neural Network, Deep Learning

*Corresponding Author: ratna.nitin.patil@gmail.com

DOI: 10.1201/9781003357346-36

1. Introduction

Human face expressions help interpret the person's state of mind much better than words and hence it has always been an interesting topic for scientists. The first documented research dates back to 1862 when a scientist named Duchenne wanted to study how facial expressions were produced by facial muscles. Another scientist named Charles Darwin studied facial and body gestures in animals. Darwin's work attracted a bigger crowd towards studying expressions. However, Paul Ekman got the first breakthrough in the field of facial emotion recognition (FER) by classifying the six basic human facial expressions which are Happiness, Fear, Disgust, Anger, Surprise, and Sadness. [1] We have trained our network to classify the image as one of the seven emotions in CK+ (six basic expressions and seventh neutral) and six emotions in FER.

Starting with traditional methods, they went on to train computers to help them study facial expressions. Many algorithms were suggested and rejected. Some famous algorithms are LBP (Local Binary Pattern), Gabor Filtering, and PCA (Principal Component Analysis) [2] [3]. After decades of work, they developed training module sets like DNN (Deep Neural Network), CNN (Convolution Neural Network), and DCNN (Deep Convolution Neural Network) that use pattern recognition to identify expressions. Galande et al. have discussed and compared various approaches to take out needless information from the images [4].

2. Literature Review

Table 36.1 depicts the comparison of previous work done by other authors.

Table 36.1   Comparison of previous works

S.No	Journal	Methodology	Result
1	Sharmeen M. Saleem Abdullah et al. (2021) [5]	MMAN (Fusion method), Speech-BERT, RoBERT Shallow fusion, Bandpass Filters (DCCN with a SincNet layer, RNN), Cross-modal attention, Fusion method, Natural language processing (NLP), Bert model LSFM	LSTM - 73.98%, RNN - 80.51%, CNN - 82.41%, SVM - 72.52%, LSTM (Bert model) - 5.77% improvement
2	Saiyed Umer et al. (2021) [6]	The techniques under statistical-based approaches are Local Binary Pattern, Histogram of Oriented Gradient, Scale Invariant Feature Transform, Bag of Words, Sparse Representation, and Coordinate descend methods	F64X64: KDEF - 75.89%; GENKI - 84.78%; CK+ - 91.87%, F128X128: KDEF - 82.79%; GENKI - 94.33%; CK+ - 97.69%
3	D. Lakshmi et al. (2021) [7]	The face is detected using Viola James. Then, the modified HOG and LBP features are extracted from the eye, nose, and mouth region. The concatenated features are fed to DSAE for dimensionality reduction. The reduced low-dimensional feature vector is passed to multi-class SVM classification	CK+: 97.66

S.No	Journal	Methodology	Result
4	Shervin Minaee et al. (2021) [8]	The network weights were randomly initialized with Gaussian variables with zero mean and 0.05 standard deviation. Adam optimizer with a learning rate of 0.005 was used for optimization and L2 regularization with a weight decay value of 0.001 was added	FER: 70.02% FERG: 99.3 JAFFE: 92.8% CK+: 98.0%
5	M. A. H. Akhand et al. (2021) [9]	Pre-trained DCNN models and TL techniques are the basis of this study. Several pre-trained DCNN models are investigated to identify the best-suited one for FER	Input Image - 360 × 360; KDEF - 73.87%; JAFFE - 91.67% Input Image - 128 × 128; KDEF - 80.81%; JAFFE - 91.67% Input Image - 64 × 64 KDEF - 69.39%; JAFFE- 83.33%

3. Data and Variables

In this study, we have used the TensorFlow Keras library to train our network with a part of our dataset and then we tested the trained network on the remaining images from the same dataset.

The first step in building a facial recognition model is to choose a suitable and diverse dataset. A dataset is a collection of images of people showing various expressions. It is better to use a dataset for training a model rather than live images as unlike datasets live images might not be able to provide the required clarity of expressions. Some of the famous datasets are mentioned in Table 36.2.

Table 36.2 List of available datasets

Dataset	No. of images	Image type	Emotion
FER 2013 [10]	32398	48 × 48 gray	6
JAFFE [11] [12]	213	256 × 256 gray	7
CK+ [13]	5876	640 × 480 gray	7
AffectNet [14]	0.4 million	Grayscale and RGB	7
SFEW [15]	700	260 × 248 RGB	7

In this study, we have worked on CK+ and FER 2013 datasets.

A. CK+ Dataset

It is the most used dataset to train FER models. It contains 593 video clips of 123 subjects with their age varying from 18 to 50 years. It covers people from different gender heritage and origins. Every clip depicts a shift in the expression from the nonpartisan appearance to a designated top demeanor; it is recorded at 30 FPS with the goal of either 640 × 490 or 640 × 480 pixels. Among these clips, 327 are named with one of the seven emotion classes: contempt, sadness, fear, anger, surprise, happiness, and disgust. This data set is considered the most widely utilized research facility-controlled look arrangement data set accessible and is utilized in most expression recognition techniques. The arrangement where we utilized this dataset had proactively removed still pictures from moving pictures in the video which we utilized for preparing and testing our model [13] [16]. An image from CK+ dataset is shown in Figure 36.1.

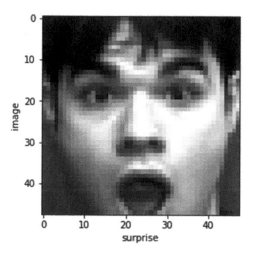

Fig. 36.1   Image from CK+ dataset

Figures 36.2 and 36.3 show the loss and accuracy graph obtained in training process of CK+ dataset for 10 and 60 epochs, respectively.

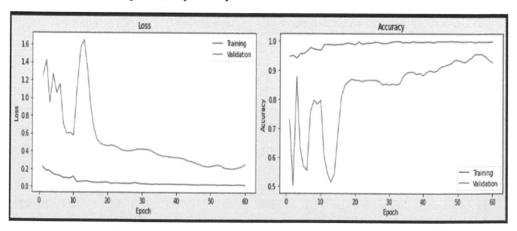

Fig. 36.2   Loss and accuracy graph for 10 epochs (CK+)

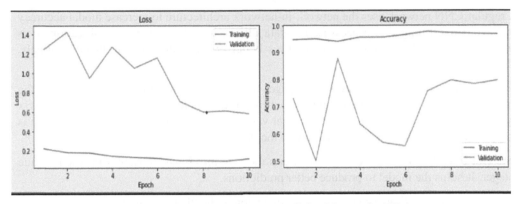

Fig. 36.3   Loss and accuracy graph for 60 epochs (CK+)

In the CK+ dataset, use of random subsets of the original database makes meta-analyses difficult. To address these and other concerns, we present the Extended Cohn–Kanade (CK+) database [13].

B. FER 2013 Dataset

This dataset consists of 48 × 48 pixel images. These are grayscale images of facial expressions [17] [18]. The images have been cropped focusing mainly on the emotion depicted by the subject. This dataset consists of 35,685 instances of 48 × 48 pixel grayscale pictures. Pictures are arranged given the inclination displayed in the looks (fear, neutral, anger, sadness, disgust, happiness, and neutral). An image from FER13 dataset is shown in Figure 36.4.

Fig. 36.4   Image from FER13 dataset

C. Methodology and Model Specification

Some of the famous methods to train a model for FER are CNN, KNN (K Neural Network), RNN (Recurrent Neural Network), LSTM (Long Short-Term Memory), PHRNN [19], and MSCNN [19].

A regular CNN network uses the network-in-network architecture to increase model accuracy and avoid overfitting. Pre-processing the data allows it to have better classification. It is observed that combining pre-processing gives better results than applying them separately. [20]

To train our model, we first convert our images to a form in which they can be passed in the training model. In the first step, we convert our image vectors to pixels and divide each matrix by 255 to get a compressed matrix with pixel values ranging from 0 to 1. After the image is compressed, we go for One Hot Encoding of the data.

One Hot Encoding is done because it converts the data into binary variables for each unique integer. It helps the model to produce better predictions.

After One Hot Encoding, we provide a 3rd dimension to the 2D image to pass it to the training model.

Figure 36.5 shows the pictorial representation of a deep neural network model.

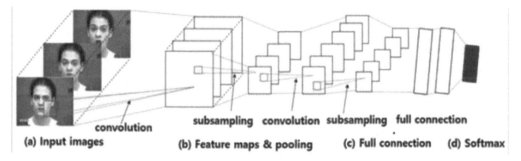

Fig. 36.5   Deep Neural Network Model

A training network model is nothing but multiple layers of CNN which work to identify the emotion. It has two mandatory layers: first, input layer which inputs the image whose emotion will be predicted and the last layer which flattens the results received from previous layers and gives the result. In our model, we have introduced two more layers between these two layers and have a total of 4 layers in our network.

In the introduced layers as well as the first layer, we included the following functions:

Base layer, Activation layer, Weight initializer, Weight regularizes, Attention layers, Locally connected layers, Weight constraints, Core layers, Convolution layers, Pooling layers, Recurrent layers, Reshaping layers, Merging layers, Pre-processing layers, Normalization layers, Regularization layers, and Activation layers. [21]

The following functions are used in this study:

(a) Activation Function: There are many activation functions available; you can use whichever best suits your datasets. We used ReLu with training layers and softmax with the output layer.

(b) Pooling Layer: It reduces the dimensions of all the variables used, thus reducing the amount of computation done in the model. Maxpooling2d is used in this project.

(c) Convolution Layer: It applies a filter to an input that activates the input. Convo2d is used in this project.

The use of these functions depends upon the dataset you are using. We have worked with various combinations and get one that best fits our dataset. Once the layer structure is ready, we can pass an image as many times as we want through the layers to get accurate results. Each time the image is passed into the layer is called Epoch. We have used 50 Epochs for our model.

The model summary in Figs 36.6 and 36.7 defines the results of the applied functions and layers in the network for CK+ and FER datasets, respectively.

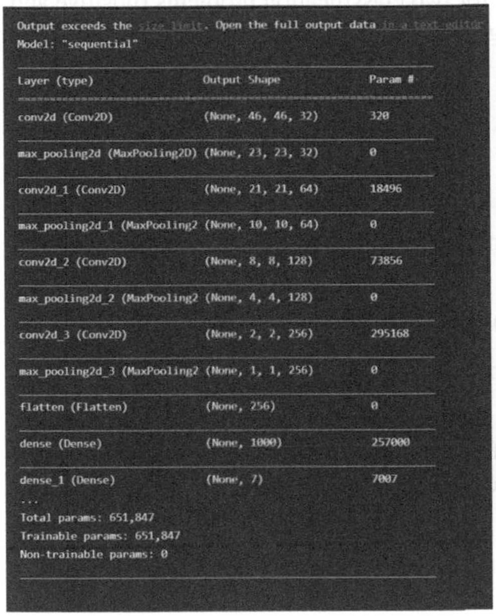

Fig. 36.6   Model summary of CK+

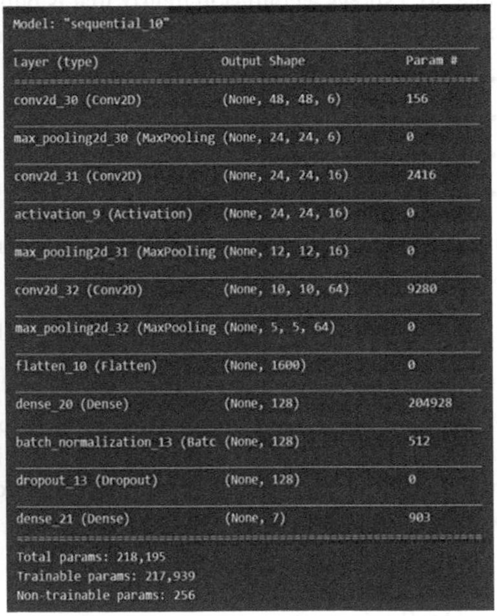

Fig. 36.7   Model summary of FER

Once we are done with the training process, we move forward to testing of data. An adequate portion of the dataset is already kept aside for testing purposes, so the model does overfit itself with training images. (Overfit: A situation where the model provides high accuracy but only with images it has worked with multiple times during training.)

After training and testing of data, we can print our results of training and testing accuracy. Our training model achieved 99 percent accuracy and the Testing set achieved an accuracy of 90 percent.

We can also plot the accuracy and loss rate of our model on a graph and print it.

D. Confusion Matrix

A confusion matrix related to a neural network model shows a matrix for the predicted and real classification. It is plotted of size n × n, where n is equal to the number of classes in which the network is classifying the subject.

Table 36.3   Confusion matrix for two classification problems

	Predictive – Negative	Predictive – Positive
Real Negative	p	q
Real Positive	r	s

Table 36.3 shows a confusion matrix that is built for two classification problems (positive and negative), hence the value of n = 2, whose passages have the following meanings:

- p: right negative predictions
- q: inaccurate positive predictions
- r: inaccurate negative predictions
- s: right positive predictions.

The Accuracy and Error of the network in determining the facial expression can be calculated using the following formulas:

$$Accuracy = (p + q)/(p + q + r + s)$$

$$Error = (q + r)/(p + q + r + s)$$

We can determine the conflict score related to a confusion network condition. As seen by applying these formulas, the disagreement is 1 if either q or r has a value equal to 0 (in such condition, the network misclassifies an instance of a single class) and is 0 if q and r are something very similar.

if q = r = 0:

$$D = 0$$

Otherwise:

$$D = |q\text{-}r|/\max\{q, r\}$$

The characteristic choice system proposed here selects attributes that have great segregation power on their own, yet more critically are reciprocal to one another. For instance, let there be two credits A and B, having comparable classification exactness. Our methodology shall consider them as a subpart of characteristics on the off-chance that they have a huge disarrangement as far as what models they misclassify. A large disagreement is indicated by D value closer to 1. [22]

Figure 36.8 displays the confusion matrix for the CNN model trained using the CK+ dataset.

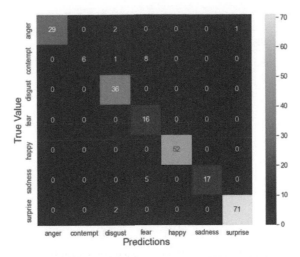

Fig. 36.8   Confusion matrix for CK+ dataset

5. Empirical Results

This model is trained and tested over CK+ and FER13 datasets.

(i) FER13

Figure 36.9 shows the test result from FER13 dataset.

Figure 36.10 shows the accuracy vs loss graph for FER13 dataset.

Test loss: 1.3659305572509766

Test accuracy: 0.5557258129119873

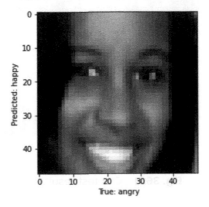

Fig. 36.9   FER13 test result

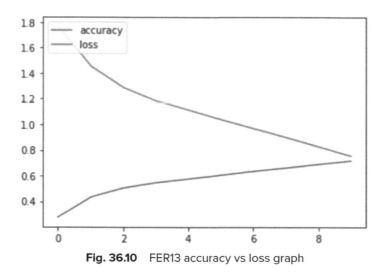

Fig. 36.10 FER13 accuracy vs loss graph

(ii) CK+:

Figure 36.11 shows the test result for CK+ dataset.

Figure 36.12 shows the accuracy vs loss graph for CK+ dataset.

Test Loss: 0.2534593641757965

Test accuracy: 0.9268292784690857

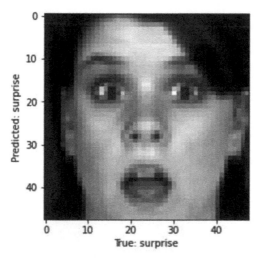

Fig. 36.11 CK+ test result

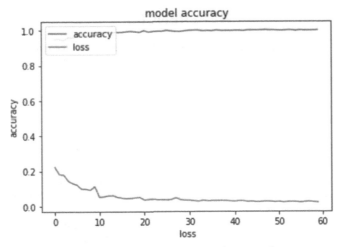

Fig. 36.12 CK+ accuracy vs loss graph

6. Conclusion

In this paper, we have worked on FER by using CNN architecture. We have tried to recognize various classes of facial expressions. In this study, we have used CNN for image classification and recognition and obtained high accuracy. Using CK+ (*Extended Cohn–Kanade dataset*) dataset, we have designed a FER model. The CNN worked in the following parts:

1. Removed the background from the picture to concentrate on the facial feature.
2. Converted the image into an array (Convolutional layer), convert negative values into zero (Activation function, Maxpooling in layers).

We have used the deep learning CNN algorithm along with Keras and TensorFlow Library on the CK+ dataset. By using these techniques and concepts, it was possible to identify the types of facial emotions of humans. The performance of our architecture reaches 99% of Training accuracy and 92% of Testing accuracy. The higher percentage indicates accurate emotion recognition, and the lower percentage of face emotions images show incorrect recognition. At last, to sum up our work, we have successfully used real-life images to create a FER model. Emotion recognition by humans is an easy task but identification of the emotion expressed by a human face is a complex and difficult task for computer architecture. In this study, we have predicted emotions expressed in images and it is still not enough for smooth interaction between humans and machines. At present, work is being done where machines can even identify real-time emotions with higher accuracy.

REFERENCES

1. W. H. Abdulsalam, R. S. Alhamdani and M. N. Abdullah, "Emotion Recognition System Based on Hybrid Techniques," *International Journal of Machine Learning and Computing,* vol. 9, 2019.

2. C. D. Wu and L. H. Chen, "FACIAL EMOTION RECOGNITION USING DEEP LEARNING," The University of Texas at Austin, 2019

3. R. N. Patil and S. C. Tamane, "A NOVEL SCHEME FOR PREDICTING TYPE 2 DIABETES IN WOMEN: USING KMEANS WITH PCA AS DIMENSIONALITY REDUCTION," *International Journal Of Computer Science And Engineering,* 2017.

4. A. Galande and R. Patil, "The art of medical image fusion: A survey,"," in *International Conference on Advances in Computing, Communications and Informatics (ICACCI),* 2013.

5. S. M. Abdullah, Y. A. Siddeeq , M. A. M. Sadeeq and S. Zeebaree, "Multimodal Emotion Recognition using Deep Learning," *Journal of applied science and technology trends,* vol. 02, no. 02, pp. 52-58, 2021.

6. S. Umer, K. R. Rout, C. Pero and M. Nappi, "Facial expression recognition with trade-offs between data augmentation and deep learning features," *Journal of Ambient Intelligence and Humanized Computing,* vol. 02, pp. 721-735, 2021.

7. D. Lakshmi and R. Ponnusamy, "Facial emotion recognition using modified HOG and LBP features with deep stacked autoencoders," *Microprocessors and Microsystems,* vol. 103834, p. 82, 2021.

8. A. Abdol, M. Minaei and S. Minaee, "Deep-Emotion: Facial Expression Recognition Using Attentional Convolutional Network," *Sensors,* vol. 21, no. 9, p. 3046, 2021.

9. M. Akhand, S. Roy, N. Siddique, M. Kamal and T. Shimamura, "Facial Emotion Recognition Using Transfer Learning in the Deep CNN," *Electronics,* vol. 10, p. 1036, 2021.

10. S. B. Musa, I. Karim, E. P. Wibowo, L. Zahara and P. Musa, "The Facial Emotion Recognition (FER-2013) Dataset for Prediction System of Micro-Expressions Face Using the Convolutional Neural Network (CNN) Algorithm based Raspberry Pi," in *Fifth International Conference On Informatics and Computing(ICIC),* 2020.

11. F. Cheng, H. Xiong and J. Yu, "Facial Expression Recognition in JAFFE Dataset Based on Gaussian Process Classification," *IEEE Transactions on Neural Networks,* vol. 21, no. 10, pp. 1685-1690, 2010.

12. K. Zhang, Y. Huang, Y. Du and L. Wang, "Facial Expression Recognition Based on Deep Evolutional Spatial-Temporal Networks," *IEEE Transactions on Image Processing,* vol. 26, no. 9, pp. 4193 - 4203, 2017.

13. P. Lucey, J. F. Cohn, T. Kanade, J. Saragih, Z. Ambadar and I. Matthews, "The Extended Cohn-Kanade Dataset (CK+): A complete dataset for action unit and emotion-specified expression," *IEEE Computer Society Conference on Computer Vision and Pattern Recognition,* pp. 94-101, 2010.

14. A. Mollahosseini, B. Hasani and M. Mahoor, "AffectNet: A Database for Facial Expression, Valence, and Arousal Computing in the Wild," *IEEE Transactions on Affective Computing,* vol. 10, pp. 18-31, 2017.

15. A. Dhall, R. Goecke, T. Gedeon and S. Lucey, "Static facial expression analysis in tough conditions: Data, evaluation protocol and benchmark," in *IEEE International Conference on Computer Vision Workshops,* 2011.

16. J. Luo, Z. Xie, F. Zhu and X. Zhu, "Facial Expression Recognition using Machine Learning models in FER2013," in *IEEE 3rd International Conference on Frontiers Technology of Information and Computer (ICFTIC),* 2021.

17. J. Luo, Z. Xie, F. Zhu and X. Zhu, ""Facial Expression Recognition using Machine Learning models in FER2013," in *IEEE 3rd International Conference on Frontiers Technology of Information and Computer (ICFTIC)*, 2021.

18. F. Cheng, Y. Jiangsheng and H. Xiong, "Facial Expression Recognition in JAFFE Dataset Based on Gaussian Process Classification," *IEEE Transactions on Neural Networks,* Vols. 1685-1690, p. 21, 2010.

19. P. Utami, R. Hartanto and I. Soesanti, "A Study on Facial Expression Recognition in Assessing Teaching Skills: Datasets and Methods," *Procedia Computer Science,* vol. 161, pp. 544-552, 2019.

20. W. Mellouk and W. Handouzi, "Facial emotion recognition using deep learning: review and insights," *Procedia Computer Science,* vol. 175, pp. 689-694, 2020.

21. C. B. Ko, "A Brief Review of Facial Emotion Recognition Based on Visual Information," *Sensors (Basel),* vol. 18, no. 2, p. 401, 2018.

22. S. Visa, A. Inoue and A. L. Ralescu, Proceedings of The 22nd Midwest Artificial Intelligence and Cognitive Science Conference 2011, Cincinnati, Ohio, USA, April 16-17, 2011, 2011.

Intelligent Systems and Smart Infrastructure – Brijesh Mishra et al. (eds)
© 2023 Taylor & Francis Group, London, ISBN 978-1-032-41287-0

CHAPTER

37

Mitigating Chemical and Biological Warfare Using Sensors: A Study

Pushpalatha S.[1]

Department of E&CE,
VTU Centre for PG Studies, Mysuru, India

K. S. Shivaprakasha[2]

Department of ECE,
N.M.A.M. Institute of Technology, Nitte, India

Abstract

Biological and chemical warfare is a threat to the human society. They are being used as weapons to destroy living organisms like humans, animals, and plant. The biological and chemical warfare are different from each other. The main aim of biological warfare is dilution and destruction of economic progress and to cause unstability in the nation. The chemical warfare uses the toxic properties of chemical substance as weapons, whereas pathogenic agents such as bacteria or viruses are used for biological warfare. The biological and chemical warfare are used in terrorist activities as well. In addition, novel and accessible technologies are being emerged for the proliferation of such weapons. Thus, the development of smart systems to facilitate the detection and mitigation of threats due to such biological or chemical warfare is the need of the day. In this paper, an attempt has been made to detail various biological and chemical warfare techniques and the systems that are developed to mitigate the effects of the same.

Keywords: Weapons of Mass Destruction (WMD), Chemical Weapons Agents (CWAs), Biological Weapons Agents (BWAs)

Corresponding author: [1]pushala.s@gmail.com, [2]shivaprakasha.ks@nitte.edu.in

DOI: 10.1201/9781003357346-37

1. Introduction

World War II has witnessed great tragedies to the mankind. At the end of World War II, many treaties were made to limit and regulate the use of Weapons of Mass Destruction (WMD). Chemical weapons have proved to be more dangerous with increased degree of destruction. The usage of bacteria and chemicals, toxins, viruses, or poisons to kill or harm militaries or citizens is named as chemical or biological warfare. Chemical Weapons Agents (CWAs) are influential deadly substances which are used as weaponries for a huge destruction. Some of the deadliest agents of chemical warfare of nerve agents are G-agents (GA, GB, GD), V-agents (VX), blister agent (SM), nitrogen mustard (NM), and Lewisite. Biological Weapons Agents (BWAs) are called as the germ warfare because various forms of agents are used in biological warfare. They use biological poisons or catching agents like bacteria, viruses, insects, and fungi to kill, damage, or injure humans, animals, or plants.

The increasing threat of CWA and BWA in the modern era has necessitated the development of smart systems using sensors to detect the possibility of such threats. In this paper, we have made an exhaustive survey on different types of biological and chemical weapons which are used in the war for destruction purpose. Also, the methods to detect the possibility of such threats and the antiagents to lessen the impact of the CWAs and BWAs are also presented.

The remainder of this paper is divided into five sections: Section II deals with the Chemical Warfare. Section III presents a review on Biological Warfare. Section IV gives a tabulation of the survey being carried out. Finally, Section V gives the concluding remarks of the paper.

2. Chemical Warfare

Chemical warfare is a process of using poisonous assets of chemical materials as weaponries. It is a strategic warfare using ingredients (such as combustible mixes, smokes, or fumes) with irritation, fiery, toxic, or suffocating assets. It has four major classes, namely, nerve, blister, chocking, and blood agents. Poisonous chemicals could be aerosolized or placed into water supplies, eventually polluting the entire region.

Sulphur mustard is a deadly agent used during the World Wars, but the terrible results of such attacks eventually led to an international agreement to ban toxic chemical weapons, the most widely used and easily multiplied weapon of mass destruction. The most critical effect of such agents is immobility of the respiratory muscles and inhibition of respiratory center.

With the advancements in the sensor technology, sensors were developed and can be used for rapid detection of the chemical or biological agents. This is an essential requirement to protect first responders so as to facilitate emergency medical personnel at local medical facilities for effective treatment of casualties.

3. Biological Warfare

Biological warfare is also called as germ warfare. Biological warfare is regarded as intentional use of disease-causing biological agents like viruses, rickettsia, bacteria, and fungi to kill or to weaken the living beings like plants and animals, including humans as a part of war strategy.

A few examples of bacteria used for biological warfare are as follows:

- Tularemia
- Anthrax
- Botulinum toxin
- Brucella species
- *Clostridium perfringens*
- *Salmonella* species
- *Shigella*
- *Staphylococcus aureus*
- *Burkholderia mallei*
- *Burkholderia pseudomallei*
- *Chalamydia psittaci*
- *Coxiella burnetii*
- *Rickettsia prowazekii*
- *Vibrio cholerae*

A few examples for viruses are as follows:

- Smallpox
- SARS
- Alphaviruses
- Hantavirus
- HIV/AIDS
- Nipah virus
- H1N1, a strain of influenza

The example for toxins are

- Botulinum toxin
- Cholera toxin
- Shiga toxin
- Trichothecenes
- Volkensin
- Modescin
- Ricin
- Abrin
- Cholera toxin

Biosensors refer to powerful and innovative devices combining a physical & biological sensing elements and with wide range of applications. BWAs are infectious microorganisms or toxins with the capability to harm or kill humans.

4. Comparative Study

In this section, an attempt is made to study various chemical and biological warfare agents used and the methods to counter the effects of the same. Table 37.1 presents a detailed study of the same.

Table 37.1 Comparative Analysis of Chemical and Biological Warfare

Paper	Chemicals/Germs	Methods	Antiagents/Agents	Parameters	Sensors Used	Remarks
[1]	Dimethyl Methyl Phosphonate (DMMP), Tri-Methyl Phosphate (TMP) and Di-Isopropyl Methyl phosphonate (DIMP)	Sample preparation and data processing and statistics	High-Performance Liquid Chromatography, Isopropyl Methyl Phosphonic Acid (IMPA), Isobutyl Methyl Phosphonic Acid (IBMP)	Thermo Fisher Orbitrap Elite, Prosocial, methanol: carbon tetrachloride, X Cali bur v. 2.2 Trace Finder v. 3.3.	Insects as chemical sensors	Establishing quickly detecting CWA and hydrolysis material in clinical biofluids using paper spray spectrometry
[2]	Triethyl phosphate (TEP).	M tube furnace of Lindberg Blue with outside diameter crystal tube	Triethyl phosphate (TEP) and Tri-Methyl Pentane (TMP), Solid Sorbent Tubes (SST)	Pyrolysis Ingredients	Molecularly imprinted polymer sensors	Triethyl phosphate by custom pyrolizers with waste gas. Analysis of substance by TD/GC/MS
[3]	G series agents like GA, GB, GD, and VX agents	Distant or deadlock charging, flame photometry by Infrared and Raman spectroscopy, recognition of photoionization and electrochemical, and on-the-spot testing of carbon nanotube gas using ionization sensors	Polymer-graphene nanoplatelet (GNP).	Di-isopropyl methyl-phosphonate (DIMP)	Graphene nanoplatelet-polymer chemiresistive sensor	This is the identification of other analyst classes that are multipurpose and flexible for the development of the developed chemiresitive sensor. Optimization of polymer-GNP blends and machine learning algorithms
[4]	Nitrogen mustard (HN3), Sulfur mustard, Lewisite, Tabun, sarin	Analytical displacement analysis for SM. Chemodosimeter method for SQ dy and SM, Acridine-based chromo-fluorogenic, Luminal identification of SM in context temperature	Ditopic receptors for tabun detection	–	Chromo-fluorogenic sesors	The development of responsive probe and accurate detection and differentiation using chromo-fluorogenic sensors in real time

Paper	Chemicals/Germs	Methods	Antiagents/Agents	Parameters	Sensors Used	Remarks
[5]	Soman (GD) and 4-nitrophenyl phosphate	Solvothermal recipe and procedure, themes by cross-sectional method, TEM images by cross-sectional method, Mapping of EDX and TOF-SIMS (time-of-flight secondary ion mass spectroscopy)	UiO-66 particles, atomic layer deposition (ALD) by TiO2. Pinacolyl methyl phosphonic acid (PMPA)	Nanofiber kebab assembly of MOF, Polyamide-6 (PA-6) of nanofiber mats		To detoxify the CWAs, MOF-non-fibers are very good reactive materials. Using these MOF-nanofiber, half-lives of GD and DMNP are compared
[6]	Numerous components of sulfide-holding CWA simulants	High Internal Phase Emulsion. Fabrication method	Poly DCPD (poly-di-cyclo-penta-diene)	M2 catalyst		Considerable amounts have been collected and reacted with polydispersed, hydroperoxides with several components of sulfide-containing CWA simulators
[7]	Both H- (mustards) & V- (nerve) type of CWAs with stoichiometric and catalytic destruction	NHC complexes are destructed by group 11 metals and vanadium, destruction by complicated tethered-NHCs, catalytic CWA destruction	Monodentate N-heterocyclic carbenes, NHC complexes of silver and alkali metals by aryloxide. Iron–NHC	Nuclear magnetic resonance spectroscopy, Ultraviolet-visble spectroscopy and Gas chromatography mass spectrometry		Use of NHC of stable air and moisture shows the annihilation of CWA stimulants. Imidazoline/Imidazolinium silver NHC complexes were compared
[8]	Sarin (GB), Tabun (GA), VX and blistering agents like mustard gas, lewisite	By heating dicyandiamide in a horizontal furnace, graphitic carbon nitride was prepared. Hummer's oxidation method used	Cotton, Cu-BTC MOF and carbon nitrite, graphitic, nanospheres	Graphitic carbon nitride (gCN)		The ability to simultaneously absorb, damage, and perceive toxic poster dews through vapors or fabric can be desgnated as "smart textile"

Paper	Chemicals/Germs	Methods	Antiagents/Agents	Parameters	Sensors Used	Remarks
[9]	Dimethyl 4 Nitrophenyl Phosphate (DMNP) and GD and VX	Hydrolysis monitoring of DMNP. Hydrolysis assay of GD and VX nerve agents	Carbon, mesoporous silica, geolites, and surfactants/metallo surfactants	Polyethyleneimine (PEI), N-ethyl Morpholine, NU-1000, Zirconium MOF		Hydrolyzing the simulator DMNP in water neural agents by various systems
[10]	Sulfur- (yperite and VX stimulants) OR and V-series of Organopho Sphorus Nerve Agents	CEES in an Oxone of oxidation and neutralization of the CWA stimulant packed-bed reactor	With oxone or aqueous form (nerve agent detoxification)	Syringe pumps, Air-tight plastic syringe, ethanol/ trifluoroacetic acid	Flow sensor	Detoxification and neutralization of CW agents, yperite by oxone and V-series OPNA simulators
[11]	Lethal	The SAPTO Analysis. RDG Analysis QTAIM Analysis UV-Vis Analysis	Lung, Blood, Blistering, and Bravery Agents	Symmetrical Limits, Dynamisms, and Geometric and Electric Possessions	A-series molecular sensors	Energy, the trend of stability of the complexes based on the adsorption
[12]	Acyclic acid Acrylamide Ethyl methacrylate Methacrylic acid	Open Force-Field (OFF), Polymer Consistent Force-Field (PCFF)	Blood, Choking, Nerve, Blister Agents	Detection and identification of risky materials	Imprinted monomers	Detection of tiny materials and consider the mechanisms of physical processes on nuclear
[13]	Organo-Phosphorus	Air-chromatography HPLC MS, and NMR Methods	Swelling Agents, Plasma Agents, and Organophosphorus Bravery Agents	Sensing and higher trivialness towards selected analyte with respect to other curious particles	OP CWAs sensors	Detection of chemical war agents is critical, in particular due to the recent international setup
[14]	Activated carbon	1. Samples, 2. SEM-EDS, 3. FESEM, 4. SR µ-XRF	Blister and Nerve Agents	Gas adsorber	Activated carbon sensors	The results provided by FESEM show impregnated metals present in the carbon-based structures. Further, XAS provides the evaluation of advanced works

Paper	Chemicals/Germs	Methods	Antiagents/Agents	Parameters	Sensors Used	Remarks
[15]	Cyanide, Phosgene, Chlorine, Sulphur Mustard	Pharmacological treatment	Swelling, Bravery, Harsh, Incapacitating/ Behaviour Altering, and Asphyxiants/ Blood Agents	Gas Poisoning; Decontamination; Drug Therapy	MOS sensors	CWAs which have been used for a long time are still a source of hazard for people and environment despite all the prohibitions
[16]	Nitrogen rich in atmosphere	InAs planar APD	Nerve Agents	Slide photodiodes, ultraviolet detectors, particle establishment	Metal organic framework sensor	In the conditions where operating biases are high, in order to avoid edge break down in APD structure, one method is the incorporation of graded p-type region
[17]	Soman and Sulfur mustard	AVLAG swatch test	Nerve Agents and Blister Agents	Allowing for better fiber-precursor contact	Chemo-sensor	The modified fibres covered by healthy grown MOF particles
[18]	Carbon nanoparticles	Synthesis of CNPs-C2-OH	Mustard, Lewisite, Nerve Agents, etc.	Optical response of the ppm level is sensitive and reversible sensoristics	Nano-structured sensor	tiny structured sensors are notices of nerve agents in air and liquid
[19]	Mustard gas	Oxidative neutralization CEES with oxone	Nerve agents and Blister agents	Oxone may be used in hard procedure or in water form	Gas Sensor	Oxone is a common possible promoter for cleansing of environmentally persistent chemical warfare mediators
[20]	Isopropyl methyl phosphate	HCPs Using 2 or 3 equivalent of crosslinker	Nerve agents	Gas capture storage and heterogeneous catalysis	Polymer-based sensors	Quality testing of polymers in contradiction of CWAs and their simulators

Paper	Chemicals/Germs	Methods	Antiagents/Agents	Parameters	Sensors Used	Remarks
[21]	Blood test	Well-equipped lab and skilled laboratory people are mandatory for the diagnosis of tularemia as Francisella tularemia is a biohazardous pathogen	Bacterium Francisella tularemia	Francisella tularensis, Tularemia, Vector-borne infection		Tularemia is obtained through the consumption of infected animals, Contact with aquatic environments, Vector reduction and inhalation of aerosols
[22]	A positive serological test results, presence of amplifiable viral RNA sequences in blood/tissue	No specific methods or treatment	Orthohanta virus	Hanta virus, Hemorrhagic fever with renal syndrome, Nephropathiaepidemica		Hanta virus shows the symptoms only after infecting the tissues. Icatibant is an antiviral drug which is used to counteract the leakage in vascular tissues
[23]	Growth on selective media, lack of hemolysis, lack of motility, capsule staining, culture	Biomedical countermeasures and diagnostic tools	Bacillus anthracis	Anthrax bacillus anthracis, Detection, diagnosis, Infection therapy		Presently, the top BWAs are microorganisms because of lethality, easy dissemination, long-term stability, and production of spores
[24]	Isolation of the virus from the blood or lesions	The speed combined with high sensitivity and specificity are the main things required for laboratory testing. They have published several real-time combinations for the identification of OPVs	Variola virus (VARV), a member of the Orthopoxvirus (OPV) genus of the Poxviridae	Smallpox, variola virus, vaccination, antiviral, eradication		Smallpox virus certified by the WHO. Viable variola virus (VARV), the causative agent of smallpox

Paper	Chemicals/Germs	Methods	Antiagents/Agents	Parameters	Sensors Used	Remarks
[25]	Finding bacteria in samples of blood, bone marrow, or other bodily fluids	Blood samples of fever patients were collected and brucellosis was suspected	Brucella	Brucellosis - Brucella melitensis - PCR - PCR-RFLP - sequencing		1. Commonly transmitted to human through the consumption of raw milk 2. To avoid the infection by monitoring the infected pets
[26]	Polymerase chain reaction, chest CT scan	—	—			—
[27]	RT-PCR is a bodily fluid and antibody detection by ELISA	No specific method or treatment	The antiviral treatment remdesivir has been effective in nonhuman primates when given as post-exposure prophylaxis	Nipah virus, Paramyxo viruses, Pteropus bat, Kozhikode		Nipah virus present in a wide variety of bat reservoirs; they can be isolated and propagated
[28]	Polymerase chain reaction is the preferred laboratory test given its accuracy and sensitivity	Genetic Methods, Phenotypic Methods, Immunological Methods, Electron Microscopy	Orthopoxvirus genus	Poxviridae, Orthopox viruses, host and tissue tropism, signaling in orthopox virus infection, host immune response to MPXV		Implementation of SORMAS is the correct way to combat MPXV
[29]	Q fever	Collection of Samples, Staining and Related Measures, Isolation and Cultivation, Serology, Molecular Method	Coxiella burnetiid	Coxiella bunetiid, qfever, In Small Ruminants, In Humans		The dwelling place of pets and animals should be in proper hygiene condition
[30]	Polymerase chain reaction, chest CT scan	—	—	Birth of COVID-19, biosecurity and biosecurity, Biological warfare		For the sake of the common good of humanity, developing a much more healthy globa power of biosecurity and biosafety

5. Conclusion

Biological and chemical warfare are becoming threats to the mankind. Attempts have been made to control the conduction of warfare and the development of weapons using harmful substances such as poisons and disillusioned weapons. Development of sophisticated sensors is playing an important role in the process of mitigating the effects of these harmful chemicals and germs. In this paper, an attempt has been made in summarizing various types of chemicals and germs used as CWAs and BWAs. Different types of mechanisms and sensors that are used to mitigate the effects of these warfare are also detailed in the paper. It is unfortunate that the developments in the fields of science and technology are also being misused for causing threats in the society. Nevertheless, technology has also enabled us to detect and nullify the effects of these threats. In summary, one has to understand that the technological advancements are to be used for the betterment of the mankind and not for the destruction purposes.

REFERENCES

1. Josiah McKenna, Elizabeth S, Dhummakupt, Theresa Connell, Paul S. Demond, Dennis B. Miller, J. Michael Nilles, Nicholas E. Manicke and Trevor Glaros. 2017. Detection of chemical warfare agent simulants and hydrolysis products in biological samples by paper spray mass spectrometry. *Analyst* 142: 1442–1451.
2. Daniel J. Van Buren , Thomas J. Mueller, Christopher J. Rosenker, John A. Barcase , Kelly A. Van Houten.2021. Custom pyrolyzer for the pyrolysis of chemical warfare agents. *Journal of Analytical and Applied Pyrolysis* 154 : 105007
3. Michael S Wiederoder, Eric C Nallon, Matt Weiss, Shannon K McGraw, Vincent P Schnee, Collin J Bright, Michael P Polcha, Randy C Paffenroth, and Joshua R. Uzarski.2017. Graphene Nanoplatelet Polymer Chemiresistive Sensor Arrays for the Detection and Discrimination of Chemical Warfare Agent Simulants. *American Chemical Society.*
4. Vinod Kumar. 2021.Chromo-fluorogenic sensors for chemical warfare agents in real-time analysis: journey towards accurate detection and differentiation. *Chem. Commun* 57: 3430–3444.
5. Junjie Zhao, Dennis T. Lee, Robert W. Yaga, Morgan G. Hall, Heather F. Barton, Ian R. Woodward, Christopher J. Oldham, Howard J.Walls, Gregory W.Peterson and Gregory N.Parsons. 2016.Ultra-Fast Degradation of Chemical Warfare Agents Using MOF– Nanofiber Kebabs. *Angew. Chem. Int. Ed.* 55: 1–6.
6. Christopher L. McGann, Grant C. Daniels, Spencer L. Giles, Robert B. Balow, Jorge L. Miranda-Zayas, Jeffrey G. Lundin, and James H. Wynne. 2018. Air Activated Self-Decontaminating Polydicyclopentadiene PolyHIPE Foams for Rapid Decontamination of Chemical Warfare Agents. *Macromol. Rapid Commun.,* 1800194.
7. Catherine Weetman,a Stuart Notmanb and Polly L. Arnold. 2018. Destruction of chemical warfare agent simulants by air and moisture stable metal NHC complexes. *The Royal Society of Chemistry.*
8. Dimitrios A. Giannakoudakis,a,b Yuping Hu,a,c Marc Florent,a and Teresa J. Bandosz.2013. Smart textiles of MOF/g-C3N4 nanospheres for chemical warfare agent rapid detection/detoxification. *J. Name* 00: 1–3.
9. Su-Young Moon, Emmanuel Proussaloglou, Gregory W. Peterson, Jared B. DeCoste, Morgan G. Hall, Ashlee J. Howarth, Joseph T. Hupp and Omar K. Farha. 2016. Detoxification of Chemical Warfare Agents Using a Zr6-Based Metal–Organic Framework/Polymer Mixture. *Chem. Eur. J.* 22: 1–6.

10. Antonin Delaune,a Sergui Mansour,a Baptiste Picard,a Philippe Carrasqueira,a Isabelle Chataigner, a Ludovic Jean,a Pierre-Yves Renard, a Jean-Christophe M. Monbaliu b and Julien Legros. 2021. Flow neutralisation of sulfur-containing chemical warfare agents with Oxone: packed bed vs. aqueous solution. *Green Chemistry* 23: 2925–2930.

11. Hasnain Sajid, Sidra Khan, Khurshid Ayub, Tariq Mahmood. 2021. Effective adsorption of a Series chemical warfare agents on Graphydiyne nano flake. *Research Square.*

12. Dumıtru Pavel, Jolanta Lagowski, Carmela Jackson Lepage.2006. Computationally monomers for molecular imprinting of chemical warfare agents. *Polymer* 47: 8389–8399.

13. Ester Butera, Agantino, Andrea, Giuseppe Trusso Sfrazzetto. 2021. Supramolecular sensing of chemical warfare agents. *ChemPlusChem* 86: 681–695.

14. Krit Won-in, Chatdanai Boonruang, Nichada Jearanaikoon. 2019. X-rays spectroscopy study of impregnated carbon used as gas adsorber for chemical warfare agents in military mas. IEEE *6th Asian Conference on Defence Technology* .

15. Seyhan Polat, Hakan Parlakpinar, Mehmet Gunata. 2018. Chemical warfare agents and Treatment strategies. *Annals of medical research* 25(4): 776–82.

16. Benjamin S, White, Ian C. Sandall, Xinxin Zhou, Andrey Krysa. 2016. High gain InAs Planar avalanche photodiodes. *IEEE*, Vol 34. No,11.

17. Min-Kun Kim, Sung Hun Kin, Myungkyu Park.2018. Degradation of chemical warfare agents over cotton fabric functionalized with Uio-66-NH2. *RSC Advances* 8: 41633.

18. Nunzio Tuccitto, Lorenzo Riela, Agatino Zammataro, Luca Spitaleri. 2020. Functionalized Carbon nano-particle based sensors for chemical warfare agents. *ACS Appl. Nano Mater* 3: 8182–8191.

19. Antonin Delaune, Sergui Mansour, Philippe Carrasqueira.2021. Flow neutalisation of sulfurContaining chemical warfare agents with oxone; packed bed vs. aqueous solution. *HAL.* https://hal.archieves-ouvertes.fr/hal-03353823.

20. Craig Wilson, Marcus J. Main, Nicholas J. Cooper, Michael E, Briggs. 2017. Swellable Functional hyper cross linked polymer networks for the uptake of chemical warfareagents. *Polymer Chemistry*: 8-1914.

21. Derya Karataş Yeni & Fatih Büyük , Asma Ashraf , M. Salah ud Din Shah. 2021. Tularemia: a re-emerging tick-borne infectious disease. *Folia Microbiologica* 66: 1–14.

22. Vaheri, A., Henttonen, H.,Mustonen, J. 2021. Hantavirus Research in Finland: Highlights and Perspectives. *Viruses* 13:1452. https:// doi.org/10.3390/v13081452.

23. Pohanka M. 2020.Bacillus anthracis as a biological warfare agent: infection, diagnosis and countermeasures. *Bratisl Med J*: 121 (3).

24. Hermann Meyer , Rosina Ehmann and Geoffrey L. Smith.2020. Smallpox in the Post-Eradication Era. *Viruses*12: 138

25. Anita Barua, Ashu Kumar, Duraipandian Thavaselvam, Smita Mangalgi, Archana Prakash, Sapana Tiwari, Sonia Arora& Kannusamy Sathyaseelan.2016. Isolation & characterization of Brucella melitensis isolated from patients suspected for human brucellosis in India. *Indian J Med Res* 143. pp. 652–658.

26. Robert Skopec, Researcher-Analyst, Dubnik, Slovakia.2020. Coronavirus is a Biological Warfare Weapon. *Auctores Publishing* – Volume 1(2)–026.

27. Dr. Prarthana M. S. 2018.Nipah virus in India: past, present and future. *IJCMPH* 3653-3658.

28. Emmanuel Alakunle, Ugo Moens, Godwin Nchinda and Malachy Ifeanyi Okeke.2020.Monkeypox Virus in Nigeria: Infection Biology, Epidemiology, and Evolution. *Viruses* 12: 1257.

29. Tolera Tagesu Tucho.2019.Q Fever in Small Ruminants and its Public Health Importance. *Dairy and Vet Sci J* 9(1): JDVS.MS.ID.555752.

30. Jing-Bao Nie. 2020. In the Shadow of Biological Warfare: Conspiracy Theories on the Origins of COVID-19 and Enhancing Global Governance of Biosafety as a Matter of Urgency. *Journal of Bioethical Inquiry Pty Ltd.*

Intelligent Systems and Smart Infrastructure – Brijesh Mishra et al. (eds)
© 2023 Taylor & Francis Group, London, ISBN 978-1-032-41287-0

CHAPTER

38

Workload Prediction Model for Autonomic Scaling of Cloud Resources with Machine Learning

Sanjay T. Singh[1], Mahendra Tiwari[2]

Department of Electronics and Communication,
J. K. Institute of Applied Physics and Technology,
University of Allahabad

Anchit Sajal Dhar[3]

Department of Computer Science & Information Technology,
Sam Higginbottom University of Agriculture,
Technology and Sciences

Abstract

Cloud computing enables clients with on-demand access to software, platform, and infrastructure, in the form of services through the Internet. Client applications are executed over the cloud which is backed by Virtual Machines (VMs) and these VMs are hosted on top of physical servers. The amount of workload traffic received by the cloud changes over time. To fulfil these fluctuating workload needs, VMs must be automatically scaled up and down to guarantee that the quality of service (QoS) to the client is maintained, which, in turn, must be achieved by ensuring that the Service-Level Agreement (SLA) criteria are not breached. To achieve this goal of automatic scaling (also known as auto-scaling), the important task is to predict the future workload demands for cloud resources so that appropriate numbers of VMs must be made ready in advance so that the requirements of clients are met. The prediction is done on the basis of the past resource usage trends. In this paper, we propose an autonomic

Corresponding author: [1]sanjayt.singh@gmail.com, [2]mahendra@allduniv.ac.in, [3]anchit.dhar@shiats.edu.in

DOI: 10.1201/9781003357346-38

resource management model, based on a Machine Learning (ML) technique called Random Forest, that attempts to solve the problem of predicting the cloud resource demands. We have evaluated the proposed classifier, for its prediction accuracy by utilizing an actual datacenter dataset containing thousands of records. Further, the classifier is compared with another classifier based on Naïve Bayes algorithm and found that our proposed algorithm gives better performance. The results obtained indicate the accuracy and efficiency of Random Forest Classifier.

Keywords: Cloud Computing, Workload Prediction, Auto-scaling, Machine Learning, Random Forest Classifier

1. Introduction

In cloud computing, workloads (user requests) keep on arriving as users access the various services provided by the cloud. To ensure that the Service-Level Agreement (SLA) parameters agreed upon with the customers are upheld and users' quality of service (QoS) is preserved (Singh & Chana, 2015a), proper distribution of cloud workloads is of utmost importance. Workloads are an abstraction of actual operations of the computer (Singh & Chana, 2015b).

Cloud computing (Armbrust, A. Fox, and R. Griffith, 2009) is a pay-as you-go computing model that delivers on-demand services such as infrastructure, platform, and software, through the Internet. It gives users access to a pool of resources that may include computing infrastructure such as processing elements, storage space, development platforms, user-specific applications, and networks. Cloud computing follows a multi-layered architecture (Buyya et al., 2011; Bashar, 2013), that can be seen in Fig. 38.1, to make these services available to the customers.

Looking at Fig. 38.1, it can be easily observed that the workloads generated by various users are received at the SLA Management Layer, which is responsible for: (i) monitoring, in real time, the incoming workload traffic, (ii) managing the SLA parameters, and (iii) monitoring applications and VM performance for ensuring the QoS. In the Virtualisation Layer, the various VMs are deployed over whom the various workloads are placed for their execution. Considering the limitation of hardware resources which make up the Infrastructure Layer, virtualisation is a key technology which presents the resources to users as though they are unlimited. At the infrastructure layer, we have the actual hardware resources over which the virtualisation layer is deployed.

Scalability (Puthal et al., 2015) is an interesting feature as well as a challenge (Dillon et al., 2010), to be dealt with, of cloud computing, which supports matching the capacity of the cloud against the incoming workload. This gives rise to the notion of Dynamic Resource Provisioning in response to workload needs. Dynamic provisioning, on the other hand, should consider the QoS and SLA limits that are positioned to guarantee that users receive the promised services.

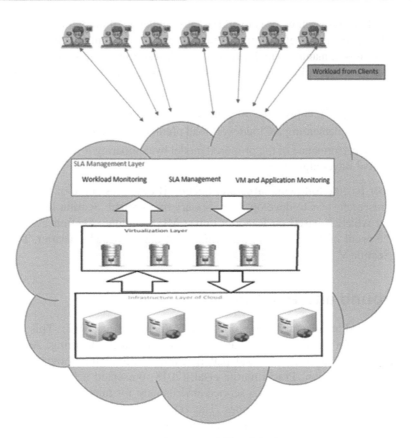

Fig. 38.1 Multi-Layered Cloud Architecture

The decision for scaling up or down VMs is taken at the SLA Management layer, depending on the workload trend. Virtualisation layer is where the VMs are deployed or removed. If the amount of workload is increasing, then new VMs must be instantiated in advance as switching on a VM takes some fraction of time. Once VMs are ready, they can be quickly assigned the newly admitted workload. On the other hand, if the workload is having a receding trend, then these VMs can be shut down so as to reduce energy consumption of the physical servers. This necessitates the presence of a prediction component in order for an effective dynamic provisioning solution to estimate demand in advance and make the resource accessible to the clients with minimal or no delay.

To address the problem of accurate and efficient prediction of future workload demands for dynamic allocation of resources, Machine Learning (ML) can be of great help. Over a period of time, researchers have successfully utilized the predictive capabilities of ML (Alpaydin, 2014) approaches. These approaches can learn from historical data of workload trends and forecast future workload trends using the learnt system model.

The intent behind this research work is to address the challenge of predicting the future cloud workload trends accurately and efficiently so that dynamic provisioning of cloud resources can be done in an optimised manner which eventually will result in savings for client as well as the cloud service provider. Following are the contributions of our work to the area of research in context:

- Review of existing ML approaches for autoscaling of Cloud resources.
- Proposing and implementing a Cloud workload prediction model.
- Evaluating the performance of the proposed model by comparing it with another classifier based on Naïve Bayes algorithm.

Rest of the paper has the following structure: Section II describes some of the related work in this area of research so as to develop the necessary background. Section III elaborates our proposed model which is based on Random Forest ML technique. Section IV is where we demonstrate the proof of the predictive performance of our proposed classifier. The paper is concluded in Section V.

2. Background and Related Work

A microservice-based architecture was proposed in Alipour & Liu (2017). This architecture consists of features for metrics monitoring and workload pattern learning. After this, the architecture predicts the future workload pattern using Multinominal Logistic Regression as well as Linear Regression. Prachitmutita et al. (2018) presented a framework for Auto-Scaling for use in an infrastructure as a service environment for microservices, with a focus on service-level agreement and cost savings. This framework is built upon Artificial Neural Network and Recurrent Neural Network. Moreno-Vozmediano et al. (2019) recommended utilising Support Vector Machine regression in order to estimate the computing load which will be on the server using the past data, and then determining the appropriate number of resources using a queuing model, to match the projected server demand and meet the SLAs. Marie-Magdelaine & Ahmed (2020) suggested a proactive autoscaling architecture that uses a forecasting model which is based on learning the system behaviour to change the resource pool dynamically, both horizontally and vertically. To improve end-to-end delay for native apps of cloud environment, their system employs a proactive autoscaling method using Long Short-Term Memory. The problem of auto-scaling is visualised as a sequence model in the work (Golshani & Ashtiani, 2021), and Convolutional Neural Networks are used to anticipate the future demand for the services provided by the cloud. The authors also do a plotting between the projected workload and the real-time as well as future quantities of necessary resources by employing neural networks.

According to this study, the bulk of research has focused on applying machine learning to the cloud datacenter resource scaling function. Utilising a supervised learning technique, we now examine the use of Random Forest Classifier to implement and improve the performance of dynamic resource provisioning function in a Cloud Computing environment.

3. Proposed Model

The goal of this work was to investigate cloud node resource utilisation based on workload factors and predict the future workload so that decisions for autoscaling the cloud resources can be taken. The challenge will be handled through obtaining a sufficiently populated dataset of labelled workload variables from various nodes of cloud after which this data will be used to develop a Random Forest Classifier. Consequently, the random forest classifier would be able to predict the future workload.

A. Random Forest Classifier

We define Random Forests, as in Breiman (2001), "A random forest is a classifier consisting of a collection of tree-structured classifiers $\{h(x, \Theta k), k = 1,...\}$ where $\{\Theta k\}$ are independent identically distributed random vectors and each tree casts a unit vote for the most popular class at input x." At training stage, random forests create a large number of individual decision trees. To produce the final forecast, the predictions acquired from all trees are combined; for regression the mean prediction and for categorization the mode of classes. This technique is called an Ensemble technique since they make a final conclusion based on a group of outcomes. To determine the significance of feature, the possibility of reaching at that node weighs towards the lessening in node impurity. The node probability is the result of division of the number of samples that arrive at the node by the overall number of samples. Higher value indicates higher feature significance.

B. Features for Characterizing Workload

For the prediction purpose, we use nine features (Al-Faifi et al., 2018b) that are used for measuring the performance of a cloud node. These features may be easily retrieved from any system. These features are presented in Table 38.1.

Table 38.1   Workload Features

S. No.	Name of the Feature
1.	No. of jobs in 1 minute
2.	No. of jobs in 5 minutes
3.	No. jobs in 15 minutes
4.	Memory capacity
5.	Disk capacity
6.	Number of CPU cores
7.	CPU speed per core
8.	Average Receive for Network Bandwidth in Kbps
9.	Average Transmit for Network Bandwidth in Kbps

Source: Modified from Al-Faifi et al. (2018b).

The features mentioned in Table 1 are the indicator of the performance of a cloud node. The number of incoming jobs has an impact on the performance of a cloud node since a big number of jobs slows down the provisioning of services. Memory capacity is the amount of physical memory available. The higher the memory capacity, the better the performance. Disk capacity implies the amount of disk drive capacity available. Higher disk drive capacity means better performance. Number of CPU cores implies the number of processing cores per CPU. CPU speed per core is the measure of speed per CPU core. We also plotted a correlation matrix, shown in Fig. 38.2, to find out the coefficients of correlation between the various features. We avoid highly correlated elements while determining the correlation so that the model does not deviate to one side or the other.

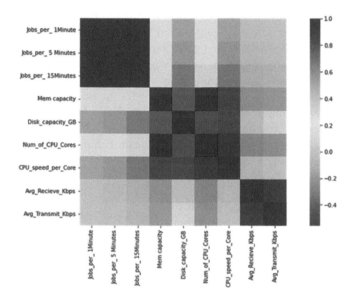

Fig. 38.2 Correlation Matrix for Dataset Features

4. Experimental Setup

A. Dataset Description

For this work, we have used an actual datacenter dataset as provided by (Al-Faifi et al., 2018a). The dataset contains 9 features and the number of nodes used for obtaining this dataset are 13. The dataset file we used for our work, amongst all the files in (Al-Faifi et al., 2018a), contains 25,697 records. The class label for the dataset are four ordinal classes: Very High (76%–100%), High (51%–75%), Low (26%–50%), and Very Low (0%–25%).

Out of 25,697 records, 75% of records were chosen randomly for training the classifier and 25% records were chosen randomly for testing the classifier.

B. Tools Used

We used Anaconda Notebook for running Jupyter Notebook and used Python for coding the classifiers. The hardware used for running the simulation was a laptop with Core i5-1035G1 processor with frequency up to 3.6 GHz, 8 GBs of RAM, and 2 GBs of Graphics Card. The operating system running on the laptop was Windows 10 64-bit.

C. Measures for Classifier Assessment

For evaluating the classifier, we have used the following measures:

- Confusion Matrix
- Balanced Accuracy Score
- Mean Absolute Error
- Mean Squared Error
- Root Mean Squared Error
- Classification Score

5. Results and Discussion

At first, we trained the Random Forest Classifier with the training dataset. Subsequently, the dataset for testing and evaluating the classifier was used to test the performance of the trained classifier. The number of outcomes for the proposed classifier is counted to create confusion matrix, which is shown in Fig. 38.3. This matrix represents true positive and true negative against predicted positive and predicted negative values.

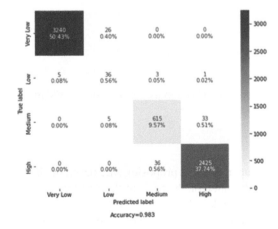

Fig. 38.3 Confusion Matrix for Random Forest Classifier

From Fig. 38.3, it can be seen that out of 3266 records for the class Very Low, the classifier is able to classify 3240 (50.43%) records correctly, and out of 2461 records for the class High, the classifier is able to classify 2425 (37.74%) records correctly. The overall accuracy of all the items predicted is 0.983. We created another classifier with Naïve Bayes algorithm and using the same dataset for prediction and found out the overall accuracy to be 0.868. So, it can be seen that our proposed classifier performs better than Naïve Bayes classifier.

The values for Balanced Accuracy Score, Mean Absolute Error, Mean Squared Error, and Root Mean Squared Error of the proposed classifier are 0.918, 0.017, 0.018, and 0.134, respectively. With Naïve Bayes classifier, the values are 0.520, 0.140, 0.178, and 0.422, respectively. Here again, we can see that our proposed model performs better.

Figure 38.4 is a graph that clearly identifies that as the amount of workload is increasing on the cloud, the number of CPU core and Memory capacity will also be required to be increased and vice versa. This is an indication that the VMs are needed to be increased and decreased accordingly to meet the variation in the amount of workload.

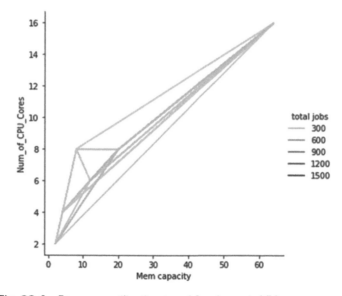

Fig. 38.4   Resource utilization trend for dynamic VM management

6. Conclusion

The work presented in this paper aimed at proposing ML-based model for predicting the future workload trends, of cloud datacenter, on the basis of the past trend of workload. Consequently, a prediction model based on Random Forest technique was proposed. This model is trained and evaluated with a real dataset. We have evaluated the model against various measures for evaluating the performance of our proposed model and found that it gives sufficient prediction accuracy. We also compared our proposed model with another classifier based on Naïve Bayes algorithm and found that our proposed classifier performs better. As for future work, it can be compared with other ML techniques and also the proposed model can be modified by merging it with a density estimation function to further investigate a scope to improve its performance.

REFERENCES

1. Al-Faifi, A. M., Song, B., Hassan, M. M., Alamri, A., & Gumaei, A. (2018a). Data on performance prediction for cloud service selection. *Data in Brief, 20,* 1039–1043. https://doi.org/10.1016/j.dib.2018.08.108

2. Al-Faifi, A. M., Song, B., Hassan, M. M., Alamri, A., & Gumaei, A. (2018b). Performance prediction model for cloud service selection from smart data. *Future Generation Computer Systems, 85,* 97–106. https://doi.org/10.1016/j.future.2018.03.015

3. Alipour, H., & Liu, Y. (2017). Online machine learning for cloud resource provisioning of microservice backend systems. *Proceedings - 2017 IEEE International Conference on Big Data, Big Data 2017, 2018-Janua,* 2433–2441. https://doi.org/10.1109/BigData.2017.8258201

4. Alpaydin, E. (2014). *Introduction to Machine Learning.*

5. Armbrust, A. Fox, and R. Griffith, M. (2009). Above the clouds: A Berkeley view of cloud computing. *University of California, Berkeley, Tech. Rep. UCB,* 07–013. https://doi.org/10.1145/1721654.1721672

6. Bashar, A. (2013). Autonomic scaling of cloud computing resources using BN-based prediction models. *IEEE 2nd International Conference on Cloud Networking (CloudNet),* 200–204.

7. Breiman, L. (2001). Random Forests. *Springer Link, 45,* 5–32.

8. Buyya, R., Garg, S. K., & Calheiros, R. N. (2011). SLA-oriented resource provisioning for cloud computing: Challenges, architecture, and solutions. *Proceedings - 2011 International Conference on Cloud and Service Computing, CSC 2011, Figure 1,* 1–10. https://doi.org/10.1109/CSC.2011.6138522

9. Dillon, T., Wu, C., & Chang, E. (2010). Cloud computing: Issues and challenges. *Proceedings - International Conference on Advanced Information Networking and Applications, AINA,* 27–33. https://doi.org/10.1109/AINA.2010.187

10. Golshani, E., & Ashtiani, M. (2021). Proactive auto-scaling for cloud environments using temporal convolutional neural networks. *Journal of Parallel and Distributed Computing, 154,* 119–141. https://doi.org/10.1016/j.jpdc.2021.04.006

11. Marie-Magdelaine, N., & Ahmed, T. (2020). Proactive Autoscaling for Cloud-Native Applications using Machine Learning. *2020 IEEE Global Communications Conference, GLOBECOM 2020 - Proceedings.* https://doi.org/10.1109/GLOBECOM42002.2020.9322147

12. Moreno-Vozmediano, R., Montero, R. S., Huedo, E., & Llorente, I. M. (2019). Efficient resource provisioning for elastic Cloud services based on machine learning techniques. *Journal of Cloud Computing, 8*(1). https://doi.org/10.1186/s13677-019-0128-9

13. Prachitmutita, I., Aittinonmongkol, W., Pojjanasuksakul, N., & Supattatham, M. (2018). Auto - scaling microservices on IaaS under SLA with cost - effective Framework. *2018 Tenth International Conference on Advanced Computational Intelligence (ICACI),* 583–588.

14. Puthal, D., Sahoo, B. P. S., Mishra, S., & Swain, S. (2015). Cloud computing features, issues, and challenges: A big picture. *Proceedings - 1st International Conference on Computational Intelligence and Networks, CINE 2015,* 116–123. https://doi.org/10.1109/CINE.2015.31

15. Singh, S., & Chana, I. (2015a). Q-Aware: Quality of service based cloud resource provisioning. *Computers and Electrical Engineering, 47,* 138–160. https://doi.org/10.1016/j.compeleceng.2015.02.003

16. Singh, S., & Chana, I. (2015b). QRSF: QoS-aware resource scheduling framework in cloud computing. *Journal of Supercomputing, 71*(1), 241–292. https://doi.org/10.1007/s11227-014-1295-6

Intelligent Systems and Smart Infrastructure – Brijesh Mishra et al. (eds)
© 2023 Taylor & Francis Group, London, ISBN 978-1-032-41287-0

CHAPTER

39

Novel Algorithm to Identify Whether the Medicine is Expired or Not through OCR

Ankit Kumar[1], Mamta[2], Bhawna Mallick[3]

Department of Computer Science and Engineering,
Meerut Institute of Engineering and Technology, Meerut, India

Abstract

We can leverage the power of Internet technology, which allows individuals to assist one another with only a single click from their phone, to fulfil the growing need for improvement in healthcare institutions. This method for giving medicine intends to give those in need access to an online forum for doing so. It aims mainly on reduction of expiry medicines those are being sold in market. As we all know, nowadays, each and every individual is more worried about their health due to the global pandemic that we all faced and came out of it strongly as unit, and people know that their immunity is the key factor for them to stay healthy. There is one thing that we all can agree to "India as a country can move faster towards the healthy life if we all will work together as a unit."

Keywords: Expiry date, Donation, Technology, OCR, Internet, Text extraction

1. Introduction

This algorithm is designed to allow user to donate and manage the medicine donation system with just simple clicks with an understandable interface to those who have minimum knowledge of technology. Although researches were already made for the detection of expiry date through scanning the wrapper of medicine, quality is not ensured, but our algorithm ensures the quality

Corresponding author: [1]ankit.kumar.cs.2018@miet.ac.in, [2]mamta.jagbir.cs.2018@miet.ac.in,
[3]bhawna.mallick@miet.ac.in

DOI: 10.1201/9781003357346-39

with the help of standard API (application programming interface). With the help of this API, space and time complexity is reduced.

2. Ease of Use

For those who are willing to get a medicine prescribed by the doctor for free can easily get through our application by just uploading their requirement.

A. Maintaining the Specification Mention by Donor

This application is designed in such a way that the donor has to upload the minimum information manually from his end as most of the information will be detected automatically through the application by using text detection which in turns reduces the chances of error happening to the minimum.

3. Background/Literature Review

Some of the important work that has been previously done in this field is mentioned here. A lot of work has been done in this field but we are referring to the particular application i.e. GIVMED app helps you donate your extra medicine easily, quickly, and reliably, without even stepping out of your door. The welfare organizations that we co-operate with are social pharmacies, retirement homes, and NGOs that cover the medical needs of socially fragile groups and employ a pharmacist who controls the medicine that is donated. GIVMED does not gather medicine but helps in networking through the app. Medicines that cannot be donated at all are as follows: any type of pharmaceutical product those that are be prescribed on red line and those that need to be refrigerated. Medicine that cannot be donated after having been opened are syrups, creams, inhaler, and drops. Our Application ensures to overcome some of the minor drawback of GIVMED such as better interactive interface specific for the donation of medicine, and OCR scanning open for the detection of expiry date from the wrapper of the medicine.

4. Proposed Methodology

Admin site android application. As we know that administrator is the super user of this android application. Only the admin has the access to all the source of data of all the activity-participating members.

Admin is the one who will have the access to all the data, request made by NGOs and User. It is the job of admin to allocate the medicine to NGOs which is donated by the user. The admin can also make a note of the feedback provided by the NGOs and look forward to improve the application.

The one who is willing to donate the medicine to NGOs first have to register himself/herself through google account mail verification. After successful verification, he/she can login to the application and create his/her profile by adding profile picture, user name, and other basic details. Afterward, he/she can move to the section where he/she can donate the medicine by typing some of the basic details and capture the wrapper of medicine to extract salt name and expiry date. Based on the required salt and validation of medicine, the message of acceptance and rejection will be prompted out.

The NGOs first need to register themself to the application through google account verification and also need to create the profile of NGOs by adding the name of NGOs, profile picture, address, and more basic details. Then, after the registration and profile creation, the NGOs can request for the medicine to the admin and one whose medicine is accepted by the validation process of medicine is donated to the requested NGOs. The NGOs will then get the medicine that is allotted by the admin. The NGO can also give the feedback on the quality of medicine donated.

5. Algorithm

The steps that are used in creating this application are mentioned below:

1. The person (user) who is willing to donate the medicine first has to capture the image through the application.
2. The image consists of the wrapper of medicine where the necessary detail about medicine is available.
3. First, the input image is scanned through OCR (optical character reader) to get the list of salt as an output.
4. Then again, the wrapper of medicine acts as an input for OCR (optical character reader) and gets the expiry date as output.
5. When the wrapper of medicine is scanned to extract the expiry date, the application also provides the option to crop the input image.
6. The dependencies to perform OCR and crop operation are added to build Gradle module to ensure complete functioning.
7. The output of expiry date is now compared with the present date with the help of inbuilt library of java.
8. If Present date is less than Expiry date, then the medicine is "ACCEPTED".
9. Else "REJECTED".
10. On the basis of donation history, the user will get reward.
11. The ACCEPTED medicine is now donated to the requested NGOs.

6. Design and Implementation of Application

The application is designed in such a way that it is easy to use as we use as similar to any other application those are being normally used.

1. Api'com.theartofdev.edmodo:android-imagecropper:2.8.+'

2. implementation'com.google.android.gms:playservices-vision:19.0.0'

3. androidTestImplementation'androidx.test.espresso:espresso-core:3.4.0'

4. testImplementation'junit:junit:4.+'androidTestImplementation'androidx.test.ext:junit:1.1.3'

5. implementation'androidx.appcompat:appcompat:1.4.0'

6. implementation'com.google.android.material:material:1.4.0'

7. implementation'androidx.constraintlayout:constraintlayout:2.1.2'

8. testImplementation'junit:junit:4.+'androidTestImplementation'androidx.test.ext:junit:1.1.3'

7. Implementation of Expiry Date Detection

The java file was used to construct the business logic for processing client requests, listing items from the database, and storing data in the database. Java also incorporates the process translation from one operation to another. Image to text translation was done with the help of OCR and Google crop APIs. These dependencies are added in built Gradle module.

8. Result Discussion

The correct evaluation of this project based on the trials is calculated using three measures: recall, precision, and F-measure. Recall tends to be the proportion of properly recognized values for the project over all correctly identified and unidentified values for the project, whereas precision is the percentage of successfully identified project values over all values present in the repository. The precision of the methodology is given by the expression mentioned below:

$$P = C/(C + W) \text{ cluster for a set of values,}$$

where the number of relevant recognized values for the project is: C,

The number of incorrectly identified values for the project is: W,

and the number of unrecognized values for the project is: M.

1. Remember R : $R = C/(C + M)$.

2. F-measure combines recall and accuracy. $F = 2PR/(P + R)$ yields the F-measure.

3. Recall R and precision P are equally weighted. Table 39.1 shows the calculated value for F-measure which indicated the accuracy of the values scanned through the application designed:

Table 39.1

Number of Inputs	C	W	M	P	R	2PR	P+R	F
5	3	0	2	1	.60	1.2	1.60	.75
10	6	0	3	1	.66	1.32	1.66	.79
15	8	0	5	1	.01	1.22	1.61	75
20	10	1	6	.90	.62	1.11	1.52	.73
25	10	1	8	.90	.55	.99	1.45	.68
30	10	1	9	.90	.52	.93	1.42	.65
35	12	2	10	.857	.54	.92	1.39	.66
40	14	3	12	.823	.53	.87	1.35	.64
45	17	2	12	.894	.58	1.03	1.47	.70
50	18	2	12	.90	.60	1.08	1.50	.72
55	19	3	13	.863	.59	1.01	1.45	.69
60	20	3	14	.869	.58	1.00	1.44	.69

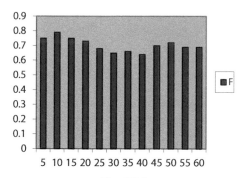

Fig. 39.1

For the above example, the application can also scan the wrapper from google image.

9. Conclusion

As we all know, nowadays, each and every individual is more worried about their health due to the global pandemic that we all faced and came out of it strongly as unit, and people know that their immunity is the key factor for them to stay healthy. More than ever people are taking their health issues more seriously and are taking regular immunity boosters and multivitamins in order for their body to stay healthy but as we all are aware of the fact that India is a country where there are lots of people those lies below the poverty line, these are the people who cannot afford this kind of medication. Covid situation has shown us the real situation of our healthcare department as well as the health conditions of the people residing in our country. It is the responsibility of the group of people like us who have the privilege to afford those medications that instead of throwing away the residual medicined donate those to the people who could use them in order to stay alive and healthy so they can continue contributing towards the country.

REFERENCES

1. Fourkanul Islam, M., Bin Zaman, S., Nazrul Islam, M., & Islam, A. (2021). Design and development of a gaming application for learning recursive programming. In Algorithms for Intelligent Systems (pp. 285–296). Springer Singapore.
2. Distribution among poor People: 2017 IEEE Region ten Humanitarian Technology
3. Conference (R10-HTC) twenty one - twenty three Dec 2017, Dhaka, Bangladesh
4. Jung, M.-J., Kim, J., Lee, S. H., Whon, T. W., Sung, H., Bae, J.-W., Choi, Y.-E., & Roh, S. W. (2022). Role of combined lactic acid bacteria in bacterial, viral, and metabolite dynamics during fermentation of vegetable food, kimchi. Food Research International (Ottawa, Ont.), 157(111261), 111261. https://doi.org/10.1016/j.foodres.2022.111261
5. Uzayr, S. B. (2022). Android Studio Tools. In Mastering Android Studio (pp. 133–178). CRC Press.
6. K. Elissa, "Title of paper if known," unpublished.
7. R. Nicole, "Title of paper with only first word capitalized," J. Name Stand. Abbrev., in press

Intelligent Systems and Smart Infrastructure – Brijesh Mishra et al. (eds)
© 2023 Taylor & Francis Group, London, ISBN 978-1-032-41287-0

CHAPTER

40

New Framework for Implementation of Decision Tree Classifier

Raghvendra Singh[1], Mahesh K. Singh[2] and Dharmendra Kumar Jhariya[3]

Department of Electronics and Communication Engineering
National Institute of Technology Delhi, Delhi 110040, India

Abstract

In Machine Learning (ML), a classifier is an algorithm that automatically orders or categorizes data into one or more of a set of "classes". This paper offers a unique structure that allows us to automate the formation of trained ML classifiers IC chips from raw datasets. The unique structure creates the trained model by performing different processing steps on the datasets that it accepts in a comma separated value (CSV) form. As the trained model is created, the unique structure develops ML-classifier on the basis of a tree structure, *viz* decision tree classifier in 2-forms; Extensible-Markup-Language (XML), and Verilog. The XML format displays tree hierarchy once generated and the code in Verilog is the HDL presentation of generated model. The code produced is used as input for a Field-Programmable-Gate-Array for prototype. The proposed work is implemented on Xilinx Artix-7 FPGA.

Keywords: Machine learning, Classifier, Decision tree, Integrated Circuit, Verilog.

1. Introduction

Machine learning (ML) brings out unique methods to exploit the potential of data. This prime technology helps the binary machines to grasp knowledge and enhance it, to ultimately form an expertise by forging programs that can easily approach data and do work via predictions and detections, the presentations of these calculation should be better. The improvement of this

Corresponding author: [1]imraghvv@gmail.com, [2]ksmahesh@nitdelhi.ac.in, [3]dharmendra.jhariya@nitdelhi.ac.in

DOI: 10.1201/9781003357346-40

kind is specially required in real-time utilizations [1], such as real NLP or Image-Recognition in video. To enhance the overall performance of ML classifiers is via imposing them through Application-Specific-Integrated-Circuit (ASIC) chips which carry out classifying job. Flow starting from raw dataset to layout files is packed with difficulties in design and implementation phase. The proposed work presents unique processing steps that can build up flow to produce trained ML classifier-integrated chip from a comma separated file (CSV).

The flow began with raw data in CSV format, which then undergo different data processing steps like data formatting and feature selecting; then this dataset is trained under ML libraries which produce a trained model in a tree-based mode. The output of this trained model is used to drive two different formats, namely, XML and Verilog. While the XML format is positional representation of tree output, tree representation is used as the guide to write the Verilog code for classifier. This code is then used in ASIC implementation. The proposed work has performed an analysis using a case study of one such dataset, Playing tennis Dataset [2] & Spect [3] dataset. The research in the field of improving the performance of ML classifiers for real-time applications are being focused on three routes as of now, namely, Graphics Processing Units (GPU), Field-Programmable Gate Array (FPGA), and ASIC design that focuses on a certain application but there was no such unique and all-inclusive flow that can direct the implementation process from dataset to HDL implementation [4]–[7]. This came to light that ASIC implementation of the chips are way better in performance than their FPGA and GPU counterparts. The proposed work aims to bridge the gap between the ML classifiers and its HDL implementation.

2. Design Automation Flow

The flow starts with the generation of the trained decision tree classifier through Python program using ID3 Algorithm. The python program uses different ML libraries and data processing steps to produce this decision tree classifier, as shown in Fig. 40.1. The steps are as follows:

- Raw dataset from the CSV file are being read and binarization of the data takes place. The binarization of data is conversion of values in datasets into binary format, as shown in Fig. 40.3. The Playing Tennis Dataset (shown in Fig. 40.2) is being used for analysis for this proposed work. The process plays an important role in the HDL implementation phase.

Fig. 40.1 Machine Learning Classifier Flow Chart

- The follow-up step involves tree formation using Iterative Dichotomiser 3 (ID3) algorithm [8]. As the name suggests, the algorithm divides features over and over in 2 or more groups at every step. The information gain (IG) is calculated for every feature. Then, maximum IG feature is used for root-node and branches are split for the binary tree; this

OUTLOOK	TEMP	HUMIDITY	WIND	PLAY
Sunny	Hot	High	Weak	No
Sunny	Hot	High	Strong	No
Overcast	Hot	High	Weak	Yes
Rain	Mild	High	Weak	Yes
Rain	Cool	Normal	Weak	Yes
Rain	Cool	Normal	Strong	No
Overcast	Cool	Normal	Strong	Yes
Sunny	Mild	High	Weak	No
Sunny	Cool	Normal	Weak	Yes
Rain	Mild	Normal	Weak	Yes
Sunny	Mild	Normal	Strong	Yes
Overcast	Mild	High	Strong	Yes
Overcast	Hot	Normal	Weak	Yes
Rain	Mild	High	Strong	No

Fig. 40.2   Playing Tennis Dataset

outlook = SUNNY	outlook = OVERCAST	outlook = RAIN	temp = HOT	temp = MILD	temp = COOL	HIGH humidity	WEAK wind	NO play
1	0	0	1	0	0	1	1	1
1	0	0	1	0	0	1	0	1
0	1	0	1	0	0	1	1	0
0	0	1	0	1	0	1	1	0
0	0	1	0	0	1	0	1	0
0	0	1	0	0	1	0	0	1
0	1	0	0	0	1	0	0	0
1	0	0	0	1	0	1	1	1
1	0	0	0	0	1	0	1	0
0	0	1	0	1	0	0	1	0
1	0	0	0	1	0	0	0	0
0	1	0	0	1	0	1	0	0
0	1	0	1	0	0	0	1	0
0	0	1	0	1	0	1	0	1

Fig. 40.3   Binarization of Playing Tennis Dataset

goes on repeatedly for the rest of features until all features are checked or the decision tree got all leaf-nodes. Tree formation steps are as follows:

1. For the root-node, we select dataset A.
2. At every repetition, the algorithm repeatedly goes through all attributes in set A (unused) and calculates Entropy (H) and IG of them.
3. Then, it selects the highest IG attribute among them.
4. The decided attribute divides set A to develop subset of data.
5. On every subset, the same procedure is followed, picking the attributes not selected before.

- The output of the interpreted Python program is shown in Fig. 40.4. The decision tree for playing tennis dataset with maximum IG which in our case is OVERCAST (entropy: 0.94), then we move to 0, HIGH HUMIDITY (entropy: 1.0), WEAK WIND (entropy: 0.722), SUNNY (entropy: 1.0), and finally class 1 and class 2. And then we move to branch 1, and move up to the tree. This is done until all the nodes are found. The comprehensible version of the decision tree is also shown in Fig. 40.4, which one can look for better understanding of the results of interpreted Python program.

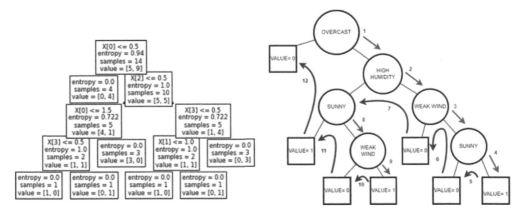

Fig. 40.4   Decision Tree Classifier of Playing Tennis Dataset

3. Tree Hardware Implementation

After the binary tree formation, this resultant tree is used to form the XML file for the representation of tree structure and also the binary tree helps to build hardware description in Verilog format using nested if-else format. The XML file helps in data visualization and the Verilog code helps in the ASIC implementation flow. The XML code and Pseudo Verilog code for the given data set are shown in Figs 40.5 and 40.6, respectively.

```
<root name = "Overcast">
        <child0 name="highHumidity">
                <child0 name= "weakWind">
                        <child0 name= "sunny">
                                <child0> class1 </child0>
                                <child1> class0 </child1>
                        </child0>
                        <child1> class0 </child1>
                </child0>
                <child1 name= "sunny">
                        <child0 name= "weakWind">
                                <child0> class1 </child0>
                                <child1> class0 </child1>
                        </child0>
                        <child1> class1 </child1>
                </child1>
        </child0>
        <child1> class0 </child1>
</root>
```

Fig. 40.5   XML Representation of the Decision Tree

```
module Playtennis( input sunny, input overcast,input highHumidity,
                input weakWind, output reg noplay);
    always@(*)begin
        if (overcast == 0) begin
            if (highHumidity == 0) begin
                if (weakWind == 0) begin
                    if (sunny == 0)
                        noplay <= 4'b0001;
                    else
                        noplay <= 4'b0000;
                    end
                else
                    noplay <= 4'b0000;
                end
            else begin
                if (sunny == 0) begin
                    if (weakWind == 0)
                        noplay <= 4'b0001;
                    else
                        noplay <= 4'b0000;
                    end
                else
                    noplay <= 4'b0001;
                end
            end
        else
            noplay <= 4'b0000;
    end
endmodule
```

Fig. 40.6   Verilog Pseudo Code for the Decision Tree Classifier

The outputs of Python program are taken as the attributes for the Verilog template, which are then arranged with the help of nested if-else statement to form a Decision Tree Classifier for the given dataset. Elements like the module name, i/p ports, and o/p port are added to code, and the description for tree classification program is added as always block. The output of Verilog implementation of classifier is shown in Fig. 40.7.

Fig. 40.7   Output Instance 1 & 2 of Implemented Decision Tree

We can verify the results with the binarize dataset in our step two of design flow. The result matches with the instances in Fig. 40.3 from our binarized data. We evaluate the performance in terms of accuracy. The Playing Tennis Dataset implemented through our Verilog code has 100% accuracy with that of Python implementation. The dataset description along with number of features instances and depth of tree is presented in Table 40.1.

Table 40.1   Description of Datasets

Dataset Name	No. of Instance s	No. of Features	No. of Classes	No. of Binary Features
Playing Tennis	14	6	2	9
Spect	267	22	2	22

The elaborated design of Verilog code of playing tennis data set is shown in Fig. 40.8. The elaborated design consists of 6 MUXs cascaded to form a tree-like structure with features like OVERCAST, HIGH HUMIDITY, WEAK WIND AND SUNNY as the select lines and NOPLAY as the output of MUX tree. The synthesis result of the implemented Verilog code for Playing Tennis Dataset is shown in Table 40.2.

Table 40.2   Synthesis Result for Playing Tennis Dataset

Resources	Utilization	Utilization %	Available
LUT	1	0.01%	20800
IO	5	4.72%	106

The playing tennis dataset overall utilization is very less as it only uses 1 LUT, 0 Flip flops and latches and only 5 IOs.

Performance of the datasets is being evaluated in terms of accuracy, that they show during and after the implementation as shown in Table 40.3, the trained model accuracy is compared in the table.

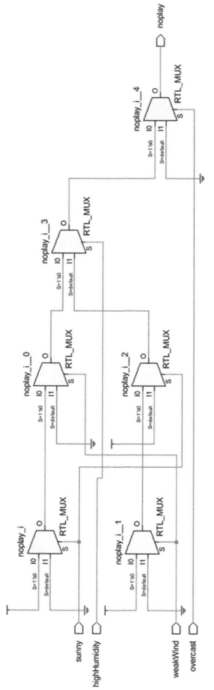

Fig. 40.8 Elaborated design of implemented decision tree

Table 40.3 Proposed Work in Terms of Accuracy

S. No	Dataset	Proposed Work Accuracy	Others
1.	Playing Tennis	100%	100%
2.	Spect	98%	71.72%

4. Conclusion

We used Playing Tennis and Spect dataset to implement the proposed unique framework. The output of the work, Verilog Designs, were produced as the final-product for the trained ML Classifier. The HDL implementation output fully resembles to that of our standard Decision Tree along with the Elaborated Design and XML representation. FPGA prototyping of the Verilog code is required to evaluate other performance metrics for the dataset like throughput and latency. The average accuracy of generated classifier is matched approximately based on trained model testing.

To further improve the proposed design flow, we intend to study other ML classifiers and algorithms like Support Vector Machine, Random Forest, Artificial Neural Networks, etc.; also we intend to examine the frameworks' response to these algorithms and analyze their performance for real-time application.

REFERENCES

1. M. A. Talib, S. Majzoub, Q. Nasir, and D. Jamal, "A systematic literature review on hardware implementation of artificial intelligence algorithms," J. Supercomput., early access, May 28, 2020. [Online]. Available: https://link.springer.com/article/10.1007/s11227-020- 03325-8#citeas, doi: 10.1007/s11227-020-03325-8..

2. Play Tennis. Accessed: Jan. 25, 2020. [Online]. Available: https://www. kaggle.com/fredericobreno/play-tennis

3. Archive.ics.uci.edu, 2021 [online] Available: http://archive.ics.uci.edu/ml.

4. A. Hernandez, H. Fabelo, S. Ortega, A. Baez, G. M. Callico, and R. Sarmiento, "Random forest training stage acceleration using graphics processing units," in Proc. 32nd Conf. Design Circuits Integr. Syst. (DCIS), Nov. 2017, pp. 1–6.

5. M. Owaida, H. Zhang, C. Zhang, and G. Alonso, "Scalable inference of decision tree ensembles: Flexible design for CPU-FPGA platforms," in Proc. 27th Int. Conf. Field Program. Log. Appl. (FPL), pp. 1–8, Sep. 2017.

6. F. Saqib, A. Dutta, J. Plusquellic, P. Ortiz, and M. S. Pattichis, "Pipelined decision tree classification accelerator implementation in FPGA (DTCAIF)," IEEE Trans. Comput., vol. 64, no. 1, pp. 280–285, Jan. 2015.

7. A. Elkanishy, D. T. Rivera, A.-H.-A. Badawy, P. M. Furth, Z. M. Saifullah, and C. P. Michael, "An FPGA decision tree classifier to supervise a communication SoC," in Proc. IEEE High Perform. Extreme Comput. Conf. (HPEC), pp. 1–6, Sep. 2019.

8. H. Zhang and R. Zhou, "The analysis and optimization of decision tree based on ID3 algorithm," in Proc. 9th Int. Conf. Model., Identificat. Control (ICMIC), pp. 924–928, Jul. 2017.

Intelligent Systems and Smart Infrastructure – Brijesh Mishra et al. (eds)
© 2023 Taylor & Francis Group, London, ISBN 978-1-032-41287-0

Big Data Analytics: Integrating Machine Learning with Big Data Using Hadoop and Mahout

Pooja Rani[1]

MMICTBM, Maharishi Markandeshwar (Deemed to be University), Mullana, Ambala, Haryana, India

Rohit Lamba[2]

Department of ECE, MMEC, Maharishi Markandeshwar (Deemed to be University), Mullana, Ambala, Haryana, India

Ravi Kumar Sachdeva[3]

Chitkara University Institute of Engineering and Technology, Chitkara University, Punjab, India

Rajneesh Kumar[4]

Department of Computer Engineering, MMEC, Maharishi Markandeshwar (Deemed to be University), Mullana, Ambala, Haryana, India

Priyanka Bathla[5]

Department of Computer Science & Engineering, Chandigarh University, Gharuan, Mohali, Punjab, India.

Abstract

As technology is advancing, data generated by users is increasing day by day. Traditional database management systems are not capable of storing and processing large sets of data. Big data is used to store such larger datasets and it can store structured, unstructured, and semi-structured data. However, only storing data has no use if it is not processed to get some useful results. Big data analytics is a field concerned with the analysis of Big data and making useful decisions and conclusions. Machine learning deals with providing the ability to machines to automatically learn to provide intelligence to machines just like humans. In Big data, large

Corresponding author: [1]pooja.rani@mmumullana.org, [2]rohitlamba14@mmumullana.org, [3]ravisachdeva1983@gmail.com, [4]drrajneeshgujral@mmumullana.org, [5]priyankabathla85@gmail.com

DOI: 10.1201/9781003357346-41

amount of data is available, and through machine learning, this data is processed to provide useful decisions. This paper introduces the concepts of Big data and machine learning. It also introduces the approach to integrating machine learning with Big data using the Hadoop and Mahout tools.

Keywords: Big Data, Machine Learning, Hadoop, Mahout, HDFS, MapReduce

1. Introduction

Big data term is used to refer to those datasets whose size is very large, and due to its size, typical database applications cannot be used to manage such data. The need for Big data arose because data generated per day is continuously increasing due to the advancement of technology. Storing and analyzing such a large amount of data was not possible with the traditional database management system. Therefore, concept of Big data came into existence to handle such a large amount of data. In Big data, different types of data are stored. So, the schema of such data cannot be defined (Tsai et al. 2015). Big data has five main characteristics (Aburawi and Albaour 2021):

(i) *Volume:* The amount of data is very large. So we cannot store it on a single system. Big data is stored on a clustered system. It uses commodity hardware for storage.

(ii) *Variety:* Data such as audio, video, images, and text can be stored. Data stored can be structured, semi-structured, or unstructured. Due to different types of data, the schema of data cannot be defined.

(iii) *Velocity:* Data is generated at a high speed. For example, whenever we add a post to Facebook, data is generated. Whenever we purchase any item on Flipkart, data is generated. So a huge amount of data is generated per second.

(iv) *Value:* Data should be helpful for any type of decision-making. It is only possible if data can be analyzcd.

(v) *Veracity:* There can be inconsistencies in the data. Data is uncertain and we have to deal with those inconsistencies to properly analyze the data.

The main advantage of Big data comes from its size. A large amount of data means there will be a vast amount of knowledge we can discover from that data. By discovering the knowledge from Big data, we can use this knowledge to make useful decisions. Big data analytics is used to analyze the data and extract knowledge from Big data (Elgendy and Elragal 2014). Machine learning is used to perform data analytics on Big data. Machine learning helps the machine to analyze the data, learn from the data, and take decisions automatically. Data analytics can be done with the help of supervised or unsupervised learning. In supervised learning, desired outputs are given for inputs (Kumar and Rani 2020). By analyzing some given data of inputs and outputs, machine learns. In unsupervised learning, only input is given and patterns in the data have to be identified by the system. There is no training data for the system (Lamba et al. 2022). When we combine machine learning with Big data, we take advantage of the huge amount of data to make automatic decisions (Al-Sai and Abdullah 2019). Machine learning and Big data can be used in a variety of applications. For example, we can use machine learning and Big data in automotive industries (Do Nascimento 2021). We can use machine learning and Big data to suggest users items to purchase on ecommerce websites. However Big data analytics has certain challenges associated with it due to Big data's characteristics as shown in Fig. 41.1.

Fig. 41.1   Challenges of Big data

Challenges of Big data are as given below (Rawat and Yadav 2021):

(i) *Incomplete Data:* Data stored can be incomplete. Analyzing such incomplete data is difficult.

(ii) *Heterogeneous Data:* Schema for data cannot be defined because different types of data are stored. Data can be structured, unstructured, or semi-structured. Analyzing such heterogeneous data is difficult.

(iii) *Security:* Data stored should be secure to avoid incorrect use. Security of data is a major challenge.

(iv) *Size:* Huge amount of data is stored in Big data and size of this data continuously increases. Handling such a large amount of data is a major challenge.

2. Literature Survey

Hassan et Al. (2017) described Big data characteristics consisting of volume, variety, velocity, and veracity. The authors also described security, size, incomplete, and heterogeneous data as Big data challenges. Jain and Kumar (2015) explained that it is expensive to store and process Big data because of the requirement of a lot of space; therefore, cloud computing can be used as a solution to this problem. In cloud computing, cluster of computers can be used for the storage and processing of huge amounts of data. Bhosale and Gadekar (2014) in their paper presented that there are various problems in Big data processing due to its size and heterogeneous nature and Hadoop can be used to deal with those problems. Prakash and Aloysius (2018) described various technologies of Big data. The authors also explained the architecture of Hadoop as including various components. Among these components, the Map

Reduce component is a programming framework that is required for processing data stored on clustered systems. Manikandan and Ravi (2014) explained in their paper the implementation of HDFS (hadoop distributed file system) clusters within Hadoop. In a HDFS cluster, there is one name node and more than one data node. Data is divided among data nodes and name node acts as a master to monitor all data nodes. Alam and Ahmad (2014) described the architecture of Hadoop consisting of two main components: HDFS and Map Reduce. They described that two functions map and reduce are programmed according to the processing required. Chen and Zhanga (2021), in their survey paper, described that machine learning can be used as a tool to discover knowledge from Big data. L'heureux et al. (2017) described various challenges that occur when we use machine learning with Big data, including dirty data, data locality, data uncertainty, data availability, real-time processing, etc. Dhyani and Barthwal (2014) described that Big data can help in enhancing the efficiency of any process if it is analyzed. Authors also presented various projects related to Hadoop such as Pig Latin, Pig Runtime, Hive, and Mahout. Acharjya and Ahmed (2016), in their paper, described Mahout as a tool for Big data analysis which provides various algorithms for machine learning. Madasamy and Parameswari (2018) described that Mahout can be used for providing machine learning in automotive industries. The authors also described machine learning algorithms such as classification, recommendation, and clustering implemented on Mahout. Esteves and Rong (2011), in their paper, described clustering algorithms as hierarchical and partitioned. The authors described k-means clustering implemented with Hadoop and Mahout. In k-means clustering, k-clusters are formed. Initially, random data points are chosen as the centroid of each cluster. After that, data points are placed in each cluster depending upon the distance between the cluster centroid and data point. A data point is placed in the cluster for which distance will be minimum. After deciding the cluster of each data point, new centroid of each cluster is calculated. Then the cluster of each data point is changed depending upon the distance between the data point and new centroid. This process is repeated until there are no improvements. The authors concluded that running the K-means algorithm in a distributed manner enhances its performance. Eluri et Al. (2016) described that canopy clustering can be used before hierarchical or k-means clustering if clustering is to be done on larger data sets to increase speed. Daoping et Al, (2016) in their paper gave implementation of parallel clustering algorithm using Mahout.

3. Hadoop Architecture

Hadoop is a framework that is used for storing Big data and performing Big data analytics. Hadoop provides distributed storage for data. Hadoop stores files on commodity hardware. Data is distributed across many nodes. Applications run in a distributed manner across many nodes. So if any node fails, task can be assigned to another node. So, it provides fault-tolerant behavior. Hadoop has many components among which HDFS and Map Reduce are two main components (Anuradha 2015). Architecture of Hadoop is shown in Fig. 41.2.

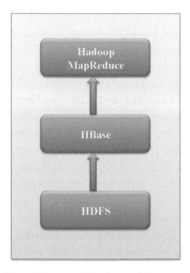

Fig. 41.2 Architecture of Hadoop

A. HDFS

HDFS component is responsible for storing Big data in distributed manner. Data is broken into blocks and distributed across nodes. There is also redundancy while distributing data blocks. Data blocks are duplicated across many nodes. The number of nodes in which blocks are duplicated is known as the replication factor which can be configured by the user. Its default value is 3 (Masadeh, Azmi and Ahmad 2020).

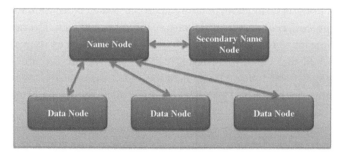

Fig. 41.3 Architecture of HDFS

This duplication makes the hadoop a fault-tolerant system. Figure 41.3 shows the architecture of HDFS. There are three components in HDFS:

(i) *Name Node:* There is one name node in HDFS. It will not contain data blocks. It will contain information about which data blocks are stored in which data nodes.

(ii) *Data Node:* There are many data nodes in HDFS. Data blocks are distributed across data nodes.

(iii) *Secondary Name Node:* It provides backup to name node when name node fails. If name node fails, then after restarting gets the information about the previous state from the secondary name node.

B. Map Reduce

Map Reduce is a programming model to deal with the processing of Big data. It can process unstructured and structured data. It is responsible for retrieving the Big data and performing the processing. Map Reduce distributes the task among nodes using two functions Map() and Reduce(). Map () function distributes the task among nodes by dividing a module into sub modules. There is a Master node that performs this allocation of tasks. Slave nodes can further divide the sub-modules. So, a task runs in distributed manner by dividing among various nodes. Reduce () function is responsible for collecting the results from different nodes and forming a final output. Map() is responsible for dividing the task and reducing is responsible for aggregating the results (Demirbaga 2021). In Map Reduce framework, there are two types of nodes: salve nodes and master nodes. Master nodes divide the task and assign the task to slave nodes. Master nodes are also known as JobTracker because they do the task of job scheduling. Task is divided among various slave nodes, and slave nodes implement the assigned task. (Rajendran et al. 2021). Figure 41.4 shows the architecture of Map Reduce.

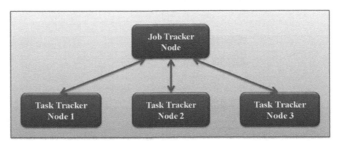

Fig. 41.4 Architecture of Map Reduce

4. Integrating Machine Learning with Big Data

Machine learning can be integrated with Big data using Mahout tool. Mahout is a Big data analysis tool and part of the Hadoop Framework (Anil et al. 2020). It uses Map Reduce in combination with Hadoop to implement machine learning. It runs on the top of Map Reduce. With the help of Mahout, various machine learning techniques as given below can be implemented (Allam 2018):

(i) *Recommendation:* In this technique, by analyzing the action user has performed, such as items which the user has purchased, recommendations can be done.

(ii) *Categorization:* In this technique, some sample data in with both input and output are provided. It is a supervised learning technique. The system classifies data into various categories.

(iii) *Clustering:* In this technique, objects are combined into various groups. Similar objects are combined into one group which is known as a cluster. Mahout implements various clustering algorithms such K-Means and Canopy Clustering. In clustering algorithms, machine learns by observations. It is unsupervised learning. There are two types of clustering algorithms: partitioned and hierarchical. In hierarchical clustering algorithms, data is decomposed into groups hierarchically (Manikandan and Ravi 2014). Machine learning can be combined with Big data using Hadoop Architecture in which HDFS of Hadoop will take care of providing storage to Big data and Mahout tool will be used for machine learning (Naureen 2021). Figure 41.5 shows the roles of different parts of the Hadoop framework for storage, processing, and machine learning associated with Big data. Figure 41.6 shows that Mahout runs on the top of Hadoop Map Reduce.

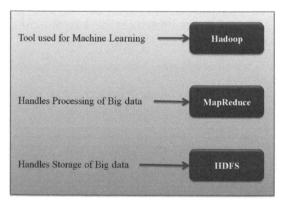

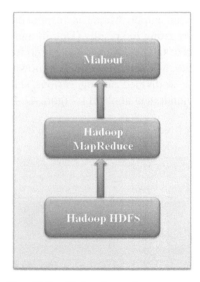

Fig. 41.5 Roles of different parts of Hadoop framework to handle storage, processing, and machine learning with Big data

Fig. 41.6 Mahout running over Hadoop Map Reduce

5. Conclusion

Big data provides way of storing a large amount of data. It also gives flexibility to store any type of data. Schema of Big data cannot be defined because of the variety in the data. Storing data is useless if we don't analyze the data and extract some useful information from it. Big data analytics help in analyzing the data for knowledge discovery. Areas of machine learning and Big data can be integrated to get efficient use of data. In this paper, the mechanism of integrating machine learning with Big data is discussed. Mahout tool used with Hadoop framework can easily implement machine learning with Big data. When we run Mahout over Map Reduce, we get the efficient implementation of machine learning algorithms because algorithms run on clustered systems. When we use more than one node to run the algorithm, performance increases. So using Mahout over Hadoop is an efficient way to integrate machine learning and Big data.

REFERENCES

1. Tsai, C. W., Lai, C. F., Chao, H. C., Chao and Vasilakos, A. V. (2015). Big data analytics: a survey. Journal of Big Data. 2(21). https://doi.org/10.1186/s40537-015-0030-3.

2. Aburawi, Y. A. and Albaour, A. (2021). Big Data: Review Paper. International Journal of Advance Research and Innovative Ideas in Education. 7(1).

3. Elgendy, N., and Elragal, A. (2014). Big data analytics: a literature review paper. In Industrial conference on data mining (pp. 214–227). Springer, cham.

4. Kumar, R., and Rani, P. (2020). Comparative Analysis of Decision Support System for Heart Disease. Advances in Mathematics: Scientific Journal. 9(6): 3349–3356. https://doi.org/10.37418/amsj.9.6.15.

5. Lamba, R., Gulati, T. and Jain, A. (2022). A Hybrid Feature Selection Approach for Parkinson's Detection Based on Mutual Information Gain and Recursive Feature Elimination. Arabian Journal for Science and Engineering. https://doi.org/10.1007/s13369-021-06544-0.

6. Al-Sai, Z. A., and Abdullah, R. (2019). Big data impacts and challenges: a review. In IEEE Jordan International Joint Conference on Electrical Engineering and Information Technology (pp. 150–155). IEEE.

7. Do Nascimento, I. J. B., Marcolino, M. S., Abdulazeem, H. M., Weerasekara, I., Azzopardi-Muscat, N., Goncalves, M. A. and Novillo-Ortiz, D. (2021). Impact of big data analytics on people's health: Overview of systematic reviews and recommendations for future studies. Journal of medical Internet research. 23(4): e27275.

8. Rawat, R., and Yadav, R. (2021). Big data: Big data analysis, issues and challenges and technologies. In IOP Conference Series: Materials Science and Engineering 1022(1): 012014. IOP Publishing.

9. Hassan, R., Manzoor, R. and Ahmad, M.S. (2017). BIG DATA & HADOOP: A Survey. International Journal of Computer Science and Mobile Computing. 6(7): 64–68.

10. Jain, V.K. and Kumar, S. (2015). Big Data Analytic Using Cloud Computing. In International Conference on Advances in Computing and Communication Engineering (pp. 667–672). IEEE. https://doi.org/10.1109/ICACCE.2015.112.

11. Bhosale, H.S. and Gadekar, D.P. (2014). A Review Paper on Bigadata and Hadoop. International Journal of Scientific and Research Publications. 4(10): 1–6.

12. Prakash, A.A. and Aloysius, A. (2018). Architecture Design for Hadoop No-SQL and Hive. International Journal of Scientific Research in Computer Science, Engineering and Information Technology. 3(1): 1069–1073. https://doi.org/10.32628/CSEIT1831245.

13. Manikandan, S.G. and Ravi, S.. (2014). Big Data Analysis using Apache Hadoop. In International Conference on IT Convergence and Security (pp. 1–4). IEEE. https://doi.org/10.1109/ICITCS.2014.7021746.

14. Alam, A. and Ahmad, J. (2014). Hadoop Architecture and its issues. In International Conference on Computational Science and Computational Intelligence (pp. 288–291). http://dx.doi.org/10.1109/CSCI.2014.140.

15. Chen, C. L., and Zhanga, C. Y. (2021). Data-intensive applications, challenges, techniques and technologies: A survey on Big Data. Information Sciences. 275: 314–347. https://doi.org/10.1016/j.ins.2014.01.015.

16. L'HEUREUX, A., Grolinger, K., Elyamany, H. F. and Capretz, M. A. M. (2017). Machine Learning With Big Data: Challenges and Approaches. IEEE Access 5: 7776–7797. https://doi.org/10.1109/ACCESS.2017.2696365.

17. Dhyani, B., and Barthwal, A. (2014). Big Data Analytics using Hadoop. Journal of advanced Computer Science and Applications. 108(12): 1–5.

18. Acharjya, D.P., and Ahmed, P.K. (2016). A Survey on Big Data Analytics: Challenges, OpenResearch Issues and Tools. International Journal of Advanced Computer Science and Applications. 7(2): 511–518. https://dx.doi.org/10.14569/IJACSA.2016.070267.

19. Madasamy, J. and Parameswari, R. (2018). Data Mining and Machine Learning Techniques (Mahout on Hadoop) For Automotive Industries: A Survey. International Journal of Pure and Applied Mathematics 119(10): 1–5.

20. Esteves, R.M., and Rong, C. (2011). K-means clustering in the cloud - A Mahout test. In International Conference on Advanced Information Networking and Applications (pp. 514–519). IEEE. https://doi.org/10.1109/WAINA.2011.136.

21. Eluri, V.R., Ramesh, M., Al-Jabri, A.S.M. and Jane, M. (2016). A Comparative Study of Various Clustering Techniques on Big Data Sets using Apache Mahout, In MEC International Conference on Big Data and Smart City (pp. 1–4). IEEE. https://doi.org/10.1109/ICBDSC.2016.7460397.

22. Daoping, X., Alin, Z. and Yubo, L. (2016). A parallel Clustering algorithm implementation. In International Conference on Instrumentation & Measurement, Computer, Communication and Control (pp. 790–795). https://doi.org/10.1109/IMCCC.2016.9.

23. Anuradha, J. (2015). A brief introduction on Big Data 5Vs characteristics and Hadoop technology. Procedia computer science. 48: 319–324. https://doi.org/10.1016/j.procs.2015.04.188.

24. Masadeh, M.B., Azmi, M.S. and Ahmad, S.S.S. (2020). Available techniques in hadoop small file issue. International Journal of Electrical and Computer Engineering. 10(2): 2097. http://doi.org/10.11591/ijece.v10i2.pp2097-2101.

25. Demirbaga, U. (2021). HTwitt: a hadoop-based platform for analysis and visualization of streaming Twitter data. Neural Computing and Applications. https://doi.org/10.1007/s00521-021-06046-y.

26. Rajendran, S., Khalaf, O.I., Alotaibi, Y. and Alghamdi, S. (2021). MapReduce-based big data classification model using feature subset selection and hyperparameter tuned deep belief network. Scientific Reports 11: 24138. https://doi.org/10.1038/s41598-021-03019-y.

27. Anil, R., Capan, G., Drost-Fromm, I., Dunning, T., Friedman, E., Grant, T. and Yilmazel, O.. (2020). Apache Mahout: Machine Learning on Distributed Dataflow Systems. Journal of Machine Learning Research. 21(127): 1–6.

28. Allam, S. (2018). Usage of hadoop and microsoft cloud in big data analytics: an exploratory study. International Journal of Innovations in Engineering Research and Technology. 5(10): 27–32.

29. Naureen, A. (2021). Big Data Analytics with Hadoop. International Journal of Engineering Research & Technology. 9(5): 33–37.

Intelligent Systems and Smart Infrastructure – Brijesh Mishra et al. (eds)
© 2023 Taylor & Francis Group, London, ISBN 978-1-032-41287-0

CHAPTER 42

Detection of Fake News Using Machine Learning

Smriti Agarwal[1], Simran Goel[2], Shubhi Agarwal[3], Amit Kumar Saini[4] and Vimal Kumar[5]

Department of Computer Science and Engineering,
Meerut Institute of Engineering and Technology,
Meerut 250005, U.P., India

Abstract

In this paper, we have proposed that in the modern era, faux news is a key point in the lives of political world, followed by social media. Fake news detection is a key investigation to be done for its diagnosis but it has some summons too. With the increase in the number of social media, platform such as Instagram and Telegram, the false news is propagating at a very high speed to billions of users in a very short time span and spreading its bad effects to the readers. This widely spreading of fake news is creating a big problem in society and has far-reaching results as the biased behavior done in election for profiting some candidates or political parties. With the help of this paper, we hereby aim to present binary categorization of variety of news reports and article in the society. We compare different machine learning classification techniques and Natural Language Processing. With the help of this paper, we collate two different machine learning classification approaches. We the authors hereby come up with the idea and abilities to differentiate and determine whether the news is fake or true in very manner it can be.

Keywords: Machine Learning, NLP, Fake News, Classification

Corresponding author: [1]smriti.agarwal.cs.2018@miet.ac.in, [2]simran.atul.cs.2018@miet.ac.in, [3]shubhi.agarwal.cs.2018@miet.ac.in, [4]amit.cs@miet.ac.in, [5]vimal.kumar@miet.ac.in

DOI: 10.1201/9781003357346-42

1. Introduction

As we all know, now a days the surfing on the internet and connectivity throughout the globe via social networking websites increasing day by day and our most of the times and lives is associated with the internet today. And presently the society and its people are hereby habited to believe in online news rather than the traditional news medium like newspaper. In order with some inspection studies that are controlled to chase out the influences of any false and fabricated message on returning via such fake news information (T. Granskogen and J. A. Gulla, 2017). The finest example for fake news is during the situation of pandemic which occurred in the entire world in the last two years. There are a variety of news articles till now that are false and fabricated and used solely to create confusion and uncertainty in the minds of discrete people and this all things tends to mislead their brains to trust that false news. Though, does anyone consider if it's real or fake?

Without discussing about the goodies of social networking websites, the Online provided articles and stories are always less than the conventional news firms. The results of analysis and survey have found that the millions of news from platforms like Twitter are generated as faux news for illegal purposes. This spreading and transmitting of false news worldwide in couple of seconds has caused a negative and strong influence on public and throughout the globe as well. While throughout the country, artiodactyl we with lime was bombarded as a preservation against the COVID-19, false information on Indian social media platforms like Indian news channels led to voters drinking and eating cow poo in order to prevent illness. The scientists placed together the rumors and looked into completely different buzz, like uptaking garlic drink, sportive hot socks, and lay out goose's fat on one's chest, as remedies for this dangerous virus. The fake account and social networking sites have been created by Russian countries to spread the false stories and reports. Fake news is causing some significant changes since some of its misleading reports aim to upset social cohesion and, secondly, to make it difficult for the general public to distinguish between the real news and the fake news.

Rest of the paper is structured as follows. Section 2 reviews the related work. Section 3 explains the research methodology. Section 4 discusses the implementation and result analysis. Section 5 summarizes the paper.

2. Related Work

With the thought of detecting the feasible reliability of the global knowledge, the paper aims to collect variety of linked work on the field of the social network of false news detection and use a simple technique for detection of fake news by using Passive Aggressive Classifier and Naive Bayes classifier. This method is being used to create software systems and is being tracked against a data set. They achieved classification accuracy of approximately **93.69%.**

Gupta (Madisetty and Desarkar, 2018) had given a structure in which to handle thousands of tweets in nano or micro seconds, and we build a variety of ML techniques that trade with different complications, including accuracy repository, hour lag with high processing time. First of all, in this, they collected 4 lakhs tweets from twitter dataset. Further, they will differentiate between spam and non-spam tweets. Then they will obtain some traits. And finally, they tend to gain a rightness which is 93.45% and then transcend the desired result by approximately 18%.

3. Proposed Methodology

This project is concerning building a false news detection model using the two machine learning algorithms. The focus of this project is on model development in a machine learning using a colab notebook, it isn't always developing various conventional package systems. Machine learning generally needed a good amount of time for training and testing of model, and also an immense and good value of dataset. In summary, using the model to approximate foresight entails choosing good predictions for the future. If the model involves news projections, the projections must result in a degree of accuracy and consistency when the predictions are compared to fake and true news (Dhruv, K., Jaipal Singh, 2019).

The whole paper is divided into two different categories.

The first category consists of machine learning classifier. The algorithms used for creating the classifier are Naïve Bias and Passive Aggression (Granik and Mesyura, 2017). Both the algorithms could be used for training the above model but out of two, the best is Passive Aggressive classifier. The model is trained using fake or real news dataset taken from Kaggle. The second category consists of the authentication model checking the authenticity of the input URL entered by the user. We have used Python programming language and various utilities are imported from the inbuilt libraries provided by Python. The most common library used in our project provided by Python is Sklearn. The front-end of the project is designed by HTML, CSS, or Bootstrap. The whole interface is created using DOM (Document Object Model) providing the structure to the webpage and making the code easy to read and more understandable. This also makes the code easily maintainable (Gilda, 2018).

The backend of the application is created using Flask framework, which supports Python programming language.

A. System Architecture

(i) Static Search

The fake news detector of the static search architecture and the system is pretty good and used the primary process of machine learning flow. The design for this search is shown in Fig. 42.1.

(ii) Dynamic Search

In the dynamic search, it asks for particular reserve words to be searched and prepared. The results our machine learning process consist of the following steps and shown in Fig. 42.1.

- We load the data set from references.
- Once the data set is loaded, we need to preprocess the data set by cleaning the text applying some tokenization techniques, i.e., removing stop word and drop the duplicate.

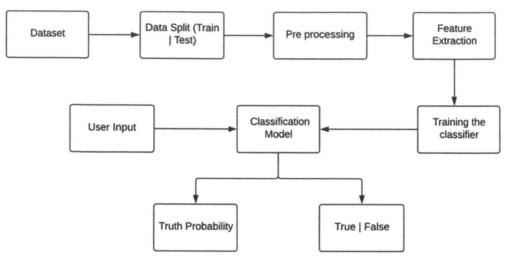

Fig. 42.1 Architecture of Static search

- Then we have a set of pre-processed data.
- We need to represent them into a numeric format, and we call this step as vectorization, where we extract the features of the data by dividing it into different groups and taking the best out of all.
- The next step is to create a trained model from the extracted data and text and evaluate the model for providing the result from the dataset.
- The final result of our model it may be true opinion or it may be fake opinion.

4. Implementation and Result Analysis

A. Implementation Steps

(a) Static Search Implementation

The first search implementation part of our model trained our dataset by using 2 algorithms out of 4 algorithms for classification process. These algorithms are Naïve Bayes and Passive Aggressive Algorithm (Ray, 2017).

Step 1: The first step is that our preprocessed dataset is extracts different features from it. These features are CountVectorizer and Tf-Idf Features.

Step 2: Then, for estimating the fake news detection, we have built the classification system and the extracted features are to be fall into different classifiers. Naïve Bayes and Passive Aggressive from skleam are used. This feature of extracted data used in each classifier.

Step 3: We calculate the accuracy from our classifier and build a confusion matrix.

Step 4: After all the training is done, we select the best model for our project fake news classification based upon the accuracy.

Step 5: Our project gives the best model, i.e., Passive Aggressive Classifier, which is based upon the accuracy of our data and then will predict the data as either true or faux news.

(b) Dynamic Search Implementation

This type of implementation contains a search field which is as follows:

(i) Dataset content are to be searched upon.

In searching, firstly, to fix the correct problem statement, the dynamic search field, Natural Language Processing (NLP), is used, and for that reason, we have tried to produce a model which can categorize the news as fake news as stated by the designation used in the newspaper or magazines articles. To pass it through a Passive Aggressive Classifier, we have used approaches of NLP like TF-IDF Vectorization and Count Vectorization ahead to yield the originality as a percentage probability of an article (Dong and Victor, 2019).

Working

Working is divided into 3 different assertions:

(i) For the news article, the project checks the reliability using the NLP.
(ii) Whenever the user is in a confusion regarding the news, then the user can directly go to the website and check for the news and would get a good response as an output.
(iii) Authenticity of a news source will be checked in this third step.

B. Confusion Matrix

In this, matrix named confusion matrix is likely easy to evaluate and recognize, but the associated phraseology can be complicating. In terms of its definition, it is a table that is often used to represent the effectiveness of a classification model based on known values for a set of test data. It is helpful because they give relatively simple values. Confusion matrix allows the inspection of the efficiency of an algorithm. It is an abstract of estimating result on a classification system in confusion matrix.

Precision: In this, we predict the truly positive values among all. The value generally lies between 0 and 1.

$$PR = \frac{TP}{TP + FP}$$

Recall: It is also known as true positive rate. In this, we determine predicted positive out of total positive values

$$RE = \frac{TP}{TP + FN}$$

Accuracy: It is defined as the number of total true prediction upon total number of required datasets.

$$CA = \frac{TP + TN}{TP + TN + FP + FN}$$

F1-score: In this, we take both false positive and negative values, and as the result, it is the mean (harmonic) of precision and recall.

$$F1 - \text{score} = \frac{2TP}{2TP + FP + FN}$$

Matrices of this kind are frequently used in the section of machine learning and authorize us to estimate the representation of a model from distinct outlook.

C. Results

The project working was completed by different algorithms with Vector features – Count Vectors and Tf-Idf Vectors. Accuracy was calculated for all algorithms.

(a) Static search confusion matrix

The various features are extracted (Count Vectorizer, Tf-Idf) on two different classifiers (Naïve Bayes, Passive Aggressive Algorithm); by this, the confusion matrix gives the result and detect whether the news is fake or real.

Table 42.1 Comparison of two classifiers used

Classification	Precision	Recall	F1-Score	Accuracy
Naïve Bayes	0.89	0.89	0.89	89.34%
Passive Aggressive	0.93	0.93	0.93	93.45%

For our project, the best model is Passive Aggressive Classifier with an accuracy of 93.45%.

5. Conclusion

Fake news detection research has never been more than it is currently, mainly during a pandemic time when the whole world is fighting against it. The approaches discussed in this paper aspect from the real news. So many techniques and criteria for fake news detection are there. Accuracy of fake news detection tasks can also be affected by dataset online via social media platform. Earlier, newspapers were favored as hard-copies are now being swapped by applications like Instagram, Twitter, WhatsApp, Facebook and news articles to be studied online. Hence, in direction to control the situation, we have built our faux news detection model. In this, the user grasps the input and categorize it to be correct/incorrect. For execution, we have used different Machine Learning algorithms and NLP. The model is trained of dataset utilizing known dataset, and using the performance calculation, a variety of performance estimations are done. The system gives the finest accuracy for the best model which is being used to categorize the report. As notified above, our most valuable system appears to be Passive Aggressive Classifier, with an accuracy of 93.45%. The exactness for dynamic search is 93.45%, and with each and every repetition, it is growing. A compound of automated and human techniques gives hike to a hybrid technique. We desire this report summons you to unite the dispute against faux news by producing better, greater, and best results.

REFERENCES

1. S. Helmstetter and H. Paulheim, "Weakly supervised learning for fake news detection on Twitter," Proc. 2018 IEEE/ACM Int. Conf. Adv. Soc. Networks Anal. Mining, ASONAM 2018, pp. 274–277, 2018.
2. T. Granskogen and J. A. Gulla, "Fake news detection: Network data from social media used to predict fakes, vol. 2041, no. 1, pp. 59–66, 2017.
3. S. Gilda, "Evaluating machine learning algorithms for fake news detection," IEEE Student Conf. Res. Dev. Inspiring Technol. Humanit. SCOReD 2017 - Proc., vol.2018–January, pp. 110–115, 2018
4. Dhruv, K., Jaipal Singh, G., Manish, G., Vasudeva, V.: Mvae: multimodal variational autoencoder for fake news detection. In: Proceedings of the 2019
5. Dong, X., Victor, U., Chowdhury, S., & Qian, L. (2019). Deep Two-path Semi-supervised Learning for Fake News Detection. arXiv preprint arXiv:1906.05659.
6. H. Gupta, M. S. Jamal, S. Madisetty and M. S. Desarkar, "A framework for real-time spam detection in Twitter," 2018 10th International Conference on Communication Systems & Networks (COMSNETS),Bengaluru, 2018, pp. 380–383
7. https://flask.palletsprojects.com/en/2.0.x/
8. M. Granik and V. Mesyura, "Fake news detection using naive Bayes classifier," 2017 IEEE 1st Ukr. Conf. Electr. Comput. Eng. UKRCON 2017 - Proc., pp. 900–903, 2017.
9. RAY,S.https://www.analyticsvidhya.com/blog/2017/09/common-machine-learning-algorithms/2017, September.

Intelligent Systems and Smart Infrastructure – Brijesh Mishra et al. (eds)
© 2023 Taylor & Francis Group, London, ISBN 978-1-032-41287-0

CHAPTER

43

Enhancing the Identification and Design of Token Smart Contracts on the Ethereum Blockchain

Annie Silviya S. H.[1], B. Sriman[2], Pranshu Jha[3]

Department of Computer Science and Engineering
Rajalakshmi Institute and Technology (RIT)
Chennai, India

Abstract

In the world of cryptocurrency, the word "token" simply refers to digital assets owned by individuals. Although, contextually, it can have different meanings, the term token usually refers to cryptocurrency denominations. Tokens utilise other blockchains, which are not native cryptocurrency. Consider ERC-20 as an example. ERC-20 utilises the Ethereum blockchains but it is not Ethereum's native cryptocurrency. Tokens have multiple use-cases like in the form of governance tokens, in decentralised financial (DeFi) sector, as NFTs and many more. Ethereum, being the first programmable blockchain and the second largest cryptocurrency, has the most popular tokens used by millions everyday be it for DeFi, governance, or in the form of NFTs. Since the tokens are built upon the robust and decentralised infrastructure of blockchain, they are automatically secure and privacy-oriented. Through this paper, we present a detailed and structured study on classification of these standard Ethereum tokens.

Keywords: Cryptocurrency, Tokens, Ethereum, NFT, blockchain

Corresponding author: [1]anniesilviya.sh@ritchennai.edu.in, [2]sriman.b@ritchennai.edu.in, [3]pranshujha.2021.cse@ritchennai.edu.in

DOI: 10.1201/9781003357346-43

1. Introduction

Ethereum is one of the most popular platforms in use today for building tokens and decentralized applications that make use of the blockchain infrastructure. Ethereum currently has 10 different token standards, each bringing forth their own advantages and disadvantages. The 10 token standards currently used by the Ethereum community are ERC 20, ERC 165, ERC 621, ERC 777, ERC 827, ERC 884, ERC 721, ERC 223, ERC 865, and ERC 1155. Through this study of these Ethereum tokens, we have portrayed the use-cases of all the standards, highlighting their pros and cons and thus concluding with the best standard to be used for a specific use case.

A. Token Basics

The existence, and the role, of differential ownership structure and types of equity ownership fuel the debate of determining the stock return. It is observed that developed countries like USA, UK, and Canada mostly witness dispersed ownership structure, while emerging countries like India have concentrated ownership (Laporta et al., 1999). Further, it found institutional investors as one of the key owners in the corporate ownership structure, where they influence the management through their holdings and monitoring (McNulty and Nordberg, 2016). Due to the existence of differential ownership, the debate on the effect of ownership holdings on stock return becomes very intense. Here, this paper is mostly based on the effect of differential ownership holdings.

B. Benefits

There are three main benefits to using tokens on a blockchain.

1. *Programmability:* The contracts allow the administrators to manage important conditions like enforcements of rules and regulations.
2. *Evidence Tampering:* A major concern with maintaining any kind of record is the possibility of it being tampered with. But since blockchain is an immutable record of transfers, hence any sort of tampering can be easily detected and dealt with.

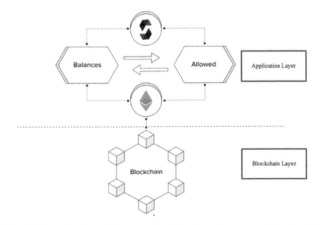

Fig. 43.1 Differentiation between Application Layer and Blockchain Layer

3. *Liquidity:* Using tokens, the ownership can be fragmented or divided into multiple parts. This increases the liquidity of certain assets which are otherwise considered indivisible.

2. Literature Survey

Although the Ethereum standards have multiple uses, like every technology, they also come with multiple flaws, which has made tokens unappealing to a certain section of people.

Through our research, we have found that token [2], [3] standards have many disadvantages like ERC20 having a very high gas cost and slow transaction speed (as shown in Fig. 43.2) but taking a look at the ERC721 token standard, we can see that it is much more advanced than previous proposed standards but lacks some features like efficient transfer capability, a good design, or the ability to get token IDs, which make the user experience complicated.

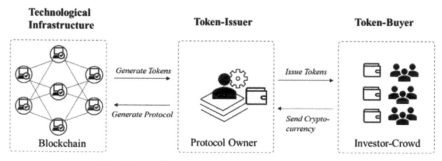

Fig. 43.2 Exchange ERC Token Infrastructure

When we look at ERC223, it provides a better user experience while providing 50% less gas fee than ERC20 so we can think it as one of the best standards to use but the disadvantage ERC223 has, is that it's still not widely accepted by the Ethereum community and integrated with many wallets. As we saw these problems still do exist as shown in Fig. 43.3, we need to devise a solution to make the tokens more efficient and user-friendly.

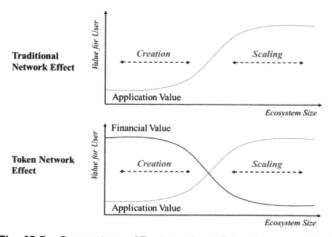

Fig. 43.3 Comparison of Traditional and Token Network Effect

Interface Standards for Tokens

As we know, there are 10 Token [3] Standards currently proposed and we can have a sketch about it here. The Token Standards are:

- ERC 20
- ERC 165
- ERC 721
- ERC 621
- ERC 777
- ERC 827
- ERC 223
- ERC 884
- ERC 865
- ERC 1155

ERC-20

ERC-20 token has been used by multiple Initial coin offering (ICO)to issue their tokens on the Ethereum blockchain [4]. ERC-20 is a fungible token standard [2][3], which in layman's terms means that any two coins built on this standard will be having the same values in any point at a time.

This token is also significant because it provides a set of basic rules that all Ethereum tokens [4] must adhere to and comply with, thus ensuring compatibility between multiple types of tokens. The basic rules defined by ERC-20 include how the transactions are approved as shown in Table 43.1, the data that can be accessed by the user about the token, the total supply of tokens, etc.

Table 43.1   ERC20 Hash and Signatures

ERC 20 Token Classification Signature	First 4 Byte Keecak Hash
total supply ()	18160ddd
balanceOf(address)	70a872b1
transfer(address,uint256)	awm928k
transferFrom(address,address,uint256)	098d81n9
approve(address,unit256)	m231o82
allowance(address,address)	96d89212
transfer(address,address,uint256)	S2323sw5

```
event Approval(address indexed _owner, address indexed _spender, uint256 _value
```

Fig. 43.4   Event Transfer

```
event Transfer(address indexed _from, address indexed _to, uint256 _value)
```

Fig. 43.5 Event Approval

The ERC-20 smart contract has a certain set of functions defined into itself that allows it to replicate the functionalities of a crypto token, allowing for transfer, balance checking, etc. of the tokens. The functions are as follows:

- Function balance of
- Function OwnerOf
- Function safeTransferFrom
- Function transferFrom
- Function approve
- Function set ApprovalForAll
- Function fetApproved

ERC 165

More than a standard for a token, ERC 165 is the standard for a method. This method standardizes detection of the interface used by a smart contract and publishing such smart contracts. Such a standard method is required because smart contracts require specific interfaces to interact with other standard tokens like ERC 721. Therefore, with this standard, the Ethereum development community can identify how they can interact with a particular smart contract. ERC 165 also forms the base for other token standards like ERC 721.

ERC 721

ERC 721 is a non-fungible token (oftentimes abbreviated as NFT), which means that each token is unique and has its own different value. These non-interchangeable data units are stored on the blockchain and can be traded or sold. Non-fungible tokens like ERC-721 globally have unique contract address, uint256, and token id which allows for the creation of virtual collectibles, tickets, real estate, school and degrees, the authenticity of medicines, etc.

A very popular example of usage of ERC-721 is cryptokitties, where users can buy, sell, or breed virtual kittens. Another example of the ERC-721 use case would be the popular BoredApe NFTs, which are being sold at over a million dollars.

ERC 223

ERC 223 is a superset of the ERC 20 standard [5], which fixes a very crucial issue that has resulted in the loss of $3,000,000 as of 31 Dec 2017. If ERC 20 tokens are sent to a contract address that can't handle the tokens, the tokens get burnt (destroyed) and can't be recovered. ERC 223 was programmed in such a way that developers can specify whether their smart contract accepts or declines tokens sent to it thus, preventing it from being burnt. Although ERC 223 provides a crucial update, it is not being used at present by any tokens.

ERC 621

ERC 621 is an extension to the ERC 20 standard, which allows contract owners to decrease as well as to increase the overall token supply using two built-in functions which are 'decreaseSupply' and 'increaseSupply'. The ERC 621 standard is currently a draft and has not been implemented on the Ethereum blockchain.

ERC 777

ERC 777 [6] is a token standard that attempts to reduce friction in crypto transactions. It gets rid of the double transaction verification used by ERC 20 and lowers the transaction overload. ERC 777 also allows users to reject incoming tokens from a blocklisted address. This ability to decline incoming payments from blocklisted addresses improves the security of Ethereum [11] DApps. However, ERC 777 [6] has not been implemented yet and is still in the EIP phase as of now.

ERC 827

ERC 827 is an extension to ERC 20 which aims to solve similar limitations that ERC 223 attempted to solve in a much more advanced manner. ERC 827 allows for token holders for transferring their tokens while also allowing 3rd parties to spend the tokens. This token standard will essentially [12] allow the wallet to reuse the tokens since both receiver and sender parties can agree for some reason to spend some amount by 3rd party.

ERC 884

ERC 884 is a standard created with the purpose of tokenizing stocks. A Recent Legislation in Delaware, USA, allows companies to use blockchain as a digital ledger to maintain share registries. ERC 884 intends to follow this legislation and it assigns each token as a share of a company which is registered in Delaware [13]. To follow the proposed legislation, ERC 884 also allows for the following:

- Whitelisting and identity verification of token holders
- Records of information maintained by regulators
- Only the whole value of tokens to be transferred
- Corporations need to prepare a list of shareholders as per the regulations ERC 884 is still a draft proposal as of now and hasn't been implemented.

ERC 865

ERC 865 is a beginner-friendly token standard that aspires to promote and help new users in the crypto world. It achieves this by proposing the token to self-pay the miner fees as well so that new users would not have to pay the miner fees themselves. ERC 865 is still a draft proposal and has not been implemented yet.

ERC 1155

ERC 1155 [7] is the most advanced Ethereum token standard [10], which comes with multiple improvements. However, ERC 1155 is still in the proposal and is referred to as EIP 1155. First

Price Auction is when a user performing a transaction pays the gas fees for the miner to solve it but only for a higher price. Instead, when someone offers lower gas prices, the transactions become slow, and to counter this problem, ERC 1155 [8] makes the transaction fees more predictable, reduces delays in transaction confirmation, and automates the fee bidding system.

ERC 1155 also increases the network capacity by changing the maximum gas limit per block to 25 million gas from the original amount of 12.5 million, thereby almost doubling the block size [14]. It also introduces a 'miner tip' which is directly paid to the miner who is prioritizing the transaction. ERC 1155 also brings in a few disadvantages like low profit for miners, wallet changes, base fee burning, etc.

3. Token Basics: FT and NFT

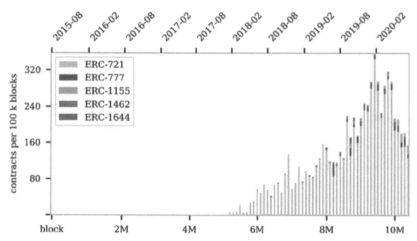

Fig. 43.6 Creation of ERC tokens (except ERC 20) (The upper horizontal axis shows the years and the lower horizontal axis shows the number of Ethereum blocks [9]. From the graph, each bar represents a bin of 10,000 blocks in the span of 2 weeks.)

The abbreviations FT and NFT stand for 'Fungible Tokens' and 'Non-Fungible Tokens', respectively. The term 'fungible' simply means that the value of such tokens is not unique and they are equivalent to each other. On the other hand, Non-Fungible means that each token has a unique value that makes it different from all other tokens.

4. Conclusion and Future Work

The theme of our research is identifying and classifying Ethereum-based token contracts [14]. Our study, unlike others, has considered all ERC standards, including the non-compliant ones. We have also focused on the contract code than activity; thus, our study is not only about the popular but also the unused tokens as well. Our study tries to understand the concept of token using readily accessible data and to explore the extent to which this study can be done.

- Compliant Tokens: 97% as shown in Fig. 43.4 of the Ethereum blockchain-based ERC tokens utilize and follow the ERC-20 standard, which is also the base standard for the fungible tokens. The remaining 3% is composed primarily of ERC-721 standard, used by non-fungible tokens, and a very small percentage of other a few hundred contracts utilizing a token standard other than ERC-20 and ERC-721. Notably, contracts following the security token standards are non-existent which could be because of the legal regulations. At the same time, about 200 tokens are being deployed every day leaving us vexed as to what their purpose might be.

- Non-compliant Tokens: 25% of the token contracts are occupied by Non-compliant Tokens. They show the same activity levels as their compliant counterparts. This adds up to a total of 272k contracts on the Ethereum blockchain taken up to 10.5 M blocks.

- Token Type: Due to legal reasons, it is necessary to have a distinction between various token types, viz. security, utility, and payment tokens [15]. A token's contract interface functions can be distinguished as token-related or neutral or any other functionality. Through this classification, we can say that a token is 'pure' if it has only token-related and neutral functions. 70% of the compliant tokens are pure tokens. They are more likely to be security tokens, while on the other hand, non-pure tokens have greater chances of being utility tokens. This can be confirmed through the SEC filings.

To better assess the tokens, we need to understand in detail about individual contracts and their current use-cases. This analysis requires tools that can better study the bytecode present in the token contract and find data structures as well as code present in it. Another requirement is of methods that can detect whether the components that form an application are reflected on the blockchain (also called on-chain) or not (off-chain) and, whether they should be considered as interrelated or not.

REFERENCES

1. Yaga D., Mell P., Roby N. 2018. "Blockchain Technology Overview", *NIST - National Institute of Standards and Technology and IR- Internal Report. https://doi.org/10.6028/NIST.IR.8202.*
2. Raskin M, Seleh F, Yermack D. 2019. "How Do Private Digital Currencies Affect Government Policy?", *NYU Stern school of business; NYU Law Research paper, Pg No 20-05. https://ssrn.com/abstract=3437529.*
3. 2020. "Bitcoin Wiki". *https://en.bitcoin.it/wiki/Script.*
4. Buteriin V., Vogelsteller F. 2020. "Ethereum.Org" Ethereum Virtual Machine". *https://ethereum.org/en/developers/docs/evm.*
5. Buteriin V., Vogelsteller F. 2015. "EIP 20: ERC-20 Token Standard". *https://eips.ethereum.org/EIPS/eip-20.*
6. Dafflon J., Baylina J., Shabai T. 2017. "EIP 777: ERC-777 Token Standard". *https://eips.ethereum.org/EIPS/eip-777.*
7. Entriken W., Shirley D., Evans J., Sachs N. 2018. "EIP 721: ERC-721 Non Fungible Token Standard", *Available at https://eips.ethereum.org/EIPS/eip-721*
8. Bellflamme P., Lambert. T, Schwienbacher A. "Crowdfunding Tapping the right crowd", *Journal of business venturing, Vol 29(5), Pg 585-609.*

9. Buterian.V. *Aug 8, 2014.* "Ether sale: A Statistical overview Ethereum blog" *Available at https:// blog.ethereum.org/.*

10. Buterian.V. *Aug 8, 2014.* "A Next level generation smart contract and decentralized application platform" *White paper Ethereum Available at https://www.weusecoins.com/ assets/pdf/ libraryEthereum_ .*

11. Lin. L. *2017.* "Why ICOs should want to be securities Coin Desk" *Available at https://www. coindesk.com/ icos-want-securities/.*

12. Massey R., Dalal D., Dakshinamoorthy A. *2017.* "Initial coin offering: A new Paradigm Delotte", *Available at https:// www2.deloitte.com/content/dam/Deloitte/us/Documents/process-and- operations/us-cons-new-paradigm.pdf.*

13. Wood G. *2014.* "Ethereum: Secure decentralized generalized transaction ledger", *.Available athttp://gavwood.com/paper.pdf.*

14. Tex. *2017.* "Tenx Payment platform whitepaper " , *Available at https://www.tenx.tech/whitepaper/ tenx_whitepaper_final.pdf.*

15. Parker G., Van Alstyne M., Choudary S. "Platform revolution: How networked markets are transformation the economy and How to make them work for you", *New York, WW Norton and Company.*

Intelligent Systems and Smart Infrastructure – Brijesh Mishra et al. (eds)
© 2023 Taylor & Francis Group, London, ISBN 978-1-032-41287-0

CHAPTER

44

Machine Learning Approaches for Cardiac Disease Prediction

Awadhesh Kumar[1]

Department of Computer Science MMV,
Banaras Hindu University, Varanasi, India

Manoj Kumar Mishra[2]

Department of Computer Science,
RGSC, Banaras Hindu University, Varanasi, India

Akhilesh Kumar[3], Sumit Gupta[4]

Department of Computer Science,
Institute of Science, Banaras Hindu University, Varanasi, India

Abstract

Cardiac disease prediction is one of the most popular research challenges for medical professionals. Cardiac disease is also a significant cause of death worldwide as the disease interferes with the functioning of the heart. It is categorized into different types of heart disease, and each type of disease has conditions that can affect the heart. These categories include congestive heart failure, pericardial disease, cardiac arrest, coronary artery disease (CAD), heart arrhythmias, etc. Lifestyle changes and daily medication can help to prevent certain types of heart disease, but for others, surgery is required to keep your organs running properly. As per a recent data survey, obesity, cholesterol, triglycerides, high blood pressure, an unhealthy diet, smoking, and tobacco use are all contributing to an increase in the severity of heart disease and mortality among people. Machine learning and data science techniques can help in early disease detection and can reduce the mortality rate. In this paper, various approaches to machine learning (ML) such as Support Vector Machines (SVMs), Decision

Corresponding author: [1]bhuipsbhu@gmail.com, [2]mmcsbhu@gmail.com, [3]akhilesh.kumar17@gmail.com, [4]sumitkugp123@gmail.com

DOI: 10.1201/9781003357346-44

Trees (DT), K-nearest Neighbors (KNN), Random Forests (RF), Extreme Gradient Boosting Classifiers (XGBs), and Logistic Regression (LR) are used to aid in the early detection of cardiac disease and identify behavioral patterns in massive quantities of data, which will aid clinicians in strategic planning and early diagnosis. This could minimize the chances of patient deaths and injuries.

Keywords: Logistic Regression, Cardiac Disease, CAD, KNN, Decision Tree, SVM, XGB, Random Forest, Machine Learning.

1. Introduction

Due to interferences in the functioning of the heart, cardiac disease is the main significant cause of global mortality but is often largely preventable by maintaining healthy lifestyles and daily medication. In the human body, cardiac disease affects the functioning of the heart, blood circulatory system, and increases the risk of blood clots (El-Hasnony et al., 2022). Due to this, there is a possibility of damage to the arteries in organs like kidney, heart, eyes, and brain. Some of the most popular cardiac diseases are described as follows (No Title, n.d.-a): In coronary heart disease, heart muscles are confined due to not reaching the oxygen-rich blood, which causes an additional burden on the heart that causes angina, heart attacks, and heart failure. A stroke is a medical emergency in which blood supply is interrupted or blocked in some part of the brain, which results in brain damage and sometimes death. A TIA is a little bit different from a stroke, in which the flow of blood is only slightly disturbed. FAST stands for Face, Arms, Speech, and Timing, and it can be used to identify the primary indicators of a stroke or TIA. Peripheral artery disease's common indicators are dull or aching in the leg, numbness, discomfort in the legs, persistent ulcers, etc., and this disease happens when blood supply is disturbed due to blockage in the arteries to the organs, usually the legs. Aortic disease refers to a group of factors that influence the aorta. It is the body's biggest blood channel, delivering blood from the heart to all other parts. An aortic aneurysm has the potential to burst and cause a life-threatening hemorrhage. As per a recent data survey, obesity, cholesterol, triglycerides, high blood pressure, an unhealthy diet, smoking, and tobacco, ethnic background, gender, alcoholism, age, family history, etc., are all contributing to an increase in the severity of heart disease and mortality among people. Regular exercise, a balanced diet, decreasing alcohol, maintaining a healthy weight, a full diet, medicine, and other measures can all help to avoid heart disease in its early stages (No Title, n.d.-a; Swathy & Saruladha, 2022), etc. There are various laboratory tests such as blood tests, coronary angiography, cardiac catheterization, electrocardiogram (ECG), echocardiography, electroencephalogram (EEG), etc., and imaging studies such as chest X-ray, cardiac MRI, and electron-beam computed tomography (EBCT) used to diagnose different types of cardiac diseases. Through physical examination and static analysis, we coordinate these findings and predict disease from outcomes and procedures (Kumar & Kumar, 2022b; Muhammad et al., 2020). Computers are taught to recognize patterns and diagnose disease by exploiting innovations in electronic health records (EHRs) and with the help of machine learning and data science approaches. This will aid in pre-diagnosis and informative decisions by the practitioner for patient care (F.Y et al., 2017;

Swathy & Saruladha, 2022). Several ML approaches such as KNN, RT, DT, LR, SVM, and XGB are used in this paper to predict future cardiac disease and identify behavioral patterns in massive amounts of data, which will aid health-care professionals in strategic planning and early diagnosis, reducing the risk of patient injuries and fatalities (Kumar & Kumar, 2021, 2022a). The following is the framework of the remaining paper: The remaining framework of the paper contains literature-related work, datasets, methodology and model specifications, implementation and experimental results, and conclusions.

2. Literature Review

For the past few years, healthcare has been a prominent focus of research. Several researchers and medical practitioners are progressively paying attention to identifying the pattern of disease from the various risk factors and medical and family history of the patients. Many types of research on the manipulation of heart disease diagnosis and prognosis show that heart defects cause more mortality around the globe (Elzeki et al., n.d.; F.Y et al., 2017). In Khourdifi et al. (n.d.), to filter out redundant information and enhance the effectiveness of heart disease diagnosis, the Fast Correlation-Based Feature Selection (FCBF) method was used. In Nakano et al. (2020), active learning was investigated in the context of hierarchical multi-label classification (HMC). Because datasets often contain a large number of identifiers and inherent characteristics that establish association among classes, HMC presents new obstacles to active learning. The author also proposed a new active learning algorithm tested on 14 datasets of different domains that are suitable for HMC. In edu/ml & 2010 (n.d.) and Khanna et al. (n.d.), the authors used a variety of heart disease classifiers and improved the accuracy of weak algorithms by combining numerous classifiers. They also proposed various machine learning-based strategies for detecting cardiac disease. El-Hasnony et al. (2022) compare and assess various machine learning algorithms for predicting angiographic disease status using a heart disease dataset, and then compare and evaluate their capacity to predict cardiac disease. In the articles (Banu & Gomathy, 2013; Related Papers, n.d.), precompiled data is grouped in a database using clustering techniques such as K-means. The findings suggest about developed prognosis strategies may accurately diagnose heart attacks. Various models discussed by various researchers employ heart disease prediction from numeric small-sized datasets, but in this paper, we discuss the several machine learning algorithms for the newly released (2020) dataset on the Kaggle platform named 'Personal Key Indicators of Heart Disease', which consists of categorical and object type attribute values and compare them with the small-sized numeric heart disease dataset.

3. Datasets and Variables

In this study, we used two datasets (Personal Key Indicators of Heart Disease | Kaggle, n.d.). The first dataset (Personal Key Indicators of Heart Disease) is part of the Behavioral Risk Factor Surveillance System (BRFSS) available on Kaggle and comes from the CDC (Centers for Disease Control and Prevention). The BRFSS is the globally biggest comprehensive patient assessment system, consulting nearly 400,000 adults each year. This data set has 401,958 rows and 18 columns, and many different questions influenced heart disease directly or indirectly. Cleveland, Hungary, Switzerland, and Long Beach V are among the datasets included in the heart.csv data package. The target characteristics of the data set describe the disease that occurs in the patients, and the dataset comprises 1025 rows and 14 columns, where each attribute has an integer value of either 0 (no disease) or 1 (disease). Each dataset has been carried out in the test and training datasets in the ratio of 3:7, indicating that the data used for training and testing purposes is 70% and 30%, respectively.

4. Methodology and Model Specifications

One of the branches of artificial intelligence (AI) is machine learning (ML). ML and AI are primarily distinguished as ML focuses on boosting accuracy while AI focuses on increasing the potential to succeed. ML approaches are used for interpreting data and learning from it to determine future predictions. These models have the potential to understand on their own based on previous experience or statistical data. Machine learning algorithms are grouped either by learning styles or by similarity and functioning and are categorized as supervised, unsupervised, and semi-supervised learning. Representation, evaluation, and optimization are the three components of every machine learning algorithm (Swathy & Saruladha, 2022). In the proposed model, first, take a collection of data from various sources, then apply various data preprocessing techniques which remove duplicate data, null values, outlier representation, and low correlated attributes from the dataset, change the object data types attribute of a dataset into a numerical array, and when all the attributes of the dataset are transformed into numeric values, the dataset is partitioned as a training and testing dataset. For predicting cardiac disease, several ML algorithms such as DT (Charbuty & Abdulazeez, 2021), RF (Iwendi et al., 2020), KNN (Isnain et al., 2021), LR (Zhu et al., 2019), SVM (Devikanniga et al., 2020), XBG classifier (Ma et al., 2020), etc. are used for training, testing, and validating the data.

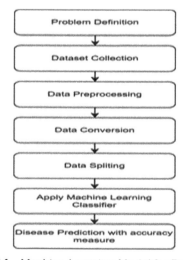

Fig. 44.1 Machine Learning Model for Prediction

5. Implementation Platform and Empirical Results

The results have been implemented in a Google Research Laboratory named "Colab". The platform is well suited for machine learning, deep learning, and data science applications and allows one to execute Python programs via a web browser and an internet connection. In this paper, small and large datasets related to heart disease are tested using machine learning algorithms such as DT, RF, KNN, LR, SVM, and XBG classifier to predict the accurate heart disease (the first dataset is large, consisting of 319,795 rows and 18 columns, and the second dataset is of small size, consisting of 1025 rows and 14 columns). The results are analyzed in terms of accuracy (using the input, or training, data, evaluate which model is best at recognizing relationships and patterns between variables in a dataset) and mean absolute error (MAE), which refers to the mean absolute value of each prediction fault in all instances of data.

Table 44.1 Before and after preprocessing dataset2 [Small Size]: execution time, accuracy, and mean absolute error of different machine learning algorithms

Sr. No.	Algorithm Name	Before pre-processing			After pre-processing		
		Execution time (in second)	Accuracy (%)	MAE	Execution time (in second)	Accuracy (%)	MAE
1.	Decision Tree	0.821	96.10	0.39	0.785	84.06	0.159
2.	Random Forest	0.958	100.00	0.000	0.88	84.06	0.159
3.	KNN	1.223	72.07	0.279	1.035	63.76	0.362
4.	Logistic Regression	0.954	85.06	0.149	0.621	84.06	0.159
5.	SVM	0.636	71.53	0.282	0.013	66.67	0.333
6.	XBG	0.212	94.16	0.058	0.767	85.59	0.145

Table 44.2 Before and after pre-processing the dataset1 [Larger Size]: execution time, accuracy, and mean absolute error of different machine learning algorithms

Sr. No.	Algorithm Name	Before pre-processing			After pre-processing		
		Execution time (in second)	Accuracy (%)	MAE	Execution time (in second)	Accuracy (%)	MAE
1.	Decision Tree	1.26	96.19	0.038	1.179	95.42	0.046
2.	Random Forest	31.48	95.80	0.042	25.998	94.81	0.052
3.	KNN	383.094	96.06	0.039	293.738	95.25	0.047
4.	Logistic Regression	3.223	96.16	0.038	2.796	96.19	0.038
5.	SVM	1146.313	96.19	0.038	789.51	71.53	0.046
6.	XBG	15.507	96.20	0.037	14.311	95.44	0.046

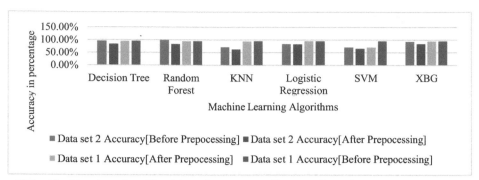

Fig. 44.2 Machine learning Algorithms Accuracy on Data set 1 and Data set2 before and after preprocessing the data

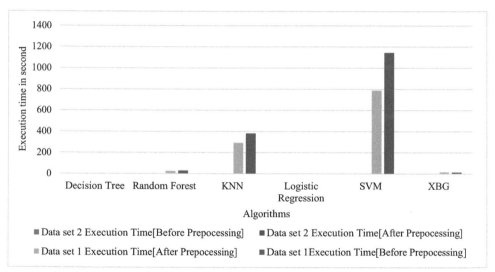

Fig. 44.3 Execution time of Algorithms Before and After data preprocessing in data set1 and dataset 2

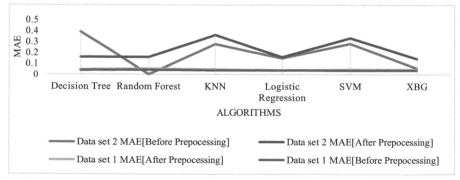

Fig. 44.4 Mean Absolute Error after and before preprocessing of dataset1 and dataset2

6. Summary Results

From Figs 44.2 and 44.3 and Tables 44.1 and 44.2, we found that the execution time taken by most of the ML algorithms is less when the algorithm is applied after preprocessing the datasets, and it usually takes more time before any preprocessing in both small-size and large-size datasets. Results showed that the KNN algorithm took more time and the XGB boost took less time, both before and after preprocessing dataset2. In the case of larger size dataset1, SVM & KNN algorithms take more execution time in both preprocessed and without preprocessing, and DT, LR, and XGB boost take very little time even in the execution of larger size dataset1 and small size dataset2. Finally, we can say that XGB boost, decision tree, and logistic regression take very little execution time to execute both the larger size dataset1 and the smaller size dataset2. The accuracy of the different algorithms is higher without preprocessing the dataset and less after preprocessing dataset2 because after preprocessing, dataset 2 contains only 228 records and 13 attributes, which is very less compared to the dataset without preprocessing, which contains 1025 records and 13 attributes. Since the size of dataset2 is small and there are very few records available for training and testing after preprocessing dataset2, the accuracy is low compared to the accuracy of various machine learning algorithms in unprocessed dataset2. We also found that the accuracy of the random forest algorithm is 100% and 84.06%, respectively, in the cases of without preprocessing and with preprocessing dataset2. The XGB Boost algorithm has good accuracy of about 96.16% and 85.59%, respectively, in both preprocessed and unprocessed dataset2. So we can say that the Random Forest and XGB Boost algorithms perform better in small datasets with and without preprocessing as compared to the rest of the algorithms.

The results from Tables 44.1 and 44.2, and Figs 44.2 and 44.3 also show that the accuracy of different algorithms is higher without preprocessing and less after preprocessing in both dataset1 and dataset2. After preprocessing, dataset1 consists of 77,224 records and 18 attributes, and dataset2 contains 228 records and 13 attributes, which is less as compared to without preprocessing records of dataset1 and dataset2, which consist of 319,995 records and 18 attributes, and 1025 records and 13 attributes, respectively. The results showed that the Radom Forest and XGB Boost Algorithm achieve greater accuracy in smaller size preprocessed and without preprocessed dataset2 as compared to the rest of the algorithms. The Decision Tree, Logistic Regression, SVM, and XGB Boost all perform with greater than 95% accuracy in larger size dataset1 without preprocessing, of which XGB Boost and SVM perform best with an accuracy of 96.19% and 96.20%, respectively. The logistic regression, XGB Boost, and decision tree algorithms perform best with an accuracy greater than 95% in larger size preprocessed dataset1, out of which Logistic Regression, Decision Tree, and XGB Boost perform best with an accuracy of 96.19%, 95.44%, and 95.42%, respectively.

According to Tables 44.1 and 44.2 and Fig. 44.4, the mean absolute error of different algorithms is lower without preprocessing and slightly higher after preprocessing of both smaller dataset2 and larger dataset1. We also discovered that when we compared a larger dataset1 to a smaller dataset2, MAE is lower in dataset2 and higher in dataset1 due to its larger size.

Finally, we can say that the Random Forest and XGB Boost algorithms are best in terms of high accuracy and low mean absolute error for smaller dataset2 while Logistic Regression, XGB Boost, and SVM algorithms are best for larger size dataset1.

7. Conclusion

A large number of people in various nations throughout the world have been affected by various forms of cardiac disease, and this number is expected to rise in the future. It is critical to discover methods for early illness identification and treatment. Machine learning has the potential to make heart disease therapy more convenient and cost-effective. Machine learning accuracy is determined by the quantity and quality of the data used to train the model. In this paper, a small-sized numeric dataset2 and a larger-sized object type dataset1 are used to test the accuracy in both datasets without and with preprocessing. It is discovered that all algorithms have higher accuracy when applied to smaller size datasets without preprocessing and lower accuracy when applied to preprocessed datasets, whereas most algorithms have higher accuracy in both without and with preprocessed larger size datasets. The implemented results show that the Random Forest and XGB Boost algorithms outperform the Logistic Regression, XGB Boost, and SVM algorithms for smaller datasets in terms of high accuracy and low mean absolute error, while the Logistic Regression, XGB Boost, and SVM algorithms outperform for larger datasets1. From the results, it is also revealed that when the dataset is large, then after and before preprocessing the dataset, most algorithms achieve good accuracy as there are enough records available for training and testing, whereas without preprocessing, small-sized datasets have achieved higher accuracy, but with preprocessing yields less accuracy than larger sized datasets as there are fewer records available for training and testing. Patients' lives are saved and medical practitioners are more efficient when heart disease and other associated diseases are diagnosed early.

8. Acknowledgments

This study is funded by a seed grant from BHU [grant number. R/Dev/D/IoE/SEED GRANT/2020-21/Scheme No. 6031] under the IoE.

REFERENCES

1. Banu, M. A. N., & Gomathy, B. (2013). *DISEASE PREDICTING SYSTEM USING DATA MINING TECHNIQUES. 1*(5), 41–45.
2. Charbuty, B., & Abdulazeez, A. (2021). Classification Based on Decision Tree Algorithm for Machine Learning. *Journal of Applied Science and Technology Trends, 2*(01), 20–28. https://doi.org/10.38094/jastt20165
3. Devikanniga, D., Ramu, A., & Haldorai, A. (2020). Efficient diagnosis of liver disease using support vector machine optimized with crows search algorithm. *EAI Endorsed Transactions on Energy Web, 7*(29), 1–10. https://doi.org/10.4108/EAI.13-7-2018.164177

4. edu/ml, A. F. ics. uci., & 2010, undefined. (n.d.). UCI machine learning repository. *Ci.Nii.Ac.Jp.* Retrieved April 15, 2022, from https://ci.nii.ac.jp/naid/10031099995/

5. El-Hasnony, I. M., Elzeki, O. M., Alshehri, A., & Salem, H. (2022). Multi-Label Active Learning-Based Machine Learning Model for Heart Disease Prediction. *Sensors*, *22*(3). https://doi.org/10.3390/s22031184

6. Elzeki, O., Elfattah, M. A., ... H. S.-P. C., & 2021, undefined. (n.d.). A novel perceptual two layer image fusion using deep learning for imbalanced COVID-19 dataset. *Peerj.Com.* Retrieved April 15, 2022, from https://peerj.com/articles/cs-364.pdf

7. F.Y., O., J.E.T, A., O, A., J. O, H., O, O., & J, A. (2017). Supervised Machine Learning Algorithms: Classification and Comparison. *International Journal of Computer Trends and Technology*, *48*(3), 128–138. https://doi.org/10.14445/22312803/ijctt-v48p126

8. Isnain, A. R., Supriyanto, J., & Kharisma, M. P. (2021). Implementation of K-Nearest Neighbor (K-NN) Algorithm For Public Sentiment Analysis of Online Learning. *IJCCS (Indonesian Journal of Computing and Cybernetics Systems)*, *15*(2), 121. https://doi.org/10.22146/ijccs.65176

9. Iwendi, C., Bashir, A. K., Peshkar, A., Sujatha, R., Chatterjee, J. M., Pasupuleti, S., Mishra, R., Pillai, S., & Jo, O. (2020). COVID-19 patient health prediction using boosted random forest algorithm. *Frontiers in Public Health*, *8*(July), 1–9. https://doi.org/10.3389/fpubh.2020.00357

10. Khanna, D., Sahu, R., Baths, V., of, B. D.-I. J., & 2015, undefined. (n.d.). Comparative study of classification techniques (SVM, logistic regression and neural networks) to predict the prevalence of heart disease. *Ijmlc.Org.* Retrieved April 15, 2022, from http://www.ijmlc.org/vol5/544-C039.pdf

11. Khourdifi, Y., Intelligent, M. B.-I. J. of, & 2019, undefined. (n.d.). Heart disease prediction and classification using machine learning algorithms optimized by particle swarm optimization and ant colony optimization. *Researchgate.Net.* Retrieved April 15, 2022, from https://www.researchgate.net/profile/Youness-Khourdifi/publication/331382781_Heart_Disease_Prediction_and_Classification_Using_Machine_Learning_Algorithms_Optimized_by_Particle_Swarm_Optimization_and_Ant_Colony_Optimization/links/5c76ac6da6fdcc4715a11f1d/Heart-Disease-Prediction-and-Classification-Using-Machine-Learning-Algorithms-Optimized-by-Particle-Swarm-Optimization-and-Ant-Colony-Optimization.pdf

12. Kumar, A., & Kumar, A. (2021). *DEEPHER: Human Emotion Recognition Using an EEG-Based DEEP Learning Network Model*. 32. https://doi.org/10.3390/ecsa-8-11249

13. Kumar, A., & Kumar, A. (2022a). Analysis of Machine Learning Algorithms for Facial Expression Recognition. *Communications in Computer and Information Science*, *1534 CCIS*, 730–750. https://doi.org/10.1007/978-3-030-96040-7_55

14. Kumar, A., & Kumar, A. (2022b). Human Sentiment Analysis on Social Media through Naïve Bayes Classifier. *Journal of Scientific Research*, *66*(01), 350–357. https://doi.org/10.37398/jsr.2022.660137

15. Ma, B., Meng, F., Yan, G., Yan, H., Chai, B., & Song, F. (2020). Diagnostic classification of cancers using extreme gradient boosting algorithm and multi-omics data. *Computers in Biology and Medicine*, *121*(January), 103761. https://doi.org/10.1016/j.compbiomed.2020.103761

16. Muhammad, Y., Tahir, M., Hayat, M., & Chong, K. T. (2020). Early and accurate detection and diagnosis of heart disease using intelligent computational model. *Scientific Reports*, *10*(1), 1–17. https://doi.org/10.1038/s41598-020-76635-9

17. Nakano, F. K., Cerri, R., & Vens, C. (2020). Active learning for hierarchical multi-label classification. In *Data Mining and Knowledge Discovery* (Vol. 34, Issue 5). Springer US. https://doi.org/10.1007/s10618-020-00704-w

18. *No Title*. (n.d.-a). http://www.nhs.uk/conditions/cardiovascular-disease/Pages/Introduction.aspx

19. *No Title*. (n.d.-b). https://www.kaggle.com/datasets/johnsmith88/heart-disease-dataset

20. *Personal Key Indicators of Heart Disease | Kaggle*. (n.d.). Retrieved April 16, 2022, from https://www.kaggle.com/datasets/kamilpytlak/personal-key-indicators-of-heart-disease

21. *Related papers*. (n.d.). https://doi.org/10.1109/ICICA.2014.36

22. Swathy, M., & Saruladha, K. (2022). A comparative study of classification and prediction of Cardio-Vascular Diseases (CVD) using Machine Learning and Deep Learning techniques. *ICT Express*, *8*(1), 109–116. https://doi.org/10.1016/j.icte.2021.08.021

23. Zhu, C., Idemudia, C. U., & Feng, W. (2019). Improved logistic regression model for diabetes prediction by integrating PCA and K-means techniques. *Informatics in Medicine Unlocked*, *17*(April), 100179. https://doi.org/10.1016/j.imu.2019.100179

Intelligent Systems and Smart Infrastructure – Brijesh Mishra et al. (eds)
© 2023 Taylor & Francis Group, London, ISBN 978-1-032-41287-0

CHAPTER

45

Trust-Aware Mitigation of Various Security Threats for Internet of Things

Renu Mishra[1], Inderpreet Kaur[2]

Dept of CSE, Galgotia College of Eng. and Technology,
Gr Noida, India

Sandeep Saxena[3]

Dept of IT, IMS Unison University,
Dehradun, India

Raghwendra Mishra[4]

Dept of Mathematics,
Govt. PG College, Rudrapur, India

Tanu Shree[5]

Chandigagh University

Meena Sachdeva[6]

Dept of MCA, Galgotia College of Engineering and Technology,
Gr Noida, India

Abstract

Nowadays people want to enjoy the shift from "always on" to "always connected" communication environment. Internet fulfills that wish with full of capabilities to become an important part of next-generation wireless network. The similarity between any dynamic network and Internet of Things (IoTs) environment opens spick-and-span ways for providing different services in such environments and also focusses on various issues in its networking

Corresponding author: [1]renutrivedi11@gmail.com, [2]kaur.lamba@gmail.com, [3]sandeep.research29@gmail.com, [4]meetgirdhar@gmail.com, [5]tanu.shree29@gmail.com, [6]meena.sachdeva@galgotiacollege.edu

DOI: 10.1201/9781003357346-45

constraints as well. Ad hoc network is a dynamic and temporary network, which is settled on the fly without any skeleton. Usually they are used in military operations, emergency rescue, disaster recovery, wireless sensor network, and commercial multimedia communication. Due to open transmission medium and absence of secure boundaries, IoT became more susceptible to attacks with malicious intent to cripple the network. This chapter has the objective to highlight the benefits of soft security-based solution to provide the security in IoT environment during routing. The proposed trust-aware routing arranges all available routes in the descending order of trust value (TV) and ascending order of hop-counts.

Keywords: IoT, routing, trust, security, throughput

1. Introduction

Mobile Internet of Thing (IoT) is an ever-changing and non-permanent network, which is settled on the fly without any skeleton. Usually, they are used as a fast and short-lived communication network in hostile environments. The basic characteristics to provide flexibility and scalability are the uniqueness of this class of networks [1]. Similar to the traditional networks, security is a paramount concern in ad hoc networks also, with major security service requirements like confidentiality, authentication, authorization, and tamper proofing. Major challenges are faced while implementing any security solution for Mobile Ad Hoc Network in Internet of Things. Distributed operation is one of the main reasons to make the network open for criminals and attackers because no central controller is here; network controlling responsibilities of the network are distributed among all. The communication among nodes is cooperation based and each node can work as a relay, to implement routing and security. Multihop routing makes the network more vulnerable to various attacks caused by selfish and malicious nodes. A non-cooperative node in data transmission is called a selfish node, which saves the battery power for its operation. Data is forwarded either directly or via some intermediate nodes (if is not in its communication range). Since nodes can move arbitrarily, this makes the network topology very much uncertain. The medium is open to all nodes without any restriction. In most cases, IoT devices may be mobile, with limited CPU computation, low battery, and limited memory. Trust should be computed and evaluated between two neighbours in IoT's for security and reliable data transmission. It is also necessary to quantify the network behaviour in terms of "Trust", to improve security services. A comparison between Cryptographic and Trust-based Methods for MANET Routing Security is presented [2]. A set of parameters for the trust evaluation process can be defined to compute the overall trust to filtrate internal attacks and dishonest recommendations [28]. A node having less trust value (TV) is said to be malicious nodes that can drop the packet in between the network. Neighbour nodes are acrophobic to send the data even in the existing shortest path [30].

2. Literature Review

In any ad hoc type of networks, routing first does route discovery and then route maintenance. All prior routing methods presumptively presume that nodes are reliable and cooperative. This thought opens the door for vulnerability in the routing protocols. Because the nodes are not so powerful in terms of resources and infrastructure are barriers for high power-consuming cryptographic algorithms, so many crypto-based schemes are proposed to protect routing information but these approaches may not be suitable for real IoT. The power capacity of a mobile node affects network survivability in IoT since nodes will be disconnected if the battery is exhausted. An energy-efficient security should guarantee the long life of the network. An energy-efficient security protocol avoids downloading huge tables and limited calculations are preferred. We need a balanced approach that must be developed for secure computation and lifetime of the node [11]. Hard security protocols are not easy to implement and light security protocols can be easily attacked. Various mechanisms and protocols have already been advised for preserving energy and securing ad hoc networks. Researchers introduced trust-aware security for gaining confidentiality and authentication with Attack-tolerance, Compatibility, and Scalability. A comparison is presented [2] between Cryptographic and Trust-based security. Earlier, several issues like compromise node, computational overload, and energy preservation are highlighted. A lot of work was contributed to "lightweight" security mechanisms using trust [80]. They provide general ideas for trust evaluation in networks by applying different approaches. Some researchers proposed a trust model to establish trust in pure IoT [12]. The trust computation is based on monitoring data delivery in the network for secure routing evaluation in MANET. A new way to compute trust relationship to identifying malicious nodes in IoT was given in [13]. The trust-based mechanism includes the notion of friends, acquaintances, and strangers. These algorithms/protocols are not suitable for MANET with less power, storage, and processing. TSAODV [14] proposal came, in which information regarding routing having the highest trust value among all. One paper [15] utilized queuing theory as Trust Evaluation Factor; each node has k trust evaluation matrices which have many trust evaluation factors like paper link quality, distance, and mobility. Trust evaluation is being used in different new paradigms of networks like MANET, e-commerce, and other multiagent systems with different requirements. Different researchers contributed and presented various models to compute trust. Author M. Branchod gave CONFIDENT named contribution [16] to check the node's Fairness is the capital work for watchdog, trust/reputation manager. Trust ratings are computed and utilized in the routing process to increase the probability of detecting malicious nodes. In this area, researchers contributed a lot [17] [18] [19] but still we can't expect one all-round perfect solution that covers all fields. We can choose suitable features from multiple models to design the solution for our area. Various existing trust management schemes involved in major areas like routing and group communication and key management are investigated with their merits and demerits & findings. We identified some work [17, 20] in multi-criteria trust evaluation. Additionally, energy is included as an important QoS trust metric[21,22] to improve the performance of the network. In the literature review, we found that integration of different dimensions of trust is essential in the composition of a trust metric which would provide better performance. Taken all these facts into the account, we modified our early trust-based routing scheme [22]. Previously we proposed a trust-based model to identify misbehaviour of the node by comparing the value threshold. However, this model was based single trust evaluation dimension to quantify and predict reliability among nodes. This single measure is not enough satisfactory in many scenarios (selfish behaviour, malicious intent, the lack of fixed infrastructure, limited resources, physical failures, etc.) of dynamic MANETs. Some modifications in the route discovery, trust update, and trust recommendation procedures are done to adjust the trust-aware communication. Lightweight trust-based routing protocol is proposed for mobile ad hoc networks, which consumes limited computational resource and suitable for blackhole and grey hole attack and specially to target denial-of-service attacks [23], [24]. Various attempts are made as an extension of AODV routing protocol with the help of direct trust and indirect trust. Direct trust is calculated from the number of packets received and forwarded, whereas Indirect trust is based on the reputation of the node, observed by other neighbour nodes. In ad hoc networks, securing routing protocols is one of the fundamental challenges.

It is simply an activity to shunt the legal policies on a system. An attacker may modify, release, insert false data, or obtain illegitimate access to disrupt network operation [2]. Since no central coordinating authority is present, the medium shared in the IoT makes it more vulnerable than wired networks. The apprehension of possible attacks will ever be the first step in the direction of designing a good security policy. External and internal attacks are the two types of attacks. An outsider can cause congestion or spread misleading routing information in an external assault. On the other hand, the internal attacks are committed by compromised member nodes, which may gain access and pretend to be authorized node [3].

A. Popular Attacks in IoT

Here are some popular attacks on the routing protocols:

1. *Black Hole Attack:* In this, the attacker node publicizes itself for having the shortest route to any desired node in the network. Normal innocent nodes rely on the received reply as they follow cooperation-based forwarding. Malicious node takes advantage of this and replies to the request, claiming for having the shortest path [4]. Source node has to trust that reply in the absence of verifying mechanism. The network can be targeted by a single black hole node or a group of attacker nodes that work together to degrade the network reliability.

2. *Gray Hole Attack:* It is a special case of blackhole attack by dropping a few packets with a set of probability [5]. The node may drop some or all the packets for some time and later behave very normal.

3. *Rushing Attack:* A malicious node rising the speed (Rush) of the routing process. It accepts the Route Request packet and forwards to its neighbours sooner as compared to others. The packet from the attacker will reach first and will be accepted and other RR will be discarded with source sequence numbers.

4. *Wormhole Attack:* Wormhole attack catches the packet from one location and sends it over the tunnel to the other location. The tunnel is planned to give the impression of having the optimized path to the destination. It happens with the help of multiple

malicious nodes, which may create choke points [6]. A wormhole attack may equally harm to proactive and reactive protocols both.

5. *Sybil Attack:* Here attacker node controls multiple identities by assuming arbitrary identities or may spoof legitimate nodes. This attack can be launched either to erase the proofs of its earlier malicious activities or to disrupt the network.

3. Security Countermeasures in IoT

Designing the adequate security framework is very hard in IoT because no such strong boundary exist to separate insider nodes from outside network. An idiosyncratic security solution is not enough due to no stability of nodes and is incapable of physical protection to catch security threats [29]. Additionally, because the ad hoc network is distributed and infrastructure-less networks, it might be best to implement security strategies at the individual node level in below two dimensions [22].

A. Cryptography (Hard Security)

Cryptography is just an art of hiding information. It works as an important security tool to provide authentication, confidentiality and other services [7]. There are two popular approaches to implement cryptography. First is a symmetric type where the same key supports encryption and decryption, while the public/asymmetric approach is based on different keys to encrypt and decrypt the data [8], [26], [27]. Although asymmetric cryptography is versatile (authentication, integrity, and confidentiality) and simple to use for key distribution, it is not without flaws. Single key cryptographic algorithms have lighter computation than the public-key approach but suffer from a key compromise problem. Any cryptosystem trusts on some inherent efficient key management system.

B. Trust Evaluation (Soft Security)

Various cryptographic algorithms are proposed to provide secure solutions but often seem unfeasible because they assume that nodes are cooperative and trustworthy [9]. The importance of trust management is realized and followed by society to design better security protocols. It is an approved tool to mitigate attacks and filter out misbehaving nodes based on social properties, each node is going to be assessed with the threshold value [10], and the isolation of node is performed by trust value. Any trust-based security solution aims to provide a performance guarantee through the evaluation of node behavior. Current routing algorithms aim only to find optimal routes but not cover performance guarantee. Widespread use of IoT creates the need for a system to rank out the behavior of the network. Here, multidimensional trust evaluation scheme is designed by including current attributes (Direct Trust) of node and the past behavior (Indirect Trust) with others to improve Quality of Service (QoS) [25].

4. Proposed Routing for Trust-based Security

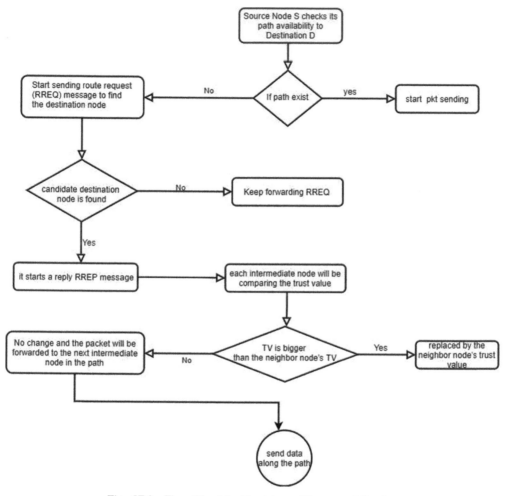

Fig. 45.1 Flow Chart for Trust-based Proposed Routing

Stage 1: Source node A will send RREQ packet to B and C, which are neighbors of it and it will continue till reaching the destination. After reaching the destination, the destination sends Rtrep packet back to source A. Rtrep packets are broadcasted from various paths over a specified time.

Go get trust-aware route; all available routes are arranged as per the descending order of TV and ascending order of hop-counts. Whenever a fresh route is found, in this stage, it could be ensured that the chosen routes are with minimum hop-count and highest TV.

Stage 2: Now, the first route is chosen and the first part of the message is routed. Similarly, the next route is chosen with a similar assumption. If all the message parts securely routed, the real routing is accomplished by chosen paths.

Stage 3: If more paths are chosen over possible eligible paths, set all these paths in their energy that need to transmit the packets. Then, choose the lowest energy path and so on.

Stage 4: Repeat until secure routes are obtained.

Stage 5: The algorithm is repeated from Stage 2 by choosing an alternate route if no secure routes are available.

Stage 6: This mechanism works until all the paths are drained. Moreover, the mechanism halts for another route if no secure route is obtained. Also, it can be assumed that the algorithm could fail if all routes are available, or a specific time interval is no longer valid.

5. Experimental Setup

On the basis of the following metrics, we compared our proposal to the normal DSR and normal AODV using the ns 2.34 simulator. The maximum node speed is set to 10 m/s, and the percentage of malicious nodes in the network is set at 10% of all nodes. We experiment with different network sizes to see how they affect the results.

As shown in Fig. 45.2, a total of 19 IoT nodes are participating in such environment and device 10 wants to send any routing packet to receiver device 19. Each node has its trust value; on the basis of the predefined threshold, only few nodes have qualified. Next step is to update the routing information with only qualified nodes.

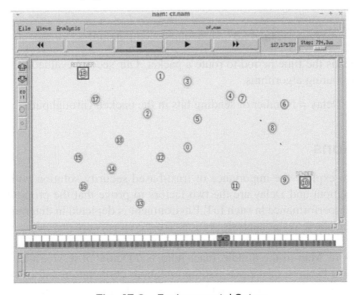

Fig. 45.2   Environmental Setup

Throughput: We evaluated the throughput of the proposed scheme with the conventional AODV routing.

$$\text{Throughput} = \text{Total packets received/Total packets sent}$$

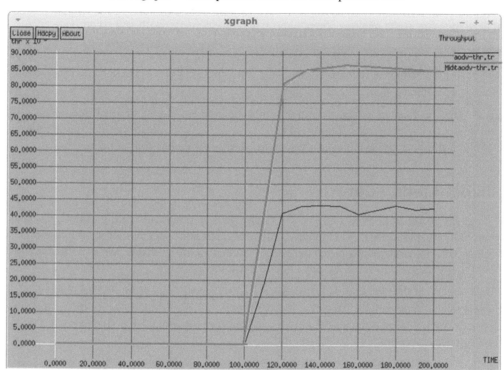

Fig. 45.3 Throughput Comparison

Delay: Delay means the time period to route a packet. Our second evaluation is done on the delay in both the routing algorithms.

$$\text{Delay} = \text{Number of sending bits in the packet/Throughput}$$

6. Conclusions

The paper tries to explore the importance of trust-based security solution with the token of proof. The Throughput and Delay are the two factors to prove that the proposed trust-based AODV gave better performance in such IoT. Environment is depicted in diagrams 5.2 and 5.3, respectively, in the above result analysis section. Trust concepts are proved as a better way to achieve security in various operations related to network communication like routing, data collection, and more. Man-in-the-middle, black hole, and Denial of service attacks are accrued very frequently just because of the pre-assumption about cooperation and trustworthiness of nodes. The paper firstly discusses IoT and how Trust works in case of such ad hoc networks.

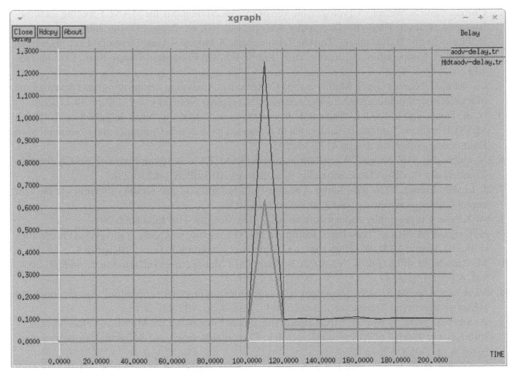

Fig. 45.4 Delay Comparison

In various sections, we identified the most possible attacks for MANETs and discuss their countermeasures. Trust-based security solutions are reviewed and declared as one of the best security solutions for such dynamic and modern environment. It investigates in detail the management of Trust through related works. Trust-based schemes are attack-tolerant, cooperative, flexible, lightweight, and scalable as well as compatible to the rapidly growing network size.

7. Acknowledgment

First of all, we would thank the Almighty GOD who has always blessed me by granting his kindness to complete the task. I would like to state my gratitude to Dr. Niraj Kumar Shukla for providing such platform.

REFERENCES

1. A. Gupta, P. Verma, and G. Sambyal, "An Overview of MANET: Features, Challenges and Applications," International Journal of Scientific Research in Computer Science, Engineering and Information Technology, vol. 4, no. 1, pp. 122–126, 2018J.

2. Cordasco and S. Wetzel, "Cryptographic versus trustbased methods for MANET routing security," ElectronicNotes in Theoretical Computer Science, vol. 197, no. 2, pp. 131-140,2008.

3. R. J.Lewicki and B.B. Bunker, "Trust in Relationships: A Model of Trust Development and Decline, In Conflict, Cooperation and Justice, B. Z. Rubin, Ed. San Francisco: Jossey-Bass, 1995,pp.133-173.

4. C. Panos, C. Ntantogian, S. Malliaros, and C. Xenakis, "Analyzing, quantifying, and detecting the blackhole attack in infrastructure-less networks," Computer Networks, vol. 113,pp.94-110,2017.

5. N. A. Funde and P. Pardhi, "Detection & prevention techniques to black & gray hole attacks in MANET: a survey," International Journal of Advanced Research in Computer and Communication Engineering, vol. 2, no. 10, pp.4132-36,2013.

6. N. Gupta and S. N. Singh, "Wormhole attacks in MANET," in 2016 6th International Conference-Cloud System andBig DataEngineering (Confluence),2016:IEEE, pp.236-239.

7. J. Lindley-French, "The Revolution in security affairs: hard and soft security dynamics in the 21st century," Europeansecurity,vol.13,no.1-2,pp.1-15,2004.

8. A. Kahate, Cryptography and network security. Tata McGraw-Hill Education,2013.

9. K. Garg and M. Misra, "Trust based security in MANET routing protocols: a survey," in Proceedings of the 1st Amrita ACM-W Celebration on Women in Computing in India,2010,pp.1-7.

10. M. Branchaud and S. Flinn, "xTrust: A Scalable Trust Management Infrastructure," in PST, 2004, vol. 4, pp. 207-218.

11. M. A. Morid and M. Shajari, "An enhanced ecommerce trust model for community based centralized systems," Electronic Commerce Research, vol. 12,no. 4, pp. 409-427,2012.

12. A. Sharma and D. N. Kumar, "Trust Based Theoretical Framework for Mobile Ad-Hoc Networks," International Journal of Advanced Research in Computer Science and Software Engineering, vol. 3, no. 6, pp. 905909,2013.

13. S. Sridhar, R. Baskaran, and P. Chandrasekar, "Energy supported AODV (EN-AODV) for QoS routing in MANET," Procedia-Social and Behavioral Sciences, vol. 73,pp.294-301,2013.

14. D. K. Prasadh and R. Senthilkumar, "Nonhomogeneous Network Traffic Control System Using Queueing Theory," International Journal of Computer Engineering & Technology (IJCET), vol. 3, no. 3, pp. 394405,2012. are using a style other than Chicago Manual of Style, 16th edition so we can ensure it is retained).

15. A. Verma and M. S. Gujral, "Trust Oriented Security Framework for Ad Hoc Network," Journal of Computer Science & Information Technology, vol. 5, pp. 19-26,2012.

16. M. Branchaud and S. Flinn, "xTrust: A Scalable Trust Management Infrastructure," in PST, 2004, vol. 4, pp. 207-218.

17. K.S.Ramana,A.Chari,andN.Kasiviswanth,"Asurveyontrustmanagementformobileadhocnetworks," International Journal of Network Security & Its Applications(IJNSA),vol.2,no.2,pp.75-85,2010.

18. K. S. Ramana, A. Chari, and N. Kasiviswanth, "Trust based security routing in mobile adhoc networks," IJCSE) International Journal on Computer Science and Engineering,vol. 2,no.02,pp.259-263,2010.

19. N. K. Nehra, M. Kumar, and R. Patel, "Neural network based energy efficient clustering and routing in wireless sensor networks," in 2009 First International Conference on Networks & Communications, 2009: IEEE, pp.34-39.

20. A. Ahmed, P. Kumar, A. R. Bhangwar, and M. I. Channa, "A secure and QoS aware routing protocol for Wireless Sensor Network," in 2016 11th International Conference for Internet Technology and Secured Transactions(ICITST),2016:IEEE,pp.313-317.

21. N. C. Fernandes and O. C. M. B. Duarte, "An efficient group key management for secure routing in ad hoc networks," in IEEE GLOBECOM 2008-2008 IEEE Global elecommunicationsConference,2008:IEEE,pp.1–5.

22. R. Mishra, I. Kaur, and S. Sharma, "New trust based security method for mobile ad-hoc networks," International Journal of Computer Science and Security (IJCS), vol. 4, no. 3, p. 346, 2010.
23. I. Kaur, "A Survey to Improve the Network Security with Less Mobility and Key Management in MANET,"2018.
24. S. Kumar, M. Goyal, D. Goyal, and R. C. Poonia, "Routing protocols and security issues in MANET," in 2017 International Conference on Infocom Technologies and Unmanned Systems (Trends and Future Directions)(ICTUS), 2017: IEEE, pp. 818–824.
25. M. Maleki, K. Dantu, and M. Pedram, "Power aware source routing protocol for mobile ad hoc networks," in Proceedings of the 2002 international symposium on Low power electronics and design, 2002, pp. 72–75.
26. S. Taneja and A. Kush, "Energy efficient, secure and stable routing protocol for MANET," Global journal of computer science and technology, 2012.
27. T. Singh, J. Singh, and S. Sharma, "Energy efficient secured routing protocol for MANETs," Wireless Networks, vol. 23, no. 4, pp. 1001–1009, 2017.
28. S. Hammer, M. Wißner, and E. André, "Trust based decision-making for smart and adaptive environments," User Modeling and User-Adapted Interaction, vol. 25, no. 3, pp. 267–293, 2015.
29. M. Cukier and S. Panjwani, "A Comparison between Internal and External Malicious Traffic," in The 18th IEEE International Symposium on Software Reliability (ISSRE'07), 2007: IEEE, pp. 109–114.
30. A. Lamba, S. Garg, and R. Kumar, "A Literature Review of MANET's Routing Protocols Along With Security Issues," 2016.
31. D. Geetha and S. Sakthivel, "Service Orient Stream Cipher Based Key Management Scheme for Secure Data Access Control Using Elliptic Curve Cryptography in Wireless Broadcast Networks," American-Eurasian Journal of Scientific Research, vol. 11, no. 1, pp. 63–71, 2016.
32. H. Kojima, N. Yanai, and J. P. Cruz, "ISDSR+: Improving the Security and Availability of Secure Routing Protocol,"IEEE Access,vol. 7, pp. 74849–74868, 2019.
33. K. Zhang, C. Wang, and C. Wang, "A secure routing protocol for cluster-based wireless sensor networks using group key management," in 2008 4th international conference on wireless communications, networking and mobile computing, 2008: IEEE, pp. 1–5.
34. K. K. Chauhan and A. K. S. Sanger, "Securing mobile Ad hoc networks: key management and routing," arXivpreprintarXiv:1205.2432,2012.
35. N. Bißmeyer,S. Mauthofer, K. M. Bayarou, and F. Kargl, "Assessment of node trustworthiness in vanets using data plausibility checks with particle filters," in 2012 IEEE Vehicular Networking Conference (VNC), 2012: IEEE, pp. 78–85.
36. R. Chen, J. Guo, F. Bao, and J.-H. Cho, "Integrated social and quality of service trust management of mobile groups in ad hoc networks," in 2013 9th International Conference on Information, Communications &Signal Processing, 2013: IEEE, pp. 1–5.
37. R. Menaka, V. Ranganathan, and B. Sowmya, "Improving performance through reputation based routing protocol for manet," Wireless Personal Communications, vol. 94, no. 4, pp. 2275–2290, 2017.
38. Rao, PV Venkateswara, and S. Pallam Setty. "Investigating the Impact of Black Hole Attack on AODV Routing Protocol in MANETS under Responsive and Non Responsive Traffic." International Journal of Computer Applications 120.22(2015)
39. S. Ba and P. A. Pavlou, "Evidence of the Effect of Trust Building Technology in Electronic Markets: Price Premiums and Buyer Behavior, MIS Quarterly, Vol. 26, pp. 243–268, 2002.
40. D. H. McKnight, L. L. Cummings, and N. L. Chervany, "Initial Trust Formation in New Organization Relationships, Academy of Management Review, Vol. 23, pp. 473–490, 1998 Common reference uses for the Chicago Manual of Style, 16th Edition are below. If you require information on a different style, please contact your Project Coordinator.

Intelligent Systems and Smart Infrastructure – Brijesh Mishra et al. (eds)
© 2023 Taylor & Francis Group, London, ISBN 978-1-032-41287-0

CHAPTER

46

Performance Analysis of Wind Energy Conversion System Using Multilevel Inverters

Asha Singh[1], Pallavi Choudekar[2], Ruchira[3]

Amity University, Noida,
Uttar Pradesh, India

Abstract

Wind turbines transform the kinetic energy of the wind into usable electrical energy. Wind turbines that are both reliable and efficient are essential for meeting the world's ever-increasing energy demand. Off-shore installations have grown much more popular as a result of the increased need for wind energy generation, and new wind turbine designs, as well as upgraded generator and converter designs, are expected in the near future. WECS (Wind Energy Conversion System) is made up of a number of non-linear components that contribute significantly to harmonic disturbances. Harmonic disturbances harm the quality of power generation. As a result, harmonic analysis and mitigation are now part of the WECS multilayer inverters, which have minimal switching stress and high voltage capabilities, and have effectively decreased harmonics. The goal of this study is to examine the total harmonic distortion and power of WECS when applied with various multilevel inverters. Additionally, the design and analysis of WECS and three different types of multilevel inverters to highlight the variations in the inverter topologies used with WECS are presented.

Keywords: Wind Turbine, Perturb and Observe, Multilevel Inverter, Capacitors

Corresponding author: [1]ashasingh.as63@gmail.com, [2]pallaveech@gmail.com, [3]er.ruchiragarg@gmail.com

DOI: 10.1201/9781003357346-46

1. Introduction

Electricity demand is on the rise these days. Faced with this dilemma, several countries have resorted to new kinds of energy known as "renewable" in order to minimise pollution. Wind is certainly in a good place among these, not as a replacement for traditional energy sources, but as a complementing energy enhancer. [1]

In comparison to wind turbines with a fixed speed, the variable speed wind turbine can improve energy efficiency, minimise mechanical stresses, and increase the pace at which electrical energy is generated. The quantity of power output of the converters is limited by IGBT properties such as maximum voltage and current. Commercial turbines have already surpassed 7.5 MW, and developing technologies are on their way to exceeding 10 MW [2]. Multilevel converters and medium voltage operation become appealing at this power level due to enhanced power quality and efficiency. This is why multilevel converters are used in many modern setups. A two-stage conversion consists of an AC–DC rectifier and a DC–AC inverter is the most typical technique of constructing a PMSG Wind Energy Conversion System with adjustable speed and adjustable frequency that is grid-connected [3]. A three-phase bridge rectifier or thyristors converter converts AC to DC, after which a DC capacitance filter is used. The second stage is an inverter created on IGBTs that employs a variety of control methods. AC–DC converters are the most competent means of extracting AC power from a variable speed generator. However, when these converters are used, the system receives non-sinusoidal current with sinusoidal voltage.

Multilevel converters are utilised to handle high voltage and high-power requirements. Multilevel converters boost output voltage, reduce voltage and current harmonics, and approximate the sine wave output waveform. [4] Furthermore, by removing low-frequency harmonics, AC inductance is reduced as a result of varying AC voltages, lowering the overall cost of the system. Multilevel converters manage the output frequency, voltage, and phase angle, as well as the phase angle, enabling for quick reaction and self-regulation. A two-level back-to-back converter construction is used in this article, with a converter on both the generator and the load side.

The purpose of this research is to compare the THD and power of WECS when they are implemented utilising various multilevel inverters and to complete the design and analysis of WECS and three different types of multilevel inverters in order to demonstrate the differences in the inverter topologies utilised with WECS. The system's core architecture includes a 5-level cascaded H-bridge multilevel inverter, a 3-level diode clamped inverter, and a 3-level flying capacitor inverter for grid-connected wind energy systems, as well as maximum power extraction for wind energy resources utilising the P&O algorithm. Section 1 provides the introduction. Wind energy conversion system is discussed in Section 2. PMSG modelling is discussed in Section 3 followed by system architecture in Section 4. The results are discussed and presented in section 5 and the work is concluded in Section 6.

2. Wind Energy Conversion System

Wind energy supplies mechanical energy to a WECS, which is then transmitted to an electrical generator for electricity generation. The connection of a WECS is depicted in Fig. 46.1. The generator receives wind energy from the wind turbine. A pulse width modulation converter controls the generator's spinning speed to get the greatest power out of the WECS. [5]

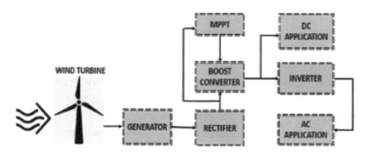

Fig. 46.1 Wind Energy Conversion System

The wind power analysis conversion setup includes an IGBT converter, a DC-Link, and a multilevel inverter, as well as a link to the DC application and a synchronous permanent magnet generator, as shown in Fig. 46.1. The IGBT inverter is a three-phase rectifier controlled by PWM. [6,7] The fact that this approach provides for total reversibility of instantaneous power generation as well as control of electromechanical parameters like electromagnetic torque and generator speed justifies it. The multilevel inverter modifies the voltage within the DC-Link as well as the active and reactive power that is exchanged with the grid [8].

A. PMSG Modelling

The direct-drive PMSG wind generator systems do not require a gearbox, making them excellent for improving overall efficiency, dependability, and availability. They are frequently larger; but, where there is no land or space constraint and adequate wind speed, this may not be a disadvantage. [9] To model PMSG-type electrical equipment, utilise the equations below, which are shown by the d-q reference frame.

$$F1 - \text{score} = \frac{2TP}{2TP + FP + FN} \tag{1}$$

$$V_q = R_s i_q + L_q \frac{di_q}{dt} - w_e L_d i_d + w_e \phi_m \tag{2}$$

Equations (1) and (2) represent the mathematical relationship for stator voltage (2). Vd and Vq are the d and q components of voltages (V). Equations 3 and 4 are used to calculate electrical torque. B is the rotor friction, J is the rotor inertia, ω_r is the rotor speed, and Tm is the mechanical torque.

$$T_e = \frac{2}{3} p\{\phi_m i_q + (L_d - L_q)i_d i_q\} \tag{3}$$

$$T_m - T_e = B\omega_r + \frac{Jd\omega_r}{dt} \tag{4}$$

B is rotor friction, J is rotor inertia, ω_r rotor speed and T_m is mechanical torque. The parameters for a permanent magnet synchronous generator are listed in Table 46.1.

Table 46.1 PMSG Parameters

PMSG	
Inductances Ld, Lq (H)	0.395e-3
Flux linkage	0.1194
Stator phase resistance	0.0485
No. of phases	3
Rotor type	Salient-pole
Inertia	0.0027
Viscous damping	0.0004924
Pole pairs	4
Rotor flux position	90 degrees behind phase A axis (modified park)

A device that converts level of voltage in DC systems is DC-to-DC converter. A boost converter is a DC-to-DC converter that outputs a higher voltage than the input. Fig. 46.2 shows the boost converter circuit diagram.

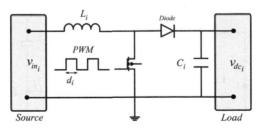

Fig. 46.2 Boost Converter

Changing the PV parameters (voltage and current) and looking for a change in output power in P&O is how the MPP is done. When it comes to obtaining MPP, choosing the right step size is crucial.[10] When making a direction choice, averaging the PV power value is required to reduce the reaction to noise. While the system is not working at MPP, it is not operating at peak performance.[11] This method perturbs the voltage and checks divergence in power by altering the duty cycle in a certain way. If dP is less than zero, the duty cycle is altered more in the opposite route until MPP is obtained. However, if dP > 0, the duty cycle will alter in the same way., consequently the MPP may be traced by following this technique.

3. System Architecture

Multilevel inverters are capable of attaining a staircase output voltage waveform with reduced harmonic content. The increase in level of inverters reduces the harmonic content in the output. Based on waveform there are many advantages of multilevel inverter such as better quality in terms of THD, better harmonic profile with reduced filter requirements, lesser dv/dt stress on the load and possibility of modulation with low and high switching frequencies. Advantages of multilevel inverter are represented in Fig. 46.5. The different topologies of multilevel inverters that are used in this work are cascaded H bridge MLI (Fig. 46.3), diode clamped MLI (Fig. 46.4) and flying capacitor MLI as shown in Fig. 46.5 [12].

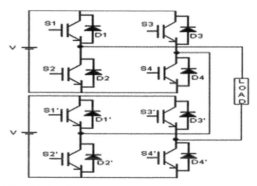

Fig. 46.3   5-Level Cascaded H-bridge Inverter

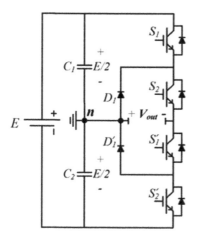

Fig. 46.4   3-Level Diode Clamped Inverter

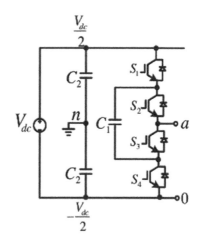

Fig. 46.5   Flying Capacitor Multilevel Inverter

Wind energy conversion system is modeled using MATLAB / Simulink. The wind energy system is integrated with three different inverter topologies namely 3-level Diode clamped inverter, a 5-level cascaded h-bridge inverter, and a 3-level flying capacitor inverter. The proposed inverter's performance dynamics is verified using the simulation. A 12.3 kW wind turbine is employed in this work, and the parameters of PMSG are analysed using equations 1-4.

4. Results

The performance comparison of three different inverter topologies integrated with the wind energy conversion system is tabulated in Table 46.2. The power quality analysis of all three inverter topologies has been done. Harmonics introduced by the inverters plays an important role in their performance. So THD comparison is shown here for all the three topologies. Fig. 46.6, 46.7 and 46.8 shows the THD obtained with 3-level Diode clamped inverter, a 3-level flying capacitor inverter and 5-level cascaded h-bridge inverter, and a respectively.

Table 46.2   Comparative Analysis of Wecs Using Multilevel Inverter Topologies

Parameters	5-Level Cascaded Inverter	3-Level Diode Clamped Inverter	3-Level Flying Capacitor Inverter
THD	88.67%	114.2%	188.3%
Output power	20 KW	40 KW	25.16 KW
Output voltage	5 KV	6 KV	4 KV
Power semiconductor switches	8	4	4
Clamping diode per phase	0	2	0
DC bus capacitor	1	1	2
Voltage unbalancing	Average	High	Very small

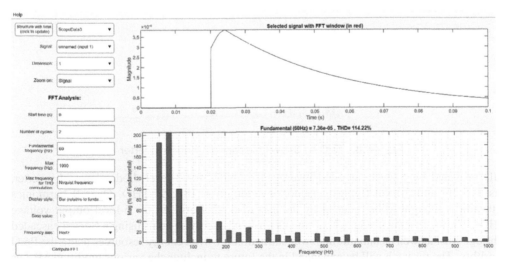

Fig. 46.6   THD by 3-level diode clamped MLI

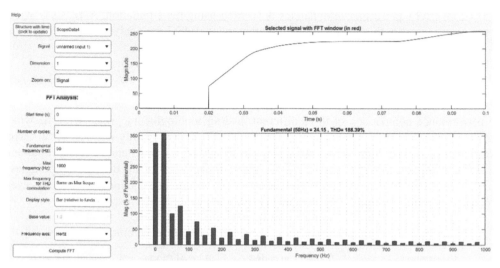

Fig. 46.7 THD by flying capacitor 3-level MLI

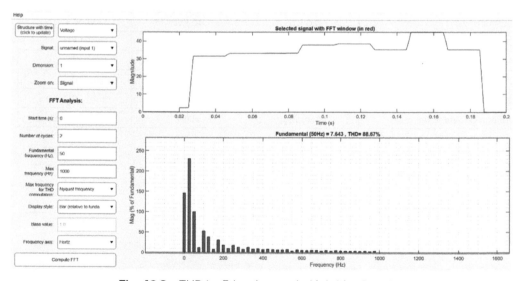

Fig. 46.8 THD by 5-level cascaded h-bridge inverter

THD provided by 5-level cascaded inverter is 88.67%, for 3-level diode-clamped inverter, it is 114.2% and 3-level flying capacitor inverter gives 188.3% THD. Overall harmonics distortion is thought to be reduced by the cascaded MLI utilised for 5-level output voltage (THD). The result shows reduced harmonics for wind energy systems using a 5-level cascaded H-bridge MLI as well as maximum power extraction for wind energy resources using the P&O algorithm. The research will be expanded in the future to include the various modified MLI as well as other renewable energy sources.

5. Conclusion

The work analyses that the P&O approach used for MPPT of wind-generating systems is a preferable methodology for dealing with large wind speed fluctuations sensitivity. Three types of multilevel inverters are used in the study for the production and absorption of reactive power. A 3-level diode-clamped inverter and a 3-level flying capacitor inverter have higher overall THD. When adopting the SPWM approach, the cascaded multilevel inverter employed for 5-level output voltage may be proven to reduce total harmonics distortion (THD). According to the findings, this scheme is clearly more suited for cascaded multilevel inverters, as the output signal quality has been increased, making it more acceptable for both independent and grid-connected systems.

REFERENCES

1. R. Esmaili, and L. Xu, D. K. Nichols: A new control method of permanent magnet Generator for Maximum Power Tracking in wind Turbine Application, vol.3, IEEE Power Engineering Society General Meeting, (2005), 2090–2095.

2. M. A Abdullah, A. H. M. Yatim, C. W. Tan: Maximum Power Point Tracking Algorithm for Wind Energy System: A Review, Vol. 2, International Journal of Renewable Energy Resources, (2012).

3. Pena, J. C. Clare: Doubly fed induction generator using back-to-back PWM converters and its application to variable speed wind-energy generation, Vol. 143, No 3, IEEE Puoc.-Electr. Power Appl., (1996), 231–241.

4. Pena, R.; Clare, J. C.; Asher, G. M.: Doubly Fed Induction Generator Using Back-to-Back PWM Converters and its Application to Variable-Speed Wind-Energy Generation, Vol. 143, No. 3, IEEE Proceedings on Electric Power Applications, May (2006).

5. Pinto, S. F.; Aparicio, L. Esteves, P.: Direct Controlled Matrix Converters in Variable Speed Wind Energy Generation Systems", Proc.POWERENG´07, International Conference on Power Engineering, Energy and Electrical Drives, Setúbal, Portugal, May 2007).

6. A. Kurella, R. Suresh: Simulation of incremental conductance MPPT with direct control method using Cuk converter, Volume 2, International Journal of Research in Engineering and Technology, (2013)

7. Abhijeet Awasthi, Ritesh Diwan, Dr. Mohan Awasthi, "Study for Performance Comparison of SFIG and DFIG Based Wind Turbines", vol. 2, issue-4, ISSN: 2278-621X, July (2013), 1–10.

8. L. Ackermann: Wind energy technology and current status: a review, Renewable and Sustainable Energy Review, (2000), 315–375.

9. Z. Lubosny: Wind turbine operation in electric power systems, Springer, (2003).

10. J. Hui, A. Bakhshai, P.K. Jain: An adaptive approximation method for maximum power point tracking (MPPT) in wind energy systems, Energy Conversion Congress and Exposition (ECCE), IEEE, (2011).

11. RuiMelicio, V. M. F Mendes : Doubly Fed Induction Generator Systems for Variable Speed turbine, ISEL, DEEA, 1950–062 Lisboa, Portugal, (2006).

12. A. Singh, P. K. Nayak, R. Ruchira and P. Choudekar: Power Quality Enhancement of Cascaded H Bridge 5 Level and 7 Level Inverters, 2021 8th International Conference on Signal Processing and Integrated Networks (SPIN), (2021), 544–5

Intelligent Systems and Smart Infrastructure – Brijesh Mishra et al. (eds)
© 2023 Taylor & Francis Group, London, ISBN 978-1-032-41287-0

CHAPTER

47

A Hybrid Approach for Prediction of Type 2 Diabetes Using Birch Clustering and Artificial Neural Network

Yogendra Singh[1], Mahendra Tiwari[2]

Department of Electronics and Communication,
University of Allahabad, Prayagraj, India

Abstract

Diabetes is a long-term disorder in which a patient's blood sugar level is abnormally high, resulting in serious effects such as heart disease, stroke, renal failure, and high blood pressure. Machine learning might aid in the early identification and diagnosis of type-2 diabetes, perhaps lowering the severity of the disease's consequences. This study aims to develop a hybrid model to predict diabetes more accurately, tackle the problems that occur from datasets, and increase the performance of the diabetes predictive model. The proposed methodology includes missing value imputation, outlier identification and imputation, random sampling, normalization, BIRCH clustering, and artificial neural network classification. The proposed methodology includes missing value imputation, outlier identification and imputation, random sampling, normalization, BIRCH clustering, and artificial neural network classification. We utilized the Frankfurt Hospital (Germany) diabetes dataset in our study, which is publicly available on Kaggle. With 0.977 precision, 0.980 recall, 0.979 f-measure, and 0.970 ROC, the suggested model obtained 97.25% accuracy. We also compared our results to those of earlier research. The results showed that our strategy is successful in getting the desired results. The suggested technique significantly increases prediction performance, according to this study. The findings indicated that BIRCH clustering and Artificial Neural Networks might accurately identify the diabetic illness.

Keywords: Diabetes, BIRCH clustering, Artificial neural network, Hybrid model, Predictive model, Outlier, Class imbalance problem

Corresponding author: [1]sachin12345jan@gmail.com, [2]mahendra@allduniv.ac.in

DOI: 10.1201/9781003357346-47

1. Introduction

Diabetes is the deadliest metabolic disease caused by glucose levels that are higher than usual due to abnormalities in insulin production, insulin action, or both (P *et al.* 2020). Diabetes is generally classified into type-1 and type-2. When the body's immune system is targeted, the beta cells in the pancreas (which generates insulin) destroy, resulting in type 1 diabetes. On the other side, type 2 diabetes, in particular, is related to insulin resistance, a condition in which cells react poorly to insulin, reducing glucose absorption from blood circulation (Han Cho 2019).

Diabetes mellitus is a worldwide public health problem. According to the International Diabetes Federation, diabetes and its complications are expected to take the lives of 4.2 million persons aged 20 to 79 in 2019. Every eight minutes, one person dies as a result of this disease. Diabetes is responsible for 11.3% of all fatalities among adults in this age range worldwide. Furthermore, one undiagnosed individual is expected to exist for every diagnosed person with diabetes (Han Cho 2019).

Early detection and medication of type 2 diabetes are one of the most important steps in preventing complications such as neuropathy (Aparicio *et al.* 2021). Early identification reduces the absolute and relative risk of heart disease and mortality (Herman *et al.* 2015).

In the health sector, several research projects have been conducted in the areas of early detection, forecasting, categorization, and diagnosis. The popularity of data mining in the medical field is steadily growing due to its ability to uncover complicated and hidden knowledge while developing powerful prediction models. Hence, various data mining techniques have been developed to support the treatment of diseases in the medical field. These techniques can be improved by reducing the risks of bias. The biasness of the model depends upon the dataset, which indicates a data imbalance. A poor dataset may lead to a biased and poor predictive model, which means the dataset has a lot of missing values, outliers, class imbalance, and high variance data points. Therefore, we proposed a hybrid approach to tackle these problems to enhance the effectiveness of the diabetes predictive model. The major purpose of this study is to enhance the accuracy of type 2 diabetes mellitus prediction utilizing hybrid approach. This model has been developed by combining BIRCH clustering and artificial neural networks. The accuracy, precision, recall, F1-score, and ROC evaluation parameters were utilized to evaluate the suggested model's performance.

The following parts comprise this research paper: Section 2 discusses the related work done in this area. Section 3 explains the Material and methods. Section 4 addresses the Results and discussion of the proposed methodology. The proposed methodology's results have been compared and analysed with conventional ML algorithms and state-of-art methods. Finally, Section 5 clarifies the conclusion of the study.

2. Related Works

AK Srivastava et al. developed a hybrid framework for diabetes prediction that can accurately forecast diabetes disease. It is claimed that combining data imputation and outlier identification approaches increases the performance of the LS-SVM. The authors' proposed model include a missing value imputation approach based on K-Mean++, an outlier identification method based on ABC, and a classification methodology based on sparse SVM. Their proposed model achieved 96.57% overall accuracy (Srivastava *et al.* 2021).

W. Chen et al. introduced a hybrid model for predicting type 2 diabetes patients that combined k-means clustering with a decision tree. In their model, K-means clustering was employed for data reduction, while the J48 decision tree was used as a classifier for classification. Their model outperformed prior research cited in their literature in terms of accuracy. Overall accuracy achieved by their proposed model was 90.04% (Chen *et al.* 2018).

B. P. Patil et al. constructed a hybrid prediction approach for t2dm patients. The authors employed a basic K-means clustering approach to confirm the selected class label of provided data (incorrectly categorized data points were isolated, i.e., patterns obtained from the original data) and then applied the classification procedure to the result set in their suggested Hybrid Prediction Model (HPM). Using the k-fold cross-validation procedure, the C4.5 algorithm created the final classifier model. For the PID dataset, their suggested HPM had a classification accuracy of 92.38% (Patil *et al.* 2010).

V. B. Kumar et al. presented a hybrid data mining technique for diabetes prediction and categorization. Their technique involved missing value imputation, data clustering, dimension reduction, and Bayesian regularized neural network classification. The data clustering approach assists in grouping diabetic variables based on their commonalities. The misclassification error was minimized using these cluster groups during dataset training and testing. Attribute selection was deemed the crucial component that greatly affects the prediction and classification with data mining methods. The authors applied a Bayesian regularized neural network for classification where the chosen characteristics were trained. The accuracy of the proposed model was 94.5% (Kumar *et al.* 2019).

B. S. Kumar and R. Gunavathi introduced a diabetic prediction model to optimize the classification accuracy by excluding the irrelevant characteristics. As a result, it is vital to use a strong attribute selection method that produces better accuracy in health forecasting than previous studies. Hence, innovative approaches, Improved Firefly (IFF) as an attribute selection and a hybrid Random Forest algorithm as a classification were introduced to construct these hybrid model. The current research delivers a superior outcome with 96.3% accuracy. The Firefly method was enhanced utilizing the weighted intensity value for attribute selection. Additionally, the RF was modified utilizing the back-propagation strategy to eliminate the undesired trees to increase the resultant performance (Senthil Kumar and Gunavathi 2019).

R. Howsalya Devi et al. proposed an innovative hybrid method for detecting diabetes mellitus that used the farthest first and SVM algorithms. Their investigation offered a combined technique for diagnosing DM using the Farthest First (FF) clustering algorithm and the Sequential Minimal Optimization (SMO) classifier algorithm. The data was divided into various groups using FF clustering. Due to the reducing size of the dataset, the computing time was considerably decreased. The authors used SMO (improved support vector classifier) to accurately identify whether a patient has diabetes or not, i.e., tested positives for diabetic patients and negatives for the non-diabetic patient. According to their findings, the suggested hybrid technique has a classification accuracy of 99.4% for predicting diabetes mellitus (Howsalya Devi *et al.* 2020).

3. Material and Methods

The proposed model's phases are described in detail below, along with their respective diagrams and tables.

A. Dataset and Software Tool

The dataset that has been considered for this experiment was taken from the hospital Frankfurt, Germany (publicly available at the Kaggle) (Diabetes | Kaggle 2022). The purpose of utilizing this dataset was to anticipate whether or not a patient has diabetes based on the attributes presented in the dataset. The dataset contains 2000 female patients aged 21–81 years, with eight attributes and one class attribute. The total number of diabetic and non-diabetic patients is 684 and 1316, respectively. Table 47.1 provides an overview of the dataset.

Table 47.1   Details description of diabetes dataset

Features	Description	Missing values	Outliers
Pregnancies	The number of times a woman has been pregnant	0	0
Glucose	Among of glucose in blood	13	0
Skin Thickness	The thickness of the triceps' skin folds (mm)	573	11
Blood Pressure	Diastolic blood pressure reading (mmHg)	90	35
Insulin	2-h Serum insulin	956	62
Diabetes Pedigree Function	The possibility of affecting by disease depending on the patient's ancestors' history.	0	68
BMI	Body Mass Index (kg/m2)	28	28
Age	Age of diabetic patient	0	0
Outcome	Binary (0/1)	0	0

The implementation of the proposed model was done on a Jupyter notebook v6.3.0 and Python 3.8.8 available with Anaconda Navigator. The Jupyter notebook is a powerful open-source, web-based application that allows you to create and edit documents for data science and scientific computing and supports dozens of programming languages.

B. Data Pre-processing

Data pre-processing is one of the most crucial processes for transforming data to develop a more accurate machine learning model (Khanam and Foo 2021). Most healthcare data has missing information and other abnormalities that might diminish the dataset's efficacy (Soni and Varma 2020). Pre-processing, which includes missing value imputation, outliers' detection and imputation, random sampling, and normalization, has been done to improve the data quality. These techniques are described in detail in the subsections below.

Missing value imputation

Missing values may result from data corruption or inability to record data. Missing data must be addressed during dataset preparation since many machine learning algorithms do not tolerate missing values. We retrieved the number of missing data points in BMI, insulin, glucose, blood pressure, age, and skin thickness. The number of missing values in the experimental dataset is illustrated in Table 47.1. The missing number was replaced with the appropriate median value, calculated by grouping the diabetic and non-diabetic. The median value is used because extrinsic and extreme values have less of an impact on it than the mean and is typically used as the preferred statistical method when the distribution is asymmetrical.

Outlier identification

Outliers are observations that deviate from the rest of the data substantially. In other terms, they are unusual numbers in a dataset. Many statistical studies are concerned about outliers because they can lead to irrelevant or biased test results. We looked for outliers based on the interquartile range in the dataset using Jupyter Notebook. Table 47.1 clearly shows that 204 outliers' values are presented in the considered dataset. Unfortunately, there are no clear statistical criteria to overcome this problem. As a result, all outlier values in the dataset above the 95th percentile and below the 5th percentile were replaced with their respective percentile values. Outlier values were found using equations (1) and (2), while extreme values were discovered using equations (3) and (4).

$$Q1 - EVF * IQR \leq x < Q1 - OF * IQR \tag{1}$$

$$Q3 + OF * IQR < x \leq Q3 + EVF * IQR \tag{2}$$

$$x < Q1 - EVF * IQR \tag{3}$$

$$x > Q3 + EVF * IQR \tag{4}$$

Where Q1 represents the 5% quartile, Q3 represents the 95% quartile, IQR demonstrates the interquartile range, the difference between Q1 and Q3, OF denotes the outlier factor, and EVF indicates the extreme value factor.

Random sampling

There are 2000 entries in the dataset, including 1316 for non-diabetic patients and 684 for diabetes patients. If this dataset is used to train a predictive model, the final output will always be biased since the model will be more trained for non-diabetics than people with diabetes. This is because the number of non-diabetic patients in the dataset is much higher than that of diabetes patients. This is known as the "class imbalance problem". We used random sampling to solve this issue. There are two kinds of random sampling: random under-sampling and random over-sampling.

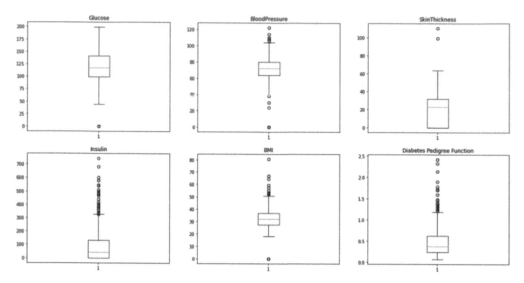

Fig. 47.1 Boxplot of identified outliers in PID dataset

In our suggested model, we utilized random oversampling. The main purpose of using this technique was to have a small dataset and avoid any information loss.

Normalization

In our research study, the Z-score normalization method was used to transform the data into the same interval that allows effective data processing. The mean and standard deviation were used to normalize the data. The mean of all the data is 0, and the standard deviation is 1. The mathematical formula for normalization is shown in Eq. (5), where Z is the normalized attribute value, x_i is the initial value of the attribute, μ is the mean of data, and σ is the standard deviation of data.

$$Z = \frac{x_i - \mu}{\sigma}$$

(5)

C. BIRCH Clustering

The BIRCH algorithm (Balanced Iterative Reducing and Clustering utilizing Hierarchies) is one of the most powerful unsupervised learning-based integrated hierarchical clustering algorithms over a particularly large dataset (Osmond *et al.* 2016). In general, clustering features (CF) and cluster feature trees (CF Tree) are two notions used to describe clusters. The clustering tree extracts information in a much smaller space than the large dataset stored in memory, which may increase the algorithm's speed and scalability while using huge data sets for clustering. It's ideal for interacting with situations involving discrete and continuous attribute data clustering. A CF is a node in the BIRCH tree. It is a short illustration of an underlying set of one or more data points. The main objective behind BIRCH's construction is that points that are significantly closer should always be treated as a group. This level of abstraction is provided by CFs. CFs are represented as a three-valued vector: CF = (N; LS; SS). The linear sum (LS), the square sum (SS), and the number of points it encloses (N) are simple to compute:

$$LS = \sum_{P_i \in N} \bar{P}_i \tag{6}$$

$$SS = \sum_{P_i \in N} |\bar{P}_i|^2 \tag{7}$$

$$C = LS/N \tag{8}$$

The centroid of the cluster is determined by dividing the linear sum by the number of points in the cluster. Both of these numbers, as the formulae show, may be derived iteratively. Any CF in the tree may be estimated by combining the CFs of its children (Kaur and Kaur 2015):

$$CF_1 + CF_2 = (N_1 + N_2, LS_1 + LS_2, SS_1 + SS_2) \tag{9}$$

A CF tree is a height-balanced tree with two parameters: B for branching factor and T for threshold. A non-leaf node is represented by the expression {CFi, childi}, where, $i = 1, 2, ...,$ B, Childi is a pointer to its child node and CFi is the CF of the sub-cluster represented by the ith child.

D. Artificial Neural Network

An artificial neural network (ANN) is a mathematical and computational approach inspired by the human nervous system. It allows the model to learn by the sample from appropriate data representing a physical event or selection process. The ANN establishes an actual relationship between dependent (output) and independent (input) attributes, and extracts subtle information and complicated and hidden knowledge from representative datasets, which is one of its distinctive features (Sadiq *et al.* 2019).

A layer of input nodes, one or more layers of hidden nodes, and a layer of output nodes are the three types of layers in an ANN (Wang 2003). Input layer nodes fire activation functions to transmit information to hidden layer nodes, whereas hidden layer nodes either fire or remain dormant depending on the information provided. The weight functions are applied to the data in the hidden layers, and when the value of a hidden layer's nodes crosses a certain threshold, a value is dispatched to the layer of output nodes (Sadiq *et al.* 2019). Eq. (10) represents the output of the j_{th} node in the hidden layer:

$$h_i = \sigma \left(\sum_{j=1}^{N} V_{ij} x_j + T_i^{hid} \right) \tag{10}$$

In the above equation, activation function is represented by, number of inputs nodes is denoted by N, V_{ij} are the weights, x_j are the inputs feature to the input nodes, and T_i^{hid} are the threshold terms of the hidden nodes.

In this study, ANN is employed as a classifier to determine whether the patient is diabetic or not, with the cluster of pre-processed dataset obtained from the BIRCH clustering which is fed into the ANN.

For this experiment, we have tuned the ANN to achieve excellent results. The following are the tuned hyperparameters: The optimizer's name is adam, the loss function is binary_ crossentropy, the batch size is set to 20, the number of epochs to 500, and the learning rate is the default learning rate which has the value of 0.01. Four hidden layers with ReLU activation function and 0.3 dropouts were used in this experiment, while the sigmoid activation function was used for the output layer. The number of neurons used in the 1st, 2nd, 3rd, and 4th hidden layers is 128, 64, 64, and 32, respectively.

E. Optimal BIRCH-ANN-based Hybrid Predictive Model

The author has proposed a hybrid predictive model for predicting diabetes more accurately. The proposed model is organized into three primary stages: data pre-processing, clustering, and classification. Data pre-processing is used to clean and enhance data quality; clustering is applied to create clusters of pre-processed datasets and classification to predict diabetic or non-diabetic patients. Table 47.2 describes constituent stages, and Fig. 47.2 depicts an overview of the proposed hybrid predictive models.

Table 47.2   Constituent stages of the hybrid proposed model

Algorithm Hybrid Diabetes Prediction Model based on BIRCH-ANN
Input: Frankfurt hospital (Germany) Diabetes Dataset
Output: Accurate and unbiased prediction
Prediction Steps
Step1: Read the Frankfurt hospital (Germany) dataset
Step2: Data Pre-processing
Step3: Create clusters of dataset using BIRCH clustering
Step4: Each cluster trained by artificial neural network
Step5: Each trained sub-models are combined to make complete hybrid model
Step6: Evaluate distance between test data point and Centroid of cluster generated by BIRCH
Step7: Test data fetch to that respective sub-trained model for which distance between test data and centroid is minimum
Step8: Hybrid Model evaluated by evaluation matrices and by plotting ROC

F. Performance Evaluation Metrics

Five performance evaluation metrics were used to evaluate the proposed model. These evaluation metrics are derived from the confusion matrix, and the ROC is a graph of the true-positive rate vs. the false-positive rate that is used to illustrate the proposed model's potential (Osmond *et al.* 2016). In the confusion matrix, True-Negative (TN) data points show the total number of non-diabetes patients correctly identified, True-Positive (TP) data points express the total number of diabetes patients correctly identified, False-Positive (FP) data points indicate the total number of non-diabetes patients classified as diabetes patients, and False-

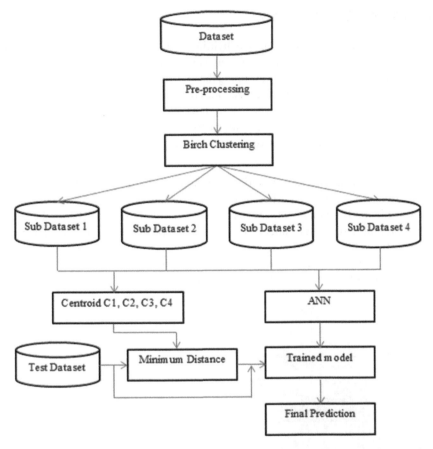

Fig. 47.2 A hybrid model using BIRCH clustering and artificial neural network

Negative (FN) data points show the total number of diabetes patients classified as non-diabetes patients (Kulkarni *et al.* 2020). The evaluation metrics computed from the confusion matrix are described in Table 47.3.

Table 47.3 Description and formulation of performance metrics

Metrics	Description	Formula
Accuracy	Percentage of adequately categorized cases among all occurrences	$\dfrac{TP + TN}{TP + TN + FP + FN}$
Precision	It is defined as the rate of correct predictions	$\dfrac{TP}{TP + FP}$
Recall	Used to evaluate classifier completeness	$\dfrac{TP}{TP + FN}$
F-Measure	Weighted average of precision and recall	$2 \times \dfrac{\text{Precision} \times \text{Recall}}{\text{Precision} + \text{Recall}}$
ROC	Receiver operating characteristic curve	—

4. Results and Discussion

The experimental findings of the proposed hybrid predictive model are presented in this section. In the pre-processing phase, 1660 missing values were identified and replaced with the appropriate median value, calculated by grouping the diabetic and non-diabetic. In the next step, a total of 204 outliers' values were recognized based on the interquartile range and replaced with their respective percentile value. After these steps, the random oversampling method was applied to generate 632 records of diabetes patients to balance non-diabetic patients' class with diabetic patients' class. To allow effective data processing, z-score normalization transformed the attributes' values into the same interval.

After pre-processing data, BIRCH clustering was used to divide the dataset into four sub-datasets or clusters with their similarities. Then, these subsets of the dataset are trained on the ANN with four hidden layers and 500 epochs to make four sub-models. The predictive model was built by combining these sub-models to make our proposed hybrid model. The proposed hybrid model achieved 97.25% accuracy with 0.977 precision, 0.980 recall, 0.979 F-measure, and 0.970 ROC. Figure 47.3(a) depicts the BIRCH-ANN model's confusion matrix. The BIRCH-ANN model's ROC curve is shown in Fig. 47.3(b).

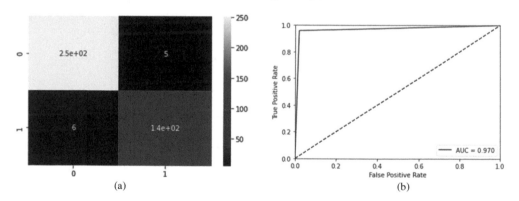

(a)   (b)

Fig. 47.3   (a) Confusion matrix, (b) ROC curve

The four traditional diabetes predictive models based on ML techniques are compared with the proposed hybrid model in Table 47.4. The table clearly shows that the proposed hybrid model outperformed all other ML models. As a result, our suggested diabetes prediction approach may be able to detect people with diabetes more effectively. Figure 47.4 shows the results of the diabetes prediction models that have been developed.

We also compared our results to other prior research in Table 47.5, and it demonstrated that our method is successful in getting the desired results and surpasses all others.

The proposed model achieved 97.25% accuracy, while other machine learning-based models failed to achieve more than 90.25% accuracy. Hence, we can conclude that our proposed model obtains promising results.

Table 47.4   Experimental observations of five ML model

Algorithms	Accuracy	Precision	Recall	F-Measure	ROC
SVM	90.25%	0.951	0.907	0.928	0.880
NB	82.46%	0.908	0.832	0.868	0.820
k-NN	87.01%	0.931	0.879	0.904	0.864
J48	87.01%	0.958	0.850	0.901	0.882
BIRCH-ANN	97.25%	0.977	0.980	0.979	0.970

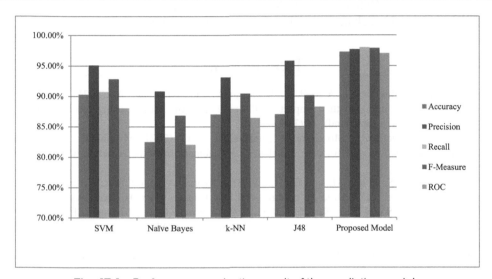

Fig. 47.4   Performance evaluation result of the predictive models

Table 47.5   Comparative analysis using state-of-the-art methods

Year	ML Algorithm	Accuracy
2021 (Kumari *et al.* 2021)	Soft voting classifier	79.08%
2019 (Mahabub 2019)	Ensemble Voting classifier	85.71%
2021 (Ahmad *et al.* 2021)	RF	88.27%
2018 (Sisodia and Sisodia 2018)	NB	76.30%
Our Study	Proposed BIRCH-ANN hybrid model	97.25%

5. Conclusion

In this paper, a novel computational model based on symptom analysis is developed to predict diabetes disease. The proposed approach is characterized by a detailed description of each intermediate step. The suggested model obtains 97.25% when pre-processing, BIRCH, and ANN are applied. The suggested model performed better than existing models and methodologies, according to the comparative research. Under the supervision of a diabetic specialist, the suggested hybrid model might be used as an expert system application to assist clinicians in making early illness detection decisions.

The suggested approach can substantially and effectively support clinicians in their decision-making for their patients. This approach might help clinicians identify diabetes illness in its early stages, reducing the chances of diabetes harming a patient's life. Furthermore, the suggested method may produce fascinating patterns from any real-world dataset containing male and female data. Hybridization of datasets may be used in the future to construct new prediction models with high accuracy rates and lower the risk of biased results.

REFERENCE

1. Ahmad, H. F., Mukhtar, H., Alaqail, H., Seliaman, M., and Alhumam, A., (2021). Investigating health-related features and their impact on the prediction of diabetes using machine learning. Applied Sciences (Switzerland), 11 (3), 1–18.

2. Aparicio, L. F., Noguez, J., Montesinos, L., and García, J.A.G., (2021). Machine learning and deep learning predictive models for type 2 diabetes : a systematic review. Diabetology & Metabolic Syndrome.

3. Chen, W., Chen, S., Zhang, H., and Wu, T., (2018). A hybrid prediction model for type 2 diabetes using K-means and decision tree. Proceedings of th1. W. Chen, S. Chen, H. Zhang, and T. Wu, Proc. IEEE Int. Conf. Softw. Eng. Serv. Sci. ICSESS 2017-Novem, 386 (2018).e IEEE International Conference on Software Engineering and Service Sciences, ICSESS, 2017-Novem (61272399), 386–390.

4. Diabetes | Kaggle [online], (2022). Available from: https://www.kaggle.com/c/diabetes/data [Accessed 14 Jan 2022].

5. Han Cho, N., (2019). IDF Diabetes Atlas. 2019th ed. International Diabetes Federation.

6. Herman, W. H., Ye, W., Griffin, S. J., Simmons, R. K., Davies, M. J., Khunti, K., Rutten, G. E. H. M., Sandbaek, A., Lauritzen, T., Borch-Johnsen, K., Brown, M. B., and Wareham, N. J., (2015). Early Detection and Treatment of Type 2 Diabetes Reduce Cardiovascular Morbidity and Mortality: A Simulation of the Results of the Anglo-Danish-Dutch Study of Intensive Treatment in People With Screen-Detected Diabetes in Primary Care (ADDITION-Europe).

7. Howsalya Devi, R. D., Bai, A., and Nagarajan, N., (2020). A novel hybrid approach for diagnosing diabetes mellitus using farthest first and support vector machine algorithms. Obesity Medicine, 17.

8. Kaur, J. and Kaur, G., (2015). Clustering Algorithms in Data Mining: A Comprehensive Study. International Journal of Computer International Journal of Computer International Journal of Computer International Journal of Computer Science.

9. Khanam, J. J. and Foo, S. Y., (2021). A comparison of machine learning algorithms for diabetes prediction. ICT Express, (xxxx).

10. Kulkarni, A., Chong, D., and Batarseh, F. A., (2020). Foundations of data imbalance and solutions for a data democracy. Data Democracy: At the Nexus of Artificial Intelligence, Software Development, and Knowledge Engineering, 83–106.

11. Kumar, V. B., Vijayalakshmi, K., and Padmavathamma, M., (2019). A hybrid data mining approach for diabetes prediction and classification. Proceedings of the World Congress on Engineering and Computer Science, 2019-Octob, 298–303.

12. Kumari, S., Kumar, D., and Mittal, M., (2021). An ensemble approach for classification and prediction of diabetes mellitus using soft voting classifier. International Journal of Cognitive Computing in Engineering, 2, 40–46.
13. Mahabub, A., (2019). A robust voting approach for diabetes prediction using traditional machine learning techniques. SN Applied Sciences, 1 (12).
14. Osmond, D., Giftson, N., Priyadarshini, R.J., and Arockiam, L., (2016). A Novel Algorithm Using Birch with K-Means Clustering Using Dataset in R Data Mining Tool, 9 (26), 315–323.
15. P, B.M.K., R, S.P., R K, N., and K, A., (2020). Type 2: Diabetes mellitus prediction using Deep Neural Networks classifier. International Journal of Cognitive Computing in Engineering, 1 (July), 55–61.
16. Patil, B. M., Joshi, R. C., and Toshniwal, D., (2010). Hybrid prediction model for Type-2 diabetic patients. Expert Systems with Applications, 37 (12), 8102–8108.
17. Sadiq, R., Rodriguez, M. J., and Mian, H. R., (2019). Empirical Models to Predict Disinfection By-Products (DBPs) in Drinking Water: An Updated Review. Encyclopedia of Environmental Health, 324–338.
18. Senthil Kumar, B. and Gunavathi, R., (2019). An enhanced model for diabetes prediction using improved firefly feature selection and hybrid random forest algorithm. International Journal of Engineering and Advanced Technology, 9 (1), 3765–3769.
19. Sisodia, D. and Sisodia, D. S., (2018). Prediction of Diabetes using Classification Algorithms. In: Procedia Computer Science. Elsevier B.V., 1578–1585.
20. Soni, M. and Varma, D. S., (2020). Diabetes Prediction using Machine Learning Techniques. International Journal of Engineering Research & Technology, 9 (9).
21. Srivastava, A. K., Kumar, Y., and Singh, P. K., (2021). Hybrid diabetes disease prediction framework based on data imputation and outlier detection techniques. Expert Systems, (July), 1–17.
22. Wang, S., (2003). Artificial Neural Network. Interdisciplinary Computing in Java Programming, 81–100.

Intelligent Systems and Smart Infrastructure – Brijesh Mishra et al. (eds)
© 2023 Taylor & Francis Group, London, ISBN 978-1-032-41287-0

CHAPTER

48

Energy Scheduling of Industrial, Commercial, and Residential Appliances Using Time of Use and Real-Time Pricing Demand Response Techniques

Abhinesh Kumar Lal Karn[1], Sandeep Kakran[2]

NIT Kurukshetra, Haryana 136119

Abstract

Demand-side management is becoming a popular field of study as every customer in every industry wants to save money on power. However, traditional methods would make it impossible to reduce costs because electricity is now supplied from both ends, i.e., generation and distribution, and old methods cannot cut costs because they do not focus on power generation from the distribution side. As a result, demand response is used to cut costs and schedule various appliances. Every industry employs more or less similar appliances but with varying power ratings. Previously, the majority of energy-scheduling work was limited to the residential and commercial sectors. However, now industrial appliance scheduling is also critical because industry consumes most of the electricity. The scheduling of various appliances in the residential, commercial, and industrial sectors is done in this study. To compare the impact of price rates for different sectors on weekdays and weekends, a Time of Use and Real-Time Price-based approach is employed with different rates on weekdays and weekends. GAMS software is used to schedule appliances, and the CPLEX solver is used to solve the mixed-integer linear programming.

Keywords: Demand Response, Time of Use, Real-Time Price

Corresponding author: [1]abhineshlal2894@gmail.com, [2]skakran@gmail.com

DOI: 10.1201/9781003357346-48

1. Introduction

With the evolution of the smart grid, it is necessary to schedule the industrial appliances with commercial and household appliances in order to decrease costs and meet consumer demand. Previously, most of the research was focused on scheduling of the household appliances, but as the industry load has reached the level of 45% of total electrical loads, so to reduce the burden on utility company, research for energy scheduling of industry controllable appliance is also required. With the improvement in the technology employed in the smart grid, controllable appliances can be scheduled when other sectors' demand is low. The pressure on utilities can be reduced with the use of renewable energy, such as solar power, wind power, biomass, and others, to meet the demand of customers in various industries. However, due to the intermittent nature of PV and wind turbines, some form of energy storage is required. In recent years, a lot of research has been done on renewable energy sources and their intermittency (Mudie and Essah, 2016).

Demand response (DR) in smart grids is a unique feature that helps to reduce power consumption. Consumers benefit from DR because it decreases their costs and provides an incentive. It can also be used to control client demand by shifting some of the load. Load response and pricing response are two types of DR. Direct load control, interruptible load, curtailable load, and scheduled load are all types of load responses. Time of use (TOU), real-time pricing (RTP), critical peak pricing, extreme day pricing, and extreme day critical peak pricing are all types of price responses (Das, 2019). For the industrial customers, the modified approach of DR, i.e., interactive time of use, has been considered to obtain the best performance (Kholerdi and Marzbali, 2021). The energy scheduling of domestic appliances has been studied by McKenna and Keane (2014) using the bottom-up load model, and the Monte–Carlo technique has been used to predict the probability of consumption considering appliance operation. In the smart grid, a stable and efficient communication infrastructure is provided via cognitive radio technology, while the game model has been used for distributed storage energy planning (Wang. et al., 2018). To reduce the energy consumption and peak to average ratio, particle swarm optimization was employed to reschedule the appliances (Babu. et al., 2018). Tabassum and Shastry (2020) present various types of appliances and their consumption patterns for successful user participation in demand side management. DR techniques such as RTP and TOU have been used to schedule energy for thermostatically controlled devices (Bhati and Kakran, 2018). RTP-based DR has been used to minimize the electricity cost of industrial facilities (Lu.et.al, 2020). The energy scheduling of a three-level integrated energy system was done using the Stackelberg game approach. The electricity utility company, natural gas utility company, and several identically designed smart energy hubs make up a three-level integrated energy system (Luo. et al.,2021). The particle swarm optimization method (PSO) has been used to plan energy consumption, reduce costs, and save energy in the residential sector (Bouakkaz. et al.,2021). In addition to three other types of algorithms, such as cuckoo search, adaptive cuckoo search, hybrid genetic algorithm, and PSO (Goyal and Vadhera, 2021), an artificial intelligence-based approach employing multi-interval programming has been employed for energy management.

In earlier work, industrial, residential, and commercial price variation during weekdays and weekends has not been done using different price-based DR. In this study, different pricing-based techniques, i.e., TOU and RTP, are used during weekdays and weekends to compare the price for different sectors, i.e., residential, industrial, and commercial. The utilization of that appliance for that particular sector has been taken into account while determining the equipment's operating hours. The novelty of this paper is to look at the variation in cost during weekdays and weekends for different sectors. GAMS software is utilized for energy scheduling of various appliances, and with the help of the CPLEX solver, the problem is solved.

2. Problem Formulation

A. Objective Function

The objective function is designed to minimize the cost of appliances in the residential, industrial, and commercial sector

$$\text{Cost} = \sum_{h=1}^{24} \sum_{app=1}^{9} E_{app}^h \times \text{Price}_g \tag{1}$$

$$E_{app} = \left[E_1^1, E_2^2, \dots E_9^{24} \right] \tag{2}$$

$$\sum_{t=1}^{24} E_{app}^h = E_{app}^{tot} \tag{3}$$

Cost: The total cost of appliances

E_{app}^h: Hourly energy consumption of each appliance

Interruptible appliance energy consumption is calculated using (4) and (5):

$$E_{app}^h = l_{h,app} \times M_{app}^{max} + (1 - l_{h,app}) \times M_{ap}^{min} \tag{4}$$

$$\sum_{t=1}^{24} l_{h,app} \times W_{app,h} = n_{app} \tag{5}$$

$l_{h,app}$: ON/OFF time of an appliance is decided using the binary variable

$M_{app,h}$: ON/OFF time of an appliance is decided using the binary variable

n_{app}: Total ON duration of the appliance

Uninterruptible appliance energy consumption is calculated using (6):

$$\sum_{n \in K} x_{n,app} = 1 \tag{6}$$

x is a binary variable, and 'K' denotes the total number of possible schedules of appliance 'app' during the user-preferred duration, and it is calculated as K = Total number of user-preferred duration – Total duration during which the appliance is ON + 2

$$E_{app}^{h} = \sum_{n \in K} x_{n,app} \times E_{n,h} \tag{7}$$

$E_{n,h}$: Energy consumption during the hth hour of schedule n

Power consumption of must run appliances is calculated as

$$E_{app}^{h} \times W_{app,h} = E_{app}^{tot} \tag{8}$$

E_{app}^{tot}: Total Energy required for appliance 'app'

Interruptible appliance energy consumption in user-preferred duration is calculated using the constraint

$$0 \leq E_{app}^{t} \times W_{app,h} \leq E_{app}^{tot} \tag{9}$$

Constraints used for energy scheduling

$$M_{app}^{min} \leq E_{ap}^{t} \times W_{app,h} \leq M_{app}^{max} \tag{10}$$

$$\sum_{a=1}^{9} E_{ap}^{t} \leq B^{max} \tag{11}$$

B^{max}: the limit put by the utility to consume energy at each hour

3. Results

The weekday and weekend pricing rates are shown in Fig. 48.1 (Mckenna and Keane, 2014; https://hourlypricing.comed.com/live-prices/). Tables 48.1–Table 48.3 show different residential, commercial, and industrial appliances, the power rating of those appliances, and their operating hours during the weekdays and weekends (Babu et al., 2018; Venkatesh et al., 2022).

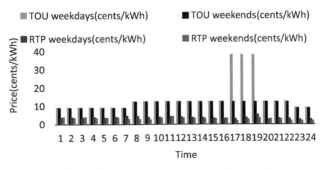

Fig. 48.1   Weekday and Weekend Pricing Rate

Table 48.1 Residential appliance's power rating and their operating hour during weekdays and weekends

Appliance	Power rating (kW)	Total Operating hours	
		Week-days	**Week-ends**
TV	0.1	4	7
Iron	1.2	2	4
Motor	0.74	2	2
Washing Machine	0.4	3	6
Dishwasher	1.2	2	4
Hair Dryer	1.6	1	2
Fan	0.1	8	12
Light	0.8	8	10
Refrigerator	0.5	24	24

Table 48.2 Commercial appliances' power rating and their operating hour during weekdays and weekends

Appliance	Power Rating (kW)	Total Operating hours	
		Week-days	**Week-ends**
Steamer	1.1	10	13
Dryer	0.23	4	6
Kettle	0.65	4	6
Oven	0.75	4	6
Coffee Maker	1.8	6	8
EV charging	0.7	5	7
Fan	0.3	10	14
AC	0.5	8	12
Light	2	10	12

Table 48.3 Industrial appliances' power rating and their operating hour during weekdays and weekends

Appliance	Power Rating (kW)	Total Operating Hour	
		Week-days	**Week-end**
Water Heater	0.5	6	3
Welding Machine	1.6	5	3
Induction Motor	1.5	7	4
DC Motor	1	7	4
Water Pump	0.72	5	2
Boiler	0.47	7	3
Chilling Plant	1.2	5	2
Fan	0.75	9	5
Air Furnace	1.1	4	2

Figure 48.2 and Fig. 48.3 show the operation of different residential, industrial and commercial appliances during weekdays and weekends, considering the flat price. The flat price for weekdays is 16.5 cents/kWh, and for weekends it is 13.5 cents/kWh.

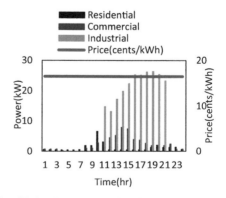

Fig. 48.2 Operation of different appliances of residence, commercial, and industry during weekdays using a flat pricing scheme

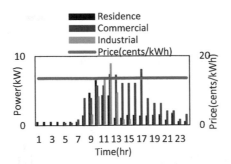

Fig. 48.3 Operation of different appliances of residence, commercial, and industry during weekends using a flat pricing scheme

Figures 48.4 and 48.5 show the operation of different residential, industrial, and commercial appliances during weekdays and weekends considering TOU. From Figs. 48.4 and 48.5, it can be seen that with the use of TOU, appliances of the different sectors have been scheduled during the minimum price as compared to the flat price rate.

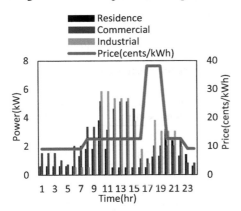

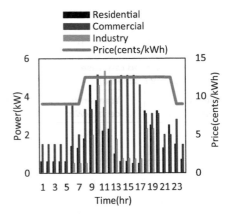

Fig. 48.4 Operation of different appliances of residence, commercial and industry during weekdays using TOU

Fig. 48.5 Operation of different appliances of residence, commercial and industry during weekends using TOU

Figures 48.6 and 48.7 show the operation of different residential, industrial, and commercial appliances during weekdays and weekends considering RTP. From Figs. 48.6 and 48.7, it can be seen that with the use of RTP, appliances of the different sectors have been scheduled during the minimum price as compared to the flat price rate.

Table 48.4 shows the cost of commercial appliances during weekdays and weekends without DR and with DR. In without DR, the flat price is considered. For weekdays, the rate assumed is 16.5 cents/kWh, and for weekends, the rate assumed is 13.5 cents/kWh. With TOU, the cost

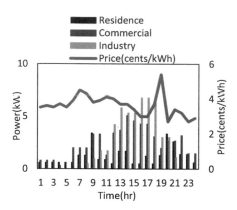

Fig. 48.6 Operation of different appliances of residence, commercial, and industry during weekdays using RTP

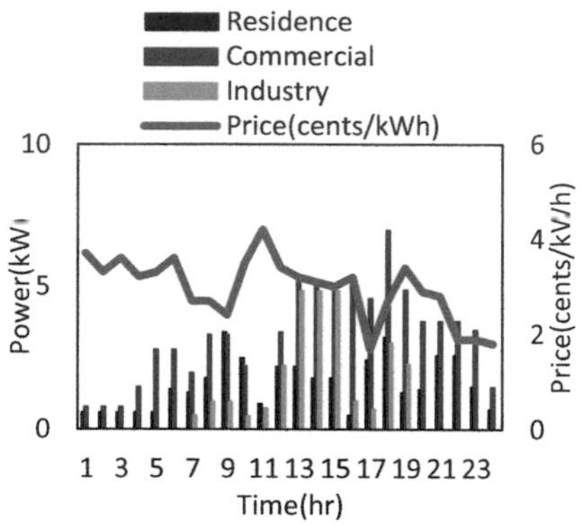

Fig. 48.7 Operation of different appliances of residence, commercial, and industry during weekends using RTP

is reduced by 12.96% during the weekdays and by 12.48% during the weekends. With RTP, the cost is reduced by 78.27% during weekdays and by 78.98% during the weekends.

Table 48.4 Cost of residential appliances during weekdays and weekends

Without demand response (Flat price scheme)		With demand response		
Weekdays (cents/kWh)	Weekends (cents/kWh)	Pricing Scheme	Weekdays (cents/kWh)	Weekends (cents/kWh)
473.22	519.48	TOU	411.870	454.600
		RTP	102.808	109.166

Table 48.5 shows the cost of commercial appliances during weekdays and weekends without DR and with DR. In without DR, the flat price is considered. For weekdays, the rate assumed is 16.5 cents/kWh, and for weekends, the rate assumed is 13.5 cents/kWh. In comparison with the flat rate tariff, with TOU, the cost is reduced by 12.109% during the weekdays and by 13.857% during the weekends, and with RTP, the cost is reduced by 77.474% during weekdays and by 78.849% during the weekends.

Table 48.5 Cost of commercial appliances during weekdays and weekends

Without demand response (Flat Price Scheme)		With demand response		
Weekdays (cents/kWh)	Weekends (cents/kWh)	Pricing Scheme	Weekdays (cents/kWh)	Weekends (cents/kWh)
937.53	1047.33	TOU	824.000	902.200
		RTP	211.184	221.512

Table 48.6, shows the cost of industrial appliances during weekdays and weekends without DR and with DR. In without DR, the flat price is considered. For weekdays, the rate assumed is 16.5 cents/kWh, and for weekends, the rate assumed is 13.5 cents/kWh. In comparison with the flat rate tariff, with TOU, the cost is reduced by 2.467% during the weekdays and by 8.35% during the weekends, and with RTP, the cost is reduced by 78.336% during weekdays and by 77.30% during the weekends.

Table 48.6   Cost of industrial appliances during weekdays and weekends

Without demand response (Flat Price Scheme)		With demand response		
Weekdays (cents/kWh)	Weekends (cents/kWh)	Pricing Scheme	Weekdays (cents/kWh)	Weekends (cents/kWh)
866.91	371.25	TOU	845.520	340.250
		RTP	187.803	84.245

4. Conclusions

This work presents the energy scheduling of different appliances in residential, industrial, and commercial sectors. The techniques used for energy scheduling of different appliances of the different sectors are TOU and RTP with different rates on weekdays and weekends. From the results, it can be noticed that the cost is significantly reduced after applying pricing-based DR techniques. It is also found that the cost of the residential and commercial sector is more on weekends in comparison to weekdays because the operating hours on weekdays are extensive, whereas in the industry sector, the price is less on weekends in comparison to weekdays. Further, from the results, it can also be concluded that most of the appliances are scheduled during the minimum price time. In the future, more interruptible appliances of the different sectors can be considered for scheduling, and different pricing schemes can be used to calculate the cost.

REFERENCES

1. S. Mudie, E.A. Essah, A. Grandison, R. Felgate, Electricity use in the commercial kitchen, International Journal of Low-Carbon Technologies, Volume 11, Issue 1, March 2016, Pages 66–74.
2. Nilima R. Das," Finding Optimal Operation Schedule for Electrical Appliances using Demand Response in Residential Sector", International Journal of Engineering Technology and Exploring Engineering (IJITEE) ISSN: 2278–3075, Volume-8 Issue-8, June 2019.
3. K. McKenna and A. Keane, "Discrete elastic residential load response under variable pricing schemes," IEEE PES Innovative Smart Grid Technologies, Europe, 2014, pp. 1–6.
4. K. Wang, H. Li, S. Maharjan, Y. Zhang and S. Guo, "Green Energy Scheduling for Demand Side Management in the Smart Grid," in IEEE Transactions on Green Communications and Networking, vol. 2, no. 2, pp. 596–611, June 2018, doi: 10.1109/TGCN.2018.2797533.

5. N. R. Babu, S. Vijay, D. Saha and L. C. Saikia, "Scheduling of Residential Appliances Using DSM with Energy Storage in Smart Grid Environment," 2nd International Conference on Power, Energy and Environment: Towards Smart Technology (ICEPE), 2018.

6. Venkatesh, B.; Sankaramurthy, P. Chokkalingam, B., Mihet-Popa, L. Managing the Demand in a Micro Grid Based on Load Shifting with Controllable Devices Using Hybrid WFS2ACSO Technique. Energies 2022, 15, 790.

7. Zahira Tabassum, Chandrasekhar Shastry "Analysis of Load Consumption" International Journal of Innovative Science and Research Technology, 2020.

8. Neelam Bhati, Sandeep Kakran ," Optimal Household Appliances Scheduling Considering Time Based Pricing Scheme" International Conference on Power Energy , Environment and Intelligent Control, 2018.

9. Renzhi Lu, Ruichang Bai, Yuan Huang, Yuting Li, Junhui Jiang, Yuemin Ding,Data-driven real-time price-based demand response for industrial facilities energy management,Applied Energy, Volume 283, 2021, 116291, ISSN 0306–2619.

10. D. Liu, Y. Xu, Q. Wei and X. Liu, "Residential energy scheduling for variable weather solar energy based on adaptive dynamic programming," in IEEE/CAA Journal of Automatica Sinica, Jan. 2018.

11. Xi Luo, Yanfeng Liu, Jiaping Liu, Xiaojun Liu,Energy scheduling for a three-level integrated energy system based on energy hub models: A hierarchical Stackelberg game approach,Sustainable Cities and Society, 2020.

12. Christian Artigues, Pierre Lopez, Alain Haït,The energy scheduling problem: Industrial case-study and constraint propagation techniques,International Journal of Production Economics, 2013.

13. Lin Gu, Jingjing Cai, Deze Zeng, Yu Zhang, Hai Jin, Weiqi Dai, Energy efficient task allocation and energy scheduling in green energy powered edge computing, Future Generation Computer Systems, 2019.

14. Abderraouf Bouakkaz, Antonio J. Gil Mena, Salim Haddad, Mario Luigi Ferrari, Efficient energy scheduling considering cost reduction and energy saving in hybrid energy system with energy storage,Journal of Energy Storage, 2021.

15. M. H. Amini, J. Frye, M. D. Ilić and O. Karabasoglu, "Smart residential energy scheduling utilizing two stage Mixed Integer Linear Programming," North American Power Symposium (NAPS), 2015.

16. Real time electricity prices Available at: https://hourlypricing.comed.com/live-prices/ [last accessed April,2022].

17. Govind Rai Goyal, Shelly Vadhera, Multi-interval programming based scheduling of appliances with user preferences and dynamic pricing in residential area, Sustainable Energy, Grids and Networks, Volume 27, 2021, 100511, ISSN 2352–4677.

18. Somayeh Siahchehre Kholerdi, Ali Ghasemi-Marzbali,Interactive Time-of-use demand response for industrial electricity customers: A case study,Utilities Policy, Volume 70, 2021, 101192, ISSN 0957–1787

Intelligent Systems and Smart Infrastructure – Brijesh Mishra et al. (eds)
© 2023 Taylor & Francis Group, London, ISBN 978-1-032-41287-0

CHAPTER

49

A Comprehensive Study on Energy Competent Clustering Hierarchy for Wireless Sensor Networks

Kumari Ritika[1], Arvind Kumar Pandey[2], and Arun Kumar Mishra[3]

Department of Electronics and Communication,
Buddha Institute of Technology, Gorakhpur - 273209,
Uttar Pradesh, India

Abstract

It is possible to create wireless sensor networks (WSNs) using erratic deployment of sensor nodes (SNs) to monitor an area. Efficiencies in power consumption are especially important in WSNs, because there is no external power source to rely on. The base station receives data from sensors deployed on SNs. The WSN's lifespan is determined by the SNs energy/battery life, and a greater battery life means a longer network lifespan. For a long time, WSN has been able to run on SNs energy efficiently. To increase the lifespan of a wireless sensor network, it is suggested that a SN from a WSN cluster be used as an aggregator or cluster head (CH). WSNs leverage clustering as a way to significantly reduce the energy consumed by sensor nodes. Clustering is one of the approaches used in WSNs to assist sensor nodes conserve energy. This manuscript examines a variety of WSN clustering algorithms and approaches based on distinct clustering features.

Keywords: WSN, energy efficiency, network lifetime, clustering, sensor node

Corresponding author: [1]ritikapandey545@gmail.com, [2]arvind415@bit.ac.in, [3]akmishra298@bit.ac.in

DOI: 10.1201/9781003357346-49

1. Introduction

WSNs are incorporated with a great number of independent electronic devices (known as nodes or sensors), with feasible mechanic elements, which are capable of remote sensing, processing the signal and transmission in an ad-hoc manner. The primary idea is that these independent nodes, which are capable of sensing the environment would be dispersed over a definite physical area and the nodes would have to be competent to convey the information through mesh networking to accomplish some purpose. Independent nodes interact with one other to form a multi-hop mesh network in a decentralized WSN model. WSN frameworks comprise of sensor nodes, application servers, and gateways. The job of sensor nodes incorporates detecting proximate data, handling information, and communicating with adjoining nodes. In certain cases, sensors make a job for the actuators in the computerization systems. Sensor nodes are installed in a designated region where information is gathered continuously or intermittently. Sensors are programmable, effectively controllable, and simultaneously are utilized for connecting with other smart appliances. Because of its small size, low power, minimal expense and negligible user intervention, it is broadly utilized in numerous viable applications. Sensor networks may be put to a variety of different uses according to the fore-mentioned described characteristics. Figure 49.1 shows the principal application area for a remote sensor network.

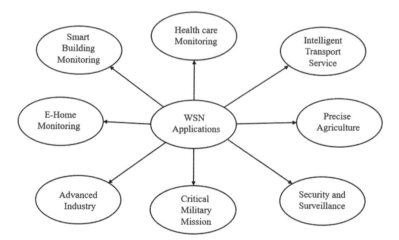

Fig. 49.1 Wide-range application area of WSN

The market of sensor network was valued about \$ 34.26 billion in 2017 and is expected to reach \$ 93.86 billion by 2023, according to a recent analysis. A CAGR of 18.55% is expected through 2023, according to this estimation (Amarlingam et al. (2018)). Efficient energy coverage (EEC) has a considerable impact on performance of WSN. Most of the WSNs utilize batteries to power their sensors, and replacing or recharging each of those batteries is not practicable. Energy conservation and the network's lifespan have been addressed in several ways in the past decade. An important one of them is sensor scheduling, which reduces power consumption by coordinating the actions of the sensors. These scheduling solutions need those devices be placed in an area of interest. Finally, certain devices can be put into a rest state while the remaining can be used to do the sensing operation. As a result of the EEC problem, several different scheduling techniques have been presented in Dwivedi and Sharma (2021). Multi-hop transmission, which uses less energy, it replaced the earlier method which aims at direct sending of data to the master unit i.e., base station (BS). With limited battery and energy requirements, clustering is the best and most productive structure that divides the nodes into specific groups of clusters.

2. Relevant Work

Many algorithms have been proposed for routing and clustering approaches recently for WSNs. They primarily aim to minimize the energy depletion of the sensor nodes to maximize the lifetime of the network. Here, authors have taken some approaches to study and put some light on it. The approaches discussed in the paper are cluster based models; fuzzy based models and nature-inspired models.

A. Cluster Based Models

Setting up the cluster-based algorithm is critical for communication and routing. Here, the formation/distribution of nodes is the sole component of network communication that is automatically optimized if the node scheme gives superior outcomes. Some of the earlier clustering-based research studies are shown here to demonstrate the significance of cluster formation and node locations in the network structure. When it comes to WSN clustering, there is only one widely acknowledged methodology, called "Low Energy Adaptive Clustering Hierarchy" (LEACH) mentioned in Agarwal, Agarwal, and Muruganandam (2018). An aggressive protocol that utilizes arbitrary, homogeneous, stationary nodes is employed in this system. Using this protocol with CH nodes as routers with sink, the clusters have been formed upon receipt of an energy signal. This protocol implements the aggregation and combining of surrounding information through the transfer of information. CH may be selected at any time by SNs with a high degree of certainty. LEACH technique has few advantages, the cluster heads merge the entirety of the information, resulting into a decrease in network traffic (Dwivedi and Sharma (2020)) and there is no need to acknowledge the exact position of each node in order to form a cluster in this case. In addition to the positives, there are some of the drawbacks too. LEACH does not tell about the number of cluster heads in the network. If for some reason the cluster head fails, then the cluster is rendered worthless since the data stored in its sink unit, the base station, would never reach it. This is a major flaw in LEACH's design.

To alleviate the drawback of LEACH's single-hop communication, 'Energy Efficient Hierarchical Clustering' (EEHC) approach has been proposed. Distributed hierarchical algorithm was utilized to increase network lifetime in this algorithm. At start, each SN is designated as CH with the possibility (P) of relaying to all SNs in the vicinity. In order to receive this message and become a cluster member, any node that is not a cluster head will be designated as such. However, when a node receives the message in a stimulated instance, it becomes a 'forced' cluster head, and the cycle is repeated. Subramani et al. (2022) proposed 'Hybrid Energy-

Efficient Distributed- Clustering' (HEED), a distributed hierarchical method for clustering. Within the cluster, one-way communication is assumed. Between a CH and a BS transmission over many hops is allowed. Consideration is given to intra-cluster communication and residual energy in determining who will serve as the cluster's head node. As a result, CH nodes were not randomly selected by HEED, and CH was distributed throughout the connected network. The clustering method involves three stages. First of all, an algorithm declared an initial percentage of CHs among all the SNs at the beginning. In second stage, except for the CHs that could communicate for less money, all SNs proceeded through the numerical iterations. In the end, each SN chose either a CH with a lower cost or announced itself to be the CH.

The 'Linked Clustering Method' (LCA) is a preliminary algorithm for clustering that relies on an exclusive property to choose CHs. Static and single hop are also features of this clustering approach. 'Threshold Sensitive Energy Efficient Sensor'(TEEN) is a hierarchical architecture. The system relied heavily on instantaneous execution. The technique suggested a 'two-tier' clustering method that employed a dual 'threshold' and was both data-centric and hierarchical. Hard threshold (HT) determines the detected accredit values, whereas a little change in those values is determined by a soft threshold (ST). CH's members were given both 'ST' and 'HT' values to help them limit the amount of data they are transmitting. Normal and advanced nodes are introduced in SEP, which is a two-level heterogeneous routing protocol. There is more energy in advance nodes than in conventional nodes both regular and advance nodes have a weighted likelihood of becoming the cluster head under stable-election protocol (SEP). SEP does not guarantee the efficient deployment of nodes. The enhanced stable Election Protocol (E-SEP) was suggested for hierarchies with three levels. Intermediate nodes were designed to bridge the gap between standard nodes and advanced nodes. The nodes in a cluster choose a leader based on their degree of energy. ESEP has the same downside as SEP. Heterogeneity may be shown in the distributed energy efficient clustering protocol (DEEC). Cluster heads in DEEC are more likely to be high-energy nodes than low-energy nodes. TEEN uses hard and soft thresholds to choose which cluster head to use to reduce the number of transmissions and save nodes' energy. As a result, the network's overall lifespan and stability are extended. Normal and advance nodes are distributed at random in SEP. Stability and throughput are impacted if most typical nodes are positioned distant from the base station, which uses more energy while delivering data. In this way, the effectiveness of SEP is reduced. Priyadarshini and Sivakumar (2021) came up with the K-Mean clustering algorithm (K-MEAN) based on partitioning. This data was divided into K equal-sized clusters using K-means method. Each observation was assigned a unique index value for the collection of nodes that comprised it. For large data sets, K-means clustering is more suited than other clustering methods since it provides true interpretation and splits clusters into single levels.

B. Fuzzy-based Protocol to Enhance the Lifetime of WSN

In CHEATS protocol, for CH selection, it's also known as a Takagi-Sugeno fuzzy system. CHEATS is 10% better than LEACH. CHEATS selects CH based on the Sugeno model and only employs two boundary variables, namely the distance of the sensor node from the sink and the residual energy of the nodes. In terms of performance, it's better than LEACH, but it doesn't address some of the issues, such as the failure of CH to pick the best appropriate CH for a given network. In CHEF protocol, clusters are created in every round, in order to pick the CH, energy and local frames are used as fuzzy output variables, just as LEACH. If compared to LEACH, it yields a comparatively better performance. If two sensor nodes have the same probability of becoming CH, then this method does not have the proper plan to determine which sensor node of these two is more acceptable for CH use. If there is a significant concentration of CH at one end of the n/w, then the number of CH is reduced due to the cluster radius condition, increasing the likelihood of CH failure. It manages to overcome the drawbacks of a revolutionary election method. Whereas FBECS cluster heads are elected by measuring the "Eligibility Index" of each sensor node in a wireless sensor network using the residual energy, density of a node's neighboring nodes, and distance from a node to its destination. It outperforms the LEACH & BCSA protocol, for example. A wide range of data may be carried to the sink while also balancing network load and extending the network's lifespan.

FLEACH protocol, the construction of WSN clusters should always be done to support the idea that it reduces energy consumption. A variety of strategies have been developed to reduce energy usage during cluster formation, however they all proven to be prohibitively expensive. A two-phase FLEEC strategy designed to maximize energy efficiency is described in this research. First, the sink node calculates the communication radius for all sensor nodes using two fuzzy inputs: Node-density and distance to sink. In order to calculate the probability of becoming CH, the second step makes use of the previous stage's leftover energy and total distance. This procedure has been shown in experiments to be more energy efficient than LEACH & EFCH. In FUCA, unbalanced clustering, on the other hand, is used to extend the life duration of WSNs. According to FLECH Protocol it has two functioning phases. Firstly, the data collection phase, in this members of the cluster will get a TDMA schedule from the newly elected CHs, the member nodes only convey data to their CHs during the designated time periods. To conserve energy, they go into a slumber mode at other times. After receiving data from all of its members, the CH nodes combine it into a single message. CHs will eventually send the compiled data to BS. Only once throughout a frame can a member node deliver data to CH. Unlike the member nodes, all the CHs ought to be in a wake-up state at all times. All of CH's functions result in an increase in energy use. Clusters get smaller and closer to the sink in the event of uneven clustering. Using the uneven clustering notion, this fuzzy based clustering method distributes the energy equally. For the fuzzy input, this technique takes into account residual energy, node density, and the distance from a sensor node to its sink. Fuzzy logic and the Mamdani approach were both employed in the selection of the CHs. Compared to LEACH and another energy-aware uneven clustering fuzzy-based protocol, this approach has proven to be a more preferred and enhanced variant. Finally in SCHFTL protocol, when it comes to reducing energy consumption in the WSN, selecting the best approach for CH selection is a crucial step. The LEACH method, which picks the CH based on a minimal threshold value, is one such algorithm. Using the LEACH method, each sensor node sends data to its CH, and the CH then sends that data to the sink as a whole. The SCHFTL protocol is described in this protocol. The Mamdani interference engine is used to pick the super cluster head from all the possible cluster heads. When compared to FMCHEL fuzzy-based master CH election leach and CHEF protocol, it performs better.

C. A Common Network Assumptions for WSN Protocols

There are few CN assumptions for WSN protocols which includes, all the nodes in a homogeneous network have the same sensing area, processing power and other characteristics. During deployment, all nodes have the same energy. While after deployment, the nodes are anchored to the ground. The 'Received Signal Strength Indicator' (RSSI) calculates the distance between nodes. All nodes are designed to broadcast a message that says HELLO that includes their ID within the range of their communication radius Rc when they are installed. Only the number of neighbors and the distance between them are counted by each node in this HELLO message. The BS also transmits a Hello message that includes its position. The main reason behind node death is energy depletion.

D. Various Input Variable for WSN

In residual energy, choosing the right node to be the CH is extremely important since it requires the node to put in more effort than any other member node. BS receives data from CH nodes that have collected data from their members. In order to do the fore-mentioned tasks, a CH must have sufficient energy. Whereas, in node centrality (NC), node degree is the total number of one-hop nearby nodes within Rc of a node. The centrality of a node is defined by its position in relation to its neighbors. The lower the NC value, the greater the likelihood that a node would be chosen as CH.

$$NC = \frac{\sqrt{\sum_{i=1}^{ND} dist_i x^2 / ND}}{Ntk_{Dimension}} \tag{1}$$

For example, in a 100 m × 100 m field, $Ntk_{Dimension}$ is 100m and 200m indicates distance from the neighboring node in Equation 1. In this case, ND (node degree) refers to the number of neighbors inside the communication radius RC of a node distance. In this case, the $Ntk_{Dimension}$ is 200 m × 200 m, and the field area is 200 square meters.

$$Distance\ to\ BS = \frac{d_i}{\alpha Ntk_{Dimension}} \tag{2}$$

$$\alpha = \frac{d_{max}}{Ntk_{Dimension}} \tag{3}$$

Distance to BS, more energy is needed to transfer data as the source goes away from the sink node. It is preferable to maintain the distance between CH and BS as short as possible to conserve energy. In Equation 2, d_{max} is the maximum distance between a node I and the BS. 'α' is the network dimension specific constant and d_{max} is the distance between node in the network and the BS.

E. Nature-inspired Models for Energy-adaptive Routing in WSN

Meta-heuristic or Nature-Inspired optimization technique is also a way to support EEC problem solving. Simulated annealing and Tabu search are single solution-based approaches. For the most part, population-based approaches fall into swarm intelligence based and evolutionary based category. ACO, PSO, FA, and evolutionary-based techniques include differential evolution and GA under the swarm intelligence category. In ant colony optimization (ACO), for dynamic routing in WSNs, an adaptive technique was proposed known as ACO. The wireless sensor network's unpredictability was taken into account when developing this. There are two ways to install sensor nodes: sparsely or densely. The ACO determines the best network configuration in order to reduce data redundancy and increase data aggregation. ACO was also used to construct an adaptive routing scheme. Routing decisions are made based on both the amount of residual energy in the nodes and their location. Clusters were not employed to organize nodes in this example. In a cross- layer WSN protocol stack, fuzzy logic was employed in addition to ACO to optimize WSN routing. In the WSN, neural networks were employed in conjunction with ACO to aid in routing. The cluster head is selected using a neural network, while the optimum path is determined using ACO. Another technique which is Particle-Swarm Optimization (PSO) is the most well-known meta- heuristic optimization method. Swarm optimization takes its cues from this approach, which has a plethora of uses. Swarming birds like flocking and shoaling are the inspiration for this design. PSO is based on a natural phenomenon: the shoaling behavior of fish. After multiple research projects, it was discovered in 1990 that fish shoaling is a social activity that evolved inside a group. They travel wherever they can find food and eat it whenever they can. Animals communicate with one another by exchanging information about their current location and whereabouts of food. In firefly algorithm (FA) Vasanthi et al. (2020) initially devised the firefly algorithm (FA), another form of optimization. This plan's basic idea is to use a spotlight to enlist the help of as many flies as possible. Fire-fly flashes have an unusual appearance that varies by species. Prey attraction and the mating behavior of fireflies are the two most important aspects of this algorithm. Female fireflies have a distinct flashing pattern in response to male fireflies, which is used for mating. If the light intensity decreases, the distance between two fireflies increases. It indicates that the intensity of light emitted by the firefly has a direct effect on the attraction between the two insects. When it comes to FA, the same results are achieved when a random set of solutions are evaluated.

Genetic algorithm (GA) is very popular and commonly used algorithm, genetic algorithms are another meta-heuristic optimization tool (GA). Reproduction, biological evolution and Darwinian theory all play a role in its development. Holland (1973) was the first to formulate this theory, which proposes that genes are passed down from parents to children via mutation, selection and crossover. In the selection phase, several characters are chosen for crossover and mutation. In the crossover phase, genes are exchanged to make offspring. This is when new characters are introduced. In order to optimize network energy and load balancing. Wang et al. (2020) presented a clustering strategy with three main properties. The method 'Mobile Sink Based Adaptive Immune Energy Efficient Protocol' (MSIEEP) is proposed to reduce energy consumption during transmission and regulate packet overhead. In order to save energy, it employs the adaptive immune algorithm (AIA). For better network performance, this research offers a memetic algorithm-based data collection routing system. GA can be helpful in clarifying search and optimization problems. 'Artificial Bee Colony' (ABC), a different kind of scheme is inspired by honey bees natural search behavior is called ABC. It was first drafted by Mann and Singh (2017) and depicts the bees foraging behavior. Artificial bees with colonies include onlookers, scouts and employed bees. Bees are utilized to seek for food or the location of a food supply. They dance with nectar quantity if the bee employed wishes to provide information about food or the location of the meal. Onlooker's bee is waiting for food in the dancing zone. The amount of nectar determines the timing of their dance. There is a direct correlation between the amount of nectar and the number of honeybees attracted. In the ABC method, solution employed bees move at random, and the quality of the solution is dependent on the fitness-value. Similar activities indicates that the protocol employed delivers a longer network lifespan by consuming less energy. There has also been a proposal to use CPMA, a clustering protocol-based meta-heuristics technique. As a result of its use of the harmony search algorithm, it seeks to lower overall power consumption while also distributing power evenly across the network. To lower the network's energy consumption, increase network life span, and enhance network robustness, authors suggested MWSN, an innovative error-tolerant algorithm for routing (ABS-PSO) that uses path coding (Liao, Kuai, and Lin (2015)).

3. Energy Consumption Model

Figure 49.2 depicts the radio hardware energy consumption model. The transmitter controls the radio electronics and the power amplifier, while the receiver controls the radio electronics at destination end. Both of these actions are energy-intensive. Consumption of energy is calculated by using the below equation for a free space channel with a distance of d:

$$E_{TX} = m \times E_{elec} + m \times \varepsilon_{fs} \times d_2, \quad d \le d_o \tag{4}$$

$$m \times E_{elec} + m \times \varepsilon_{mp} \times d_4, \quad d > d_o \tag{5}$$

where, d_o can be estimated by

$$d_o = \sqrt{\frac{\varepsilon_{fs}}{\varepsilon_{mp}}} \tag{6}$$

which is the electronic energy, relies on the coding done in digital form, modulation, filtering the unwanted signal and spreading the signal. The specified function of the receiver's energy

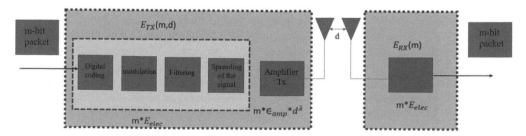

Fig. 49.2   Energy dissipation model

is the converse of the transmitter, as depicted in Fig. 49.2. The parameters f_s and ε_{mp} are the amplification factors of the transmitting circuit when $d \leq d_o$ and $d > d_o$, here, d_o is the minimum threshold. To obtain a message of m bits, the receiver expends energy as shown in Fig. 49.2.

4. Conclusion

WSN is a successful potential technology that is currently used in a variety of applications that require little to no human intervention. Authors reviewed many clustering algorithms used for making WSN energy efficient in this paper. A review of various energy efficient clustering models for WSNs has been discussed in this article. The article also mentions a few areas of cluster- based, nature-inspired, and fuzzy-logic based WSN clustering algorithms. Authors looked at the possibility of a number of different models for calculating WSN optimum routing and clustering. In present study, authors have found that it is possible that the optimal results are not always achieved by a single optimization procedure so multiple algorithms can be put out together to obtain a new efficient algorithm and is termed as hybrid protocol. Fuzzy-logic and nature-inspired techniques can be combined in numerous ways depending on the situation. Indeed, WSN has a great scope in future but still it is dwelling on energy efficiency issues which needs to be addressed. Researchers are still trying to get new findings which are WSN optimization oriented. In this article authors are intended to show different protocols which are framed for energy efficient WSN. Clusterization of the SNs is one such way to achieve this. This study can be a motivation for other researchers in this field.

5. Acknowledgement

Authors are thankful to the department of electronics and communication engineering, Buddha Institute of Technology, Gorakhpur - 273209, Uttar Pradesh, India for providing the necessary resources to carry out the research work.

REFERENCES

1. Agarwal, Karan, Kunal Agarwal, and K Muruganandam. 2018. "Low energy adaptive clustering hierarchy (leach) protocol: Simulation and analysis using matlab." In *2018 International Conference on Computing, Power and Communication Technologies (GUCON)*, 60–64. IEEE.

2. Amarlingam, M, Pradeep Kumar Mishra, Pachamuthu Rajalakshmi, Mukesh Kumar Giluka, and Bheemarjuna Reddy Tamma. 2018. "Energy efficient wireless sensor networks utilizing adaptive dictionary in compressed sensing." In *2018 IEEE 4th World Forum on Internet of Things (WF-IoT)*, 383–388. IEEE.

3. Dwivedi, Anshu Kumar, and A Sharma. 2020. "FEECA: Fuzzy based Energy Efficient Clustering Approach in Wireless Sensor Network." EAI Endorsed *Transactions on Scalable Information Systems* 7 (27).

4. Dwivedi, Anshu Kumar, and Awadesh K Sharma. 2021. "I-FBECS: improved fuzzy based energy efficient clustering using biogeography based optimization in wireless sensor network." *Transactions on Emerging Telecommunications Technologies* 32 (2): e4205.

5. Holland, John H. 1973. "Genetic algorithms and the optimal allocation of trials." *SIAM journal on computing* 2 (2): 88–105.

6. Liao, Wen-Hwa, Ssu-Chi Kuai, and Mon-Shin Lin. 2015. "An energy-efficient sensor deployment scheme for wireless sensor networks using ant colony optimization algorithm." *Wireless Personal Communications* 82 (4): 2135–2153.

7. Mann, Palvinder Singh, and Satvir Singh. 2017. "Energy efficient clustering protocol based on improved metaheuristic in wireless sensor networks." *Journal of Network and Computer Applications* 83:40–52.

8. Priyadarshini, R Raj, and N Sivakumar. 2021. "Cluster head selection based on minimum connected dominating set and bi-partite inspired methodology for energy conservation in WSNs." *Journal of King Saud University-Computer and Information Sciences* 33 (9): 1132–1144.

9. Subramani, Neelakandan, Prakash Mohan, Youseef Alotaibi, Saleh Alghamdi, and Osamah Ibrahim Khalaf. 2022. "An Efficient Metaheuristic-Based Clustering with Routing Protocol for Underwater Wireless Sensor Networks." *Sensors* 22 (2): 415.

10. Vasanthi, V, et al. 2020. "Fractional Gaussian Firefly Algorithm and Darwinian Chicken Swarm Optimization for IoT Multipath Fault-Tolerant Routing." *International Journal of Computer Networks and Applications* 7 (6): 167–177.

11. Wang, Chuhang, Xiaoli Liu, Huangshui Hu, Youjia Han, and Meiqin Yao. 2020. "Energy-efficient and load-balanced clustering routing protocol for wireless sensor networks using a chaotic genetic algorithm." *IEEE Access* 8: 158082–158096.

Intelligent Systems and Smart Infrastructure – Brijesh Mishra et al. (eds)
© 2023 Taylor & Francis Group, London, ISBN 978-1-032-41287-0

Movie Recommendation System Using Combination of Content-based and Collaborative Approaches

Shubham Sharma[1], Chetan Sharma[2], Abdul Aleem[3]

School of Computing Science and Engineering,
Galgotias University

Abstract

Recommendation systems are a kind of auto suggestion system. These systems have minimal amount of human interaction and everything here is done by machine learning (ML) and AI In today's life, ML and AI play a big role in our daily life; for example, social medias and e-commercial websites. In the proposed Movie Recommendation System, we have used ML for recommendation. It helps users to get suggestion of movies based on movies watched earlier or liked previously. For example, let us suppose Ram watches a movie of genre of sci-fi; so by the type of genre Ram watched, we are going to recommend the movie of the same genre to Ram. To do so, we have taken help from ML and build an algorithm based on content-based and collaborative-based approaches. We have combined content and collaborative approach for model building. Here ML made work easy and less time-consuming without much human interaction. Therefore, the aim of the research is to give user better results in the form of recommendation system by utilizing user's past experience of movies watched.

Keywords: Collaborative filtering, Content-based filtering, Clustering, Hybrid approach

Corresponding author: [1]0shivamsharma10@gmail.com, [2]Chetansharmaa.cs733@gmail.com, [3]Abdul.aleem@galgotiasuniversity.edu.in

DOI: 10.1201/9781003357346-50

1. Introduction

In today's world, we have seen the progress of Internet. We have now entered the Web 2 era, and trillions of data has been generated in a minute on internet. Generating a relevant data amid from a sea of data has been a prominent topic of investigation. The picture is one of the most unearthly entertainments, but it also has the problem of data overload. To address this problem, this study proposes a personalised film recommendation system. Customized suggestions try to understand the client's characteristics and preferences by gathering the information and develop a meaningful insight from the data is a main purpose. Performing the task on the basis of client what he is going to like or dislike and understanding the type of taste he/she have and then recommending him things accordingly can be only done by Deep Learning.

A customised framework is a type of data that distinguishes origination from the competition. It is a structured framework that integrates a range of specific mining algorithms and user-related data in order to meet clients' real or predicted needs. There are three types of framework named as cooperative separating proposal framework, substance-based proposal framework, and a crossbreed proposal framework, which are all part of the standard proposal framework. Because each proposal calculation has a particular use reach and usage requirement, numerous proposal evaluations are used for the same data suggestion. When the proposed framework is used in practise, it will primarily be an edge recommendation structure. That is, to incorporate the positive aspects of each proposition evaluation into the defined cycle in order to sustainably expand the proposal's influence. In this paper, we will look upon.

2. Related Work

We researched the recommendation mechanism and found three major divisions: collaborative filtering Recommend System, content-based Recommend System, and hybrid Recommend System, as described in the abstract part. We have additionally explained these studies because contemporary CFRS research has focused on metaheuristic. We have compiled a list of the studies next.

A. CF Approach

CF known as Collaborative Filtering is the most common method for recommendation system. Basically it works as a connection between two users. Suppose user1 likes comedy movie and he gives 5 rating to that movie and user2 gives 3 rating to the same movie. Thereby, it is predicted that user1 and user2 have the same taste of movie if user1 also likes thrill movies, so it is assumed that user2 will also like that movie. Hence, in CF, it's a relation between User and Items. Ma et al. [1] introduced a model variable investigation approach, which helps in organizing data and rating record.

Kleinberg [2] introduced Social Poisson Factorization (SPF), which is a model known as a probabilistic model that deals with data from interpersonal organisations to typical factorization process; SPF connects the social and computational worlds. Kleinberg [2] presented a probabilistic recommender system based on the grouping strategy. Their software discovers co-inclination patterns in real-world data and recommends remedies as they appear. The research we divide it into two groups named as Connection Forecasting and lattice factorization approaches. A method for minimising the number of factors in a grid is called grid factorization.

Xia et al. [3] developed Improved Network-based inference, INBIW, a recommendation computation that extended the first-weight organization-based deduction by incorporating an adjustable border and reducing the impression of significant level axis. In order to evaluate INBIW, the posting rate and hit rate are calculated.

Execution of the proposal

In the executive business process, Xia et al. [3] suggested a technique for aiding asset distribution. Basically it deals with multi-choice problems, which is to resolve with the approach called entropy-based classification strategy. It is mainly introduced for asset distribution to view multi-choice problem.

Tuzhilin et al. [4] proposed various interface-based forecasting models: the conventional (non-Bayesian) model which separates a bunch of elements to prepare a double characterization model the probabilistic model models the joint likelihood among the elements in an organization utilizing Bayesian graphical models; the straight logarithmic model registers the comparability between the hubs in an organization by rank-decreased likeness grids. The models have been found to be effective for prediction to be utilized for recommendation.

Based on hybrid lattice factorization methods, Zhang et al. [5] created a model for medical care; it proposed the same working as recommendation system but for medical purpose. Doctors differ from earlier studies in the following ways: (i) Client inclination and expert highlighted can be extracted by regular lattice factorization by latent Dirichlet; (ii) client inclination and expert highlighted can be extracted from latent Dirichlet allocation and joined to lattice factorization.

Jalili et al. [6] introduced a model recommendation using lattice factorization process and the system is added into trust proliferation; it is based on trust-based recommendation. One more model was introduced by Roth et al. [7], which state a recommendation between friend–friend on social network for example mutual friends on Facebook

Ma et al. [1] introduced the novel probabilistic factor analysis technique that takes into account the clients' preferences as well as their friends' blessings. Aleem et al. [8] computations summarise connect expectation, emphasising commitments based on existing lookouts and techniques, for example arbitrary walk-based procedures, highest probability strategies. They also introduced some common applications: recreating organisations, valuing network progressing components, and rearranging partially named networks.

B. Clustering Approach

In the machine learning model, it is important to classify the data into sub-classes. In data science, we use to call it a clustering. In clustering approach, we used to divide the data into classes; for example, a bank company wants to know how many bank customers will pay back the loan. For such analysis, we have to divide the data into cluster of users who have previously paid the loan by their age group. Developers will look up to how many age group users had paid the loan, suppose age group of 20–25 have paid the loan previously. So the developer will create a cluster of age group by building the model accordingly.

Clustering can be done by five methods. Methods are named as Hierarchical, Density-Based, Partitioning, and Fuzzy. But in this project, we have used Partitioning. It is also known as k-Mean Clustering in which k means a number of groups of cluster. Its divides the unlabelled data into clusters and from the partition of cluster in the way so that distance between the centroid of partition is minimum as compare to another cluster.

In the same way, we have divided our datasets into cluster; for example, User1 liked the movie of genre-type comedy and gives 4–5 rate so we are going to form the cluster of every user liked the same type of movie and by algorithm and feature engineering we are going to suggest User the same type of Movie he/she like from there past experience. In our dataset, we had used previous movie rating given by them.

So, accordingly, we had grouped the rating by the Python in-build method called as group by, and after that, we have ordered the data by the Order by Python method. But before performing feature engineering, we have cleaned the data using the clustering approach.

For information sparsity problem in collaborative filtering, Xia et al. [3] proposed a SOM grouping collective clarifying computation based on Singular Value Decomposition (SVD), which putrefies the item and client inert variable and elements of first matrix are decreased. A novel CF technique is used for exactly imagining missing assessments. Their proposed strategy, TCFACO (Trust-conscious Collaborative Filtering Ant Colony Optimization), uses belief articulations as fun data source.

C. Content-based Approach

Mooney et al. introduced a text classification paradigm based on content-based book recommendations. The technique recognises the benefits of being able to recommend previously underappreciated information to customers with new interests and to amplify recommendations. Substance-based recommender system combines a method for breaking down video content and focusing on a number of agent-sensitive features (lighting, shading, and movement) trapped in current Applied Media Theory organisations. When the idle elements can't be gathered from user data, apply a latent variable model for the proposal and expect them from music.

Ma et al. [1] wished for a film proposal framework that was based on the clients' ratings. The consequences of access media security and access control are broken down using the film valuation framework, and cross-breed circulation storage is the solution announced. The Portable Edge Computing (PEC) technology is employed for the public cloud to meet the high productivity demands for interactive media broadcasting. Deployment, user login, job task, data encoding, and data decryption are all part of the framework's sequence.

D. Hybrid Approach

Hybrid approach is a combination of different machine learning models and rule engine. In the rule engine, we do the feature engineering to train the model. The rule engine can be a function or only one line of code into the programming. Basically, rule engine is used for modification in the data. So that while training the model, we can get a better precision, F1 Score. Apart from the rule engine, we use different kind of machine learning models to get a better accuracy score.

Moreover, Li et al. [7] introduced the three methods of CF suggestion models. It helps model building in different aspects such as deducing, profiling, and foreseeing while building an accuracy score of model. Li et al [8] there methods combine two machine learning models, strategies i.e. case-based, self-organizing map. The authors converted unsupervised clustering into inclination thinking.

HMRS is an easy way for predicting what client will like and his/her demand, collecting the data from previous record of user. Gather all the data and perform some EDA for better analysis of user interest. In this approach, we used to make the clusters of similar data and train the model accordingly. While gathering the similar data, we use rule engine so that data can be classified in similar cluster manner. While gathering the data, we use one in-build method to do that, such as group by, count and order by. These methods are in-build methods. Apart from these, we used correlation method to find out the correlation between two columns or more.

For clustering approach, it is necessary to divide the data into specific manner so that the model can be trained to make model more predictable and works properly in real-world problem; therefore, we have used these approaches and some in-build method. And the since we have Simple data so we have split the data Randomly and trained accordingly.

3. Proposed HMRS-RA

There are two phases inb HMRS-RA: online and offline. The data is pre-processed in the offline phase and then the suggestion is made in the online phase. HMRS-RA filters users in the offline phase based on the basis of their gender (female and male) and age (different ranges 20–30, 35–50). According to the initial evaluating grid, which contains every client's rating for noticed motion pictures, four rating matrices are constructed in this phase, where (25–50) represents the rating of users with more than 25 years of experience but less than 50 years of expertise. The ratings range from 1 to 5, with 1 being the most terrible intense dislike and 5 representing the deepest affection for a film. The Self-Organizing Map (SOM) is a clustering method used by HMRS-RA to identify sets of comparable customers based on movie ratings, and N groups are used to construct a project to analysis on the internet-based stage, the rating of previously unnoticed motion of films for versatile client.

4. Evaluation of HMSRS-RA

The Movie Lens informational collection, which includes 845 clients, 1580 modules, and 100,000 users' assessments for motion movies, is used to test the viability of HMRS-RA. The ratings are from sets 1, 2, 3, 4, 5, which show how much people enjoy or detest the films mentioned. The five-overlay cross-approval calculation is used to evaluate the HMRS-RA. The cross-approval strategy is divided into five cycles, with 60 percent of the information considered prepared data and the remaining information (40 percent) considered test data in each cycle. The SOM grouping technique's underlying weight esteems thought randomly; therefore, with autonomous running, every single edge is repeated eight times. Using various cross-approval cycles, the average value of the Mean Absolute Error (MAE) was determined.

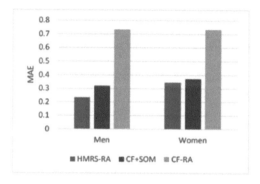 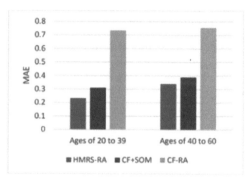

$$MAE = \frac{\sum_{i=1}^{n} |\hat{r}_{u,i} - r_{u,i}|}{n}$$

Fig. 50.1 MAEs values for various recommendations

The initial stage is to determine the true rating value for each category by calculating the average rating regard for distinct films in each category for dynamic consumers. The MAE was then calculated based on the class's true rating of greater than esteem 4 as determined by HMRS-RA. The most pessimistic scenario and the best situation, respectively, were determined by the class with the highest and lowest MAE system. We used HMRS-RA to predict how films would be rated in the worst-case and best-case scenarios, and then computed the MAE based on that underlying rating grid.

In the final step, MAE augmentation determines the MAEs for the worst and best circumstances. The HMRS-RA was linked to the CF-RA and the RS in the MAE regulations, which cemented the old CF and SOM methods. The results show that our predicted computation is accurate, as well as that HMRS-suggestions are becoming more accurate. The MAE measures demonstrate the RA's the evaluation of HMRS-RA to the most recent work in each and every strategy in suggested frameworks listed in them on recommender framework.

5. Conculsion

We delivered a Hybrid Movie Recommendation system in this paper. A context that combines public unscrambling with content-based sorting to address the new thing's chill start problem. HMRS-RA lessens the error start issue for new gesture images by taking into account appropriate data, such as genre. The proposed technique (HMRS-RA) addresses the issue of flexibility by reducing the extension of the data by grouping. By treating the asset component as a load for recognising the similarity among consumers in each group, we were able to demonstrate the suggestion in comparison to several best-in-class and most recent works in recommender frameworks. The MAEs of our suggested calculation are 0.13, 0.24, 0.13, and 0.24 for males, females, age 20–39, and age 40–60, respectively, according to the exploratory data. In this way, HMRS-RA expanded the exactness of suggestions. Later on, we might want to characterize the client's dependence on profound learning approaches, for example, convolutional neural organizations on account of an enormous dataset.

REFERENCES

1. H. Ma, and H. Liu, "Design of Clothing Clustering Recommendation System on SOM Neural Network," In 8th International Conference on Social Network, Communication and Education, Atlantis Press, 2018.
2. D. Liben-Nowell, and J. Kleinberg, "The link-prediction problem for social networks", Journal of the American Society for Information Science and Technology, Vol. 58, No. 7, pp. 1019–1031, 2007.
3. Z. Wu, and Y. Li, Xia , "Link Prediction Based on Multi-steps Resource Allocation", in Proceedings of the 2014.
4. T. Zhou, L. Lü, and Y. Zhang, "Predicting missing links via local information," The European Physical Journal B, Vol. 71, No. 4, pp. 623–630, 2009.
5. A. Javari, J. Gharibshah, and M. Jalili, "Recommender systems based on collaborative filtering and resource allocation," Social Network Analysis and Mining, Vol. 4, No. 1, pp. 234, 2014.
6. M. Roth, A. Ben-David, D. Deutscher, G. Flysher, I. Horn, A. Leichtberg, and R. Merom, "Suggesting friends using the implicit social graph," in Proceedings of the 16th ACMSIGKDD international conference on Knowledge discovery and data mining, ACM: Washington, DC, USA., pp. 233–242, 2010.
7. M. Li, BM. Dias, I. Jarman, W. El-Deredy, and P.J. Lisboa, "Grocery shopping recommendations based on basket sensitive random walk," in Proceedings of the 15th ACM SIGKDD international conference on Knowledge discovery and data mining, ACM: Paris, France, pp. 1215–1224, 2009.
8. A. Aleem, and A. Kumar, and M. M. Gore, "A Study of Manuscripts Evolution for Perfection". In Proceedings of the 2nd International Conference on Advanced Computing and Software Engineering (ICACSE-2019), pp: 278–282, 2019.

Intelligent Systems and Smart Infrastructure – Brijesh Mishra et al. (eds)
© 2023 Taylor & Francis Group, London, ISBN 978-1-032-41287-0

CHAPTER

51

Pareto-Based Differential Evolution Algorithm Using New Adaptive Mutation Approach

Shailendra Pratap Singh[1], Gyanendra Kumar[2]

SCSE, Galgotias University,
Greater Noida,
UP, India

Vibhav Prakash Singh[3]

Computer Science and Engineering,
MNNIT Allahabad, Prayagraj,
UP, India

Abstract

This paper proposes an extended version of differential evolutionary (DE) to deal the multi-objective optimization issue by incorporating a new adaption-based mutation operator. We have proposed a new mutation method with a Pareto-based differential evolution (PBDE) algorithm for non-dominating sorting. The process provides more diversity among candidate solutions. The new mutation operator provides more bandwidth concerning diversity, which enhances the convergence rate. In addition, this study employs a non-dominated sorting method to lower the temporal complexity of Pareto dominance. The proposed work is evaluated comparatively with the latest variants of the standard multi-objective-based algorithm on six benchmark functions, such as the 2-objective and 3-objective test functions.

Keywords: Evolutionary Algorithm, DE, Multi-objective, Pareto Strategy

Corresponding author: [1]shail2007singh@gmail.com, [2]maurya.gyanendra@gmail.com, [3]vibhav@mnnit.ac.in

DOI: 10.1201/9781003357346-51

1. Introduction

In the literature, many multi-objective differential evolutionary (MODE) algorithms and Pareto-based methodologies have been developed [1-8]. The authors of [7] presented a MODE-based method to implement mutation and crossover depending on the parent population. The researcher proposed the dominance notion in [12] to solve non-dominated problems. Deb et al. [9] suggested a non-dominated sorting-based technique in 2002. This approach saved time and would be utilised for the sorting algorithm, which takes $O(mn^2)$ for sorting based on (non-dominating) the search space. On several benchmark functions, this approach outperforms Pareto fronts. The authors proposed the NSGA2 algorithm, which is based on non-dominant sorting strategies.

Furthermore, it also provides the ranking methodology. A non-dominating sorting algorithm follows this method. This method found the first rank of non-dominating algorithms and the second rank of proposed methods. This method was checked for the performance of ZDT and DTLZ. [8–12] is the authors' proposed method for a new operator in the mode algorithm. This operator aids in increasing diversity for problems with two or three objectives. The proposed method was checked for the performance of ZDT and DTLZ multi-objective functions. But this algorithm does not have sufficient diversity and conversion speed for multi-objective complex problems. The authors recommended changing the non-dominating sorting strategy for the best ranking method in [13-16]. The temporal complexity of the non-dominant operation is reduced by using this strategy. This method was checked for many-objective problems. This algorithm has better performance on some functions. But it did not provide sufficient conversation speed. In [4], the authors proposed the multiobjective-based evolutionary algorithm. This method was used in the adaptation-based strategy for solving multi-objective problems (MOP). This concept is explained in Algorithm 1.

Algorithm 1 PBDE-NAB
Set the Input variable of the algorithm:
1 $fun_{obj}, obj = 1, 2, ...n$. Multiobjective problem with obj is a objectives
2 $Search\_space\_of$ D $decision\_variables$
3 Number of generation according to fitness functions
Set the Output: find the better value
step1: Using the nondominated sorting algorithm
Begin Procedure of algorithm
step1.1: for $(i = 1; i < n; i++)$
step1.2: Set the each solution according to fitness function
step1.3: Apply the non-dominated sorting algorithm
step1.4: Apply mutation operator
step1.5: Apply crossover and using the tuning parameter
step1.6: Apply selection strategy and ranking of best solutions according to fitness for $fun_{obj}, obj = 1, 2, ...n$
step1.6: End for
End Procedure of algorithm

Algorithm 2 PBDE-NAB
Set the Input variable of the algorithm:
1 $fun_{obj}, obj = 1, 2, ...n$. Multiobjective problem with obj is a objectives
2 $Search\_space\_of$ D $decision\_variables$
3 Number of generation according to fitness functions
Set the Output: Optimum value
step1: Using the nondominated sorting algorithm and generate the new population
Begin Procedure of algorithm
step1.1: for $(i = 1; i < n; i++)$
step1.2: Set the each solution according to IE_i randomly vector with in feasible search space. Apply the non-dominated sorting algorithm and generate their Pareto front number with assign ranks to the solutions
step1.3: Apply new Internal adaption based operator
step1.4: Apply crossover and using the tuning parameter
step1.5: Apply selection strategy and ranking of best solutions according to fitness for $fun_{obj}, obj = 1, 2, ...n$
step1.6: End for
End Procedure of algorithm

[6] is the authors' proposed approach for application in software cost estimation. This approach was used as a homeostasis factor to increase the diversity of NASA 93 projects. This algorithm provides better performance than multiobjective-based software cost estimation problems. This algorithm has better performance indicators like spacing, generation distance, and inverted generation distance. In [18], a hybrid algorithm was proposed for many objectives. This method manages two different operators. These operators enhanced the diversity of the hybrid algorithm for many-objective problems. This method has better performance than many other many-objective algorithms. Being motivated, we have introduced a novel PBDE algorithm for MOP. The authors' contribution are as follows:

- A novel adaption-based strategy is devised and incorporated with Pareto-based differential evolution (PBDE) algorithm.
- The devised mutation operator considers the new adaption-based environment (NA) is designed and multiplied with each current vector of the better pool as a mutant strategy.
- The proposed method aims to retain the diversity from the starting generation to the last. In the proposed algorithms, increment in diversity is considered a new adaption-based vector.

The remaining structure of the paper is as follows: The first suggested PBDE employing a novel adaptation-based operator is addressed in Section 2. Section 3 presents the results and outcome. Finally, Section 4 brings the paper to a close end.

2. Proposed Approach: New Adaptive Mutation Operator for Pareto-Based Differential Evolution Algorithm (PBDE-NAB)

This work introduces the Novel Adaptive Mutation Operator for Pareto-Based Strategy, a new form of the algorithm that combines the PBDE method's operators. The working of the internal environment vector (IE1) is explained in Eq. (1).

$$IE1 = (current_{vector}) * EF1 \tag{1}$$

where $current_{vector}$ denotes the current population, and EF1 denotes environment factor of search space. The environment factor (EF1) is used to stabilise the searching space's environment both locally and globally.

The proposed adaption operator (NAB) is implemented with multi-objective operators. This operator produces a variety of solutions. The algorithm is used as an internal environment vector which provides sufficient diversity for MOPs. Therefore, we have applied new operators to generate the candidate solutions. After that, we applied the sorting algorithm to find the dominating and non-domination solutions. These solutions, which are at the first rank, are called "non-domination," and the second rank (remaining) solution is called "dominant."

Therefore, we have created new formations of vectors of mutation strategy according to environmental factors. This factor is incorporated into the mutation strategy of the DE algorithm. The process is explained in the form given in Eq. (2). As a result, we begin by employing non-dominating-based Pareto front solutions, as per Eq. (2).

$$\vec{\gamma_i}A = \overrightarrow{\alpha_{best}}, G + \delta_1.(\vec{IE1}_{r_1^i} - \vec{IE1}_{r_2^i}, G) \tag{2}$$

As a result, Euclidean distance from the first front solution is calculated to determine which solution is closest to the suppressed solution. The solution with the least distance is treated as α_{best}, G. This uses the best individuals for a non-dominating sorting algorithm. For the mutation operator, two solutions will be chosen from the population.

3. Result Analysis and Discussions

The setting of the experimental environment and parameters is taken from [4, 6]. In this paper, the proposed method applies the multi-objective benchmark functions (MOPs). The MOPs problems are analyzed in the performance indicator and also checked for satisfactory indicator performance from the multi-objective function. We have applied the spacing performance indicator to two or more functions. This indicator checked the performance of the non-dominating sorting algorithm. Furthermore, it checks the conversion speeds and diversification of the multiobjective-based algorithms. We have incorporated the spacing indicator into our proposed method. This process is explained in equation 3.

Where m is the number of objectives based on internal environment vectors, and distance is calculated on the Pareto front between a single objective function and two or more objective functions. Here k denotes the solution sets of the non-dominating ranking. EDabc_i represents the distance between two or three objective functions. The spacing value is minimum, which is the better convergence speed of the MOPs problems.

$$Spacing = \sqrt{1/(mabc-1)\sum_{i=1}^{mabc}(E\bar{D}abc - EDabc_i)^2}$$

$$EDabc_i = minimum_i, j \neq i(\sum_{k=1}^{k}|fabc_i^k - fabc_j^k|) \tag{3}$$

4. Experimental Results and Discussion

The proposed method compares the results of several scenarios, such as ZDT and DTLZ. Furthermore, each goal function was performed 30 times in succession, and the statistics were calculated. The proposed method is used for the different generations like 100, 200, 500, and 1000 for testing the spacing indicator.

The diversity and convergence speed are used for optimum Pareto fronts to compare with other sophisticated PBDE-NAB algorithms. The non-dominating sorting algorithm is based on the Pareto front strategy. This technique is used to find a ranking-based approach, which is to find the best rank. The best rank means selecting the first rank from a circumstance area like the global search area. Therefore, its first rank is a feasible solution to the search space. So, we have found the best rank in the global search area. This Pareto front strategy checks the rank of one solution between two or more objectives. These objectives lie in the one rank Pareto front, which is the best optimization solution. Otherwise, these solutions are not feasible for search purposes.

The suggested technique is compared to conventional Pareto-based or multi-objective algorithms. For all mean and standard deviation values, the suggested method outperforms the referenced methods in terms of conversion speed and variety as shown in Figs 51.1 and 51.2.

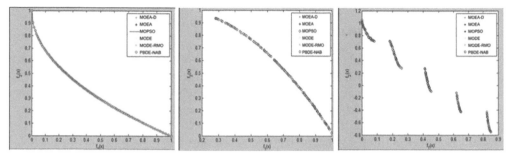

Fig. 51.1 PBDE-NAB compared with global optimization algorithm on ZDT family

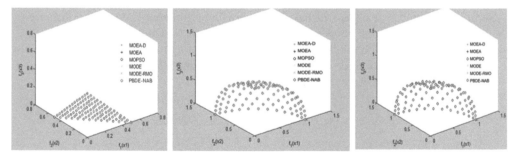

Fig. 51.2 PBDE-NAB compared with global optimization algorithm on DTLZ family

5. Conclusion

This work developed a novel mutation operator variation that offers enough variety for the PBDE method and improves the convergence rate of the optimum Pareto front. Furthermore, integrating the non-dominating sorting method reduces the computational complexity of the Pareto front. Also, the presented technique is assessed on multi-objective problems for different generations, and the cost is precisely anticipated by optimising the tuning parameters. The observed results on several benchmark functions show that the performance of the suggested technique is much better than the maximal variations of the DE algorithm.

REFERENCES

1. Storn Rainer, and Kenneth Price, Differential Evolution-A Simple and Efficient Adaptive Scheme for Global Optimization Over Continuous Spaces, International

2. Swagatam Das, Sankha Subhra Mullick, P. N. Suganthan, Recent advances in differential evolution-An updated survey, Swarm and Evolutionary Computation (27) 2016, pp. 1–30, 2016.

3. Chen, Bili, Yangbin Lin, Wenhua Zeng, Defu Zhang, and Yain-Whar Si, Modified differential evolution algorithm using a new diversity maintenance strategy for multi-objective optimization problems, Applied Intelligence 43, no. 1 (2015), pp. 49–73, Harvard, 2015.

4. Singh, S.P., Dhiman, G., Tiwari, P. and Jhaveri, R.H., 2021. A soft computing based multi-objective optimization approach for automatic prediction of software cost models. Applied Soft Computing, 113, p.107981.

5. Shailendra Pratap Singh. "Improved based Differential Evolution Algorithm using New Environment Adaption Operator", Journal of The Institution of Engineers (India): Series B, 2021.

6. Singh P. Shailendra, and Kumar Anoj , 'Pareto Based Differential Evolution with Homeostasis Based Mutation', Journal of Intelligent \& Fuzzy Systems, vol. 32, no. 5, pp. 3245–3257, 2017.

7. XuChen, Wenli Du , Feng Qian, Multi-objective differential evolution with ranking-based mutation operator and its application in chemical process optimization, Chemometrics and Intelligent Laboratory Systems 136 (2014), pp. 85–96, 2014.

8. Zhang X, Tian Y, Cheng R, Jin Y., An efficient approach to nondominated sorting for evolutionary multi-objective optimization, Evolutionary Computation, IEEE Transactions on. 2015 Apr, 19(2):201-13, pp. 201–13, 2015.

9. Deb, K., Pratap, A., Agarwal, S., Meyarivan, T., A fast and elitist multi-objective genetic algorithm: NSGA-II, IEEE Transactions on Evolutionary Computation 6 (2), pp. 182–197, 2002.

10. Bader, J., Zitzler, E., HypE: An algorithm for fast hyper volume-based manyobjective optimization, TIK Report 286, Computer Engineering and Networks Laboratory (TIK), ETH Zurich, 2008.

11. Deb, L. Thiele, M. Laumanns, E. Zitzler, Scalable multi-objective optimization test problems, Proceedings of the Congress on Evolutionary Computation (CEC-2002), (Honolulu, USA), 2002, pp. 825–830, 2002.

12. Merah, Ahmed \& Adjabi, Mohamed, 'Multi-objective Optimization in Power Systems Including UPFC Controllerwith NSGA III', International Journal of Modeling and Optimization, Vol. 8, pp. 318–325, 2018.

13. Z. Wang, Q. Zhang, A. Zhou, M. Gong, and L. Jiao, Adaptive Replacement Strategies for MOEA/D, IEEE Transaction. Cyber., vol. 46, no. 2, pp. 474486, 2016.

14. Leandro L. Minku and Xin Yao, 'Software Effort Estimation as a Multi-objective Learning Problem, ACM Transactions on Software Engineering and Methodology (TOSEM), Volume 22 Issue 4, October 2013.

15. SP Singh, A Kumar, 'Software cost estimation using homeostasis mutation based differential evolution, IEEE Conference 11th International Conference on Intelligent Systems and Control (ISCO), 2017.

16. Singh SP, Singh, VP, Mehta AK, 'Differential evolution using homeostasis adaption based mutation operator and its application for software cost estimation', Journal of King Saud University-Computer and Information Sciences, 2021.

17. Tan, Y.Y., Jiao, Y.C., Li, H. and Wang, X.K., A modication to MOEA/D-DE for multi-objective optimization problems with complicated Pareto sets, Information Sciences, 213, pp. 14–38, 2012.

18. Romit Beed , Arindam Roy,Sunita Sarkar, and Durba Bhattacharya, A hybrid multi-objective tour route optimization algorithm based on particle swarm optimization and artificial bee colony optimization, Computational Intelligence, Vol. 36, 2020.

Intelligent Systems and Smart Infrastructure – Brijesh Mishra et al. (eds)
© 2023 Taylor & Francis Group, London, ISBN 978-1-032-41287-0

CHAPTER

52

Stroke Prediction and Analysis Using Machine Learning

Kausthubh Priyan[1], Pavan Kumar K. N.[2], Manish H. R.[3]

Computer Science and Engineering,
Global Academy of Technology, Bangalore

Snigdha Sen[4]

Global Academy of Technology,
Bangalore

Abstract

Among many diseases, Stroke (derived from the ancient Greek term '*Apoplexy*' meaning "struck down with violence") is also a ruling cause of death over the past few years. Stroke is mostly caused to a clot in an artery or a blockage of an artery supplying blood to the brain. It is also caused due to brain hemorrhage. It is said that in their lifetime, 1 in 4 adults above the age of 25 will suffer from a stroke. Stroke has already become a worldwide epidemic. Now with COVID-19 introduced into the world, the plausibility of a person getting a stroke is extremely high. This paper pre-dominantly targets analyzing stroke prediction dataset by utilizing multiple Machine Learning (ML) algorithms and evaluating the performance of the best algorithm. In this paper, we have experimented with algorithms such as Logistic Regression (LR) and Support Vector Machine (SVM). Apart from the aforementioned algorithms, we have also experimented with strategic algorithms such as ANN, lazy learner algorithm (KNN), Random Forest, and tree structure algorithms (Predictive Decision Tree Algorithm) to learn and analyze the prediction of stroke. To achieve this, we have considered a stroke dataset. In addition, We have also presented the results acquired from multiple methods/algorithms and determined the best algorithm to use to predict the occurrence of stroke.

Corresponding author: [1]Kausthubh.priyan.work@gmail.com, [2]Pavankumbadi2@gmail.com, [3]Manishhvr2@gmail.com, [4]snigdha.sen@gat.ac.in

DOI: 10.1201/9781003357346-52

Keywords: Stroke, Machine Learning, Decision Tree, Prediction, Deep Learning, Neural Network (NN)

1. Introduction

Machine learning has proved to be very beneficial in the health care department over the past few years and have also yielded superior predictive value. The occurrence of a stroke can be predicted using ML techniques. However, it is not widely used in different clinics or hospitals and less known of risk scores for predicting stroke either. The algorithm first makes prediction and then the operator corrects them. This process is continued until the algorithm reaches a high level of accuracy. Its high level of accuracy gives us even more reason to use ML. Ancient Persia and Mesopotamia were one of the first places to witness stroke. 'Apoplexy', is a Greek word which translates to 'struck down with violence' was first describes in Hippocratic writings. While the word 'stroke' was first coined in the year 1599, the term 'brain attack' was used by American Stroke Association in 1990. Around 5.5 million people die every year from stroke and half of them are left impaired for the rest of their lives. It also the second deadliest epidemic in the world. In 2010, around seventeen million people suffered a stroke, and 33 million people had a stroke previously and were fortunate enough to be alive. We, as computer engineers, are determined to predict stroke, using ML algorithms and date analytics to prevent this deadly cause of death. We use different ML algorithms and evaluate their performance and analyze the dataset. These algorithms can also be applied to different datasets, and flask can also be implemented at a further stage.

The manuscript is organized as follows: Section 1 discusses related work in this field, Dataset is described in Section 2, multiple ML algorithms used in this work is explained in Section 3, we report results in Section 4 and finally, we conclude with possible future work.

2. Literature Review

Recently, AI, ML, and DL has been used extensively to predict stroke at it early stages resulting in its prevention along with other domains [1-5]. [6] discusses about stroke prediction using machine learning as a whole. In [7], the predictive algorithm, decision tree, helps in selecting/ choosing features, the prominent component analysis algorithm assists in dimension reduction, and the backpropagation NN classification algorithm is assists in building a classification design. Using [8], basics of how to create a stroke prediction model can be understood. [9] draws an analogy between Cox statistical models, predictive analytics, and ensemble models merging both approaches to predict risk of stroke in a study done by the people of China. Authors [10] displays the usage of the Random Forest Algorithm to anticipate cryptogenic stroke. [12] exhibits the usage of Decision Tree with the help of a greedy algorithm (CART) to encounter splits.

3. Description of the Dataset Used

According to the World Health Organization, the second most leading cause of death is stroke. Approximately 11% of the deaths worldwide are caused by stroke. In this paper, we used a Kaggle dataset [12] that predicts whether a patient is likely to have a stroke. This works on the basis of inputs i.e., gender, high blood pressure, age, other diseases, and smoking habits. Each row of data has information about the patient.

4. ML Algorithms Used

A. SVM Algorithm

SVMs work on the principle of partitioning an n-dimensional space via a decision line into classes based on our requirements. The new data points will be categorized into the respective classes hereby [13].

B. LR Algorithm

Since LR is a type of classification algorithm, the outcome is only 2 values, either a true (1) or a false (0) / yes (1) or a no (0). [13]

C. KNN Algorithm

K-Nearest Neighbor (KNN) is a type of machine learning algorithm that works on both classification as well as regression problems. It is one of the supervised learning algorithms [13] [14].

D. ANN Algorithm

ANN consists of a huge number of interconnected processing units that collaborate to interpret data and produce meaningful outputs [13].

E. Decision Tree Algorithm

Here, the aim is to construct a series of if-else like tree structure. Each node and child node is a series of if-else-like statements [13][15].

F. Random Forest Algorithm

Random forest algorithm is also used to resolve both classification and regression problems [16][17].

5. Implementation and Result

Data analysis was done using Python in Google in collaboration with libraries like Matplotlib, NumPy, Pandas for visualization.

Comparative Analysis of ML algorithms: Different algorithms have been used on the dataset such as SVM, LR, ANN, KNN, Random Forest and tree structure algorithm (Decision Tree). Synthetic Minority Oversampling Technique (SMOTE) has been availed to balance the imbalanced dataset. In KNN, the value of 'k' is considered to be 5. In ANN, we have considered Rectified Linear Unit Activation Function (ReLU) activation function and sigmoid activation function using TensorFlow, and Network was trained in 100 epochs. Fig. 52.1 presents the comparative study.

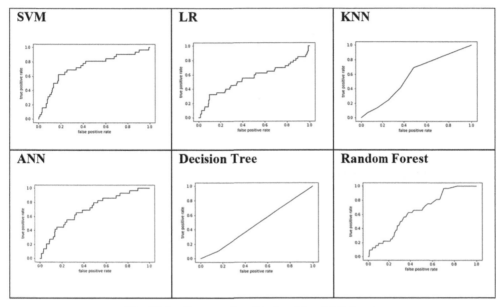

Fig. 52.1 Comparative Study of the ROC Graph

Figures above are the comparative study that shows us the ROC Graph for each of the algorithms used in Stroke Prediction. On X-axis, we have the false positive rate and on y-axis, we have the true positive rate. A **Receiver Operating Characteristic (graph plotting) curve** outlines a classification model's achievement across all levels. The percentage of true positive and false positive are mapped on this curve. Table 52.1 describes the results of all algorithms.

Metrics Used

Accuracy = TP + TN/TP + FP + FN + TN; TP = True Positives, TN = True Negatives, FP = False Positives, FP = False Negatives.

Precision = TP/(TP + FP)

Recall Score = TP/(TP + FN)

F1 Score = 2 ∗ ((precision ∗ recall)/(precision + recall)).

Confusion Matrix: The performance of a classification algorithm is condensed using a confusion matrix.

Table 52.1 Accurateness, Precision Score, Recall & F1 scores when different algorithms are employed

Parameters	SVM	LR	KNN	ANN	Decision Tree	Random Forest
Accuracy	88.0%	71.0%	73.0%	74.0%	83.0%	83.0%
Precision Score	0.098214	0.055351	0.028455	0.073077	0.039216	0.041667
Recall Score	0.343750	0.348837	0.241379	0.655172	0.206897	0.187500
F1 Score	0.152778	0.095541	0.050909	0.131488	0.065934	0.068182

Table 52.2 Confusion Matrices for all algorithms used (3 × 3 Matrix)

SVM

n = 982	Anticipated: NO	Anticipated: YES
Original: NO	TN = 849	FP = 101
Original: YES	FN = 21	TP = 11

LR

n = 982	Anticipated: NO	Anticipated: YES
Original: NO	TN = 683	FP = 256
Original: YES	FN = 28	TP = 15

KNN

n = 982	Anticipated: NO	Anticipated: YES
Original: NO	TN = 714	FP = 239
Original: YES	FN = 22	TP = 7

ANN

n = 982	Anticipated: NO	Anticipated: YES
Original: NO	TN = 712	FP = 241
Original: YES	FN = 10	TP = 19

Decision Tree

N = 982	Anticipated: NO	Anticipated: YES
Original: NO	TN = 806	FP = 147
Original: NO	FN = 23	TP = 6

Random Forest

n = 982	Anticipated: NO	Anticipated: YES
Original: NO	TN = 812	FP = 138
Original: YES	FN = 26	TP = 6

Based on the parameters above, from Table 52.1 and 52.2, we can see that SVM works the best compared to the other algorithms where the accuracy is the most with the highest precision value.

6. Conclusion

Predicting a human being's risk of having a stroke has been a popular study topic for many different authors throughout the world since it is a common illness with good evidence that early detection of the risk can help with prevention and treatment. Through this paper, we have tried to point out the accurateness using different algorithms in Stroke Prediction. We have also considered confusion matrices for each of the algorithms used which is represented in a 3×3 matrix. From the algorithms used, SVM fared far better compared to the other algorithms with an accuracy of 88.0%. Further, we would also be implementing this by providing it with a sophisticated front-end via Flask. Collecting more information and experimenting with alternative deep learning methods might be part of a future research project to improve prediction even further.

REFERENCES

1. Sen, Snigdha, et al. "Astronomical big data processing using machine learning: A comprehensive review." Experimental Astronomy (2022): 1-43.
2. Sen, Snigdha, et al. "Implementation of neural network regression model for faster redshift analysis on cloud-based spark platform." International Conference on Industrial, Engineering and Other Applications of Applied Intelligent Systems. Springer, Cham, 2021.
3. Sandeep, V. Y., Snigdha Sen, and K. Santosh. "Analyzing and Processing of Astronomical Images using Deep Learning Techniques." 2021 IEEE International Conference on Electronics, Computing and Communication Technologies (CONECCT). IEEE, 2021.
4. Monisha, R., Sen, S., Davangeri, R.U., Sri Lakshmi, K.S., Dey, S. (2022). An Approach Toward Design and Implementation of Distributed Framework for Astronomical Big Data Processing. In: Udgata, S.K., Sethi, S., Gao, XZ. (eds) Intelligent Systems. Lecture Notes in Networks and Systems, vol 431. Springer, Singapore. https://doi.org/10.1007/978-981-19-0901-6_26
5. Khasnis, Namratha S., Snigdha Sen, and Shubhangi S. Khasnis. "A Machine Learning Approach for Sentiment Analysis to Nurture Mental Health Amidst COVID-19." Proceedings of the International Conference on Data Science, Machine Learning and Artificial Intelligence. 2021.
6. Akash, Kunder & Shashank, H & .S, Srikanth & A.M, Thejas. (2020). *Prediction of Stroke Using Machine Learning.*
7. M. S. Singh and P. Choudhary, 2017 "*Stroke prediction using artificial intelligence*" 8th Annual Industrial Automation and Electromechanical Engineering Conference (IEMECON), 2017, pp. 158-161, doi: 10.1109/IEMECON.2017.8079581.
8. **Blogs:** how to create a stroke prediction model: https://www.analyticsvidhya.com/blog/2021/05/how-to-create-a-stroke-prediction-model/
9. Matthew Chun, Robert Clarke, Benjamin J Cairns, David Clifton, Derrick Bennett, Yiping Chen, Yu Guo, Pei Pei, Jun Lv, Canqing Yu, Ling Yang, Liming Li, Zhengming Chen, Tingting Zhu, the China Kadoorie Biobank Collaborative Group, August 2021, *Stroke risk prediction using machine learning: a prospective cohort study of 0.5 million Chinese adults,* Journal of the American Medical Informatics Association, Volume 28, Issue 8, Pages 1719–1727, https://doi.org/10.1093/jamia/ocab068
10. Bai, J., Yang, J., Song, W. *et al.,* 2022, "Intelligent Prediction of Cryptogenic Stroke Using Patent Foramen Ovale from TEE Imaging Data and Machine Learning Methods". *Int J Comput Intell Syst* **15,** 13.

11. S. Peñafiel, N. Baloian, H. Sanson and J. A. Pino, 2021, "Predicting Stroke Risk With an Interpretable Classifier," in IEEE Access, vol. 9, pp. 1154-1166, doi: 10.1109/ACCESS.2020.3047195.

12. **Dataset from:**
https://www.kaggle.com/code/tegarridwansyah/Stroke-prediction-logistic-regression-96-acc/notebook

13. Uddin, S., Khan, A., Hossain, M. et al. 2019. *Comparing different supervised machine learning algorithms for disease prediction.* BMC Med Inform Decis Mak 19, 281.

14. P. Soucy and G. W. Mineau, 2001, *"A simple KNN algorithm for text categorization,"* Proceedings 2001 IEEE International Conference on Data Mining, pp. 647-648, doi: 10.1109/ICDM.2001.989592.

15. Harsh H. Patel, Purvi Prajapati, 2018, *"Study and Analysis of Decision Tree Based Classification Algorithms"*, International Journal of Computer Sciences and Engineering, Vol.6, Issue.10, pp.74-78.

16. *Arun D. Kulkarni, Barrett Lowe, March 16, "Random Forest Algorithm for Land Cover Classification"*, International Journal on Recent and Innovation Trends in Computing and Communication (IJRITCC), ISSN: 2321-8169, PP: 58 - 63, Volume 4 Issue 3.

16. https://www.datasciencecentral.com/stroke-prediction-using-data-analytics-and-machine-learning/

Intelligent Systems and Smart Infrastructure – Brijesh Mishra et al. (eds)
© 2023 Taylor & Francis Group, London, ISBN 978-1-032-41287-0

CHAPTER

53

Bitcoin Price Prognosis Using Different Machine Learning Techniques: A Survey

Vipul Kadam[1], Shiva Tyagi[2]

AKGEC Ghaziabad,
Dr. A.P.J. Abdul Kalam Technical University

Abstract

Nowadays, crypto-currency such as Bitcoin is the top searched word on Google. Bitcoin becomes the bull's eye for investors; as also, it is the trending topic for researchers. Bitcoin gets its exposure due to a massive gain in its price from 2011 to 2018. Many giant investors have shifted their portfolios from other investments to Bitcoin. Traders also got a new opportunity for trading and multiplying their funds. So, there arises a huge race for the prediction of the price of Bitcoin and other cryptocurrencies. So, machine learning techniques of prediction get into the story. Many researchers approach different prediction techniques. In the upcoming section, we will review the different techniques of machine learning through which we can predict the price of Bitcoin. We will encounter the various factors that affect the price of Bitcoin. We will also look at data classifier techniques, and feature selection techniques. And in the end, we will look at the future scope and possibilities.

Keywords: Bitcoin, Data Classifier Techniques, Time Series Data, Machine Learning

1. Introduction

Bitcoin's Transaction is the first application of Blockchain and Bitcoin is a game-changer in the money market. It is an alternative to the stock market. There are also ups and downs

Corresponding author: [1]vip27may@gmail.com, vipul2010017m@akgec.ac.in; [2]tyagishiva@akgec.ac.in

DOI: 10.1201/9781003357346-53

in the price of Bitcoin. Due to its highly volatile nature, various factors affect the Bitcoin price, i.e. (Supply & Demand, Regulations, and Sentimental Factors). In case the demand for Bitcoin is high and the supply is limited, then the price of Bitcoin surges high and conversely. In the case of regulation, if the adoption rate is high, then the price of Bitcoin is high. In the case of Sentimental Factors, if the sentiments lead in a positive direction, the price increases; otherwise it decreases. To gain more profit, the prediction of the Bitcoin price is much more important. There are many machine learning techniques through which we can predict the price of Bitcoin. Examples include Auto-Regression Integrated Moving Average (ARIMA), Linear Regression Model, Latent Source Model (LSM), Binomial Generalized Linear Model (BGLM), Generalised Auto-Regression Conditional Heteroskedasticity Model (GARCH), etc. For Using Data Classifier Techniques, some data about the cryptocurrency is essential. So much data can be collected from quandl.com, CoinMarketCap [14], Binance, [15], etc.

2. Organization of Paper

The remaining paper is organized as follows. The second part looks into the literature review, which summarizes techniques, limitations of some techniques, and advantages of the techniques and classifiers over the different datasets. The result and analysis are represented in

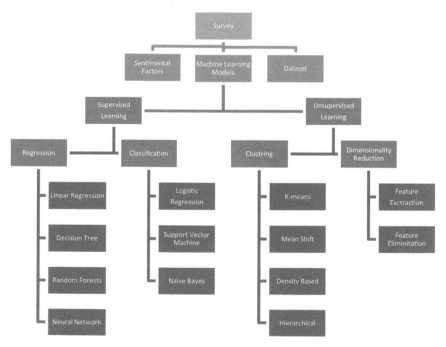

Fig. 53.1 Hierarchical structure of survey with machine learning models [20], dataset, and sentimental factors

the third section. Section four is the brief discussion and future scope of the paper. Finally, all the references are presented.

3. Literature Review

Azari et. al [1] use the Auto-Regression Integrative Moving Average (ARIMA) for the prediction of Bitcoin price. The authors train ARIMA for a 3-year-long period of data, in which ARIMA is unable to predict the sharp fluctuation in Bitcoins price. But for a short period of time, i.e., 1-day price prediction ARIMA outperforms Recurrent Neural Network (RNN). The author further uses Residual Sum of Square (RSS) for preprocessing of data to make stationary prices and also uses Mean Square Error (MSE) to reduce the error in the predicted price of Bitcoin. Hence, the authors conclude that the ARIMA model is feasible for price prediction in sub-periods of time span to several time spans. Also, they conclude that MSE helped to reduce errors in predicted prices from 118,000 to 16,000.

Pant et. al [2] have divided the process into various steps. At first, the authors collect the tweets from various tweeter accounts such as BitcoinNews@BTCTN, Cryptoyoda, etc. Further, they preprocess the tweeter data to remove irrelevant data from the tweets by using Regex and Weighed Search. And for feature selection, they use two methods: (1) WordzVector and (2) Bag of Words. Then the authors performed sentiment analysis with the help of five different algorithms (Naïve Bayes, Bernouli Naïve Bayes, Multinominal Naïve Bayes, Linear Support Vector Classifier, and Random Forest), from which they obtain the result of sentiment analysis in terms of variables, i.e. ($p = 0.26$ and $n = 0.41$), where p denotes positive and n denotes negative. Further Sentimental Analysis and Historical Prices are fed to a Recurrent Neural Network (RNN) to predict the new price for Bitcoin. The authors also claim to obtain 77%–82% accuracy.

Rane et. al [3] describe various methods and techniques used in machine learning to predict the price of Bitcoin. The authors have mentioned 9 techniques for prediction. The authors gave a brief introduction to each type of technique and classified them according to their accuracy. The following techniques with their accuracy are mentioned by the authors, i.e., ARIMA – 53.74%, Linear Regression Model – 56%, Binomial Generalized Linear Model(BGLM) – 51.6%, Generalized Auto-regression Conditional Heteroskedasticity Model (GARCH), SVM – 54%, long short-term memory (LSTM) – 52.78, NARX – 52%, Multilayer Perception MLP – 55%, The LSM in combination with Bayesian Regression examines the existing pattern in the system. And the authors claim that this algorithm gives a 200% return within 60 days. It is also concluded that the accuracy of NARX is 60%–70%.

Velankar et. al [4] have divided the investigation work into two phases. In the First Phase, the authors collect the data for 5 years from two well-known sources, i.e., Quandl and CoinMarketCap. After collecting the data, the authors process the data for its normalization, so as to eliminate insignificant, duplicate data, and hence redundancy can be removed. So, for the normalization, the authors have used five algorithms. They have also given a brief introduction to all the five algorithms:

- Log Normalization
- Standard Deviation Normalization
- Boxcox Normalization
- Z Score Normalization
- In-Built MATLAB Method

So, after the first phase, the authors will feed up with the selected features like Block Size, Total Bitcoins, Number of Transactions, Trade Volume, Day High, and Day Low to the predictive network. So the variation in the value of Bitcoin will generate patterns that are useful for the prediction of the Bitcoin price. The authors have used Bayesian Regression and GLM/Random Forest for the implementation of the model. And also they give a brief introduction to both the algorithms.

Aggarwal et. al [5] use various Deep Learning Approaches to predict the price of Bitcoin. The authors also studied the effect of gold prices on Bitcoin prices. The authors give a brief introduction about Root Mean Square Error (RMSE), CNN, LSTM, and gated recurring unit (GRU). At first, the authors collected the Bitcoin dataset from Poloniex for a 5-minute interval. And also collected Gold Dataset from datahub.io. The authors have also performed a Sentiment Analysis by using Valence Aware Dictionary for Sentiment Reasoning (VADER). In the next step, the authors will feed the data into various Deep Learning Algorithms that are mentioned above. The authors also checked the RMSE of the mentioned algorithm, and found LSTM has better accuracy than CNN and GRU. In addition, the authors do not find any correlation between Bitcoin price and gold price. The authors also conclude that there is no effect of gold price on Bitcoin price.

Paulyshenko et. al [6] assume that the pattern can be predicted by experts. But the authors implement a fusion of the Regression Model and Expert Opinion. The authors use the Bayesian Regression approach infusion with expert opinion. The authors give a brief introduction to Bitcoin, Time Series, and the Bayesian Regression Technique. The authors have implemented the numerical modeling using Python. The authors also give knowledge about various Python Packages like Panda, Numpy, Scipy, Matplotlib, Seaborn, Sklearn, qundl, and MWViews for getting time series of Wikipedia pages, with the help of Google Search and Wikipedia time series data. The authors show the pattern formulation/oscillation in Bitcoin charts. It is assumed that these patterns can be recognized by experts. It claims expert opinion and regression models give more accuracy in the prediction for a short period of time.

Rathan et. al [7] have compared the two techniques of Bitcoin price prediction, i.e. Decision Tree and Linear Regression. So, first, the authors give a brief introduction to the decision tree and linear regression. The authors collected the dataset from Quandl.com. The authors have also highlighted the features of the data just like "Time_stamp, Open, High, Low, Close, Volume_btc, Volume_currency, Weighted_price". In the second step, the authors will train the dataset and feed them to the classifier techniques. And then they compare the result of both the algorithms. Decision trees are claimed to provide an accuracy of 95.88% while linear regression is claimed to provide an accuracy of 97.59%. Finally, the authors conclude that Linear Regression outperforms the decision tree in Bitcoin price prediction.

Aniruddha Dutta et. al. [8] used sequence model, RNN, LSTM, and suggested the GRU model with recurrent dropout performs better than the popular existing model. LSTM has historically been used to examine time series in the deep learning literature. The GRU architecture, on the other hand, appears to outperform the LSTM model in our tests. Adding a recurrent dropout increases the GRU architecture's performance; nonetheless, more research into the dropout phenomena in GRU architectures is needed. Our trained GRU architecture has been used to execute two different sorts of investing strategies. The findings suggest that when machine learning models are deployed correctly, they may benefit the investing business in terms of financial gains and portfolio management. Recurrent machine learning models outperformed traditional ones in price prediction in this example, making the investing methods more lucrative.

Suhwan et. al [9] used sequence models, RNNs, and LSTM, and showed that a GRU model with recurrent dropout performs better than a popular existing Bitcoin price prediction model, such as deep neural networks (DNN), LSTM, convolutional neural networks, deep residual networks, etc. The LSTM-based models outperformed all other prediction models in terms of price prediction (regression), but the DNN-based models outperformed all other prediction models in terms of price ups and downs prediction. Additionally, a simple profitability study showed that classification models performed better than regression models for algorithmic trading.

Shiva et. al [10] used an approach of LSTM with a recurrent neural network for Bitcoin price value from 2019 to 2021 between the real price and predicted price values. The forecasting and analysis was done using the tool jupyter notebook.

Aljojo et. al [11] used Nonlinear Autoregressive Exogenous (NARX) to predict the influence of Bitcoin timestamp on Bitcoin transactions. And it is founded that the performance of currency is highly influenced and the prediction accuracy is 96%. The authors use a dataset of 4 years from 1 Jan 2012 to 31 Jan 2016.

Sujatha et. al [12] use a Non-linear Autoregressive Network with exogenous inputs to select seven attributes from thirteen attributes that help in forecasting the future trend of Bitcoin. Further, the result is fed to Network Training Algorithm which shows the following result (i.e., Comparatively BR possesses minimum testing error with a value lies between $-4.9e + 09$ and $5.9e + 09$).

Jagannath et. al [13] perform co-relation analyses and classifies user, miner, and exchange activity. To accurately predict the price of Bitcoin, the author uses the jSO optimization algorithm to find the best values of these parameters. Further, the authors compared the results with the LSTM model, and found jSO outperforms LSTM with high accuracy and a minimum error rate. The authors made a comparative analysis of the Mean Absolute Error of LSTM and LSTM-jSO (For a Train Size of 0.1, MAE of LSTM is 4.0567 and LSTM-jSO is 2.06088, For a Train Size of 0.5, MAE of LSTM is 3.0542 and LSTM-jSO is 2.04965, For Train Size of 0.9, MAE of LSTM is 2.9035 and LSTM-jSO is 1.89352).

Shahbazi et. al [16] focus on two cryptocurrencies, i.e. Litecoin and Monero. Using the Reinforcement Learning Algorithm (RLA), the authors analyze and predict prices. Additionally, RLA is found to outperform ARIMA, SARIMA, etc. with better reliability and less error rate.

Tanwar et. al [17], as part of the study, two cryptocurrencies are considered: Litecoin and Zcash. The authors propose a price prediction model using two deep learning algorithms, namely GRU and LSTM. They also perform a comparative analysis with the proposed model. Litecoin MSE losses derived from the proposed model for the 1-day and 3-day periods are 0.02038 and 0.02103, respectively, and Zcash MSE losses are 0.00461 and 0.00483. The proposed model predicts Litecoin prices accurately for 7-day and 30-day periods. It however follows the stochastic nature of Zcash.

4. Result and Analysis

Year and Reference No.	Dataset Used	Algorithm Used	Result	Remark
[1] (2019)	3-years long data with 1-day variation	ARIMA, RSS, MSE	ARIMA outperforms other algorithms, MSE reduce error from 118,000 to 16,000	ARIMA can be used for short variation of time
[2] (2018)	Tweets from tweeter, 3-year data of Bitcoin price	RNN, GRU, LSTM, Sentiment Analysis	Accuracy of prediction lies between 77% and 82%	The combination of sentimental analysis & LSTM gives a fair prediction
[3] (2019)	Bitcoin Time series data is used	ARIMA, LSM, BGLM, GARCH, SVM, LSTM, NARX	NARX gives the highest accuracy of 60%–70%	Combination of two or more algorithms gives better accuracy.
[4] (2018)	5-year daily record of Bitcoin price & payment network	Bayesian Regression, GLM	Normalization of data is done	Normalization of data helps in gaining more accuracy
[5] (2019)	5-minute interval prices of Bitcoin and gold are used	RMSE, CNN, LSTM, GRU	No correlation with gold price whereas correlation with sentiments	Correlation between Bitcoin and its factors is proved good for price prediction
[6] (2019)	1-Year monthly data is used	Bayesian Regression Model	Pattern recognition is done with the help of Bayesian Regression	Pattern recognisation is different approach for sentimental analysis
[7] (2019)	5-Days Dataset is used	Decision Tree, Linear Regression	Decision Tree gives 95.88% and Linear Regression gives 97.59% accuracy	Further, the combination of linear and exponential regression can give better result

Year and Refer-ence No.	Dataset Used	Algorithm Used	Result	Remark
[8] (2020)	Bitcoin Time-series data, factors driving Bitcoin are used	GRU, RMSE, RNN, LSTM	RNN models like GRU and LSTM perform better than machine learning models	It is still premature to solely use for price prediction
[9] (2019)	Bitcoin time-series data from 29-Nov-2011 to 31-Dec-2018	DNN, LSTM	Deep Learning models seem to predict very well than regression models	There is no clear winner. Results of deep learning models were compared with each other
[10] (2022)	Bitcoin price value from 2019 to 2021.	RNN, ANN, LSTM	Real-time value of Bitcoin price are higher than predicted values	LSTM & RNN are not individually sufficient for more accuracy
[11] (2021)	4-Years data from 1-Jan-2012 to 31-Jan-2016 is used	NARX	The accuracy of proposed work is 96%	The influence of Bitcoin time stamp on Bitcoin transaction is a new technique of sentimental prediction.
[12] (2021)	1-Year data of 2019 is used	NARX, Bayesian Regularization, Scaled Conjugate Gradient algorithm	BR NN shows good performance with minimum error and regression value 0.9984	BR NN is a unique technique for the price prediction
[13] (2021)	Three types of data are used i.e. user activity, miner activity and exchange activity.	JSO, LSTM	JSO outperforms LSTM with minimum average error	The correlation between users, miners & exchanges fills up the loop holes and gives high accuracy
[16] (2021)	4-Years data from 2016 to 2020 with 1276 data points are used	RLA	RLA outperforms ARIMA & SARIMA with high accuracy and low error	RLA proves to be a better model for prediction if used with sentimental analysis
[17] (2021)	For LSTM data from 24-Aug-2016 to 26-May-2021 is used. For GRU data from 29-Oct-2016 to 26-May-2021 is used.	GRU, LSTM	Predicted price of Litecoin is similar to actual price. But there is difference between predicted and actual price of ZCash	The performance of GRU is better than LSTM as it is easy to modify

In Fig. 53.2, a comparative analysis of the Data Classifier has been done on the bases of how many times it is used by the authors in the above-mentioned research papers. Most of the authors use LSTM for the prediction of Bitcoin price. Here, other refers to the other algorithms used in the above-mentioned research papers.

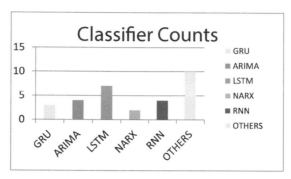

Fig. 53.2 Graphical Representation of Data Classifier Counts

5. Conclusion

As part of our study, we have examined the introduction to Bitcoin, how its value varies, and why it is important to predict its value. We have also seen the sentimental factors that affect the value of Bitcoin. We have outlined the dataset sources that feed into machine learning algorithms. We have examined the classification of machine learning techniques used for prediction. We have reviewed the work done by other researchers using deep learning and machine learning to predict prices. We have tracked the preprocessing of data using algorithms like Regex and weighed search, Wordz vector, and Naïve Bayes. This value is fed to different classifier techniques to increase the accuracy of the predictions. As a result, machine learning has a high prediction efficiency. In the next section, we discuss the future scope of Bitcoin price prognosis.

6. Future Scope

In the future, we can use Support and Resistance [18] levels in combination with sentimental analysis for price prediction. We can also use the Relative Strength Index (RSI) [19] and pattern recognition technique for the prediction of Bitcoin price. We can also use the Moving Average Method as well as the combination of the above-mentioned techniques for the prediction of Bitcoin price.

REFERENCES

1. A. Azari, "Bitcoin price prediction: An ARIMA approach", *arXiv preprint arXiv:1904. 05315*, 2019.
2. D. R. Pant, P. Neupane, A. Poudel, A. K. Pokhrel, en B. K. Lama, "Recurrent neural network based bitcoin price prediction by twitter sentiment analysis", in *2018 IEEE 3rd International Conference on Computing, Communication and Security (ICCCS)*, 2018, bll 128–132.

3. P. V. Rane en S. N. Dhage, "Systematic erudition of bitcoin price prediction using machine learning techniques", in *2019 5th International Conference on Advanced Computing & Communication Systems (ICACCS)*, 2019, bll 594–598.

4. S. Velankar, S. Valecha, en S. Maji, "Bitcoin price prediction using machine learning", in *2018 20th International Conference on Advanced Communication Technology (ICACT)*, 2018, bll 144–147.

5. A. Aggarwal, I. Gupta, N. Garg, en A. Goel, "Deep learning approach to determine the impact of socio economic factors on bitcoin price prediction", in *2019 Twelfth International Conference on Contemporary Computing (IC3)*, 2019, bll 1–5.

6. B. M. Pavlyshenko, "Bitcoin Price Predictive Modeling Using Expert Correction", in *2019 XIth International Scientific and Practical Conference on Electronics and Information Technologies (ELIT)*, 2019, bll 163–167.

7. K. Rathan, S. V. Sai, en T. S. Manikanta, "Crypto-currency price prediction using decision tree and regression techniques", in *2019 3rd International Conference on Trends in Electronics and Informatics (ICOEI)*, 2019, bll 190–194.

8. A. Dutta, S. Kumar, en M. Basu, "A gated recurrent unit approach to bitcoin price prediction", *Journal of Risk and Financial Management*, vol 13, no 2, bl 23, 2020.

9. S. Ji, J. Kim, και H. Im, 'A comparative study of bitcoin price prediction using deep learning', *Mathematics*, τ. 7, τχ. 10, σ. 898, 2019.

10. M. K. S. Kumar, S. Pathapati, και P. M. J. Bhaskarini, 'BITCOIN PRICE PREDICTION'.

11. N. Aljojo, A. Alshutayri, E. Aldhahri, S. Almandeel, και A. Zainol, 'A Nonlinear Autoregressive Exogenous (NARX) Neural Network Model for the Prediction of Timestamp Influence on Bitcoin Value', IEEE Access, τ. 9, σσ. 148611–148624, 2021.

12. R. Sujatha, V. Mareeswari, J. M. Chatterjee, A. A. A. Mousa and A. E. Hassanien, "A Bayesian Regularized Neural Network for Analyzing Bitcoin Trends," in IEEE Access, vol. 9, pp. 37989-38000, 2021, doi: 10.1109/ACCESS.2021.3063243.

13. N. Jagannath et al., "A Self-Adaptive Deep Learning-Based Algorithm for Predictive Analysis of Bitcoin Price," in IEEE Access, vol. 9, pp. 34054-34066, 2021, doi: 10.1109/ACCESS.2021.3061002.

14. Coinmarketcap.com. 2022. [online] Available at: <https://coinmarketcap.com/> [Accessed 10 April 2022].

15. Binance. 2022. *Buy/Sell Bitcoin, Ether and Altcoins | Cryptocurrency Exchange | Binance*. [online] Available at: <https://www.binance.com/en> [Accessed 10 April 2022].

16. Z. Shahbazi and Y. -C. Byun, "Improving the Cryptocurrency Price Prediction Performance Based on Reinforcement Learning," in IEEE Access, vol. 9, pp. 162651-162659, 2021, doi: 10.1109/ACCESS.2021.3133937.

17. S. Tanwar, N. P. Patel, S. N. Patel, J. R. Patel, G. Sharma and I. E. Davidson, "Deep Learning-Based Cryptocurrency Price Prediction Scheme With Inter-Dependent Relations," in IEEE Access, vol. 9, pp. 138633-138646, 2021, doi: 10.1109/ACCESS.2021.3117848.

18. Investopedia. 2022. Support and Resistance Basics. [online] Available at: <https://www.investopedia.com/trading/support-and-resistance-basics/#:~:text=Technical%20analysts%20use%20support%20and,to%20a%20concentration%20of%20demand.> [Accessed 19 April 2022].

19. Investopedia. 2022. Relative Strength Index (RSI). [online] Available at: <https://www.investopedia.com/terms/r/rsi.asp> [Accessed 19 April 2022].

20. 2022. [video] Available at: <https://www.youtube.com/watch?v=yN7ypxC7838&t=111s> [Accessed 7 May 2022].

Intelligent Systems and Smart Infrastructure – Brijesh Mishra et al. (eds)
© 2023 Taylor & Francis Group, London, ISBN 978-1-032-41287-0

Defected Ground Inset Fed Multi-Slot-Loaded Multiband Patch Antenna for WiFi/LTE/C & X Band Applications

Navendu Nitin[1], J. A. Ansari[2]

Department of Electronics & Communication,
University of Allahabad,
Prayagraj, India

Neelesh Agrawal[3]

Department of ECE, SHUATS,
Prayagraj, India

Abstract

A defected ground inset fed multi-slot-loaded multiband antenna is proposed and fabricated to develop multiband resonance for applications like LTE, etc. The resonating frequencies of the simulated proposed antenna are 5.5 GHz, 6.1 GHz, 6.7 GHz, 7.6 GHz, 8.7 GHz, 9.4 GHz, and 10.1 GHz while resonating frequencies of the proposed fabricated antenna 5.5078GHz, 6.1326 GHz, 7.0072 GHz, 8.1318 GHz, and 9.7561 GHz. The obtained frequency bands are useful for Wi-Fi, LTE, satellite communication, radar communication, and C & X band applications. Frequency band like 5.5 can be suitable for Wi-Fi & LTE applications. The simulation of the projected antenna is done through HFSS software.

Keywords: LTE, Multi-Slot, Return Loss (RL), Microstrip Antennas (MSA), Defected Ground, Inset Feeding

Corresponding author: [1]navendunitin09@gmail.com, [2]neelesh.agrawal@shiats.edu.in, [3]jaansari@rediffmail.com

DOI: 10.1201/9781003357346-54

1. Introduction

Due to unmatched requirement of different wireless communication structures and applications like satellite, LTE, WIFI, WiMAX etc., the need for the multiband patch antennas is increasing day by day. Some configurations to achieve the need that were projected before are as follows: Lizzi and Mass (2011) reported a monopole patch antenna having a shape of Sierpinski fractal to get dual band for LTE applications. The two bands obtained were 700 MHz and 2.6 GHz. Then, Moosazadeh and Kharkovsky (2014) reported a monopole patch antenna consisting of U- and L- shaped slotted patch. The antenna was able to get tri band. The antenna was useful for WLAN & WiMAX. Further, Chen et al. (2014) introduced a rectangular slotted patch antenna with defected ground plane, a fork & L-shaped strips to produce dual wide-band behaviour. The antenna was applicable for LTE and WLAN. After that, Wang et al. (2014) used a Yagi patch antenna having three director arrays, a folded dipole, and a ground plane to generate broad band resonance, whereas Khan et al. (2016) used a dielectric resonator patch antenna to produce multiband behaviour for LTE applications. Further, Chou et al. (2015) designed a monopole MSA having coupled ground to obtain five bands for GSM & LTE applications. Then, Wong and Huang (2015) used a rectangular open-slotted patch antenna along with an L-shaped strip to get two different open-slotted resonating paths for LTE applications. However, Feng et al. (2015) designed a cross-magneto electric dipole dual-layer structured patch antenna to get dual wide-band resonance. The antenna was useful for WLAN, LTE, 3G, and 2G applications. Afterwards, Ulibarri and Bertuch (2016) used a complementary Yagi Uda-printed antenna to get resonance at 2.44 GHz. The antenna was having U-shaped slotted reflector and radiator elements while the rectangular slot was represented as a director. Huang and Gong (2016) configured a patch antenna consisting of two sets of folded dipoles as a lower band element and two crossed dipoles as an upper band element to get dual polarized behaviour for LTE & GSM applications. Further, Zhang and Wang (2016) used a two-port MIMO-printed antenna having single ring covering the range from 1.85 to 2.62 GHz for LTE applications. Then, Moorefield et al. (2016) configured a frequency-tuneable printed antenna having a liquid metal actuated slot to get frequency tuning from 1.85 to 2.07GHz. However, Sarkar and Srivastava (2017) used an inverted L-shaped monopole MIMO patch antenna along with a split-ring resonator for LTE, Wi-Fi, and other applications. Then, Al-Bawri et al. (2018) configured a Y-shaped slot-loaded dipole patch antenna to get resonance at 3.5 GHz, 2.4 GHz, and 5.5Ghz frequencies for LTE and WLAN applications. Further, Morsy and Morsy (2018) used a MIMO patch antenna with two sections of meander line antennas for LTE & UMTS applications.

Here, the proposed patch antenna contains multi-slots on to a rectangular patch, which has been empowered (energized) by inset feeding. A defected ground plane has been used to get the desired results. The antenna shows good results for different applications like satellite, Wi-Fi, LTE, and C & X band applications.

2. Antenna Design

The simulated and fabricated design of the proposed defected ground inset fed multi-slot-loaded multiband patch antenna is indicated in Fig. 54.1. The dielectric substrate of the proposed structure is FR4 with relative permittivity (ε_r) and thickness (h) as 4.4 & 1.6 mm, respectively. The feeding in the discussed antenna is done by 50 Ω microstrip line. The length (L_F) & width (W_F) of the microstrip feeding line are revealed in Fig. 54.1(a). In order to have a proper impedance matching, an inset feeding has been incorporated along with the line feeding. The length (L_{IF}) and width (W_{IF}) of the inset feeding are mentioned in Fig. 54.1(a). The proposed antenna has slots with multiple characters that are "JESUS". The parameters of the multiple slots are given in Fig. 54.1(a). Equations have been acquired from Garg et al. (2001) & Balanis (2005) for the designing of the proposed structure.

Different parameters regarding the antenna structure which have been optimized are revealed in Fig. 54.1(a). In order to get the desired results, a defected ground plane has been used. The optimized value of the defected ground plane is shown in Fig. 54.1(a).

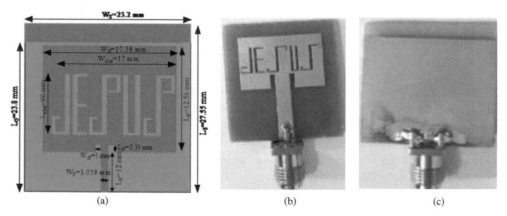

(a) (b) (c)

Fig. 54.1 (a) The proposed antenna. (b) Front view of the fabricated antenna. (c) Back view of the fabricated antenna

3. Parametric Study

A. Change in Ground Plane Length

Figure 54.2 shows the effect on return loss (S11) if the length of the ground plane is varied from $L_G = 6.8$ mm to $L_G = 27.55$ mm. For ground length $L_G = 23.8$ mm which is less than substrate length (L_S), the objective to get an antenna which could radiate multiband useful for LTE, C band & X band applications is fulfilled. Since selected ground plane $L_G = 23.8$ mm is less than substrate length (L_S), the term defected ground structure (DGS) has been used in the proposed paper. As discussed in Khandelwal et al. (2017) that because of DGS there will be a

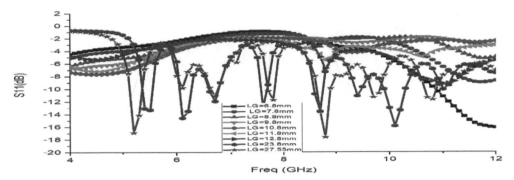

Fig. 54.2 Simulated return loss versus frequency for different ground length L_G

transformation in current distribution around the ground plane proves to be useful to tune the antenna according to the desired frequency bands.

B. Change in Inset Feed Length (LIF)

Figure 54.3 shows the variation in return loss (RL) if inset feed length is changed from $L_{IF} = 0.5$ mm to $L_{IF} = 3.35$ mm. The desired result obtained is for inset feed length $L_{IF} = 3.35$ mm.

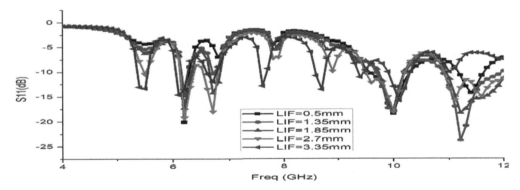

Fig. 54.3 Simulated return loss (RL) vs frequency for different inset feed length L_{IF}

C. Change in Inset Feed Width (WIF)

Figure 54.4 shows the variation in return loss (RL) if inset feed width is changed from $W_{IF} = 1$ mm to $W_{IF} = 6$ mm. The desired result obtained is for inset feed width $W_{IF} = 1$ mm.

D. Variation in Number of Slots

A variation in number of slots has been done to analyse the changes in return loss. Figure 54.5 indicates the return loss (RL) for different number of slots, i.e., for slot "J", for slot "JE", for slot "JES", for slot "JESU", and for slot "JESUS" Therefore, from Figure 54.5, it is evident that for slot "JESUS", a better result has been realised by having more number of resonating frequencies than rest of the variations.

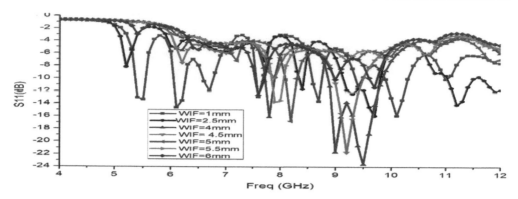

Fig. 54.4 Simulated return loss vs frequency for different inset feed width WIF.

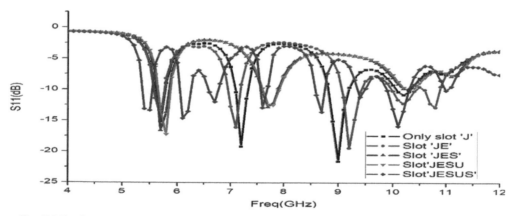

Fig. 54.5 Simulated return loss for slot "J", slot "JE", slot "JES" ,slot "JESU" & slot "JESUS"

4. Antenna Performance

The graph of simulated and fabricated return loss (RL) versus frequency is shown in Fig. 54.6. The simulated outcomes show multiple resonating frequencies like 5.5 GHz, 6.1 GHz, 6.7 GHz, 7.6 GHz, 8.7 GHz, 9.4 GHz, 10.1 GHz having return losses (RL) −13.3169, −14.5924, −11.9419, −12.8604, −13.5555, −11.1121, and −15.8465 dB respectively, while fabricated measured result also shows multiple resonating frequencies like 5.5078GHz, 6.1326GHz, 7.0072GHz, 8.1318GHz, 9.7561GHz with RL as −16.2853, −17.6983, −14.3985, −11.7594, and −20.0473 dB respectively. The fabricated measured result also shows that the resonating frequency 9.7561 GHz with 1.0601 GHz (−10 dB RL) bandwidth is covering almost two simulated resonating frequencies that are 9.4 GHz and 10.1 GHz.

The simulated (−10 dB RL) bandwidth for resonating frequencies 5.5 GHz, 6.1 GHz, 6.7 GHz, 7.6 GHz, 8.7 GHz, 9.4 GHz, 10.1 GHz are 184.1 MHz (5.5446–5.3605 GHz),

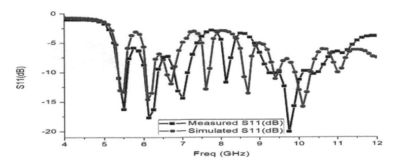

Fig. 54.6 Simulated & fabricated return loss (RL) vs frequency

212.2 MHz (6.2608–6.0486 GHz), 181.2 MHz (6.7660–6.5848 GHz), 118.4 MHz (7.6818–7.5634 GHz), 173.4 MHz (8.7569–8.5835 GHz), 119.8 MHz (9.4795–9.3597 GHz), 401.3 MHz (10.2992–9.8979 GHz), whereas measured (−10 dB RL) bandwidth for resonating frequencies 5.5078GHz, 6.1326GHz, 7.0072GHz, 8.1318GHz, 9.7561GHz are 236.2 MHz (5.5988–5.3626 GHz), 347.3 MHz (6.3723–6.0250 GHz), 355.6 MHz (7.1366–6.7810 GHz), 82.7 MHz (8.1844–8.1017), 1.0601 GHz (10.3028–9.2427 GHz).

Figure 54.7 reveals the (measured & simulated) VSWR. For frequencies 5.5GHz, 6.1GHz, 6.7GHz, 7.6GHz, 8.7GHz, 9.4 GHz, 10.1GHz, the simulated VSWR has magnitudes 1.5505, 1.4581, 1.6769, 1.5890, 1.5317, 1.7709, and 1.3847 respectively, while for frequencies 5.5078GHz, 6.1326GHz, 7.0072GHz, 8.1318GHz, and 9.7561GHz, the measured VSWR has magnitudes 1.3623, 1.2998, 1.4709, 1.6963, and 1.2209 respectively. The simulated and the measured outcomes have shown better agreement towards each other. The variances in some areas like different number of resonating frequencies, etc. may be because of some unwanted problems in designing (simulation & fabrication) like simulation of radiation box, itching, soldering, etc. of the proposed antenna.

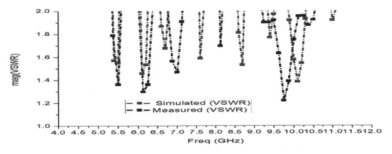

Fig. 54.7 Simulated & measured VSWR versus frequency

Simulated 2D & 3D gain for resonating frequency 5.5 GHz are shown in Fig. 54.8. For resonating frequencies 5.5 GHz, 6.1 GHz, 6.7 GHz, 7.6 GHz, 8.7 GHz, 9.4 GHz, 10.1 GHz, the discussed antenna generates the highest gain of around 2.7420 dB, 1.5073 dB, 2.1361 dB, 0.38588 dB, 2.3227 dB, 1.7197 dB, and 1.8819 dB respectively.

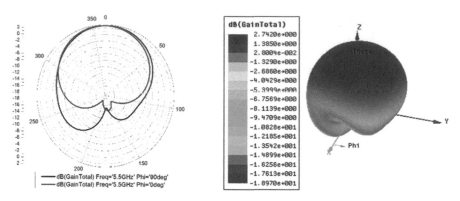

Fig. 54.8 Simulated 2D & 3D gain for selected frequencies 5.5 GHz

5. Conclusion

The proposed designed patch antenna contains multi-slots (i.e., "JESUS") on to a rectangular patch having an inset feed line along with defected ground plane. Results illustrate that the designed antenna can generate multi-resonating frequencies for LTE and other applications. At frequency 10.1GHz, the antenna has a simulated maximum (−10dB RL) bandwidth of 401.3 MHz (ranging from 9.8979 GHz to 10.2992 GHz), whereas measured maximum (−10 dB RL) bandwidth for frequency 9.7561GHz is 1.0601 GHz (ranging from 9.2427 GHz to 10.3028 GHz). For frequency 10.1 GHz, 1.3847 is the minimum simulated VSWR while 1.2209 is the minimum measured VSWR for frequency 9.7561GHz. 2.7420 dB is the maximum simulated gain for frequency 5.5GHz with broad radiation pattern. After analysing the results, it is concluded that the proposed defected ground inset fed multi-slot-loaded multiband patch antenna may be valuable for Wi-Fi, LTE, and C & X band applications.

REFERENCES

1. Lizzi, L. and Mass, A. (2011). Dual-band printed fractal monopole antenna for LTE applications. IEEE Antennas and Wireless Propagation Letters.10(1): 760–763. doi: 10.1109/LAWP.2011.2163051.
2. Moosazadeh, M., and Kharkovsky, S. (2014). Compact and small planar monopole antenna with symmetrical L- and U-shaped slots for WLAN/WiMAX applications. IEEE Antennas and Wireless Propagation Letters 13: 388–391. doi: 10.1109/LAWP.2014.2306962.
3. Chen, S., Dong, D., Liao, Z., Cai, Q., and Liu, G. (2014). Compact wideband and dual-band antenna for TD-LTE and WLAN applications. Electronics Letters 50(16): 1111–1112. doi:10.1049/el.2014.1576,
4. Wang, Z., Liu, X. Yin, Y., and Wu., J. (2014). Dual-element folded dipole design for broadband multilayered Yagi antenna for 2G/3G/LTE applications. Electronics Letters 50(4): 242–244. doi:10.1049/el.2013.4146.
5. Khan, R., Jamaluddin, M. H., Kazim, J. U. R., Nasir, J., and Owais, O. (2016). Multiband-dielectric resonator antenna for LTE application. IET Microwaves, Antennas and Propagation 10(6): 595–598. doi: 10.1049/iet-map.2015.0535.

6. Chou, Y. J., Lin, G. S., Chen, J. F., Chen, L. S., and Houng, M. P. (2015). Design of GSM/LTE multiband application for mobile phone antennas. Electronics Letters 51, (17): 1304–1306. doi: 10.1049/el.2015.1839.

7. Wong, K. L., and Huang, C. Y. (2015). Triple-Wideband Open-Slot Antenna for the LTE Metal-Framed Tablet device. IEEE Transactions on Antennas and Propagation 63(12): 5966–5971. doi: 10.1109/TAP.2015.2491321.

8. Feng, B., An, W., Yin, S., Deng, I.., and Li, S. (2015). Dual-wideband complementary antenna with a dual-layer cross-ME-dipole structure for 2G/3G/LTE/WLAN applications. IEEE Antennas and Wireless Propagation Letters 14:626–629. doi: 10.1109/LAWP.2014.2375338.

9. Ulibarri, P. R., and Bertuch, T. (2016). Microstrip-fed complementary Yagi-Uda antenna. IET Microwaves, Antennas and Propagation 10(9): 926–931. doi: 10.1049/iet-map.2015.0734.

10. Huang, H., Liu, Y., and Gong, S. (2016). A novel dual-broadband and dual-polarized antenna for 2G / 3G / LTE base stations. IEEE Transactions on Antennas and Propagation 64(9): 4113–4118.

11. Zhang, Y., and Wang, P. (2016). Single ring two-port MIMO antenna for LTE applications. Electronics Letters 52(12): 998–1000. doi: 10.1049/el.2016.0857.

12. Moorefield, M. R., Gough, R. C., Morishita, A. M., Dang, J. H., Ohta, A.T., and Shiroma, W. A. (2016). Frequency-tunable patch antenna with liquid-metal-actuated loading slot. Electronics Letters 52(7): 498–500.

13. Sarkar, D., and Srivastava, K. V. (2017). Compact four-element SRR-loaded dual-band MIMO antenna for WLAN/WiMAX/WiFi/4G-LTE and 5G applications. Electronics Letters 53(25): 1623–1624. doi: 10.1049/el.2017.2825.

14. Al-Bawri, S. S., Jamlos, M. F., Soh, P. J., Junid, S. A. A. S., Jamlos, M. A. and Narbudowicz, A. (2018). Multiband slot-loaded dipole antenna for WLAN and LTE-A applications. IET Microwaves, Antennas and Propagation 12(1):63–68. doi: 10.1049/iet-map.2017.0008.

15. Morsy, M. M. and Morsy, A. M. (2018). Dual-band meander-line MIMO antenna with high diversity for LTE/UMTS router. IET Microwaves, Antennas and Propagation 12(3): 395–399. doi: 10.1049/iet-map.2017.0802.

16. Garg, R., Bhartia, P., Bahl, I., Ittipiboon, A. (2001). Microstrip Antenna Design Handbook, Artech House London.

17. Balanis, C. A. (2005). Antenna Theory Analysis and Design, John Wiley and Sons,3rd edn., 811–872.

18. Khandelwal, M. K., Kanaujia, B. K., and Kumar, S. (2017). Defected ground structure: Fundamentals, analysis, and applications in modern wireless trends. International Journal of Antennas and Propagation 2017. doi: 10.1155/2017/2018527.

Intelligent Systems and Smart Infrastructure – Brijesh Mishra et al. (eds)
© 2023 Taylor & Francis Group, London, ISBN 978-1-032-41287-0

CHAPTER **55**

Implementation of RAM Feeder Cycles and Their Steps Logic in Waste to Energy Plant

Vivek Kumar Srivastava[1], Poonam Syal[2]

Department of Electrical Engineering,
National Institute of Technical Teachers Training & Research
Chandigarh, India

Abstract

In this paper, we describe the methods of implementing the RAM feeder cycles/hour and also the step inclusion in the cycle. The logic is implemented in India's largest integrated waste to energy facility situated in New Delhi. The logic is prepared in plant DCS. The method is calculating the waiting time of the feeder. The results analyzed for the applied logic the cyclic operation success with steps in DCS. RAM feeder is mainly responsible for pushing the waste into furnace of incinerator. Dumping of municipal solid waste (MSW) in landfills is not a proper solution for handling the MSW as the land resource is limited. In MSW incineration, the volume of the waste is reduced significantly. The waste to energy (WTE) concept is a very good solution for the problem of MSW.

Keywords: Waste To Energy (WTE), Incineration, Municipal Solid Waste (MSW), Waste Combustion, RAM feeder, Waste handling, Waste management.

1. Introduction

Waste is a problem if it is dump in landfill, but if we use waste for energy generation, then it can become fuel and the drastic problem of dumping the waste can be reduced. Although use

Corresponding author: [1]vivekosho786@gmail.com, [2]poonamsyal@nitttrchd.ac.in

DOI: 10.1201/9781003357346-55

of waste as a fuel for energy generation is increasing across the globe, in India, it is in initial stage. In foreign countries, waste to energy (WTE) is a fully accepted concept. India has very few operating WTE plants at present. The waste generated mostly reached to landfills, which is not a permanent solution. Energy generation from waste is a good option. In India, WTE is growing now. It is very good solution for Municipal Solid Waste (MSW) to use as fuel for generating energy. MSW or refused derived fuel (RDF) can be used as a fuel in waste incinerator for generating steam.

In Indian context, Banerjee, Srivastava, and Hung (2013) evaluated the best way to handle the waste. They also mentioned that with the economic development, life standards improve and the waste generation also increases. So with the developments, the waste handling focus is also mandatory for sustainable development. Bhat et al. (2018) reviewed the MSW generation and its management in India. They analyzed the waste characteristics in India such as the content of moisture and the calorific value of waste. The authors found that the waste management in India needs to improve. The Indian waste characteristics are highly variable due to lack of awareness for segregation of waste as per properties. Charles Rajesh Kumar, J. et al. (2019) described the waste management through WTE in India opportunities and the impact on environment. They mentioned that the advancement in WTE technologies in India economically provides an optimal solution for the restoration of power and heat and helps in fighting the rising energy demand. These technologies decrease waste volumes, environmental influence, threats to public health, and dependence on fossil fuels for power production. India has an estimated potential of WTE of about 2.554 GW from MSW and about 1.683 GW from urban and industrial wastes.

The waste is very heterogeneous in nature. The formation of uniform layer on the grate improves the burning of waste inside the furnace. Implemented logic controls the ram feeders by cycles/hour with steps inclusion in the cycles for uniform distribution of waste inside the furnace.

2. Components of WTE Plant

In WTE plant, waste is used as fuel just like coal in thermal power plant. Although there are various technologies for generating heat, we focus on incineration technology. MSW is used as fuel either after segregation in process and disposal plant or directly. WTE plant has the following sections:

A. Waste Pit

In waste pit, waste is stored for the aging for feeding. Pit waste is stored and aged waste is fed into hopper of the boiler for burning. Feeding is by grab crane.

B. Grab Crane

Grab crane is used for feeding waste into boiler hopper. It is also used for shuffling the stored waste in pit for making it dry so that the moisture in the waste reduced. By shuffling, the waste becomes dry because moisture goes down to the pit in the form of leachate.

C. Fuel Feeding System

Fuel feeding has ram feeders. The waste put into hopper through grab crane is fed into boiler by ram feeders. Ram feeders are main feeder by which required quantity of waste is fed into the boiler for incineration. The number of ram feeders can be different as per design of plant. The plant in which logic implemented, has four RAM feeders for one boiler.

D. Grate System

There are two grates called upper grate and lower grate. Upper grate has four parts and lower grate also has four parts. The design of grate may vary as per plant design. In upper grate, two zones are there, namely drying zone and burning zone. The waste pushed by ram feeder comes to upper grate. When the upper grate is in backward position, the waste drying is performed by warm primary air. By the upper grate movement, waste comes to the burning zone. In burning zone, the waste burning occurs above 800°C temperature. After burning, the material comes to lower grate, and from here, it goes to ash disposal.

E. Incinerator Furnace

The burning of waste occurs in furnace. In the furnace, there is water wall. In water wall, the heating of water occurs. The vaporization occurs.

F. Turbine-generator (TG)

The superheated steam generated in boiler goes to turbine for rotating it. The generator coupled with turbine rotate and thus electricity generated.

G. Steam Soot Blowing System

Steam soot blowing system is used for removing the ash from evaporator and super heater zones. In steam soot blowing system steam of 14–16 kg/Cm^2 and above, 280°C is used for the removal of ash.

H. Pulse Soot Blowing System

Pulse soot blowing system is used for economiser zone ash removal. Pulse soot blowing system creates the high resonance in the zone and deposited ash removed from the tubes.

I. Flue Gas Treatment System

Flue gas is mixture of various gases generated from the burning. It also has toxic gases (NO, SO_2, etc.), which are not be went out to atmosphere, so the treatment of flue gas is mandatory before it goes to the stack of plant.

3. Process Description of WTE Plant

The energy generation process in waste to energy plant is shown in Fig. 55.1.

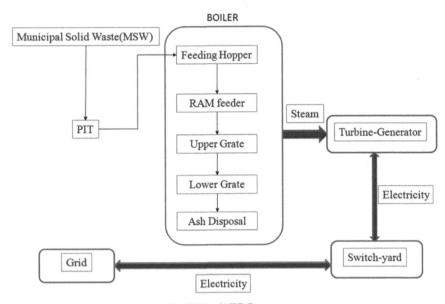

Fig. 55.1 WTE Process

The MSW has been collected from various locations. The collected waste unloaded into pit for aging because aging is very important for good combustion. Grab cranes shuffle the waste and put the waste into the hopper. The appropriate level of waste in the hopper is must because waste provides the sealing of the pit and the hopper from the furnace. In the lower point of hopper, RAM feeders are present for pushing the waste into various zones of the furnace. The zones are dry zone, combustion zone, and ash zone, respectively, from up to down in the furnace. From Ram, feeder waste goes to upper grate where primary air is present; here drying takes place and also the waste readiness for combustion. From upper grate, it comes to lower grate, and after complete combustion, it goes to ash disposal.

Due to waste combustion, steam is generated in boiler. The super-heated steam goes to TG for rotating the turbine. From TG, electricity is generated and goes to switch yard of the plant. Plant runs in synchronization with grid and supplying energy to the grid as per power purchase agreement (PPA). Plant can also run in isolation mode in case of any problem and fulfill the auxiliary power required for equipment.

4. Implementation

Arena et al. (2018) developed a combustion control for waste incineration plant. The flame image observation by the cameras is used for the corrections of output. The authors mentioned that the feed rate to the grate should be varying as per fire position on the grate visible in furnace camera. Falconi et al. (2020) describe the control strategy for combustion optimization in WTE incineration plant with assumption of homogenous distribution of waste on grate. In the described strategy, the accurate waste composition information availability (Qin et al. 2008) gives the design of combustion control system, which is implemented in DCS. The steam flow is set point in the given design and the grate speed varied as per error. The only grate speed variation in practical may not be as effective because the waste load on grate plays very important role. Magnanelli et al. (2020) presented a comprehensive dynamic model of moving grate WTE plant using MATLAB Simulink. They mentioned that various parameters response time of WTE plant largely depend upon waste characteristics. The steam response time is slower than the oxygen. The air response time is more as compared to the waste load and grate speed. Leckner and Lind (2020) compare the fluidised bed (FB) and grate combustion technology for waste combustion. They found that grate technology is more used than the fluidised bed. In fluidised bed, waste preparation is required, but in grate, firing direct MSW feeding. Dan et al. (2022) characterized the MSW incineration and flue gas emission in Tibet Plateau. They found that temperatures effect is significant on flue gas emission. The furnace draft and the oxygen concentration have also the effect on the flue gas emission. The furnace draft, oxygen level, and furnace temperature are necessary to be maintained.

India's largest commercial running integrated waste to energy facility situated in New Delhi has the incineration capacity of 1200 tonnes per day (TPD) with 24 MW generation capacity. In the running plant, we analyzed that RAM feeder is the main feeder for pushing the waste into the grate. There are a total of 4 RAM feeders for a boiler. The cyclic operation implementation with step inclusion is based on waiting time calculation.

The total travelling length of RAM feeder is 1400 mm. We reduced the travel length and also the travel of reduced length is in step operation as the full length travel exposes the RAM table in high temperature. A forward and backward speed command goes to proportional valve for operating the RAM feeder. We analyzed the time taken for going a fixed distance with a fixed speed and then generalize the same. Forward command of fixed speed goes for a fixed time after that command reverse in auto mode of operation. The operator gives the set point of cycles/hour and number of steps as per the furnace condition visible in the control room in the monitor of furnace camera. By this, waste layer adjusted on the grate inside the furnace.

Suppose speed percentage is "s" and forward distance travel in time "t" sec is "d" mm. So forward command is given for time t, backward command is also given for the same time t at same speed percentage s for returning the feeder to home position. So the total time for one cycle will be 2t sec.

Suppose number of cycles/hour set point "n", then we calculate the waiting time (time of feeder in home position) for number of steps = 1 and number of steps > 1. Number of steps means that forward movement of the RAM feeder in single shot or the number of shots (steps) selected by operator. Waiting time calculations are as follows:

$$\text{Time/Cycle, } t_c = 3600/n \qquad (1)$$

where n is the set point number of cycles/hour.

For number of steps = 1

$$\text{Waiting time, } t_w = t_c - 2t \tag{2}$$

where t is the time taken in forward movement.

For number of steps > 1, waiting time for each step is equally divided and waiting time in home position will reduce accordingly.

If number of steps > 1, then the forward command time is divided according to number of steps. Suppose number of steps = n1 > 1, then first forward command goes for time

$$t_{n1} = t/n1 \tag{3}$$

After elapse of waiting time in forward direction, again forward command goes for time t_{n1}. Waiting time in forward direction will be

$$t_{wf} = t/n1 \tag{4}$$

where n1 is the number of steps >1.

For steps > 1, the waiting time in home position will be

$$t_{wn1} = t_w/n1 \tag{5}$$

where t_w is the waiting time in home position for number of steps = 1.

The above described logic with different selections graphics implemented in plant DCS. The logic is prepared in Function Block Diagram (FBD). VFFBDBUILDER software is used for logic preparation. VFHMICfg is used for graphics design to provide the interface of system for operators in control room. VFExplorer is used for linking the logic to graphics and loading the program to controller. These all are part of installed plant Supcon DCS Visual field (V4.20.00.01-180728-M). Hierarchy of logic software and the graphics software tree is shown in Fig. 55.2.

5. Results

The logic is loaded in the controller and tested for the operation and the command from the controller is as per implementation, as shown in Fig. 55.3. The testing is in two steps and the running is as expected.

Figure 55.4 shows the command in forward and backward direction in graphics. All Ram feeders are running in auto mode with the same number of cycles and steps as visible in figure. From the graphics, operator can continuously watch of the operation of the RAM feeder if there be any deviation in any time. Clear observations can be done.

Figure 55.5 shows the home position when the speed command is zero and also the selection face plates for all feeders.

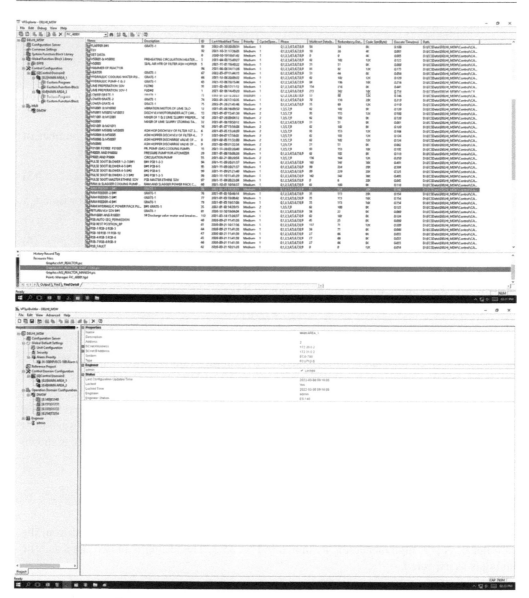

Fig. 55.2 DCS Logic Software & Graphics Software Tree

In graphics, by selection, face plates operator can select operation mode (auto/manual). For starting all RAM feeders with same set point in one shot, a push button available in graphics.

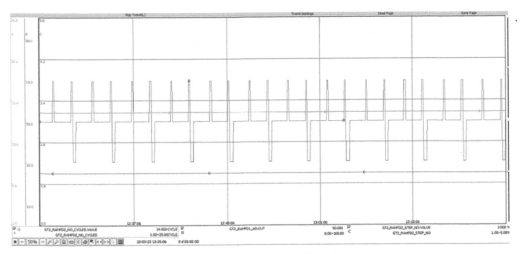

Fig. 55.3 Command Trend

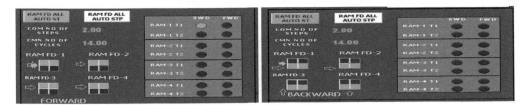

Fig. 55.4 Forward & Backward Command

Fig. 55.5 Home Position Zero Command & Selection Face Plates

Different combinations of operation of feeders can be selected by operator as per plant requirement. The selections that can operator select are given in Table 55.1. All feeders can be the same number of cycles and steps and also in different number of cycles and steps as required.

The furnace view in the control room is shown in Fig. 55.6 by observing it and feedback from field operator desk person can take the decision. In manual mode, the feeder is hold in home position and no waste input till the operator gives the command.

Table 55.1 Mode Selections

Serial No.	RAM-1 Mode	RAM-2 Mode	RAM-3 Mode	RAM-4 Mode
1	Auto	Auto	Auto	Auto
2	Manual	Auto	Auto	Auto
3	Auto	Manual	Auto	Auto
4	Auto	Auto	Manual	Auto
5	Auto	Auto	Auto	Manual
6	Manual	Manual	Auto	Auto
7	Manual	Auto	Manual	Auto
8	Manual	Auto	Auto	Manual
9	Auto	Manual	Manual	Auto
10	Auto	Manual	Auto	Manual
11	Auto	Auto	Manual	Manual
12	Auto	Manual	Manual	Manual
13	Manual	Auto	Manual	Manual
14	Manual	Manual	Auto	Manual
15	Manual	Manual	Manual	Auto
16	Manual	Manual	Manual	Manual

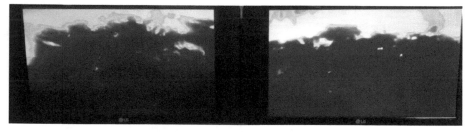

Fig. 55.6 Furnace view in control room

6. Conclusion

A new waiting time calculation-based cyclic operation implementation is done commercially running WTE. In this logic, the combination for the operator as per plant point of view is vast. By those combinations and also by the different number of cycles/hour set point for different feeders, the waste loaded on the grate can be well adjusted and the burning improved as the layer of waste on grate is maintained. The logic is totally software-based, so accuracy is high.

7. Acknowledgment

The authors are thankful to Delhi MSW Solutions Limited for providing the plant observations and agree for implementing the logic. Also we thanks to Control & Instrumentation team of plant. Special thanks to Mr. J. S. Thakur and Mr. S. S. Parashar from Delhi MSW Solutions Limited for guiding about the practical operational aspects.

REFERENCES

1. Arena, Umberto, J Rajavel, Eva Thorin, Emilia Den Boer, Olga Belous, Han Song, Emerging Countries, et al. 2018. "Developments in Combustion Control for Waste Incineration Plants." *Waste Management* 2 (1): 37–41. https://doi.org/10.1016/j.wasman.2017.08.046%0Ahttp://dx.doi.org/10.1016/B978-0-444-64083-3.00012-9%0Ahttp://dx.doi.org/10.1016/j.wasman.2017.06.049%0Ahttp://dx.doi.org/10.1016/j.energy.2010.12.070%0Ahttps://doi.org/10.1016/j.biteb.2019.03.003%0Ahttp://dx.

2. Banerjee, Tirthankar, Rajeev Kumar Srivastava, and Yung Tse Hung. 2013. "Plastics Waste Management in India: An Integrated Solid Waste Management Approach." *Handbook of Environment and Waste Management: Volume 2: Land and Groundwater Pollution Control*, 1029–1060. doi: 10.1142/9789814449175_0017.

3. Bhat, Rouf Ahmad, Shabeer Ahmad Dar, Davood Ahmad Dar, and Gowhar Hamid Dar. 2018. "Municipal Solid Waste Generation and Current Scenario of Its Management in India," no. April.

4. Dan, Zeng, Wenwu Zhou, Peng Zhou, Yuechi Che, Zhiyong Han, A. Qiong, Bu Duo, et al. 2022. "Characterization of Municipal Solid Waste Incineration and Flue Gas Emission under Anoxic Environment in Tibet Plateau." *Environmental Science and Pollution Research* 29 (5). Environmental Science and Pollution Research: 6656–6669. doi: 10.1007/s11356-021-15977-x.

5. Falconi, Franco, Hervé Guillard, Stefan Capitaneanu, and Tarek Raïssi. 2020. "Control Strategy for the Combustion Optimization for Waste-to-Energy Incineration Plant." *IFAC-PapersOnLine* 53 (2): 13167–13172. doi: 10.1016/j.ifacol.2020.12.125.

6. J, Charles Rajesh Kumar, Mary Arunsi B, R Jenova, and M A Majid. 2019. *Sustainable Waste Management Through Waste to Energy Technologies in India-Opportunities and Environmental Impacts*. Vol. 9.

7. Leckner, Bo, and Fredrik Lind. 2020. "Combustion of Municipal Solid Waste in Fluidized Bed or on Grate – A Comparison." *Waste Management* 109 (2020). The Authors: 94–108. doi: 10.1016/j.wasman.2020.04.050.

8. Magnanelli, Elisa, Olaf Lehn Tranås, Per Carlsson, Jostein Mosby, and Michael Becidan. 2020. "Dynamic Modeling of Municipal Solid Waste Incineration." *Energy* 209. Elsevier Ltd: 118426. doi: 10.1016/j.energy.2020.118426.

9. Qin, Yufei, Yan Bai, Zhongli Shen, and Keming Zhang. 2008. "Design of Combustion Control System for MSW Incineration Plant." *Proceedings - International Conference on Intelligent Computation Technology and Automation, ICICTA 2008* 1: 341–344. doi: 10.1109/ICICTA.2008.309.

Intelligent Systems and Smart Infrastructure – Brijesh Mishra et al. (eds)
© 2023 Taylor & Francis Group, London, ISBN 978-1-032-41287-0

CHAPTER

56

A Comparative Performance Analysis of QCA Full Adder

Ayushi Kirti Singh[1], Subodh Wairya[2], Divya Tripathi[3]

Department of Electronics and Communication Engineering,
Institute of Engineering and Technology, Lucknow, A.K.T.U
Lucknow, India

Abstract

Quantum-Dot cellular automata (QCA), which is a latest and fast-growing technology, is thought to be the most potent alternative for CMOS. Major challenge faced by CMOS technology is that, we have reached a point where, in order to increase transistor density further, we must fabricate transistors at atomic sizes; this poses a significant issue because we must now think about the impact of the quantum tunnelling phenomenon. QCA is a Quantum Physics application that seems to be used for the implementation of various digital circuits. With the use of Quantum tunnelling phenomenon, information is passed on from one end to the other even when there is no actual current flow, resulting in low value power consumed and extremely small size which is of order of nanometer. It is important to note that QCA has substantial advantages over CMOS technology in terms of size, speed, and total power consumed. In this research paper, a comparative analysis of the performance of various previous QCA 1-bit full adder circuits is done on QCADesigner Software, which proves that all the circuits perform well and comparison of various parameters like area, clock latency, and quantum cost are also presented. When compared to previous QCA-based full adder designs, it was seen that there is 90% reduction in the total area covered by the cell and 70% improvement in the speed of QCA circuit.

Keywords: QCA (Quantum Dot cellular automata), QCA Designer, CMOS Technology, 1-bit Full Adder

Corresponding author: [1]2000521195004@ietlucknow.ac.in, [2]swairya@gmail.com, [3]divyatripathi.cest@gmail.com

DOI: 10.1201/9781003357346-56

1. Introduction

Nowadays, researchers are very much interested in nanotechnology because of the limitations of existing silicon based transistor technology, which is popularly known as CMOS Technology. QCA is a novel nano method that provides a fresh approach to data transformation and processing. Primary building blocks of which a QCA circuit is made include the 3-Input Majority gate and basic Inverter. Compared to CMOS technology, it attracts attention because of its smaller size, faster speed, i.e., up to 1 THz, higher switching frequency, and reduced power consumption (Afrooz & Navimipour, 2017). Herein, previous QCA full adder design is explained and their parameters were verified. The paper is categorized into 4 sections. Section 1 deals with the introduction of QCA, considered as an alternate solution to CMOS technology, and also deals with QCA design schemes. Section 2 deals with a review of few existing QCA-based full adder designs like using 59 cells, 37 cells, 26 cells, 19 cells, 15 cells, and 14 cells. Furthermore, Section 3 elaborates discussion on simulation results, and Section 4 deals with the conclusion.

A. Basic QCA Cells

A basic QCA cell is a square having 4 quantum dots that are present in all the corners. Free electrons tunnel in these dots and occupy diagonally opposite sites; it is important to notice here that the electrons don't occupy adjacent positions due to insufficient energy present with them, which forms the basis of quantum-dot cellular automata (QCA) (Moustafa, 2019).

Coulombic repulsion determines a localized polarization. Two polarizations are attainable since the two electrons are in opposing corners, as seen in Fig. 56.1. As a result, there will be two binary states that can be expressed.

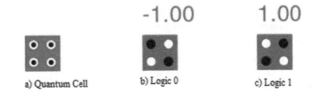

a) Quantum Cell   b) Logic 0   c) Logic 1

Fig. 56.1 Layout of (a) Quantum cell, (b) Logic 0, (c) Logic 1

B. Clock Zones

The clock has a game-changing role in QCA technology as the flow of signals is controlled by the clock itself. The overall circuit area of QCA is divided into 4 different/distinct clock zones named as clock-1 also known as switch phase, clock-2 as hold phase, clock-3 as release phase, and clock-4 as relax phase. Each clock is lagging behind the preceding clock by 90°; this is clearly shown in Fig. 56.2. The following are four clock phases:

Switch Phase: In this phase, actual computation is performed, i.e., during this phase, the QCA cell settles down to one of its states either as logic "1" or as logic "0" as affected by its neighbouring cell.

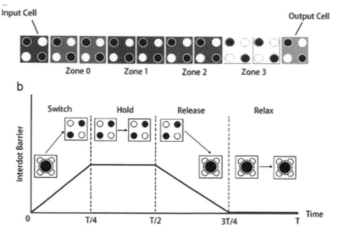

Fig. 56.2 Clocking Scheme in QCA

Hold Phase: Here, QCA cell maintains its current state as it is; in other words, the cell polarity is paused.

Release Phase: In this phase, boundary potential is diminished bit by bit and the cell loses its polarity.

Relax Phase: In this phase, the cell remains unpolarized because of the absence of an inter-dot barrier.

C. QCA Wires

Binary QCA cables, like other traditional wires, are available. QCA wire is used for transferring data from mother-cell to sister-cells in order to generate output. The mother cell is the one that accepts input, while the sister cells are the ones that transmit data (Joy et al., 2021). In the QCA design process, there are two types of wires, which are described as 90° wire and 45° wire, as shown in Fig. 56.3(a) and 56.3(b). Positive and negative polarizations can be applied in both wires (Sayedsalehi et al., 2011).

Fig. 56.3 (a) Basic QCA wire 90°, (b) Basic QCA Wire 45°

D. 3-Input Majority Gate

A Majority gate circuit consist of 5 cells (having 3 inputs and 1 acting as output). As illustrated in Fig. 56.4, Majority Logic gate comprises three cells which act as input, one cell is present in the centre, and one acts as output. This gate (Three-I/P majority) can be used to implement basic AND or OR logic simply by setting the value of an input cell to either logic 1 or logic 0 level, and that input cell is known as a control input. Equation (1) can be used to express the input–output of a majority

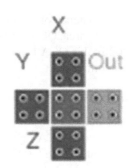

$$\text{Maj } (X, Y, Z) = XY + YZ + XZ \qquad (1)$$

Fig. 56.4 3-Input Majority Gate

2. Review of QCA-Based Full Adders

In this section, a comparative study of QCA 1-bit FA has been done by using different number of cells. QCA-based adders are better when compared with conventional CMOS-based adders because of small size, i.e., they occupy less area when compared to CMOS and energy dissipation associated with QCA-based adders is less than that of CMOS. The reason why so much emphasis is being laid on FA is that, it acts as a stepping stone for designing various other complex digital circuits. A 1-bit full adder, which is a combinational circuit, performs a 3-bit Sum. There are three inputs namely (X, Y, and Z) and two outputs that make up the circuit (Sum and Carry). Further, the numerous full-adder QCA layout using 59 cells, 37 cells, 26 cells, 19 cells, 15 cells, and 14 cells were designed and their respective output waveform was also verified using QCADesigner software.

A. Full Adder Circuit Using 59 Cells and 37 Cells

QCA-based circuit diagram for 1-bit Full adder using 59 cells is shown in Fig. 56.5(a) (Abedi et al., 2015); this QCA layout, the total area covered by the cells, was found to be 0.043 μm^2

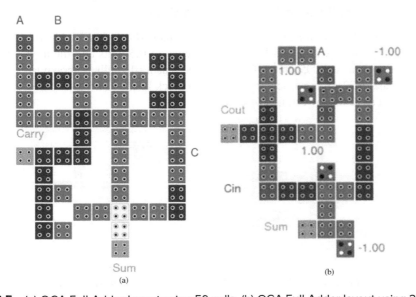

(a)   (b)

Fig. 56.5 (a) QCA Full Adder layout using 59 cells, (b) QCA Full Adder layout using 37 cells

with a clock latency of 1 unit. Out of all the layouts discussed in this paper, this one occupies the maximum area. QCA Full Adder layout using 37 cells is shown Fig. 56.5(b) (A. H. Majeed et al., 2020); this QCA layout has clock latency of 0.5 and area covered by the cell was found to be 0.04 μm^2.

B. Full Adder Circuit Using 26 Cells and 19 Cells

QCA-based circuit diagram for Full Adder using 26 cells is shown in Fig. 56.6(a) (Alkaldy et al., 2020); this QCA layout has a clock latency of 0.25 units and the area covered by the cells was found to be 0.023 μm^2. QCA-based adder using 19 cells is shown in Fig. 56.6(b) (Salimzadeh & Heikalabad, 2021); this layout has a clock latency of 0.5 and area covered was found to be 0.012 μm^2.

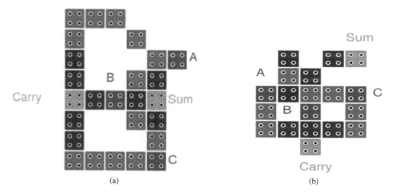

(a)   (b)

Fig. 56.6 (a) QCA Full Adder layout using 26 cells, QCA Full Adder layout using 19 cells

C. Full Adder Circuit Using 15 Cells and 14 Cells

QCA-based circuit diagram for Full Adder using 15 cells is shown in Fig. 56.7(a) (A. Majeed & Alkaldy, 2022); this QCA layout has a clock latency of 0.5 units and the area covered by the cells was found to be 0.005 μm^2. QCA-based adder using 14 cells is shown in Fig. 56.7(b) (Gassoumi et al., 2022); this particular layout has a clock latency of 0.5 and area covered by the cell is found to be the lowest, i.e., 0.0035 m^2.

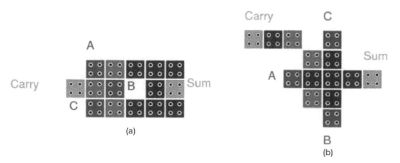

(a)   (b)

Fig. 56.7 (a) QCA Full Adder using 15 cells, (b) QCA Full Adder using 14 cells

3. Result and Discussion

All the QCA-based layouts have been verified with the help QCA Designer tool. Simulation parameters like cell count, total area covered by the cells, clock latency (delay), and quantum cost have all been measured and verified as simulation parameters. Cell count is defined as the total number of cell that are used to form the circuit. Lower cell count is generally preferred as it helps in reducing the overall area. The simulation result of a QCA-based full adder with 14 cells is shown in Fig. 56.8, and the waveform clearly demonstrates a latency of 0.5 units. The technological cost of bringing a QCA design into effect is known as quantum cost.

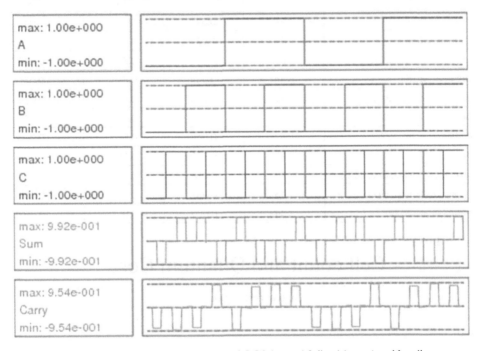

Fig. 56.8 Simulation waveform of QCA based full adder using 14 cells

From Table 56.1, it can be observed that Full Adder using 14 cells is the best performing QCA layout circuit till date as it provides a significant improvement in the cell count, area, and clock latency. The comparison between Cell count and Quantum cost for various Full adder designs is represented in Fig. 56.9, and Fig. 56.10 represents the comparison between area and clock latency.

4. Conclusion

QCA technology is best suited for the designing of various combinational and sequential circuits due to its high speed of operation and computation. Mostly, the researches which are

Table 56.1 Performance comparison of QCA-based 1-bit full adder

Circuit	Cell count	Area (μm^2)	Clock Latency	Quantum Cost
(Taherkhani et al., 2017)	228	0.28	1.75	111.7
(Lakshmi & Athisha, 2011)	192	0.2	2	76.8
(Wang, 2022)	124	0.04	1	4.96
(Kianpour, Moein,2014)	105	0.14	1.25	18.38
(Poorhosseini & Hejazi, 2018)	96	0.1	2	19.2
(Sayedsalehi et al., 2011)	75	0.09	0.75	5.063
(Abedi et al., 2015)	59	0.04	1	0.04
(Abedi et al., 2015)	82	0.07	1	0.07
(Labrado & Thapliyal, 2016)	63	0.05	0.75	0.037
(Babaie et al., 2019)	44	0.034	0.5	0.071
(Jaiswal & Sasamal, 2018)	40	0.033	0.5	0.075
(Momenzadeh et al., 2005)	38	0.032	0.5	0.075
(A. H. Majeed et al., 2020)	37	0.040	0.5	0.020
(Alkaldy et al., 2020)	26	0.023	0.25	0.006
(Salimzadeh & Heikalabad, 2021)	19	0.012	0.5	0.006
(A. Majeed & Alkaldy, 2022)	15	0.010	0.5	0.005
(Gassoumi et al., 2022)	14	0.007	0.5	0.0035

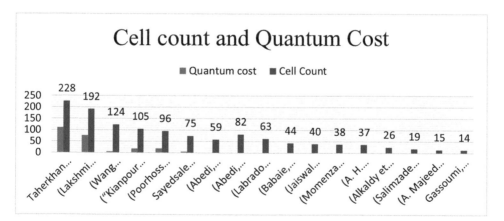

Fig. 56.9 Comparison between Cell Count and Quantum cost

being done in this field (QCA) focuses on two factors; firstly, it focuses on the designing of QCA cell, and secondly, optimization of the area of the digital logic circuits using QCA. In this paper, various high-performance QCA-based Full adder circuits have been studied and different parameters like the total number of cells used, the total area covered by these cells,

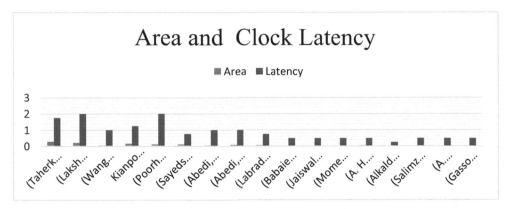

Fig. 56.10   Comparison between Area and Clock Latency

clock latency, and quantum cost of QCA-based 1-bit Full Adder circuit are compared. It has been demonstrated in the paper that out of all the adder circuits, QCA adder using 14 cells results in highest degree of optimization due to lower cell count as compared to previous existing designs and it was observed that there is a 90% reduction in the total area covered by the cell and a 70% improvement in the speed of the circuit.

REFERENCES

1. Abedi, D., Jaberipur, G., & Sangsefidi, M. (2015). Coplanar Full Adder in Quantum-Dot Cellular Automata via Clock-Zone-Based Crossover. *IEEE Transactions on Nanotechnology*, *14*(3), 497–504. https://doi.org/10.1109/TNANO.2015.2409117
2. Afrooz, S., & Navimipour, N. J. (2017). Memory Designing Using Quantum-Dot Cellular Automata: Systematic Literature Review, Classification and Current Trends. *Journal of Circuits, Systems and Computers*, *26*(12). https://doi.org/10.1142/S0218126617300045
3. Alkaldy, E., Majeed, A. H., Shamian Bin Zainal, M., Bin, D., & Nor, M. D. (2020). Optimum multiplexer design in quantum-dot cellular automata. *Indonesian Journal of Electrical Engineering and Computer Science*, *17*(1), 148–155. https://doi.org/10.48550/arxiv.2002.00360
4. Babaie, S., Sadoghifar, A., & Bahar, A. N. (2019). Design of an Efficient Multilayer Arithmetic Logic Unit in Quantum-Dot Cellular Automata (QCA). *IEEE Transactions on Circuits and Systems II: Express Briefs*, *66*(6), 963–967. https://doi.org/10.1109/TCSII.2018.2873797
5. Gassoumi, I., Touil, L., & Mtibaa, A. (2022). An efficient QCA-based full adder design with power dissipation analysis. *https://Doi.Org/10.1080/21681724.2021.2025440*. https://doi.org/10.1080/21681724.2021.2025440
6. Jaiswal, R., & Sasamal, T. N. (2018). Efficient design of full adder and subtractor using 5-input majority gate in QCA. *2017 10th International Conference on Contemporary Computing, IC3 2017*, *2018-January*, 1–6. https://doi.org/10.1109/IC3.2017.8284336
7. Joy, U. B., Chakraborty, S., Tasnim, S., Hossain, M. S., Siddique, A. H., & Hasan, M. (2021). Design of an Area Efficient Quantum Dot Cellular Automata Based Full Adder Cell Having Low Latency. *International Conference on Robotics, Electrical and Signal Processing Techniques*, 689–693. https://doi.org/10.1109/ICREST51555.2021.9331135

8. Kianpour, Moein, Reza Sabbaghi-Nadooshan, and Keivan Navi. *"A novel design of 8-bit adder/subtractor by quantum-dot cellular automata." Journal of Computer and System Sciences 80, pp: 1404-1414, 2014 - Google Search.* (n.d.). Retrieved May 12, 2022

9. Labrado, C., & Thapliyal, H. (2016). Design of adder and subtractor circuits in majority logic-based field-coupled QCA nanocomputing. *Electronics Letters*, *52*(6), 464–466. https://doi.org/10.1049/EL.2015.3834

10. Lakshmi, S. K., & Athisha, G. (2011). Design and Analysis of Adders using Nanotechnology Based Quantum dot Cellular Automata. *Journal of Computer Science*, *7*(7), 1072–1079. https://doi.org/10.3844/JCSSP.2011.1072.1079

11. Majeed, A., & Alkaldy, E. (2022). High-performance adder using a new XOR gate in QCA technology. *Journal of Supercomputing*, 1–16. https://doi.org/10.1007/S11227-022-04339-0/TABLES/4

12. Majeed, A. H., Zainal, M. S. bin, Alkaldy, E., & Nor, D. M. (2020). Full Adder Circuit Design with Novel Lower Complexity XOR Gate in QCA Technology. *Transactions on Electrical and Electronic Materials*, *21*(2), 198–207. https://doi.org/10.1007/S42341-019-00166-Y/TABLES/5

13. Momenzadeh, M., Huang, J., & Lombardi, F. (2005). Defect characterization and tolerance of QCA sequential devices and circuits. *Proceedings - IEEE International Symposium on Defect and Fault Tolerance in VLSI Systems*, 199–207. https://doi.org/10.1109/DFTVS.2005.26

14. Moustafa, A. (2019). Quantum Information Review Efficient Quantum-Dot Cellular Automata for Half Adder using Building Block. *Quant. Inf. Rev*, *7*(1), 1. https://doi.org/10.18576/qir/070101

15. Poorhosseini, M., & Hejazi, A. R. (2018). A Fault-Tolerant and Efficient XOR Structure for Modular Design of Complex QCA Circuits. *Https://Doi.Org/10.1142/S0218126618501153*, *27*(7). https://doi.org/10.1142/S0218126618501153

16. Salimzadeh, F., & Heikalabad, S. R. (2021). A full adder structure with a unique XNOR gate based on Coulomb interaction in QCA nanotechnology. *Optical and Quantum Electronics*, *53*(8), 1–9. https://doi.org/10.1007/S11082-021-03127-Z/TABLES/4

17. Sayedsalehi, S., Moaiyeri, M. H., & Navi, K. (2011). Novel efficient adder circuits for quantum-dot cellular automata. *Journal of Computational and Theoretical Nanoscience*, *8*(9), 1769–1775. https://doi.org/10.1166/JCTN.2011.1881

18. Taherkhani, E., Moaiyeri, M. H., & Angizi, S. (2017). Design of an ultra-efficient reversible full adder-subtractor in quantum-dot cellular automata. *Optik*, *142*, 557–563. https://doi.org/10.1016/J.IJLEO.2017.06.024

19. Wang, X. (2022). Designing digital circuits based on quantum-dots cellular automata using nature-inspired metaheuristic algorithms: A systematic literature review. *Optik*, 169251. https://doi.org/10.1016/J.IJLEO.2022.169251

Intelligent Systems and Smart Infrastructure – Brijesh Mishra et al. (eds)
© 2023 Taylor & Francis Group, London, ISBN 978-1-032-41287-0

Towards Deep Learning based Wilt Disease Detection System

Ruchilekha[1]

DST-CIMS, Institute of Science,
Banaras Hindu University, Varanasi, India

Manoj Kumar Singh[2]

Department of Computer Science, Institute of Science,
Banaras Hindu University, Varanasi, India

Rajiv Singh[3]

Department of Computer Science,
Faculty of Mathematics and Computing,
Banasthali Vidyapith, Rajasthan, India

Abstract

Wilt disease detection is one of the major issue of remote sensing. Here, we have used wilt dataset available at University of California Irvine (UCI) Repository that includes five features of multispectral images acquired by Quickbird satellite sensor. The purpose of the study is to advance the efficiency and performance of the algorithm for wilt disease detection system. In this paper, we have developed a method for wilt disease detection system. The proposed algorithm involves two steps: (1) the preprocessing of dataset that includes standardization of dataset using z-score normalization and oversampling of minority-class training data using Synthetic Minority Over-Sampling Technique (SMOTE), and (2) classification of data into wilt or normal class using deep learning feed-forward networks designed with H2O Automated Machine Learning (AutoML). The performance of dataset is measured in terms of accuracy and area under curve (AUC). The accuracy and AUC of the proposed method are obtained as 88.8 percent and 0.9553 respectively, which are quite higher in comparison to Support Vector

*Corresponding author: manoj.dstcims@bhu.ac.in

DOI: 10.1201/9781003357346-57

Machine (SVM), Extreme Gradient Boosting (XGBoost), and Deep feed-forward networks algorithms.

Keywords: Wilt disease, SMOTE, deep learning and H2O AutoML.

1. Introduction

Our ecosystem is generally composed of two significant factors: biotic & abiotic. Biotic factor encompasses living parts of the ecosystem, for example plants, animals, fungi, algae, and bacteria whereas abiotic factor encompasses non-living parts of the ecosystem, such as rocks, minerals, soil, and water. Forest is the most obvious feature of our ecosystem. It can be affected either by biotic factor or abiotic factor. Today, deforestation is one of the major problems in our ecosystem. It can be considered as one of the factors contributing to global climate change including greenhouse effect and desertification. Deforestation occurs due to the action of cutting of trees, removal of vegetation and permanent destruction of the forest.

In this paper, we have considered the case of deforestation due to biotic factors such as insects and beetles. The vast areas of forest are destroyed due to wilt disease spreading in many regions which mainly affects the vascular system of plants. It occurs due to attack by organisms such as nematodes, fungi, and bacteria. This results into the abrupt killing of plants, tree branches or even entire trees. The purpose of the study is the rapid detection of wilt disease, so that the immediate treatment can be done to prevent the trees from further destruction.

The various methods and techniques have developed in the field of remote sensing for the wilt disease detection. In paper [1], the detection of diseased trees, Japanese pine wilt and Japanese oak wilt, is done using high-resolution multispectral images acquired by using Quickbird [2] (a space-borne sensor). In proposed method, the pansharpening [2, 3] approach is performed on these images using intensity-huesaturation smoothing filter-based intensity modulation (IHS-SFIM) [1, 4]. Then segmentation is done on pansharpening images by utilizing multiresolution segmentation [1, 5]. SMOTE technique is applied to oversample the training data that belongs to wilt class. At last, for classification purpose multiscale object-based image classification approach is used.

In paper [6], an algorithm is developed for positive and unlabeled data. This algorithm trains a classifier and generates binary predictions for test samples using a calibrated threshold. In paper [7], author has proposed a new optimization task which is built on the top of SVM based genetic programming and it gives more improved results. In paper [8], a work is performed on Unmanned Aerial Vehicle (UAV) based wilt detection using Convolutional Neural Networks (CNN). In this, aerial images of radish fields are captured using a camera attached to UAV devices. These high-quality images are analyzed using color and texture features. Then k-means clustering algorithm is applied to cluster image data separately into four classes: healthy radish, fusarium wilt radish, bare ground, and mulching film. At last, the fusarium wilt radish data is used for classification purpose using CNN model (GoogleNet). Recently, the area of remote sensing is presented in paper [9] to distinguish healthy pine trees from sick trees on the basis of spectral analysis. The two different approaches were considered for metrics filtering. For classification purpose, random forest based approach was designed for both data of multispectral and hyperspectral images separately for early detection of decaying trees.

From previous papers of remote sensing, the basic idea is taken. Here, we have used the dataset available at UCI Machine Learning Repository in which the training data and testing data are collected separately for two classes: Wilt and Normal. The leading theme of our work is to develop a technique that improves the performance of wilt disease detection system. It is observed that deep learning model designed using H2O AutoML gives better results when it is compared to other existing methods such as XGBoost, SVM, and Deep feed-forward networks algorithms.

2. Methodology

The proposed methodology is divided into two steps. Firstly, the preprocessing of data is done to transform the data into simplified and balanced form. The preprocessing is basically a combination of two folds: (i) Standardization of data using z-score normalization that involves rescaling of data, and (ii) Oversampling of training data that belongs to minority class to balance imbalanced data using SMOTE mechanism. Secondly, these preprocessed training data are trained using various algorithms and are compared with our proposed method. Further for classification, we have used deep learning feed-forward networks designed using H2O AutoML to predict test data into their respective classes. Figure 57.1 depicts the process of the proposed methodology.

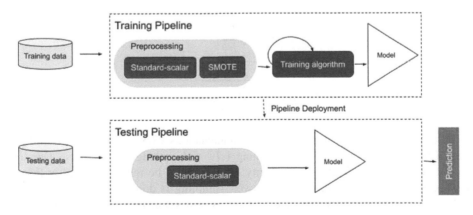

Fig. 57.1 Proposed methodology

A. Dataset Description

In this study, the dataset is fetched from the site of UCI Repository [10]. In this dataset, features extracted from high-resolution multispectral satellite images acquired using Quickbird sensor are provided. It has the records of five features namely: average green value, average red value, average near infrared (NIR) value, standard deviation of panchromatic band, and average gray level co-occurrence matrix (GLCM) of panchromatic band. Further, this dataset is partitioned into two categories: (i) training data & (ii) testing data; each set is divided into two classes, wilt and normal. The training data consists of 4339 data that comprises 4265 normal and 74 wilt data. Similarly, the test data consists of 500 data which comprises 313 normal and 187 wilt data. Now we have used this training and testing dataset in our paper for implementation and evaluation of the proposed method.

B. Preprocessing of Dataset

The preprocessing of the dataset is done to prepare our data to the best expose to the problems of machine learning algorithms. The obtained dataset is imbalanced and inaccurate, so to resolve this problem preprocessing of data is required. In this paper, preprocessing of data is performed in two steps: (i) rescaling of data using z-score normalization, and (ii) resampling of data using SMOTE technique [1, 11]. Standardization of data is a technique to obtain attributes with a Gaussian distribution. The standardization is done to transform the dataset (x_j) of various units which are on different scales to the same scale with mean value as 0 and standard deviation as 1. Each value in the dataset is subtracted by sample mean value (ii) and then divided by the standard deviation (σ). After this process, a new set of standard data is achieved. The formula for standardization (z) is given below:

$$z = \frac{(x_j - \mu)}{\sigma} \tag{1}$$

With mean:

$$\mu = \frac{1}{N} \sum_{j=0}^{N} x_j \tag{2}$$

And standard deviation:

$$\sigma = \sqrt{\frac{1}{N} \sum_{j=1}^{N} (x_j - \mu)^2} \tag{3}$$

Since the available training dataset is highly imbalanced as 98.3% of the total training data consists of normal class data and the remaining 1.7% only belongs to wilt class. Due to this reason data belongs to wilt class or minority class will not be mapped correctly that is of our interest. So to overcome this problem, over- sampling is done using SMOTE. SMOTE [11] takes the data of minority class and introduces new synthetic samples by interpolating in between minority class sample and its closed neighborhoods. SMOTE consists of two parameters: the number of closed neighborhoods and the percent of new synthetic samples. In this process, for each selected sample of minority class m_j, one of its k-closed neighborhoods n_j was randomly chosen and the scaling factor is also randomly obtained in between 0 to 1. A new synthetic sample of minority class m'_j gets synthesizes in addition to the existing dataset.

This process is repeated for entire training dataset which doubles the size of training data for the minority class.

$$m'_j = m_j + scaling * (n_j - m_j) \qquad (4)$$

C. Various Learning Networks & Learning Mechanisms

There are so many architecture and learning mechanisms are available. In this paper, a deep feed-forward architecture using H2O AutoML is proposed method for classification. This proposed method is compared with other learning mechanisms such as deep feed-forward network, XGBoost, and SVM. The brief descriptions of these learning techniques are given below.

Deep Learning Feed-Forward networks. In deep learning models [12, 13, 14], the deep feed-forward network is the most basic classifier. This network basically consists of three layers named as input layer, hidden layer, and output layer. Any network having two or more hidden layers is called Deep Neural Networks. Figure 57.2 depicts the architecture of the network. In this figure, the five input nodes (X1, X2, X3, X4, X5) represent the five features followed by one hidden layer that consists of 3 nodes and an output layer with one node. The detail of this network is briefly shown in Table 57.2. The output layer can be classified either into wilt or normal class. The net input is calculated at each node with net output using an

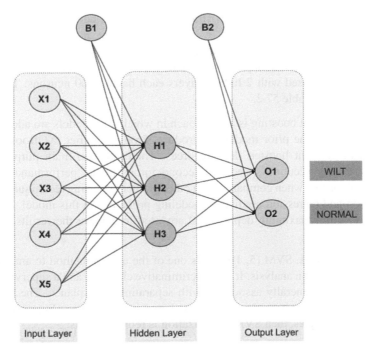

Fig. 57.2 Architecture of the deep feed-forward network

activation function for each layer. This process is repeated for each node till last output node. This phase is called the feed-forward phase. After the feed-forward phase, the obtained result is matched with target output. In case of mismatch, the error is calculated by subtracting the obtained output from the targeted output. This error is propagated backward to minimize the error. This phase is called back-propagation phase. This process continues till the error is reduced, and then weights and biases are updated. In this manner, the network is trained by iterating it with known outputs until convergence. Finally, the trained network is used for prediction and classification of the test data.

Deep Learning Feed-Forward networks using H2O AutoML. The idea of AutoML [15, 16, 17] has emerged from automating the entire pipeline of machine learning that can automatically build high-quality custom models including task detection without human assistance. It is an open-source machine learning platform which trains and tunes various models. This user friendly tool is designed with two stopping criteria such as specific restricted time and/ or number of models to be trained. It widely uses statistical and machine learning algorithms. The goal of H2O AutoML is to obtain the best fit model by training and cross-validating various algorithms. In this automating tool, two stacked ensemble models are trained in which one ensemble has optimized models, and the other ensemble gives the best performing model. Then it returns a leaderboard of all models along with their results.

In our paper, we achieved that deep learning feed-forward networks (from DNNs class/family) performs better among all other models. So we have selected this network as our preferred model and have compared it with other existing state-ofart techniques. This proposed model automatically selects various parameters by itself, such as the number of layers, number of neurons, type of activation functions, etc. in such a way that it gives improved and best-performing result. It is observed that the best results are obtained when H2O AutoML model with 5 input nodes processed with 2 hidden layers each having 200 neurons. The details of parameters are shown in Table 57.2.

XGBoost Model. Gradient boosting is an approach in which new models are added to correct the errors originated by the prior models to produce strong classifier. XGBoost [18] is the implementation of gradient boosted decision trees. There are two main purpose of using XGBoost: first for its execution speed and second for its model performance. Generally, XGBoost is fast algorithm when compared with other gradient boosting techniques. It supports both classification and regression predictive modeling problems. In this model, 322 numbers of trees patterns with a maximum depth of 20 are used to achieve the results as shown in Table 57.2.

Support Vector Machines. SVM [5, 19, 20] is one of the easiest method to analyze data for classification and regression analysis. It is a discriminative classifier and a supervised machine learning algorithm that generally associated with separating hyperplanes. The basic idea of SVM is based on the concept of decision planes and decision boundaries. Suppose, there are N numbers of training samples, then SVM optimization is represented as:

$$f(x) = \sum_{j=1}^{N} \alpha_j y_i K(x_j, x) \tag{5}$$

With kernel:

$$K(x_j, x) = \exp\left(-\frac{\|x_j - x\|^2}{2\sigma^2}\right) \qquad (6)$$

where x_j is the jth input vector, $K(x_j, x)$ is the kernel, and α_j is SVM optimized value. SVM optimization maps data from input space to feature space. Here, we have used radial basis function as a kernel and penalty parameter (C) & kernel coefficient are set to their default value as 1.0 & auto respectively. The details of parameters are given in Table 57.2.

3. Result and Discussion

The overview of the dataset used in this study is shown in Table 57.1. Initially, the standardization of dataset is done and minority data are balanced by oversampling them. These preprocessed data are used to train the various models and corresponding to that testing is done to analyze the performance of various algorithms. The parameters used in various classifiers are shown in Table 57.2.

Table 57.1   Total instances of the dataset belong to each class

	Normal Class	Wilt Class
Training set	4265	74
Testing set	313	187

Table 57.2   Details of various networks and mechanisms

Deep feed-forward network using AutoML (Epochs:10)				
Parameters	**Input layer**	**Hidden layer**	**Hidden layer**	**Output layer**
No. of neuron Activation fn. Kernel	5 Relu Auto	200 tanh Auto	200 tanh Auto	1 Sigmoid Auto
Deep feed-forward network (Epochs:50)				
Parameters	**Input layer**	**Hidden layer**	**Output layer**	
No. of neuron Activation fn. Kernel	5 Relu Normal	3 Relu Normal	1 Sigmoid Normal	
XGBoost				
No. of trees Depth of tree				322 20
Support Vector Machine (SVM)				
Kernel Kernel coefficient Penalty parameter, C				rbf Auto 1.0

Here, we have considered two cases, one is evaluated with preprocessing and other is evaluated without preprocessing. The results of Deep Learning feed-forward networks using H2O AutoML, XGBoost, SVM, and Deep Learning feed-forward are obtained for both case. Table 57.3 below describes the experimental results and performance of each algorithm for both the cases. In this paper, accuracy and AUC are considered as parameters for evaluation. From the results, we can conclude that our proposed method obtains 88.8% of accuracy and 0.9553 AUC which gives better results in comparison to other methods.

Table 57.3 Comparison and performance of various methods.

Model	Preprocessing	Accuracy	AUC
Deep Learning Feed-Forward (H2O AutoML)	SMOTE + Standard-scaler	**88.8**	**0.9553**
Deep Learning Feed-Forward (H2O AutoML)	NO	72.2	0.9403
XGBoost	SMOTE + Standard-scaler	87.2	0.9547
XGBoost	NO	79.2	0.9412
Deep Learning Feed-Forward	SMOTE + Standard-scaler	87.8	0.9367
Deep Learning Feed-Forward	NO	78	0.9515
SVM	SMOTE + Standard-scaler	88.6	0.953
SVM	NO	80.8	0.9501

The performance of each method is evaluated on the basis of overall accuracy and AUC [21]. The basic difference between overall accuracy and AUC is that overall accuracy depends on a particular cut-point, while AUC tests all the cut-points and plots the graph between sensitivity and specificity. In general, the accuracy gives the percentage of the correct classifications at a specific threshold whereas AUC gives the overall performance of the classifier. AUC is the area under the ROC (Receiver Operating Characteristic) curve which is obtained by modifying the threshold from 0 to 1 in small steps, also specificity and sensitivity are measured for each value of the threshold. The figures on next page shows the comparison between four described methods on the basis of accuracy and AUC. Figure 57.3 shows the comparison of accuracy and Fig. 57.4 shows the comparison of AUC among various methods.

4. Conclusion and Future Work

From the previous section, it can be concluded that Deep Learning feed-forward network using AutoML gives better result in comparison to XGBoost, SVM, and Deep learning feed-forward networks. The accuracy and AUC of the proposed method are determined as 88.8% and 0.9553 respectively. The important fact about this model is that it automatically checks all the thresholds and selects the threshold that gives the best result. Similarly, it checks the number of neurons, a number of hidden layers, activation functions, etc and corresponds to the best suitable criteria that enhance the performance of the model.

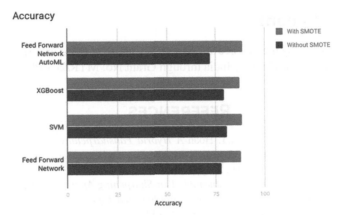

Fig. 57.3 Comparison of various methods using their accuracy

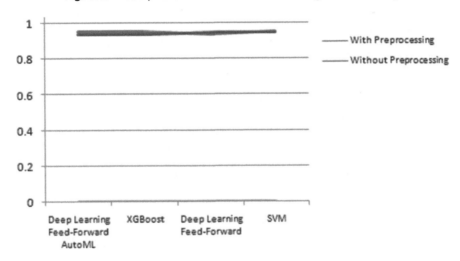

Fig. 57.4 Comparison of various methods using AUC

This field of remote sensing has a wide scope in future research. We can improve this work in various ways. Firstly, we can use aerial images besides satellite images that can be captured at lower altitudes with the help of drones or other devices. This will be helpful in many areas such as forest-fire monitoring, traffic monitoring, agricultural monitoring areas, search & rescue operations, number of trees counts in a particular region, etc. Secondly, we can use different preprocessing methods and deep learning models which help in training the networks and hence improves the performance and gives more accurate results in comparison to previous studies.

5. Acknowledgment

This work was supported by the Science and Research Engineering Board (SERB), Department of Science and Technology, Govt. of India through Grant No. MTR/2018/001103.

REFERENCES

1. Johnson B. A., R. Tateishi, and N.T. Hoan *A Hybrid Pansharpening Approach and Multiscale Object-Based Image Analysis for Mapping Diseased Pine and Oak Trees, International Journal of Remote Sensing*, (2013) Vol. 34, No. 20, pp. 6969–6982.
2. Tu T., C. Hsu, P. Tu, and C. Lee *An Adjustable Pan-Sharpening Approach for IKONOS/Quickbird/ GeoEye-1/WorldView-2 Imagery, IEEE Journal of Selected Topics in Applied Earth Observations and Remote Sensing*, 2012 Vol. 5, No. 1, pp. 125–134.
3. Johnson B., R. Tateishi, and N. T. Hoan *Satellite Image Pansharpening Using a Hybrid Approach for Object-based Image Analysis, ISPRS International Journal of Geo-Information*, (2012), Vol. 1, No. 3, pp. 228–241.
4. Tu T., S. Su, H. Shyu, and P. Huang *A New Look at IHS-Like Image Fusion Methods, Information Fusion*, (2001), Vol. 8, No. 2, pp. 177–186.
5. Trimble (2012) eCognition Developer 8.7.1, (2012) Reference book, pp. 34–38.
6. Li W., Q. Guo, and C. Elkan *A positive and Unlabeled Learning Algorithm for One-class Classification of Remote-sensing Data, IEEE Transactions on Geosciences and Remote Sensing*, (2011) Vol. 49, No. 2, pp. 717–725.
7. Muhammad S. M. P., Md Nasir S., N. Mustapha, and T. Perumal A New Classification Model for a Class Imbalanced Dataset using Genetic Programming and Support Vector Machines: Case Study for Wilt Disease Classification, (2015) *Remote Sensing Letters*, Vol. 6, No. 7, pp. 568–577.
8. L. Minh Dang, Syed I. Hussan, Im S., A. k. Sangai, Irfan M., S. Rho. S. Seo, H.Moon *UAV Based Wilt Detection System via Convolutional Neural Networks, Sustainable Computing: Informatics and Systems*, (2018) http://hdl.handle.net/10454/17186
9. Marian-D., Vasco, Elsa, Klaas and Nicolas, *A Machine Learning Approach to Detecting Pine Wilt Disease Using Airborne Spectral Imagery, Remote Sensing*, (2020) Vol. 12, No. 2280.
10. Johnson B., *UCI Machine Respiratory*, (2014) http://archive.ics.uci.edu/ml/datasets/wilt
11. Chawla, N., K. Bowyer, L. Hall, and W. Kegelmeyer, *SMOTE: Minority Over-Sampling Technique, Journal of Artificial Intelligence Research*, (2002) Vol. 16, pp. 321–357.
12. I. Goodfellow, Y. Benjio, and Aaron C., Deep Learning, *The MIT Press*, (2016) http://www.deeplearningbook.org
13. Y. LeCun, Y. Benjio, and G. Hinton Deep Learning, *Nature*, (2015) Vol. 521, No. 7553, pp.436–444
14. Min Ji, L. Liu and Manfred B., *Identifying Collapsed Buildings Using Post-Earthquake Satellite Imagery and Convolutional Neural Networks: A Case Study of the 2010 Haiti Earthquake, Remote Sensing*, (2018) Vol. 10, pp. 2072–4292.
15. AutoML, http://www.automl.org
16. Automated ML, https://www.ml4aad.org/automl/
17. AutoML: A Survey of the State-of-the-Art, https://arxiv.org/abs/1908.00709v5
18. XGBoost, https://xgboost.readthedocs.io
19. Rukshan B. and V. Palade Class Imbalance Learning Methods for Support Vector Machines, *Singapore-MIT Alliance for Research and Technology Centre, University of Oxford.*

20. J. Munoz Mari, J., F. Bovolo, L. Gomez-Chova, L. Bruzzone, and G. C. Walls (2010) *Semisupervised one-class support vector machines for classification of remote sensing data, IEEE Trans. Geoscience Remote Sensing*, Vol. 48,No. 8 pp. 3188–3197.

21. Jin Huang and C. X. Ling (2005) *Using AUC and Accuracy in Evaluating Learning Algorithms, Transactions on Knowledge and Data Engineering*, Vol. 17, No. 3, pp. 299–310.

Intelligent Systems and Smart Infrastructure – Brijesh Mishra et al. (eds)
© 2023 Taylor & Francis Group, London, ISBN 978-1-032-41287-0

CHAPTER

58

Mathematical Analysis of Self-Organizing Maps for Clustering

Femy N. S.[1], Sasi Gopalan[2]

Department of Mathematics,
Cochin University of Science and Technology,
Cochin-22, Kerala, India

Oscar Castillo[3]

Tijuana Institute Technology,
Tijuana, Mexico

Abstract

In this paper, mathematical analysis of an unsupervised machine learning technique called Self-Organizing Maps (SOM) is done for clustering. The publicly available data on COVID-19 in various countries worldwide is used for analysis. The countries with similar characteristics are clustered together, and spatial analysis of coronavirus cases is done using Self-Organizing Maps. This will make it easier to apply the same pandemic-control strategy to all nations with similar characteristics. The mathematical analysis of SOM is a significant contribution to this paper, as it enables similar group countries to use the same tactics to combat viral transmission. For the experimental study, the available data on COVID-19 is taken from Humanitarian Data Exchange.

Keywords: Self-Organizing Maps, Coronavirus, Random sample initialization, Linear initialization.

Corresponding author: [1]femyleny2009@gmail.com, [2]sgcusat@gmail.com, [3]ocastillo@tectijuana.mx

DOI: 10.1201/9781003357346-58

1. Introduction

Coronavirus disease is a communicable respiratory disease that has spread throughout the world (Shereen et al. 2020). COVID-19 had a significant influence on the health system, particularly the patients with cardiovascular diseases, diabetes, and kidney diseases. The presence of pre-existing cardiac conditions increases the chance of complications, including mortality, in coronavirus-affected patients (Bandyopadhyay et al. 2020). Diabetes and chronic kidney diseases are factors for the prognosis and progression of COVID-19 (Guo et al. 2020, Askari et al. 2021). As it is now essential to take protective measures to control the virus, in our study, we have done clustering of countries with coronavirus cases using Self-Organizing Maps (SOM).

The paper's major contribution is the mathematical analysis of SOM, which helps in the spatial analysis of coronavirus cases. This will group the countries with similar characteristics so that the same strategy can be applied to the same group of countries against the spread of virus.

The paper is organized as follows. A brief review of the literature on SOM is done in Section 2. Section 3 explains the fundamentals of SOM. Section 4 deals with the mathematical analysis of SOM that has been utilized for clustering countries in the world. The spatial analysis of coronavirus cases using Self-Organizing Maps is done in Section 5. Section 6 provides results and discussions. Conclusion and future works are discussed in Section 7.

2. Literature Review

This section has discussed different machine learning techniques applied for COVID-19 and related works on Self-Organizing Maps. The study in the paper of Melin et al. (2020) gives an idea about the spread of coronavirus pandemic in the world using SOM. Another analysis is made by developing the Gaussian regression model for predicting coronavirus by Ketu and Mishra (2021). A COVID-19 forecasting model is developed by Saqib (2021). The other cases are: a convolutional neural network for classifying coronavirus in chest X-ray images by Abbas et al. (2021), the Susceptible-Infected-Removed model to investigate the spread of COVID-19 inside a community by Cooper et al. (2020), a literature survey on the application of artificial intelligence in fighting against coronavirus cases by Mohammad and Tayarani (2021). The other works related to deep learning are by Rachna and Meenu (2021), Elkornany and Elsharkawy (2021). Also, soft computing techniques are applied in this area by Bhardwaj and Bangia (2020).

In Clarka et al. (2020), Self-Organizing Maps are used in water resources research and engineering. It is also applied by Dong et al. (2021) for image segmentation and by Neisari et al. (2021) for Spam review detection, Chen et al. (2020) for high-impact malicious tasks submission in mobile, Ruiz-del-Solar (1998) for texture segmentation, and by Liukkonen and Hiltunen (2018) for recognizing the systematic Spatial patterns in Silicon Wafers. The other applications of SOM related to COVID-19 are done by Yu et al. (2021) and Souza et al. (2021). In this work, we have used SOM to cluster countries worldwide according to groups of coronavirus cases.

3. Preliminaries

A. Self-Organizing Maps

SOM is an unsupervised artificial neural network that helps visualize data of higher dimensions in a lower dimensional space (Kohonen 2001). In the input space of SOM, the data vectors, $x_i, i = 1, 2, ..., N_1$; where N_1 is the number of input data vectors, are vectors in the n-dimensional space $\mathbb{R}^n$ In the output space, the weight vector for each node is $u_j, j = 1, 2, ..., m_1$, where m_1 is number of nodes in output space. The distance from data vector to weight vector corresponding to each node in the output space is computed. The node having minimum distance will be the winner of the competition.

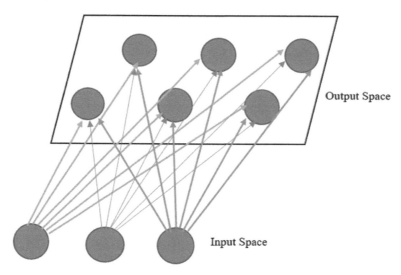

Fig. 58.1 Structure of SOM

4. Methodology

In this section, the mathematical analysis is done on the SOM implementation using the Minisom package in Python 3.8.

Result 1: Let $\vec{x}_i$ and $\vec{u}_j$ be two vectors in $\mathbb{R}^n$ where $\vec{x}_i$ and $\vec{u}_j$ are the input vectors and weight vectors in SOM. Then there exists a metric $d: \mathbb{R}^n \times \mathbb{R}^n \to \mathbb{R}$ such that $d(\vec{x}_i, \vec{u}_j) = \|\vec{x}_i - \vec{u}_j\|$, $i = 1, 2, ..., N_1, j = 1, 2, ..., m_1$.

Remark 1: By result 1, the minimum distance from input vector to weight vector in the output space is $\min_j \|\vec{x}_i - \vec{u}_j\|$ and the index of winning neuron is $Arg \min_j \|\vec{x}_i - \vec{u}_j\|$.

Result 2: Let $\mathbb{R}^n$ be a vector space and W be the set of all whole numbers. Then, corresponding to SOM, the map $f: \mathbb{R}^n \times \mathbb{R}^n \to W \times W$, defined by $f(\vec{x}_i, \vec{u}_j) = arg \min_j \|\vec{x}_i - \vec{u}_j\|$, $i = 1, 2, ..., N_1, j = 1, 2, ..., m_1$, is not one-to-one.

While considering the map, the input vectors with similar characteristics are clustered in one neuron. Hence the mapping is not one-to-one.

The number of nodes in output space of SOM and number of iterations in SOM are taken as in the literature.

Remark 2: In the output layer of SOM the number of nodes is $5 \times \sqrt{\text{number of input data vectors}}$ (Vcsanto and Alhoniemi 2000).

Remark 3: The number of iterations in SOM should be at least 500 times the number of output neurons, according to a rule of thumb (Kohonen 2001).

The different parameters used in SOM implementation are map size, input length, learning rate, neighborhood radius, and neighborhood function.

Definition 1: (Kohonen 2001) The learning rate $\alpha: \mathbb{N} \to [0, 1)$ is defined by $\alpha(t) = \dfrac{\alpha(0)}{1 + \dfrac{t}{T}}$, where $\alpha(0)$ is the initial learning rate, t is the current iterations, T is the $\dfrac{\text{number of iterations}}{2}$,

which is a decreasing function of time.

The learning rate is a training parameter that controls the size of weight vectors in the learning of SOM.

Definition 2: (Kohonen 2001) The neighborhood radius of SOM $\sigma: \mathbb{N} \to [0, \infty)$ is defined by $\sigma(t) = \dfrac{\sigma(0)}{1 + \dfrac{t}{T}}$, where $\sigma(0)$ is the initial radius, t is current iteration, T is the $\dfrac{\text{number of iterations}}{2}$.

Definition 3: (Kohonen 2001) The neighborhood function of SOM $h_{jc}: \mathbb{N} \to (0, 1]$ is defined as $h_{jc}(t) = e^{-\dfrac{d_{jc}^2(t)}{2\sigma^2(t)}}$, where $\sigma(t)$ is the neighborhood radius, and $d_{jc}(t)$ is the horizontal distance between the indices of winner neuron v_c and neighboring neuron v_j.

h_{jc} satisfies the following properties:
(i) h_{jc} is symmetric and $h_{jj} = 1$.
(ii) h_{jc} depends only on the distance d_{jc} between neurons v_c and v_j and decreases with the increasing distance.

By definitions 1, 2, and 3, the weight updating is done using (1)

$$\vec{u}_j(t+1) = \vec{u}_j(t) + \alpha(t)h_{jc}(t)(\vec{x}_i(t) - \vec{u}_j(t)) \tag{1}$$

where $\vec{x}_i(t)$ is the input vector, $\vec{u}_j(t)$ and $\vec{u}_j(t+1)$ are the weight vectors at time t and $t + 1$ respectively, $\alpha(t)$ is the learning rate, $h_{jc}(t)$ is neighborhood function.

Weight initialization is significant in the implementation of SOM. The two important weight initialization methods in SOM are Random Sample Initialization (RSI) and Linear Initialization (LI) (Attik et al. 2005, Akinduko et al. 2016). In RSI, the weight vectors are taken randomly from input vectors, whereas in LI, the weight vectors are taken from the linear combination of two eigenvectors corresponding to the biggest two eigenvalues of input data vectors. In this work, LI is used in clustering countries around the world. The covariance matrix of the input data is computed in LI. The data dispersion and direction are determined by the covariance matrix. The linear transformation of input data is closely related to the covariance matrix of input data. As a result, if $A = cov(X)$, then $A\vec{v} = \lambda\vec{v}$ denotes λ the eigenvalue and $\vec{v}$ the eigenvector. The eigenvector depicts the data's biggest variance directions, while the eigenvalues depict the amount of variance in these directions. In LI of SOM, two eigenvectors, $\vec{v}_1$ and $\vec{v}_2$ related to two biggest eigenvalues of covariance matrix, are considered.

Lemma 1: Let $\vec{v}_1$ and $\vec{v}_2$ be two eigenvectors related to two biggest eigenvalues of covariance matrix for input vectors in $\mathbb{R}^n$, then $W = \text{Span}\{\vec{v}_1, \vec{v}_2\}$ is a subspace of $\mathbb{R}^n$ in linear initialization of self-organizing maps.

Lemma 2: $\{\vec{v}_1, \vec{v}_2\}$, stated in Lemma 1 form an orthogonal basis for $W = \text{Span}\{\vec{v}_1, \vec{v}_2\}$, in $\mathbb{R}^n$

The proof trivially follows as the eigenvectors of the covariance matrix are orthogonal.

A projection of input vectors from input layer to output layer is proven by orthogonal decomposition theorem (Limaye 2013), and by the Best approximation theorem (Limaye 2013), it is shown that in the output space, this projection is nearest to input vector. The following are the theorems using the orthogonal projection theorem, and best approximation theorem.

Theorem 1: Let $\vec{v}_1$ and $\vec{v}_2$ be two eigenvectors related to two biggest eigenvalues of covariance matrix for input vectors in $\mathbb{R}^n$, and let $W = \text{Span}\{\vec{v}_1, \vec{v}_2\}$, is a subspace of $\mathbb{R}^n$. Then, each input vector $\vec{x}$ in $\mathbb{R}^n$ can be expressed uniquely in the form $\vec{x} = \hat{x} + \vec{z}$ where $\hat{x}$ is in W and $\vec{z}$ in $W^{\perp}$.

If $\{\vec{v}_1, \vec{v}_2\}$ is an orthogonal basis of W, then $\hat{x} = \left(\dfrac{\langle \vec{x}, \vec{v}_1 \rangle}{\langle \vec{v}_1, \vec{v}_1 \rangle}\right)\vec{v}_1 + \left(\dfrac{\langle \vec{x}, \vec{v}_2 \rangle}{\langle \vec{v}_2, \vec{v}_2 \rangle}\right)\vec{v}_2$ and $\vec{z} = \vec{x} - \hat{x}$.

The orthogonal projection of $\vec{x}$ onto W is vector $\hat{x}$.

Theorem 2: Let $W = \text{Span}\{\vec{v}_1, \vec{v}_2\}$ be a subspace of $\mathbb{R}^n$, $\hat{x}$ is orthogonal projection of $\vec{x}$ onto W, where $\vec{x}$ is an input vector in $\mathbb{R}^n$. Then, $\hat{x}$ is the nearest point to $\vec{x}$ in W, such that $\|\vec{x} - \hat{x}\| < \|\vec{x} - \vec{y}\|$ for all $\vec{y}$ in W that are different from $\hat{x}$.

5. Simulation Results

This section uses SOM to group countries according to high, medium, low, and very low death, confirmed, and recovered COVID-19 cases. The database for analysis is taken from Humanitarian Data Exchange (HDX, 2022). It contains time-series data of coronavirus-affected countries from March 1, 2020, to August 3, 2021. The dataset contains the death, recovered, and confirmed cases of countries for 560 days. Hence, an average of 14 days is taken as one column, and a new dataset is formed with 40 columns for the analysis. Each row of the dataset represents data from one country, and it is 40-dimensional. This high-dimensional data is visualized in a 2-dimensional plane using SOM.

A. Experimental Analysis of Clustering Countries

The analysis of the coronavirus death, confirmed, and recovered cases are made using the Minisom package in Python (Vettigli 2018). Since the dataset consists of 274 countries, by Remark 2, the number of output neurons is $5 \times \sqrt{274} \approx 83$. Hence, 81 neurons are considered in the SOM implementation, and the map size is taken as 9×9. By Remark 3, the number of iterations is $500 \times 81 = 40500$. The neighborhood radius in the Minisom package is by default 1; the learning rate is 0.5, and the neighborhood function is a Gaussian function. The different parameters used in SOM are shown in Table 58.1.

Table 58.1   Parameters used in SOM implementation

Dataset	N_1	M	α	σ	T	n
Covid confirmed	274	9×9	0.5	1	40500	40
Covid death	274	9×9	0.5	1	40500	40
Covid recovered	259	9×9	0.5	1	40500	40

(a) N_1-Number of inputs
(b) M-Map size
(c) α-Learning rate
(d) σ-Neighborhood radius
(e) T-Number of iterations
(f) n-dimension of input data

Analysis of confirmed, death, and recovered coronavirus cases globally

The SOM implementation of confirmed, recovered, and death coronavirus cases is shown in Fig. 58.2(a)-(c). For each pattern in the dataset, the corresponding winning neuron is marked

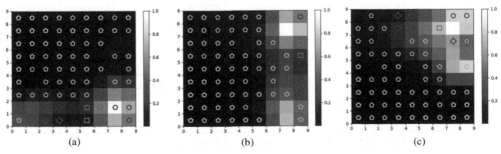

   (a)   (b)   (c)

Fig. 58.2   SOM implementation for (a) confirmed, (b) recovered, and (c) death coronavirus cases

Marker color represents different groups of COVID-19 cases:
 (a) Red-high   (b) Green-medium   (c) Blue-low   (d) Yellow-very low

Link for colored figure: https://drive.google.com/file/d/1P3aK18OiAyKNPrg3AB_kGAxLBRDk0XDk/view?usp=sharing

using a marker. Each type of marker represents a cluster of high (red), medium (green), low (blue), and very low (yellow) confirmed coronavirus cases. The mean interneuron distance of weights of the winning neuron and its neighbors is used in the background (the values are shown in the color bar on the right). The white boxes represent high interneuron distances (countries with higher coronavirus cases), and the black represents low interneuron distances (countries with low coronavirus cases).

Based on this implementation of SOM, the result is applied to group countries around the world according to high, medium, low, and very low coronavirus cases, which are indicated by the decreasing order of intensity of brown color in Fig. 58.3. Table 58.2(a) shows that the countries US, India, and Brazil have high confirmed coronavirus cases, the Russia, France, UK, and Turkey have medium confirmed cases, etc. Table 58.2(c) shows that India and Brazil have high recovered coronavirus cases, Russia, and Turkey have medium recovered cases, etc. Table 58.2(b) shows that the US and Brazil have high death cases; India and Mexico have medium death cases, etc. This is indicated by dark and light brown colors, respectively, in Fig. 58.3(a)-(c). The cluster of countries formed using the SOM method, clearly showing the total confirmed, recovered, and death cases from July 21, 2021, to August 3, 2021, is visualized in Fig. 58.3(a)-(c).

Table 58.2 Number of (a) confirmed, (b) death, and (c) recovered coronavirus cases (up to August 3, 2021)

(a)

Cluster	Country	Cases
High	US	34808698
	India	31510967
	Brazil	19775608
Medium	Russia	6105000
	France	5963713
	UK	5757495
	Turkey	5659149
Low	Argentina	4882261
	Colombia	4749721
	Spain	4371864
	Italy	4330098
	Iran	3776769
	Germany	3768927
	Indonesia	3259246
Very Low	Poland	2882466
	Mexico	2786095
	....	....

(b)

Cluster	Country	Cases
High	US	611596
	Brazil	552801
Medium	India	422369
	Mexico	239436
Low	Peru	196062
	Russia	153251
	UK	129372
	Italy	128003
	Colombia	119578
	France	110730
	Argentina	104609
Very Low	Germany	91603
	Iran	89652
	....	....

(c)

Cluster	Country	Cases
High	India	30679041
	Brazil	17678476
Medium	Russia	5469776
	Turkey	5435485
Low	Colombia	4527556
	Argentina	4523427
	Italy	4128258
	Germany	3648806
	Iran	3303009
Very Low	Poland	2653463
	Indonesia	2618722
	....	....

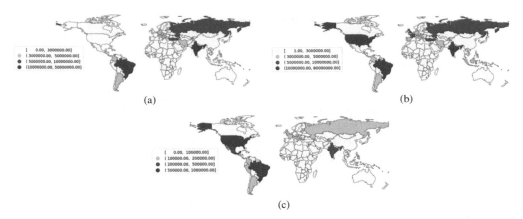

Fig. 58.3 Total (a) recovered, (b) confirmed, and (c) death coronavirus cases from 7/21/2021 to 8/03/2021

Link for colored figure: https://drive.google.com/file/d/1Rbm8IX50M0H5bY18pFZKdTz4RCkkCS0n/ view?usp=sharing

6. Results and Discussion

From Fig. 58.2(b) and Table 58.2(a), it is found that the US, India, and Brazil have high confirmed cases; Russia, France, UK, and Turkey have medium confirmed cases, and so on. From Fig. 58.2(a) and Table 58.2(c), India and Brazil have high recovered cases; Russia and Turkey have medium recovered cases, and so on. (Data from US was not available). From Fig. 58.2(c) and Table 58.2(b), US and Brazil have high death cases; India and Mexico have medium death coronavirus cases, and so on. India has high confirmed and high recovered cases but medium death cases. Brazil has high confirmed, high death, and high recovered cases. Russia has medium confirmed, and medium recovered cases, but low death cases. Mexico has very low confirmed cases, very low recovered cases, but medium death cases. This analysis assists each country's government in taking the necessary steps to prevent the epidemic from spreading.

7. Conclusion

The mathematical analysis of an unsupervised artificial neural network known as Self-Organizing Maps for clustering is presented in this paper. This is used in spatial analysis of coronavirus-affected countries. Countries with similar features are grouped together so that similar countries can use the same pandemic control strategies. The available data from Humanitarian data exchange is taken for experimental analysis. In the future, other methods such as fuzzy logic, evolutionary algorithms, particle swarm optimization, etc., can be used to handle this complex problem. Also, other works involving fuzzy models (Zhou et al. 2021, Castillo et al. 2021) can be used to deal with the problem.

REFERENCES

1. Shereen, M. A., Khan, S., Kazmi, A., Bashir, N., & Siddique, R. (2020). COVID-19 infection: Emergence, transmission, and characteristics of human coronaviruses. *Journal of advanced research*, *24*, 91–98.

2. Bandyopadhyay, D., Akhtar, T., Hajra, A., Gupta, M., Das, A., Chakraborty, S., ... & Naidu, S. S. (2020). COVID-19 pandemic: cardiovascular complications and future implications. *American Journal of Cardiovascular Drugs*, *20*(4), 311–324.

3. Guo, W., Li, M., Dong, Y., Zhou, H., Zhang, Z., Tian, C., ... & Hu, D. (2020). Diabetes is a risk factor for the progression and prognosis of COVID-19. *Diabetes/metabolism research and reviews*, *36*(7), e3319.

4. Askari, H., Sanadgol, N., Azarnezhad, A., Tajbakhsh, A., Rafiei, H., Safarpour, A. R., ... & Omidifar, N. (2021). Kidney diseases and COVID-19 infection: causes and effect, supportive therapeutics and nutritional perspectives. *Heliyon*, *7*(1), e06008.

5. Melin, P., Monica, J. C., Sanchez, D., & Castillo, O. (2020). Analysis of spatial spread relationships of coronavirus (COVID-19) pandemic in the world using self organizing maps. *Chaos, Solitons & Fractals*, *138*, 109917.

6. Ketu, S., & Mishra, P. K. (2021). Enhanced Gaussian process regression-based forecasting model for COVID-19 outbreak and significance of IoT for its detection. *Applied Intelligence*, *51*(3), 1492–1512.

7. Saqib, M. (2021). Forecasting COVID-19 outbreak progression using hybrid polynomial-Bayesian ridge regression model. *Applied Intelligence*, *51*(5), 2703–2713.

8. Abbas, A., Abdelsamea, M. M., & Gaber, M. M. (2021). Classification of COVID-19 in chest X-ray images using DeTraC deep convolutional neural network. *Applied Intelligence*, *51*(2), 854–864.

9. Cooper, I., Mondal, A., & Antonopoulos, C. G. (2020). A SIR model assumption for the spread of COVID-19 in different communities. *Chaos, Solitons & Fractals*, *139*, 110057.

10. Tayarani, M. (2020). Applications of artificial intelligence in battling against covid-19: A literature review. *Chaos, Solitons & Fractals*.

11. Jain, R., Gupta, M., Taneja, S., & Hemanth, D. J. (2021). Deep learning based detection and analysis of COVID-19 on chest X-ray images. *Applied Intelligence*, *51*(3), 1690–1700.

12. Elkorany, A. S., & Elsharkawy, Z. F. (2021). COVIDetection-Net: A tailored COVID-19 detection from chest radiography images using deep learning. *Optik*, *231*, 166405.

13. Bhardwaj, R., & Bangia, A. (2020). Data driven estimation of novel COVID-19 transmission risks through hybrid soft-computing techniques. *Chaos, Solitons & Fractals*, *140*, 110152.

14. Clark, S., Sisson, S. A., & Sharma, A. (2020). Tools for enhancing the application of self-organizing maps in water resources research and engineering. *Advances in Water Resources*, *143*, 103676.

15. Dong, B., Weng, G., & Jin, R. (2021). Active contour model driven by Self Organizing Maps for image segmentation. *Expert Systems with Applications*, *177*, 114948.

16. Neisari, A., Rueda, L., & Saad, S. (2021). Spam review detection using self-organizing maps and convolutional neural networks. *Computers & Security*, *106*, 102274.

17. Chen, X., Simsek, M., & Kantarci, B. (2020). Locally reconfigurable self organizing feature map for high impact malicious tasks submission in mobile crowdsensing. *Internet of Things*, *12*, 100297.

18. Ruiz-del-Solar, J. (1998). TEXSOM: Texture segmentation using self-organizing maps. *Neurocomputing*, *21*(1-3), 7–18.

19. Liukkonen, M., & Hiltunen, Y. (2018). Recognition of systematic spatial patterns in silicon wafers based on SOM and K-means. *IFAC-PapersOnLine*, *51*(2), 439–444.

20. Yu, Z., Arif, R., Fahmy, M. A., & Sohail, A. (2021). Self organizing maps for the parametric analysis of COVID-19 SEIRS delayed model. *Chaos, Solitons & Fractals*, *150*, 111202.
21. Souza, A. A. D., Almeida, D. C. D., Barcelos, T. S., Bortoletto, R. C., Munoz, R., Waldman, H., ... & Silva, L. A. (2021). Simple hemogram to support the decision-making of COVID-19 diagnosis using clusters analysis with self-organizing maps neural network. *Soft Computing*, 1–12.
22. Kohonen, T. (2001). Self-organizing Maps.-Springer Series in Information Sciences, V. 30, Springer.
23. Attik, M., Bougrain, L., & Alexandre, F. (2005, September). Self-organizing map initialization. In *International Conference on Artificial Neural Networks* (pp. 357–362). Springer, Berlin, Heidelberg.
24. Akinduko, A. A., Mirkes, E. M., & Gorban, A. N. (2016). SOM: Stochastic initialization versus principal components. *Information Sciences*, *364*, 213–221.
25. Limaye, B. V. (2013). *Functional analysis*. New Age International.
26. Vesanto, J., & Alhoniemi, E. (2000). Clustering of the self-organizing map. *IEEE Transactions on neural networks*, *11*(3), 586–600.
27. The Humanitarian Data Exchange, [Online]. [Accessed March 1, 2022]. Available: https://data.humdata.org/dataset/novel-coronavirus-2019-ncov-cases.
28. Giuseppe Vettigli. (2018). MiniSom: minimalistic and NumPy-based implementation of the Self-Organizing Map. https: //github.com/JustGlowing/minisom/
29. Zhou, J., Pedrycz, W., Yue, X., Gao, C., Lai, Z., & Wan, J. (2021). Projected fuzzy C-means clustering with locality preservation. *Pattern Recognition*, *113*, 107748.
30. Castillo, O., & Melin, P. (2021). A new fuzzy fractal control approach of non-linear dynamic systems: The case of controlling the COVID-19 pandemics. *Chaos, Solitons & Fractals*, *151*, 111250.

Intelligent Systems and Smart Infrastructure – Brijesh Mishra et al. (eds)
© 2023 Taylor & Francis Group, London, ISBN 978-1-032-41287-0

CHAPTER

59

Green Computing:
An Ecofriendly Technology

Rupal Singh[1], Niharika[2], Pooja Dehraj[3]

Department of Computer Science and Engineering,
Noida Institute of Engineering and Technology, Noida, India

Abstract

Currently, computer systems are the basic need for various purposes like study, business, job, storage, and other purposes. Now, computers are not only used in offices for business purposes but they are used in homes also. As the usage of computers rapidly increases, the energy consumption through computers is also increasing and due to more energy consumption, the emission of carbon footprints is increasing in environment, which is harmful for ecosystem. Due to this concern, many IT product manufacturing industries start using the green computing technology. Green computing is an ecofriendly and sustainable computing. The main objective of green computing is to save the environment and maintain sustainability.

Keywords: Green Computing, Green Technology, Carbon Footprints, IT Products

1. Introduction

The revolution of industries leads to the exploitation of fossil fuels and this also becomes the main reason for polluting the environment because burning of fossil fuels emits carbon footprints and water footprints. It took a long time to realize that the industrial revolution harming the environment and then the green technology approaches are the need [1].

It is common knowledge that "Go Green" is trending all over the world to make our environment pollution-free. The main purpose of "Go Green" is to make aware people of the

Corresponding author: [1]rupalsinghriet999@gmail.com, [2]niharikayad23@gmail.com, [3]drpoooja.cse@niet.co.in

DOI: 10.1201/9781003357346-59

rising pollution and to save the earth from pollution [1]. If we talk about electronic devices, radiations are emitted which are very harmful to the ecosystem, that's why green computing is in existence to use electronic devices mainly computer-related resources in an eco-friendly manner. Sometimes, the term green computing is also known as green IT or green technology and this technology is the study or practice of environmentally sustainable computing. Green computing says that computers and their other resources should be used in an eco-friendly manner. Green computing is the technology that uses the eco-friendly approach to protect the environment from harmful radiations and gases like carbon dioxide that are emitted through computers [2].

The green computing technology can be characterized as the methods of utilizing computer resources and arrangement of servers in an efficiently way with very less or almost negligible impact on environment [3]. Many companies are using the approach of green computing to save the environment from carbon footprints released through computers and its resources. If talking about laptops, "Dell" company uses the approach of green computing. Dell laptops are automatically going into sleep mode when they are not in use which results in less consumption of energy and avoid unnecessary energy consumption and Dell outlet refurbishes the products like laptops to maintain sustainability because the sustainability of IT products is the concerned subject of green computing so its works on it. If talking about operating systems, the "Linux" operating system has used the approach of green computing [4].

A. History of Green Computing

In 1992, an American agency named Environment Protection Agency (EPA) launched a program named Energy Star. This Energy Star is a labeling program that awarded products that succeed in maximizing efficiency by minimizing the use of energy. The Energy Star program is applied to electronic products like printers, televisions, computers, etc., and applied to temperature control devices used like A.C., refrigerators, etc. Energy Star promotes energy efficiency products [5]. Energy Star created the sleep mode for devices by which they consumed less energy because the maximum energy consumption results in maximum emission of carbon gas, and carbon gases are very harmful to the ecosystem, and it affects global warming. Later on, the green star program is known as green computing [6]. Green computing mainly works on IT products to maintain the sustainability and to save the energy.

The Environment Protection Agency (EPA) is still providing energy-efficient methods for products and it also plays an active and important role for consumers by cost-effective methods. EPA establishes a way to save U.S. households and business money in 2006. EPA from programs like Energy Star expects to prevent the emissions of greenhouse gases from electronics products. Some organizations get the certifications for the leading groups in green computing and two of them are European Union organizations and TCO [7].

B. Objectives of Green Computing

There are various reasons behind the initiative of green technology but the main reason is the concern of pollution that increased by using electronic and digital devices. Due to this polluted nature, many manufacturers started to make biodegradable IT products and electronic devices to save the environment. There are other various objectives of green computing and some of them are mentioned below:

- To reduce the implementation of hazardous materials.
- To maximize the energy efficiency of electronic devices.
- To maintain the sustainability of IT products and other electronic devices.
- To promote and implement the biodegradable IT and electronic devices [6].
- To promote the recyclability of the e-wastes or electronic wastes to maintain the sustainability of products.

C. Importance of Green Computing

Everyone was aware that our environment is getting destroyed or damaged by the resources which we are using and that it is due to the resources which are not renewable; then humans started realizing that we have to do something so that we can protect our environment for the future [7]. Here comes the green computing:

- It's always difficult to renew or recycle computers or any electronic device so that's why green computing helps electronic devices to stay for a long time with the help of virtualization.
- Green computing also financially helps us, when we are using green computing techniques, we can save more money with the help of energy and resources which are applied to the device.
- Green computing also helps in reducing the carbon footprints for the business organizations and also helps the program to get implement easily and with more energy efficiency.
- For saving the environment from pollution, we start using green techniques [7].

2. Various Types of Green Computing

Green computing addresses a way to save the earth from global warming by using natural resources that do not emit harmful radiation like carbon dioxide. By using the natural resources, the green technology implements for generating electricity, and these implementations are mentioned below as the types or can say categories of green computing:

A. Solar Power System

The solar power system is the technique to transfer solar radiation into electrical energy. In a solar panel, the sun radiations are absorbed by the photovoltaic cells and then this generates electric charges and through that, the electric current is generated. The solar panel is made up of silicon cells, a glass case, a metal frame, and a wire connect to it which is used to transfer the electric current [8]. By using solar power system, generating electric current results in reducing the use of coal and fuel, and due to this, environment pollution can be reduced.

Fig. 59.1 Solar panels to generate electricity by absorbing the sunlight [10]

The solar power system can be installed for commercial purposes, business purposes, and also for personal purposes because it is available on market. The solar power system is the one and most widely used category of the green computing technique.

B. Geothermal Power

Geothermal energy is the technique in which the heat (thermal energy) stored in the earth's crust is used to generate electricity. It is common knowledge that thermal energy exists on the earth for millions of years and the geothermal energy technique uses a rich amount of thermal energy that is underground to generate electricity. Geothermal energy plants are specifically used on the site. The implementation of geothermal energy plants is quite cheap, especially when it comes to generating direct heating. The implementation of geothermal energy plants is quite challenging because the temperature of underground heat is very high [11].

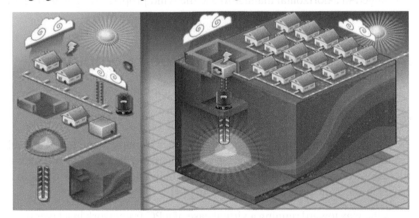

Fig. 59.2 Geothermal plant is implemented to generate electricity [11]

Geothermal energy plants come under the category of green technology because it does not emit any harmful gases and produces very less amount of carbon dioxide that does not pollute or harm the environment system.

C. Wind Energy

Wind energy is one of the categories that come under green computing technology because it does not harm the environment and it is renewable energy. The wind is simply defined as the air that is moving in the environment. The wind is formed when the earth is heated by the sun rays, which means it's a balanced reaction to control the heat of the earth. Wind energy or wind power is the method by which the wind is used to generate electricity. Wind energy is the method in which kinetic energy is converted into mechanical energy with the help of wind turbines by friction force used in wind power implementation [11].

Fig. 59.3   Wind turbines in plant to generate electricity using wind energy [12]

The wind energy plant consists of wind turbines, as shown in Fig. 59.3. These wind turbines are machines that consist of a rotor and three pro-parallel blades. These wind turbines are implemented in the place where the speed of the wind is high. The blades used in the wind turbines are arranged in a horizontal manner to use the wind properly for generating electricity [13].

3. Approaches for Implementing Green Computing

There are several approaches for implementing green computing:

A. By Virtualization

Virtualization is one of the main strategies and approaches to implement the green technology in which the data centre power consumption can be reduced and the use of IT products reduced. To reduce the use of computers-related resources, virtualization allows the users to work on virtual machines in the same host operating system [14].

Virtualization is the way toward running a virtual case of a PC framework in a layer preoccupied from the genuine equipment. It runs the various operating system on a single machine, and at the same time that helps to reduce the overall time. Virtualization helps in green computing as it helps in saving the power of the system by performing multiple tasks. Virtualization is the response to settling the force utilization of server farms. The main task of virtualization is to save energy or reuse the energy for various tasks. When the number of workers is diminished, it likewise implies that server farms can reduce the structure size too.

Some assets of the virtualization which directly hit on productivity and add to the climate consist the following[14]:

- Arranged personal time as drop by moving a virtual machine starting with one actual work and then by the other workers.
- Strongly adjusted jobs across a worker bunch and give natural failover to virtualized applications.
- Asset assignment is better overseen and kept up with.
- Virtualization dramatically increases a worker gathering's capacity to share utility.
- Worker usage rates can be incremented by up to 80% rather than an underlying 10% to 15%.

B. Use ACPI

ACPI stands for Advanced Configuration and Power Interface. ACPI has been started from Windows 98 and further in all upcoming operating systems. This gives considerably more steady and productive force to the board and makes it workable for the working framework to kill chosen gadgets, for example, a screen or CD-ROM drive, when they are not being used. ACPI should assist with taking out PC lockup on entering power-saving or rest mode. This will consider further developed force the board, particularly in compact PC frameworks where decreasing force utilization is basic for broadening battery life [14].

ACPI is embedded in all the operating systems nowadays which helps devices to save the battery life of the devices and we can use the devices for a large period. ACPI also has existing plug-and-play BIOS to add the new hardware devices easily within the operating system. It also helps to put some hardware in sleep mode at the time when it is not in use.

C. Use Star Lablled Products

Energy `labelling' is quite possibly the savviest strategy device for further developing energy productivity and bringing down the energy cost of apparatuses/hardware for the purchasers. Energy names can be utilized independently or supplement energy guidelines. As well as giving data that permits customers who care to choose productive models, labels additionally give a typical energy-proficiency benchmark that can work in relationship with other approach measures. The products that are registered with the bureau get a star rating ranging from 1 to 5 on the basis of their energy efficiency. An endorsement label is also provided for some products` [15].

D. Use LCD Reather Then CRT

LCD stands for liquid-crystal display and CRT stands for a cathode-ray tube. Using LCD is better than using CRT because it uses less power and it helps to reduce carbon dioxide. Today most people have replaced CRT with LCD. There are many advantages of LCD like space, power consumption, brightness, and screen flicker.

- LCD covers less space as compared to CRT because LCD has flat plane.
- Energy utilization of LCD screens increments as screen size increments, yet at the same time remains essentially lower than that of CRT screens.
- LCD has more brightness than CRT because energy utilization of LCD screens increments as screen size increments, yet at the same time remains essentially lower than that of CRT screens.
- LCD screens are prepared to do a lot higher revive rates, with paces of 75 and 85 Hertz being normal [16].

4. Conclusion

It has been seen that green computing is becoming very important for today's world according to the pollution situation increasing day by day. As in today's world, every person needs their individual computer or any electronic device so here comes the use of green computing. It helps us in keeping the environment neat and clean. It helps in global warning issues and helps the devices to stay for a longer time. Nonetheless, on the grounds that processing improvements can create awareness and inspire people and organizations to receive and use greener ways of life and work styles, as far as the ecological discussion, figuring is certainly both part of the issue and part of the arrangement. By embracing green registering rehearses, business pioneers can contribute emphatically to ecological stewardship – what's more, ensure the climate while likewise diminishing energy and paper costs.

REFERENCES

1. Debnath, Biswajit, Reshma Roychoudhuri, and Sadhan K. Ghosh. "E-waste management–a potential route to green computing" Procedia Environmental Sciences 35 (2016): 669–675.
2. Green Computing, a contribution to save the environment https://www.lancaster.ac.uk/data-science-of-the-natural-environment/blogs/green-computing-a-contribution-to-save-the-environment [Accessed on May 2021]
3. Jeba, Jenia Afrin, et al. "Towards green cloud computing an algorithmic approach for energy minimization in cloud data centers." Research Anthology on Architectures, Frameworks, and Integration Strategies for Distributed and Cloud Computing. IGI Global, 2021, 846–872.
4. Vikram, Shweta. "Green computing." *2015 International Conference on Green Computing and Internet of Things (ICGCIoT)*. IEEE, 2015.
5. Lakshmi, S. V. S. S., I. S. L. Sarwani, and M. Nalini Tuveera. "A study on green computing: the future computing and eco-friendly technology." International Journal of Engineering Research and Applications (IJERA) 2.4 (2012): 1282–1285.
6. What is Green Computing, Advantages | Disadvantages | Examples https://digitalthinkerhelp.com/what-is-green-computing-advantages-disadvantages-examples/ [Accessed on May 2021]
7. Saha, B., 2014. "Green computing," International Journal of Computer Trends and Technology (IJCTT), 14(2), pp. 46–50.
8. Anam, Ayesha, and Anjum Syed. "Green Computing: E-waste management through recycling." International Journal of Scientific & Engineering Research 4.5 (2013): 1103.

9. Yao, Leether, and Teng-Shih Tsai. "Novel hybrid scheme of solar energy forecasting for home energy management system" 2016 IEEE International Conference on Internet of Things (iThings) and IEEE Green Computing and Communications (GreenCom) and IEEE Cyber, Physical and Social Computing (CPSCom) and IEEE Smart Data (SmartData). IEEE, 2016.

10. File:Solar panels in Ogiinuur.jpg
    https://en.wikipedia.org/wiki/File:Solar_panels_in_Ogiinuur.jpg

11. GeothermalEnergy:Introduction *https://www.tutorialspoint.com/renewable_energy/geothermal_ energy_introduction.htm* [Accessed on June 2021]

12. Wind energy https://openei.org/wiki/Wind_energy [Accessed on May 2021]

13. John, Sibu Sam, et al. "Nanotechnology for solar and wind energy applications recent trends and future development," International Conference on Green Computing and Internet of Things (ICGCIoT) IEEE, 2015.

14. Anwar, Sidra, et al. "E-waste reduction via virtualization in green computing." American Scientific Research Journal for Engineering, Technology, and Sciences (ASRJETS) 41.1 (2018): 1–11.

15. Kern, Eva. "Green computing, green software, and its characteristics: Awareness, rating, challenges." From Science to Society. Springer, Cham, 2018. 263–273.

16. The Advantages of LCD Monitors Over Traditional CRT Monitors *https://www.techwalla.com/ articles/the-advantages-of-lcd-monitors-over-traditional-crt-monitors* [Accessed on July 2021]

Intelligent Systems and Smart Infrastructure – Brijesh Mishra et al. (eds)
© 2023 Taylor & Francis Group, London, ISBN 978-1-032-41287-0

Defected Ground Structure with Four Band Meander-Shaped Monopole Antenna for LTE/WLAN/WIMAX/ Long Distance Radio Telecommunication Applications

Chandan[1], Ashutosh Kumar Singh[2], R. P. Mishra[3]

Department of ECE, Institute of Engineering & Technology,
Dr. Rammanohar Lohia Avadh University, Ayodhya, U.P., India

Ratneshwar Kumar Ratnesh[4]

Department of ECE, Meerut Institute of Engineering and Technology,
Meerut, U.P., India

Parimal Tiwari[5]

Research Scholor, Department of Physics and Electronics,
Dr. Rammanohar Lohia Avadh University, Ayodhya, U.P., India

Abstract

A meander-shaped monopole microstrip-fed antenna with the defect in the ground plane is presented in this paper. The antenna generates four bands with wide coverage of the wireless applications. The frequencies covered are 1.5 GHz, 2.6 GHz, 3.5 GHz, and 6.6 GHz, which are suitable for LTE/WLAN/WiMAX/Long distance radio telecommunication applications. The respective return losses are −14.64dB, −18.59 dB, −22.64 dB, and −13.41 dB. The bandwidth of 10%, 21.66%, 37.14% and 4.5% are achieved, respectively, with respective gain of 6.02 dBi,

---

Corresponding author: [1]chandanhcst@gmail.com, [2]aksinghelectronics@gmail.com, [3]director.rpm@gmail.com, [4]ratnes123@gmail.com, [5]parimal.tiwari1@gmail.com

DOI: 10.1201/9781003357346-60

5.33 dBi, 2.70 dBi, and 3.91 dBi. This meander-shaped antenna presents the figure of 8-type radiation pattern and the results simulated are demonstrated using HFSS simulation tool. This meander-shaped antenna presents the figure of 8-type radiation pattern and the results simulated are demonstrated using HFSS simulation tool.

Keywords: Meander-Shaped, LTE, Microstrip-fed, Defected Ground Structure and Omni-Directional.

1. Introduction

The microstrip antenna recently has gained attention because of its light weight, low-profile, easily fabricated nature, low cost, polarization patterns, resonant frequency, and impedance among the researchers and academicians when a particular shape and mode are selected. The microstrip antenna consists of a radiating patch of any geometry placed on a ground plane and separated by the dielectric substrate. Many printed monopole antennas are designed in the recent past for the wireless applications. The microstrip antennas with meander line design techniques conclude in the small size and wideband performance [1]. The rectangular tri-band with inverted L-slot monopole antenna for a WLAN and WiMAX application is bigger than the proposed antenna [2]. A low-profile triple band antenna for applications of WiMAX and WLAN is discussed in [3], also having the size greater than the proposed antenna. In [4], a metamaterial based tri-band antenna for WiMAX and WLAN applications is presented [4] that is also big in size than the proposed antenna. The monopole U-slot antenna with truncated ground structure for multiband applications discussed in [5] also has large size than the proposed antenna. In references [6-19], the literature review has been carried out and has gone through different design techniques.

The performance and design of the prototype antenna in terms of radiation pattern, gain and current distribution is best. The antenna has the dimensions of $26 \times 38 \times 0.8 = 790.4$ mm^3 and covers the frequency ranges from (1.5–1.65GHz)/1.5 GHz, (2.4–2.92 GHz)/2.6 GHz, (3.4–4.7 GHz)/3.5 GHz, and (6.4–6.7 GHz)/6.6 GHz, which are much suitable for LTE/WLAN/WiMAX/Long distance radio telecommunication applications. All the simulations and parametric studies are carried out on HFSS simulation tool.

2. Antenna Structure

Assisted by the ANSYS HFSS tool, the simulated geometry of the proposed antenna with the top and the bottom view is fully described in Fig. 60.1(a) and (b), respectively. The radiating patch has three number of turns fed with feeding technique of type microstrip to support 50Ω input port. The ground is on the reverse side of the radiating patch having defect in the structure to produce numerous resonant frequencies. The geometry is proposed on the FR4 substrate with the relative permittivity of 4.4 and loss tangent 0.2.

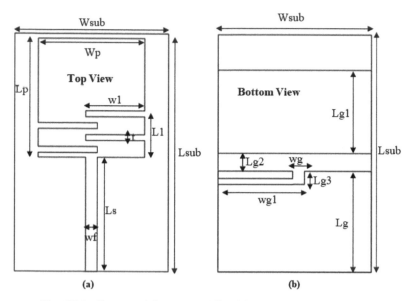

Fig. 60.1 Proposed Antenna (a) Top View and (b) Bottom View

In the geometry described in Fig. 60.1, the length and width of the microstrip line are denoted as Ls and wf, respectively. A rectangular patch is slotted at the bottom side with four equally spaced, with width w1, thickness t, slots making this design as meander-shaped. Lp is the length and Wp is width of the rectangular patch. The dielectric substrate has the dimensions of Lsub as length and Wsub as its width.

Moving on to the bottom side, a cut is made on the ground first with width wg and length Lg3 and then extended after towards the left with width wg1. This ground has the length of Lg. Just after Lg with the spacing of Lg2, one more ground structure is designed with its height as Lg1 as shown in Fig. 60.1(b). This concludes the antenna design.

Table 60.1 provides the whole dimensions with nomenclatures used in the geometry design of the antenna. All the dimensions are given in mm.

Table 60.1 Dimension of the Proposed Antenna (in mm)

Wsub	Lsub	Wp	Lp	h	Ls	wf	t
26	38	18	20	0.8	19	2	1
L1	Lg	Lg1	Lg2	Lg3	wg	wg1	
5	17	14	2	1	12	1	

Figure 60.2 demonstrates the return losses (S11 $\leq$ −10 dB) at all the four frequencies obtained after the simulation presented in between S_{11} versus Frequency. It gives quad-band with enhanced bandwidth and good impedance matching with improved gain.

The frequency range this antenna covers are from (1.5–1.65GHz)/1.5 GHz, (2.4–2.92 GHz)/2.6 GHz, (3.4–4.7 GHz)/3.5 GHz, and (6.4–6.7 GHz)/6.6 GHz, which are much suitable for LTE/ WLAN/WiMAX/Long distance radio telecommunication applications. The bandwidth of 10%, 21.66%, 37.14%, and 4.5% is achieved.

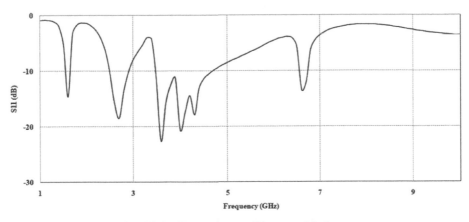

Fig. 60.2   Return Loss of Proposed Antenna

3. Antenna Configuration

In Fig. 60.3, the different antennas configurations are presented so as to achieve the final proposed antenna. Antenna 4 is the proposed antenna. Antenna 1 is the simple rectangular patch with anticlockwise rotated L-slit cut in the ground plane. This antenna generates two bands with very low return losses at the obtained frequencies. After that, two slots of equal dimensions are cut on the patch at the lower part as shown in the figure as Antenna 2. This antenna now generates dual-band too but with very low return losses; it again doesn't achieve the target. Now, two more slots are added with the same dimensions as of the previous slots on the rectangular patch as shown in Antenna 3. This antenna generates triple bands but with less bandwidth and low return losses. Finally, in the previous design of Antenna 3, one more ground structure of height Lg1 is added at Lg2 space from the lower ground part as shown in Antenna 4. This provides the quad-band with enhanced bandwidth and improved gain and impedance matching.

4. Result and Discussion

The antenna dimensions are $26 \times 38 \times 0.8 = 790.4$ mm^3, which is of compact size. The antenna provides good impedance matching with high return loss of −22.64 dB. This antenna gives omnidirectional type of radiation; 10%, 21.66%, 37.14% and 4.5% are achieved.

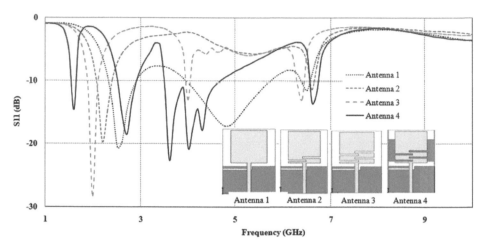

Fig. 60.3   Configuration of Different Antennas

Table 60.2   Comparison chart of the prototype antenna with other reference antennas

Reference of Antenna	Size of the Antenna	Bandwidth
[2]	36 × 26 × 1.6 = 1498 mm	25.8%, 6.9% & 15.3%
[3]	30 × 40 × 0.8 = 960 mm	6.9%, 5.1% & 28%
[4]	35 × 35 × 1 = 1225 mm	10.7%, 21.7% & 12.7%
[5]	70 × 52 × 1 = 3640 mm	28.4%, 26.45%, 14.67% & 12.84%
[6]	29 × 36 × 0.8 = 835.2 mm	13.7%, 4.2% &18.2%
Proposed Antenna	26 × 38 × 0.8 = 790.4 mm	10%, 21.66%, 37.14% and 4.5%

Table 60.2 demonstrates and illustrates the comparison between the reference antennas and the antenna proposed. It can be seen from the table that the proposed antenna is smaller than the antennas discussed and compared in the references. The comparison chart shows the better impedance bandwidth of the proposed antenna than the reference antennas and size is also compact. Pattern is the same as that of the monopole antenna with the ground plane of $\lambda/4$. This antenna covers (1.5–1.65GHz)/1.5 GHz, (2.4–2.92 GHz)/2.6 GHz, (3.4–4.7 GHz)/3.5 GHz and (6.4–6.7 GHz)/6.6 GHz, which are much suitable for LTE/WLAN/WiMAX/Long distance radio telecommunication applications. The bandwidth of 10%, 21.66%, 37.14% and 4.5% are achieved.

The return loss curves for the successive values patch length varying from 19 mm to 21 mm is shown in Fig. 60.4. It can be viewed from the given figure that optimized value of lp = 20 mm provides the quad-band with high return losses when compared to the Lp = 19 mm and Lp = 21 mm. In Fig. 60.4 S11 parameter in terms of length of patch is shown in which Lp = 20 mm shows the optimum result when compared to Lp = 19 mm and Lp = 21 mm, which provide low impedance matching at the corresponding resonant frequencies. Hence, Lp = 20 mm provides the optimum value.

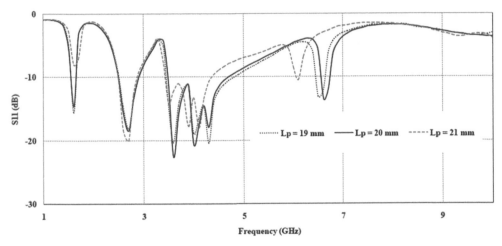

Fig. 60.4 Length of Patch Varying from 19-21 mm

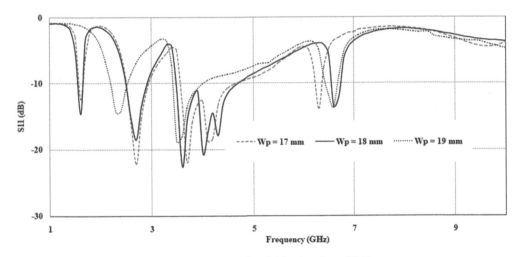

Fig. 60.5 Width of Patch Varying from 17-19 mm

Figure 60.5 shows the curves for return losses due the variation in the width of patch Wp from 17 mm to 19 mm. When Wp is taken 17 mm, the antenna resonates four bands but with low return losses at some frequencies. When the value is taken as 19 mm, then antenna resonates three times only, that too with less return losses. But when the value of Wp is taken as 18 mm, quad-band is generated by the proposed antenna with good impedance matching and high return losses. Hence, Wp with the value 18 mm provides the optimum results. The reflection coefficient curves for the variation in the height of the substrate are shown in Fig. 60.6. It is clearly evident from the figure that antenna produces four bands when the height of the substrate is taken as h = 0.8 mm than the value h = 1.6 mm, which only gives out three bands that too with very less return losses. When h = 0.8 mm is taken, the results

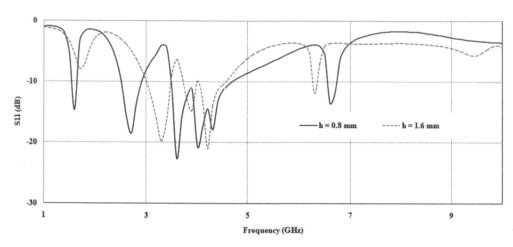

Fig. 60.6   Height of Substrate Varying from 0.8-1.6mm

are coming out optimum and good return losses and impedance matching are obtained. Figure 60.7 illustrates the return losses curves when the length Lg of the ground is varied from 16 mm to 18 mm successively. When the value is given 16 mm, only three bands are generated from the proposed antenna, which doesn't fit. Again, when the value to the Lg is given as 18 mm, then it generates only dual-band, which doesn't fulfil the requirements. But finally, when the value to Lg is given as 17 mm, quad-band is generated by the antenna with better return losses and impedance matching; hence, Lg = 17 mm is the optimum value.

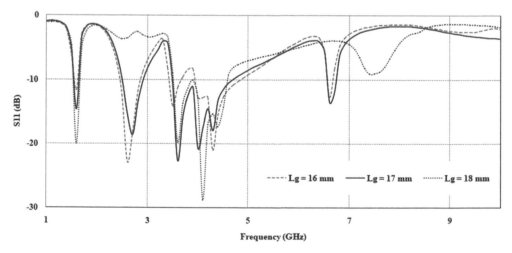

Fig. 60.7   Length of Ground Plane Varying from 17-19 mm

The gain plot of the propose antenna is shown in Fig. 60.8, which shows that improved gain has been achieved through the designed antenna. The gain at (1.5–1.65GHz)/1.5 GHz is

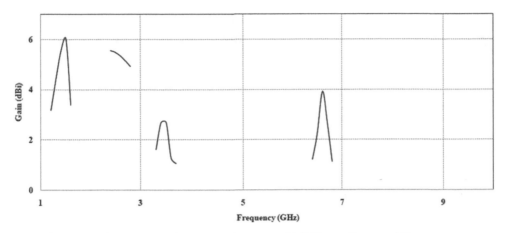

Fig. 60.8 Total Gain of Proposed Antenna with Different Resonant Frequency

6.02 dBi, (2.4–2.92 GHz)/2.6 GHz is 5.33 dBi, (3.4–4.7 GHz)/3.5 GHz is 2.70 dBi and (6.4–6.7 GHz)/6.6 GHz the gain is 3.91 dBi.

Figures 60.9, 60.10, 60.11 and 60.12 show the surface current distribution of antenna at 1.5 GHz, 2.6 GHz, 3.5 GHz, and 6.6 GHz, respectively. The red indication provides the highest distribution of current through that surface

Figures 60.13, 60.14, 60.15 and 60.16 demonstrate the radiation patterns of the proposed antenna at four different frequencies obtained after the simulation at 1.5 GHz, 2.6 GHz, 3.5 GHz, and 6.6 GHz. It is evident from the figures that the proposed antenna behaves totally like the omnidirectional antennas just like Fig. 60.8. The radiation pattern is taken at phi = 0 degree and phi = 90 degrees at all the resonant frequencies.

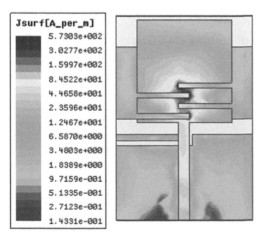

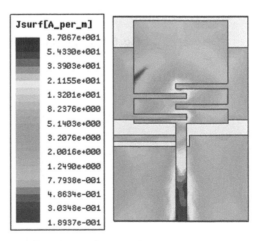

Fig. 60.9 Current Distribution of Antenna at 1.5 GHz

Fig. 60.10 Current Distribution of Antenna at 2.6 GHz

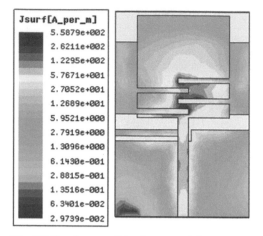
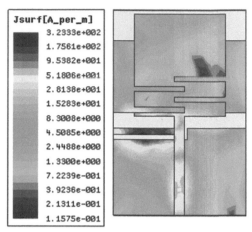

Fig. 60.11 Current Distribution of Antenna at 3.5 GHz

Fig. 60.12 Current Distribution of Antenna at 6.6 GHz

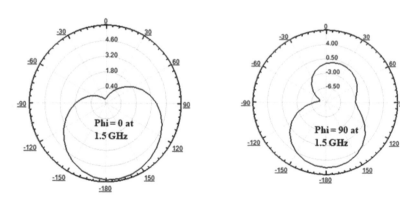

Fig. 60.13 Radiation Pattern of Antenna at 1.5 GHz

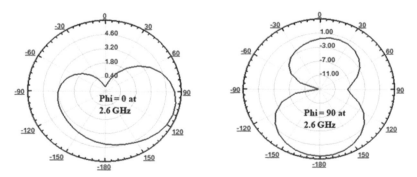

Fig. 60.14 Radiation Pattern of Antenna at 2.6 GHz

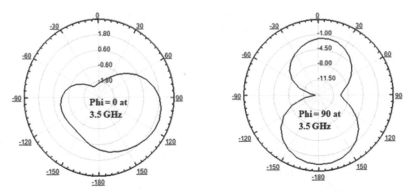

Fig. 60.15   Radiation Pattern of Antenna at 3.5 GHz

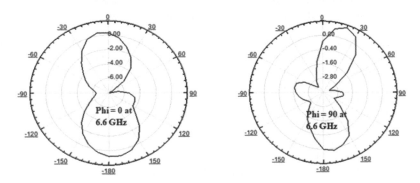

Fig. 60.16   Radiation Pattern of Antenna at 6.6 GHz

5. Conclusion

The size of this compact antenna is $26 \times 38 \times 0.8 = 790.4$ mm^3. This antenna covers the frequency ranges from (1.5–1.65GHz)/1.5 GHz, (2.4–2.92 GHz)/2.6 GHz, (3.4–4.7 GHz)/3.5 GHz, and (6.4–6.7 GHz)/6.6 GHz, which are much suitable for LTE/WLAN/WiMAX/Long distance radio telecommunication applications. The bandwidth of 10%, 21.66%, 37.14%, and 4.5% is achieved. The gains at the respective frequencies are 6.02 dBi, 5.33 dBi, 2.70 dBi, and 3.91 dBi, respectively. The antenna behaves the same like omnidirectional antenna with the same type of radiation patterns. This antenna provides the applications like LTE, WLAN, WiMAX, and other wireless applications. The simulation is done on ANSYS HFSS with FR4 material.

REFERENCES

1. Ramesh Garg, Prakash Bharti and Inder Bahl, "Microstrip Antenna Design Handbook," Artech House INC. Boston, London, 2001.

2. Chen H, et al, "Tri-Band Rectangle Loaded Monopole Antenna with Inverted-L Slot for WLAN/WiMAX Applications," Electron Lett., Vol. 49, pp. 1261–1262, 2013.

3. Verma S and Kumar P., "Compact Triple-Band Antenna for WiMAX and WLAN Applications," Electron Lett. Vol. 50, pp. 484–486, 2014.

4. Pushkar P and Gupta VR., "A Metamaterial Based Tri-band Aantenna for WiMAX/WLAN Application," Microwave Opt Techno Lett., Vol. 58, pp. 558–561, 2016.

5. Chandan, Toolika Srivastava and B.S.Rai, "Multiband Monopole U-Slot Patch Antenna with Truncated Ground Plane," Microwave and Optical Technology Letters, vol. 58, No. 8, pp. 1949-1952, 2016.

6. Chandan,"Truncated Ground Plane Multiband Monopole Antenna for WLAN and WiMAX Applications," IETE Journal of Research, pp. 1–6, 26 January 2020.

7. Chandan, RK Ratnesh, Amit Kumar, "A Compact Dual Rectangular Slot Monopole Antenna for WLAN/WiMAX Applications," Springer Cyber Physical Systems, Lecture Notes in Electrical Engineering book series (LNEE) Volume 788, pp. 699–705, 2021.

8. Chandan, G. D. Bharti, P. K. Bharti and B. S. Rai, "Miniaturized Pi(π)Slit Monopole Antenna for 2.4/5.2.8 Applications," AIP Conference Proceedings, American Institute of Physics pp. 200351–200356, 2018.

9. Chandan, G. D. Bharti, Toolika Srivastava and B. S. Rai, "Dual Band Monopole Antenna for WLAN2.4/5.2/5.8 with Truncated Ground," AIP Conference Proceedings, American Institute of Physics pp. 200361–200366, 2018.

10. Chandan, G. D. Bharti, Toolika Srivastava and B. S. Rai, "Miniaturized Printed Kshaped Monopole Antenna with Truncated Ground Plane for 2.4/5.2/5.5/5.8 Wireless LAN Applications," AIP Conference Proceedings, American Institute of Physics pp. 200371-200377,2018.

11. Chandan, Toolika Srivastava and B. S. Rai, "L Slotted Microstrip Fed Monopole Antenna for Triple Band WLAN and WiMAX Applications," Springerin Proceedings Theory and Applications, Vol. 516, pp. 351–359, 2017.

12. Chandan and B. S. Rai, "Dual Band Monopole Patch Antenna Using Microstrip Fed for WiMAX and WLAN Applications," Springer Proceedings Information Systems Design and Intelligent Applications, Vol. 2, pp. 533–539, 2016.

13. Chandan and B. S. Rai, "Bandwidth Enhancement of Wang Shape Microstrip Patch Antenna for Wireless System," IEEE Fourth International Conference on Communication Systems and Network Technologies (NITTRBHOPAL), PP. 11–15, 2014.

14. Chandan and B. S. Rai, "Dual Band Wang Shaped Microstrip Patch Antenna for GPS and Bluetooth Application," IEEE Sixth International Conference Computational Intelligence and Communication Networks (Udaipur), pp. 69–73, 2014.

15. Sparsh Singhal, Pranjal Sharma, Chandan, "A Low-Profile Three-Stub Multiband Antenna for 5.2/6/8.2 GHz Applications," Springer Cyber Physical Systems, Lecture Notes in Electrical Engineering book series (LNEE) Volume 788, pp. 2021.

16. Siddharth Vashisth, Sparsh Singhal and Chandan, "Low-Profile H Slot Multiband Antenna for WLAN/Wi-MAX Application," Springer Cyber Physical Systems, Lecture Notes in Electrical Engineering book series (LNEE) Volume 788, pp. 2021.

17. Shivam Choudhary, Yash Sharma, Shubham Kumar, Chandan, "Dual Circular-Inverted L Planar Patch Antenna for Different Wireless Applications," Springer Cyber Physical Systems, Lecture Notes in Electrical Engineering book series (LNEE) Volume 788, pp. 2021.

18. K. P. Singh and Chandan, "A Compact Tri Band Monopole Antenna with LStub for WLAN/WiMAX Applications," 3rd IEEE International Conference, CIPECH-18, KIET, Ghaziabad, pp. 249–253, 2018.

19. Missi Bhasker, Chandan and R. K. Prasad "Design of Monopole Antenna with defected Ground for WLAN Antenna," IEEE International Conference on Emerging Trends in Electrical, Electronics & Sustainable Energy Systems (ICETEESES-16, KNIT Sultanpur), Vol. 2, pp. 187–190, 2016.

Intelligent Systems and Smart Infrastructure – Brijesh Mishra et al. (eds)
© *2023 Taylor & Francis Group, London, ISBN 978-1-032-41287-0*

CHAPTER

61

Herbal Plants Leaf Image Classification Using Machine Learning Approach

Gaurav Kumar[1], Vipin Kumar*, and Aaryan Kumar Hrithik[2]

Department of Computer Science and Information Technology,
Mahatma Gandhi Central University, Motihari, Bihar 845401, India

Abstract

Herb refers to any part of a medicinal plant. Medicinal plants are those kinds of plants that can be used for medicinal purposes to cure and prevent humans or animals from diseases. Identifying these herbs plays a crucial role in using them as a medicine. So, the identification and classification of herbs are critical. Only an expert or a person who has prior knowledge can identify it. To make use of herbs by people at large, the problem of identification must be solved, and methods should be proposed that are easy to use and cost-effective. Any plants can be identified by their leaves' images. In this paper, six classical ML algorithms have been applied on collected 25 kinds of herbal plant leaves, and a comparative analysis of the results of these six algorithms has been done. An accuracy of 82.51% is achieved by using Multi-layer Perceptron (MLP) classifier.

Keywords: Classification, Herbal plant leaf, Machine learning, RGB image.

1. Introduction

Herb refers to any part of the medicinal plant like stigma, flower, leaf, fruit, bark, seed, stem, or a root, as well as a non-woody plant. Medicinal plants are those kinds of plants that can be used for medicinal purposes to cure and prevent humans or animals from diseases. The herbal

*Corresponding author: rt.vipink@gmail.com
[1]viagrv@gmail.com, [2]aaryanhrithik7@gmail.com

DOI: 10.1201/9781003357346-61

plants are also used for other purposes other than medicinal plants. It is used in the perfume manufacturing industry as a flavonoid, eaten as a food, and used in many spiritual activities. The primary goal of this paper is the classification of herbs for medicinal purposes, but it can be used for identification and classification for other purposes as well. Indian AYUSH system has codified 8000 kinds of herbal remedies in India (Mahtab Alam Khan 2016). As the population grows, people are shifting more towards Ayurveda, which mainly depends on herbal plants. There are different side effects of synthetic drugs nowadays. People are becoming more and more resistant to currently used drugs in the case of many infectious diseases. These are the few reasons to shift people towards herbal remedies rather than modern synthetic drugs. According to WHO, about 80% of the population worldwide use herbs as their primary healthcare needs (WHO Monographs on Selected Medicinal Plants Volume 4, 2022).

Medicinal plants are classified generally by manual identification, i.e., a person who is an expert of it by experience is only able to identify the herbs. So, we are people's knowledge dependent. The primary and pervasive problem that arises is identifying these herbal plants by anyone. If the identification problem is solved, people living in rural areas lacking primary healthcare facilities can quickly identify and use herbal plants in their neighborhoods. With the advancement of image processing through Machine learning (ML) and Deep learning (DL), computer-based image recognition is possible nowadays. The leaf of a plant has many features useful in uniquely identifying the different classes of plants' leaves. The shape of the leaf is the most used feature in automated training via different machine learning algorithms. Apart from the shape, colors, textures, and veins are also commonly used by researchers.

The novelty of the proposed research work:

- To the best of our knowledge, this is the first multiple herbal plant leaves image dataset that has been created for 25 categories of herbs with a total of 6628 RGB images.
- The comparative study of the classification performance of the classifiers has been done where the best performer has been identified successfully based on accuracy, precision, recall, F-measures, and support performance measures (Dj Novakovi et al. 2017).
- The highly misclassified leaf has been analyzed based on textual information. It has been observed that visual images of misclassified leaf categories are similar.

2. Literature Review

The authors described a method for identifying and classifying medicinal plants based on their texture and color (Anami et al., 2022). Various plant species such as Papaya, Tulsi, and Garlic are studied for their medicinal properties. The system's accuracy for distinguishing color and texture features is better than 80%. It is also more accurate than standard methods for identifying herbs and shrubs. Kan, Jin, and Zhou (2017) introduced an improved classification method to identify different types of medicinal plants. The proposed method allows the automatic recognition of plants based on their leaf images. It achieves a 93.3% success rate and can also be used with SVM. The paper considers the texture and the shape features of leaves. Various models are then used to study the effectiveness of the proposed method. The paper's findings support the development of a classification system for medicinal plants.

This study aimed to identify the different types of herbal plant leaves collected from Pakistan's Agriculture Department (Naeem et al., 2021). Some of these include Basil, Catnip, Basil, and Lavinia. The data collected by the system was then analyzed using a computer vision laboratory. Five machine learning classifiers were deployed on the data. One of them achieved a 99.10% accuracy in classifying six different kinds of medicinal plant leaves. One paper proposes a mobile application called MedLeaf, designed to identify medicinal plants based on their leaf image (Prasvita and Herdiyeni 2013). For the study, the authors used 30 species of medicinal plants in Indonesia. The data collected by the app were used for evaluation, which will help manage nature reserves and botanical gardens. It uses image processing and computer vision techniques to identify different types of plants and classify them into edible/poisonous categories. This paper analyzes the benefits of various automated procedures for leaf pattern recognition. A computer vision approach is proposed that can entirely ignore the context of the image and provide an accurate and fast leaf recognition process. It is possible to identify algae and plants without having a definite form.

3. Proposed Methodology

The workflow of the proposed methodology is shown in Fig. 61.1. Each step of the methodology part will be discussed in the subsequent subsections.

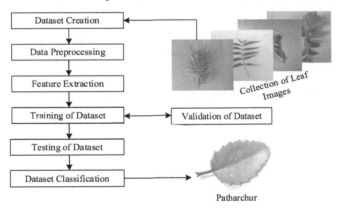

Fig. 61.1 The Workflow of Herbal Plants Leaf Classifications

Dataset preparation: The leaf images, captured in RGB format, are read by a machine and converted adequately into a numerical form. This dataset must be in the tabular format, where each row represents a particular instance or example, here in our case, an image, and each column to be a specific feature. If the total pictures in the dataset are N and indexed with lowercase n = 1, 2, ...N. So, each row referred to as a data point, or an example is denoted by n. Our problem of leaf classification is a supervised learning problem where we have a level , the category name of a leaf associated with each example of x_n. The dataset is written as a set of data point-label pairs $\{(x_1, y_1), ..., (x_n, y_n), ..., (x_N, y_N)\}$. The table of data points

$\{x1, x2, ..., xn\}$ can be written as $X \in R^{N \times D}$. The required resized dataset $\{x_n^r, y_n\}$ obtained from resized function $Re(.)$ is shown in eq. 1.

$$X^r = Re(\{x_n^r, y_n\}_{n=1}^N, p \times q) \tag{1}$$

Feature extraction: Since we have a vector representation of image data, which is high-dimensional data, we can rearrange data to represent it better. Finding a principal component via Principal Component Analysis (PCA) is related to eigenvalues, and PCA is also related to singular value decomposition. PCA is an algorithm for linear dimensionality reduction. X represents the data matrix, where the rows represent the different images. The principal component is the data covariance matrix as shown in eq. 2:

$$S = \frac{1}{N} \sum_{n=1}^N x_n x_n^T \tag{2}$$

Furthermore, we assume there exists a low-dimensional compressed representation (code): $z_n = B^T x_n \in R^M$ of x^n, where we define the projection matrix as $B := [b_1, ..., b_M] \in R^{D \times M}$. It aims to search for z_n and the vectors $b_1, ..., b_M$ to maintain them as the original vector x_n And the data loss can be reduced due to compression (Deisenroth, Faisal, and Ong 2020).

Divide dataset into train, validation, and test set: The resized image dataset is represented by X^r. This dataset is divided into Training set, Validation set, and the Test dataset, namely T_r, V_a, and T_e, respectively, i.e., $T_r \subset X^r$, $V_a \subset X^r$, and $T_e \subset X^r$, where $X^r = T_r \cup V_a \cup T_e$. The height (H), width (W), and color channel (C) of the RGB image may be represented in tensor T. The ith image of the dataset $X_i^r \in X^r$ can be represented as $X_i^r = [H, W, C, i]$, where i = 1, 2, ..., b, where b is the batch size.

Applying ML Algorithms: A predictor function f is a function that, when given to training data as input, shows the output as a predicted level. Suppose the output is a single real-valued number. Then, it can be represented as $f(x) : R^D \to R$. Here the input has D-dimensions or D-levels. For N data points $x_n \in R^D$ and corresponding scalar level $y^n \in R$. For the supervised learning setting, the pairs are $\{(x_1, y_1), (x_2, y_2), ..., (x_N, y_N)\}$. The predictor function will be as $f(., \theta) : R^D \to R$, where θ is a parameter of each classifier. Our target is to optimize parameter θ^* in such a way that it fits in the dataset well, which is shown as $f(x_n, \theta^*) \approx y_n$ for all $n = 1, 2, ..., N$. The output of the predictor function will be denoted by $\hat{y}_n$ corresponding to level y_n and hence look at eq. 3:

$$\hat{y}_n = f(x_n, \theta^*) \tag{3}$$

Our target is to minimize the average loss function shown in by finding a suitable parameter θ^* on the set of N-training data points. The average loss function is given in eq. 4:

$$L_{avg} = \frac{1}{N} \sum_{n=1}^N l(y_n, \hat{y}_n) \tag{4}$$

4. Experiments

Dataset descriptions: Authors have collected the 25 classes (6628 RGB images) of herbal plant leaves images, where each class has more than 250 images (shown in Table 61.1), which are 1. *Calotropis gigantea* (Akvana): 262; 2. *Emblica officinalis* (Amla): 278; 3. *Justicia adhatoda* (Arusa): 267; 4. *Cannabis sativa* (Bhang): 269; 5. *Clerodendrum infortunatum* (Bhat): 259; 6. *Eclipta prostrata* (Bhringraj):260; 7. *Ageratum houstonianum* (Bluemink): 254; 8. *Capsella bursa-pastoris* (Capsella): 266; 9. *Tylophora indica* (Dambel): 260; 10. *Datura stramonium* (Dhatura):256; 11. *Nyctanthes arbor-tristis* (Harsingar): 289; 12. *Cirsium arvense* (Kantaiya): 272; 13. *Pongamia pinnata* (Karanja): 264; 14. *Murraya koenigii* (Meethaneem):256; 15. *Plumeria pudica* (Nagchampa): 272; 16. *Azadirachta indica* (Neem): 289; 17. *Vitex negundo* (Nirgundi): 252; 18. *Coleus barbatus* (Patharchur): 254; 19. *Mentha arvensis* (Pudina): 277; 20. *Putranjiva roxburghii* (Putrajeevak): 261; 21. *Catharanthus roseus* (Sadabahar): 256; 22. *Vernonia cinerea* (Sahadevi): 257; 23. *Asparagus racemosus* (Shatawari): 260; 24. *Ocimum sanctum* (Tulsi): 267; and 25. *Verbesina virginica* (Virginica): 271 (details as S. No. followed by Scientific/Botanical name (Common name): Number of images in that category).

Table 61.1 Sample images of all 25 categories of herbal plant leaf

Experiment design: Figure 61.1 is the experimental flowchart, which shows the whole process from data collection by the authors to the classification of herbs. The first phase is creating the

dataset by collecting leaves images. More than 250 images of leaves in every 25 categories of herbal plants are gathered in this phase. Afterward, the collected images were cropped, resized, and organized based on their classes in the pre-processing data phase. Cropped images were resized into 250 × 250 sizes, and each category of images was stored in separate folders with their names. Once the data is pre-processed, it is ready for the ML program to do the execution, and the features of these leaves' images have been extracted in the feature extraction part. After feature extraction, the algorithm divides each class into the train, validation, and test sets. The validation set and test set consist of 50 images each, and the remaining more than 150 images in each category are being used for the training set (Raschka 2018). Now, different classifiers have been applied to get the performance in the form of accuracy, precision, recall, etc. Lastly, the authors can do the classification based on the results obtained from the previous stage.

Hardware and Software: A standard smartphone is used to capture RGB images of the leaves. The pre-processing and execution of the ML algorithms have been done over the workstation: Intel® Xeon® Silver 4210 CPU @2.20Ghz 2.1 Hz, RAM-64 G.B. along with 64bit window operating system. The scikit-learn package of ML library is used over JupyterLab 3.2 with Python programming language.

5. Result and Analysis

A. Result

The Bar Plot shown in Fig. 61.2 shows the classification performance based on the accuracy of the six classifiers, namely, L.R., KNN, SVM, D.T., MLP, and N.B. In this Figure, the x-axis represents the name of the classifiers, and the y-axis represents their respected accuracy. For a clear view, the accuracy of each classifier has been shown just above their bars. Figure 61.3 shows the Box Plot of all the same six classifiers. Here also, the classifier's name is along the x-axis, and the accuracy is along the y-axis. The classification performance (Precision, Recall, f1-score, support) of the best classifier, which is the MLP classifier, is shown in Table 61.3. Fig. 61.4 is the confusion matrix using a heat map for our herb's dataset. It shows the classification and misclassification amongst different categories of herbs (Visa et al. 2011; DeBoer, 2015). The x-axis and y-axis both contain the name of the herbal plants, of which classification has been done in this paper. The color range of this heat map varies from 0 to 4.5. For the diagonal elements, which start from left-top to right bottom, the brighter (closer to 4.5), color of a cell shows that the category has that much better correlation with itself, which must be, i.e., the classification has been done by the classifier that much successfully, but if these diagonal cells' color is darker (closer to 0), much misclassification has occurred. But for the cells except for this diagonal cell, the story will be just the opposite. That is darker (closer to 0). The color shows the least the correlation of that class with corresponding other herb classes.

B. Analysis of Result

Classification Comparison using Bar Plot: As the bar plot in Fig. 61.2, Linear Regression (L.R.) has 73.65%, k-Nearest Neighbors (KNN) has 68.83%, Support Vector Machine (SVM)

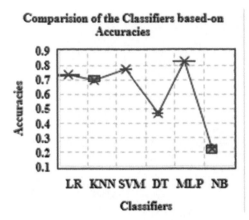

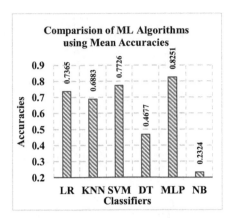

Fig. 61.2 Box plot of ML algorithms classification performance based on accuracy measure

Fig. 61.3 Bar plot of ML algorithms classification performance based on accuracy measures

Table 61.2 Performance of the MLP classifier based on Precision, Recall, F1-Score, and Support

S.N.	Precision	Recall	F1-Score	Support	S.N.	Precision	Recall	F1-Score	Support
1.	0.95	0.92	0.93	75	13.	0.97	1	0.99	70
2.	1	0.99	0.99	72	14.	0.76	0.56	0.65	55
3.	0.74	0.73	0.73	70	15.	0.71	0.8	0.75	69
4.	0.87	0.83	0.85	64	16.	1	1	1	78
5.	0.69	0.78	0.73	69	17.	0.82	0.78	0.8	60
6.	0.96	0.98	0.97	66	18.	0.7	0.74	0.72	70
7.	0.87	0.92	0.89	59	19.	0.93	0.94	0.94	69
8.	0.65	0.7	0.67	56	20.	0.68	0.7	0.69	61
9.	0.92	0.89	0.91	65	21.	1	0.98	0.99	53
10.	0.84	0.74	0.78	69	22.	0.9	0.92	0.91	65
11.	0.68	0.7	0.69	74	23.	0.89	0.89	0.89	64
12.	0.86	0.82	0.84	67	24.	0.94	0.91	0.92	74
					25.	0.65	0.65	0.65	68
					Accuracy			0.84	1662
				Macro Avg.	0.84	0.84	0.84	1662
				Weighted Avg.	0.84	0.84	0.84	1662

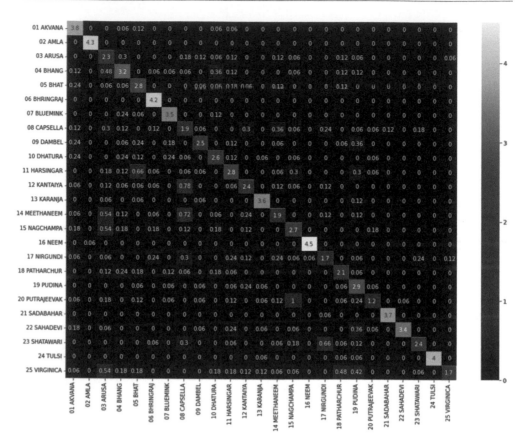

Fig. 61.4 Heatmap showing the correlation amongst different classes of leaves based on classification performance

has 77.26%, Decision Tree (D.T.) has 46.77%, Neural Network (MLP) has 82.51%, and the Naïve Bayes (N.B.) has 23.24% classification accuracy. So, based on classification accuracy, if we choose MLP, which has the maximum precision, this classifier for the classification of herbal plants will give the best result compared to other classifiers. And the N.B. will provide the worst classification report because it has the least accuracy.

Classification Comparison using Box Plot: The evaluation criteria for analyzing a box plot are max maximization, min maximization, size of a box plot, and the median of the box plots. From Fig. 61.3, MLP has the maximum max, the maximum min, the median too has a maximum value, and the smallest box size compared to other classifiers. So again, the MLP classifier is performing best based on all four evaluation criteria of a box plot.

Classification Comparison based on Performance: The performance of the MLP classifier is shown in Table 61.3. As we can see, S. No. 16 of herbal plant leaf has the best performance because it has a maximum value of precision, recall, f1-score, and support. At the same time, the category-20 leaf has a minor version due to the lowest deal of the accuracy, recall, and f1-score. Based on support, the category-21 leaf has a minimum performance evaluation. These results are validated clearly from the heatmap in Fig. 61.4.

Performance Analysis of Best and Worst Performers based on Confusion Matrix: From Fig. 61.4, we can see that the best classification is happening for Category-16, *Azadirachta indica* (Neem). When we go to the cell corresponding to category-15 along the x-axis and category-20 along the y-axis, the misclassification is maximum, and its value is 1. It is due to the visual similarity between their leaf images, as shown in Table 61.1.

6. Conclusion

This paper classified 25 categories of herbal plant leaves images. Firstly, the shape feature is extracted to do the classification. Secondly, various ML algorithms have been applied to get the classification report. Based on these results, the analysis has been drawn through box plot, bar plot, and heat map. The accuracy of up to 82.38% is achieved by using MLP Classifier. In the future, the Deep Learning model will be implemented on the same 25 kinds of herbs to get better accuracy. More advanced methods will be used for highly misclassified. More categories of herbs can be collected for classification.

REFERENCES

1. Alpaydin, Ethem. 2020. *Introduction to Machine Learning*. MIT press.
2. Anami, BS, SS Nandyal, A Govardhan - International Journal of, and undefined 2010. 2022. "A Combined Color, Texture and Edge Features Based Approach for Identification and Classification of Indian Medicinal Plants." *Academia.Edu*. Accessed March 13. https://www.academia.edu/download/73139079/A_Combined_Color_Texture_and_Edge_Featur20211019-23901-eqqy68.pdf.
3. DeBoer, Mike. 2015. "Understanding the Heat Map." *Cartographic Perspectives*, no. 80: 39–43.
4. Deisenroth, MP, AA Faisal, and CS Ong. 2020. *Mathematics for Machine Learning*. https://books.google.co.in/books?hl=en&lr=&id=pFjPDwAAQBAJ&oi=fnd&pg=PR9&dq=mathematics+for+machine+learning&ots=VLkm1JG9Ai&sig=IFpO5yPObG1EexIwuOlpuUr-aHM.
5. Dj Novakovi, Jasmina, Alempije Veljovi, Siniša S Ili, ˇ Zeljko Papi, and Milica Tomovi. 2017. "Evaluation of Classification Models in Machine Learning." *Theory and Applications of Mathematics & Computer Science*. Vol. 7.
6. Dr. Mahtab Alam Khan. 2016. "Introduction and Importance of Medicinal Plants and Herbs." *Zahid*. May 20.
7. Erickson, Bradley J., Panagiotis Korfiatis, Zeynettin Akkus, and Timothy L. Kline. 2017. "Machine Learning for Medical Imaging." *Radiographics* 37 (2). Radiological Society of North America Inc.: 505–15. doi:10.1148/rg.2017160130.
8. Kan, H. X., L. Jin, and F. L. Zhou. 2017. "Classification of Medicinal Plant Leaf Image Based on Multi-Feature Extraction." *Pattern Recognition and Image Analysis 2017 27:3* 27 (3). Springer: 581–87. doi:10.1134/S105466181703018X.

9. Alpaydin, Ethem. 2020. *Introduction to Machine Learning*. MIT press.

10. Anami, BS, SS Nandyal, A Govardhan - International Journal of, and undefined 2010. 2022. "A Combined Color, Texture and Edge Features Based Approach for Identification and Classification of Indian Medicinal Plants." *Academia.Edu*. Accessed March 13. https://www.academia.edu/download/73139079/A_Combined_Color_Texture_and_Edge_Featur20211019-23901-eqqy68.pdf.

11. DeBoer, Mike. 2015. "Understanding the Heat Map." *Cartographic Perspectives*, no. 80: 39–43.

12. Deisenroth, MP, AA Faisal, and CS Ong. 2020. *Mathematics for Machine Learning*. https://books.google.co.in/books?hl=en&lr=&id=pFjPDwAAQBAJ&oi=fnd&pg=PR9&dq=mathematics+for+machine+learning&ots=VLkm1JG9Ai&sig=IFpO5yPObG1EexIwuOlpuUr-aHM.

13. Dj Novakovi, Jasmina, Alempije Veljovi, Siniša S Ili, ˇ Zeljko Papi, and Milica Tomovi. 2017. "Evaluation of Classification Models in Machine Learning." *Theory and Applications of Mathematics & Computer Science*. Vol. 7.

14. Dr. Mahtab Alam Khan. 2016. "Introduction and Importance of Medicinal Plants and Herbs." *Zahid*. May 20.

15. Kan, H. X., L. Jin, and F. L. Zhou. 2017. "Classification of Medicinal Plant Leaf Image Based on Multi-Feature Extraction." *Pattern Recognition and Image Analysis 2017 27:3* 27 (3). Springer: 581–87. doi:10.1134/S105466181703018X.

16. Mitchell, Tom Michael. 2006. *The Discipline of Machine Learning*. Vol. 9. Carnegie Mellon University, School of Computer Science, Machine Learning.

17. Naeem, Samreen, Aqib Ali, Christophe Chesneau, Muhammad H. Tahir, Farrukh Jamal, Rehan Ahmad Khan Sherwani, and Mahmood Ul Hassan. 2021. "The Classification of Medicinal Plant Leaves Based on Multispectral and Texture Feature Using Machine Learning Approach." *Agronomy 2021, Vol. 11, Page 263* 11 (2). Multidisciplinary Digital Publishing Institute: 263. doi:10.3390/AGRONOMY11020263.

18. Prasvita, Desta Sandya, and Yeni Herdiyeni. 2013. "MedLeaf: Mobile Application for Medicinal Plant Identification Based on Leaf Image." *International Journal on Advanced Science, Engineering and Information Technology* 3 (2). Insight Society: 103. doi:10.18517/ijaseit.3.2.287.

19. Raschka, Sebastian. 2018. "Model Evaluation, Model Selection, and Algorithm Selection in Machine Learning." *ArXiv Preprint ArXiv:1811.12808*.

20. Tan, Pang-Ning, Michael Steinbach, and Vipin Kumar. 2016. *Introduction to Data Mining*. Pearson Education India.

21. Verma SS, Prasad A, Kumar A. CovXmlc: High-performance COVID-19 detection on X-ray images using Multi-Model classification. Biomedical Signal Processing and Control. 2022 Jan; https://doi.org/10.1016/j.bspc.2021.103272

22. Visa, Sofia, Brian Ramsay, Anca L Ralescu, and Esther van der Knaap. 2011. "Confusion Matrix-Based Feature Selection." *MAICS* 710: 120–27.

23. "WHO Monographs on Selected Medicinal Plants Volume 4." 2022. Accessed March 17. https://www.who.int/medicines/areas/traditional/SelectMonoVol4.pdf.

Intelligent Systems and Smart Infrastructure – Brijesh Mishra et al. (eds)
© 2023 Taylor & Francis Group, London, ISBN 978-1-032-41287-0

CHAPTER

62

Wrench-Shaped Triple-Port MIMO Antenna for Modern Radar Applications

**Aditya Kumar Singh*, Utkarsh Sharma, Amrees Pandey,
Vandana Yadav, R. S. Yadav**

Department of Electronics & Communication,
University of Allahabad, Prayagraj, India

Abstract

A triple-port MIMO antenna is constructed & studied in a miniaturized size of $20 \times 20 \times 1.6$ mm^3, which is acceptable for X band & Ku band applications & particularly for modern radar & satellite applications presented. The simulation results are shown (A-1, A-2, & A-3) in terms of simulated peak gain, reflection coefficient, radiation pattern, ECC, & DG. The Ringing phenomenon and its effect on the performance of the micro-strip antenna are explained using coaxial and micro-strip line feeding. The specifications of proposed antenna have been improved for preferred antenna parameters & operation. Simulated isolation is less than −15 dB, ECC is less than 0.18, and DG in the range between 9.940 and 9.999 (dB) is obtained.

Keywords: DG, ECC, Radiation Efficiency, Ringing Phenomenon, Triple-MIMO Antenna

1. Introduction

In recent years, a remarkable development is noticed in indoor and outdoor wireless communication systems due to multiple input multiple output technology. The work on MIMO MSA is reported for the wireless local area network (WLAN), Word Wide interoperability for microwave access (WiMax), long-term evolution (LTE), & 5G/6G communication systems. A

*Corresponding author: aditya08129@gmail.com

DOI: 10.1201/9781003357346-62

very less MIMO antenna system is reported for the X, Ku, K bands. A faster transmission rate along with better communication signal quality enables the MIMO system and is desirable for the bands like X, Ku, & K also. There are only a few MIMO antennas for three ports reported in literature. A small-scale triple port MIMO antenna for the WLAN applications, laptop computers, & systematic design systems of mobile multiple antenna system triple port by utilizing the theory of characteristics modes is reported [1] [2] [3]. An ultra-compact three-port multiple input multiple output antenna directional radiation pattern for 2.4 GHz WLAN & with high isolation applications is reported in [4]. A triple-port MIMO dielectric resonator antenna (DRA) applying decoupled techniques for X-band application is reported [5].

In this paper, we considered the triple-port with compact size MIMO antenna of the frequency range (11-15) GHz on FR-4 substrate. We also investigated the performance of the MIMO antenna by using coaxial feeding and micro-strip line feeding. The overall compact dimension of antenna $20 \times 20 \times 1.6$ mm^3, printed on FR4 epoxy substrate ($\varepsilon_r = 4.4$, tan $\delta = 0.02$ and thickness of 1.6 mm) with simulated by ANYSYS HFSS 18 electromagnetic are used to model, design, and construct the Wrench-shaped MIMO antenna.

2. Systematic Layout and Evolution of Antenna Structure

The evolution of the considered triple-port Wrench-shaped MIMO antenna is presented in Figs 62.1 & 62.2. The analytical growth for A1, A2, & A3 (proposed) is introduced in Fig. 62.1 and its geometric representation and dimensions value (cf. Fig. 62.2) are tabulated in Table 62.1. The red color shows the radiating element (front-view), blue color shows ground (back view) (L × W) mm^2, and black color shows the feeding point of the antennas (cf. Figs. 62.1 & 62.2). A three circular-shaped radiation elements has associated with coaxial feed at port-1 and micro-strip line feed at port 2 and port 3 respectively. A-2 is designed through A-1 by using three circular cut in which diameter of the circular slot D = 6 mm^2 of the radiating element (cf. Fig. 62.1).

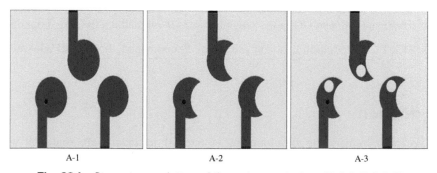

A-1   A-2   A-3

Fig. 62.1 Step-wise evolution of the antenna designs (A-1, A-2 & A-3)

Similarly, A-3 (proposed) is designed through Ant-2 by giving three radii of the circular slot (r = 5 mm) of the radiating element (cf. Figs. 62.1 and 62.2). The designed Wrench-shaped antenna has simple structure, printed on FR4 substrate ($\varepsilon_{rsub} = 4.4$, tan $\delta = 0.0009$ & thickness of 1.6 mm). The suggested antenna is simulated, designed, and implemented using an ANSYS HFSS 18 electromagnetic simulator.

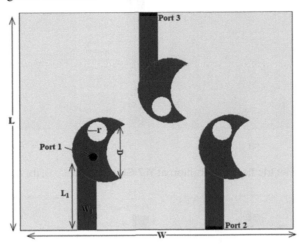

Fig. 62.2 Geometrical specification of triple-port Wrench-shaped MIMO antenna (A-3)

Table 62.1 Specifications of the Proposed MIMO Antenna (A-3)

Geometric Specifications	Values (in mm)
W (Width of the A-3)	20
L (Length of the A-3)	20
W1 (Width of the rectangular slot)	02
L1 (Length of the rectangular slot)	05
D (Diameter of the circular slot)	06
r (Radius of the circular slot)	0.5

3. Results and Discussion

In this appraisal, we have discussed the result of anticipated triple- port MIMO antenna in brief. The surface current distribution of Ant-3 (proposed) is resonating on 11.7 GHz at the phi (0 & 90), which is represented in Fig. 62.3(a) & (b), respectively; for this distribution, maximum value of surface current is 58 A/m. Similarly, the surface current distribution on 13.9 GHz at phi (0 & 90) is represented in Fig. 62.4(a) & (b), respectively; for this distribution, maximum value of surface current is 53 A/m.

In Ant-1, we have found the two bands, that is, 11.2 GHz to 12.2 GHz & 13.2 GHz to 14.4 GHz, at both the bands are showing "ringing effect" phenomenon. The ringing effect is known to be a phenomenon in which an antenna exhibits two resonant frequencies in the same band [6]. Due to which the gain in the same band is falling at particular resonance frequency.

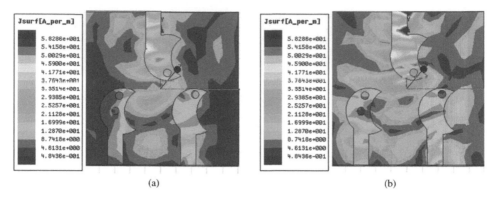

Fig. 62.3 Electric field distribution at 11.7 GHz (a) at phi = 0, (b) at phi = 90

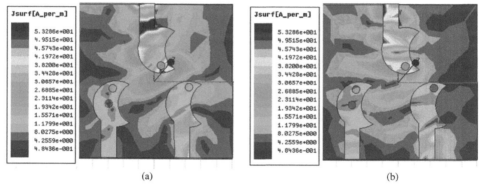

Fig. 62.4 Electric field distribution at 13.9 GHz (a) at phi = 0, (b) at phi = 90

The Ant-1 has a maximum gain of 3.6 dBi, but has seen a drop in gain in the band (11.2–12.2) GHz at 11.7 GHz and in the band (13.2–14.4) GHz at 13.9 GHz, respectively. Furthermore, we focus on port management to remove this unwanted ringing effect. (cf. Fig. 62.5).

In A-2, we obtained "ringing effect" only for the first operating band (11.2-12.2), and for second operating band (13.2-14.4) GHz, it is removed and further peak gain increases up to 4.6 dB (cf. Fig. 62.6).

In A-3, "ringing effect" at both operating bands, i.e. (11.2- 12.2) GHz & (13.2-14.4) GHz, is removed and further peak gain increased up to 5.0 dB (cf. Fig. 62.7). A-3 shows the better gain as compared to Ant-1 & Ant-2. Ant-3 (proposed) obtained two bands (11.2-12.2) GHz and (13.2-14.4) GHz without causing any "ringing effect" phenomenon.

Figure 62.8 shows the reflection coefficient for all three ports at frequency range (11-15) GHz. In port-1, we found the two operating bands (11.2-12.2) GHz and (13.2-14.4) GHz and their bandwidth (BW) is 1 GHz and 1.2 GHz, respectively. At port-2, we obtained only single operating band (13.5-14.2) GHz with BW of 0.7 GHz. In port-3, (11.3-12.0) GHz operating band and its BW 0.7 GHz are obtained.

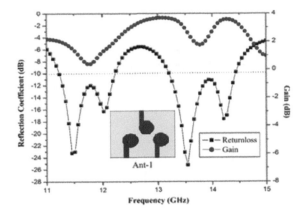

Fig. 62.5 Return loss & peak gain plot of A-1

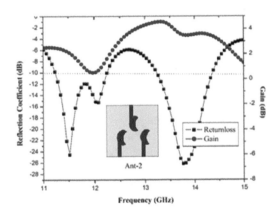

Fig. 62.6 Return loss & peak gain plot of A-2

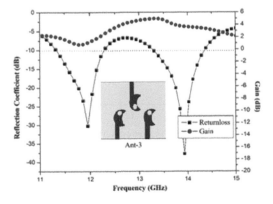

Fig. 62.7 Return loss & peak gain plot of A-3

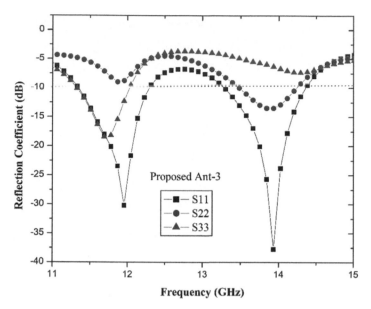

Fig. 62.8 |S11|, |S22| & |S33| versus gain plot of A-3 (proposed)

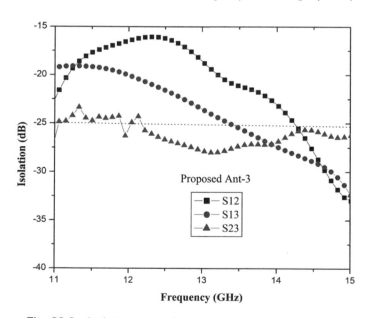

Fig. 62.9 Isolation versus frequency plot of A-3 (proposed)

Peak gain of the proposed A-3 in the entire operating band at port 1, port 2, & port 3 is above the zero and peak gain up to 5.0 dB is observed (c.f. Fig. 62.7). Isolation for the proposed A-3 is < −15 dB, as shown in Fig. 62.9.

In this triple-port MIMO antenna, ECC, isolation, and diversity gain are obtained with the help of S-parameter and interpreted in equations (1) and (2). ECC value for the MIMO antenna system is suggested to be < 0.5 and its value should be near to zero, so that mutual coupling effect of the antenna elements would be better. Fig. 62.10 shows the maximum value of ECC is < 0.16 at the entire operating frequency range (11–15) GHz. Diversity gain can be calculated with the help of ECC and the formula is shown in equation (2). Diversity gain of the proposed Ant-3 varies between 9.84 and 10 (cf. Fig. 62.10).

$$Diversity\ Gain = 10\sqrt{1 - ECC^2} \tag{1}$$

$$ECC = \frac{\left| \iint_{4\pi} F1(\theta,\varnothing).F_2(\theta,\varnothing)d\Omega] \right|^2}{\iint_{4\pi} |F1(\theta,\varnothing)| \dagger d \iint_{4\pi} |F2(\theta,\varnothing)|^2 d\Omega} \tag{2}$$

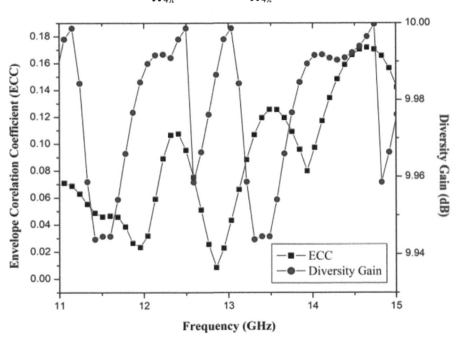

Fig. 62.10 ECC and diversity versus frequency plot of A-3

Radiation efficiency of a proposed MIMO antenna is shown in Fig. 62.11, which varies between 64% and 70% at entire frequency range of 11 to 15 GHz. Radiation pattern in E & H is represented in Fig. 62.12, at resonating on 11.9 GHz and 13.9 GHz; both frequencies show the omni-directional radiation pattern, respectively (Fig. 62.12).

The comparative analysis is in terms of antenna size, type of antenna, impedance BW, gain and application. The proposed work in Table 62.2, is compared with practically available antennas for X and Ku band applications. Antenna(A-3) occupies a lesser size as compared to antennas reported in references [7, 8, 9 & 10].

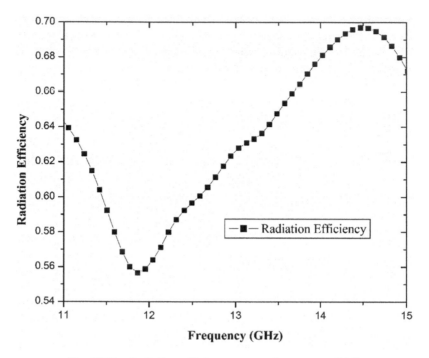

Fig. 62.11   Radiation efficiency versus frequency of A-3

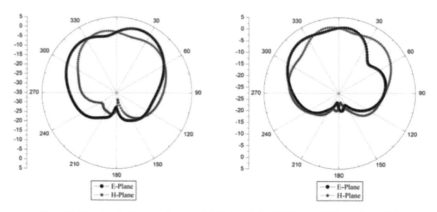

Fig. 62.12   Radiation pattern of A-3 (a) at 11.9 GHz and (b) at 13.9 GHz

4. Conclusions

In this paper, we investigated the ports analysis with the help of coaxial & micro-strip line feeding, and ports will play a significant role on the antenna efficiency. Multiple resonant frequencies in a single operating band are named as "ringing effect", which is also the causal of gain declination. Antenna 3 shows the better gain as compared to Ant 1; Ant 2 & Ant 3

Table 62.2 Comparison of considered antenna with previous antenna

Antenna size (mm³)	Antenna type	Operating frequency band (GHz)	Impedance BW (%)	Peak gain	Application
$28.8 \times 37.21 \times 1.6$[7]	Wideband	10.6-15.9	40.01	9.2	X and Ku band
$40 \times 40 \times 1.6$ [8]	Dual-band	8.12-16,01, 17.03-18.91	6.66 5.61	5.1 2.2	X and Ku band
$50 \times 50 \times 1.52$ [9]	Dual-band	11.44-12.48 13.47-14.39	8.7 6.6	3.71	X and Ku band
$36 \times 36 \times 0.76$[10]	Wide-band	9.89-17.55	55.83	NR	X and Ku band
$20 \times 20 \times 1.6$ (Proposed Work)	Dual-band	11.2-12.2, 13.2-14.4	8.5 8.7	5	X and Ku band

(proposed) obtained the two operating bands (11.2-12.2) GHz and (13.2-14.4) GHz without causing any "ringing effect" phenomenon. So we conclude that the Ant-3 is suitable for the X and Ku band applications and in particular for modern Radar and Satellite applications.

REFERENCES

1. Deng, C, Shi, B, Liu, D.: Compact omnidirectional three-port MIMO antenna with the same vertical polarization for WLAN applications. Microw Opt Technol Lett. Vol. 62, pp. 800–805, 2020.
2. Chen, W- S, Lin, R- D.: Three- port MIMO antennas for laptop computers using an isolation element as a radiator. Int J RF Microw Comput Aided Eng. Vol. 31, pp., 2021.
3. Martens, Robert, Manteuffel, Dirk.: Systematic design method of a mobile multiple antenna system using the theory of characteristic modes.IET Microwaves, Antennas & amp; Propagation, Vol. 8, pp. 887–893, 2014.
4. Wang, H, Liu, L, Zhang, J, Li, Y and Feng Z.: Ultra-Compact Three- Port MIMO Antenna With High Isolation and Directional Radiation Patterns. IEEE Antennas and Wireless Propagation Letters, Vol. 13, pp. 1545–1548, 2014.
5. A. Abdalrazik, A. S. A. El-Hameed and A. B. Abdel-Rahman.: A Three- Port MIMO Dielectric Resonator Antenna Using Decoupled Modes. IEEE Antennas and Wireless Propagation Letters, Vol. 16, pp. 3104–3107, 2017.
6. Singh, V., Mishra, B., Singh, R.: Dual-wideband semi-circular patch antenna for Ku/K band applications, Microwave and Optical Technology Letters, Vol. 61, pp. 323–329, 2017.
7. Yadav, N , Hu, G and Yao, Z.: Parallel Notch and H Shape Slot Loaded Compact Antenna for X and Ku Band Applications. Open Journal of Antennas and Propagation, Vol. 7, pp. 13–21, 2019.
8. Ravi Kumar, KS, Prasad, L, Ramesh, B, Vinay, KP, Mallikarjuna Rao P.: Broad Band Comb Shaped Micostrip Patch Antenna For X-Band & Ku-Band Applications. American International Journal of Research in Science, Technology, Engineering & Mathematics Vol. 23(1), pp. 52–54, 2018.
9. Thi, TN, Hwang, KC, Kim, HB.: Dual-band circularly-polarised Spidron fractal microstrip patch antenna for Ku-band satellite communication applications. Electron Lett.; 49(7), pp. 444–445, 2013.
10. Khandelwal MK, Dwari S, Kanaujia BK, Kumar S. Design and analysis of microstrip DGS patch antenna with enhanced bandwidth for Ku band applications. In: International Conference on Microwave and Photonics. 2013.

Intelligent Systems and Smart Infrastructure – Brijesh Mishra et al. (eds)
© 2023 Taylor & Francis Group, London, ISBN 978-1-032-41287-0

CHAPTER

63

A Review on Plant Disease Detection Using Machine Learning Techniques

Waris Abid Ahrari[1], Shreyansh Yadav[2],
Manish Kumar Gupta[3], Ranjeet Singh[4]

Computer Science & Engg Deptt
Buddha Institute of Technology, GIDA
Gorakhpur, UP, India

Abhishek Kumar Pandey[5], Ashutosh Pandey[6]

Computer Science & Engg Deptt
Shambhunath Institute of Engineering and Technology
Prayagraj, UP, India

Abstract

Plant and crop diseases is a major risk when it comes to the quantity and quality of farming; however, due to a lack of essential systems in certain regions of the world, early detection of plant and crop diseases is still very challenging. The advances in computer vision technologies with the help of deep learning, along with the increase in global smart phone penetration rate, have made it easier for smart phone-assisted disease diagnosis. This paper talks about the different Deep Learning Algorithms along with their accuracy scores and various image processing techniques for pre-processing of images that can be used for the classification of different diseases in plants and crops. Overall, the advancement of computer vision technologies and the use of freely available plant and crop disease datasets paves the way for crop and plant disease detection using a smart phone.

Keywords: Deep Learning; image processing; plant disease detection

Corresponding author: [1]waris.ahrari@gmail.com, [2]shreyansh.yadav44@gmail.com, [3]manish.testing09@gmail.com, [4]ranjeetsingh370@bit.ac.in, [5]abhi_007cs@yahoo.co.in, [6]pandeyashutosh33@gmail.com

DOI: 10.1201/9781003357346-63

1. Introduction

Agriculture has become a major factor of economic development in India. The appropriate crop is chosen by the farmer depending upon the soil, local climatic conditions, its economic worth, etc. Because of rising population, changing weather, and political instability, agriculture companies began looking for innovative ways to enhance food output. As a result, researchers are looking for new efficient and precise technologies to increase food quality and quantity.

In a successful farming system, accurate disease identification in plants and crops is essential. A farmer, in general, identifies disease symptoms in plants and crops through naked eye observations, which necessitates constant monitoring. However, when it comes to the implementation of this manual process on large plantation, it is very expensive and inefficient. In the developing countries like India, experts are hired for the identification of diseases in crops, which makes it even more expensive for the farmers.

Plants are attacked by several diseases, which usually target the specific plant parts like stem, fruit, leaf and seed (Singh et al. 2019). Leaf symptoms are an important origin of knowledge to detect the diseases in multiple types of plants and therefore it has to be contemplated in identifying the disease. This paper reports about the revelation of leaf disease using Image processing strategies, as it is a trendy process being used in agriculture.

A. Plant Disease

Diseases in plants are of two types: Biotic (infectious) and Abiotic (noninfectious). Living organisms like bacteria, fungus, viruses, insects, nematodes, and animals cause biotic diseases (infectious diseases). Non-infectious diseases, often known as abiotic diseases, are caused by non-living factors such as harsh environmental conditions (freeze injury, wind injury, chemical injury, drought stress, sunscald, nutrient deficiency, etc.). Black spot is a fungal disease common in rose leaves and it causes round black spots on the leaves. Like this, there are a lot of plant and crop diseases that affect the quality, quantity and growth of the plants and crops. Both biotic and abiotic diseases have some type of visual effect on the plants. The effects of diseases in plants vary based on the types of pathogens, infections, infected part, duration of the infection, environmental conditions and Diseases that affect the leaves of plants include leaf spot and leaf blight, among other things. Similarly, root rots affect root; fruit rots and fruit spots affect the fruits, etc.

Plant and crop diseases are causing low productivity of the goods and financial losses for farmers all around the world. Before adequate control strategies can be offered, a rapid and precise identification of the disease is necessary. It is the initial step to treat any disease. Therefore, a fast and reliable method for detecting the plant disease is very important.

B. Deep Learning

Based on the working of hidden layers, neural networks are classified into different types. Convolutional Neural Network (CNN or ConvNet) is the most common technique when it comes to the problems related to image classification using Deep Learning, as in CNN, every pixel acts as the input nodes in input layer. The hidden layers in CNN consist of filters or kernels for features or patterns to be searched in the images; these filters then assign scores to each pixel based on how closely they match with the filters, and these set of filters or kernels are called convolutional layer. The initial convolutional layer extracts low-level features of the images while later layers capture more abstract high-level features like objects and shapes.

Many study areas, such as medical diagnostics, bioinformatics, and satellite imagery applications, have been mastered by deep learning. Deep learning has recently been implemented in agricultural applications and it solves various problems such as it is helpful for classifying different crops and plant diseases, for identifying plant type using leaves and for counting fruits on a tree using color as a feature.

For detecting and classifying images in agriculture as well as in variety of other fields (like medical, self-driving cars, etc.), there are vast number of supervised learning models that are being developed nowadays and CNN (or Convolutional Neural Network) is mostly used for creating these models as it automatically extracts relevant characteristics.

2. Related Works

A wide range of surveys have been conducted to compare different Machine Learning algorithms for plant and crop disease detection and classification. There is a lot of work done in the domain to develop algorithms for classification, identification and recognition of plant and crop diseases. We have looked into several research papers to compare efficiency of different algorithms for identification and classification of plant disease.

A. ANN Classifier

Bashish, Braik, and Bani-Ahmad (2010), tested a proposed approach for plant disease recognition using a feed-forward back-propagation algorithm and found that it functioned effectively, with a precision of roughly 93 percent. Ramakrishnan. M and Anselin Nisha (2015) proposed a work on identification of the disease Cercospora (leaf spot) in groundnut plant using Backpropagation neural network method. Their proposed model classified 4 different disease and shows the accuracy of around 97 percent using 100 diseased leaves images sample for training and testing purpose. And in another paper, the author used a back-propagation algorithm to identify four different diseases of pomegranate plants with an accuracy score of around 90% (Dhakate and Ingole A. B. 2015).

B. KNN Classifier

K-Nearest Neighbors (or KNN) is used for both classification and regression purposes. It is a supervised learning algorithm. There is a lot of papers on classification of plant disease using K-Nearest Neighbor classifier. Eaganathan et al. (2014) used a KNN (K-Nearest Neighbor) algorithm for classification Leaf scorch, which is a sugarcane disease with the accuracy score of around 95 percent. Parikh et al. (2016) proposed an algorithm to classify Grey Mildew disease, which is a common cotton plant disease that weakens the plant and its productivity. The accuracy score for the proposed algorithm is around 83 percent using just 40 sample images for training and testing purposes.

C. SVM Classifier

Masazhar and Kamal (2017) used SVM classifier to detect two different diseases of palm oil and have achieved the accuracy of 97% and 95%, respectively.

Using SVM classifiers, Hossain et al. (2018) proposed an algorithm to recognize tea plant diseases. The proposed algorithm was able to detect three diseases and achieved an accuracy around 90%. They used 200 sample images of tea leaves, 150 sample images for training purpose, and 50 sample images for testing purpose. In another paper, the authors used a total of 160 sample images of three diseases combined for the training purpose of an SVM classifier and the classifier achieved the accuracy score of around 90 percent (Agrawal, Singhai, and Agarwal 2017).

D. FUZZY Classifier

Majumdar et al. (2015) proposed an algorithm to detect and classify four different diseases present in wheats using the Fuzzy C-Means Classifier, where the value of C is 2 and 4; 2 here means the algorithm classifies only healthy and unhealthy leaves, and also when C is 4, then R, MR, MS and S are the four clusters that were considered for the classification. This algorithm is used for classification between healthy and unhealthy leaf samples and also for categorizing four different diseases. Accuracy of 88 percent was achieved when only healthy and unhealthy leaves are classified, and for the classification of a particular diseases, this algorithm gives the accuracy of 56%.

E. Convolutional Neural Network (CNN)

In 2016, Sladojevic et al. proposed a model to detect disease of apple, pear, peach, grapevine and cherry using CNN classification technique. The dataset of approx. 31,000 sample images was used for the training and testing purposes and it achieved accuracy score is in range 91%–98% for separate class test and with the average accuracy score of 96.3% (Sladojevic et al. 2016). Another paper proposed a classification algorithm for detection of diseases in plants and crops using CNN technique and trained it over a dataset that contains around 88,000 sample images of 25 variety of plants and they combined 58 diseases and produced an accuracy score of 99.53%, which is very impressive result over such large dataset (Ferentinos 2018). Mohanty Sharada P. et al. trained a Deep Convolutional Neural Network for detecting disease in plants and crops using publicly available dataset that contained around 54,300 images of 14 different crops and plants and combination of 26 diseases and the algorithm yields the accuracy score of 99.35% when 20% of sample image data is used for testing and 98.2% when 80% of sample image data is used for testing (Mohanty, Hughes, and Salathé 2016).

Table 63.1 Comparison of different classification techniques

[Reference]	Disease	Total Images	Classifier	Accuracy (%)
(Sladojevic et al. 2016)	13 Diseases	30880	Deep convolutional NN	96.3
(Francis and Deisy 2019)	Healthy or Unhealthy	3663	CNN	87
(Mohanty, Hughes, and Salathé 2016)	26 Disease	54306	CNN	99.35
(Ferentinos 2018)	58 Disease	87848	CNN	99.53
(Majumdar et al. 2015)	Healthy or Unhealthy	310	Fuzzy C-means (C = 2)	88
	4 Disease		Fuzzy C-means (C = 4)	56
(Hossain et al. 2018)	3 Disease	200	SVM	90
(Agrawal, Singhai, and Agarwal 2017)	Black rot, Esca & Leaf Blight	160	SVM (HSI and LAB color model)	90
		160	SVM (LAB color model)	82.5
(Eaganathan et al. 2014)	Leaf scorch disease	-	k-NN	95
(Parikh et al. 2016)	Grey Mildew	40	k-NN	82.5
(Bashish, Braik, and Bani-Ahmad 2010)	4 Disease	-	ANN	93
(Ramakrishnan. M and Anselin Nisha 2015)	Cercospora (leaf spot)	100	ANN	97.41
(Dhakate and Ingole A. B. 2015)	4 Disease	500	ANN	90

3. System Architecture

The most popular system architecture for detection of plant diseases using deep learning and image processing is presented in Fig. 63.1. The different phases for detecting diseases in plants using image processing and classification are the following.

A. Acquisition

Image acquisition is a very necessary phase as the efficiency of the system is highly depended on the sample images used for training. For this paper, the images of the plant leaf (both diseased and healthy) are from publicly available dataset, i.e., Plant Village Images. Images in the dataset are in RGB form.

B. Pre-processing of Sample Image

In this step, different techniques are used to improve the quality of the input image that later is fed into the training phase. Different pre-processing techniques include deblurring, color space conversion, smoothing, cropping and enchantment. Based on the quality of the sample

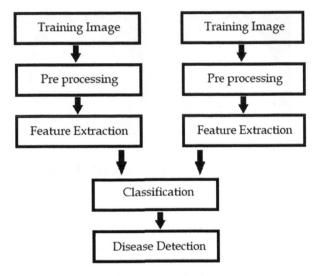

Fig. 63.1 System Architecture

image used, this phase functions differently. Commonly, Enhancement and Filtering is done after color space conversion. Since it closely resembles properties of humans in sensing color, HSV is widely used as color space.

Basically, this step enhances some of the image features necessary for further processing. Cropping of image is performed to get rid of the unnecessary region in the image, which could act as noise while training the model. For increasing the contrast of the image, we need to perform image enrichment.

Equation (1) is used for converting RGB image to gray image as follows:

$$F(x) = r * 0.2721 + g * 0.5782 + b * 0.094 \tag{1}$$

C. Image Segmentation

Image Segmentation is the process of dividing images into smaller parts, which have exactly the same or similar features.

The identification of diseased or healthy images becomes very easy when the sample image is very well segmented based on its features (S. Phadikar, J. Sil, and A. K. Das 2012). Some known segmentation techniques like edges, thresholds, locality or color-based, etc. is found to work better for systems detecting plant or crop diseases. There are different techniques to perform segmentation like k-means clustering, Otsu's method, conversion RGB into HIS (Hue, Saturation and Intensity), etc. For this paper, the segmentation process is demonstrated using K-means clustering technique and Otsu Threshold Algorithm.

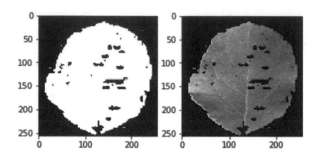

Fig. 63.2 Otsu Threshold Algorithm for Image Segmentation

D. Feature Extraction

In Feature extraction, initial set of large data set is reduced without losing any important information which helps to build more efficient and less noisy model. Hence, further processing is much easier and efficient. Since feature extraction removes the redundant data from the image, which is not necessary for classification, it plays an important part for the object identification. To detect plant disease, features such as color, texture, edges, morphology, etc. can be considered. However, in paper (Jhuria, Kumar, and Borse 2013), Monica Jhuria et al. consider texture, color and morphology of leaves as a feature for detecting diseases in plant. In their paper, they concluded that morphological features like shapes and edges of plants leaves give better results than any other features.

E. Classification of Sample Image

Classification is the final and the most important module of the plant disease detection system. The process of recognizing or detecting plant and crop diseases based on the morphological features of the leaves or other parts of plants is known as classification. Classifier is trained using the image dataset that are pre-processed using earlier module, then the trained classifier can recognize or classify the leaves in the testing phase.

There are various machine learning algorithms that can be used to identify the disease; some of them are discussed briefly in this paper.

4. Experiments and Results

A total of 1709 image datasets (805 healthy leaf images and 904 diseased leaf images) are used for training purposes. These data are fitted with different algorithms and the accuracy of each algorithm is calculated and represented visually using Box Plot with the help of matplot library. Table 63.2 shows the algorithms that are considered for this paper, along with their accuracy scores.

Figure 63.3 presents a statistical analysis that integrates the image dataset of 805 healthy and 904 diseased leaf, grouped by the performance of seven different Machine Learning Algorithms.

Table 63.2 Accuracy Score of Different Algorithm

Algorithms	Accuracy	Standard Deviation
Linear Regression	0.920312	0.021195
LDA	0.903122	0.025958
K-Nearest neighbor	0.922656	0.025303
CART	0.923438	0.028556
Random Forest	0.959375	0.015149
Naïve Bias	0.861719	0.020086
SVM	0.774219	0.043561

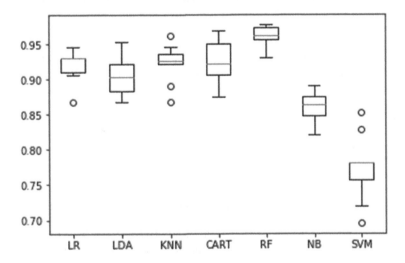

Fig. 63.3 Machine Learning Algorithm Comparison

These stats show that among different machine learning algorithms, Random Forest has the highest accuracy of 95.93 %. Other algorithms like Linear Regression, K-Nearest Neighbor, and CART have slightly less accuracy of around 92%.

5. Conclusion

For efficient and effective farming, the plant and crop diseases must be detected with accuracy and treated in the early stages to avoid widespread of the disease. This can be achieved with the help ever-improving computer vision technologies. This paper discussed various techniques to segment all the diseased part of the plant, along with various techniques for feature extraction and disease classification to extract the features of infected leaf and the classification of plant diseases. For this review, a total of seven different machine learning algorithms were studied for the detection and classification of plant and crop diseases. Detection and treatment of plant diseases in the initial stage can prevent the wide spread of the diseases in the fields, which result in improvement of quality and quantity of crops, hence increasing the efficiency of farmers.

REFERENCES

1. Agrawal, Nitesh, Jyoti Singhai, and Dheeraj K. Agarwal. 2017. "Grape Leaf Disease Detection and Classification Using Multi-Class Support Vector Machine." In *2017 International Conference on Recent Innovations in Signal Processing and Embedded Systems (RISE)*, 238–44. IEEE. doi: 10.1109/RISE.2017.8378160.
2. Bashish, Dheeb al, Malik Braik, and S Bani-Ahmad. 2010. "A Framework for Detection and Classification of Plant Leaf and Stem Diseases." *2010 International Conference on Signal and Image Processing*, 113–18.
3. Dhakate, Mrunmayee, and Ingole A. B. 2015. "Diagnosis of Pomegranate Plant Diseases Using Neural Network." In *2015 Fifth National Conference on Computer Vision, Pattern Recognition, Image Processing and Graphics (NCVPRIPG)*, 1–4. IEEE. doi: 10.1109/NCVPRIPG.2015.7490056.
4. Eaganathan, Umapathy, Jothi Sophia, Vinukumar Luckose, Feroze Jacob Benjamin, Vels University-India, Csi JACON-India, and Nilai University-Malaysia. 2014. "Identification of Sugarcane Leaf Scorch Diseases Using K-Means Clustering Segmentation and K-NN Based Classification." In .
5. Ferentinos, Konstantinos P. 2018. "Deep Learning Models for Plant Disease Detection and Diagnosis." *Computers and Electronics in Agriculture* 145 (February): 311–18. doi: 10.1016/j.compag.2018.01.009.
6. Francis, Mercelin, and C. Deisy. 2019. "Disease Detection and Classification in Agricultural Plants Using Convolutional Neural Networks — A Visual Understanding." In *2019 6th International Conference on Signal Processing and Integrated Networks (SPIN)*, 1063–68. IEEE. doi: 10.1109/SPIN.2019.8711701.
7. Hossain, Selim, Rokeya Mumtahana Mou, Mohammed Mahedi Hasan, Sajib Chakraborty, and M. Abdur Razzak. 2018. "Recognition and Detection of Tea Leaf's Diseases Using Support Vector Machine." In *2018 IEEE 14th International Colloquium on Signal Processing & Its Applications (CSPA)*, 150–54. IEEE. doi: 10.1109/CSPA.2018.8368703.
8. Jhuria, Monika, Ashwani Kumar, and Rushikesh Borse. 2013. "Image Processing for Smart Farming: Detection of Disease and Fruit Grading." In *2013 IEEE Second International Conference on Image Information Processing (ICIIP-2013)*, 521–26. IEEE. doi: 10.1109/ICIIP.2013.6707647.
9. Majumdar, Diptesh, Arya Ghosh, Dipak Kumar Kole, Aruna Chakraborty, and Dwijesh Dutta Majumder. 2015. "Application of Fuzzy C-Means Clustering Method to Classify Wheat Leaf Images Based on the Presence of Rust Disease." In , 277–84. doi: 10.1007/978-3-319-11933-5_30.
10. Masazhar, Ahmad Nor Ikhwan, and Mahanijah Md Kamal. 2017. "Digital Image Processing Technique for Palm Oil Leaf Disease Detection Using Multiclass SVM Classifier." In *2017 IEEE 4th International Conference on Smart Instrumentation, Measurement and Application (ICSIMA)*, 1–6. IEEE. doi: 10.1109/ICSIMA.2017.8311978.
11. Mohanty, Sharada P., David P. Hughes, and Marcel Salathé. 2016. "Using Deep Learning for Image-Based Plant Disease Detection." *Frontiers in Plant Science* 7 (September). doi: 10.3389/fpls.2016.01419.
12. Parikh, Aditya, Mehul S Raval, Chandrasinh Parmar, and Sanjay Chaudhary. 2016. "Disease Detection and Severity Estimation in Cotton Plant from Unconstrained Images." In *2016 IEEE*

International Conference on Data Science and Advanced Analytics (DSAA), 594–601. doi: 10.1109/DSAA.2016.81.

13. Ramakrishnan. M, and Sahaya A Anselin Nisha. 2015. "Groundnut Leaf Disease Detection and Classification by Using Back Probagation Algorithm."

14. S. Phadikar, J. Sil, and A. K. Das. 2012. "Classification of Rice Leaf Diseases Based on Morphological Changes." *International Journal of Information and Electronics Engineering* 2 (3): 460–63.

15. Singh, Uday Pratap, Siddharth Singh Chouhan, Sukirty Jain, and Sanjeev Jain. 2019. "Multilayer Convolution Neural Network for the Classification of Mango Leaves Infected by Anthracnose Disease." *IEEE Access* 7: 43721–29. doi: 10.1109/ACCESS.2019.2907383.

16. Sladojevic, Srdjan, Marko Arsenovic, Andras Anderla, Dubravko Culibrk, and Darko Stefanovic. 2016. "Deep Neural Networks Based Recognition of Plant Diseases by Leaf Image Classification." Edited by Marc van Hulle. *Computational Intelligence and Neuroscience* 2016. Hindawi Publishing Corporation: 3289801. doi: 10.1155/2016/3289801.

Intelligent Systems and Smart Infrastructure – Brijesh Mishra et al. (eds)
© 2023 Taylor & Francis Group, London, ISBN 978-1-032-41287-0

CHAPTER

64

3D Convolution for Driver Yawning Detection

Sandeep Mandia[1], Ashok Kumar[2]

Dept. of Electronics and Communication Engineering,
Govt. Mahila Engineering College, Ajmer, Rajasthan, India

Jitendra Kumar Deegwal[3]

Dept. of Electronics Instrumentation & Control Engineering,
Government Engineering College, Ajmer, Rajasthan, India

Karan Verma[4]

Dept. of Computer Science & Engineering,
National Institute of Technology, Delhi, Delhi, India

Abstract

Yawning is the main indicator of drowsiness in driving. The yawning detection well on time can save lives by alerting drivers. In this study, the yawning is modeled as a spatiotemporal activity. A 3D convolutional neural network model is presented for yawning detection in this paper. It is assessed on the yawning detection dataset (YawDD). We also worked on the dataset for making it more robust. The proposed methods exhibit 83.53% and 85.65% accuracies for 8 and 16 temporal length inputs, respectively. The proposed model having 16 temporal length input exhibits precision, recall, and F1-score values of 84.09%, 87.50%, and 82.80%, respectively.

Keywords: Yawning, 3D convolution, deep learning.

Corresponding author: [1]smandia20@gmail.com, [2]kumarashoksaini@gmail.com, [3]jitendradeegwal@gmail.com, [4]karanverma@nitdelhi.ac.in

DOI: 10.1201/9781003357346-64

1. Introduction

To improve road safety, prior warning detection, assisting vehicle control, and monitoring are integral parts of intelligent driving for research (Marina Martinez et al. 2018). Thousands of people every year have been killed or gravely wounded when drivers fall asleep while driving. Driver fatigue is a severe threat to road safety. Therefore, the research on automatic driver drowsiness detection is vital for improvement in road safety. To assist drivers to drive safely, many driving drowsiness detection methods have been demonstrated over the past few years (Kholerdi et al. 2016; Mandal et al. 2017; McDonald et al. 2014; Sun et al. 2017; Yang et al. 2020). Blinking of eyes, eyes closing, nodding, and yawning are the behavioural attributes of driver in fatigue driving. The yawning is considered one of the foremost causes of fatigue for drivers (Li, Chen, and Li 2009).

In the genuine driving environment, an accurate and robust yawning detection method becomes challenging as the drivers have trivial facial activities and expressions. Various yawning detection methods have been proposed (Abtahi, Hariri, and Shirmohammadi 2011; Haque et al. 2016; Ibrahim et al. 2015; Li, Chen, and Li 2009; Pan, Ma, and Huang 2015; Yang et al. 2020). Most of the existing papers use single static frame technique and detect yawning based on mouth opening. These techniques detect the yawning in accordance with differences in frame attributes through open and closed mouths. Several deep learning-based methods have been also applied on static frames to detect opening and closing of mouth. The approaches accordant with single static frame face challenges of confusing yawning with other facial actions and expressions.

This paper addresses the driver yawning detection problem using a deep learning model. The YawDD dataset contains examples of all normal, talking, and yawning activities in the same sample which is processed and divided into two classes namely yawning and not yawning.

The contributions of this work are as follows:

1. A 3D convolutional model is proposed for driver yawning detection.
2. The YawDD dataset is pre-processed to make it more robust for yawning detection.
3. The 3D convolutional model is evaluated on YawDD dataset and compared with other cited methods.

The remainder of the paper is organised in following order. The subsequent section describes the literature review, next to that section presents the results and discussion, and finally last section concludes the work.

2. Literature Review

In this section, the cited yawning detection approaches or methods will be presented.

Preliminary approaches use histogram features to extract the geometric and colour information from the mouth area of static image frame. However, these approaches have low accuracy and robustness. A mouth aspect ratio based yawning detection method is proposed by (Ibrahim et al. 2015) and they conducted experiment to find the threshold for the aspect ratio. The state is classified yawning when the aspect ratio is less than the threshold identified. Facial keypoint-based method for yawning detection is proposed in Haque et al. (2016). The distance between midpoints of lower and upper lips has been used for the detection of yawning. The threshold for the distance is calculated by performing the experiments with tester images. The main challenge in facial keypoint detection is the rotation of subject more than a critical angle. These methods are computationally efficient; however, these methods need to calculate the threshold that is responsive to external attributes and anti-noise capacity of these methods is poor. The other limitation associated with these approaches is not robust to the camera location for data collection and inference, and camera jitter may result in misclassification.

Subsequently, to eliminate the problems associated to preliminary methods, researchers have started to develop the feature-based methods (Abtahi, Hariri, and Shirmohammadi 2011; Ibrahim, Soroghan, and Petropoulakis 2013; Omidyeganeh, Javadtalab, and Shirmohammadi 2011; Pan, Ma, and Huang 2015). The feature-based methods have been improved through classification accuracy and algorithm robustness in several environments. The AdaBoost algorithm has been used to classify fatigue based on local binary pattern (LBP) features of face out of various different LBP features (Y. Zhang and Hua 2015). This method achieved good performance metric in the most states; however, this algorithm has not been robust to lighting variations. Another method that does not utilize any classifier is demonstrated by Anitha, Venkatesha, and Adiga (2016), which lowers the time and space complexity. It utilizes histogram of vertical projection of the lower face part. For yawning detection in this approach, a binary image of black blob in the mouth area is considered. There might be manifold blobs near face due to non-skin region that leads to false detection. Also, it requires to keep the mouth area accurately in the region of interest. A method that uses curve fitting of mouth inner corner points to detect yawning has been presented by (Ding, Zhang, and Chen 2013). For yawning detection, a dual-stage mouth inner contour mathematical discrimination model is considered. The spatio-temporal descriptors approach is used to detect yawning for monitoring driver fatigue in a nonstationary and non-linear environment (Akrout and Mahdi 2016). Du et al. (2011) extracted seven features such as mouth width, height of the lip gap, gray-level co-occurrence matrix, etc. and evaluated these features through rough set method of kernelized fuzzy, trained decision tree and support vector machine (SVM) to categorize the yawn. For yawning detection, the image sequences techniques are also considered. It provides the paramount performance between the conventional methods. Using multiple features, it might not solve the problem as many human mouth actions, namely shouting, are analogous to yawning. Therefore, these steps might be falsely detected as yawning.

The advancements in memory and compute capacity have enabled deep learning frameworks to solve the problem. The yawning detection has also been explored using deep learning by researchers (Choi, Hong, and Kim 2016; Han and Chong 2016; Husain et al. 2022; Yang et al. 2020). To enhance detection accuracy and lower the computation time, Zhang et al. (2015) demonstrated a framework for yawning detection that combines network features and traditional features.

Most of the studies have utilized image-based approach for yawning detection. However, yawning is a spatio-temporal activity and we have modelled it in spatio-temporal space.

3. Proposed Methodology

As yawning is spatiotemporal activity, a 3D convolutional neural network (Tran et al. 2015) model for yawning detection is proposed. Fig. 64.1 shows the proposed methodology for yawning detection and architecture of 3D convolution model. Initially, the images from the video sequences are extracted. Further, the image frames with a fixed temporal length passed to the 3D convolutional model. The model is responsible for detection of yawning; if the yawning is detected, an alert signal is initiated by the system. This work addresses the yawning detection using the 3D convolution deep neural network architecture. The particulars of the architecture are summarized in Table 64.1.

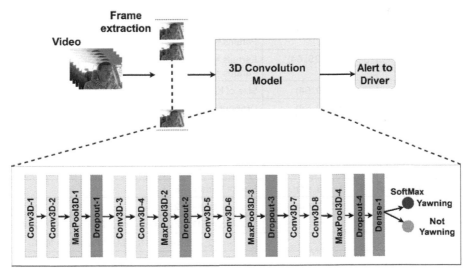

Fig. 64.1 Proposed methodology for yawning detection and model architecture

The 3D convolutional network contains the three main layers, namely Conv3D, MaxPool3D, and dropout; these layers will be discussed one by one. To start with, the Conv3D layer is similar to the Conv2D layer; however, unlike image data, the video has a four-dimensional tensor. The Conv3D layer has a three-dimensional kernel that takes care of spatial as well as temporal changes. The MaxPool3D layer is also similar to the MaxPool2D but the kernel of this layer is a three-dimensional. The dimension reduction happens in spatial and the temporal dimensions. Furthermore, the dropout layer is responsible to avoid the overfitting; it automatically turns off randomly some of the neurons decided by dropout rate from the output and make the model to predict based on the remaining outputs.

The proposed architecture is able to achieve higher accuracy than other methods proposed on the same dataset.

Table 64.1 Details of the architecture

Layer Name	Output Dimensions	Layer Attributes
Conv3D-1	(16, 224, 224, 32)	Kernels = 32, Size of Kernel = (3, 3, 3), Activation = relu, Params = 2624
Conv3D-2	(16, 224, 224, 32)	Kernels = 32, Size of Kernel = (3, 3, 3), Activation = relu, Params = 27680
MaxPool3D-1	(6, 75, 75, 32)	Size of Kernel = (3, 3, 3)
Dropout-1	(6, 75, 75, 32)	Dropout rate = 0.25
Conv3D-3	(6, 75, 75, 64)	Kernels = 64, Size of Kernel = (3, 3, 3), Activation = relu, Params = 55360
Conv3D-4	(6, 75, 75, 32)	Kernels = 64, Size of Kernel = (3, 3, 3), Activation = relu, Params = 110656
MaxPool3D-2	(2, 25, 25, 64)	Size of Kernel = (3, 3, 3)
Dropout-2	(2, 25, 25, 64)	Dropout rate = 0.25
Conv3D-5	(2, 25, 25, 64)	Kernels = 64, Size of Kernel = (3, 3, 3), Activation = relu, Params = 110656
Conv3D-6	(2, 25, 25, 64)	Kernels = 64, Size of Kernel = (3, 3, 3), Activation = relu, Params = 110656
MaxPool3D-3	(1, 9, 9, 64)	Size of Kernel = (3, 3, 3)
Dropout-3	(1, 9, 9, 64)	Dropout rate = 0.25
Conv3D-7	(1, 9, 9, 64)	Kernels = 64, Size of Kernel = (3, 3, 3), Activation = relu, Params = 110656
Conv3D-8	(1, 9, 9, 64)	Kernels = 64, Size of Kernel = (3, 3, 3), Activation = relu, Params = 110656
MaxPool3D-4	(1, 3, 3, 64)	Size of Kernel = (3, 3, 3)
Dropout-4	(1, 3, 3, 64)	Dropout rate = 0.25
Dense	512	Neurons = 512, Activation = relu, Params = 295424
SoftMax	2	Neurons=2, Activation = SoftMax, Params = 1026

4. Experimental Results

The proposed model is implemented using TensorFlow framework on the Google Colaboratory platform. In this study, the model is trained using categorical cross-entropy loss function, and Adam optimizer having 0.001 learning rate is considered. The training ran for 50 epochs.

A. Dataset

The experiment is performed on the yawning detection dataset (YawDD) (Abtahi, Shabnam et al. 2020), which consists of video data. The videos are captured in varying lighting conditions. All videos are recorded at 30 frames/second and stored in an AVI format which contains 640×480 resolution and RGB color. The videos have different activities, namely normal driving, talking, and yawning. This dataset has two subsets in which the first subset is recorded from the car dashboard with and without subject wearing glasses or sunglasses. The first subset has a total of 29 videos of both male and female subjects. Conversely, the second subset is recorded using the camera mounted on the rear-view mirror of the car. This subset also includes subjects with and without sunglasses or glasses. In the second subset, there are a total of 322 videos.

The videos are divided into yawning and not yawning classes; the yawning videos are manually clipped to the length that only captures yawning from onset to peak to end. This resulted into a total of 217 videos from yawning category and 253 videos of not yawning category.

B. Results and Discussion

The prepared data is divided into the train and test set with 80:20 ratio, respectively. The experiment is performed at two different temporal lengths. To start with, the temporal length 8 is taken for the first experiment.

After first experimentation, the test set accuracy of about 83.53% is observed, which is better than many other cited methods. Since the yawning activity starts with mouth opening, then mouth opening increases and finally the mouth is closed. To capture whole of this into the same sample, the temporal dimension has been increased from 8 to 16. This results into an increase in the accuracy about two percent and observed accuracy of about 85.65%. Table 64.2 presents the comparison of the proposed work with the cited works.

Table 64.2 Comparison of the proposed work with the cited works

Method	Accuracy (%)
(Yu et al. 2019)	76.2
(Puja Seemar * 2017)	79
(Abtahi et al. 2014)	60
(Yang et al. 2020)	83.4
Proposed method with temporal length- 8	83.53
Proposed method with temporal length- 16	85.65

The proposed method is compared with the existing methods on the accuracy evaluation metric. Furthermore, the accuracy gives the limited information of the model, the precision gives the information about the predicted positive class actually belong to true positive class, the recall gives the information about the out of total positive class samples how many are predicted as positive, and the F1-score combines the precision and recall into same metric, and is the geometric mean of these two. Hence, the precision, recall, and F1-score of the proposed model having temporal length 16 are also calculated for better understanding. Table 64.3 presents their values.

Table 64.3 Precision, recall, and F1-score of proposed model

Precision (%)	Recall (%)	F1-Score (%)
84.09	87.50	82.80

5. Conclusion

In this paper, the yawning is modelled as spatiotemporal activity and proposed a 3D convolutional network for the yawning detection. The YawDD dataset is considered to categorise within yawning and not yawning categories. In the not yawning category, the talking and normal samples are also considered, which makes the data more robust. The proposed method provided better accuracies than cited works. In future, the model will be optimized for edge devices to deploy in the real-time scenario.

REFERENCES

1. Abtahi, Shabnam, Behnoosh Hariri, and Shervin Shirmohammadi. 2011. "Driver Drowsiness Monitoring Based on Yawning Detection." In *2011 IEEE International Instrumentation and Measurement Technology Conference*, 1–4. https://doi.org/10.1109/IMTC.2011.5944101.

2. Abtahi, Shabnam, Mona Omidyeganeh, Shervin Shirmohammadi, and Behnoosh Hariri. 2014. "YawDD: A Yawning Detection Dataset." In *Proceedings of the 5th ACM Multimedia Systems Conference on - MMSys '14*, 24–28. Singapore, Singapore: ACM Press. https://doi.org/10.1145/2557642.2563678.

3. Abtahi, Shabnam, Omidyeganeh, Mona, Shirmohammadi, Shervin, and Hariri, Behnoosh. 2020. "YawDD: Yawning Detection Dataset." IEEE DataPort. https://doi.org/10.21227/E1QM-HB90.

4. Akrout, Belhassen, and Walid Mahdi. 2016. "Yawning Detection by the Analysis of Variational Descriptor for Monitoring Driver Drowsiness." In *2016 International Image Processing, Applications and Systems (IPAS)*, 1–5. https://doi.org/10.1109/IPAS.2016.7880127.

5. Anitha, C., M. K. Venkatesha, and B. Suryanarayana Adiga. 2016. "A Two Fold Expert System for Yawning Detection." *Procedia Computer Science*, 2nd International Conference on Intelligent Computing, Communication & Convergence, ICCC 2016, 24-25 January 2016, Bhubaneswar, Odisha, India, 92 (January): 63–71. https://doi.org/10.1016/j.procs.2016.07.324.

6. Choi, In-Ho, Sung Kyung Hong, and Yong-Guk Kim. 2016. "Real-Time Categorization of Driver's Gaze Zone Using the Deep Learning Techniques." In *2016 International Conference on Big Data and Smart Computing (BigComp)*, 143–48. https://doi.org/10.1109/BIGCOMP.2016.7425813.

7. Ding, Wu Yang, Ling Zhang, and Yun Hua Chen. 2013. "Yawning Detection Based on Mouth Feature Points Curve Fitting." *Advanced Materials Research* 605–607: 2227–31. https://doi.org/10.4028/www.scientific.net/AMR.605-607.2227.

8. Du, Yong, Qinghua Hu, Degang Chen, and Peijun Ma. 2011. "Kernelized Fuzzy Rough Sets Based Yawn Detection for Driver Fatigue Monitoring." *Fundamenta Informaticae* 111 (1): 65–79. https://doi.org/10.3233/FI-2011-554.

9. Han, Hyungseob, and Uipil Chong. 2016. "Neural Network Based Detection of Drowsiness with Eyes Open Using AR Modelling." *IETE Technical Review* 33 (5): 518–24. https://doi.org/10.1080/02564602.2015.1118362.

10. Haque, Mohammad A., Ramin Irani, Kamal Nasrollahi, and Thomas B. Moeslund. 2016. "Facial Video-Based Detection of Physical Fatigue for Maximal Muscle Activity." *IET Computer Vision* 10 (4): 323–30. https://doi.org/10.1049/iet-cvi.2015.0215.

11. Husain, Syed Sameed, Junaid Mir, Syed Muhammad Anwar, Waqas Rafique, and Muhammad Obaid Ullah. 2022. "Development and Validation of a Deep Learning-Based Algorithm for Drowsiness Detection in Facial Photographs." *Multimedia Tools and Applications*, March. https://doi.org/10.1007/s11042-022-12433-x.

12. Ibrahim, Masrullizam Mat, John J. Soraghan, Lykourgos Petropoulakis, and Gaetano Di Caterina. 2015. "Yawn Analysis with Mouth Occlusion Detection." *Biomedical Signal Processing and Control* 18 (April): 360–69. https://doi.org/10.1016/j.bspc.2015.02.006.

13. Ibrahim, Masrullizam Mat, John S Soroghan, and Lykourgos Petropoulakis. 2013. "Mouth Covered Detection for Yawn." In *2013 IEEE International Conference on Signal and Image Processing Applications*, 89–94. https://doi.org/10.1109/ICSIPA.2013.6707983.

14. Kholerdi, Hedyeh A., Nima TaheriNejad, Reza Ghaderi, and Yaser Baleghi. 2016. "Driver's Drowsiness Detection Using an Enhanced Image Processing Technique Inspired by the Human Visual System." *Connection Science* 28 (1): 27–46. https://doi.org/10.1080/09540091.2015.1130019.

15. Li, Lingling, Yangzhou Chen, and Zhenlong Li. 2009. "Yawning Detection for Monitoring Driver Fatigue Based on Two Cameras." In *2009 12th International IEEE Conference on Intelligent Transportation Systems*, 1–6. https://doi.org/10.1109/ITSC.2009.5309841.

16. Mandal, Bappaditya, Liyuan Li, Gang Sam Wang, and Jie Lin. 2017. "Towards Detection of Bus Driver Fatigue Based on Robust Visual Analysis of Eye State." *IEEE Transactions on Intelligent Transportation Systems* 18 (3): 545–57. https://doi.org/10.1109/TITS.2016.2582900.

17. Marina Martinez, Clara, Mira Heucke, Fei-Yue Wang, Bo Gao, and Dongpu Cao. 2018. "Driving Style Recognition for Intelligent Vehicle Control and Advanced Driver Assistance: A Survey." *IEEE Transactions on Intelligent Transportation Systems* 19 (3): 666–76. https://doi.org/10.1109/TITS.2017.2706978.

18. McDonald, Anthony D., John D. Lee, Chris Schwarz, and Timothy L. Brown. 2014. "Steering in a Random Forest: Ensemble Learning for Detecting Drowsiness-Related Lane Departures." *Human Factors* 56 (5): 986–98. https://doi.org/10.1177/0018720813515272.

19. Omidyeganeh, M., A. Javadtalab, and S. Shirmohammadi. 2011. "Intelligent Driver Drowsiness Detection through Fusion of Yawning and Eye Closure." In *2011 IEEE International Conference on Virtual Environments, Human-Computer Interfaces and Measurement Systems Proceedings*, 1–6. https://doi.org/10.1109/VECIMS.2011.6053857.

20. Pan, Renlong, Lihong Ma, and Yue Huang. 2015. "A Local Posture Manifold with Lagrangian Parallel Constraint for Subtle Action Discrimination." In *TENCON 2015 - 2015 IEEE Region 10 Conference*, 1–5. https://doi.org/10.1109/TENCON.2015.7372826.

21. Puja Seemar *, Anurag Chandna. 2017. "Drowsy Driver Detection Using Image Processing," July. https://doi.org/10.5281/ZENODO.823118.

22. Sun, Wei, Xiaorui Zhang, Srinivas Peeta, Xiaozheng He, and Yongfu Li. 2017. "A Real-Time Fatigue Driving Recognition Method Incorporating Contextual Features and Two Fusion Levels." *IEEE Transactions on Intelligent Transportation Systems* 18 (12): 3408–20. https://doi.org/10.1109/TITS.2017.2690914.

23. Tran, Du, Lubomir Bourdev, Rob Fergus, Lorenzo Torresani, and Manohar Paluri. 2015. "Learning Spatiotemporal Features with 3D Convolutional Networks." *ArXiv:1412.0767 [Cs]*, October. http://arxiv.org/abs/1412.0767.

24. Yang, Hao, Li Liu, Weidong Min, Xiaosong Yang, and Xin Xiong. 2020. "Driver Yawning Detection Based on Subtle Facial Action Recognition." *IEEE Transactions on Multimedia* 23: 572–83. https://doi.org/10.1109/TMM.2020.2985536.

25. Yu, Jongmin, Sangwoo Park, Sangwook Lee, and Moongu Jeon. 2019. "Driver Drowsiness Detection Using Condition-Adaptive Representation Learning Framework." *IEEE Transactions on Intelligent Transportation Systems* 20 (11): 4206–18. https://doi.org/10.1109/TITS.2018.2883823.

26. Zhang, Weiwei, Yi L. Murphey, Tianyu Wang, and Qijie Xu. 2015. "Driver Yawning Detection Based on Deep Convolutional Neural Learning and Robust Nose Tracking." In *2015 International Joint Conference on Neural Networks (IJCNN)*, 1–8. https://doi.org/10.1109/IJCNN.2015.7280566.

27. Zhang, Yan, and Caijian Hua. 2015. "Driver Fatigue Recognition Based on Facial Expression Analysis Using Local Binary Patterns." *Optik* 126 (23): 4501–5. https://doi.org/10.1016/j.ijleo.2015.08.185.

Intelligent Systems and Smart Infrastructure – Brijesh Mishra et al. (eds)
© 2023 Taylor & Francis Group, London, ISBN 978-1-032-41287-0

CHAPTER

65

Electric Vehicle EMS with FOPID

Gaurav Singhal[1], Shujaat Husain[2], Haroon Ashfaq[3]

Department of Electrical Engineering,
Jamia Millia Islamia,
New Delhi, India

Abstract

In order to enhance their performance and satisfy the demands of drivers, additional information on the power management utilised in electric vehicles is required. This also reduces pollution, fossil fuel depletion, and carbon emissions. A FOPID controller was proposed for an energy management system. Optimization-based methodologies are the subject of this study. A state equation is used to put the theory into action. Based on optimization-based methodologies, this work is focused on. To put this theories into practice, you'll need to apply a state equation. There are multiple inputs and outputs to the linked fifth-order state-space model being investigated. Using a state-space model to generate transfer matrices for decoupling the system, and then fine-tuning the system using these transfer matrices in combination with a FOPID controller. By employing a particle swarm optimization (PSO) technique, the adjusted parameters are reduced to their minimum. System modelling and de coupling techniques are also used in this research for efficient performance of electric vehicle.

Keywords: FOPID, PSO, Energy Management System.

1. Introduction

When a vehicle is pushed between one or more electric voltage motors, it is known as an electric vehicle (EV). In certain cases, it may be fueled by a main reservoir that draws energy from

Corresponding author: [1]Gaurav15796@gmail.com, [2]shujaathusain12@gmail.com, [3]hashfaq@jmi.ac.in,

DOI: 10.1201/9781003357346-65

the local environment, while in others, it may be powered by something like a battery. Using solar panels, turning gasoline into electricity using energy storage, or a generator, amongst many other techniques, it may be charged (M. VERMA and S. SRIVASTAVA, 2021). Railway line transit, surface and underwater watercraft, electric aircraft, and electric spaceships are all examples of electrified transportation. As a result of having minimal moving parts to maintain, electric cars have minimal operating costs and are particularly ecologically beneficial owing to the fact that they need little or no biodiesel (petrol or diesel) (A. Sing and S. Suhag, 2019). While some battery-electric vehicles (EVs) used sponsor or nickel-metal hydride batteries in the past, fuel cells really are the guideline for modern battery-electric vehicles, despite the fact that they have a greater chance of survival and are more energy-efficient, with a monthly subconscious rate of only 5 percent (Lu et al., 2018). Despite this increased efficiency, these batteries continue to raise concerns since they are susceptible to flashover, as has happened in the burning of the Tesla Model S, despite attempts to enhance battery safety in recent years. Fully refueling an electric vehicle at home might cost as little as £7.80, and in public parking spaces, it could even be free.

There are three categories of electric automobiles (EVs), according to L. Abualigah et al. (2021): battery electric vehicle, cord hybrid vehicles, and hybrid electric vehicles.

A. Battery Electric Vehicles (BEVs)

BEVs, often known as electric cars, are self-driving, networked vehicles that run on batteries rather than a mechanical drivetrain. The charging case, which is recharged from the grid, is responsible for providing the vehicle with all of its energy (B. Yang et al., 2020). Since BEVs generate no hazardous exhaust emissions and do not contaminate the environment in the same manner that typical gasoline-powered cars do, they are considered zero-emission vehicles.

B. Hybrid Plug-in Electric Vehicles (PHEVs)

PHEVs, or plug-in hybrid electric cars, are vehicles that have both a battery and an internal combustion engine (P. Gupta et al., 2021). They, like classic hybrids, can recharge their batteries via regenerative braking. They differ from standard hybrids in that vehicles have a much larger battery and can be recharged by connecting into the grid. At moderate speeds, conventional hybrids may drive 1-2 miles that before single cylinder kicks in (O. Lozynskyy et al., 2017). Before the powertrain comes in, PHEVs may go somewhere around 10 to 40 miles. When the all range is exhausted, PHEVs revert to a regular hybrid mode, which allows them to go hundreds of km through tank of gasoline. And while most PHEVs are not suitable of quick charging, an EV go L2 connector can charge any PHEV (T. Zhang, X. Wang and Z. Wang, 2018). Hybrid electric cars, rather of depending only on an electric engine, combine rechargeable batteries with conventional diesel (or diesel) to provide power. As a result, they are suited for long commute since they can run on conventional fuel rather than charging stations. The same drawbacks as gasoline-powered cars exist with PHEVs, including greater maintenance requirements, engine noise and pollution, as well as increased operating and fuel expenses. PHEVs have bigger battery packs, which implies they have a shorter range than conventional vehicles (P. P. Dey et al., 2020).

C. Hybrid Electric Vehicles (HEVs)

HEVs, or Hybrid Electric, are vehicles that are powered by both a power supply and a hydrocarbon internal combustion engine (A. Wong, 2016). Regen braking, which adapts or else frittered power in braking to aid the propulsion system all through acceleration, provides all of the battery capacity. In a normal internal combustion engine automobile, much of this braking energy is dissipated as heat throughout the brake pads and rotors. Regular hybrids are unable to use EV go to charge or connect to the electricity to recharge (B. Benlahbib et al., 2020). The energy management system (EMS) of an electric vehicle ensures that energy is supplied smoothly from of the power drive to the vehicle's wheels. Electric vehicle batteries use a lot of energy during acceleration and deceleration (R. Items et al., 2018).

Fossil fuel engines including gasoline, petrol, diesel, and others are responsible for 25% of worldwide CO_2 emissions. An internal combustion engine (ICE) powered by fossil fuels is not only harmful, but it also has the weakest electrical efficiency of just 20%. With a number of other aspects in mind (X. Zhang et al., 2021), recent years have seen a significant increase in interest in research on electric cars that are powered either implicitly or expressly by electricity, such as hybrid automobiles. It is predicted that through 2050, all commercial cars will be zero-emission vehicles, thanks to excellent power conversion efficiency of up to 60 percent.

D. Problem Section

It is important to understand the capability of energy storage (ESS), as well as control techniques and micromanagement (EM) methodologies will now be crucial in the event that electric vehicles have a hopeful future as a source of transportation. Because among the most effective and high energy storage systems, batteries exposed to several short-term charging/discharging cycles during acceleration and DE acceleration, their service life is frequently restricted; these operating pressures are relatively severe. Furthermore, because of their low power density, batteries have a hard time responding fast to significant transient currents but superconducting magnetic storage system (SMES) has the benefit of high power density as well as a higher transient response time but lacks in the energy density.

As a consequence, battery/SMES hybrids energy storage devices (BSM-HESS) have become popular as a cost-effective way to integrate the functional advantages of batteries as well as SMES systems (see Fig. 65.1), allowing for the optimal performance of a broad range of energy storage demands (H. Li et al., 2020).

A HESS system with BSM control, on the other hand, is more difficult to construct than a single EESS system since varied attributes of numerous energy storage devices must be considered. When power electronic devices are utilized, however, many variables, severe during operation, nonlinearities including coupling, along with BSM-HESS modelling errors, occur.

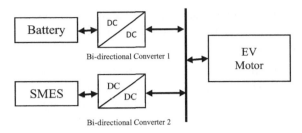

Fig. 65.1 Battery/SMES hybrids energy storage system block diagram

Energy management strategy (EMS), which encompasses both online and offline EMS, is often utilized to achieve an optimum power assignment across several energy storage devices. RBS, FBS, and FLS are examples of semi-empirical tactics that may be used online, whereas DP and Pontryagin's minimal principle are examples of offline procedures that use global optimization strategies. An EMS must establish an adequate underlying controller to successfully regulate the BSM-HESS references, as well as a considerable improvement in robustness, after creating the BSM-HESS references (H. Jiang et al.,2019).

Because of its high reliability and structural simplicity, the most often employed control technique for BSM-HESS is a simple linear control system, such as proportional-integral-derivative (PID) control. Because one-point linearization determines its control gains, it is unable to maintain globally consistent control performance when operating circumstances change, and in extreme cases, the closed-loop system stability may even collapse due to inappropriate control gains. Several innovative control strategies have been presented to address such a difficult problem. The use of port-controlled Hamiltonian (PCH) model-based energy shaping (ES) control for BSMHESS was also investigated, allowing the closed-loop system's control performance to be significantly enhanced. The BSM-HESS has a new droop control that regulates charging/discharging priority and protects batteries from sudden cycling and excessive power demand. Innovative synergistic management of firm voltage fluctuations was also shown using SMEs and batteries with considerable power/energy capabilities within the effective working range of BSM-HESS.

Because of the large range of modelling errors, the BSM-HESS is indeed a highly nonlinear system, with the utilisation of power converters, as well as the close coupling of various components (L. F. de Mingo López et al., 2021).

By using BSM-HESS and NRFOC (nonlinear robust fractional-order control) in EV applications, the authors want to demonstrate a novel strategy to prolonging the battery life span by first using RBS to provide a current reference for the battery through optimal power assignment. Nonlinearities, parameter fluctuations, and nonlinear dynamics are then combined into a perturbation, which is successfully estimated by using the 5th averaged BSM-HESS model by such a high-gain disturbance observer (HGPO).

Furthermore, a fractional-order PID control scheme totally compensates for the estimate's disturbance, allowing for high resilience while simultaneously improving transient responsiveness and smooth tracking. NRFOC does not require a precise BSM-HESS model since it just needs to monitor battery current as well as DC bus voltage. A large set of case studies from a hardware-in-the-loop (HIL) research on the D Space platform exhibits the control system's implementation viability.

In this study, the importance of electric vehicle EMS with fractional order proportional integral derivatives is highlighted. This paper discusses the fifth-order system state-space equation and its derivation, as well as some fundamental facts about coupling and particle swarm optimization (PSO).

The remainder of the paper is formatted in this manner that section 2 deals with related work, section 3 demonstrates the proposed methodology, section 4 displays results and section 5 concludes the work.

2. Related Work

For a completely active battery/supercapacitor HESS, a novel L2-ARC technique based upon passivity-based L2-gain adaptive robust control (L2-ARC) is proposed. To build a terminal Hamilton model including dissipation for HESS, internal structural characteristics are used and IDA-PBC is built to recognise its underlying control with regulation power management used to make the current competitive. IDA-PBC is an intercommunication as well as damping assignment passive controller design. As opposed to lowering time and placing the L2-ARC on a filled HESS system, however, the study's author is not interested (X. Zhang et al., 2021).

For an electric vehicle (EV) hybrid energy storage device, a nonlinear strong fractional-order control (NRFOC) as well as a rule-based strategy (RBS) for effectively distributing power needs are being developed. The D Spaces platform's practicality was shown through a HIL test. To apply NRFOC, the author of this research did not employ a BSM-HESS with Electric Automobiles (B. Yang et al., 2020).

The author will address the latest advances in rule-based and optimization-based EMS and power system requirements, as well as their merits and limitations, in this review-based study. Pontryagin's Minimum Principle (PMP) is another method that employs Markov theory to investigate the driver's intent and the procedure state identification of HEVs, optimising power control platform in the event of a change in the actual operational issue by considering the influence of people, vehicles, as well as roads on the power supply system (M. VERMA and S. SRIVASTAVA, 2021) .

A hybrid power system for an electric car with a fuel cell stack and a Lithium-ion battery is constructed and power converters for power transmission are simulated in this article. A better state machine technique is also being developed to partition the power generated by power sources. The regulations are intended to prevent battery overcharging in high SOC and overdischarging in low SOC, as well as frequent charging and discharging in the standard SOC range, all while reducing hydrogen usage. The improved state machine technique is more trustworthy, helpful to the battery, and utilises less hydrogen (meaning it is less costly) than the original state machine strategy, according to a MATLAB simulation (H. Li et al., 2020).

Fuel cell, battery, and supercapacitor powertrain energy usage and reliability models for simply a bus are proposed in this research. To reduce energy consumption and system deterioration, 2D Dynamic Programming (DP) was developed. Fuel cell, battery, and supercapacitor powertrain energy usage and reliability models for simply a bus are proposed in this research. To reduce energy consumption and system deterioration, 2D Dynamic Programming (DP) was developed. (H. Jiang et al.,2019).

We show that the mean square error is not necessarily the most accurate indication of a problem's solution using a mathematical concept of proportionate integrative co product controllers. The optimal performance of the system is calculated using the mathematical foundation of proportional integrative derivative controllers. A PID controller's gains may be optimised by using a new objective function that is derived from a variety of computational intelligence methodologies. The issue can be solved using some of the most common computational intelligence techniques (L. F. de Mingo López et al., 2021).

To efficiently manage EV speed, the author provided a fuzz fractional-order PID (FOPID) controller based on the Ant Colony Optimization (ACO) technique. All controller settings as well as membership functions may be updated using the ACO approach in real time. The suggested controller's speed tracking capabilities are tested in the MATLAB-Simulink environment utilising a new European driving cycle (NEDC). Fuzzy integer-order PID (IOPID) controllers based on IOPID, FOPID, and ACO outperform the currently available controllers. This fuzzy-based controller might be turned into an ANFIS, which combines fuzzy inference with neural networks. It is difficult to improve learning and adaptive capabilities without professional knowledge (M. A. George et al., 2021).

3. Methodology

In this research, electric vehicle EMS with fractional order proportional integral derivatives is emphasized. In this fifth-order system, state-space equation and its derivation are discussed and some of the basic facts on coupling and PSO are also derived.

A. System Modelling

Fifth-order system state-space equation and its derivation

Define the state vector as

$$r = (r_1, r_2, r_3, r_4, r_5)^T = (s_1, s_2, i_1, i_2, v_0)^T \tag{1}$$

$$\text{Output } p = (p1, p2)^T = (i_1, v_0)^T \tag{2}$$

$$\text{Control input } l = (l_1, l_2)^T = (D_1, D_2)^T \tag{3}$$

$$\text{Tracking error } e = [e_1, e_2]^T = [i_1 - i_1^*, v_0 - v_0^*]^T \tag{4}$$

You'll obtain the following result if you separate the path loss e from the express control input u

$$\begin{cases} e_1 = \dfrac{v_1}{L_1} + \left(\dfrac{R_{on2} - R_{on1}}{L_1} D_1 - \dfrac{R_{L_1} + R_{on2}}{L_1} \right) i_1 + (D_1 - 1)\dfrac{v_0}{L_1} - \overset{\centerdot}{i_1^*} \\[4mm] e_2 = (1 - D_1)\dfrac{i_1}{C_0} + (1 - D_2)\dfrac{i_2}{C_0} - \dfrac{v_0}{RC_0} - \overset{\centerdot}{v_0^*} \end{cases} \tag{5}$$

$$\begin{bmatrix} \dot{v}_1 \\ \dot{v}_2 \\ \dot{i}_1 \\ \dot{i}_2 \\ \dot{v}_0 \end{bmatrix} = \begin{bmatrix} -\dfrac{v_1}{R_e C_1} - \dfrac{i_1}{C_1} + \dfrac{E}{R_e C_1} \\[3mm] \dfrac{I_{SC} - i_2}{C_2} \\[3mm] \dfrac{v_1 - v_0}{L_1} - \left(\dfrac{R_{L_1} + R_{on_2}}{L_1} \right) i_1 \\[3mm] \dfrac{v_2}{L_2} + \left(\dfrac{R_{on_4} - R_{on_3}}{L_2} D_2 - \dfrac{R_{L_2} - R_{on_4}}{L_2} \right) i_2 + (D_2 - 1)\dfrac{v_0}{L_2} \\[3mm] (1 - D_1)\dfrac{i_1}{C_0} - \dfrac{v_0}{RC_0} + \dfrac{i_2}{C_0} \end{bmatrix} + \begin{bmatrix} 0 & 0 \\ 0 & 0 \\ \dfrac{R_{on_2} - R_{on_1}}{L_1} i_1 + \dfrac{v_0}{L_1} & 0 \\ 0 & 0 \\ 0 & -\dfrac{i_2}{C_0} \end{bmatrix} \begin{bmatrix} D_1 \\ D_2 \end{bmatrix} \tag{6}$$

As illustrated by the state-space equation above, the system is a multi-input multi-output coupled system. We estimated the transfer matrix from the state-space model to decouple the state-space model. This transfer matrix is used to achieve decoupling (M. A. George et al., 2021).

B. Decoupling

To gain a better understanding of disturbance decoupling, the transfer function system is examined here.

$$\Sigma \begin{cases} z = Ad + Bu, \\ y = Cd + Du, \end{cases} \tag{7}$$

In all four appropriate rational matrices, disturbance (d), control (u), output (z), and measurement are included (y). All four of these properties are also present in the proper rational matrices (as shown in the preceding list) as well as in the strictly proper rational matrices (D).

The measurement feedback may be used to regulate the system (also called compensator)

$$l = Cp, \tag{8}$$

Using C as an appropriate rational matrix, we may create a closed-loop computer with such a transfer matrix.

$$A + B(1 - CD)^{-1} CC \tag{9}$$

If we refer to the input from the measurements as Σ

Are there compensators that can be used to make the transfer matrix in a closed-loop system equal to zero if C is a valid rational matrix (M. A. George et al., 2020), i.e.

$$A + B(I - CD)^{-1} CC = 0 \tag{10}$$

Compute the highest order of a matching in G from U to Z say r.

Compute the highest order of a matching in G from D to Y, say q.

Compute the minimum weight of an order r matching in G from U to Z, say μ.

Compute the minimum weight of an order q matching in G from D to Y, say v.

C. Calculations

Compute the maximum order of a matching in G from U $\cup$ D to Z, say r'

Compute the maximum order of a matching in G from D to Y $\cup$ Z. Say q'

We recall that r and q equal the generic ranks of L and M, respectively. Furthermore, it is immediate that r' and q' equal the generic ranks of $[A, B]$ and $\begin{bmatrix} A \\ C \end{bmatrix}$, respectively. We see that if or , the DDPM is not generically solvable and we can stop checking its solvability here.

Compute a size-X matching from U to Z with weight equal to μ. Assume that the matching links u_1, u_2, … …, u_r to t_1, t_2, … …, t_r. Denote by the square matrix made up of the first rows and columns of B (A. Karki et al., 2020)

Compute a size-q matching from D to Y with weight equal to v. Assume that the matching links d_1, d_2, … d_q. to p_1, p_2, … p_q. Denote by M' the square matrix made up by the first q rows and columns of M. Both identify and represent a matrix with rows and columns at its beginning Only when the initial system with transference matrices are useful for providing DDPM solution for this specific system.

After decoupling the state-space model, a FOPID controller is used for tuning purpose. The R_p, R_i, R, λ and μ are pieces of gains that work to correct or reduce the error.

In order to achieve ITAE, a PSO technique is used (Integral time absolute error). The formula of the ITAE is given below.

$$ITAE = \int_0^{t_1} |err| * t \tag{11}$$

where,

$err \rightarrow$ Error, $t \rightarrow$ Time and $t_s \rightarrow$ Simulation time

D. Particle Swarm Optimization

PSO is indeed a simple bio-inspired approach for determining the optimum solution in a solution space. It is distinguished from other optimization techniques either by fact that it just requires the objective function and is unaffected by gradients or distinguishable forms (M. S. H. Lipu *et al*, 2022)

Computational science problems may be solved using a computer-aided approach called PSO, in which candidates are continually improved on quality metrics. It uses a basic mathematical formula based just on particle's location and velocity to remedy a problem by creating a population of alternative solutions, which are referred to as "nanoparticles" in this context. The best-known site in each particle's local proximity influences its course, but it is also guided toward the best-known sites in the search space, which also are updated as better regions are discovered by other particles. As a consequence, the swarm's attention should be focused on the most effective options.

There are few, if there are any, assumptions about the problem at hand by PSO, which makes it a metaheuristic (Z. Lu and X. Zhang, 2022) Like the PSO, traditional optimization methods such as linear regression and semi methods don't need a differentiable optimization issue since they don't employ the problem's gradient. PSO, on the other hand, does not guarantee that the best solution will be discovered. To avoid divergence, the inertia weight must be less than one. Using the constriction approach, the two additional parameters may be computed or freely selected, albeit the study suggests convergence places to confine them. Swarm-based stochastic optimization, PSO, is used to find the best solution. For the purpose of re-creating their social behaviour, insects, mammals, birds, and fish all employ the PSO algorithm. These swarms use a cooperative food-finding technique, with each member of the swarm modifying the search pattern based on their own and others' learning experiences (X. Zhou et al.,).

4. Results

From Fig. 65.2, it is observable that the FOPID controller rapidly regulates the load power to its reference point, compared to PID controller. The raising time, overshoot, and settling time of the FOPID controller are 0.12, 0, and 0.185, respectively. The raising time, overshoot and settling time of the PID controller are 0.16, 0, and 0.32, respectively. And from Fig. 65.3, it is observable that the FOPID controller rapidly regulates the DC bus voltage to its reference point, compared to PID controller. The raising time, overshoot and settling time of the FOPID controller are 0, 0 and 0.26. The raising time, overshoot and settling time of the PID controller are 0, 0, and 0.28.

From Fig. 65.4(a), it is observable that the FOPID controller rapidly regulates the battery current to its reference point, compared to PID controller. The raising time, overshoot, and settling time of the FOPID controller are 0.11, 0, and 0.15, respectively. The raising time, overshoot, and settling time of the PID controller are 0.13, 0, and 0.23, respectively. From Fig. 65.4(b), it is observable that the FOPID controller rapidly regulates the SMES current to its reference point, compared to PID controller. The raising time, overshoot, and settling time of the FOPID controller are 0.11, 0, and 0.2. The raising time, overshoot, and settling time of the PID controller are 0.134, 1.2%, and 0.28, respectively.

From Fig. 65.5(a), it is observable that the FOPID controller rapidly regulates the duty cycle of battery-side converter, compared to PID controller. The raising time, overshoot, and settling time of the FOPID controller are 0.13, 0, and 0.18, respectively. The raising time,

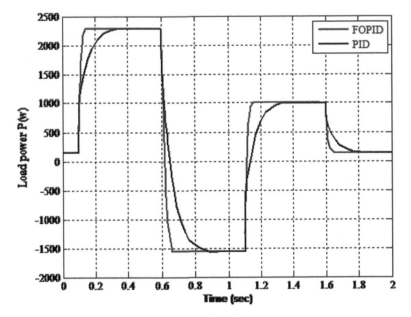

Fig. 65.2 Load power

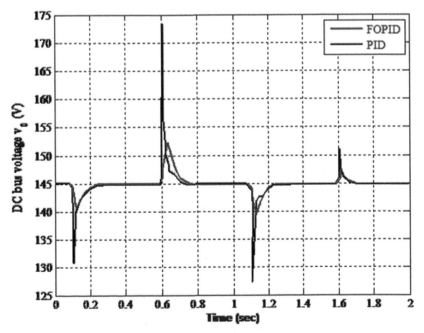

Fig. 65.3 DC bus voltage

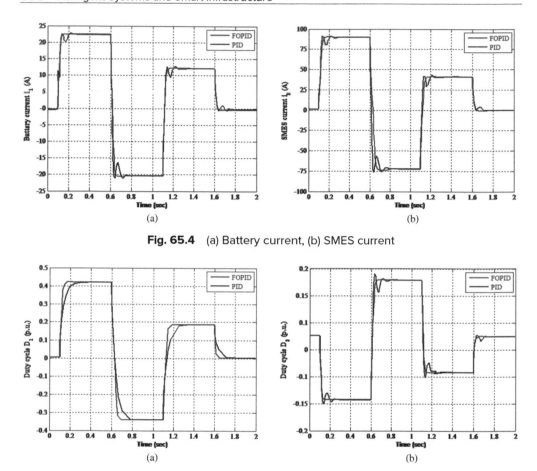

Fig. 65.4 (a) Battery current, (b) SMES current

Fig. 65.5 (a) Battery converter duty cycle, (b) SMES converter duty cycle

overshoot, and settling time of the PID controller are 0.155, 0, and 0.36, respectively. And from Fig. 65.5(b), it is observable that the FOPID controller rapidly regulates the duty cycle of SMES-side converter, compared to PID controller. The raising time, overshoot, and settling time of the FOPID controller are 0, 0, and 0.68. The raising time, overshoot, and settling time of the PID controller are 0, 2.1%, and 0.85, respectively.

5. Conclusion

The main aim of this work is to build an energy management system in this research; the authors used a FOPID controller, which is a fractional-order proportional integral derivative. To put the concept into action, state-space equations are employed. The fifth-order state-space model under investigation is a connected multi-input multi-output system. We calculated a transfer matrix using the state-space model for decoupling the system, and then used these transfer matrices to decouple the system and to control the system we use the FOPID controller, we obtain from result that FOPID controller rapidly regulates the load power, dc bus voltage, battery, and SMES CURRENT to its reference point, compared to PID controller. The tuned parameters are minimised using a PSO technique with state-space model, and decoupling techniques are used for this research to evaluate and control the efficiency of the electric vehicles.

REFERENCES

1. X. Zhang, Z. Lu, X. Yuan, Y. Wang, and X. Shen, "L2-Gain Adaptive Robust Control for Hybrid Energy Storage System in Electric Vehicles," *IEEE Trans. Power Electron.*, vol. 36, no. 6, pp. 7319–7332, 2021, doi: 10.1109/TPEL.2020.3041653.

2. B. Yang *et al.*, "Design and implementation of Battery/SMES hybrid energy storage systems used in electric vehicles: A nonlinear robust fractional-order control approach," *Energy*, vol. 191, no. xxxx, p. 116510, 2020, doi: 10.1016/j.energy.2019.116510.

3. M. VERMA and S. SRIVASTAVA, "A Review on Effectiveness of Real time optimization techniques in Power Management System in Hybrid Electric Vehicle," *Int. J. New Technol. Res.*, vol. 7, no. 4, 2021, doi: 10.31871/ijntr.7.4.13.

4. H. Li, D. Zhao, Y. Zhang, Z. Liang, H. Dang, and Y. Liu, "An Improved State Machine Strategy for Fuel-Cell-Based Hybrid Electric Vehicles," *IECON Proc. (Industrial Electron. Conf.*, vol. 2020-October, pp. 3957–3962, 2020, doi: 10.1109/IECON43393.2020.9254705.

5. H. Jiang, L. Xu, J. Li, Z. Hu, and M. Ouyang, "Energy management and component sizing for a fuel cell/battery/supercapacitor hybrid powertrain based on two-dimensional optimization algorithms," *Energy*, vol. 177, pp. 386–396, 2019, doi: 10.1016/j.energy.2019.04.110.

6. L. F. de Mingo López, F. S. García, J. E. Naranjo Hernández, and N. G. Blas, "Speed Proportional Integrative Derivative Controller: Optimization Functions in Metaheuristic Algorithms," *J. Adv. Transp.*, vol. 2021, 2021, doi: 10.1155/2021/5538296.

7. M. A. George, D. V. Kamat, and C. P. Kurian, "Electronically Tunable ACO Based Fuzzy FOPID Controller for Effective Speed Control of Electric Vehicle," *IEEE Access*, vol. 9, pp. 73392–73412, 2021, doi: 10.1109/ACCESS.2021.3080086.

8. M. A. George, I. V. L. Durga Bhavani, and D. V. Kamath, "EX-CCII based FOPID controller for electric vehicle speed control," *2020 IEEE Int. Conf. Distrib. Comput. VLSI, Electr. Circuits Robot. Discov. 2020 - Proc.*, pp. 47–51, 2020, doi: 10.1109/DISCOVER50404.2020.9278055.

9. A. Karki, S. Phuyal, D. Tuladhar, S. Basnet, and B. P. Shrestha, "Status of pure electric vehicle power train technology and future prospects," *Appl. Syst. Innov.*, vol. 3, no. 3, pp. 1–28, 2020, doi: 10.3390/asi3030035.

10. [M. S. H. Lipu *et al.*, *Power Electronics Converter Technology Integrated Energy Storage Management in Electric Vehicles: Emerging Trends, Analytical Assessment and Future Research Opportunities*, vol. 11, no. 4. 2022. doi: 10.3390/electronics11040562.

11. Z. Lu and X. Zhang, "Composite Non-Linear Control of Hybrid Energy-Storage System in Electric Vehicle," *Energies*, vol. 15, no. 4, p. 1567, 2022, doi: 10.3390/en15041567.

12. X. Zhou, F. Sun, C. Zhang, and C. Sun, "Stochastic predictive co-optimization of the speed planning and powertrain controls for electric vehicles driving in random traffic".

13. A. Singh and S. Suhag, "Frequency Regulation in AC Microgrid with and without Electric Vehicle Using Multiverse-Optimized Fractional Order- PID controller," Int. J. Comput. Digit. Syst., vol. 8, no. 4, pp. 375–385, 2019, doi: 10.12785/ijcds/080406.

14. X. Lü, Y. Qu, Y. Wang, C. Qin, and G. Liu, "A comprehensive review on hybrid power system for PEMFC-HEV: Issues and strategies," Energy Convers. Manag., vol. 171, no. April, pp. 1273–1291, 2018, doi: 10.1016/j.enconman.2018.06.065.

15. L. Abualigah, M. Shehab, M. Alshinwan, S. Mirjalili, and M. A. Elaziz, "Ant Lion Optimizer: A Comprehensive Survey of Its Variants and Applications," Arch. Comput. Methods Eng., vol. 28, no. 3, pp. 1397–1416, 2021, doi: 10.1007/s11831-020-09420-6.

16. P. Gupta, V. Pahwa, and Y. P. Verma, "Switching function based inverter modeling for a gridconnected SOFC system in real time," IOP Conf. Ser. Mater. Sci. Eng., vol. 1033, no. 1, 2021, doi: 10.1088/1757-899X/1033/1/012023.

17. O. Lozynskyy, A. Lozynskyy, B. Kopchak, Y. Paranchuk, P. Kalenyuk, and Y. Marushchak, "Synthesis and research of electromechanical systems described by fractional order transfer functions," Proc. Int. Conf. Mod. Electr. Energy Syst. MEES 2017, vol. 2018-January, pp. 16–19, 2017, doi: 10.1109/MEES.2017.8248877.

18. T. Zhang, X. Wang, and Z. Wang, "A novel improved grey Wolf optimization algorithm for numerical optimization and PID controller design," Proc. 2018 IEEE 7th Data Driven Control Learn. Syst. Conf. DDCLS 2018, pp. 879–886, 2018, doi: 10.1109/DDCLS.2018.8515951.

18. P. P. Dey, D. C. Das, A. Latif, S. M. Suhail Hussain, and T. S. Ustun, "Active power management of virtual power plant under penetration of central receiver solar thermalwind using butterfly optimization technique," Sustain., vol. 12, no. 17, 2020, doi: 10.3390/SU12176979.

19. A. Wong, "TCTAP C-112 A Complex Calcified Lesion: A Lesson Learnt Twice!," J. Am Coll. Cardiol.,vol. 10.1016/j.jacc.2016.03.319.

20. B. Benlahbib et al., "Experimental investigation of power management and control of a PV/wind/fuel cell/battery hybrid energy system microgrid," Int. J. Hydrogen Energy, vol. 45, no. 53, pp. 29110–29122, 2020, doi: 10.1016/j.ijhydene.2020.07.251.

30. R. Items, W. Rose, W. Rose, T. If, and W. Rose, "This is a repository copy of Enhanced receding horizon optimal performance for online tuning of PID controller parameters . White Rose Research Online URL for this paper : Version : Accepted Version Article : Enhanced Receding Horizon Optimal Performance for On-line Tuning of PID Controller Parameters Yongling Wu Shaoyuan Li," 2018

Intelligent Systems and Smart Infrastructure – Brijesh Mishra et al. (eds)
© 2023 Taylor & Francis Group, London, ISBN 978-1-032-41287-0

Comparative Analysis of Defected Ground Goblet Shape Microstrip Patch Antennas with Slotted as well as without Slotted for Multiband Application

Manvendr[1], Anil Kumar[2]

ECE, SHUATS, UP, India

Abstract

This paper is presented as comparison of defected ground structure with slotted (as inverted question mark shape) and without slotted microstrip patch antenna and fed by a single microstrip line feed. With slotted antenna has designed for multiple frequency bands such as L, S and C bands. With and without slotted goblet shaped MPA's have been designed for S, X and Ku band application and they are also compared to each other's. These antennae are basically for different communications; it has a good bandwidth with low return loss, good gains and all other parameters that are shown in the result section. The proposed slotted goblet in shape antenna designed for multiband resonant frequencies f_1 = 1.4 GHz, f_2 = 3.5GHz, f_3 = 5.6 GHz and f_4 = 7.1 GHz. As well as without slotted goblet in shape antennas whose resonant frequencies are 3.7, 10.7, 12.0, and 16.3 GHz, the antenna is for multiband performance, reduction in size and cost-effective; the predicted compact size MPA's realised and stable radiation pattern with suitable VSWR. This paper is analysed and simulated on HFSS simulator.

Keywords: Goblet antenna, Defected ground, Slotted patch, Multiband

Corresponding author: [1]smandia20@gmail.com, [2]kumarashoksaini@gmail.com, [3]jitendradeegwal@gmail.com, [4]karanverma@nitdelhi.ac.in

DOI: 10.1201/9781003357346-66

1. Introduction

In the last few years, microstrip patch antennas (MPAs) are suitable applicants to meet the required condition for the prominent wireless technology applications such as military telemetry, Global Positioning System carriers, satellite mobile phones, air traffic control radar, weather radar, surface ship radar, and satellite communication, due to easy fabrication and bearable manufacturing cost [1]. Many techniques are reported to obtain multiband with slots and without slot. The multiband antenna is designed and compared with dual-band and multiband microstrip antennas designed for the application of wireless communication as well as fractal and defected ground structures [3]. These designs cover the requisite bandwidth for the application cellular phone, IEEE as well as wireless local area network, Wi-Fi, bluetooth. It is light weight, cost-effective, with plain configuration and multiband functionality [4]. To achieve multi-operating frequency, each dedicated to specific purpose, a multiband antenna can be explored.

In the past, many techniques are reported to obtain multiband with slotted, dual, triple, quadrant band antenna being designed [5]. Numerous microstrip patch antennas have low profile and conformal mapping and easy fabrication [6]. The microstrip line feed triple band microstrip patch antenna used for WLAN/Wi-MAX application was reported and similar structure of a multi-resonator which is suitable for multi-band application was presented [7]. There were such types of several antennas designed and utilized with microstrip line feeding was proposed [8]. For military and satellite applications, multiband as the S, C, and X bands has been reported. Multiple resonance frequencies have recently demonstrated with coaxial feeding in a disc-patch with C-shaped slots [9]. A smaller in size antenna is necessary for mobile communication, e.g. mobile phones, laptops, tablets, and other portable devices, rather than compact antennas with patch narrow slits with multi-band antenna reported [10]. It is now clear that designing a multiband microstrip patch antenna can be as simple as cutting the slots in the patch. The parametric analysis is done, and various parameters are determined as functions of frequency.

2. Antenna Structures and Design Simulation

The presented antennas are designed on high-frequency structure simulator (HFSS) with dielectric substrate FR4 epoxy $\varepsilon_r = 4.4$ and substrate height h = 1.5 mm. The patch has the dimensions of 10.71 mm × 41.50 mm. The ground has the dimensions of 4.5 mm × 50.5 mm with height of substrate 1.5 mm. Antennas excited with microstrip feed having characteristics impedance of 50 Ω. Here with microstrip inset feed is used to transfer the signal as current on the microstrip patch, as dimension of 8 mm × 4 mm on to the patch. The geometry of the goblet-shaped microstrip patch antenna is shown in Fig. 66.1.

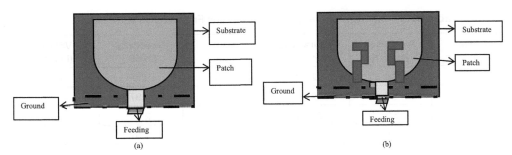

Fig. 66.1 Design structure (a) without slotted (b) with slotted

3. Result and Discussion

The antenna is designed using HFSS simulator and the result is obtained for the following parameters; for example, Return loss, Voltage Standing Wave Ratio, Gain and Radiation Pattern, which are shown in following Tables 66.2 & 66.3.

Table 66.1 Design specifications for goblet-shaped MPA without slotted

S. No.	Parameters	For goblet in shape, values are in mm	For slotted goblet in shape, values are in mm
1	Substrate Length	41.21	41.21
2	Substrate Width	50.5	50.5
3	Ground Plane Length	4.5	10
4	Ground Plane Width	50.5	50.5
5	Patch Length	10.71	10.71
6	Patch Width	41.5	41.5
7	Feed-line Length	11.0	4
8	Feed-line Width	6	8
9	Substrate Thickness (h)	1.5 mm	1.5 mm
10	Relative Dielectric Constant (ε_r)	4.4	4.4

Table 66.2 Different Characteristics at Various Frequencies without Slotted

S. No.	Frequency (in GHz)	Return Loss (in dB)	Gain (in dB)	Directivity (In dB)	VSWR (in dB)
1.	3.7	−17.02	5.03	5.09	2.46
2.	10.7	−14.25	2.27	1.50	3.41
3.	12.0	−12.02	6.33	7.07	4.44
4.	16.3	−12.48	4.42	6.48	4.21

Table 66.3 Different Characteristics at Various Frequencies with Slotted

S. No.	Frequency (in GHz)	Return Loss (in dB)	Gain (in dB)	Directivity (in dB)	VSWR (in dB)
1.	1.4	-12.02	0.78	1.10	1.66
2.	3.5	-16.50	4.76	4.87	1.35
3.	5.6	-20.35	2.45	2.69	1.21
4.	7.1	-17.86	5.02	-4.93	1.29

A. Return Loss

In telecommunications, the return loss is a loss of power in signal returned, i.e., the reflected by a discontinuity/boundary in a transmission line. The proposed antennas have been configured with low return loss. The return loss must be greater than −10 dB but at operating frequency without slotted antenna 3.7, 10.7, 12.0 and 16.3 GHz, the return loss is −17.02, −14.25, −12.02 and −12.48 dB, respectively, and for operating frequency with slotted antenna 1.4, 3.5, 5.6 and 7.1 GHz, the return loss is −12.02, −16.50, −20.35 and −17.86 dB, respectively, which is desirable for good performance of antenna

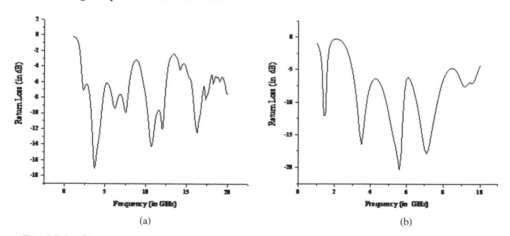

(a)   (b)

Fig. 66.2 Represents return loss at various frequencies (a) without slotted (b) with slotted

B. Gain

Gain is a signal strength as the signal is processed by the antenna at a given incident angle. The gain of the proposed antenna without slotted is varied from 5.03 dB, 2.27 dB, 6.33 dB and 4.42 dB at operating frequency 3.7, 10.7, 12.0 and 16.3 GHz, respectively, and gain of slotted antenna is 0.78, 4.76, 2.45 and 5.02 dB at operating frequency of 1.4, 3.5, 5.6 and 7.1 GHz, respectively; the gain of the proposed antenna is shown in Fig. 66.3.

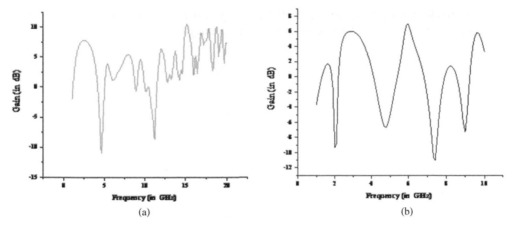

Fig. 66.3   Represents gain at various frequencies (a) without slotted and (b) with slotted

C. Directivity

The ratio of maximum radiation intensity to its average radiation intensity assigns the directivity of antenna. In this work, the directivity of without slotted antenna is 5.09, 1.50, 7.07 and 6.48 dB at 3.7, 10.7, 12.0 and 16.3 GHz, respectively, and the directivity of with slotted antenna is 1.10, 4.87, 2.69 and 4.93 dB at 1.4, 3.5, 5.6 and 7.1 GHz. In general, the value of directivity is from 1 to ∞ for various operating frequencies. Figure 66.4 represents the graph for operated frequency range and obtained directivity.

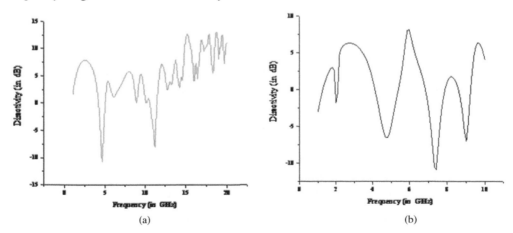

Fig. 66.4   Represents directivity at various frequency (a) without slotted and (b) with slotted

D. Voltage Standing Wave Ratio

In this work, the design without slotted antenna having 2.46, 3.41, 4.44 and 4.21 VSWR at 3.7, 10.7, 12.0 and 16.3 GHz, respectively, and with slotted antenna having 1.66, 1.35, 1.21

and 1.29 VSWR at 1.4, 3.5, 5.6 and 7.1 GHz frequency, respectively, is provided. Figure 66.5 represents the graph between VSWR and obtained frequency range.

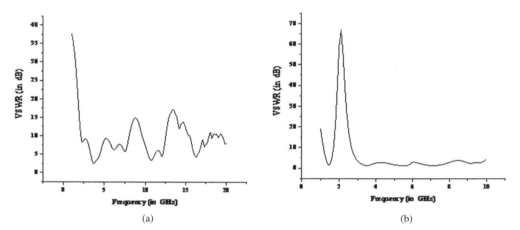

(a)   (b)

Fig. 66.5   Represents VSWR at various frequency (a) without slotted and (b) with slotted

Radiation Pattern: The radiation pattern decides the coverage area in free space and it has the ability to receive and transmit in different directions. In the present work, Fig. 66.6 represents the optimized E-plane and H-plane at various frequencies. The high directivity provides high radiation efficiency in all the possible directions.

4. Conclusion

The navel goblet in shape with slotted patch antennas as Defected Ground Structure is demonstrated well and its properties are improved as return loss, VSWR, bandwidth, gain are compared with each other for their performance. The fundamental parameters are modelled and observed with the help of HFSS simulator and the results are measured. The effects of introducing slot into the patch of antenna and parameters have been successfully investigated. The antennas operated at its corresponding frequencies of operations. These antennas i.e. With and without slotted goblet in shape patch antennas are also compared with conventional defected ground antennas, regarded its size compactness improved overall.

According to the analysed result, the four bands provided a better result with respect to return loss, gain, directivity, and radiation pattern, as shown in Tables 66.2 and 66.3. The simulated results are presented as suitable for S, C, X and Ku bands. The compared results are formatted in Table 66.1.

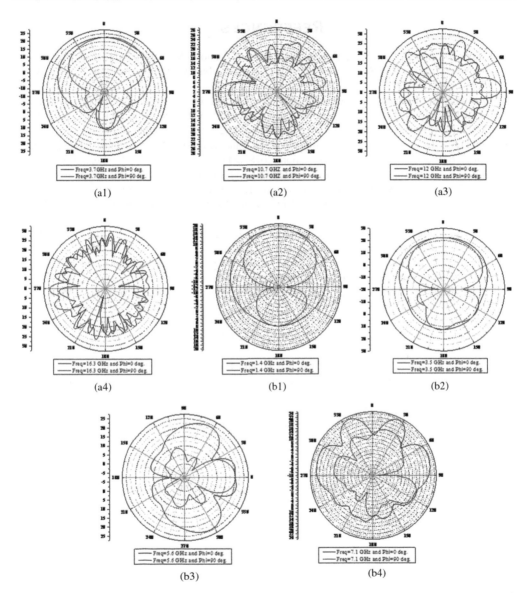

Fig. 66.6 Represents radiation pattern at various frequency (a) without slotted and (b) with slotted

5. Acknowledgement

We acknowledge to DR. Arvind Kumar Associate Professor, Dept of ECE, MNNIT for their help and support of this work.

REFERENCES

1. Kumar, M., Ansari, J. A., Saroj A., Saxena, R., Devesh, "A novel Microstrip fed L-shaped Arm Slot and Notch Loaded RMPA with mended ground plane for bandwidth improvement" PIER C, vol. 95, pp. 47–57, 2019.

2. Chandrabhan, Kumar, A., "Triple Band Notch Loaded Hexagonal Patch Stacked Antenna" International Journal of Engineering and Technology (IJET) Vol. 9, No. 3, Jun–Jul 2017, pp. 2485–2490 ISSN (Print): 2319–8613 ISSN (Online): 0975–4024; Thomsonreuter IF=0.797

3. Naji D.K., "Design of Compact Dual-band and Tri-band Microstrip Patch Antennas" International Journal of Electromagnetics and Applications 2018, 8(1): 26–34 DOI: 10.5923/j.ijea.20180801.02

4. Akkole S.,Vasudevan N. "Microstrip Fractal MultiBand Antenna Design and Optimization by using DGS Technique for Wireless Communication. International Conference on Inventive Computation Technologies [ICICT 2021] IEEE Xplore Part Number: CFP21F70-ART; ISBN: 978-1-7281-8501-9

5. Singh, A., Singh, G. P., Manvendra, Philip, P. C., Kumar, M., Saxena, R., "Analysis of V Slot Multiband Microstrip patch Antenna for S, C and X Bands

6. Rajmohan, I. J., Hussein, M. I., "A compact multiband planar antenna using modified L-shape resonator Slots"Volume 6, Issue 10, October 2020, e0528

7. Hu, W., Yin, Y. Z., Yang, X., Fei, P. "Compact Multiresonator-Loaded Planar Antenna for Multiband Operation" IEEE TRANSACTIONS ON ANTENNAS AND PROPAGATION, VOL. 61, NO. 5, MAY 2013

8. Gao, S., Hodges, R. E. "Advanced Antennas for Small Satellites" Vol. 106, No. 3, March 2018 | Proceedings of the IEEE

9. Cicchetti, R., Miozzi, E., Testa, O., "Wideband and UWB Antennas for Wireless Applications: A Comprehensive Review" Hindawi International Journal of Antennas and Propagation Volume 2017, Article ID 2390808

10. Anand, H., Kumar, A., "Design of Frequency-Reconfigurable Microstrip Patch Antenna." Indian Journal of Science and Technology, Vol 9(22), pp. 1–4. June 2016 , ISSN: (Print) : 0974-6846 ISSN (Online): 0974–5645.

Intelligent Systems and Smart Infrastructure – Brijesh Mishra et al. (eds)
© 2023 Taylor & Francis Group, London, ISBN 978-1-032-41287-0

CHAPTER

67

Understanding the Need of Reinforcement Learning for Load Balancing in Cloud Computing

Prathamesh Vijay Lahande[1], Parag Ravikant Kaveri[2]

Symbiosis Institute of Computer Studies and Research,
Symbiosis International (Deemed) University,
Pune, India

Vinay Chavan[3]

Seth Kesarimal Porwal College,
Nagpur University,
Nagpur, India

Abstract

Cloud computing provides services to users by executing tasks on the cloud Virtual Machines (VMs) by making use of resource scheduling algorithms which are statically fixed prior to task execution. Cloud performance is directly proportional to how resources are scheduled. If resources are scheduled appropriately, then the performance of the cloud will be enhanced and cloud can execute more tasks. Similarly, if resources are poorly scheduled, the cloud performance is hampered. Also, with less task size, the cloud is less loaded, but if the task size is huge in number, then cloud has to schedule resources in such a way that tasks execution and load balancing are done smoothly. Task size is never the same across time and it keeps varying and hence load balancing has always been a challenge to the cloud to keep a smooth flow of tasks execution irrespective of tasks size. The main aim of this research paper is to test performance and behaviour of resource scheduling algorithms and perform empirical

Corresponding author: [1]prathamesh.lahande@sicsr.ac.in, [2]parag.kaveri@sicsr.ac.in,
[3]drvinaychavan@yahoo.com

DOI: 10.1201/9781003357346-67

analysis of the same. This research paper is divided into three phases: the first phase includes a simulation experiment conducted on WorkflowSim environment where different task sizes are executed using Max–Min (MX–MN) and Min–Min (MN–MN) resource scheduling algorithms on a series of VMs; the second phase includes a comparative empirical analysis of how these resource scheduling algorithms behaved using a mathematical model of Linear Regression Equations and R^2 analysis; the last phase includes understanding the need of Reinforcement Learning model for enhancing this entire process of load balancing.

Keywords: Cloud Computing, Cost, Reinforcement Learning, Resource Scheduling, Time

1. Introduction

Cloud refers to a network providing a variety of services by executing tasks on Virtual Machines (VM) (Michael et al. 2010, Tharam et al. 2012). The end-user submits these tasks to the cloud with the expectation that it will be processed at a minimum cost and time. The task size varies at every given instance. Hence, the cloud has to perform load balancing by proper resource scheduling (Arulkumar et al. 2022, Nagendra et al. 2022, Zhang et al. 2021). Therefore, load balancing becomes a challenge in itself due to the unpredictability in task size (Wael et al. 2022, Shridhar et al. 2022, Gao et al. 2021, Kaur et al. 2021). The amount of cost and time required for task scheduling (Yuejuan et al. 2021, Devi et al. 2021) is directly proportional to how the resources are scheduled on the cloud (Feng et al. 2022). If the resources are scheduled in a proper way, then the productivity of task execution will be high, otherwise low (Rupali et al. 2021). Cloud uses a static resource scheduling algorithm for all the execution of tasks and hence results obtained will be limited with respect to (w.r.t.) cost and time taken. Hence, to enhance and output better results, Reinforcement Learning (RL) (David 2007, Aram et al. 2004, Richard et al. 1998 Kaelbling et al. 1996) can be used for dynamic resource scheduling to execute tasks at better time and cost requirements.

2. Resource Scheduling Algorithms

A. MX-MN Algorithm Pseudo Code [13]

Figure 67.1 shows MX–MN pseudo code.

```
Input: Task Set T: T₁, T₂, T₃, …, Tₘ; VM Set: VM₁, VM₂, VM₃, …, VMₙ
Output: Total Cost and Total Time
Step 1: Start
Step 2: For all tasks in 'T'
Step 3: For all the VMs 'VMj' in the VM Set 'VM'
Step 4: Calculate Completion Time of Task 'Ti': CTij = Execution Time ETij + VMj
Step 5: End of For
Step 6: End of For
```

Step 7: **Loop** until all tasks in Task Set 'T' are scheduled and executed by a 'VM'
Step 8: **For** all tasks in 'T'
Step 9: **Search** a Task 'Ti' with Maximum Completion Time 'CTij' and corresponding VM 'VMj' which the Task 'Ti' will get scheduled and executed to.
Step 10: **Allocate** the Task 'Ti' to that VM 'VMj'
Step 11: **Update** Start Time 'ST' of 'Ti' when task gets scheduled.
Step 12: **Update** Finish Time 'FT' of Task 'Ti' when task finishes its execution.
Step 13: **Calculate** time taken by Task 'Ti' for execution: Time = FT − ST.
Step 14: **Calculate** the Cost taken by the 'VMj' which executes Task 'Ti".
Step 15: **Remove** Task 'Ti' from Task Set 'T'.
Step 16: **End** of Loop
Step 17: **Calculate** the total cost and total time to execute 'M' tasks by 'N' VMs.
Step 18: **Stop**

Fig. 67.1 Pseudo Code of MX-MN Algorithm

B. MN–MN Algorithm Pseudo Code [13]

Figure 67.2 shows the MN–MN pseudo code.

Input: Task Set T: T_1, T_2, T_3, ..., T_m; VM Set: VM_1, VM_2, VM_3, ..., VM_n
Output: Total Cost and Total Time
Step 1: **Start**
Step 2: **For** all tasks in 'T'
Step 3: **For** all the VMs VMj in the VM Set 'VM'
Step 4: **Calculate** Completion Time of Task 'Ti': CTij = Execution Time ETij + VMj
Step 5: **End** of For
Step 6: **End** of For
Step 7: **Loop** until all tasks in Task Set 'T' are scheduled and executed by a 'VM'
Step 8: **For** all tasks in 'T'
Step 9: **Search** a Task 'Ti' with Minimum Completion Time 'CTij' and corresponding VM 'VMj' which the Task 'Ti' will get scheduled and executed to.
Step 10: **Allocate** the Task 'Ti' to that VM 'VMj'.
Step 11: **Update** Start Time 'ST' of 'Ti' when task gets scheduled.
Step 12: **Update** Finish Time 'FT' of Task 'Ti' when task finishes execution.
Step 13: **Calculate** the time taken by Task 'Ti' for execution: Time = FT − ST.
Step 14: **Calculate** the Cost taken by the VMj which executes Task 'Ti'.
Step 15: **Remove** Task 'Ti' from Task Set 'T'.
Step 16: **End** of Loop
Step 17: **Calculate** Total Cost and Total Time to execute 'M' tasks by 'N' VMs.
Step 18: **Stop**

Fig. 67.2 Pseudo Code of MN–MN Algorithm

3. Evaluation

A. Simulation and Experiment

An experimented was conducted on WorkflowSim [8] [14] environment where four scenarios were considered to test the performance of Resource Scheduling algorithms: MX–MN and

MN–MN algorithm [13]. First scenario includes executing 25 tasks using both MX–MN and MN–MN algorithm on VMs in the series: 5, 10, ..., 50 one by one. This means 25 tasks will be executed firstly on 5 VMs, then on 10 VMs, ... so on up to 50 VMs. This process is repeated for the second, third, and fourth scenarios where the environment remains the same, but the tasks size vary in each scenario. For the second scenario, the task size is doubled to 50, third scenario task size is 100, and last scenario task size is 1000. After every task execution, the total cost and total time required for its execution are recorded for performing empirical analysis and comparison in second phase. The purpose of this experiment is to understand how resource-scheduling algorithms perform under several different conditions. For conducting this experiment, resource scheduling algorithm is statically fixed prior to executing tasks.

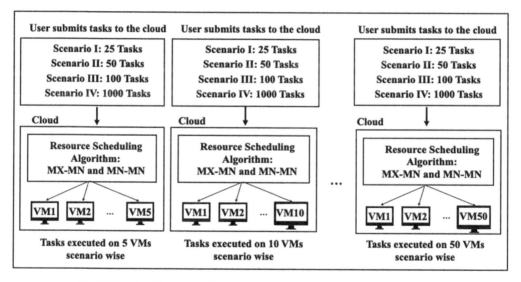

Fig. 67.3 Architecture of conducted experiment for all four scenarios

Figure 67.3 shows the architecture of the conducted experiment. Tasks are submitted to the cloud for execution purposes in all four scenarios which the cloud accepts and executes them separately using both MX–MN and MN–MN algorithm on cloud VMs ranging in the series 5, 10, ..., 50 throughout all the four scenarios.

B. Empirical Analysis of Resource Scheduling Algorithms Behaviour

Empirical analysis of resource scheduling algorithm's behaviour is done using Linear Regression Equations and R^2 analysis for every scenario.

Scenario I: Tasks Size = 25; VMs: 5, 10, ..., 50 - Scenario I consists of executing 25 tasks on cloud VMs in the series 5, 10, ..., 50 separately using both resource scheduling algorithms MX–MN and MN–MN. The total cost required and total time taken for executing a particular task on all VMs are recorded. Figure 67.4 shows the total cost comparison graph for executing these 25 tasks and Fig. 67.5 shows the total time comparison graph for executing 25 tasks.

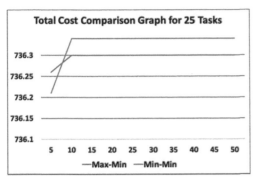

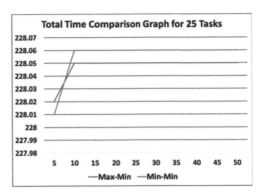

Fig. 67.4 Total Cost Comparison Graph   **Fig. 67.5** Total Time Comparison Graph

Table 67.1 Comparison w.r.t. total cost and time for executing 25 tasks

Criteria	Total Cost Comparison		Total Time Comparison	
	MX–MN	**MN–MN**	**MX–MN**	**MN–MN**
Linear Regression Equation	y = 0.0022x + 736.28	y = 0.0071x + 736.29	y = 0.0016x + 228.04	y = 0.0027x + 228.04
Regression Line Slope	0.0022	0.0071	0.0016	0.0027
Slope Sign	Positive	Positive	Positive	Positive
Regression Line Y-Intercept	736.28	736.29	228.04	228.04
Relationship	Positive	Positive	Positive	Positive
R^2 value	0.2727	0.2727	0.2727	0.2727
VM Analysis	↑VM = ↑Cost	↑VM = ↑Cost	↑VM = ↑Time	↑VM = ↑Time
Performance	MX-MN ≈ MN-MN		MX-MN ≈ MN-MN	

The following points can be observed from Table 67.1:

- ↑VM = ↑Cost & ↑VM = ↑Time: With increase in number of VMs for executing 25 tasks, the total cost and time required increase.
- MX–MN and MN–MN behavior is similar for executing 25 tasks across all VMs.

Scenario II: Tasks Size = 50; VMs: 5, 10, …, 50 - Since the behaviour of MX-MN and MN–MN was similar in first scenario, the task size is doubled to 50 with other experiment configuration same. Figure 67.6 shows the total cost comparison graph and Fig. 67.7 shows the total time comparison graph for executing 50 tasks, respectively.

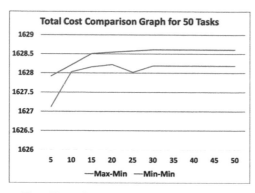

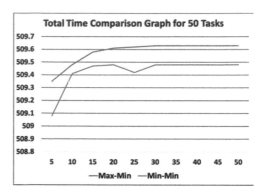

Fig. 67.6 Total Cost Comparison Graph

Fig. 67.7 Total Time Comparison Graph

Table 67.2 Comparison w.r.t. total cost and time for executing 50 tasks

Criteria	Comparison w.r.t. Total Cost		Comparison w.r.t. Total Time	
	MX–MN	**MN–MN**	**MX–MN**	**MN–MN**
Linear Regression Equation	y = 0.0589x + 1628.2	y = 0.0665x + 1627.7	y = 0.0236x + 509.45	y = 0.0255x + 509.29
Regression Line Slope	0.0589	0.0665	0.0236	0.0255
Slope Sign	Positive	Positive	Positive	Positive
Regression Line Y-Intercept	1628.2	1627.7	509.45	509.29
Relationship	Positive	Positive	Positive	Positive
R^2 value	0.5882	0.3634	0.5887	0.3834
VM Analysis	↑VM = ↑Cost	↑VM = ↑Cost	↑VM = ↑Time	↑VM = ↑Time
Performance	MX-MN > MN-MN		MX-MN > MN-MN	

The following points can be observed from Table 67.2:

- ↑VM = ↑Cost & ↑VM = ↑Time: With increase in number of VMs for executing 50 tasks, the total cost and time required increase.
- MX–MN > MN–MN: The performance of MX–MN is better than MN–MN for executing 50 tasks across all VMs.

Scenario III: Tasks Size = 100; VMs: 5, 10, …, 50 – For the third scenario, the task size is made 100 with the experiment environment same as the earlier two scenarios. Figure 67.8 shows the total cost comparison graph for executing these 50 tasks and Fig. 67.9 shows the total time comparison graph for executing 50 tasks.

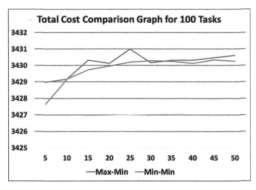

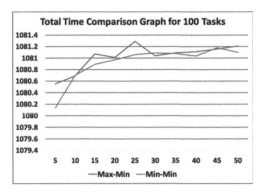

Fig. 67.8 Total Cost Comparison Graph   **Fig. 67.9** Total Time Comparison Graph

Table 67.3 Comparison w.r.t. total cost and time for executing 100 tasks

	Comparison w.r.t. Total Cost		Comparison w.r.t. Total Time	
Criteria	MX–MN	MN–MN	MX–MN	MN–MN
Linear Regression Equation	y = 0.1433x + 3429.4	y = 0.2098x + 3428.6	y = 0.0571x + 1080.7	y = 0.0795x + 1080.5
Regression Line Slope	0.1433	0.2098	0.0571	0.0795·
Slope Sign	Positive	Positive	Positive	Positive
Regression Line Y-Intercept	3429.4	3428.6	1080.7	1080.5
Relationship	Positive	Positive	Positive	Positive
R^2 value	0.4914	0.5794	0.5669	0.6139
VM Analysis	↑VM = ↑Cost	↑VM = ↑Cost	↑VM = ↑Time	↑VM = ↑Time
Analysis	MN-MN > MX-MN		MN-MN > MX-MN	

The following points can be observed from Table 67.3:

- ↑VM = ↑Cost & ↑VM = ↑Time: With increase in number of VMs for executing 100 tasks, the total cost and time required increase.
- MN-MN > MX–MN: The performance of MN–MN is better than MX–MN for executing 100 tasks across all VMs.

Scenario IV: Tasks Size = 1000; VMs: 5, 10, …, 50 – For scenario IV, tasks are 1000. Figure 67.10 shows the total cost comparison graph for executing these 50 tasks and Fig. 67.11 shows the total time comparison graph for executing 50 tasks.

The following points can be observed from Table 67.4:

- ↑VM = ↑Cost & ↑VM = ↑Time: With increase in number of VMs for executing 1000 tasks, the total cost and time required increase.
- MN–MN > MX–MN: The performance of MN–MN is better than MX–MN for executing 100 tasks across all VMs.

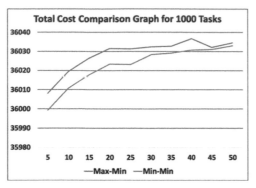

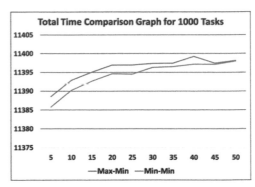

Fig. 67.10 Total Cost Comparison Graph   **Fig. 67.11** Total Time Comparison Graph

Table 67.4 Comparison w.r.t. total cost and time for executing 1000 tasks

Criteria	Total Cost Comparison		Total Time Comparison	
	MX–MN	**MN–MN**	**MX–MN**	**MN–MN**
Linear Regression Equation	y = 2.3118x + 36016	y = 3.2243x + 36005	y = 0.8558x + 11391	y = 1.135x + 11388
Regression Line Slope	2.3118	3.2243	0.8558	1.135
Slope Sign	Positive	Positive	Positive	Positive
Regression Line Y-Intercept	36016	36005	11391	11388
Relationship	Positive	Positive	Positive	Positive
R^2 value	0.6618	0.8348	0.6792	0.8189
VM Analysis	↑VM = ↑Cost	↑VM = ↑Cost	↑VM = ↑Time	↑VM = ↑Time
Performance	MN-MN > MX-MN		MN-MN > MX-MN	

C. Need of RL

From the conducted experiments, results obtained, and empirical analysis, we can say that different resource scheduling algorithms behave differently, especially when the number of tasks varies. Hence, fixing any resource scheduling algorithm statically, which will work under all conditions, will output limited results. In order to get better results, Reinforcement Learning (David 2007, Aram et al. 2004, Richard et al. 1998 Kaelbling et al. 1996) mechanism can be implemented to dynamically select the resource scheduling algorithm for execution of tasks in any number. With Reinforcement Learning, the agent, in this case the cloud, will go into a learning phase, and with trial-and-error technique and reward mechanisms, the cloud will understand to dynamically use resource scheduling algorithms while executing varying sized tasks.

4. Results and Conclusion

Table 67.5 shows the comparison of MX–MN and MN–MN algorithms across different environmental conditions. We can observe that no algorithm stands out and gives the best performance across all conditions. The major problem here is that algorithm chosen to execute the tasks is fixed throughout the execution of the tasks. In this research paper, we have also seen that different algorithms behave differently under different circumstances and scenarios. To get better results, we can use Reinforcement Learning mechanism to make the resource scheduling dynamic in nature. With Reinforcement Learning, the cloud will go into a trial-and-error phase to learn and adapt the environment using reward strategies. Once the cloud learns, then it will give better results as compared to the earlier static system.

Table 67.5 Overall comparison of MX–MN and MN–MN

Tasks Size	Algorithm Performance w.r.t. Total Cost	Algorithm Performance w.r.t. Total Time
25	MX–MN ≈ MN–MN	MX–MN ≈ MN–MN
50	MX–MN > MN–MN	MX–MN > MN–MN
100	MN–MN > MX–MN	MN–MN > MX–MN
1000	MN–MN > MX–MN	MN–MN > MX–MN

REFERENCES

1. Arulkumar V, N. Bhalaji. (2022). Load balancing in cloud computing using water wave algorithm.
2. Feng Li, T. Warren Liao, Wentong Cai. (2022). Research on the collaboration of service selection and resource scheduling for IoT simulation workflows.
3. Wael Khallouli, Jingwei Huang. (2022). Cluster resource scheduling in cloud computing: literature review and research challenges
4. Nagendra Prasad Sodinapalli, Subhash Kulkarni, Nawaz Ahmed Sharief, Prasanth Venkatareddy. (2022). An efficient resource utilization technique for scheduling scientific workload in cloud computing environment.
5. Shridhar G. Domanal, G. Ram Mohana Reddy. (2022). Optimal load balancing in cloud computing by efficient utilization of virtual machines.
6. Gao M., Li Y., Yu J. (2021). Workload Prediction of Cloud Workflow Based on Graph Neural Network.
7. Rupali, Mangla N. (2021). Resource scheduling on basis of cost-effectiveness in cloud computing environment.
8. Rajput R.K.S., Hussain R., Goyal D. (2021). Modelling and Simulation of Cloud Service Cost Analysis using Resource Scheduling.
9. Yuejuan K., Zhuojun L., Weihao O. (2021). Task scheduling algorithm based on reliability perception in cloud computing.
10. Devi K.L., Valli S. (2021). Multi-objective heuristics algorithm for dynamic resource scheduling in the cloud computing environment.
11. Kaur G., Bala A. (2021). Prediction based task scheduling approach for floodplain application in cloud environment.

12. Zhang B., Zeng Z., Shi X., Yang J., Veeravalli B., Li K. (2021). A novel cooperative resource provisioning strategy for Multi-Cloud load balancing.
13. Ibrahim A. Thiyeb, Sharaf A. Alhomdy. (2020). HAMM: A Hybrid Algorithm of Min-Min and Max-Min Task Scheduling Algorithms in Cloud Computing.
14. Weiwei Chen, Ewa Deelman. (2012). WorkflowSim: A Toolkit for Simulating Scientific Workflows in Distributed Environments.
15. Michael Armbrust, Armando Fox, Rean Griffith, Anthony D. Joseph, Randy Katz, Andy Konwinski, Gunho Lee, David Patterson, Ariel Rabkin, Ion Stoica, Matei Zaharia. (2010). A View of Cloud Computing.
16. Tharam Dillon, Chen Wu, Elizabeth Chang. (2012). Cloud Computing: Issues and Challenges.
17. David Vengerov. (2007). A Reinforcement Learning Approach to Dynamic Resource Allocation.
18. Aram Galstyan, Kristina Lerman. (2004). Resource Allocation in the Grid Using Reinforcement Learning.
19. Kaelbling L.P., Littman M.L., Moore A.W. (1996). Reinforcement learning: A survey.
20. Richard S. Sutton, Andrew G. Barto. (1998). Reinforcement Learning - An introduction.

Intelligent Systems and Smart Infrastructure – Brijesh Mishra et al. (eds)
© 2023 Taylor & Francis Group, London, ISBN 978-1-032-41287-0

CHAPTER 68

Penta Band Microstrip Patch Antenna for S, C, and X Band Applications

Abhishek Kumar, Chirag Parashar

Meerut Institute of Engineering and Technology,
Meerut (Uttar Pradesh), India

Ajay Kumar*

IMS Engineering College, Ghaziabad, Uttar Pradesh, India

Abstract

In this article a small size Penta band antenna was planned for C, S, and X band wireless applications. The multiple band operation of this antenna is achieved by meandering line. This antenna covered step-by-step process for Penta band antenna design. The antenna is planned on FR-4 material with the permittivity of 4.4- and 1.6-mm substrate thickness. The size of the antenna is $15 \times 19 \times 1.6$ mm^3 and achieved frequency bands of (2.53–2.64) GHz, (3.95–4.34) GHz, (5.53–5.89) GHz, (6.79–7.23) GHz, and (9.18–10.21) GHz that cover the operating bands of C, S, and X bands applications. The presented antenna has a great impact on various applications like airport surveillance radar, surface ship radar, satellite communication, and medical industry.

Keywords: Multiple Patches, Return Losses, Radiation Pattern, Microstrip Patch Antenna (MSA).

1. Introduction

To fulfill the requirements of current scenario, multi-band antennas are required for high-frequency wireless applications. There are different antennas utilized for wireless applications

*Corresponding author: akgangwarr@gmail.com

DOI: 10.1201/9781003357346-68*

but for compact design and economical making, the need of microstrip antenna is increased. The microstrip antenna consists a dielectric layer which is sandwiched between two conducting layers. The antenna is ease to integrate in the circuit board. The multi-band antennas also consist of the advantages like easy to integrate numerous wireless communication bands in a single-antenna system [1],[2]. For designing a high-frequency band antenna, different high-frequency antennas are studied such as in the year of 2017, Rajesh Kumar et. al. designed a multi-band antenna for C, L, and S band applications [3]. It is planned to make by cutting slots in to the bowtie radiating patch. The antenna is planned on FR4 substrate simulated using HFSS software. The antenna is resonated at 1.6, 2.8, 4.8, 5.6, and 6.6 GHz frequency band; further, in the year of 2018, Ruixing Zhi et. al. planned a compact size multi-frequency antenna for wireless applications [4]. The antenna was designed using two strips with parasitic stubs, and for achieving monopole operation, partial ground plane is used. This antenna covered the bandwidth from 2.28 to 2.57 GHz, 5.0 to 6.27 GHz, and 7.11 to 7.96 GHz bandwidth, The antenna is suitable for wireless local area network and X-frequency band. The arc with parasitic antenna provides dipole-like (omnidirectional) radiation pattern in E-plane and H plane, respectively. Following the year, Nikita Saxena et al. planned a multi-frequency band antenna for S and C wireless band [5]. The antenna is planned on RT Duroid substrate with the permittivity of 2.2. It is proposed using square radiating patch with four L-shaped slots cut in the patch for achieving multi-band characteristic. The design antenna resonates at 2.379, 2.45, 3.75, 4.46, 5.0 and 6.25 GHz resonance frequencies.

From the study of above antenns, it is found that the size of the antennas are large. Thus a new compact pentaband antenna is designed and analysis. The performances of the antenna is evaluated by returnloss, current distribution in the radiating patch and radiation performance at resonance frequency.

2. Antenna Design

The Penta band antenna is designed for five resonates at frequency operations. The proposed antenna and their dimensions size are mentioned in antennas, the size of ground has been kept $Wg \times Lg$ (15.0×19.0 mm^2) and it is designed using meander line. The width of the meander line is 2.0 mm. The antenna is designed and simulated with FR4 dielectric substrate with permittivity of 4.4, loss tangent 0.002, and height of 1.6 mm. For designing of the antenna, FR4 dielectric substrate is used because it is of low cost and sufficiently available. The multifrequency band (Penta band) antenna power is fed by a microstrip line of dimension $W_f \times L_f$ (1.0×5.0 mm^2) [6],[7]. The labeling of antenna and patch constituents has been done and prototype of the antenna is placed on a vacuum radiation box of volume $48.0 \times 56.0 \times 19.2$ mm^3.

3. Results and Discussion

The proposed five-band antenna is designed and validated using high-frequency simulation software, the S_{11} parameter is graphed, and radiation pattern is observed followed by the current distribution pattern. The different stages of antenna are illustrated in Fig. 68.1(a) to (c) and first-stage antenna and their return loss are shown in Fig. 68.2.

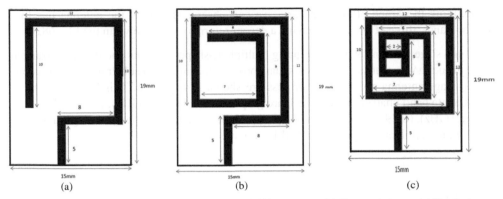

Fig. 68.1   Different stages of the antennas: (a) First stage; (b) Second stage; (c) Final stage

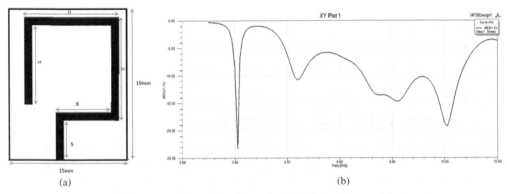

Fig. 68.2   The first-stage antenna design; (b) Return loss of the antenna

As we can see in the introductory stage, an inverted rectangular hook-shape-like patch is created on the substrate, which produces two proper frequency bands and formation of other bands can also be seen in the graph. In order to bring out multi-bands, an L-shaped patch is introduced in the second stage of antenna, and their return loss is shown in Fig. 68.3.

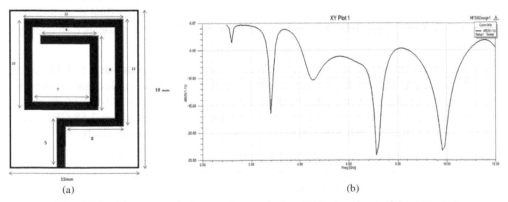

Fig. 68.3   The second-stage antenna design; (b) Return loss of the antenna

After studying the graph of second stage, we can observe that now the antenna has four frequency bands, three of which are having a decent return loss, as shown in the Fig. 68.3.

For the final stage, antenna is designed using multiple L-shaped patches and they are arranged together in such a way that they form two square boxes at the middle of substrate approximately. The final-stage antenna and their return loss are shown in Fig. 68.4. This helps antenna to generate penta band characteristics which are graphed at (2.53–2.64 GHz), (3.95–4.34 GHz), (5.53–5.89 GHz), (6.79–7.23 GHz) and (9.18–10.21 GHz) that cover the operating bands of S, C, and X bands applications provided by IEEE organisation.

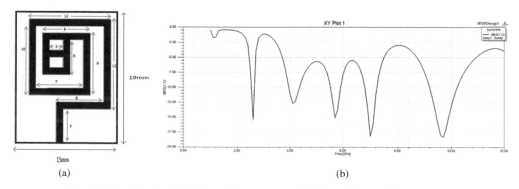

(a)   (b)

Fig. 68.4   The final design of the antenna; (b) Return loss of the antenna

In all three stages, the patches are united together and assigned electric field boundary, the port is given the excitation of lumped port, the ground is also giving perfectly E-boundary and outer boundary is gave by radiation box. The solution setup for the above analysis has solution frequency of 2.4 GHz and the value of maximum delta S is 0.02 with number of passes count as 6, and the sweep type is kept fast in frequency sweep setup.

The ground is kept in the form of simple strip (15 × 1.15 sq.mm) yet efficient. An analysis of current distribution pattern and 2D radiation pattern is done to observe the fluctuations caused by altering the parameters of antenna. The current distribution pattern in the antenna signifies the flow of charge coming through the resistance and helps to understand the radiation mechanism of the antenna. Fig. 68.5 displays current distribution of the final design of the antenna at 2.58 GHz, 4.14 GHz, 5.71 GHz, 7.01 GHz, and 9.69 GHz resonance frequency, respectively.

On doing a comparative analysis of the above results, we can observe that at 2.58 GHz, 5.71 GHz, 7.01 GHz, and 9.69 GHz, the patch is more greenish, which implies that antenna is more conductive in Fig. 68.5 (a), (c), (d), and (e) than in Fig. 68.5(b) at 4.14 GHz. To study radiations of the patch, 2-D radiation patterns are produced at frequencies at 2.58 GHz, 4.14 GHz, 5.71 GHz, 7.01 GHz, and 9.69 GHz shown in Fig. 68.6(a)–(e), respectively.

After observing all the patterns, we infer that the following antenna radiates stably in almost all direction at all the operational frequency, and we can conclude that the antenna has the good and stable omnidirectional radiation pattern.

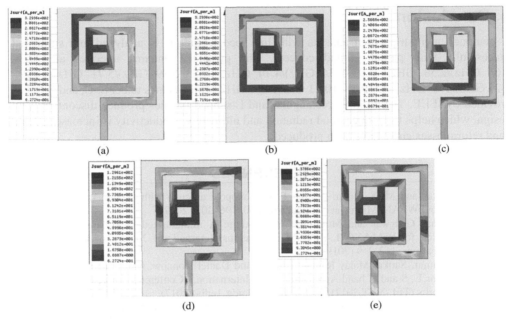

Fig. 68.5 Current distribution of the antenna: (a) 2.58 GHz; (b) 4.14 GHz; (c) 5.71 GHz; (d) 7.01 GHz; (e) 9.69 GHz

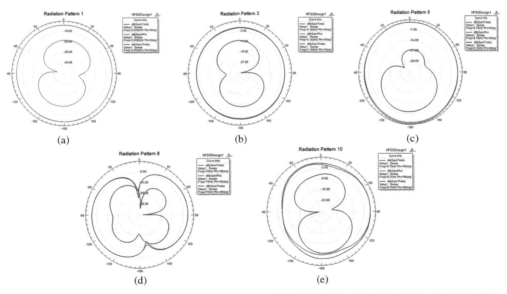

Fig. 68.6 2D radiation pattern of the antenna: (a) 2.58 GHz; (b) 4.14 GHz; (c) 5.71 GHz; (d) 7.01 GHz; (e) 9.69 GHz

4. Conclusion

A compact microstrip patch antenna with line feed technique is analyzed and simulated successfully with partial ground concept. The presented antenna operates at (2.53–2.64 GHz)/2.58 GHz, (3.95–4.34 GHz)/4.14 GHz, (5.53–5.89 GHz)/5.71 GHz, (6.79–7.23 GHz)/7.01 GHz, and (9.18–10.21 GHz)/9.69 GHz, which comes under S, C, and X bands provided by IEEE. The multiple rectangular and L-shaped patches produce discontinuity in design, which helps to produce good radiation and ultimately conductivity is increased. Penta band return losses curve have been produced.

REFERENCES

1. A. Gangwar and M. S. Alam: A high FoM monopole antenna with asymmetrical L-slots for WiMAX and WLAN. Microwave optical letter. 60, 196–202 (2017).
2. Ajay Kumar Gangwar and Muhmmad Shah Alam: A miniaturized quad-band antenna with slotted patch for WiMAX/WLAN/GSM applications. International Journal of Electronics and Communications. 112, 1-9 (2019).
3. Rajesh Kumar, Sarika, Malay Ranjan Tripathy and Daniel Ronnow: Multi-band Slotted Bowtie Antenna for L, S and C band Applications. 2nd International Conference on Telecommunication and Networks (TEL-NET 2017), pp.1-4, IEEE, Noida, India (2017).
4. Ruixing Zhi, Mengqi Han, Jing Bai, Wenying Wu, and Gui Liu: Miniature Multiband Antenna for WLAN and X-Band Satellite Communication Applications, Progress In Electromagnetics Research Letters, 75, 13-18 (2018).
5. Nikita Saxena and Asmita Rajawat: Design and Analysis of Multi Band antenna for S and C Band. International Conference on Advances in Computing, Communication Control and Networking (ICACCCN), pp. 894-898, IEEE, Greater Noida, India.
6. M. Karthikeyan, R. Sitharthan, Tanweer Ali, Sameena Pathan, Jaume Anguera, · and D. Shanmuga Sundar: Stacked T-Shaped Strips Compact Antenna for WLAN and WiMAX Applications. Wireless Personal Communications, 308 (2021).
7. Bahareh Badamchi, JavadNourinia, Changiz Ghobadi and Arash Valizade Shahmirzadi: Design of compact reconfigurable UWB slot antenna along with the switchable single/dual band notch functions. IET , Microwaves, Antennas & Propagation , 8, (2014).
8. Chao-Ming Luo, Jing-Song Hong, Muhammad Amin and Lei Zhong: Compact UWB antenna with triple notched bands reconfigurable. IEEE International Conference on Microwave and Millimeter Wave Technology (ICMMT), (2016).
9. G. AishvaryaaDevi, J. Aarthi;P. Bhargav, R. Pandeeswari, M. Ananda Reddy and R. Samson Daniel: UWB frequency reconfigurable patch antenna for cognitive radio applications. IEEE International Conference on Antenna Innovations & Modern Technologies for Ground, Aircraft and Satellite Applications (iAIM), 13,10 (2017).
10. Yingsong Li and Wenhua Yu: Design of a ultra wideband antenna with tunable and reconfigurable band-notched characteristics. IEEE 4th Asia-Pacific Conference on Antennas and Propagation (APCAP), (2015).
11. Paris Lotfi, Mohammadnaghi Azarmanesh and SandaberSoltani: Rotatable Dual Band-Notched UWB/Triple-Band WLAN Reconfigurable Antenna. IEEE Antennas and Wireless Propagation Letters, 12, (2013).
12. Shan Wang, Jian Dong and Meng Wang: A Frequency-Reconfigurable UWB Antenna with Switchable Single/Dual/Triple Band Notch Functions. Cross Strait Quad-Regional Radio Science and Wireless Technology Conference (CSQRWC), (2019).
13. Manish Sharma, Amit Kumar Goel, Naresh Kumar and Yozendra Kumar Awasthi: Reconfigurable Dual Notched Band UWB Antenna. 8th International Conference on Cloud Computing, Data Science & Engineering (Confluence), (2018).

Intelligent Systems and Smart Infrastructure – Brijesh Mishra et al. (eds)
© 2023 Taylor & Francis Group, London, ISBN 978-1-032-41287-0

CHAPTER

69

A Review on IoT-based Smart Agriculture

Vaibhav Bhatnagar[1], Tamanna[2], Rahul Kumar Sharma[3]

Department of Computer Science and Engineering,
Noida Institute of Engineering and Technology, Noida, India

Abstract

The Internet of Things (IoTs) is the present growing as well as the future technology of every agricultural land which is impacting life of everyone by making overall system intelligent in the agricultural field. Because of the fast expansion of economy of any country, agriculture sector plays a vital role. The agricultural systems which are developed through IoT are the most sustainable systems that decrease the resource consumption and thereby increasing the production of food. As we can see in the near future that the population will increase globally and so the demand of energy and food will keep on increasing and this will be a much greater challenge ever faced by humankind. The IoT technology uses sensors which are connected to cloud used in the agricultural field that helps in agriculture modernization as well as communication in savvy farming. Better and capital yielding agriculture can be achieved through IoT as it helps in improved management of crops and resources. Smart farming is a very high technology system through which clean crops can be grown which are sustainable for the population. This paper reviews various smart systems developed using IoT in the field of agriculture. This paper will also focus on existing state-of-the-art approaches for crop improvement. Investigation will also be done on various kinds of sensors used in agricultural land for making the field smart. Along this, this paper will help the present and future scientists for the advancement of savvy agriculture with collaboration to enough information.

Keywords: IoT, sensors, savvy farming, smart system, cloud

Corresponding author: [1]vaibhav.rahi1992@gmail.com, [2]tamannaveer22@gmail.com, [3]rahulsharma9045@gmail.com

DOI: 10.1201/9781003357346-69

1. Introduction

The Internet of Things (IoTs) will be the foundation for making everything smart in the future. IoTs has the capability of transforming the existing and traditional technology from offices to homes and in every sector to a whole next level or the upcoming generation everywhere computing. The tune foundation for the development and betterment of different IoTs items includes smart living, smart education, smart transportation, and smart automation, which are all aspects of smart living and more. It is being used commercially and on a wide scale in various sectors like manufacturing, business management, agriculture, and so on to uplift the capital yielding as well as for a better living.

The most researched and challenging topic is smart agriculture as most of the population in India are connected and are based on this sector for the living as well as food security. As agricultural sector is the most crucial sector, different research has been carried out in this field to make it smart.

In view of this, a system is developed which uses wireless sensor network (WSN) for tracking water management, crop management and energy management. This model will further develop a nexus model which is based on the data in real time [3].

Traditional methods of agriculture management require a massive amount of manpower and resources; this has an adverse impact on economy. To overcome this, state-of-the-art methods have been used. In this method, a hardware-based system is used which when deployed in the field, whic will give an idea of the moisture and temperature of the soil and it is done in real time [4]. An important aspect of agriculture is having a good and proper irrigation system which requires various sensors that communicate different framework of soil sensed to the cloud using Arduino Uno microcontroller (Atmega 328). The data received from the sensors is classified into various categories based on the values of threshold using machine learning algorithms and the farmer receives an email or message to take a certain type of action [5].

Smart agriculture facilitates farmers to ensure the growth of economy as well as eradicating hunger from the country. This can be achieved by developing a system of poly farm houses in which various sensors are used such as water volume, soil pH, soil moisture, motion detector, and air temperature sensors and these sensors ensure the higher yielding with minimum amount of resources used [6]. In order to calculate the water content in the soil, YL-38 sensor is used, DTH 11 sensor is used to record the environment temperature and humidity, DS18B20 sensor senses the soil temperature, 16-by-2 LCD displays the reading and a GSM module is required to connect one device to another which enables IoT connection. Using so many sensors will probably lead to an error at any node and rectification of errors is also automated without any human intervention [5, 8].

Fertilizer is one of the important ingredients in crop production as better quality and quantity of fertilizer will make a high-yielding crop. IoT-based low-cost fertilizer is designed to make the agriculture smart. A sensor known as Nitrogen–Phosphorus– Potassium (NPK) is used with LDR and LED to sense and analyze the soil nutrients. Fuzzy logic concept is applied in order to capture the deficiency of nutrients in the soil [12]. There are various research carried out in the field of smart agriculture but one problem is always there, i.e., the cost of IoT hardware and power consumption and so the farmer could not make much amount of profit, and, in fact, suffer loss. To overcome this challenge, a system is developed using WSN which will operate at a suitable cost and has a low power consumption. The system developed uses Raspberry pi 2 model B which is based on free OS Debian optimized for hardware, various sensors such as leaf wetness, wind direction and speed, LCD (zigbee transmitter), DC motor, and LM 380 as a speaker [13, 16].

In older times and even today, farmers use to plan a high-yielding crop production on prediction basis, which means farmers predict the upcoming temperature, rainfall, and various other factors which play a vital role in the agricultural sector. Now in this smart era, various prediction modules in collaboration with IoT are being used for the better production and one of the modules of prediction uses SVM (support vector machine) classifier [14]. A system is also there which uses IITH mote in which the node of sensors works on the solar power so that the cost of power consumption is cut off and the profit is high [19].

There are various proposed systems which are developed and many require further studies for the sake of agricultural industry. All these proposed systems may require different sensors but the goal is common to be achieved, i.e., smart agriculture which in turn requires IoT. A smart agriculture system is one which helps the farmers and even non-farmers to be able to work in agriculture industry and contribute in the growth of the economy of a country. In this paper, some of the smart agriculture systems are reviewed, which will enhance the future research and scientists will come up with some excellent ideas in the agricultural industry to make it savvy.

2. Literature Review

A. Ownership Concentration and Stock Return

In a system proposed by Jash Doshi et al [1], the monitoring of the farm and the monitoring of the greenhouse are both included. To keep an eye on the system, various readings generated by various sensors such as humidity sensors, temperature sensors, pH sensors, and so on must be stored. These many readings will provide farmers with different alerts in various situations. The authors created an IoTs gadget based on a plug-and-sense principle in this study. The temperature and humidity sensors in this model are DTH 11 and ESP, respectively. With jumper wires, the ESP 32S and the node MCU, which are wireless and wifi-controlled, are linked to the breadboard. Every 18 minutes, the ESP 32S will be turned off, and when it wakes up, it will upload the reading to the cloud. If the farmer misses the buzzer (SI1145IR is used) or notification, LEDs positioned around the bucket will remain on. The soil moisture sensor is positioned at the bottom of the bucket, and temperature, humidity, IR, and a buzzer are mounted at the top. The proposed system is of low power, requiring only a 6000 mAh powerbank to function.

Chayapol Kamyod [2] uses a proposed architecture to examine and test the end-to-end reliability of an IoT-based smart farm. Type 1 IoT network communication and type 2 IoT network communication are two separate architectures that are IoT-connected, according to the author. Form 1 networks examine sensor data at the IoT node, and this type of networks is appropriate for low-cost, low-complexity systems where the data supplied by the sensor is not large and computational operations are minimal. Type 2 is the polar opposite of type 1, in that it is more significant and generates large amounts of data, which is then stored on a cloud server and analysed. The suggested model uses a wireless router to facilitate communication between sensor nodes, as well as the IEEE 802.11B standard at 11 mb/s. The suggested system's authors also state that the distance between the router and the sensor node must be less than 50 metres to avoid connectivity issues. Because these three parameters can affect system reliability, this article focuses on throughput, IoT server load, and latency.

Yemeserach Mekonnen et al [3] designed a prototype of smart farm which involves field preparation associated with accommodation of crop line and preparation of soil and also have solar panel which is having floating foundation and placement of pole. This module consists of raised beds of size 4´25 ft and each bed having dispense WSN, PV panels, and smart sensors for irrigation and database. Electrical system, Wireless sensor, and Irrigation system are the three primary divisions. The electrical system unit includes a storage unit and a power supply via a PV panel with a peak power of 320W and 37.2 Vmpp. The panel placement angle is 26° as the latitude, which will boost energy output to the maximum extent possible. A 20A 12V dc MPPT is connected to the panel to prevent overcharging of the batteries. The wireless sensor unit includes a variety of sensors linked to it, and an Arduino-based node with multiple radio options is employed. The sensor board and the principal functionality module board are the two essential components. The data generated by sensors is transferred to a router gateway using the XBee-Pro S2 module. The RTC (real-time clock) on the WSU microcontroller is utilised to turn on the device when measurement is required. The WSU features a 6000 mAh battery that can be recharged using a solar panel with a voltage of 7 V and a current of 500 mA. This suggested work makes use of the Zigbee protocol to send data to a gateway. The irrigation system unit, which has a regulator, 24-volt DC pump, 24-volt solenoid valves, and relays that can activate the water pump and valves at the threshold value, is the last unit in the research.

According to Reuben Varghese et al [4], machine learning algorithms and the IoTs are two major strategies to make agriculture smart. Different crops are used in this suggested work, and a logistic regression-based model is used.

Tensorflow is a type of flow that deals with the state of the soil, as well as the temperature and moisture in the surrounding environment. SVM detects the current state of the soil and crop suggestions, and if an error occurs, the system will issue an alarm to the farmer via SMS API. If human control of the machinery is required, a web application using PHP is created in this study. For the current module, the researchers have specified the following software and hardware requirements: Amazon AWS T2.Micro EC2, which stores data and acts as a database, Tensorflow, and an Apache server are all software needs. Raspberry Pi, DTH 11 temperature sensor, FC-28 moisture sensor, MQ-135 air quality sensor, and LM-393 sensor that measures the appropriate quantity of sunshine for the crop, a 5 volt–10 A relay as a switch to operate the equipment, and Arduino MCP-3208 ADC are required as hardware needs. Because it has only the most basic sensors, this system is both inexpensive and of low cost.

Bhanu K. N et al [5] proposed a smart irrigation system based on IoT in a research paper. This approach also focuses on the proper amount of water consumption by crops in order to preserve water. The authors proposed a system in which all of the sensors work at the same time, sensing the soil's correlative properties and sending them to Arduino. It will send the information to the Thingspeak cloud server, which will store it. Moving forward, data saved in the cloud is classified using a classification algorithm and analysed at the Thingspeak cloud, and if the classed data falls below a threshold value, an email or alert is delivered to the farmer for action. A sensor branded 4L38 is utilised to calculate the current water percentage for precise water consumption. DTH 11 is the same temperature sensor that was utilised in [4]. The DS18B20 sensor is used in this system to monitor the soil temperature.

Rahul Dagar et al [6] have developed a smart farming system based on IoT, which will help the agriculture industry become more intelligent. The authors of the research created a poly home that includes a temperature sensor, humidity sensor, pH sensor, motion detector sensor, and other sensors. The polyhouse is a sturdy construction that is completely covered in polythene to protect it from harmful sun rays, heavy rainfall, storms, and other natural occurrences. According to the study, utilising IoT technology in the polyhouse will enhance production by up to two fold.

Ibrahim et al [7] propose a smart agriculture system based on the IoTs. The power module in this system is divided into four basic modules: processing (controller), processing (memory), communication, and sensing. A CPU with a microcontroller that detects the input signal is required for the proposed system. It also requires a power supply unit, a memory unit, and programming tools to store the control measures obtained by the programme. The module consists of an actuator that controls the valves rather than acting as a control centre since it merely sets the parameters or values that were to be communicated to remote nodes. In order to activate the attached devices, the remote node will now compare the actual parameter to the configured value. On the GUI control centre, the entire procedure may be observed. The alert function, which the authors have focused on in this study, is one item that plays a very vital part in GUI. This alarm feature can be customised to meet our specific demands, such as the type of alarm in various scenarios and the set point level.

Vaishali Puranik et al. [8] proposed a system that focuses on maintenance automation, pesticide control, irrigation management, and crop monitoring on a regular basis. The entire research project is focused on developing an IoT-based agriculture automation system. Various sensors are placed in the agriculture farm to achieve such a system, such as the moisture mapping sensor in the soil. Moisture-out ports on this sensor aid in the measurement of soil moisture by monitoring the potential difference between soil particles. The sensor's output is connected to the microcontroller in order to keep the soil at the proper moisture level. Water management in the field is to be carried out using irrigation sensor after reaching the exact soil moisture. The moisture content is examined first, and if it is less than the threshold value, then there is an issue with the irrigation system. If the irrigation system fails or the sensor fails to produce reliable data, an alert is issued to the farmer and the service team is called; however, if the irrigation system fails or the sensor fails to produce accurate data, the alarm is sent to the farmer and the service team is contacted. This study also suggests a two-stage irrigation automation system. In step 1, fertiliser insufficiency or efficiency must be reported, and in stage 2, which fertiliser is causing a problem with crop development must be determined. The two stages necessitate two sets of sensors, the first of which will identify the pH value and nutrients, and the second of which will monitor the excess or lack of various fertilisers.

Ramalakshmi Ramar et al. [9] designed a smart irrigation system for agricultural irrigation that employs IoTs and cloud technologies to collect and store data provided by numerous sensors put in the crop field. The suggested and implemented system can control a variety of solenoid valve implementations. This project will focus on the design of IoTs devices, software, and hardware, as well as their integration with network and cloud connectivity. The use of various sensors and a microcontroller is specified by the author. This system requires a WEMOS D1 controller (ESP-8266EX), which regulates the flow, monitors all sensors, and operates the DC motor. The soil moisture sensor 4L69, the DTH 11 temperature sensor, the water flow sensor 4F-201, and the pH sensor pH metre probe were all used in this study (0-10). The data collected by all of these sensors is saved in Thingspeak's cloud, where it will be monitored via a mobile app or a web interface.

Kushagra Agarwal et al [10] published a study in which they analysed several existing systems for smart agriculture with IoT and recommended a system that will focus on numerous agricultural elements. The authors of this study are interested in water management, crop management, pesticide control, food production, and crop safety from diverse natural and human phenomena. Various sensors are utilised at various positions in the agricultural field to achieve this.

Saurav Verma et al [11] proposed a smart agriculture system based on the IoTs and a four-layer module. Application layer, middleware layer, network layer, and perception layer are the layers of the four-layer module. The perception layer deals with the various sensors needed in the field, as well as detectors and all of the system's hardware requirements. The perception layer's components are connected to a microcontroller, which aggregates and analyses real-time input. The network layer is the next layer, which aids in data transfer. This layer connects the hardware components (sensors, actuators, etc.) to the software components (database, WebGis, etc.) as well as the middleware layer. The network's middleware layer enables network protocols, and this layer is responsible for information exchange between network sensors and the operating system. This layer connects the application layer and the network layer. The application layer is the final layer, and it consists of communication protocols that are used to host in a communication network. MQTT, AMQP, COAP, and other protocols are examples.

Lavanya G et al [12] propose an IoT-based system that monitors soil nutrients and, based on that data, alerts the user or farmer in their field on the fertiliser quantity to be used during crop cultivation. This study will be built on a lower cost fertiliser intimation system based on the

IoTs, which will be automated and incredibly beneficial to agricultural production. Agriculture will become smarter, thanks to an IoT-based low-cost fertiliser. To measure and analyse soil nutrients, a sensor known as NPK is utilised in conjunction with LDR and LED. Fuzzy logic concept is applied in order to capture the deficiency of nutrients in the soil. A user-friendly system which is also internet based and automated intimate from time to time to the farmers about smart devices.

K. Lokesh Krishna et al. [13] presented a research project involving the development of a wireless sensor network system (WSN). The proposed system will run on low power, making it energy-efficient, as well as costing the farmer significantly less to install on their farmland, making it budget-friendly. Hermo hygro sensor, soil moisture sensor, humidity sensor, UV (ultra violet) sensor, obstacle sensor, carbon dioxide sensor, and pH sensor were employed in this study. This variant runs on the Raspberry Pi 2 model B, a free Debian-based operating system that is tailored for Raspberry Pi hardware. All of the sensors mentioned above, as well as the camera used in the crop field in this proposed work, are connected to a Raspberry Pi 2 Model B, which is mounted on a wireless mobile robot. A liquid colour display (LCD) is utilised to display the data readings, which is connected to the Zigbee transmitter. A DC motor is employed as a sprayer in the field, and a speaker labelled LM 380 is fitted in this system for spraying. The activities intended for the robot may be accessed through the PC part, and the total system is powered by a solar panel that charges the batteries.

G. S. Nagaraja et al. [14] created a system that is based on the agricultural business and includes IoT. The proposed system is made up of two modules: data monitoring and crop prediction, according to the authors. The data collected from agricultural sensors is analysed using a Raspberry Pi 3 and the node MCU Devkit module. According to the research article, the sensor utilised in the data monitoring module is DTH 11, which is a temperature sensor, and a huge amount of data is aggregated from sources such as data acquired through lab tests or soil samples, or data created by sensors in the crop prediction module. In this module, a farmer or user enters data into the module and the module forecasts a crop that is suitable, using machine learning techniques. The data from the sensors is sent into Thingspeak, a cloud application for data visualisation, and it is also stored in real time on the Firebase database. When the moisture level falls below the threshold, the motor turns on.

K. A Patil et al. [15] established a concept for smart agricultural systems in which they propose a three-module system with three layers. Farm side, server side, and client side are the three modules, and the layers mentioned are perception, network, and application. The perception layer has a Ubi-sense-mote, which is a board with a temperature and humidity sensor, senses the proximity of an IR LED, and alerts through a buzzer, whereas the network layer is responsible for transactions to the application layer and consists of a ubi-mote with IEEE 802.15.4 in which ARM and SOC Cortex M3 are used with an external flash memory, and the application layer is used to collect data from various sources. WINGZ is employed as a Zigbee technology in this suggested system. The authors also implement some essential decision support models for pest and disease forewarning in this module. Image processing may be used to identify diseases in crops, and alarms can be sent to farmers in real time. This can be readily controlled via a mobile or online application.

Piyush Patil et al [16] suggest a method in which an IoT-based toolbox is used to provide smart agriculture for the benefit of both the country's economy and the farmer. The suggested system consists of a Raspberry Pi model with a DTH 11 temperature and humidity sensor, a SEN-0114 sensor for measuring moisture in the soil, and an FC-37 sensor for monitoring leaf wetness. The SEN-08942 may also be used to determine the direction and speed of the wind, and additional sensors such as a rainfall sensor and a soil pH sensor are also installed in the farm. One of the project's highlights is the identification of appropriate and superior seed for sowing in the soil. Image processing is employed for this recognition, and the authors recommend a number of approaches for doing so, including artificial neural networks, Euclidean distance methods, rule-based systems, and so on.

Gokul L. Patil et al [17] describe an IoT-based system that can be used in the agriculture industry to make it more sophisticated. The microcontroller Arduino mini pro board is employed in this suggested system, and the LM-35 is used as a temperature sensor, along with several other types of sensors such as moisture, pressure, and humidity sensors. The data created by all of these stated sensors is collected at the cloud interface, which is linked to a mobile application via which the total system is monitored, and all of these sensors and devices are integrated using an agro logger and guided through a central platform. For further investigation, the suggested system employs Xively as an IoT platform.

Prathibha SR et al [18] present a smart monitoring system based on IoT in agriculture. The suggested system comprises of a CC3200 main block that has a microcontroller, network processor, and a wifi unit. As we have seen in past research, it uses a separate temperature sensor, TMP007, which has an internal math engine and digital control, as well as a humidity sensor, HDC1010. A camera sensor MT9D11 is used to monitor the entire system, and it is connected to a booster camera pack through a PCB. The proposed system is simple to install in any crop field because it is portable, and it is also energy efficient because it runs on low power.

Soumil Heble et al [19] proposed and developed an IoT-based system with low-power operations. The authors proved the method very effectively by using a maize crop area of around 648 square metres. The field is divided into 27 plots, each having a 24 square metre area. Now, each plot has one of each type of sensor node, resulting in a total of 27 nodes and one sink in the whole field. The field sensors will generate a significant amount of data, which will be transferred to the sink for uploading to the server, where it will be analysed and consumed. IITH mote, a wireless and 802.15.4 compliant technology, is employed for the sink and sensor node. Moving on, the authors said that the communication gateway will be an Intel Edison with a 4G modem, with the IITH mote acting as the sink. In this study, the readings from several sensors were examined, and a fully functional system was created.

Based on the current state of the agricultural industry, R. Ramya Priya et al [20] presented an IoT-based solution. The suggested system uses an Atmega 8A pH sensor that does not require batteries, making it an energy-efficient solution. The meter's probe, which is attached to the sensor, is placed into the soil, and the desired reading is taken. Because the Atmega 8A sensor provides accurate readings, it prevents overwatering and underwatering, as well as water waste. The above-mentioned system is based on sensor, wireless, and IoT technology integration. The integrated strategy uses RMS (remote monitoring system) to reduce human resource requirements, lowering costs.

Courage Kpotosu et al. [21] present a smart agricultural system that assists in automating the watering process while also providing critical information to the farmer about changes in the land. To create this system, the authors employed a soil moisture sensor FC-28 and a soil pH sensor SKU:SEN0161 to manage the pH of the soil, which is a critical factor in high crop output. In this experiment, a 5-V 1-channel relay is utilised to turn on/off the water pump when the pH or moisture sensors detect a change. The gateway we come across in the system is the Raspberry Pi 3 model, and the wireless technology employed is SIM 900A GSM module with Digi Xbee Zigbee module, which has been used previously in the above-mentioned studies.

Different techniques and systems proposed by various authors along with observations and conclusion are compared in Table 69.1.

3. Conclusion

Agriculture may become more precise and productive as a result of IoT-enabled developments. The IoTs can be used in a range of agricultural applications. Water and energy are the most significant data sources, and their costs can make or break the horticultural sector. As a result of poor water system frameworks, inefficient field application techniques, and the growing of water-concentrated crops in some unwanted emerging locations, water wastage has come to an end. Its operation necessitates the use of pumps, supports, lighting, and other electrical devices. To make water technology more intelligent for agriculture, IOT should be able to monitor and change water amount, area timing, and stream span. With the assistance of the IoTs. In this area, the usage of manures, as well as pesticides based on them, is a big concern.

4. Future Scope

Although the structures shown in the previous region make the IOT idea appear to be feasible, more research is required. This section looks into specialised issues with existing IoT models. Later, a specialised concept of IoT engineering was developed to address all of the key components that are currently lacking in current designs. Before the IoTs is extensively adopted and sent around the world, a complete grasp of mechanical attributes and needs on cost, security, protection, and hazard must be examined.

Let's have a look at some of the challenges that scientists may face in the near future.

1. Maintenance of costs
2. Design of SOA (service oriented architecture) which deals with performance and costs
3. Present DBMS is not so efficient to handle vast amount of data generated
4. High number of sensor node in IoT etc.

Table 69.1 Literature review sheet

S. no	Author Name	Year	Objective	Technique Used	Observation	Dataset	Conclusion
1.	Jash Doshi, Tirthkumar Patel, Santosh kumar Bharti	2019	Smart Farming with IoT is a system for monitoring farming conditions in the most efficient way possible	Plug and Sense	Bring about messages to farmers on various platforms, easy installation, economical	Temperature from DTH11 sensor, Humidity, Soil Moisture, UV, IR from SI1145 sensor, Nutrients in the soil	Proposed an efficient, accurate, and cheap product for the farmers using hardware and materials
2.	Chayapol Kamyod	2018	An IoT-based Smart Agriculture System's End-to-End Reliability Analysis	OPNET for comparison of two IoT-based network architectures	Reliability and performance is increased, automatic control on water usage	Wireless LAN delay at various sensor nodes, Throughput at different sensors, retransmission attempts, average Ethernet delay	Proposed a paradigm for evaluating the end-to-end reliability of two major IoT-based communication networks in the context of smart agriculture.
3.	Yemeserach Mekonnen, Lamar Burton, Arif Sarwat, Shekhar Bhansali	2018	For the agriculture industry, an IoT sensor network is being developed to optimise the use of water, fertilisers, and electricity.	Zigbee protocol manages WSU data transmission through AP	Managed aquisition, smart farming is implemented, increased functionality	Temperature, humidity, air pressure, pluviometer, anemometer, solar radiation, soil temperature, soil moisture, and leaf wetness are some of the variables that can be measured.	Design and implemented a system for smart farm prototype and also energy, water and food modelling
4.	Reuben Varghese, Smarita Sharma	2018	IoT and machine learning for smart farming module using State-of-art methods	Model based on logistic regression using Tensorflow, SVM	Affordable smart farming, decreased human interference, crops future conditions can be predicted, cost-efficient	Soil-specific data from various regions, Temperature sensor data from DTH11, FC-28 sensor data for soil moisture, MQ135 sensor data for air quality, LM- 393 sensor data for plant receiving amount of sunlight	Proposed an efficient solution which is cost effective, decreasing dependency on manual labour for affordable smart agriculture

S. no	Author Name	Year	Objective	Technique Used	Observation	Dataset	Conclusion
5.	Bhanu K.N., Mahadevaswamy H.S., Jasmine H.J	2020	Agriculture IoT-based Smart System for Improved Irrigation	KNN, SVM, Naïve Bayes, Arduino receives data and send to ThingSpeak cloud	Better performance, supplies information about soil conditions to the farmer, quality crops obtained	Moisture sensor data by YL-38 sensor, temperature and humidity data through DTH11 sensor, soil temperature data by DS18B20 sensor	Proposed system avails the knowledge on various parameters of soil to ensure a good system of irrigation for crop agriculture
6.	Rahul Dagar, Subhranil Som, Sunil Kumar Khatri	2018	Smart Farming – IoT in Agriculture	Various sensors installed in Poly houses	Improved quality and quantity, appropriate decision taken, less insecticides needed	Data from a water volume sensor, a pH sensor in the soil, a motion detector sensor, and a soil moisture sensor	Concluded a system in which IoT is successfully implemented in Poly houses
7.	Ibrahim Mat, Mohamed Rawidean Mohd Kassim, Ahmad Nizar Harun, Ismail Mat Yusoff	2018	The Internet of Things in Agriculture	Power module consisting of controller, memory, transceiver, interface circuit	Reduce wastage, enhance productivity, highly efficient system, yielding high-quality varieties	Temperature sensor data, humidity sensor data, CO2 data from an IoT system	Sensors and an automated watering system are used to construct a system for crop monitoring
8.	Vaishali Puranik, Sharmila, Ankit Ranjan, Anamika Kumari	2019	Automation in Agriculture and IoT	Extension and implementation of workflow to help standalone system to thrive without human interference	Maintenance automation, pesticides and insecticides control, strategize crops according to market	pH sensor data, DTH11 temperature sensor data, soil moisture sensor data	Proposed a system which to solve various problems using IoT and automation

S. no	Author Name	Year	Objective	Technique Used	Observation	Dataset	Conclusion
9.	V. Ramachandran, R. Ramalakshmi, and Seshadhri Srinivasan	2018	An Automated Irrigation System for Smart Agriculture Using the Internet of Things	ThingSpeak cloud stores data sensed from different sensors and transferred through Wi-Fi modem using GSM automated using ARM controller through optimization model	Water utilization reduced, data availability and visualization is improved, smart irrigation system, low cost sensors,	YL-69 soil moisture sensor datasets, YF- 201 water flow sensor datasets, DTH11 temperature and humidity sensor dataset	Concluded a system to decrease water consumption by combining IoT, cloud computing as well as optimization.
10.	Kushagra Agrawal, Nikunj Kamboj	2020	Smart Agriculture Using IOT: A Futuristic Approach	Smart grid, M2M platform, clod computing, SOA technology, APSCM, Machine learning	Water conservation, farmer profit, increased production, animal damage control, monitoring irrigation system	Datasets through temperature, humidity, wind, rainfall sensors	Proposed a system focused on crop management, controlling pests, production and safety of food
12.	Lavanya G, Rani C, Ganeshkumar P	2018	An automated low-cost IoT-based Fertilizer Intimation System forsmart agriculture	NPK sensor with LDR & LED, Fuzzy logic, Mamdani Inference System	Nutrients monitoring, high crop yielding, reduce soil degradation, right time usuage of fertilizer, SMS alert system	NPK sensor dataset	Suggests an IoT-based system that analyses the soil nutrients and also alerts the farmers anout the amount of fertilizer
13.	K.Lokesh Krishna, Omayo Silver, Wasswa Fahad Malende, K.Anuradha	2017	Internet of Things Application for Implementation of Smart Agriculture System	WSN, Raspberry Pi 2 model B,	Robot-based system helps in scaring birds, pesticides spraying, devices connected through internet, change tracking in the field	Thermohygro sensor dataset, soil moisture sensor datasets, humidity sensor datasets, UV datasets, obstacle sensor datasets, CO2 sensor and pH datasets	Proposed a system for mobile robot which is low cost and lower power operated for the sake of smart agriculture

S. no	Author Name	Year	Objective	Technique Used	Observation	Dataset	Conclusion
14.	G. S. Nagaraja, Avinash B Soppimath, T. Soumya, Abhinith A	2019	IoT-based smart agriculture Management system	ThingSpeak IoT cloud platform, firebase, SVM classifier,	Helps to increase production, adopting precision agriculture such that it reduces resource wastage, efficient system	DTH11 temperature sensor dataset, hygrometer dataset, moisture sensor dataset	Proposed implementation of SAMS to help the farmer to monitor various environmental factors responsible in crop production
15.	K. A. Patil, N. R. Kale	2016	A Model for Smart Agriculture Using IoT	WINGZ Zigbee technology, Remote monitoring system(RMS), three module system,	SMS-based alerts, easy access to facilities of agriculture, monitoring weather conditions for high yield	Datasets with temperature, relative humidity, light intensity, barometric pressure, and proximity sensors	Concluded a model for agricultural industry in combination with ICT which monitors various environmental factors in real time
16.	Piyush Patil, Vivek Sachapara	2017	Providing Smart Agricultural Solutions/Techniques By Using Iot-Based Toolkit	Artificial neural network, rule based system, Euclidean distance method	Smart automated system, able to detect environment conditions, sustainability increased	SEN-0114 sensor data for soil moisture, FC-37 for leaf wetness, SEN-08942 sensor data for wind direction, temperature sensor dataset, soil pH and rainfall sensors datasets	Proposed a system which can recognise the better seeds depending on the environmental conditons and also the nature of the soil
17.	Gokul L. Patil, Prashant S. Gawande, R. V. Bag	2017	Smart Agriculture System based on IoT and its Social Impact	Xively IoT platform, Agrologger, cloud interface	Smart irrigation based on real-time data, maintaining environment issues	LM-35 temperature sensor datasets, moisture, pressure, humidity sensors datasets	17

REFERENCES

1. Jash Doshi, Tirthkumar Patel, Santosh kumar Bharti. "Smart Farming using IoT, a solution for optimally monitoring farming conditions" In Recent advances on Internet of Things: Technology and Application Approaches(IoT-T&A 2019), Elsevier, November 4-7, 2019, Coimbra, Portugal

2. Chayapol Kamyod. "End-to-End Reliability Analysis of an IoT based Smart Agriculture" The 3rd International Conference on Digital Arts, Media and Technology (ICDAMT2018), 978-1-5386-0572-1/18 ©2018 IEEE

3. Yemeserach Mekonnen, Lamar Burton, Arif Sarwat, Shekhar Bhansali. "IoT Sensor Network Approach for Smart Farming:An Application in Food, Energy and Water System" 978-1-5386-5566- 5/18 ©2018 IEEE

4. Reuben Varghese, Smarita Sharma. "Affordable Smart Farming Using IoT and Machine Learning; An AI powered cost-effective solution to improve traditional farming" Proceedings of the Second International Conference on Intelligent Computing and Control Systems (ICICCS 2018), IEEE Xplore Compliant Part Number: CFP18K74-ART; ISBN:978-1-5386-2842-3

5. Bhanu K.N, Mahadevaswamy H.S, Jasmine H.J. "IoT based Smart System for Enhanced Irrigation in Agriculture" Proceedings of the International Conference on Electronics and Sustainable Communication Systems (ICESC 2020) IEEE Xplore Part Number: CFP20V66-ART; ISBN: 978-1-7281-4108-4

6. Rahul Dagar, Subhranil Som, Sunil Kumar Khatri. "Smart Farming – IoT in Agriculture" Proceedings of the International Conference on Inventive Research in Computing Applications (ICIRCA 2018) IEEE Xplore Compliant Part Number:CFP18N67-ART; ISBN:978- 1-5386-2456-2

7. Ibrahim Mat, Mohamed Rawidean Mohd Kassim, Ahmad Nizar Harun, Ismail Mat Yusoff. "Smart Agriculture Using Internet of Things" 2018 IEEE Conference on Open Systems (ICOS) 978-1-5386-6666-1/18 ©2018 IEEE

8. Vaishali Puranik, Sharmila, Ankit Ranjan, Anamika Kumari. "Automation in Agriculture and IoT" 978-1-7281-1253-4/19 © 2019 IEEE

9. V. Ramachandran, R. Ramalakshmi, and Seshadhri Srinivasan. "An Automated Irrigation System for Smart Agriculture Using the Internet of Things" 2018 15th International Conference on Control, Automation, Robotics and Vision (ICARCV) Singapore, November 18-21, 2018, 978-1-5386-9582-1/18/$31.00 ©2018 IEEE

10. Kushagra Agrawal, Nikunj Kamboj. "Smart Agriculture Using IOT: A Futuristic Approach" in International Journal of Information Dissemination and Technology • December 2019, Researchgate, DOI: 10.5958/2249-5576.2019.00036.0

11. Saurav Verma, Rahul Gala, S. Madhavan, Sanchit Burkule, Swapnil Chauhan, Chetana Prakash. "An Internet of things (IoT) architecture for Smart Agriculture" 2018 Fourth International Conference on Computing Communication Control and Automation (ICCUBEA), 978-1-5386-5257-2/18/$31.00 ©2018 IEEE

12. Lavanya G, Rani C, Ganeshkumar P. "An automated low cost IoT based Fertilizer Intimation System forsmart agriculture" Sustainable Computing: Informatics and Systems, https://doi.org/10.1016/j.suscom.2019.01.002 2210-5379/© 2019 Elsevier Inc

13. K.Lokesh Krishna, Omayo Silver, Wasswa Fahad Malende, K.Anuradha. "Internet of Things Application for Implementation of Smart Agriculture System" International conference on I-SMAC (IoT in Social, Mobile, Analytics and Cloud) (I-SMAC 2017), 978- 1-5090-3243-3/17/$31.00 ©2017 IEEE

14. G. S. Nagaraja, Avinash B Soppimath, T. Soumya, Abhinith A. "IOT BASED SMART AGRICULTURE MANAGEMENT SYSTEM" 978-1-7281-2619-7/19/$31.00 ©2019 IEEE

15. K. A. Patil, N. R. Kale. "A Model for Smart Agriculture Using IoT" 2016 International Conference on Global Trends in Signal Processing, Information Computing and Communication, 978-1- 5090-0467-6/16/$31.00 ©2016 IEEE

16. Piyush Patil, Vivek Sachapara. "Providing Smart Agricultural Solutions/Techniques By Using Iot Based Toolkit" International Conference on Trends in Electronics and Informatics ICEI 2017, 978-1-5090-4257-9/17/$31.00 ©2017 IEEE

17. Gokul L. Patil, Prashant S. Gawande, R. V. Bag. "Smart Agriculture System based on IoT and its Social Impact" International Journal of Computer Applications (0975 – 8887) Volume 176 – No.1, October 2017, Researchgate

18. Prathibha S R, Anupama Hongal , Jyothi M P. "IOT BASED MONITORING SYSTEM IN SMART AGRICULTURE" 2017 International Conference on Recent Advances in Electronics and Communication Technology, 978-1-5090-6701-5/17 $31.00 © 2017 IEEE, DOI 10.1109/ ICRAECT.2017.52

19. Soumil Heble, Ajay Kumar, K.V.V Durga Prasad, Soumya Samirana, P.Rajalakshmi, U. B. Desai. "A Low Power IoT Network for Smart Agriculture" Indian Institute of Technology - Hyderabad, India

20. D. Betteena Sheryl Fernando, M. Sabarishwaran, R. Ramya Priya, S. Santhoshini. "Smart Agriculture Monitoring System Using Iot" International Journal of Scientific Research & Engineering Trends Volume 6, Issue 4, July-Aug-2020, ISSN (Online): 2395-566X, Researchgate

21. Courage Kpotosu and Scholastica Memusi. "Smart Agriculture Using IoT" DOI: 10.13140/ RG.2.2.17522.86726, Researchgate

Intelligent Systems and Smart Infrastructure – Brijesh Mishra et al. (eds)
© 2023 Taylor & Francis Group, London, ISBN 978-1-032-41287-0

CHAPTER

70

Machine Learning Use in Agricultural for Humidity and Rainfall Prediction

Om Mani[1], Bramha P. Pandey[2]

Department of Electronic and Communication Engineering,
Madan Mohan Malaviya University of Technology, Gorakhpur 273010, U.P., India

Abstract

Because of different climatic conditions around the world, rainfall and humidity forecasting has become increasingly significant money method developed to predict climate behavior with an accuracy of 75% with the help of ML, but in this, we get 89% accuracy this model. "It is a powerful technique to forecast the weather with more accuracy." ("Prediction of rainfall and humidity forecasting by Using Machine Learning"). A climate data is gathered and evaluated with algorithms applied to anticipate the outcome. Heavy sleet prognosis could be big flowing because it is so directly linked to human economics and existence, for earth science department. It is the source of natural disasters such as floods and droughts, which afflict people all over the world. For countries like India, where agriculture is the main source of income, the reliability of rainfall and humidity estimates is critical. Applied mathematics techniques fail to correct statements due to the dynamic nature of the atmosphere. Regression may be used in the prediction of rainfall and humidity utilizing ML techniques. The goal of this study is to give non-experts an easy access to the tools used in this study as well as ML methodology.

Keywords: Classification, Rainfall and humidity, humidity forecast with rain, ML in agriculture artificial neural network, reinforcement learning

Corresponding author: [1]ommani444783@hotmail.com, [2]bppece@mmmut.ac.in

DOI: 10.1201/9781003357346-70

1. Introduction

Rainfall and humidity use of science and technology as well as in agriculture to forecast condition at particular location as well as periods. Since the manner for thousands of year weather forecast are increasingly based on computer-based models that take into account various atmospheric parameters instead of laborious calculation based on changes in barometric pressure, current weather, sky condition or cloud cover. Human judgment is still required to the most accurate forecast model, which includes understanding of pattern recognition technologies, model performance, and model bias [1]. The instability of the atmosphere, the enormous processing power needed to address the problems that characterise the environment, the mistake involved in measuring the initial circumstances, and a lack of understanding of the projected course of events all contribute to errors in the predictions. [2] The time disparity between now and when the forecast is made is growing, so does desired range? Overall using template agreements can help in the narrowing of error and the selection of the most likely outcome. Weather forecast has a wide range of applications. Rainfall warnings are vital forecasts that can be utilized to safeguard lives and property, as well as be beneficial to farmers. Forecasts based on humidity and rainfall are crucial for agriculture, and commodities merchants rely on them as well. A prediction algorithm based on data mining was developed as a way to anticipate the greatest and lowest temperatures in order to identify how the weather is changing over time. For rain and snow, the recommend data model employs the random forest model. For extracting weather conditions observation, artificial neural network is used. Rain and humidity forecast is particularly significant because excessive and irregular rain can have a range of implications, including agricultural and farm devastation as well as property damage [1]. For early warning that reduces dangers to life and property, as well as improved management of agricultural field, a better forecasting model is necessary. Farmer's benefit from this prediction and water resource are better managed. Predicting rainfall is a difficult undertaking and the outcomes must be precise.

The following are the steps involved in machine learning:

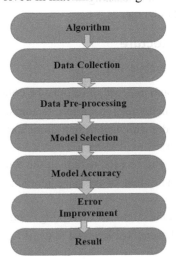

Data collection:-Data collection is a phase in terms of both quality and quantity. It determines the accuracy of our prediction model training data refers to the information gathered.

Preparation of data: the preparation of data is the next phase; this stage converts data into a tabular format, which is then loaded into an appropriate location and ready to be used in machine learning training. The data set is divided into two pieces; the testing data set is the other trainig data set. Both sections are used to enhance the model's performance.

Selecting a model: after completing the desired data process, further performing is to select a model that has been developed over time by both data scientists and researchers. The most important thing is to pick the right model for the job.

Training: A significant quantity of training data is used in this stage to assess the model's performance, and the technique comprises initialising the model with certain random variables, such as A and B, from which we may anticipate the model's output based on the prediction made by the model.

The following phase in this process compares the anticipated value to the forecast made by the model. Then, if necessary, change the values to match those anticipated previously. Some percentages denoted of split train and test data, respectively, are 70-30%, 60–40%, 65–35% and so on.

Evaluation: After training, the next phase is evaluation. This stage entails comparing the model to the data. To verify our model, we compare the data used in the training step with the new test data.

Fine-tune the Parameters: the next phase is parameter adjustment; If it was possible, we used this strategy to enhance our training. Another factor is the learning rate, which shows how far the line is shifted away with each step. The model data set and training procedure all influence how these parameters are configured.

Prediction: This is the last step in machine learning; prediction is utilized to solve queries at that moment, the benefit of machine learning is understood. Now you may fully finish the algorithm to provide the desired outcome.

A. Machine Learning Architecture

We go over the machine learning architecture

- Data of user, their behavior, and material titles must be collected.
- After the data has been obtained, transform it into features.
- After that, we'll train our model with data from training and testing. Training is also a part of the model selection process.
- A live-model serving system will be used to deliver the train model.
- The model is fed back to the output of the entertainment streaming site's to create the recommended page.
- The outcome results are sent back into the marketing channel of movie stream.
- In order to better understand user behaviour, we may utilise models to analyse a variety of movie-watching keywords.

Machine learning is well-suited to acquiring a competitive edge in digital company for the following reason

- Quickness allow for faster processing and decision-making; when the right set requirements are satisfied, ML may be used to provide useful work insights faster and more efficiently than many other analytics approaches because there is no need to programmer every scheme [4].
- The capacity to handle and analyse large amount of data ML can manage a significantly large quantity of data than the prior technique and it can do far more in-depth analytics.[4]
- More efficient than traditional analytic and programming methods in creating more models with higher accuracy. It saves time and money by allowing users to develop models and judgment without the need code.[4]
- Intelligence depends on learning abilities on its own and to provide previously unknown information.[6]

B. Form of Machine Learning

Supervised Learning: Supervised learning is a sort of machine learning in which the computer is given input and is expected to produce a certain result. Input and output are labeled for categorization purposes. The destination variable is forecasted using a collection of predictions. Other supervised learning algorithms include KHH, Decision Tree, Regression, Logistic Regression, Boosting Result Prediction and fraud detection, Classification and Regression deal filtering and classification, and Decision Tree deal with Threat management system and Risk management

Unsupervised learning: this method of learning is to use to make conclusion from data that does not have a labelled solution. We don't have s supervisor to help us out. The self-organizing map, the apriority algorithm, K-means, adaptive resonance theory, and the apriority algorithm are all examples of algorithms. Unsupervised learning examples Examples of models include cluster analysis, which deals with financial transactions, streaming analytic pattern recognition, which handles spam and fraud detection, biometrics, and identity management, and association rule, which deals with the production and assembly of bioinformatics for the internet of things.

Reinforcement learning: This is a type of learning in which you are rewarded for. Reinforcement learning is similar to supervised learning in that it employs the usage of a reward and punishment system to teach algorithms. Reinforcement learning is a method of teaching a machine by interacting with its surroundings. When a person does something right, the supervisor will praise them, and when they do something bad, they will be punished. Machines are trained here, so that it can make its own choices. Reinforcement learning allows them to learn from their previous experiences. It will capture the greatest possible knowledge in order to make an exact decision.

C. Artificial Neural Networks

The input layer, hidden layers, and resultant layer comprise Artificial Neural Network. There could be multiple hidden layers. The number of neurons will be used in each layer. The

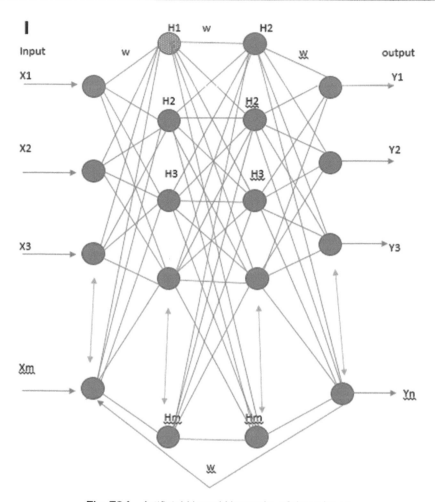

Fig. 70.1 Artificial Neural Networks of three layer

activation function associated with each neuron will be used in each layer. The activation function is responsible for adding non-linearity to the relationship. The output layer in our situation must have a constant activation function. There are two steps to Artificial Neural Networks.

- Propagation in future
- Propagation backward

The process of increasing weight with each characteristic and adding them is known as Forward Propagation. The outcome includes learning as well. The process of recalculating the model's weight is known as backward propagation.

D. Machine Learning Issues

Machine learning proponents believe that technology is a revolutionary way to combat security concerns, even halting attackers in their tracks before they can breach a network [3]. Some in the security industry regard the more sceptical, or Balanced, members as a useful tool, but not as a fix for the issues facing the sector. Problem reverse engineering machine learning models are also possible. Research showed in 2016 that by logging the outcomes of a few thousand queries, they could replicate an Amazon ML model with almost flawless accuracy. This has significant implications for how attackers use publicly available methods to exploit training data and threat information to generate new attacks that seem normal to ML-driven analysis.

Dataset

The dataset was obtained from NASA Official: Paul Stackhouse and 2017 center for Hydrometeorology and Remote Sensing. The data set is included of the following variables:

- **Temperature (°C)***:* The temperature variable has significant impact on precipitation and is related to humidity (%). The dataset has both maximum and minimum temperatures along with the mean values.
- **Cloud:** In addition to reflecting and absorbing heat from the earth's surface, the cloud variable measured in effect flattens incoming sunlight. The dataset forms of daily cloud data measured in a range from 0 to 8
- *Wind Speed***:** The wind variable measured in knots shows how quick the air is moving. The wind speed also has a direction and has various impacts on surface water and evaporation. The dataset forms of daily prevailing wind speed.
- **Rainfall (mm):** The rainfall variable is an especially important metric in weather forecasting. It helps the environment to continue to stay in its position the way it should be. Agriculture of Bangladesh mostly depends on rainfall. The dataset has daily basis of total rainfall data in millimeters.
- **Sunshine:** The sunshine variable measures the amount of sunshine at a particular place. The Sun is the basic element of our changing weather. The day–night cycles in the weather have obvious causes and effects on weather.
- **Humidity (%):** The humidity variable measured in percentage helps to calculate the amount of moisture in a given time in a particular day, which is simply the ratio of water vapor and dry air.
- **Sea Level Pressure (millibars):** The sea level pressure variable plays an important role in the formation of weather conditions in a certain area. It is a part that has mass and weight. This means a vast ocean of air inserts a huge amount of pressure. So, it is natural that the air will change the Earth's weather.

Table 70.1 Parameter use

PARAMETER	RAIN	HUMIDITY
DATE	∗	∗
TIME	∗	∗
LOCATION	∗	
SPACIFIC HUMIDITY	∗	∗
RELATIV HUMIDITY	∗	
TEMPERATURE	∗	∗
DEW/FROST		∗
SURFACE PRESSURE	∗	∗
WIND SPEED	∗	∗
WIND DIRECTION	∗	∗
WIND SPEED AT 10 (M/S)	∗	∗
RAINFALL		∗
SUNSHINE	∗	
WIND GUST SPEED	∗	
WIND GUST DIRECTION	∗	
CLOUD	∗	

NOTE: ∗ Symbol shows the used parameter.

2. Result

Rain prediction

Table 70.2 Rain Prediction

	Rain on tomorrow	Prediction rain
1	NO	NO
2	NO	NO
3	NO	NO
4	NO	NO
	NO	NO
6	NO	NO
....	....	......
1576	YES	NO
1579	YES	YES
580	NO	NO
1581	NO	NO

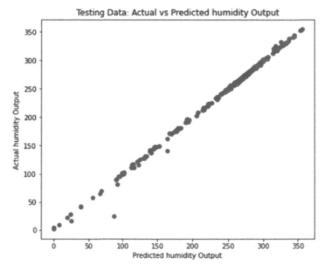

Fig. 70.2 ScatterChart of testing dataActual vs predictied Humidity Output

Humidity prediction

Table 70.3 Humidity Prediction

	Actual humidity	Predicted humidity
1	279.99	279.57
2	261.61	264.71
3	254.05	253.25
4	281.99	283.99
5	142.71	139.24
6	273.70	274.59
7	273.67	273.21

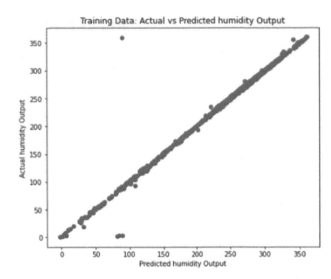

Fig. 70.3 ScatterChart of Training data Actual vs predictied Humidity Output

3. Accuracy

For rain: After applying this model on the collected data, it is found that the accuracy between the predicated result and actual result is 89.38% for the rain

For humidity: After applying this model on the collected data, it is found that the accuracy between the predicated result and actual result is 98.71% for the humidity

4. Conclusion

This is devoted to ran and humidity estimations, and it is hoped that it will be a useful as well as acceptable technique, helping the client manage the challenges pertaining to distributional characteristics of key components, information geometry, and the all-too-common model problem over fitting. For display, the choice bits ability is analytic to direct and non-straight relationship; in this model, we get the accuracy up to 89%; In order to measure model execution, we advise beginners to use straight and piece separately because I believe this is preferable to using an expectation method, which cannot detect non-linearity in data collection but still proves useful in such circumstances. I'm processing mean absolute error for both models, so get this and look at the mode's. This model is overall beneficial for farmer directly because with this help, they get the proper information of rain and humidity in their region and then they plan for the crops according to the need of this crop like water and humidity scenically required for this.

REFERENCES

1. Prediction Of Rainfall Using Machine Learninge Techniques Moulana Mohammed, Roshitha Kolapalli, Niharika Golla, Siva Sai Maturi International journal of scientific & technology research volume 9, issue 01, january 2020 issn 2277-8616 3236 (2020) Pages 5–9

2. Prediction of Weather Forecasting by Using Machine LearningN. Sri Lakshmi, P. Ajimunnis2, V. Lakshmi Prasanna, T. YugaSravani, and M. RaviTeja International Journal of Innovative Research in Computer Science & Technology (IJIRCST) ISSN: 2347–5552, Volume-9, Issue-4, July Article ID IRP1186,(2021) Pages 30–32

3. Dirmeyer, Paul A.; Schlosser, C. Adam; Brubaker, Kaye L. "Precipitation, Recycling, and Land Memory: An Integrated Analysis 2008JHM1016.1 . Dec.(2016) Pages 3–6

4. Dash, Yajnaseni, SarojK. Mishra, and Bijaya K. Panigrahi. "Rainfall prediction for the Kerala state of India using artificial intelligence approaches." Computers & Electrical Engineering (2018) Pages 66–73.

5. Forecasting using Machine Learning Amit Kumar Agarwal, Manish Shrimali, Sukanya Saxena, Ankur Sirohi, Anmol Jain International Journal of Recent Technology and Engineering (IJRTE) ISSN: 2277-3878, Volume-7 Issue-6C, April (2019) Page 38–49

6. S. B. Kotsiantis, "Supervised machine learning: A review of classification techniques," in Proceeding of the Conference on Emerging Artificial Intelligence Applictions in Computer Engineering: Real World AI Systems with Applications in Computer Egineering: Real World AI systems with Applications in eHealth,HCI Amsterdam, The Netherlands, The Netherlands: IOS Press,(2007) Pages 3–24.

7. Sardeshpande, Kaushik D., and Vijaya R. Thool. "Rainfall Prediction: A Comparative Study of Neural Network Architectures." Emerging Technologies in Data Mining and Information Security. Springer, Singapore, (2019)) Pages 19–28.

8. Ms. Ashwini Mandale, "Weather Forecast Prediction: A Data Mining Application", International Journal of Engineering Research and General Science Volume 3, Issue 2, MarchApril, 2015, ISSN 2091-2730. (2015) Pages 3–5

9. Moon, Seung-Hyun, et al. "Application of machine learning to an early warning system for very short-term heavy rainfall. "Journal of hydrology 568 (2019) Pages 1042–1054.

10. H. S. Hippert , C. E. Pedreira and R. C. Souza, "Combining Neural Network and ARIMA models for Hourly Temperature Forcast", Proceedings of the IEEE-INNS-ENNS International Joint Conference on Neural Networks, vol 12, no. 1,(2014) Pages. 57–28.

Intelligent Systems and Smart Infrastructure – Brijesh Mishra et al. (eds)
© 2023 Taylor & Francis Group, London, ISBN 978-1-032-41287-0

CHAPTER

71

Convolutional Neural Network-based Approach for Landmark Recognition

Lokendra Singh Umrao[1], Ravi Ranjan Choudhary[2]

Department of Computer Science & Engineering, Institute of Engineering and Technology,
Dr. Rammanohar Lohia Avadh University, Ayodhya, India

Ratnesh Prasad Srivastava[3]

Department of Information Technology, College of Technology,
GBPUAT Pantnagar, India

Rohit[4]

Department of Computer Science & Engineering IERT,
Prayagraj, India

Abstract

Convolutional Neural Networks (CNNs) are considered to be an essential architecture for Neural Network in order to run Deep Learning-based applications. These networks thus achieve a good accuracy in accomplishing task of processing and predicting images, sounds and videos. In Bell Lab, Prof. Yann Lecumn had developed first CNN by late 1990s. Within reasonable training time, Multilayer Perceptron (MLP) produced greater classification results over MNIST datasets. The performance goes on decreasing with increase in training time for large datasets because of parameter size in a model increases exponentially.

A picture as a Landmark can be well recognized with API of Landmark recognition supported in the package of Machine Learning. The images as input parameter is passed to the API, which detects the area it belongs to with the help of geospatial & geographical coordinates.

Corresponding author: [1]lokendra.manit@gmail.com, [2]ravir686@gmail.com, [3]ratnesh.mnnit@gmail.com, [4]rohitatiiit@gmail.com

DOI: 10.1201/9781003357346-71

This information further helps in generating image metadata by which user shares and creates his experiences. The main purpose of this article is to present an approach for the use of Convolutional Neural Network for historical landmark detection.

Keywords: Bag of Words, Landmarks, Feature Extraction, Wavelet, Gaussian Transform

1. Introduction

To reach the landmark place, navigation is provided from the staring to end point as discussed in general public landmark (Chen et al., 2009). The feature extraction of query image through bag of words and visual words is accessed to find the closest match. The image as a query is then classified to compare the query vector representation for training datasets of images. The higher number of nearest neighbor for a query image is then classified. In order to meet the desired classification result, the image background is cluttered for separating foreground and background images (Chen et al., 2011).

This paper is organized as follows: section II introduces the process of feature extraction, section III introduces the techniques associated with feature selection, section IV introduces retrieval of landmark through images, and section V finally discusses the conclusion.

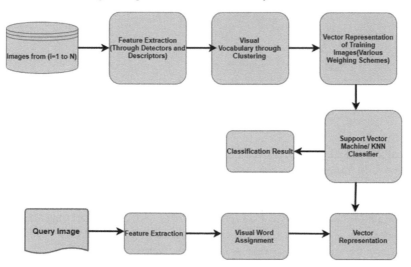

Fig. 71.1 Bag of Words Model for Land Mark Recognition

2. Feature Extraction

Initially the start point for recognizing an image is to be extracted for useful features. Features used for visual attention for historical identification are often considered as global attributes.

They are used for detecting and identifying landmarks (Lazebnik et al., 2006). The authors use Soble filter for finding image derivatives in relevance to x & y for mapping intensity. The S & V channels along with Sting Histogram are used for filter tuning and scaling, which is a combination of Sobel edge detection and two parameters a & b ranging from <1, and a + b = 1.0. The monochrome image is tested with (6 orientation, 4 scales) used in steerable pyramid. The Gabor filter also produces results mostly similar in nature (Yan Ke and Rahul Sukthankar, 2004). By this way, pixel represents global features for producing satisfied results for landmarks recognition. Key points are detected in scale-space (Nowak et Al., 2006). The (DoG) Difference of Gaussian is the reliable matching. Due to be employed, the correct discriminates key points often have a significantly closer nearest neighbor than the nearest mismatch (Bosch et al., 2008). In terms of historical identity, this does not suit well because buildings like structures tend to the edge-related ROC curve. Two stages has been adopted namely feature extraction and image enhancement for landmark recognition. In the first stage, the number of key points searched or queried for an image or picture grows larger in volumes. Therefore in order to retain key points with high information SIFT based posterior entropy is used. The best matched key points with high information is extracted through SIFT based matching mechanism (Martin et al. 1981). A polling based mechanism is adopted in order to reduce computation for deep analysis. In the second stage for color and edge feature detection, a color-edge histogram patches(CEHPs) are employed. In local color, the histogram is used in the first stage and SIFT in the detection of the features. The query represents each image stored in dataset. The five stages verification that are verified geometrically produces the best N images.

3. Landmark Retrieval based on Images

The retrieval process of images is being sped up with inverted index structure of Bag of Words model. This would also include the load classification and the second is of the most popular discriminative classification approaches used to find k nearest pixel. Since natural images can be classified into different classes, we therefore decided to use binary SVM classifier, which distinguishes different classes of landmarks from other landmarks.

4. Conclusion

For a given picture as input, in order to increase recognition rates for a landmark, algorithm uses image, location and visual classification such as historical image or ordinary image as an input. It has been tired to find significant methods in order to choose best possible classifier for landmarks. Due to contradictory nature between specificity and locality, it has been avoided to use local features which may no longer be helpful to boost per-image recognition rate.

REFERENCES

1. A. Bosch, A. Zisserman and X. Munoz (2008). "Scene Classification Using a Hybrid Generative/Discriminative Approach," in IEEE Transactions on Pattern Analysis and Machine Intelligence, vol. 30, no. 4, pp. 712–727.

2. D. M. Chen et al. (2011). "City-scale landmark identification on mobile devices," CVPR 2011, pp. 737–744, doi: 10.1109/CVPR.2011.5995610.

3. Martin A, Fischler and Robert C. Bolles (1981). Random sample consensus: a paradigm for model fitting with applications to image analysis and automated cartography. Commun. ACM 24, 6, 381–395.

4. Nowak, E., Jurie, F., Triggs, B. (2006). Sampling Strategies for Bag-of-Features Image Classification. In: Leonardis, A., Bischof, H., Pinz, A. (eds) Computer Vision. Lecture Notes in Computer Science, vol 3954. Springer, Berlin, Heidelberg.

5. O. Chum and J. Matas (2010). "Large-Scale Discovery of Spatially Related Images," in IEEE Transactions on Pattern Analysis and Machine Intelligence, vol. 32, no. 2, pp. 371–377, doi: 10.1109/TPAMI.2009.166.

6. S. Lazebnik, C. Schmid and J. Ponce (2006). "Beyond Bags of Features: Spatial Pyramid Matching for Recognizing Natural Scene Categories," IEEE Computer Society Conference on Computer Vision and Pattern Recognition, pp. 2169–2178.

7. T. Chen, K. Wu, K. Yap, Z. Li and F. S. Tsai (2009). "A Survey on Mobile Landmark Recognition for Information Retrieval," Tenth International Conference on Mobile Data Management: Systems, Services and Middleware, pp. 625–630.

8. T. Chen, Z. Li, K. Yap, K. Wu and L. Chau (2009). "A multi-scale learning approach for landmark recognition using mobile devices," 7th International Conference on Information, Communications and Signal Processing, pp. 1–4.

9. Yan Ke and Rahul Sukthankar (2004). PCA-SIFT: a more distinctive representation for local image descriptors. In Proceedings of the IEEE computer society conference on Computer vision and pattern recognition (CVPR'04). IEEE Computer Society, USA, 506–513.

Intelligent Systems and Smart Infrastructure – Brijesh Mishra et al. (eds)
© 2023 Taylor & Francis Group, London, ISBN 978-1-032-41287-0

CHAPTER

72

A Brief Review on
Outlier Detection Techniques in IoT

Preet Kamal Singh[1], Ramesh Mishra[2], Chandan[3]

Electronics & Communication Department,
IET DRMLAU, Ayodhya, India

Abstract

The Internet of Things (IoT) is already a reality, with vast number of hubs doing a wide range of tasks. The purpose of every network, from modest networks to vast-scale networks, is to transmit information from hubs to the base hub. This information is vulnerable to a various circumstances that could jeopardise the usefulness of the gathered data or the operation of network's, and thus the desired service quality (QoS). One of the primary difficulties that needs additional investigation and particular solutions in this area is outlier detection. Outliers must be identified and classified as either errors that may be ignored or serious events that demand immediate action to prevent future service degradation. We present a complete overview of contemporary outlier detection approaches utilised in the IoTs setting in this study.

Keywords: event detection, outlier detection, IoT, detection efficiency

1. Introduction

The Internet of Things (IoTs) is a set of technologies that work together to provide Internet-based services and applications. Heterogeneous sensors can gather information for process control with the use of electronic gadgets attached to real-world objects [1]. Several asset-obligated hubs are sent to detect, gather, and transport information to a base station or a server

Corresponding author: [1]aayush_btech@yahoo.co.in, [2]rameshmishra1985@gmail.com, [3]chandanhcst@gmail.com

DOI: 10.1201/9781003357346-72

farm as part of the IoT. In this approach, the right decision can be taken in a safe setting. Climate monitoring, clinical, farming, disaster warning, smart city, and assembly are all examples of where IoT is employed. Despite this, the information acquired is powerless in the face of anomalies; those are information tests that aren't your standard information tests. In the IoTs, an exception may occur owing to inbuilt characteristics of the sensor elements or due to the environmental condition in which the hubs are installed. Sensor dissatisfaction, commotion, breakdown, missing or copied information values, and other factors can all affect information quality [30, 31]. These exceptions can also apply to network data exchanged in relation to business activities (i.e., sending and getting messages). In this approach, recognising these abnormalities before any in-network treatment, such as combination of gathered information, is critical in order to maintain the limit of the flow of wrong information and moderate the gathered information's adequacy to make a sensible decision. To guarantee network QoS, the IoTs should recognise the exceptions and make the vital move to dispose of the debasement of help. Also, IoTs accompany the part of countless heterogeneous implanted gadgets that produce huge information, which makes the anomaly identification more perplexing against such enormous information [2]. Besides, with the critical headways in man-made reasoning (AI) strategies, abnormality following has additionally tended to a recent fad in information assortment and keeping a serious level of privacy for the communicated information. The force of the AI will have extra assets to safeguard weak organisations and information from defective parts. Thus, AI can then break down network exercises actually, find an example, and distinguish a wide range of deviations or abnormalities in its gathered information. With such a methodology, it is a lot simpler to rapidly distinguish network abnormalities.

As shown in Fig. 72.1, in IoTs and networks, outlier detection is frequently employed in a range of real-world applications. As example, the Internet of Things for Medical (IoMT) [3] is widely deployed in medical profession. It continuously monitors the well-being of patients and provides remote support as well as timely warnings to any changes in the circumstances stated [4]. Furthermore, in today's environment, sensors are embedded into machines to monitor assets across the board, and there are various characteristics that can be utilised to recognise occurrences and send pertinent warnings, such as temperature, moisture, and strain [5]. The Manufacturing of Internet of Things (MIoT) [6] is another approach for analysis of modern huge information. In the horticulture space, sensors are also transported into harsh

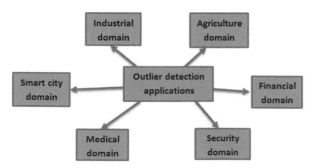

Fig. 72.1 Outlier detection in IoTs has a wide range of applications

environments to monitor and collect a variety of parameters such as temperature and stickiness over time [7]. Exceptions or odd properties in the security space could be an attack that jeopardises the organisation's security [8]. Exception location methods, like as water quality monitoring, can be used in dazzling city applications in general. It provides continuous data and alarms for water contamination control [9]. In the financial sector, evaluating the review logs for the monetary trades included in a data set, and then reporting and confirming any anomalous behaviour seen in the data [10], is crucial.

2. Related Work

Numerous techniques and calculations have been created to recognise anomalies in datasets. The great majority of these computations and techniques were developed for specific purposes and localities. Many audit and overview articles, as well as a few books, have focused on anomaly location. Patcha et al. also disseminated a research of exception location strategies for interruption recognition, as did Snyder [9,10]. Markou and Singh [11,12] reviewed anomaly detection approaches based on factual methodologies and brain organisations. Hodge and Austin [13] provided a comprehensive summary of anomaly discovery estimates created in quantifiable regions and by AI. A few unaided exception discovery calculations were evaluated by Goldstein et al. [14]. Wang et al. [15] presented the various ways for locating anomalies and discussed the advancement of exception finding calculations up to the year 2019. Tellis et al. [16] published a review that focused to show the different strategies for finding anomalies in data and information streams. The purpose of their work was to distinguish between the many methodologies that can be used in the finding of a peculiarity in applications that include information flows, as example of criminal behaviour, Mastercard misrepresentation recognition, PC interruption tracking, and calling cards. For information streams, Park et al. [17] introduced a detailed audit of concept float placement, exception recognition, and strangeness design identification [18-23] contains further audit, review and various algorithms regarding anamoly detection in the light of enormity and information deluge.

Factual-based, closest neighbour-based, grouping-based, characterisation-based, and techniques based on phantom decay are the several types of exception recognition methods for IoTs. [3] Classification-based techniques are significant meaningful methodologies in the information mining and AI communities. Grouping-based techniques acquire familiar with an order model containing a number of information cases during the preparation step, and they organise an inconspicuous sample into one of the learned (typical/exception) classes during the testing stage. SVM-based approaches are derived from group of order-based algorithms and offer three key benefits: (a) have a straightforward mathematical translation; (b) by increasing the choice limit's edge, providing an optimum answer for order; and (c) avoid dealing with the problem of dimensionality.

The way that in numerous IOTs applications, pre-grouped typical/atypical information is neither generally accessible nor simple to acquire, which suggests that solo arrangement methods suit the IOTs the best. Further, numerous interruption and anomaly location frameworks that are executed in the space of IOTs center around distinguishing network interruption as opposed to recognising gatecrashers in the actual climate. Existing discovery frameworks either utilise a factual-based recognition strategy or a multitude knowledge-based procedure. A) Architectural Structure: Existing irregularity location strategies fundamentally utilise either unified or dispersed or nearby methodology for identifying the oddities. (i) Centralised Approach: In concentrated identification, the base station is where the oddity is discovered. IoTs collect data from sensor hubs and transfer it to a base station where it is processed and broken down. This data can be used by anomaly detection tools to identify any missing information or information irregularities acquired. Banerjee et al. [7] added a gatecrasher trail monitoring component based on a subterranean insect state interruption location component. To get sensor organisations, this process can be used in conjunction with existing AI-based interruption location methods. This project investigates how anomalies are disrupted after they are discovered. Hodge and Austin [13] explained an approach that relies on scattered non-parametric inconsistency discovery and needs sensor devices to maintain a tree correspondence network geography. Every sensor device uses a fixed-width grouping calculation to group its tested estimations, then separates measurements of the groups (i.e., the centroid and number of information vectors contained), and then delivers them to respective parent or main hub. (ii) Dispersed Method: The discovery specialist is introduced at each hub in the dispersed approach. It observes the behaviour of neighbouring hubs that are in its broadcast area in order to detect any unexpected behaviour. To play out an ongoing oddity identification, some standard-based discovery methods are utilised in a hub. Hub listens indiscriminately to adjoining hubs inside its transmission reach to gather information important for peculiarity recognition. The gathered information will be broken down to identify any deviation from ordinary conduct utilising adjoining chronicled information put away in the memory. When the abnormalities have been distinguished an alert message or alarm is shipped off at base hubs and adjoining hubs.

In [4], Intrusion Detection Systems (IDS) for a sensor network based on the organisation's actions were described (e.g., number of progress and disappointment of verifications). To trace down harmful assaults from an intruder, the framework compares event data and mark records.

Ref. [5] used the discovery framework in a cluster-based sensor network, which is similar to the framework developed in this exposition. This type of location framework is capable of recognising abnormalities that it has previously encountered. In any event, the goal of this investigation is to find anomalies in rare situations where there are no unusual models for the framework to learn.

Onat and Miri [6] created interruption location conspiracies to build a model of regular traffic behaviour, and then used this model to identify unexpected traffic designs. Their methods can distinguish between assaults that have occurred lately.

In light of Bayesian Belief Networks (BBNs), [8] proposed an anomaly recognition calculation (BBN). The framework can also evaluate sensor data with missing properties. The BBNs can identify spatial transient correlations between sensor hubs as well as relationships between sensor hub features.

Patcha and Park [9] devised a method for detecting k-closest neighbour anomalies, such as directions whose distance from their K-NN exceeds an acceptable threshold limit or the top n focusing on their K-NNs. Every sensor keeps a histogram-like outline of relevant data over a sliding window of its most important informational features. The sink hub gathers these outlines and asks the organisation for any extra information required to properly assess the exceptions across the board. Synopses allow for less correspondence than would be possible with a gullible, focused approach. Their methodology differs from ours in a number of respects. For starters, they only distinguish exceptions not of one-layered information, and the difficulties of developing smaller, more comprehensive histograms will prevent any further extension. Second, they only consider the two K-NN-based exception definitions described above, whereas our methodology considers all of them, and the possibilities are endless. Third, their methodology only applies in instances where geographic proximity is not important, whereas our technology has the ability to compel spatial proximity if necessary ("semi-nearby" anomaly identification).

Snyder [10] specifies that the sensors maintain a tree correspondence geography and identify anomalies using a measure of the hidden likelihood appropriation from which the data is derived. Every sensor creates such a gauge by using an arbitrary example of its information perceptions.

Markou and Singh [11] constructed a system based on a BBNs created through the IoTs (and dispersed to every sensor). This can be used by any sensor to evaluate the likelihood of a tuple being observed and, as a result, to distinguish irregularities.

Markou and Singh [12] correct huge segregated spikes in single sensor data streams using a wavelet-based method. A powerful time travelling (DTW) distance-based procedure is additionally employed to recognise all the more consistent timespans sensor information by contrasting the information surges of geographically adjacent sensors anticipated to supply comparable streams.

Goldstein and Uchida [14] concentrated on the peculiarities in IOT, helpful properties of inconsistency recognition strategies, and break down the different oddity discovery procedures for remote sensor organisations.

Wang et al. [15] illustrated many types of IOTs as well as proposed solutions for dealing with the concerns listed and a variety of other issues. This article will include information on the IoTs and its various varieties, as well as a writing audit, so that an individual can learn more about this rapidly growing topic.

In [16], the remote sensor network was suggested as a potential expansion in the field of radio organisations, which also enables new applications with another model for detecting and broadcasting data from different scenarios, with the capacity to serve a wide range of applications at a cheap cost.

Ref. [17] is in charge of addressing directing issues and analysing steering-related issues. As a result, the writing is evaluated based on the reproduction and trial settings, as well as the Quality of Service (QoS) considerations and the sending to various programmes. Pimentel et al. [18] published a paper on IoT organisations, in which the number of organisation hubs is limited in terms of energy supply, confined computational limit, and correspondence data transmission. It examined the actual connection between the power utilisation and the connection use of remote sensor organisations.

In [24], creators proposed a dispersed shortcoming recognition (DFD) calculation for WSN, where every sensor hub helps out its neighbours to send and get information to distinguish and recognise the defective hubs. This strategy is known as the adjoining coordination method. From that point onward, a factual z-test will break down gathered information at every hub to settle on its delicate anomalies and anticipate those of its neighbours. This test utilises an ordinary conveyance, and its exactness is high in the event that more information is gathered. As a result, they have little upward correlation, with high DR and low FAR.

The creators presented a scattered technique for identifying exceptions in real-time series WSN in [25]. It takes into account the spatial-transient link in order to recognise the ordinariness of information values as well as mistakes and circumstances. In view of the Autoregressive and Moving Average (ARMA) expectation model, each hub differentiates worldwide exceptions. Then it consults with its neighbours to see if the exceptions it has discovered are also spatial abnormalities. This theory is known as TSOD (transient and spatial genuine information-based exception recognition), and it has an unavoidable upward correspondence.

In [26], the authors presented a web-based, dispersed strategy for examining exceptions in various levels of WSNs in light of a histogram, with no need for a check methodology to detect anomalies. They demonstrate through a hypothetical report that the blunder of another gauge is insignificant. Their methodology is proficient and has low intricacy.

In [27], creators proposed a guess versatile part thickness assessor (AKDE) approach. They ascertain the PDF in light of the piece thickness assessment (KDE) strategy for online anomaly location in information streams. They show that their calculation is better than KDE's.

In [28], creators projected a web-based versatile calculation in view of the ARMA model, which can recognise and supplant exceptions progressively. Also, this calculation can accomplish the interest of the ongoing radar's medical services application. Their calculation can investigate the relationship of neighbourhood data to display them utilising ARMA. Their methodology includes speed visualisation and forecasting based on SVM and brain groupings.

Based on four factual models, the creators proposed an IoT engineering for detecting mistakes and occurrences in [29]. The spatial–transient relationship is the foundation of their models. They employ the Classification and Regression Trees (CART) methodology to isolate the data and develop a forecast method for every parcel. They have a decision tree as a result of this characterisation. When accurately categorising, they also take into account the forecasting mistake. The Random Forest (RF) model is then used to collect a huge number of trees. A relapse and characterisation model is also utilised with the Gradient Boosting Machine (GBM). Finally, the Linear discriminate analysis (LDA) model is utilised as a direct classifier, distinguishing classes based on elements or borders. Overall, the quantifiable approaches work well if the circulation model is adequately specified and the data used to create the model is not necessary. Previous information on information circulation is not generally available or difficult to obtain, as previously indicated. Temporary connections can be utilised to detect anomalies in data collection utilising measurable ways. By adjusting information dispersion, this relationship will be broken and irregularities in streaming data will be easier to notice. Sensor data in IoT applications does not lend itself to parametric approaches.

3. Conclusion

In the IoTs, detecting outliers is a serious concern. Traditional data and information analysis methods are ineffective in light of the increasing use of IoT in many applications. As a result, the IoTs necessitate the development of novel, outlier detection systems that are energy-efficient and capable of overcoming certain limitations and constraints. In recent years, many outlier identification strategies for IoT have been presented. This book delves into the principles of outlier detection, as well as the various outlier sources, existing methodologies, how to evaluate an intrusion detection strategy, and the difficulties of developing such systems.

REFERENCES

1. Chandola, V.; Banerjee, A.; Kumar, V. Anomaly detection: A survey. ACM Comput. Surv. 41, 15, 2009.
2. Raja Jurdak, X. Rosalind Wang, Oliver Obst, and Philip Valencia,"Wireless Sensor Network Anomalies: Diagnosis and Detection Strategies", 2011.
3. Y. Zhang, N. Meratnia, and P. J. M. Havinga, "Outlier Detection Techniques for Wireless Sensor Network" A Survey,Technical Report, University of Twente, 2008.
4. Techateerawat, P. and Jennings, "A Energy efficiency of intrusion detection systems in wireless sensor networks", In Proceedings of the 2006 IEEE/WIC/ACM international conference on Web Intelligence and Intelligent Agent Technology (WI- IATW), pages 227–230, Washington, DC, USA. IEEE Computer Society, 2006.
5. Hai, T. H., Khan, F. and nam Huh, E. "Hybrid intrusion detection system for wireless sensor networks", Computational Science and Its Applications ICCSA, 4706:383–396, 2007.
6. Onat, I. and Miri, "An intrusion detection system for wireless sensor networks", In IEEE International Conference on Wireless And Mobile Computing, Networking and Communications, Los Alamitos, IEEE Computer Society Press, 2005.
7. Banerjee, Banerjee S., Grosan C., and Abraham, "Ideas: intrusion detection based On emotional ants for sensors", In The 5th International Conference on Intelligent Systems Design and Applications (ISDA), pages 344–349, Wroclaw, Poland, 2005.
8. Janakiram, D., Reddy, V., and Kumar A., "Outlier detection in wireless sensor Networks using Bayesian belief networks". In First International, Conference on Communication System Software and Middleware, pages 1–6, 2006.
9. Patcha, A.; Park, J.-M. An overview of anomaly detection techniques: Existing solutions and latest technological trends. Comput. Netw. 2007, 51, 3448–3470.
10. Snyder, D. Online Intrusion Detection Using Sequences of System Calls. Master's Thesis, Department of Computer Science, Florida State University, Tallahassee, FL, USA, 2001.
11. Markou, M.; Singh, S. Novelty detection: A review—Part 1: Statistical approaches. Signal Process. 2003, 83, 2481–2497.

12. Markou, M.; Singh, S. Novelty detection: A review—Part 2: Neural network based approaches. Signal Process. 2003, 83, 2499–2521.

13. Hodge, V.; Austin, J. A Survey of Outlier Detection Methodologies. Artif. Intell. Rev. 2004, 22, 85–126.

14. Goldstein, M.; Uchida, S. A Comparative Evaluation of Unsupervised Anomaly Detection Algorithms for Multivariate Data. PLoS ONE 2016, 11, e0152173.

15. Wang, H.; Bah,M.J.; Hammad,M. Progress in Outlier Detection Techniques: A. Survey. IEEE Access 2019, 7, 107964–108000.

16. Tellis, V.M.; D'souza, D.J. Detecting Anomalies in Data Stream Using Efficient Techniques: A Review. In Proceedings of the 2018 International Conference on Control, Power, Communication and Computing Technologies (ICCPCCT), Kannur, India, 23–24 March 2018.

17. Park, C.H. Outlier and anomaly pattern detection on data streams. J. Supercomput. 2019, 75, 6118–6128.

18. Pimentel, M.A.; Clifton, D.A.; Clifton, L.; Tarassenko, L. A review of novelty detection. Signal Process. 2014, 99, 215–249.

19. Chauhan, P.; Shukla, M. A review on outlier detection techniques on data stream by using different approaches of K-Means algorithm. In Proceedings of the 2015 International Conference on Advances in Computer Engineering and Applications, Ghaziabad, India, 19–20 March 2015.

20. Salehi, M.; Rashidi, L. A Survey on Anomaly detection in Evolving Data. ACM Sigkdd Explor. Newsl. 2018, 20, 13–23.

21. Chandola, V.; Banerjee, A.; Kumar, V. Anomaly detection: A survey. ACM Comput. Surv. 2009, 41, 1–58.

22. Domingues, R.; Filippone, M.; Michiardi, P.; Zouaoui, J. A comparative evaluation of outlier detection algorithms: Experiments and analyses. Pattern Recognit. 2018, 74, 406–421.

23. Safaei, M.; Asadi, S.; Driss, M.; Boulila, W.; Alsaeedi, A.; Chizari, H.; Abdullah, R.; Safaei, M. A Systematic Literature Review on Outlier Detection in Wireless Sensor Networks. Symmetry 2020, 12, 328.

24. Panda, M.; Khilar, P.M. Distributed soft fault detection algorithm in wireless sensor networks using statistical test. In Proceedings of the 2012 2nd IEEE International Conference on Parallel, Distributed and Grid Computing (PDGC 2012), lSolan, India, 6–8 December 2012; pp. 195–198.

25. Zhang, Y.; Hamm, N.A.; Meratnia, N.; Stein, A.; van de Voort, M.; Havinga, P.J. Statistics-based outlier detection for wireless sensor networks. Int. J. Geogr. Inf. Sci. 2012, 26, 1373–1392.

26. Xie, M.; Hu, J.; Tian, B. Histogram-based online anomaly detection in hierarchical wireless sensor networks. In Proceedings of the 11th IEEE International Conference on Trust, Security and Privacy in Computing and Communications, TrustCom-2012—11th IEEE International Conference on Ubiquitous Computing and Communications, IUCC-2012, Liverpool, UK, 25–27 June 2012; pp. 751–759.

27. Boedihardjo, A.P.; Lu, C.T.; Chen, F. Fast adaptive kernel density estimator for data streams. Knowl. Inf. Syst. 2015, 42, 285–317.

28. Lv, Y. An Adaptive Real-time Outlier Detection Algorithm Based on ARMA Model for Radar's Health Monitoring. In Proceedings of the 2015 IEEE AUTOTESTCON, National Harbor, MD, USA, 2–5 November 2015.

29. Nesa, N.; Ghosh, T.; Banerjee, I. Outlier detection in sensed data using statistical learning models for IoT. In Proceedings of the IEEEWireless Communications and Networking Conference, WCNC, Barcelona, Spain, 15–18 April 2018; pp. 1–6.

30. Redhwan Al-amri et. al., "A Review of Machine Learning and Deep Learning Techniques for Anomaly Detection in IoT Data", Appl. Sci. 2021, 11, 5320. https://doi.org/ 10.3390/app11125320.

31. Samara, M.A.; Bennis, I.; Abouaissa, A.; Lorenz, P, "A Survey of Outlier Detection Techniques in IoT: Review and Classification" J. Sens. Actuator Netw. 2022, 11, 4. https:// doi.org/10.3390/ jsan11010004.

Intelligent Systems and Smart Infrastructure – Brijesh Mishra et al. (eds)
© 2023 Taylor & Francis Group, London, ISBN 978-1-032-41287-0

Implementation of Deep Learning Algorithm Using Generative Adversarial Network

Ratnesh Prasad Srivastava[1]

Department of Information Technology,
College of Technology, GBPUAT Pantnagar, India

Lokendra Singh Umrao[2], Divya Singh[3]

Department of Computer Science & Engineering,
Institute of Engineering and Technology,
Dr. Rammanohar Lohia Avadh University, Ayodhya, India

Ali Imam Abidi[4]

Department of Computer Science & Engineering,
School of Engineering & Technology, Sharda University,
Greater Noida, India

Abstract

The recent advancements in Computer Science have all been made possible by advancements in AI/DL techniques. Once that could have been classified a dream, is a now a reality: Image interpolation, in-video lip-syncing, auto-text generation, sky replacement, collision/particle simulation, fixing broken art, face synthesis, just to name a few, the majority of which include use of Generative Adversarial Networks (GANs). The main challenge in objectifying personality is that no two people under similar sets of conditions behave similarly. However, similarly, no two people have the same set of characteristics that define them. The personality assessment includes, but is not limited to, using automated algorithms for data extraction and

Corresponding author: [1]ratnesh.mnnit@gmail.com, [2]lokendra.manit@gmail.com, [3]evergreendivya@gmail.com, [4]aliabidi4685@gmail.com

DOI: 10.1201/9781003357346-73

cross-validation. Previously done works on the matter involve gathering digital records and footprints for a person in concern and later creating scales which are then assessed using traditional methods. Our focus, in this paper, is on computational approach. Loads of data are generated every day, which can be used to detect, extract, and make use of patterns so found.

Keywords: Deep Learning, Generative Adversarial Networks, Personality Assessment, Big Data

1. Introduction

Personality assessment (Pradhan et al., 2020) helps any organization to bring the best out of their employees. Motivations, attitude, emotional states, and approaches to interpersonal relations are one of the few things that can be concluded from one's PA. When using tradition means, the tests either are expensive to undertake or time-consuming. GANs can effectively train a generator model using a discriminator model, which later can be used to develop a classifier model (Bleidorn & Hopwood, 2019). A GAN consists of two parts: (i) a generative network (or generator) that generates data so as to "fool" the discriminator, a deconvolutional neural network, and (ii) a discriminator, which evaluates the data, often a CNN (Ventura et al., 2017). GANs make use of unlabeled datasets to train the application for personality assessment. The better it is trained, the better is the assessment. The five traits to be focused upon in any personality assessment are openness, conscientiousness, extraversion, agreeableness, and neuroticism. Through the training process, the generator model learns the features that can be used to identify the personality type. The accuracy of the generator model is analyzed by the discriminator model. Using a Deep Neural Network, one can model mappings between different input and output spaces. Our aim here is to determine whether there is any predictable or learnable signal in input data: if personality can be determined, what aspects of an image are useful. We employ convolutional neural networks in order to understand what imputable attributes we may derive those are simpler to reason. Personality prediction (Stachl et al., 2020) is modelled as a classification task. The efficiency of GAN models is, however, hard to analyze because unlike other neural networks that try to minimize a loss function, the generator model is analyzed by another neural network. Our model solely focuses on using face crops from this subset for learning relationships between faces and predicting personality (Zar, 1972).

2. Literature Review

We want to map between input pdf files $X = \{x_1, x_2, x_3, \cdots, x_n\}$ and output space of citation count, $Y = \{y_1, y_2, y_3, \cdots, y_n\}$. For evaluating the performance of the model, we use Spearman's correlation coefficient (ρ) in both direct and SVR (Dave, 2016) prediction. It indicates correlation between prediction and practical truth. Denoted by corr(P^k, Y^k), it gives reasonable score of performance as it provides correlation between ordinal numbers. Mathematically,

$$\rho = \text{corr}(P^k, Y^k) = 1 - \frac{6\sum d_i^2}{n(n^2 - 1)} \quad \text{where } d_i = \mathbf{rg}(X_i) - \mathbf{rg}(Y_i)$$

The final evaluation of kth trait is measured using ρ. For direct, it is $\text{corr}(P_{dir}^k, Y^k)$ while for SVR, it is corr (P_{svr}^k, Y^k). The correlation between two quantities x_i and y_i is given by the relation:

$$r = \frac{\sum (x_i - \bar{x})(y_i - \bar{y})}{\sqrt{\sum (x_i - \bar{x})^2 \sum (y_i - \bar{y})^2}}$$

The closer the value of r is to 1, the more are the quantities x_i and y_i related, i.e., if Y_i increases then X_i also increases similarly decreases by decreasing Y_i. A negative value of r implies that whenever the quantity x_i decreases, y_i increases and vice versa.

   (i) Our hypothesis is that a non-zero co-relation exists between one's personality and his/ her facial features. Current prediction models are more successful in predicting some personality traits like neuroticism better over the others (N. Pal and S. Pal, 1993). For psychologists, the relationship between faces and one's personality is of active research. The works done till now have proven that the method used by N. Pal and S. Pal visualizes the results of relationship between faces and personality, which later can be interpreted qualitatively.

  (ii) Visualizes data as statistical mean (average) (Westen & Rosenthal, 2003).

The input image vector for personality prediction model happens to be of size 224×224, in our case, while the output vector is of 5×1. The output layer contains 5 neurons, each representing one distinct personality which happens to be dominantly present in the input image.

Input Vector: $X = \{x_1, x_2, \cdots, x_n\}$

Output Vector: $Y = \{y_1, y_2, y_3, y_4, y_5\} = \{$openness, conscientiousness, extraversion, agreeableness, neuroticism$\}$

The mapping from set X to set Y is found, automatically, by the neural network (GAN) (Ratner, 2017). We found out that apart from just using images, feeding metadata like text and other related information would lead in formation of a better and accurate model. On a limited data sample, k-fold cross-validation can also prove useful.

3. Methodology and Model Specifications

The evaluation of personality by ML means requires algorithms for data collection, validation, and prediction. This is done by finding any correlation that exists between the records available (digital records) and the precisely established trait groups. Digital record or footprint is any data generated as a result of a person's usage of internet services. For practical purposes, this record is represented in the form of a matrix, which usually contains a large number of rows and columns, most of which are zeroes or null because this data is often incomplete. This sparse matrix is later refined to eliminate the presence of any such data that could possibly lead to an incorrect conclusion or makes data regression difficult. This is called the cleaning step.

The main challenge lies in what we aim to predict: (i) the true personality, the behaviour what makes a person that person, or (ii) the perceived personality, the personality that is attributed to an individual by other people who interact with them. Perceived personality varies from individual to individual, a result of experiences formed because of one's interaction with the test subject. Every individual can be roughly identified as one among these Big-Five personalities:

(i) *Agreeableness:* Straight-forwardness, trustworthiness, generosity, modesty.

(ii) *Conscientiousness:* Efficiency, organizability.

(iii) *Openness:* Curiosity, inventiveness, cautiousness.

(iv) *Extraversion:* Talkatively, energy, reservedness, outgoingness.

(v) *Neuroticism:* Sensitivity, security, confidence, nervousness.

GANs can find hidden, complex, and subtle relations between one's behaviour and traits. Various NLP tools have also been developed for language-independent, unsupervised personality recognition from unlabelled text. Only the above average feature values in the input text are correspondingly mapped to personality traits. In most of the studies, the apparent or the perceived personality is modelled instead of the true personality of an individual. A better result is obtained when a combination of audio or images, multimedia apart from text input, is for personality detection. Myers–Briggs Type Indicator is currently the most widely used personality assessment method. Deep Convolutional Neural Networks (DCNNs) like GANs have often demonstrated the capability to capture even non-linear patterns in unstructured data like images and learn corresponding parameters for prediction with high accuracy.

For evaluating the performance of the model, we use Spearman's correlation coefficient (ρ) in both direct as well as SVR prediction. It indicates correlation between prediction and practical truth. Denoted by corr(P^k, Y^k), it gives reasonable score of performance as it provides correlation between ordinal numbers. Mathematically,

$$\rho = \text{corr}(P^k, Y^k) = 1 - \frac{6\sum d_i^2}{n(n^2 - 1)} \qquad \text{where } d_i = \mathbf{rg}(X_i) - \mathbf{rg}(Y_i)$$

4. Empirical Results

Our model uses face crops from client profile pictures subset for learning and forecast. Since we accept that the input information space is sensibly inconspicuous in its unpredictability, we utilize deep neural networks that can learn nearby spatial connections and elevated-level portrayals of the facial content. The number of hidden layers in the model also influences the performance of the model. For correlation purposes, we likewise present the aftereffects of preparing a model on the original uncropped crude pictures utilizing a comparable, however, more straightforward model, a modified form of AlexNet. Fine-tuning models also improve its performance as shown in Table 73.1.

Table 73.1 Result Obtained After Fine-tuning AlexNet Model

Neural Net	Feature	setting	avg.	ope	con	ext	agr	neu
P-Net	fc8	$cpi = 10,$ $fti = 140k$	0.12	0.09	0.14	0.14	0.09	0.13
P-Net	fc6 + SVR	$tb \geq 100,$ $fti = 0$	0.12	0.11	0.13	0.15	0.07	0.13
P-Net	fc6 + SVR	$tb \geq 100,$ $fti = 140k$	0.16	0.14	0.18	0.18	0.13	0.19
P-Net	fc7 + SVR	$tb \geq 100,$ $fti = 140k$	0.15	0.11	0.17	0.18	0.13	0.17

Personality analysis should be concentrated, in more profundity, from the point of view of understanding what parts of the face cause or relate with the exact character attributes of the individual. This might be concentrated regarding intrinsic and extrinsic features. An investigation of the inherent perspectives is dispensing with factors not naturally contained in the individual's facial physiognomy.

Here fc8, fc7, and fc6 are the respective layers of the modified AlexNet model. The outcomes concur with our instinct that complete images are better for indicating personalities as compared to cropped ones. Additionally, we found that calibrating and fine-tuning improves model performance. Tweaking alongside a SVR learnt on the fc6/fc7 highlights brings the normal anticipated connection score nearer to the gauge for human-appointed authorities. Also, we noticed that "agreeableness" is one of the hardest personalities to anticipate.

5. Conclusion

The ongoing AI and Deep Learning models were evaluated based on the character gauge and examined to predict personality. Thus each character gauge represents different view points. Therefore expectations of personality prediction is focused on two parts Hypothesis and Test building. There are numerous assorted uses of computerized character location which can be utilized in the business. We found out that GANs could discover hidden, perplexing, and inconspicuous relations between one's conduct and qualities. Different NLP algorithms have likewise been produced for language-free, solo character acknowledgment from unlabelled content. In any case, Generative Adversarial Networks are as incredible as the information used to prepare them. There is a desperate need of bigger, more precise, and more assorted datasets for character discovery.

REFERENCES

1. Bleidorn, W., & Hopwood, C. J. (2019). Using Machine Learning to Advance Personality Assessment and Theory. *Personality and Social Psychology Review, 23*(2), 190–203.
2. C. Stachl, F. P. Sven, H. G. M. Harari, R. Schoedel, S. Vaid, S. D. Gosling and M. Buhner (2020). "Personality Research & Assessment in the era of machine learning," *European Journal of Personality.*
3. C. Ventura, D. Masip and A. Lapedriza (2017). "Interpreting CNN Models for Apparent Personality Trait Regression," IEEE Conference on Computer Vision and Pattern Recognition Workshops (CVPRW), pp. 1705–1713.
4. Dave, A. (2016). Application of convolutional neural network models for personality prediction from social media images and citation prediction for academic papers. *UC San Diego.* ProQuest ID: Dave_ucsd_0033M_15783. Merritt ID: ark:/13030/m53v44v9. Retrieved from https://escholarship.org/uc/item/6fb9v8j0
5. N. R. Pal and S. K. Pal (1993). "A review on image segmentation techniques. Pattern recognition," 26(9): 1277–1294.
6. Ratner, B. (2017). Statistical and Machine-Learning Data Mining: Techniques for Better Predictive Modeling and Analysis of Big Data (3rd ed.). Chapman and Hall/CRC. https://doi.org/10.1201/9781315156316
7. T. Pradhan, R. Bhansali, D. Chandnani and A. Pangaonkar (2020). "Analysis of Personality Traits using Natural Language Processing and Deep Learning,"Second International Conference on Inventive Research in Computing Applications (ICIRCA), pp. 457–461, doi: 10.1109/ICIRCA48905.2020.9183090.
8. Westen, D., & Rosenthal, R. (2003). Quantifying construct validity: Two simple measures. *Journal of Personality and Social Psychology, 84*(3), 608–618.
9. Y. Mehta, N. Majumder and A. Gelbukh (2020). Recent trends in Deep Learning based Personality Detection. ArtifIntell Rev 53, 2313–2339. https://doi.org/10.1007/s10462-019-09770-z
10. Zar, J. H. (1972). Significance Testing of the Spearman Rank Correlation Coefficient. *Journal of the American Statistical Association, 67*(339), 578–580.

Intelligent Systems and Smart Infrastructure – Brijesh Mishra et al. (eds)
© 2023 Taylor & Francis Group, London, ISBN 978-1-032-41287-0

CHAPTER

74

Categorical Data Analysis of Covid-19 Data Using Machine Learning: A Comparative Analysis

Ashish Mishra[1], Bramha P. Pandey[2]

Electronics and Communication Engineering,
Madan Mohan Malaviya University of Technology, Gorakhpur, India

Abstract

The coronavirus, which was first identified in 2019, has affected almost every aspect of life. This paper aims to study the impact of COVID-19 pandemic by performing complete Exploratory data analysis (EDA). This study proposes an ML model to help and understand the impact of COVID-19 on different parameters. Comparison of accuracy is made to find best algorithm for prediction of pure categorical dataset. We used several machine learning algorithms like K-Nearest Neighbor (KNN), Logistic Regression, Decision Tree, Random Forest, and Naive Bayes, and further accuracy is calculated. The best accuracy is found to be 85.18%. The Decision Tree algorithm is found to be more precise for purely categorical data analysis. Finally, we used our own dataset which was collected through an online survey for accuracy evaluation.

Keywords: Chi-square test, Decision tree, EDA, KNN, Logistic regression, Naïve bayes, Random forest.

1. Introduction

The novel coronavirus COVID-19 was first detected in December 2019, and over past few years, it has caused great loss to lives as well as economy. Research is going on to find the

Corresponding author: [1]imashishmishra2230@gmail.com, [2]bppece@mmmut.ac.in

DOI: 10.1201/9781003357346-74

best cure of the disease. Machine learning has played a vital role in medical field. Using machine learning, we can find the best suited algorithm for prediction. This will prevent and increase awareness for the future challenges. The main goal is to develop and deploy machine learning model that could predict the status of vaccination based on various other features. To develop such model, we need to know about interdependency among different features. For this, Chi-square test is used for testing the independency of two categorical variables [1]. In the study, nominal data is converted into numerical data with the help of dummy variables. Once the data gets converted, it is used for Chi-square test. Based on Chi-square test, decision is made whether or not the two categorical variables are interdependent. Further, the accuracy comparison is done and the best suited algorithm for prediction in categorical dataset is found [5]. Much of the literature is focused on the outbreak in India in early days of 2021. It has also covered various aspects of the virus, including its effects on different parameters. The study will strengthen our fight against the deadly COVID pandemic and its mutated variants too.

2. Material and Method

A. Dataset

Table 74.1 Dataset along with features

Age	Sex	Positivity	Motivation	Status	Vaccine	Effects	State
22	Male	Yes before vaccination	Internet/Social media	Fully vaccinated	Covishield	Mild fever, fatigue, body pain	Uttar Pradesh
25	Male	Yes after vaccination	Internet/Social media	Fully vaccinated	Covishield	Mild fever, fatigue, body pain	Uttar Pradesh
29	Male	Yes before vaccination	Tv, Radio	Partially vaccinated	Covaxin	Severe illness	Uttar Pradesh
30	Female	Never infected	Newspaper	Partially vaccinated	Covishield	No effect	Uttar Pradesh
52	Male	Yes after vaccination	Newspaper	Partially vaccinated	Covaxin	Mild fever, fatigue, body pain	Uttar Pradesh

1. Features

From Table 74.1, you can see eight different variables. For performing EDA, one must have basic understanding of these variables:

- *Age:* This tells about the age of a person.
- *Sex:* This gives an idea about the gender to which the person belongs.

- *Positivity:* It explains whether the person was infected from COVID or not. They might be infected in past or never.
- *Motivation:* At the beginning, people were afraid of vaccines, so different platforms were used to motivate them. This explains what motivated the person to get vaccinated.
- *Status:* This tells whether the person is vaccinated or not and at what level, partial or fully.
- *Vaccine:* Various vaccines were launched so to know which one is preferred more for vaccination.
- *Effects:* People were suffering from adverse effects after they got vaccinated. This tells about the side effects after vaccination.
- *States:* This tells about the belongings of an individual.

B. Data Understanding

(a) Data Analysis:

Data analysis is done with the help of different plots:

1. Bar plot:

The plot usually shows quantity as their heights.

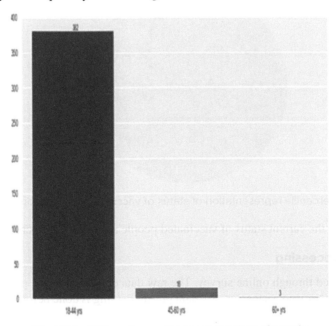

Fig. 74.1 Different age groups vs number of people

Figure 74.1 shows the bar plot representation of different age groups of people participated in the survey along with their numbers, for the proposed model in the specific range. This shows the clear majority belongs to the teenage group.

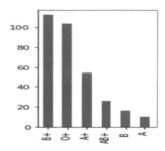

Fig. 74.2   Blood group vs number of people

Figure 74.2 shows the bar plot of the blood group with the numbers. This shows the majority people participated in the survey belongs to B positive group.

2. Pie-chart

The pie-chart has an advantage of showing numerical as well as percentile data.

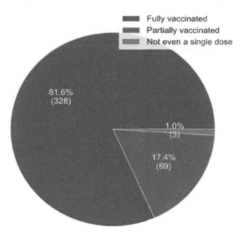

Fig. 74.3   Percentile representation of status of vaccination of the proposed model

Figure 74.3 shows the current status of vaccinated people along with the percentage.

C. Data Pre-processing

The data is collected through online survey. The raw data consists of large number of missing values and many other inconsistencies. So, data pre-processing is done before passing the data through an algorithm.

The inconsistencies present in the data will directly affect the performance of the model. It must be removed in order to increase the better training capability of the model. Data pre-processing consists of formatting data to make it standardize [7]. The columns consisting of more missing values and other inconsistencies, which turns to be insignificant, are removed. The features with no missing value is processed further.

D. Feature Selection

The aim of feature selection is to find out the best set of features for building a model with great accuracy. As we know that our dataset is categorical, Chi-square test is used to test the relation and dependency among the features. This is a hypothesis test which is used when we want to determine whether two features are interrelated or not [9]. Based on p-value, the relationship among features is determined. A greater p-value than 0.05 means that there is no relation among the features [10].

Table 74.2 Features along with p-values.

-------------	Effects	Motivation	Status
Sex	0.11	0.20	0.28
Effects	---------	---------	0.06
Motivation	0.94	-----------	0.49

From Table 74.2, it is very clear that there is no relation among the features as per the p-values ($p > 0.05$). But mild relationship is found between the set of features, Effects and Status. While testing these two features together, we find p-value to be 0.06, which is very close to the significance level of 0.05.

E. Algorithm Used

Different classifiers are used as it comes under supervised learning. Classifiers used are Logistic Regression, Naïve Bayes, Random Forest Classifier, Decision Tree Classifier, and KNN classifier. With the help of these, classifier accuracy is calculated [8]. Accuracy gives an idea about the exact prediction.

Table 74.3 Accuracy comparison of different features

	Status	Effects	Motivation	Sex
Logistic reg.	0.827	0.530	0.654	0.580
Native bayes	0.758	0.518	0.703	0.291
Random Forest	0.814	0.555	0.617	0.334
Decision tree	0.851	0.555	0.654	0.340
KNN	0.802	0.555	0.604	0.323

Table 74.3 states the different classifiers along with the accuracy of the features. The best accuracy of feature Status is found in the decision tree classifier. So, decision tree classifier can be used for the most accurate prediction of categorical data.

3. Result

Our aim was to analyze the data using different available plots, to find the relationship among the features and best suited method for the prediction of categorical data, hence the decision was made on the basis of accuracy comparison of different algorithms like Logistic Regression, Naïve Bayes, Random Forest, Decision Tree, and KNN.

The data visualization is done using bar plot and pie-chart. The mild relationship is found between the feature pair, Effects and Status, with the help of Chi-square test. Decision Tree classifier shows best accuracy for prediction of 85.1%.

We can visualize the predicted values and the true values with the help of confusion matrix.

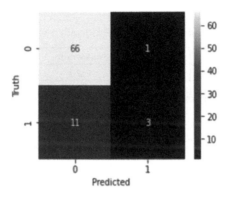

Fig. 74.4   Confusion matrix for prediction of Status feature

References

1. M. Zuvić-Butorac, "Characteristics of categorical data analysis," *Acta Med. Croatica*, vol. 60 Suppl 1, no. February 2006, pp. 63–79, 2006.
2. Worldbank, "Reference_3.pdf." 2019. [Online]. Available: https://ghr.nlm.nih.gov/condition/parkinson-disease#inheritance
3. C. Anastassopoulou, L. Russo, A. Tsakris, and C. Siettos, "Data-based analysis, modelling and forecasting of the COVID-19 outbreak," *PLoS One*, vol. 15, no. 3, pp. 1–21, 2020, doi: 10.1371/journal.pone.0230405.
4. E. J. Williamson *et al.*, "Factors associated with COVID-19-related death using OpenSAFELY," *Nature*, vol. 584, no. 7821, pp. 430–436, 2020, doi: 10.1038/s41586-020-2521-4.
5. S. D. Withers, "Categorical Data Analysis," *Int. Encycl. Hum. Geogr.*, pp. 456–462, 2009, doi: 10.1016/B978-008044910-4.00409-0.
6. F. Virgantari and Y. E. Faridhan, "in Indonesia ' s Provinces," vol. 5, no. 2, pp. 1–7, 2020.
7. A. Kumar, "Documentation on By: Ashutosh Kumar," no. June, 2020, doi: 10.13140/RG.2.2.23615.94880.
8. Nilima, V. Nayak, and V. Guddattu, "Categorical Data Analysis: Fundamentals and Perspective Applications in Health Sciences," *J. Clin. Diagnostic Res.*, no. March, pp. 0–4, 2019, doi: 10.7860/jcdr/2019/40148.12737.
9. A. C. De Young *et al.*, "COVID-19 Unmasked Global Collaboration Protocol: longitudinal cohort study examining mental health of young children and caregivers during the pandemic," *Eur. J. Psychotraumatol.*, vol. 12, no. 1, 2021, doi: 10.1080/20008198.2021.1940760.
10. A. O. Adejumo, O. O. M. Sanni, E. T. Jolayemi, and R. O. Ogedengbe, "Analysis of Categorical Panel Data," *Int. J. Stat. Appl.*, vol. 2, no. 5, pp. 56–59, 2012, doi: 10.5923/j.statistics.20120205.02.

Intelligent Systems and Smart Infrastructure – Brijesh Mishra et al. (eds)
© 2023 Taylor & Francis Group, London, ISBN 978-1-032-41287-0

CHAPTER

75

Development of a Chatbot Using LSTM Architecture and Seq2Seq Model

Adil[1], Tanmaya Garg[2], Rajesh Kumar[3]

Department of Electrical Engineering,
Delhi Technological University, New Delhi, India

Abstract

There has been an increasing need for better ways to process Natural Language, and so chatbots have been on a rise, especially commercially, with improving technology. In this paper, we explore the field of Natural Language Processing, which deals with making computers analyse, process, and work on Natural Language Text, by developing a chatbot. To achieve this, we have employed a deep Recurrent Neural Network (RNN). But since regular RNNs are plagued by issues of their own, such as exploding gradient and vanishing gradient, we've used a Long Short-Term Memory (LSTM) architecture and a sequence-to-sequence (seq2seq) model, which has a bilinear encoder–decoder architecture. To overcome some of the shortcomings of the seq2seq model, an attention mechanism was also implemented. For improved training, cross-validation was also performed, and, finally, a working chatbot was realised.

Keywords: natural language processing; recurrent neural networks; long short-term memory; sequence-to-sequence; chatbot

1. Introduction

Neural Network is a term which is quite prevalent in the field of Computer Science and Artificial Intelligence (AI) right now. Their applications are varied, including the fields of self-driving cars, image captioning, etc. One such sub-domain of AI is Natural Language

Corresponding author: [1]adil_2k18ee007@dtu.ac.in, [2]tanmayagarg_2k17ee219@dtu.ac.in, [3]rajeshkumar@dtu.ac.in

DOI: 10.1201/9781003357346-75

Processing (NLP). It deals with the process of understanding and analysing human languages. We delve into one such application of NLP, that is, chatbots, which allow us to automate human conversations. They are currently being employed in various industries such as healthcare, education, banking, customer support, and many more.

In NLP, the development of chatbots using AI and ML techniques is a field generating significant interest. Apart from research purposes, there is a strong commercial incentive, since as of now, most companies are increasingly using chatbots for customer interactions. Most of these chatbots are developed using relatively simple models such as Naive Bayes and logistic regression which have shown poor results, hence we look towards deep learning for better possibilities.

We realized that Recurrent Neural Networks have a significant edge over simple Artificial Neural Networks, in this particular application, due to their recurrent connection to their hidden state, which makes them better equipped to deal with sequential data such as texts. For developing our chatbot, we have applied a sequence-to-sequence (seq2seq) model with attention. But Recurrent Neural Networks (RNNs) have their own limitations, especially the vanishing gradient which makes them forget long-term dependencies. Hence, instead of standard Vanilla RNN, we have built our model using Long Short-Term Memory (LSTM) to tackle this issue.

RNNs, however, have their own limitations, especially the vanishing gradient, which makes them forget long-term dependencies. Hence, instead of standard Vanilla RNN, we have built our model using LSTM to tackle this issue. For developing our chatbot, we have applied a seq2seq model. But since seq2seq models depend too heavily on the final content vetor output of the encoder, it loses effectiveness over longer sequences. To deal with this, an attention mechanism can be implemented that gives a weighted context vector to the decoder instead of a simple one.

2. Literature Review

In recent years, conversation agents like chatbots have gained significant commercial acceptance in sectors such as retail, banking, and education. The continuous research in this field is focused on making the conversation between the user and the chatbot as human-like as possible.

The field of NLP has evolved from traditional techniques to Deep Learning techniques, which is now being used to model and train conversational AI bots, an example of which can be seen by the chatbot developed by Anbang Xu et al (2017) [1]. One of the models used with Deep Learning for text processing, with great results, is seq2seq, which is being deployed in areas such as Neural Network translation applications (Sutskever et al., 2014)[2] and Cold Fusion with pre-trained language models (Sriram et al, 2017) [3] .

seq2seq model can be deployed in various types of applications such as text generation and translation, summary generation, and captioning. This model has an encoder–decoder architecture, where both layers have a simple RNN architecture (Sherstinsky, 2020) [4] by default. Now, RNN has its intrinsic problems of vanishing gradient and exploding gradient, which becomes more pronounced when the dataset is large. Bengio et al., (1994) [5] discusses these problems in RNNs, possible solutions, and alternatives, as well.

One of the most widespread solutions used for vanishing gradient problems is LSTM, which was first thought of by Hochrieter and Schmidhuber (1997) [6]. LSTM cells have been found to be particularly useful in language-modelling problems. LSTM cells are an improvement and variation of RNN cells. LSTMs have been implemented with seq2seq architecture for NLP applications such as for summary generation as done by Gabor Szucs [7].

From the above, we see that LSTM and the seq2seq model can be applied to create conversational AI chatbots, with potentially better results than with more rudimentary machine learning models or even with simple ANNs or RNNs.

3. Background

In this section, we discuss some of the concepts implemented in the model.

A. RNN

RNNs are a class of Artificial Neural Networks that are said to have "memory", meaning the output of the network at any time 't' depends on both the inputs at time 't' and at 't−1'. This dependence of RNN on inputs at both present and past values of 't' is what enables it to be one of the best ways to process sequential data [8], [11]. The reason for this is that in sequential data, knowing the state of the immediate predecessor part of data is paramount.

B. LSTM

LSTM is an RNN variation that can solve the problem of long-term dependencies [13]. It is useful when dealing with long sequential data and has practical applications in the fields of time-series predictions, speech processing, language modelling, machine translation, etc. The LSTM maintains its long-term dependencies using its cell state "Ct" with only linear changes, allowing information to be passed on relatively unchanged, as shown in Fig. 75.1.

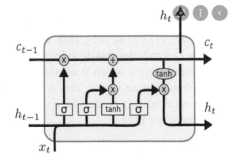

Fig. 75.1 LSTM architecture

Evaluation of new input to see what is added to the cell state and what is to be removed is done using the following processes in the cell structure.

Forget Gate: As the name suggests, the forget gate decides which information can be ignored and which needs to be retained. For example, in the sentence "Thomas is walking with Sarah", the primary subject changes from Thomas to Sarah, and it is the forget gate that is responsible for doing it. It consists of a sigmoid function. Below is the mathematical equation that describes the forget gate, with ft representing the forget gate.

$$ft = (Wf\ [ht - 1\ , xt] + bf) \tag{1}$$

Middle Layer: The next layer consists of two components: an input layer 'it', which decides which data is updated to the network, achieved by a sigmoid function; and a tanh function that makes the next candidate state 'Ct'.

$$it = (Wi\ [ht - 1, xt] + bi) \tag{2}$$

$$Ct = \ tanh\ (Wc\ [ht - 1, xt] + bc) \tag{3}$$

Based on the above two equations and the LSTM architecture, the next candidate state then can be written as:

$$Ct = ft * Ct - 1 + it * Ct \tag{4}$$

Output Layer: It is made up of two components: first, there is a sigmoid function and then a tanh function. The sigmoid function helps in filtering the output by deciding the candidate states that will be part of it, and then the tanh function is also employed on this output. The tanh function takes the final Candidate State as its input.

$$ot = (Wo\ [ht - 1, xt] + bo) \tag{5}$$

$$ht = ot * tanh\ (Ct) \tag{6}$$

C. Sequence to Sequence (seq2seq)

A seq2seq model takes as its input a series such as text, time-series, etc. and gives as its output another sequence of items, and has applications in machine translation, conversational models, speech recognition, etc [9], [10], [12]. [14], [15], [17]. The model

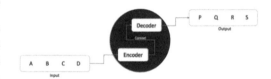

Fig. 75.2 Seq2Seq model architecture

comprises an encoder and decoder structure, as shown in Fig. 75.2. The encoder is responsible for the input series, which it converts into a form that can be analyzed by the computer. It is called a hidden state vector. This is then sent to the decoder. The decoder receives this hidden state vector and generates the final output sequence [16].

D. Attention Mechanism

A major drawback of the seq2seq model is that the output from the decoder relies heavily on the single final hidden state of the encoder. Because of this, it becomes difficult for the seq2seq to process longer input sequences. Unlike the normal seq2seq model, with an attention mechanism, the number of hidden state vectors employed is the same as the input sequence [18]. The weighted sum of these hidden state vectors is then taken to produce the context vector for the output sequences, as shown in Fig. 75.3. Use of weighted sum reduces dependency on any one hidden state, thus solving the problem faced previously.

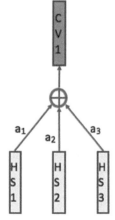

Fig. 75.3 Context Vector

This context vector is then concatenated with corresponding hidden states to give the new hidden vector with attention, called attention hidden state which is used to predict the output for any given input.

4. Model

The chatbot model was a three-layered LSTM RNN built using the tensorflow library in Python and the dataset used for training was the "Cornell Movie Dialogue Corpus", which contains conversations from movies with over 220,000 conversations.

A. Data Pre-processing

This step involved converting the two datasets into two lists of questions and answers. They were then cleaned and filtered by length and frequency. These question and answer words were then mapped onto unique integers.

B. Model Building

Functions were made for creating TF placeholders for all the hyperparameters. Then functions were created for the Encoder RNN, to create the embeddings matrix, decoding the training, validation and test sets, and then the Decoder RNN. Embedding is the process of representing words in the form of multidimensional arrays or vectors for processing. Finally, the function for the seq2seq model was created which incorporated all the above functions.

C. Training

First, all the hyperparameters were set, model inputs were loaded, and sequence length was set. The loss error function and optimizer were set up and the padding function was made.

D. Testing

First, the stored training weights were loaded and the chat loop was created where the conversations took place for testing, and the answers from the chatbots were observed.

5. Results

Since speech is so subjective, it becomes very hard to test and evaluate text-generative models, so we have done a manual check through random inputs. After

Table 75.1 Hyperparameters

Hyperparameter	Value
Epochs	300
Batch Size	64
RNN Size	512
Layers	3
Embeddings Matrix	512
Learning Rate	0.01
Learning Rate Decay	0.9
Keep Probability	0.5
Sequence Length	25
Activation Function	ReLu
Optimizer	Adam

training our model with the complete movie corpus dataset for 50 epochs the following are some of the responses we got for our given inputs (Table 75.2):

Table 75.2 Results

Input	Output
hi	hows the weekend
great thanks	youre welcome
how are you	im in nyc
how is nyc	this is so good
happy birthday	thank you
how are you feeling	the whole day is like i was just thinking about it
bye	i was thinking of that
it is really hot	it is great

6. Conclusion

An AI conversational chatbot using deep learning recurrent neural networks was developed using Tensorflow based on movie dialogues from the Cornell Movie Dialogue Dataset. This is more advanced than older and more conventional chatbots that work on rudimentary if-else conditions or simple ML models, as it is a generative model that learns sentence structure instead of relying on pre-decided static responses to inputs. It implements the seq2seq model, LSTM, as well as Attention. We see from the results that the chatbot, after training, often responds with relevant answers, but also gives irrelevant or tangential answers. We see, however, that the chatbot mostly always answers in coherent sentences, which shows that the attention model has proven effective to some degree in learning grammar and sentence structure.

7. Future Work

For future work, we see that the chatbot needs further training with more epochs and tuning in its hyperparameters to improve its performance. More relevant and exhaustive datasets can also be tried for better results. Improvements can also be made to incorporate longer sentences and their responses. The chatbot also lacks memory to continue a coherent conversation, instead of simple responses to individual questions which can be worked upon. The model needs to be evaluated with more standardized testing, which will also help compare and contrast it with other models more qualitatively for better study.

REFERENCES

1. Xu A., Liu Z., Guo Y., Sinha V. & Akkiraju R. 2017. A New Chatbot for Customer Service on Social Media. *Proceedings of the 2017 CHI Conference on Human Factors in Computing System – CH '17.* 3506-10. https://doi.org/10.1145/3025453.3025496
2. Sutskever, Ilya, OriolVinyals, and Quoc V. Le. 2014. Sequence to sequence learning with neural networks. *Advances in Neural Information Processing Systems (NIPS).* 27. 3104–3112.
3. Sriram, A., Jun, H., Satheesh, S., & Coates, A. 2017. Cold fusion: Training seq2seq models together with language models. *arXiv preprint arXiv:1708.06426.*
4. Sherstinsky A. 2020. Fundamentals of Recurrent Neural Network (RNN) and Long Short-Term Memory (LSTM) network. *Physica D: Nonlinear Phenomena.* 2020. 404(8).

5. Bengio Y., Simard P. & Frasconi P. 1994. Learning long-term dependencies with gradient descent is difficult. *IEEE transactions on neural networks.* *5*(2): 157–166.

6. Hochreiter S., Schmidhuber J. 1997. Long short-term memory. *Neural Computation.* 9 (8): 1735–1780

7. Szucs G. & Huszti D. 2019. Seq2seq Deep Learning Method for Summary Generation by LSTM with Two-way Encoder and Beam Search Decoder. *2019 IEEE 17th International Symposium on Intelligent Systems and Informatics (SISY).* 221–226. doi:10.1109/sisy47553.2019.911150

8. Pascanu R., Mikolov T. & Bengio Y. 2013. On the difficulty of training recurrent neural networks. In the *International conference on machine learning* (pp. 1310-1318). PMLR.

9. Bahdanau D., Cho K. & Bengio Y. 2014. Neural machine translation by jointly learning to align and translate. *arXiv preprint arXiv:1409.0473.*

10. Srivastava N., Hinton G., Krizhevsky A., Sutskever I. & Salakhutdinov R. 2014. Dropout: a simple way to prevent neural networks from overfitting. *The journal of machine learning research.* *15*(1): 1929–1958.

11. Karpathy A., Johnson J., & Fei-Fei L. 2015. Visualizing and understanding recurrent networks. *arXiv preprint arXiv:1506.02078.*

12. Luong M. T., Pham H. & Manning C. D. 2015. Effective approaches to attention-based neural machine translation. *arXiv preprint arXiv:1508.04025.*

13. Olah C. 2015. Understanding LSTM Networks. *Colah's Blog.* https://colah.github.io/posts/2015-08-Understanding- LSTMs/

14. Lee C., Wang Y., Hsu T., Chen K., Lee H., & Lee L. 2018. Scalable Sentiment for Sequence-to-Sequence Chatbot Response with Performance Analysis.*2018 IEEE International Conference on Acoustics, Speech and Signal Processing (ICASSP).* 6164–6168

15. Young T., Hazarika D., Poria, S., & Cambria E. 2018. Recent trends in deep learning based natural language processing[Review Article]. *IEEE Computational Intelligence Magazine.* *13*(3). 55–75.

16. Ali A. & Amin M. Z. 2019. Conversational AI Chatbot Based on Encoder-Decoder Architectures with Attention Mechanism. *Artificial Intelligence Festival 2.0, NED University of Engineering and Technology.*

17. Durga P. 2019. Attention-Seq2Seq Models. *Towards Data Science.* https://towardsdatascience.com/day-1-2-attention-seq2seq-models-65df3f49e26

18. Dhyani, M. & Kumar, R. (2021). An intelligent Chatbot using deep learning with Bidirectional RNN and attention model. *Materials today: proceedings.* *34(3).* 817–824.

Intelligent Systems and Smart Infrastructure – Brijesh Mishra et al. (eds)
© 2023 Taylor & Francis Group, London, ISBN 978-1-032-41287-0

CHAPTER

76

An Evaluation of Tree-Based Algorithms on Imbalanced Dataset: An Astronomical Case Study

Maneesh Sagar[1], Pavan Chakraborty[2]

IIIT Prayagraj, Allahabad

Snigdha Sen[3]

Global Academy of Technology, Bangalore

Abstract

The main objective of the paper is to evaluate the performance of multiple tree-based algorithms on skewed datasets. Although data in astronomical observation is very huge, still in the high redshift zone, it is very less as target values have very few observations in that area, leading to imbalanced distribution of data. For this purpose, we have chosen a redshift dataset from astronomy that contains photometric redshift data in galaxies and quasars in SDSS in the range $(0 < z < 7)$. The prime purpose of this paper is to analyze the performance of Catboost, XGboost, AdaBoost, Random Forest, and Xtratree algorithms on this imbalanced dataset. As the data in the 0–7 range is very much skewed, we first converted target values into the logarithmic values and then we compared multiple algorithms such as Catboostregressor, XGboostregressor, AdaBoostRegressor, Random Forest, ELM, LDS with ANN, and Etratreeregressor, where Extratreesregressor proved to be best in terms of performance.

Keywords: Astronomical Big data, Extratreeregressor, Catboostregressor, XGboostregressor, AdaBoostRegressor, AdaBoostRegressor, Random Forest, Redshift

Corresponding author: [1]maneeshsagar97@gmail.com, [2]pavan@iiita.ac.in, [3]snigdha.sen@gat.ac.in

DOI: 10.1201/9781003357346-76

1. Introduction

Data imbalance problems exist in many real-world datasets. Data shows uneven distributions with long tails where fewer data points can be found in some classes. The situation of a skewed dataset imposes great challenges to the machine learning models and already many techniques addressing the data imbalance problem (Cao, 2019). Most of the work done on data imbalance is a class full dataset where targets are divided into many categories but some work is also done for the continuous data targets. In recent years, many surveys like COSMOS, Dark Energy Survey (DES), and Wide-field Infrared Survey Explorer (WISE) have come into the picture in the field of astronomy and have provided a huge amount of data. The huge amount of data drew attention to the use of machine learning techniques to explore and analyze with high accuracy and speed. The study of accelerated cosmic expansion is affected by the vacuum density energy and vacuum density energy is related to the Redshift estimation, so estimation of redshift has great importance in the field of astronomy (Riess, 1998). Moreover, it is possible to trace the origin of the universe by studying distant objects. In this paper, we have analyzed the performance of various tree-based algorithms such as Extratree, Catboost, Random Forest, etc. on the SDSS dataset by estimating the redshift. We have evaluated all the models on various metrics such as MSE (mean squared error), MAE (mean absolute error), RMSE (root mean squared error), etc. First, we prepossessed the dataset and then converted the label redshift into a logarithmic domain and trained all the models. Apart from that, we have smoothed the label space using LDS (Label distribution smoothing) (Yang, 2021) and trained the ANN model.

2. Related Work

There have already been a variety of machine learning algorithms that are used for this purpose, but they are successful in forecasting low redshifts, up to 1, with a high degree of accuracy. The Support Vector Regression algorithm gives the same kind of accuracy in the range of $0<z<0.7$. There was no significant improvement in performance z [0-1] using Random Forest on 80000 galaxy data points. Deep CNN, an image-based method, was used to predict high redshift in the range $(0 < z < 4)$ (D'Isanto, 2018). The Weak Gated Experts (WGE) algorithm, which is a combination of clustering and regression techniques, yielded an MSE of 0.08; this method has been tested on a wider variety of data, and there are only a few samples in the test set with redshift values greater than 3.5. There is no ideal dataset that is not imbalanced; the performance of any machine learning model will be degraded if there exists any kind of skewness in the dataset, so skewness should be tackled before exposing in front of the model; there is much work that has been done for sake of imbalance data, and we can majorly divide the work on the basis of classification and regression dataset. CovXmlc (Kumar, 2022) is a multimodal classifier that has been used in recent years for classification. Some techniques increase the minority and decrease the majority; these solutions are for class-based problems; LDS (Label Distribution smoothing) and FDS (Feature Distribution Smoothing) come for the regression problem that learns from such imbalanced data with continuous targets, and after that deals with data that are not available or missing for the certain label, and generalize it to the complete range of target label. As we know that the volume of astronomical data is increasing day by day, so astronomical data is also being treated like big data and much research and dataset analysis is being done to help astronomers to solve multiple vital intriguing problems (Monisha, R., et al., Sen S. a., 2022). Even for the faster redshift analysis, the Neural Network was trained using Lipschitz based adaptive learning on the cloud-based spark platform (Sen S. a., 2021). Some works involved customized CNN to analyze and prepossess astronomical images (Sandeep, 2021). Scalable benchmarking for redshift estimation (Henghes B. a., 2021) and estimating the redshift using images (Henghes B. a., 2022) has been done in recent years. A comparison (Moonzarin Reza, 2020) of Extratree with some algorithms is done.

3. Description of Dataset

[1]There are many surveys like SDSS and COSMOS working for years and producing the huge amount of data that is being used by many research scholars across the world. We decided to use SDSS for our study. SDSS provides the user interface to run the SQL on the database and results get downloaded in CSV format.

The Sloan Digital Sky Survey (SDSS) is a major multi-spectral imaging and spectroscopic redshift survey, a dataset consisting of 592,312 data points; the skewness of data is 0.97 concerning the redshift. It can be seen clearly in Fig. 76.1 that data points are unevenly distributed throughout the range of 0 to 7. There are 41,185 data points in the 0–1 range, which is higher among all other ranges and 8134 is the number of data points in the range of 6–7, which is the lowest among the ranges. Table 76.1 shows the range-wise distribution of redshift over 7 classes.

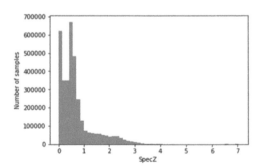

Fig. 76.1   Dataset Distribution on Redshift

4. Our Work

- We have calculated the performance of the tree-based algorithms like Extaratree, Catboost, XGboost, Gradientboost, Random Forest, etc.
- We have found out the performance metrics for ELM (Extreme Learning Machine) and LDS with ANN.
- We analyzed and compared the results of all algorithms.

---

[1]Dataset can be found on https://skyserver.sdss.org/dr14/en/tools/search/sql.aspx

Firstly, we pre-processed the dataset and scaled using Min–Max scalar keeping hypermeters to default and data points are fewer in the high redshift range so the logarithmic values of the label are used in the model instead of the actual values. GradientBoost Random Forest and Catboost result are mentioned in Table 76.1. All the algorithms are tree-based ensemble learning algorithms.

Table 76.1 Result for the Tree-Based Models

	MAE	MSE	R2 Score	RMSE	Outlier %	NMAD
XtraTree	0.0432	0.0098	0.8445	0.0990	11.7520	0.01675
Random Forest	0.0485	0.0107	0.8304	0.1034	13.5382	0.0209
Adaboost	0.1107	0.0248	0.6064	0.1576	36.2844	0.1390
XGBoost	0.0661	0.0167	0.7354	0.1292	18.0375	0.0382
Catboost	0.0599	0.0152	0.7585	0.1235	15.7279	0.03326
Gradient Boost	0.0732	0.0185	0.7058	0.1362	20.9886	0.0463

Extreme Learning Machine is a feed-forward neural network with a single layer or multiple layers of hidden nodes, where the hidden node parameters do not need to be defined and gradient-based backpropagation is not necessary. Table 76.2 shows the detailed result.

Table 76.2 Result for Extreme Learning Machine (ELM)

	MAE	MSE	R2 Score	RMSE	Outlier %	NMAD
ELM	0.0658	0.0211	0.4814	0.1452	16.0960	0.0453

LDS is a technique to smoothen the targets based on kernel density estimation that can be used in any machine learning model like ANN, etc. MSE for this algorithm is 0.258, which is higher among all the algorithms which are implemented in this paper.

5. Result & Discussions

This part of the paper represents the analysis and testing results of all the implemented algorithms.

Metrics: For evaluation of the model's performance, we are using the following metrics:

1. **Mean Absolute Error(MAE):** $\dfrac{1}{N}\sum |z - \bar{z}|$

2. **Mean Squared Error (MSE):** $\dfrac{1}{N}\sum (z - \bar{z})^2$

3. **R² Score:** $1 - \dfrac{\sum (z - \bar{z})^2}{\sum (z - \text{mean}(z))^2}$

4. **Root Mean Squared Error (RMSE):** $\sqrt{\dfrac{1}{N} \sum (z - \bar{z})^2}$

5. **Normalized median absolute deviation (NMAD)** (Geurts, 2006): $1.48 \times \text{median} \dfrac{|z - \bar{z}|}{1 + z}$

where the number of the test samples is represented by N, z is the actual value, and $\bar{z}$ is the predicted value.

As we have defined our metrics MSE, MAE, R^2 Score, etc., we analyzed the results based on the same defined metrics. We can see the result in Table 76.1. That the lowest MSE (0.0098) is of Extratree and the highest value of MSE (0.0248) is of Adaboost and the second-lowest value of MSE (0.0107) is of Random Forest. R^2 Score of the Extratree is higher among all, Random forest is having the second highest R^2 score while Adaboost is at a lower level. RMSE of the Extratree is the lowest, then Random Forest and the highest value are being held by Adaboost. NMAD is lowest for the Extratree, then for the Random forest, and the highest is for the Adaboost. The outlier percentage of Extraatree is also best because Extratree holds the lowest outlier percentage while Adaboost is having the highest outlier percentage and in terms of MAE, Extratree is having the lowest, random forest is at second lowest, and Adaboost holds the highest MAE value. So we concluded that among ensemble learning algorithms, the performance of the Extra tree is the best, Radom Forest is the second best, and Adaboost is the worst performer. In Table 76.2, we can see the performance of the ELM MAE (0.0658), MSE (0.0211), R^2 Score (0.4814), RMSE (0.1452), outlier percentage is 16 percent, and NMAD is 0.0453; if we consider all metrics at the same time, then ELM performed as good as Adaboostregressor. So we can safely conclude that some ensemble learning techniques are good for imbalanced datasets and ExtratreeRegressor is the best performing technique.

6. Conclusion

The fast growth of the data in the field of astronomy attracts more researchers to this field to produce more research work. This paper explores many machine learning techniques, mainly tree-based ensemble learning techniques on the SDSS dataset which is a skewed dataset, and the results of all techniques are compared. This paper conveys that the ensemble learning techniques are working great with the imbalanced dataset, mainly the Random Forest and Random Forest-based algorithm like Extratreesregressor. ELM is working better than ANN and LDS with ANN and as good as AdaboostRegressor. So we can conclude that Extratree is the best performer, Random Forest is the second-best among all algorithms, Adaboost is the worst performer among ensemble and tree-based algorithms while ANN with LDS is the worst among all.

REFERENCES

1. Cao, K. a. (2019). Learning imbalanced datasets with label-distribution-aware margin loss. *Advances in neural information processing systems*, 32.
2. D'Isanto, A. a. (2018). Photometric redshift estimation via deep learning-generalized and pre-classification-less, image based, fully probabilistic redshifts. *Astronomy \& Astrophysics*, A111.
3. Geurts, P. a. (2006). Extremely randomized trees. *Machine learning*, 3–42.
4. Henghes, B. a. (2021). Benchmarking and scalability of machine-learning methods for photometric redshift estimation. *Monthly Notices of the Royal Astronomical Society*, 4847-4856.
5. Henghes, B. a. (2022). Deep learning methods for obtaining photometric redshift estimations from images. *Monthly Notices of the Royal Astronomical Society*, 1696-1709.
6. Kumar, S. S. (2022). CovXmlc: High performance COVID-19 detection on X-ray images using Multi-Model classification. *Biomedical Signal Processing and Control*, 103272.
7. Monisha, R., et al. (2022). An Approach Toward Design and Implementation of Distributed Framework for Astronomical Big Data Processing. *Intelligent Systems*. Springer, Singapore, 267–275.
8. Moonzarin Reza, M. A. (2020). Photometric redshift estimation using ExtraTreesRegressor: Galaxies and quasars from low to very high redshifts. *Astrophysics and Space Science volume 365*, 50.
9. Riess, A. G. (1998). Observational evidence from supernovae for an accelerating universe and a cosmological constant. *The Astronomical Journal*, 1009.
10. Sandeep, V. a. (2021). Analyzing and Processing of Astronomical Images using Deep Learning Techniques. *2021 IEEE International Conference on Electronics, Computing and Communication Technologies (CONECCT)*, 01–06.
11. Sen, S. a. (2021). Implementation of neural network regression model for faster redshift analysis on cloud-based spark platform. *International Conference on Industrial, Engineering and Other Applications of Applied Intelligent Systems*, 591–602.
12. Sen, S. a. (2022). Astronomical big data processing using machine learning: A comprehensive review. *Experimental Astronomy*, 1–43.
13. Yang, Y. a. (2021). Delving into deep imbalanced regression. *International Conference on Machine Learning*, 11842–11851.

Intelligent Systems and Smart Infrastructure – Brijesh Mishra et al. (eds)
© 2023 Taylor & Francis Group, London, ISBN 978-1-032-41287-0

CHAPTER

77

Empirical Evaluation of Density-Based Clustering Rule Mining Models for Dynamic Data: Statistical Perspective

Jayshri Harde[1]

Computer Science and Engineering Department,
G H Raisoni University, Saikheda, India

Swapnili Karmore[2]

Data Science Department GHRIET,
Nagpur, India

Abstract

Rule mining & density-based clustering are the most widely used approaches for data processing, which involves data relationship & pattern management. These approaches are used by a wide variety of applications including pattern recognition, e-Commerce recommendations, social media analysis, etc. In order to efficiently perform data mining, experts have devised various algorithms that estimate the relationships between different data instances. These relationships are converted into rules via evaluation of support, confidence, and other scoring mechanisms. Algorithms like density-based spatial clustering applications with noise (DBSCAN), top-k rules, a priori, charm, cm-spam, etc. are proposed. Efficiency of these algorithms is measured in terms of precision, recall, computational complexity, and end-to-end delay. Algorithms need to showcase high precision performance with lower complexity & lower delay for being applicable to a wide variety of datasets. Moreover, researchers test a wide variety of rule mining approaches on their application for selecting the best performing ones, which increases system design cost & time needed for deployment. Thus, it is ambiguous to select a particular rule mining & density-based clustering algorithm when designing a new

Corresponding author: [1]jayshri.banpurkar@gmail.com, [2]swapnilikarmore@gmail.com

DOI: 10.1201/9781003357346-77

data mining application. In order to reduce this ambiguity, the underlying text reviews a wide variety of density-based rule mining approaches, and compares their performance in terms of aforementioned efficiency metrics.

Keywords: Rule mining, density clustering, a priori, charm, top-k rules, DBSCAN, machine learning, deep learning.

1. Introduction

Density-based clustering & rule mining are fields of data mining which require multiple algorithmic models to work in tandem for effective pattern analysis & clustering. These models include, but are not limited to, density-based clustering, item set mining, pattern filtering, association rule estimation, etc. A typical clustering model initializes by pre-processing data from various sources, and converting it into representable form. Models like average filtering, wavelet transform, Auto Regressive Integrated Moving Average (ARIMA), etc. are used for this purpose. Results of this model are given to an integration layer; wherein various warehousing operations are performed. The integrated data is given to a transformation layer; which assists in converting this data into a mining friendly format. This converted data is then processed by various mining algorithms, wherein metrics like support, confidence, lift, etc. are estimated. These metrics assist in estimating relative frequency patterns between datasets. For instance, the value of support between two items indicates the absolute frequency of both items appearing together in the dataset, while confidence of two items indicates their relative frequency of appearance. This can be observed from equations 1 and 2 wherein values of support & confidence are evaluated.

$$S_{x=>y} = \frac{F(x, y)}{N} \tag{1}$$

$$C_{x=>y} = \frac{F(x, y)}{F(x)} \tag{1}$$

where S, C, N, and F represent support, confidence, number of values in dataset, and frequency of these values in the dataset. Using these metrics, various algorithms are designed for solving different pattern analysis tasks. The output patterns are converted to clusters depending upon their support and confidence values, thereby making the number of clusters variable in number, which leads to dynamic clustering. A survey of some of the recently researched density-based clustering & rule mining algorithms can be observed from the next section.

2. Literature Review

A wide variety of generic & application-specific approaches are proposed for density-based clustering & rule mining. These approaches utilize data variance, deviation, kurtosis, and other statistical parameters for performing high-efficiency clustering. An example of this can be observed in [1], wherein bidirectional long-short-term-memory (BiLSTM) model is used along with hybrid DBSCAN (HDBSCAN) to perform clustering. This performance can be improved via use of nearest neighbor graphs (NNG) as observed from [2], wherein adaptive DBSCAN (ADBSCAN) model is proposed. The model utilizes 2-hop distance for graph construction, and then generates a core query interface using average & standard deviation value.

Survey of density-based clustering algorithms can be observed from [3], wherein various versions of DBSCAN are described from 1996 by highlighting key features, time complexity, and cluster extraction technique. It is observed that Mass-based clustering of spatial data with application of noise (MBSCAN), KIDBSCAN, MRDBSCAN, and Cludoop outperform other models in terms of precision & recall performance. But they are observed to have high delay due to multiple intermediate processing steps. This delay can be reduced via use of blocks DBSCAN as proposed in [4], wherein cover tree & inner block identification   processes are used for reducing redundant computations during clustering.

Another density-based clustering model that uses hybrid processing is proposed in [5], wherein class samples are divided into majority and minority samples. The minority samples are initially clustered using k means, and then cluster densities are obtained. The denser clusters are further interpolated to form minority samples, which are further clustered to form dynamic clusters. This method is named as Clustering and Density-Based Hybrid (CDBH), and is majorly used for converting any imbalanced training set into a balanced one via roulette wheel selection & redundancy removal. These algorithms have good performance when all data values are present, but the performance degrades with missing values. Thus, this models must be modified to handle missing values as observed from [6], wherein k CMM (Clustering Mixed Numerical and Categorical Data with Missing Values) model is described. The model utilizes initialization, clustering, & multiple imputation methods with decision trees & correlative modeling for clustering data with missing values. Various seep learning approaches discussed in [8] and The DenMune method utilize mutual nearest neighbour consistency principle which allows the model to have better precision, than k-nearest neighbours with an indexing ratio (NPIR).

A relative density relationship method for density peak clustering (RD2PC) is described in [9], wherein average distance-based kernel is used. The model reduces the drawbacks of empty clusters & reduced variance clusters via averaging clustering centroids for better performance. The performance of RD2PC can be improved using neighbourhood distance entropy consistency (NDEC) proposed in [10], wherein signal entropy is maximized in order to reduce intra-cluster variance, and increase inter-cluster variance. The NDEC model outperforms k means, DBSCAN, and OPTICS models, due to incorporation of sub-cluster generation.

3. Statistical Analysis

Performance evaluation of the reviewed models is done in terms of precision (P), recall (R), computational delay (D), and computational complexity (C). This is done on the Clustering basic benchmark corpus, which is available at http://cs.joensuu.fi/sipu/datasets/ with open-source licensing for research use cases. While absolute values of precision & recall are available in the literature, the values of delay & complexity are inferred from respective internal model designs. These values are categorized into low (L), medium (M), high (H), and very high (VH) ranges, depending upon delay & complexity performance of internal modules. These performance metrics are tabulated in Table 77.1, wherein comparison of different models is observed.

Table 77.1 Statistical analysis of different models

Method	P	R	D	C	Method	P	R	D	C
HDBSCAN [1]	17	39	L	M	AHC [5]	76	79	H	H
DBSCAN [1]	8	20	H	L	FRB + CHC [5]	79	80	H	VH
DBSCAN PCA [1]	5	12	M	M	kCMM [6]	89	85	M	H
kMeans [1]	14	58	M	L	kModes [6]	85	80	H	H
NNG ADBSCAN [2]	75	81	H	L	kRep. [6]	86	79	M	M
HDBSCAN [2]	71	79	M	M	AE [7]	90	89	H	VH
SNN [2]	65	61	M	M	DBN [7]	91	90	H	VH
DP [2]	66	71	M	H	VAE [7]	93	91	VH	VH
CLUB [2]	72	68	H	H	CAE [7]	91	90	VH	VH
RNN DBSCAN [2]	74	76	VH	H	CAHL [7]	90	89	VH	VH
DBSCAN [3]	61	60	L	M	KLD [7]	85	83	H	H
CLIQUE [3]	63	61	L	M	DenMune [8]	85	83	M	M
GDBSCAN [3]	65	64	M	M	NPIR [8]	83	80	M	M
IDBSCAN [3]	68	65	M	L	CBKM [8]	80	77	H	H
OPTICS [3]	69	67	M	L	RS [8]	85	83	M	M
MBSCAN [3]	75	72	M	L	FastDP [8]	79	77	L	L
KIDBSCAN [3]	76	74	M	M	FINCH [8]	83	80	M	H
MRDBSCAN [3]	79	75	M	M	HDBSCAN [8]	79	77	M	M
Cludoop [3]	74	71	H	H	RCC [8]	76	74	M	H
Block DBSCAN [4]	79	76	H	VH	kMedoids+ [8]	74	71	M	M
p-Approx DBSCAN [4]	71	70	M	M	RD2PC [9]	85	83	M	H
CDBH [5]	83	80	M	H	Cut off [9]	79	76	M	M
SMOTE [5]	74	73	M	H	Gaussian [9]	80	77	M	L
ADASYN [5]	77	79	M	M	NDEC [10]	86	83	M	M
sTL [5]	81	80	H	M	k Means [10]	75	73	L	L
sSafe [5]	80	79	H	H	DBSCAN [10]	74	72	M	M
sRST [5]	79	76	H	M	OPTICS [10]	70	67	M	M
EUSCHC [5]	74	75	H	H					

From this analysis, various observations about the performance can be evaluated. From the precision performance in Table 77.1, it is observed that VAE [7], NG [27], DBN [7], CAE [7], HA [26], and MOPSO [35] outperform other models. Similarly, the recall of VAE [7], NG [27], DBN [7], and NSDBSCAN [38] outperform other models. Thus, they must be used for real time dynamic clustering & rule mining applications. From the delay performance, it is observed that SS DBSCAN [25], DPARM [31], Fast DBSCAN [24], KD [28], and FastDP [8] have faster performance than other models. Similarly, computational complexity of these models is evaluated in Table 77.1, wherein it is observed that DBSCAN [1], FastDP [8], FDPC [12], Gaussian [9], IC [37], MOGA [34], NNG ADBSCAN [2], OPTICS [3], PLSA [17], and PWCLU [15] have the lowest complexity of deployment. Based on this statistical analysis, researchers and system designers can select the most applicable algorithm(s) for their given application, which will assist them in high-speed system deployment.

4. Conclusion and Future work

It is observed that most of the dynamic clustering models are variations of the standard DBSCAN model. Some of these models use deep learning algorithms, while some use rule mining approaches for improving efficiency of dynamic cluster creation. It is observed that some models outperform other models in terms of precision; on the other hand, some have better recall performance and lower complexity. In future, it is recommended that deep learning methods must be explored, and pre-trained models must be deployed along with incremental & transfer learning to improve performance of dynamic clustering & rule mining for multiple scenarios.

References

1. Min, Wei & Liang, Weiming & Yin, Hang & Wang, Zhurong & Li, Mei & Lal, Alok. (2021). Explainable Deep Behavioral Sequence Clustering for Transaction Fraud Detection.
2. Li, Hao & Liu, Xiaojie & Li, Tao & Gan, Rundong. (2020). A Novel Density-Based Clustering Algorithm Using Nearest Neighbor Graph. Pattern Recognition. 102. 107206. 10.1016/j.patcog.2020.107206.
3. Bhattacharjee, P., Mitra, P. A survey of density based clustering algorithms. *Front. Comput. Sci.* **15,** 151308 (2021). https://doi.org/10.1007/s11704-019-9059-3
4. Chen, Yewang & Zhou, Lida & Bouguila, Nizar & Wang, Cheng & Chen, Yi & Du, Jixiang. (2020). BLOCK-DBSCAN: Fast clustering for large scale data. Pattern Recognition. 109. 107624. 10.1016/j.patcog.2020.107624.
5. Mirzaei, Behzad & Nikpour, Bahareh & Nezamabadi-pour, Hossein. (2021). CDBH: A clustering and density-based hybrid approach for imbalanced data classification. Expert Systems with Applications. 164. 114035. 10.1016/j.eswa.2020.114035.
6. Clustering mixed numerical and categorical data with missing values, https://www.sciencedirect.com/science/article/pii/S0020025521004114
7. Md Rezaul Karim, Oya Beyan, Achille Zappa, Ivan G Costa, Dietrich Rebholz-Schuhmann, Michael Cochez, Stefan Decker, Deep learning-based clustering approaches for bioinformatics, *Briefings in*

Bioinformatics, Volume 22, Issue 1, January 2021, Pages 393–415, https://doi.org/10.1093/bib/bbz170

8. DenMune: Density peak based clustering using mutual nearest neighbors, https://www.sciencedirect.com/science/article/abs/pii/S0031320320303927

9. Chunzhong Li, Yunong Zhang, "Density Peak Clustering Based on Relative Density Optimization", *Mathematical Problems in Engineering*, vol. 2020, Article ID 2816102, 8 pages, 2020. https://doi.org/10.1155/2020/2816102

10. T. Kamali and D. W. Stashuk, "Discovering Density-Based Clustering Structures Using Neighborhood Distance Entropy Consistency," in IEEE Transactions on Computational Social Systems, vol. 7, no. 4, pp. 1069-1080, Aug. 2020, doi: 10.1109/TCSS.2020.3003538.

11. Exploring urban travel patterns using density-based clustering with multi-attributes from large-scaled vehicle trajectories, https://www.sciencedirect.com/science/article/abs/pii/S0378437120306865

12. Fast density peak clustering for large scale data based on kNN, https://www.sciencedirect.com/science/article/abs/pii/S0950705119302990?dgcid=rss_sd_all

13. Liu, Xiangyu & Niu, Xinzheng & Fournier Viger, Philippe. (2021). Fast Top-K association rule mining using rule generation property pruning. Applied Intelligence. 51. 1-17. 10.1007/s10489-020-01994-9.

14. Choi, C.; Hong, S.-Y. MDST-DBSCAN: A Density-Based Clustering Method for Multidimensional Spatiotemporal Data. *ISPRS Int. J. Geo-Inf.* **2021**, *10*, 391. https://doi.org/10.3390/ijgi10060391

15. Sharma, Krishna & Seal, Ayan. (2021). Outlier-robust multi-view clustering for uncertain data. Knowledge-Based Systems. 211. 106567. 10.1016/j.knosys.2020.106567.

16. M. D. Nguyen and W. Shin, "An Improved Density-Based Approach to Spatio-Textual Clustering on Social Media," in IEEE Access, vol. 7, pp. 27217-27230, 2019, doi: 10.1109/ACCESS.2019.2896934.

17. Djenouri, Y., Belhadi, A., Djenouri, D. *et al.* Cluster-based information retrieval using pattern mining. *Appl Intell* **51,** 1888–1903 (2021). https://doi.org/10.1007/s10489-020-01922-x

18. Li, X., Han, Q. & Qiu, B. A clustering algorithm with affine space-based boundary detection. *Appl Intell* **48,** 432–444 (2018). https://doi.org/10.1007/s10489-017-0979-z

19. Min Deng, Xuexi Yang, Yan Shi, Jianya Gong, Yang Liu & Huimin Liu (2019) A density-based approach for detecting network-constrained clusters in spatial point events, International Journal of Geographical Information Science, 33:3, 466-488, DOI: 10.1080/13658816.2018.1541177

20. Heidari, S., Alborzi, M., Radfar, R. *et al.* Big data clustering with varied density based on MapReduce. *J Big Data* **6,** 77 (2019). https://doi.org/10.1186/s40537-019-0236-x

21. Corizzo, Roberto & Pio, Gianvito & Ceci, Michelangelo & Malerba, Donato. (2019). DENCAST: distributed density-based clustering for multi-target regression. Journal of Big Data. 6. 10.1186/s40537-019-0207-2.

22. P. Lin, K. Ye, M. Chen and C. Xu, "DCSA: Using Density-Based Clustering and Sequential Association Analysis to Predict Alarms in Telecommunication Networks," 2019 IEEE 25th International Conference on Parallel and Distributed Systems (ICPADS), 2019, pp. 1-8, doi: 10.1109/ICPADS47876.2019.00010.

23. KR-DBSCAN: A density-based clustering algorithm based on reverse nearest neighbor and influence space, https://www.sciencedirect.com/science/article/abs/pii/S0957417421011374#!

24. Nanda, Satyasai & Panda, Ganapati. (2015). Design of computationally efficient density-based clustering algorithms. Data & Knowledge Engineering. 95. 23-38. 10.1016/j.datak.2014.11.004.

25. Castro Gertrudes, J., Zimek, A., Sander, J. *et al.* A unified view of density-based methods for semi-supervised clustering and classification. *Data Min Knowl Disc* **33,** 1894–1952 (2019). https://doi.org/10.1007/s10618-019-00651-1

26. Hui Liu, Yang Liu, Ran Zhang, Xia Wu, "A Clustering Algorithm via Density Perception and Hierarchical Aggregation Based on Urban Multimodal Big Data for Identifying and Analyzing Categories of Poverty-Stricken Households in China", *Scientific Programming*, vol. 2021, Article ID 6692975, 13 pages, 2021. https://doi.org/10.1155/2021/6692975

27. Sonia Setia, Verma Jyoti, Neelam Duhan, "HPM: A Hybrid Model for User's Behavior Prediction Based on *N*-Gram Parsing and Access Logs", *Scientific Programming*, vol. 2020, Article ID 8897244, 18 pages, 2020. https://doi.org/10.1155/2020/8897244

28. Chen Jungan, Chen Jinyin, Yang Dongyong, Li Jun, "A -Deviation Density Based Clustering Algorithm", *Mathematical Problems in Engineering*, vol. 2018, Article ID 3742048, 16 pages, 2018. https://doi.org/10.1155/2018/3742048

29. Malzer, Claudia & Baum, Marcus. (2019). A Hybrid Approach To Hierarchical Density-based Cluster Selection.

30. C. Chen, H. Chou, T. Hong and Y. Nojima, "Cluster-Based Membership Function Acquisition Approaches for Mining Fuzzy Temporal Association Rules," in IEEE Access, vol. 8, pp. 123996-124006, 2020, doi: 10.1109/ACCESS.2020.3004095.

Intelligent Systems and Smart Infrastructure – Brijesh Mishra et al. (eds)
© 2023 Taylor & Francis Group, London, ISBN 978-1-032-41287-0

CHAPTER

78

Design and Implementation of a Smart Taser for Women Safety

Aditi Rawat[1], Vaibhav Chaudhary[2], Vijay Kumar Tayal[3]

Amity University Uttar Pradesh, Noida, India

Abstract

Women have long been discriminated against and excluded from political and familial matters in many modern countries. Despite the enormous amount of work being performed by women on a daily basis to support their families, their voices are rarely heard and their rights are severely restricted. It is a challenging task to achieve a stand-alone taser unit as it consumes considerable power and if made powerful, its size also increases proportionally. This paper proposes a smart device that tracks the location of the victim and sends that information to concerned people. Further, a taser circuit has been implemented to protect women in such alarming situations.

Keywords: Women security, microcontroller, GPS, GSM, IPC, simulation

1. Introduction

From the time she is born, or possibly before she is born, a girl might be a victim or a target of crime. Crimes vary in nature, just as the phases do. The illustrations and tables that follow emphasize important facets of the topic. In 2012, National Crime Records Bureau published a study, according to which, the country recorded a crime rate of 46 per 100,000. Rape accounted for 2 per 100,000; dowry homicide for 0.7 per 100,000; and domestic abuse via spouses or in-laws for 5.9 per 100,000. "While the country's [India's] 85 percent prevalence of sexual assault is among the lowest in the world, it is predicted to harm 275,000 women in India [given India's

Corresponding author: [1]aditi.eddy2000@gmail.com, [2]ch.vbhv.eee@gmail.com, [3]vktayal@amity.edu

DOI: 10.1201/9781003357346-78

big population]," according to a 2014 report published in the *Lahore Journal of Public Health*. [1]

Table 78.1 Crimes Faced by Women at Different Stages of Life. [1]

Stage	Group	Description
STAGE 1	Infant	Female foeticide results due to pregnancy diagnosis techniques due to economic or cultural desire for boy as a child
STAGE 2	School going	A lot of girls are denied entry to and completion of adequate lower and upper education, and may endure prejudice from their parents and instructors, as compared to boys
STAGE 3	Adolescence	Sexual violence, both internal and external, psychological abuse, acid assaults, rape, forced marriages, and even HIV/AIDS affect many young females
STAGE 4	Wedding	After marriage, countless women face abuse emotionally, economically and psychologically by their spouses and in-laws
STAGE 5	Motherhood	During and after pregnancy, women are not constantly provided with adequate medical care and nutritious food. She is frequently forced to abort a female foetus
STAGE 6	Workplace	Trafficking, uneven remuneration for equal labour, rejection of criterion advancements, and physical, financial, and emotional mishandling are all common occurrences for women

Throughout these levels, the woman suffers silently, or if she expresses her voice, it is muted or stifled, and she is unaware of her legal rights to oppose these crimes and the legal recourse available to defend herself. It is past time for women to take action to defend themselves. Any act of cruelty or criminality should be reported. The Economist Intelligence Unit rated India 53rd out of 58 countries in gender gap scrutiny done for the first time. "The Women's Empowerment: Assessing the Global Gender Gap", was released in 2005. As per this study, examination of discrepancies between women and men in the subsequent key areas is done:

- Economic opportunity
- Economic engagement
- Access to education
- Access to birth control
- Social inclusion

The root for the study was the results of the United Nations Development Initiative on global trends of gender inequalities for women. The poor score indicates a significant gender gap in all five categories of the measure. [1]

2. Objective of Proposed Work

The persistence of this project is to develop a cost-effective Smart Taser that can be used by the general public (especially, working women) so as to save them from compromising circumstances where they force get victimized by the pathetic, cruel, creepy, and corrupt citizens of our nation. In case of crisis, first, the device can be activated in advanced so that in case of mishap, the current location can be sent to trusted personnel on a timely basis. Other than this, if there is a condition where the subject is caught by the malpractitioner, then the Taser or Stun gun can buy the subject time by stunning the bad guy with high-voltage electric shock and, meanwhile, since the taser is used, this infers that the subject is in serious threat and hence SOS alerts are floated to the registered numbers and can be updated on a server which can be accessed by nearby patrolling vehicle and women safety cell. Thus, immediate help can be sent for the victim for rescuing them from the criminals.

3. Literature Survey

A. Literature Review

In India, 93 women have been raped every day in 2014, as per the National Crime Records Bureau (NCRB). In 2014 alone, 3,37,922 incidences of violence towards women were registered [2]. In the year 2020, despite nationwide lockdown and strict regulations, 23,722 complaints were received by National Commission for Women (NCW) with all sorts of evils that one can think of and also the ones that perhaps not be heard. These are just the registered ones and there will be a lot more such cases that are never filed. Some top charters being Harassment of married women/Dowry harassment, Domestic Violence, Rape/Attempt to rape, Cyber-crimes, Sexual assault, Harassment even at workplace, and the list goes on. [3] In this article [2], author want to provide safekeeping, via an application, which is given with adequate data like humanoid behaviour, previous case learning, etc. With GPS services enabled, app could get the location access of women and look for the health condition of women, and as per these data, relief could be sent for her. Hence, technical knowledge is used to mitigate this issue. Proposed work of this paper [4] shall help to protect women from attackers. Various devices like GPS, GSM, and panic button have been implemented in this project. The proposed model is that of a band that women can carry with them so that they can stay out for work late without any hesitation. Sending information of threats via notifications to victim's relatives and bordering police station is the basis for providing security to a woman. Quite similar work is proposed in paper [5]; the model is used to locate victim as per GPS technology. In this way, the signals that have been created are sent to the board, signal processing is done, and SMS is sent to close ones, so emergency calls can be shared with the location of the coordinates to save women from harassment. In this paper [6], the author gives a brief about the condition of women in this world. They propose a smart device, which is comfortable to carry around. Further, they stated that it would be simple to apply in many regions for women's protection and monitoring. This study [7] is about ensuring women's safety when creating smart tools. This gadget aids in the detection of women's crucial situations. In treacherous situations, this will act as shielding hand by using microcontroller-interfaced GSM and GPS. Whenever a lady senses uneasy in any situation, she can push a remote key to get a position using GPS and GSM. This structure aids in the management of the risky circumstances in which women find themselves. This document also contributes to the design's ongoing development by offering fundamental analysis and technical facts like interfacing a camera with the kit too so that the assaulter can be captured later on. This paper [8] uses 8051 micro-controllers for their project and were successfully able to send the location data to a registered mobile number which they have shown in the paper itself. The "Suraksha" device is a security-system planned specifically for women in distress. It's a lightweight, easy-to-carry device, with a lot of possibilities and features. The basic strategy is to help victims by empowering them so that they can provide their immediate location along with a distress note to the cops and their relatives' phone number in order to prevent unfortunate incidents and provide real-period evidence for swift action over perpetrators of crimes against women. Currently, work is being done to make it as compact as possible so that it can be embedded in jewellery, mobile phones, and other items, making it a more versatile instrument for the general public. It has the potential to play a significant role in upcoming projects such as the CCTNS, which will digitise all police records across India and integrate all police stations. Further they did literature review which showed various software-based apps like Jivi 2010 or VithU, etc. that help in achieving this objecting; however, the authors think relying only on smartphone is not a good option as, at times, batteries may discharge or the assaulter first attacks and snatches the phone to make the victim vulnerable. In [9], the authors used 8051 microcontroller, ADC converters, clock generators, three-axis accelerometer (for fall sensing), voice recorder, body temperature sensor, and piezo-electric pressure sensor as when someone panics, body temperature upsurges, and the fight or flight response is activated that increases the heart rate, and strength of grip due to panicking may vary, thus helping more precisely to detect the emergency\critical situation. The authors [10] made smart wearable prototype that can be used for keeping the children safe when they are external and, in case of any mishap, the device keeps the parents updated about their child's whereabouts; it also have a camera to keep an eye on the surroundings; it sends updates on the app Telegram directly from the Raspberry pi, and IOT capabilities are added in the device too. This provides a basic understanding how to proceed with the proposed work. A very detailed analysis is done in this paper [11] by the author where all possible existing women safety devices are compared under various divisions. Even various new ideas are proposed for women protection. In this paper [12], a survey is conducted and the critical situation of Maharashtra and India with shocking figures have been put forth. Reference has been taken from the NCRB – National Crime Records Bureau's records; according to them, Maharashtra held third position with 2910 cases reported by women facing some sort of harassment or discrimination. The authors even stressed on the lack of timely action taken, which results in the non-identification of the victims. The authors have proposed a system but have not implemented it. Only a simple flowchart is shown in the paper with Raspberry-pi Zero W controller which is connected to a mobile-phone via wifi and bluetooth and then sends data to the phone. Neither simulation nor working model is made. In [13], Seftipin is an android application that utilises a safety score to regulate which locations are harmless and which are not. However, it is occasionally unable to provide information about the location, or the app is unavailable. Furthermore, such an app would take a lot of data to make such a forecast; therefore, this notion isn't really useful. SMARISA is a multifunction ring for ladies based on the Raspberry Pi; the IoT device functionality included in the ring is described in [14]. To activate the services, the author utilised a Raspberry Pi camera, buzzer, and button. The model roles as both a safety and an alerting device. The controller is mostly based on an Android app, with various hardware components interfacing with it. An AVR microcontroller-based wearable jacket for women's protection is presented in [15]. The authors propose a wearable gadget that may be hidden within a jacket. A swapping unit, GPS-Module, GSM, LED, Buzzer, and Module are all included. A buzzer can be used to sound an alarm with a loud volume. This study [16] proposes the Women Safety Device and Application-FEMME concept. Both application and physical units are included in the model. The GSM broadcasts the victim's location to authorities when the device's emergency button is touched once. The system is designed capturing audio and transmitting the victim's location when the alarm button is double-clicked. GSM transmits a message and makes a call to the police if any of the emergency knob is held down for a long time. The victim's position is tracked using a GPS gadget. The audio from the event is recorded using an audio recorder. A hidden camera detector that uses the RF signal as well as a hidden camera detector may identify concealed cameras in rooms. The authors [17] did a nice literature review and even expressed their views on the proposed works of the papers that they reviewed. A very compelling point is made by the authors that majority of the population in India reside in villages. Hence, the main issue is also the literacy of the users, and availability of electricity and mobile phones. They proposed a device with colour coding in spite of the instructions written there; this helps in memorizing easily the working of each button. Further, they had not addressed the issue of lack of smart devices and network anomalies. The author of this [18] paper has created an IoT grounded device called "Anti-Molestation" to ensure women and children's safety in everyday life. The device can send SMS to the victim's friends and family with his or her current location. The device can also make an on-demand call to "999" (Bangladesh's emergency number). They designed the device in such a way that it is affordable to women and people of all income levels. Furthermore, they designed the architype model to be far too small, but it can be formed into a smaller device to be cast-off in a locket, handbag, or pocket, or as a bracelet, among other things. The author's device makes the following contributions:

- It has the ability to send SMS with the provided current location.
- By pressing a single button, it can make a phone call.
- Cost-effective and energy-efficient device.
- Ability to overcome the problem of burst transmission.

The authors [19-20] propose the women safety device with a panic button. When pressed, the button alerts contacts such as a selected family member or friend, as well as the police. Being compact, it could be attached to the clothes like in jacket, suit, etc. The device will be able to send the location to the contacts as it is equipped with a GPS system. The device is equipped with a camera that captures images and videos in response to the user's commands, and these images are sent to the user's contacts and stored locally on SD card as well. Triggering signal will be from a button press, which will activate the GSM module, which would send the GPS-tracked location to the phone numbers specified. The device needs to be triggered manually by the victim, which is the area of concern. The captured images could be helpful in catching the wrongdoers as well. The authors even made a prototype of the same; however, it was not compact as claimed by the authors. However, they were successful in demonstrating the successful working of the prototype by sharing the images captured by the device and the location being shared over SMS service.

B. Findings of Literature Review

- The authors wanted to provide confidentiality, via an application, which is given with accurate data like human behaviour, previous case learning, etc., thus, providing Machine Learning capabilities.

- A further type that has been described includes multiple gadgets such as GPS, GSM, and a danger toggle.

- One gave a concept of a band that women can wear and carry with them so that they can stay out for work late without any hesitation. In case of emergency, it can sense the temperature and pulse of the subject, and then raise SOS alerts.

- Mostly, the signals that have been created are sent to the board, signal processing is done, and SMS is sent to the close ones, so emergency calls can be communal with the location of the coordinates to save women from harassment.

- These papers propose designs for apps and devices for women in emergency and in distress.

- An intelligent gadget is suggested that is convenient to transport. It would also be simple to deploy in many regions for women's security and monitoring.

- This gadget aids in the detection of women's clutch situations. It's really about ensuring women's safety when building smart devices.

- In dangerous situations, this will turn as defensive hand by using GPS 9 and GSM module 9 with Arduino device. Previously, either the shock mechanism has not been implemented, or if it did, then the means used for shock production were not viable.

- If a lady feels unsafe in any scenario, she may push a key that gives her the position over GPS and GSM.

- Literature review gave us a lot more ideas like using temperature sensor, piezoelectric pressure sensor, voice recorder, camera, and many other scopes that can be achieved in this project.

4. Component Description

A. Arduino Uno

Arduino Uno is a microcontroller board which was designed by Arduino.cc and it is Atmega328p microprocessor based open-source microcontroller. Various external boards (shields) and various other circuits could be attached to this particular board as it contains 14 digital and 6 analog I/O pins. Six of the digital pins are powered with the PWM output generation. For programming the controller, Arduino IDE–IDE is needed. The controller is connected to the pc via type B USB cable. To power up the controller, either it requires a USB cable with voltages oscillating from 7 to 20 Volts. Arduino website contains all the hardware design reference for the microcontroller panel and it has 2.5 licence. Even some of the hardware files are accessible for general public as well. "Uno" in Italian means "One" that represents the first release of Arduino Software. This board is the first of the USB board series. Altogether with 1.0 of Arduino IDE version which serves as the reference version, which is updated by

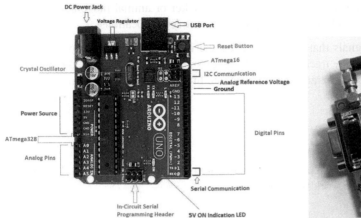

Fig. 78.1 Arduino Uno   **Fig. 78.2** GSM Module

later releases. The bootloader on the ATmega328 on the board allows it to be programmed. It includes everything users will probably need to get started with the microcontroller. Connect it to a computer through USB or use an AC-to-DC converter or battery to power it. Anyone can play about with the Uno without worrying about making a mistake. If something goes wrong, users can swap the chip for a rare buck and start over. The Arduino IDE is an unrestricted download from the Arduino Official Spot.

B. GSM Modem

A GSM modem is a particular form of modem that accepts a SIM card and functions on a mobile operator's subscription, similar to a mobile phone. It appears to a mobile operator to be identical to a mobile phone. Using this connected device, a computer could communicate via mobile network. While the majority of these GSM modems provide mobile web connectivity, several of them can also guide and receive SMS and MMS messages. A GSM modem could be either a discrete modem including a uart, USB, or bluetooth connection or a portable device with GSM network capability. In particular for this dissertation, the phrase GSM Modem is directing to any modem which is capable enough for supporting GSM's evolutionary family protocol, e.g., 2.5G technology, namely, GPRS, EDGE and 3G technology; WCDMA, UMTS, HSDPA, and HSUPA. A GSM modem presents an interface via which programs like NowSMS

are able to send and retrieve information or data through messages. The operator charges this service as charged on a smart phone as the operator is no't able to distinguish between the two.

C. GPS Module

Global Positioning System, GPS, tracking unit or tracker is a navigation device, which uses various technologies to get the location of the user, i.e., co-ordinates of the position (longitude & latitude). It

Fig. 78.3 GPS Module

can be used as vehicle tracker, or asset tracker, person tracker or animal movement tracker. This device through WGS84 UTM geographic position identify the locations.

Through unique satellite signals that are decoded by the receivers, location is pointed ovn the globe. Locations are held by tracking unit. By using cellular network or web-connected device, radio transmission, or data transmission over satellite, or wifi can be sent to concerned authorities.

5. Software Simulations

For simulating the circuits, Proteus 8 professional software is used. First, the models have been created and then the simulation work was done.

A. Taser Circuit

Authors simulated 2 different circuits, both of which are giving desired results as far as simulation output is concerned.

(a) Using 555 timer IC and transistor

(b) Using passive devices only.

Both of these circuits use voltage multiplier circuits to increase the voltage so as to attain such a higher value of voltage that will break down the air itself, hence producing strong arcs, which is the basic requirement of a Taser.

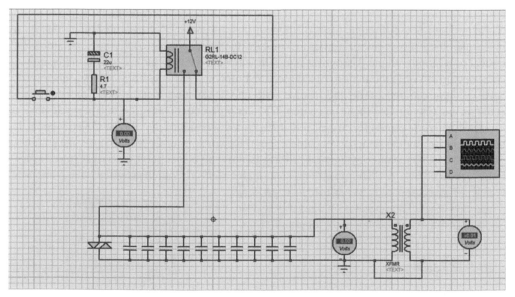

Fig. 78.4 Proteus Model for Taser Circuit using Passive Elements only

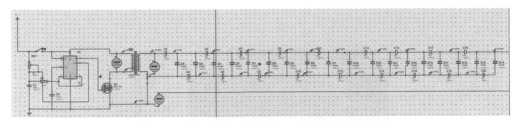

Fig. 78.5 Taser Circuit using 555 Timer and Transistor

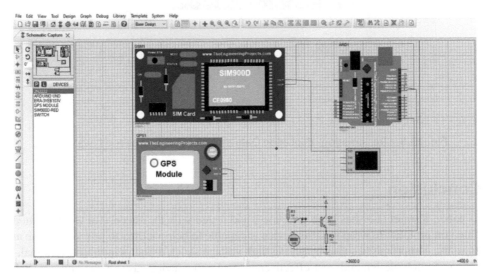

Fig. 78.6 GPS, GSM Module and a Switch Interfaced with Arduino

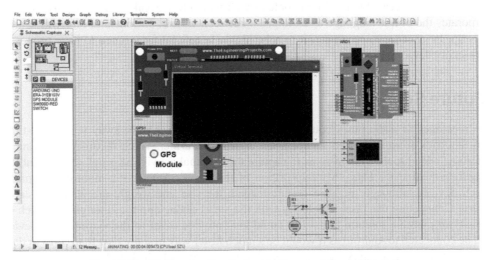

Fig. 78.7 Waiting for the Output in Terminal Box of Proteus

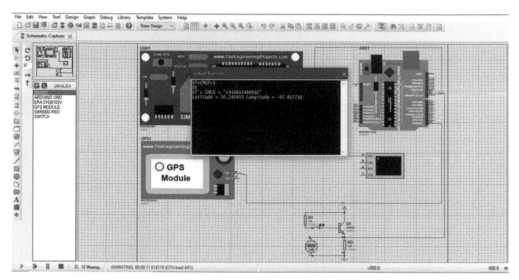

Fig. 78.8 Transmission of location Data through SMS to Desired Location

6. Conclusion

In this paper, a smart taser has been presented, and tested with the successful working model simulated on proteus. To wrap up, this device is capable to help victims to buy some time and flee from the compromised space by giving high voltage shock to the perpetrator, criminal, or wrongdoer. Simple design and cost-effectiveness make the device easy to use and accessible to all. Illiterate population can also understand the use of smart taser due to its simplicity. The taser may be accessible to blind people as well with a few tweaks only. The basic model incorporates the braille language to distinguish between various buttons and directing which side is up. In insecure environment, it may be highly beneficial for women to have proposed handy safety device.

7. Future Scope

Voice messages or audio notes could be uploaded on the server right away to get real-time audio feedback from the crime scene, when activated. Further, voice recorder and camera can also be added to the system so as to capture the criminal activity as proof so as to put the wrongdoers behind the bars. 'Miniaturization of the Taser': This is the main disadvantage of such devices; although they are feature-loaded, the size and weight are also an area of concern. Connecting the Taser wirelessly with the smart phone app too, though not much of a remedy or possible solution but still it could help provide relief to the victim if the device could be connected to Android as it would be a much faster way to share data than SMS or data updation on server through Arduino as it has limitations of itself. IOT and Machine Learning capabilities would help collecting data from such crime scenes and then further innovation could be fueled through the collected data. A new system can be added in the Taser, so that it can send distress alerts to all the smartphones within a specific radius and if someone wants to help the victim, they can do so by following the directions given by the smartphone that leads to the crime scene. However, this would require a much more superior controller, algorithms, and a far better sympathetic of the Android and IOS interfaces. It has been observed that such acts usually take place where least possible human intervention is there, or in lonely places. Hence, some cameras with IOT-based SOS detector and broadcaster could be used in such stranded places. And these devices could send data to the concerned authorities to take action immediately.

REFERENCES

1. Available on: http://www.Womenlawsindia.com/legal-awareness/crimes-against-women/. Retrieved on 14/11/2021.
2. Simon L. Cotton and William G. Scanlon (2009). Millimeter - wave soldier–to-soldier communications for covert battlefield operation. IEEE communication Magazine, 47(10): 72–81.
3. Available on: http://ncwapps.nic.in/frmReportNature.aspx?Year=2020. Retrieved on 27/08/2021.
4. Vamil B. Sangoi (2014). Smart security solutions. International Journal of Current Engineering and Technology, Vol. 4, No. 5.
5. B. Chougula (2014). Smart girls security system. International Journal of Application or Innovation in Engineering & Management, Vol. 3, No. 4.
6. Hock Beng Lim (2010). A soldier health monitoring system for military applications. International Conference on Body Sensor Networks, BSN 2010, 7–9.
7. Premkumar.P, Cibi Chakkaravarthi. R, Keerthana.M, Ravivarma. R, Sharmila (2015). One touch alarm system for women's safety using gsm. International Journal of Science Technology & Management. Volume No 04, Issue No. 01, ISSN (online): 2394–1537.
8. Nishant Bharadwaj and Nitish Aggarwal (2014). SURAKSHA, a device to help women in distress: an initiative by a student of ITM University, Gurgaon. International Journal of Information & Computation Technology, ISSN 0974-2239 Volume 4, No. 8, pp. 787–792.
9. B. Vijaylashmi, Renuka. S, Pooja Chennur and Sharangowda Patil (2015). Self defense system for women safety with location tracking and sms alerting through gsm network. International Journal Research in Engineering And Technology (IJARTET), Volume: 04 Special Issue: 05, eISSN: 2319-1163 I pISSN: 2321-7308.
10. Elsyea Adia Tunggadewi, Eva Inaiyah Agustin and Riky Tri Yunardi (2021). A smart wearable device based on internet of things for the safety of children in online transportation", Indonesian Journal of Electrical Engineering and Computer Science 22(2): 708, 2021.
11. Aradkar, A., and Sharma, D. (2015). All in one intelligent safety system for women security. International Journal of Computer Applications, Vol. 130, No. 11.
12. Mona Chaware, Dipali Itankar, Diksha Dharale, Divya Borkar, Shraveen kumar Pendyala and Mrudula Nimbarte (2020). Smart safety gadgets for women: a survey. Journal of University of Shanghai for Science and Technology, Vol. 22, No. 12.
13. Ravi Sekhar Yarrabothu and Bramarambika Thota (2015). ABHAYA: an android app for the safety of women. Annual IEEE India Conference (INDICON) ISSN: 2325–9418.

14. Navya R. Sogi, Priya Chatterjee, Nethra U. and Suma V (2018). SMARISA: a raspberry pi based smart ring for women safety using iot. International Conference on Inventive Research in Computing Applications (ICIRCA), IEEE Xplore ISBN:978-1-5386-2457-9.

15. Daniel Clement, Kush Trivedi, Saloni Agarwal and Shikha Singh (2016). AVR Microcontroller Based Wearable Jacket for Women Safety. IRJET, Volume: 03 Issue: 05, e-ISSN: 2395–0056.

16. D. G. Monisha, M. Monisha, G. Pavithra, and R.Subhashini (2016). Women safety device and application FEMME. Indian Journal of Science and Technology, Vol: 9, Issue: 10.

17. Pragna B R, Poojary Praveen Mahabala, Punith N, Sai Pranav and Shankar Ram (2018). Women safety devices and applications. International Journal of Engineering Research & Technology (IJERT), Vol. 7, No. 07.

18. Md. Imtiaz Hanif, Shakil Ahmed, Wahiduzzaman Akanda and Shohag Barman (2020). Anti-molestation: an iot based device for women's self-security system to avoid unlawful activities. International Journal of Advanced Computer Science and Applications (IJACSA), Vol. 11, No. 11.

19. S. Tayal, H. P. Govind Rao, A. Gupta and A. Choudhary (2021). Women safety system design and hardware implementation. 9th International Conference on Reliability, Infocom Technologies and Optimization (Trends and Future Directions) (ICRITO), pp. 1–3.

20. Jismi Thomas, Maneesha K J, Nambissan Shruthi Vijayan and Prof. Divya R. (2018). TOUCH ME NOT - a women safety device. International Research Journal of Engineering and Technology (IRJET), Vol. 05, No. 03.

Intelligent Systems and Smart Infrastructure – Brijesh Mishra et al. (eds)
© 2023 Taylor & Francis Group, London, ISBN 978-1-032-41287-0

CHAPTER

79

Treatment of Hepatocellular Carcinoma Using 5-Slot Microwave Ablation Antenna at 2.45 GHz

**P. Niranjan[1], T. Naveen Kumar Reddy[2],
P. Vijay Kumar Reddy[3], P. Janardhana Reddy[4],
Vivek Singh[5], Ajay Kumar Dwivedi***

Nagarjuna College of Engineering and Technology,
Bengaluru, Karnataka, India

Abstract

Every year, about 7 lakh Indians succumb to cancer, with an additional 10 lakh being newly diagnosed. In most cases, surgical removal of hepatocellular carcinoma is not a viable option. It has been shown that microwave coagulation therapy (MCT) is an effective alternative to resection in tissue and that it is safe. Other thermal ablation treatments have the advantage of being more rapid and producing an ablation region that is instantaneously hypoechoic on real-time ultrasound monitoring, as opposed to laser ablation. An intrusive technique is used in this treatment, which entails introducing a thin microwave coaxial slot antenna (operating at 5GHz) into the tumor in order to coagulate the cancer cells. When it comes to analyzing complex structures, the finite element approach is applied.

Keywords: Microwave Coagulation Therapy (MCT), Cancer Cells, Coaxial Slot Antenna, Thermal Ablation.

*Corresponding author: er.ajaydwivedi@gmail.com
[1]niranjanpothamsetty@gmail.com, [2]naveenchinnu144@gmail.com, [3]vijju2234@gmail.com,
[4]janardhanareddy5597@gmail.com, [5]vivek.10singh@gmail.com

DOI: 10.1201/9781003357346-79

1. Introduction

In Microwave ablation (MWA), interventional procedures include exposure to very high temperatures, which results in tissue necrosis. The treatment employs an antenna probe that is image-guided to the tumor's target area and eliminates it with high-frequency dielectric heating. This therapy is safe when used to treat malignancies of the kidney, liver, bone, lung, and other soft tissue. Additionally, owing to its cost-effectiveness, it has become a commonly used approach in immunology and oncology research [1].

Thermal ablation (MWA) is a minimally invasive treatment that may be done under local anesthesia and may result in same-day discharge. As a result, it is commonly used in the treatment of hepatocellular carcinoma (HCC). Each year, around 7,60,000 new HCC cases are reported globally, with India accounting for more than 50,000 instances; hence, HCC accounts for one in every six cancer patients [2]. Additionally, according to a research performed by ILBS, New Delhi (2020) on the demographic distribution of HCC patients, Uttar Pradesh accounts for 24% of the entire cases, which is a significant proportion [3]. Age, gender, and alcohol use are all prominent risk factors for HCC, and as life expectancy increases globally, the direct association between older patients and HCC is becoming an increasing source of worry [4]. Certain kinds of hepatitis may also result in HCC, and Uttar Pradesh alone accounts for 25,000 cases of viral hepatitis each year according to the study conducted in 2019, which can progress to HCC if left untreated [5]. According to some reports, surgical resection is the primary line of therapy for HCC, with thermal ablation reserved for those who are not surgical candidates. However, as stated in a recent paper [6], the death rate associated with surgical resection has risen in older patients relative to younger patients, and numerous surgical resections may impair a patient's quality of life and are a very invasive and costly procedures. Additionally, significant technological and clinical breakthroughs have been documented in the years since the microwave ablation treatment was introduced. This development comprises a matching antenna applicator, the employment of a state-of-the-art antenna structure, the optimization of the antenna applicator site, the reduction of the elongated ablation pattern, and a larger and quicker ablation zone to compensate for the thermo-regulatory impact.

The purpose of this research is to demonstrate a new minimally intrusive antenna and also to study the influence of different physical and thermal parameters of a biological system on ablation. A three-dimensional finite element model (FEM) is utilized to simulate the ablation environment. It includes thermo-physical heat properties, Penne's bioheat model, and Arrhenius kinetics.

2. Antenna Design

The Finite Element Method is an effective approach for analyzing complicated structures that allows modifying the antenna's configuration. This approach involves expressing a domain using geometrically simple forms from which approximation functions may be generated. Using COMSOL Multiphysics, a 2D finite element model calculates the absorbed power and temperature distribution around a single thin microwave coaxial antenna. In order to test the antenna model for FM analysis, the outer conductor of the narrow coaxial wire was chopped to a 1-mm-wide ring form. The antenna is encased in a PTFE sleeve (catheter) for hygiene reasons. In microwave coagulation treatment, the antenna operates at 2.45 GHz. The antenna shape is 2D modeled and studied with varied slot dimensions of 1 mm, 1.2 mm, 1.5 mm, 1.7 mm, and 1.9 mm from the tip, as shown in Fig. 79.1. The wire is shorted at the tip and a 1.5-mm-wide ring hole is cut 6 mm from the tip. In addition to the slot's axial and radial dimensions, the pin's radial dimensions provided excellent energy interaction between the microwave source and the tissue. The COMSOL Multi-physics user interface includes CAD tools for modeling 2D and 3D coaxial antennas.

3. Thermal Setup

Figure 79.2 depicts a three-dimensional arrangement with FEM boundary conditions. MWA is classified as a kind of dielectric heating. Because of the dielectric heating, polar water molecules in tissue align themselves with electromagnetic fields. The dielectric rotation of the molecules raises the kinetic energy of the molecules and, as a result, the temperature of the tissue. Because tissue is made up of many veins and arteries, a simplified assumption is required to assess the bioheat transfer model. Applying thermal boundary conditions to a 100 micro-m diameter is difficult. The impact of adding thermo-physical factor in the equation of conduction is handled by considering that the tissue area of interest has no independent blood vessel but its effect over a restriced volume is included in penne's bioheat heat model by considering the overall perfusion rate and is given by: [7]

$$\int_v \rho_t c_t \frac{\partial}{\partial t} T_t(r,t) dV = \int_v k_t \nabla^2 T_t(r,t) dV + \int_v \omega_b \rho_b c_b [T_{art}(x,t) - T_t(x,t)] + Q_m \qquad (1)$$

Where the term q_m. refers to the metabolic heat production in the tissue and the term q_h-blood. refers to the contribution of blood flow to the local tissue temperature distribution, respectively. Similarly, ρ is the tissue density [kg/], C is specific heat capacity expressed in [J/kg.K], and k is thermal conductivity expressed in W/m.K.

4. Electromagnetic Setup

Simulated design of the unique antenna was carried out, and the impact of thermoelectrical parameters on tissue models was explored. An electromagnetic-bioheat transfer model was developed that includes a temperature-dependent thermo-electrical tissue property as well as the evaluation of thermally induced tissue injury. Specifically, the following are the Maxwell's equations for a transverse electromagnetic mode, which are determined by the transverse electromagnetic mode [8]

$$E(r) = \left(\frac{A e^{j(\omega t - kz)}}{r} \right) r \qquad (2)$$

$$H(r) = \frac{Ae^{j(wt-kz)}}{r} \cdot Z_{in} \tag{3}$$

where E and H are the electric and magnetic field in r and φ directions, and the propagation direction of electromagnetic waves is represented by z.

5. Modeling Computational Domain

The antenna was numerically modeled and analyzed using finite element approach. It includes breaking up a complicated geometry into tiny pieces for a set of partial differential equations. The microwave source is at the coaxial cable tip. Table 79.1 shows the material volume put in the antenna. Several factors influence the capacity of a tissue to create heat when subjected to a time-varying electric field. These factors include conductivity, relative permittivity, and thermal conductivity.

Table 79.1   Parameters for model

S.No	Parameters	Values	Description
1	Rho_blood	1000 kg/m^3	Density of the blood
2	Cp_blood	3639 J/(kg.k)	Blood specific heat.
3	Omega_	0.0036000 1/s	The rate of blood perfusion
4	T_blood	310.15 K	Temperature of blood
5	Eps_diel	2.03	Relative permittivity
6	Eps_cat	2.6	Catheter permittivity
7	Relative blood permittivity, Catheter	2.45GHz	The frequency of microwave
8	P_in	20 W	Microwave power at the source

6. Results and Discussion

A longitudinal plane is shown in Figs 79.1 and 79.2, where 2D modelling is used to depict the antenna environment and surface power distribution in the liver tissue at the completion of the heating process (the steady state).

Figure 79.3 shows fraction of damage to the tissue with respect to time. The value of 1 indicates 100 percent of the tissue damage. The position of observation point is fixed at the longitutional direction and variable along transversal direction. It can be seen that faster rate of ablation is reached at closer observational point along transversal direction. Figure 79.4 shows temperature distribution till 10 min of time.

Figure 79.5 shows specific absorption rate (SAR) along the arc length of the antenna ablation zone. The arc length is the distance between two point along line drawn from ablation

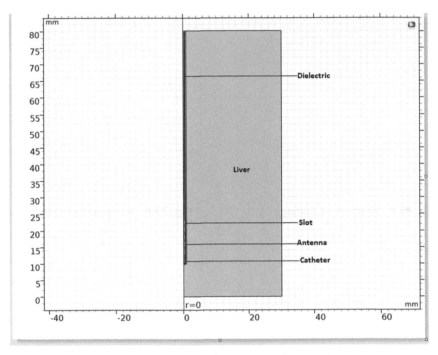

Fig. 79.1 Simulated antenna environment

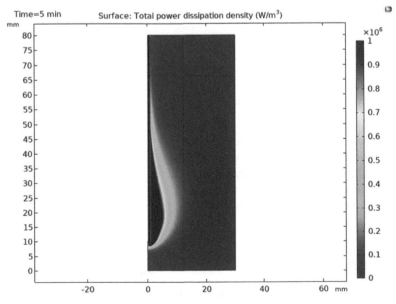

Fig. 79.2 Power dissipation

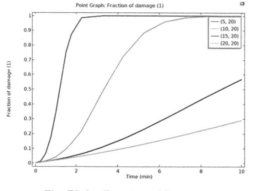

Fig. 79.3   Fraction of Damage

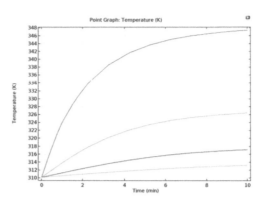

Fig. 79.4   Temperature Distribution

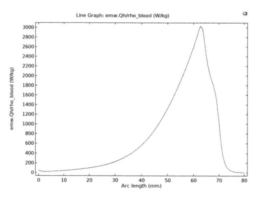

Fig. 79.5   SAR Pattern

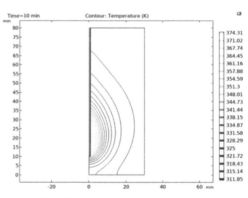

Fig. 79.6   Temperature Contour

boundary along axial direction. Figure 79.6 shows isothermal temperature contour due to dielectric heating.

Figure 79.7 shows ablation zone due to microwave ablation. Since 5 slots were introduced in the antenna, a higher amount of ablation is seen. Slots improve field distribution along the transversal direction of antenna.

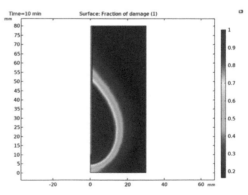

Fig. 79.7   Thermal Damage

7. Conclusion

We analyse an axisymmetric model with a narrow microwave coaxial antenna in COMSOL Multiphysics. The mesh statistics of the model, as well as the SAR pattern and temperature distribution in tissue, are all analysed in detail. The models offer temperature distribution, surface temperature on tissue, and power absorption in tissue with variable electric, thermal, and geometric variations. The effect of temperature and heat distribution on tissue at 40 W input power is studied using a steady state model. In the designed antenna, we introduced a total of 5 slots with variable diameter, which improves the overall ablation zone and more spherical ablation zone.

REFERENCES

1. Siegel L, Miller D, Jemal A, Cancer statistics, 2020. CA Cancer J. Clinics. 2020; 70 (1): 7–30.
2. Jha S, Sharma PK, Malviya R. Hyperthermia: Role and Risk Factor for Cancer Treatment. Achiev. Life Sci. 2016; 10(2): 161–167.
3. Alibakh shikenari M,Virdee B S, Shukla P, et al. Metamaterial-inspired antenna array for application in microwave breast imaging systems for tumor detection. IEEE Access. 2020; 8: 174667–174678.
4. Simon CJ,Dupuy DE, Mayo Smith WW. Microwave ablation: Principles and applications. Radiographics. 2005; 25(SPEC. ISS.): 69–84
5. Imajo K, Ogawa Y, Yoneda M, Saito S, Nakajima A. A review of conventional and newer generation microwave ablation systems for hepatocellular carcinoma. J. Med. Ultrason. 2020; 47(2): 265–277.
6. Poulou LS, Botsa E, Thanou I, Ziakas PD, Thanos L.Percutaneous microwave ablation vs ablation radiofrequency ablation in the treatment of hepatocellular carcinoma. World J. Hepatol.2015; 7
7. Glassberg MB, Ghosh S, Clymer JW, et al. Microwave ablation compared with radiofrequency ablation for treatment of hepatocellular carcinoma and liver metastases: A systematic review and metaanalysis. Onco. Targets. Ther. 2019; 12: 6407–6438.
8. Luyen H, Gao F, Hagness SC, Behdad N. Microwave ablation at 10.0 GHz achieves comparable ablation zones to 1.9 GHz in Ex vivo bovine liver. IEEE Trans. Biomed. Eng. 2014; 61(6): 1702–1710.

Intelligent Systems and Smart Infrastructure – Brijesh Mishra et al. (eds)
© 2023 Taylor & Francis Group, London, ISBN 978-1-032-41287-0

CHAPTER

80

Coordinated Control and Energy Management Strategy of Battery and Super-capacitor of Micro-grid

Vineet Kumar[1], Haroon Ashfaq[2], Rajveer Singh[3]

Department of Electrical Engineering
Jamia Millia Islamia, New Delhi, India

Abstract

Renewable energy is unreliable and intermittent, causing substantial grid issues. An energy storage micro-grid may effectively decrease its volatility. This study suggests an energy management strategy for micro-grids using lithium batteries and supercapacitors. A photovoltaic system may protect the battery from overcharging in two ways: A virtual resistor and virtual capacitance are employed in hybrid energy storage devices to share power. However, the battery offers steady power changes amid rapid changes in load and PV output, resulting in lower transitional strain and extended battery life. By contrast, a battery delivers constant power, whereas a supercapacitor only supports variable power. A virtual impedance control solution is offered to improve power-sharing. Adding a high-pass filter to the voltage control loop restores the super-capacitors terminal voltage to its original value. How the hybrid energy storage system works depends on the state of renewable energy storage devices. This stabilizes renewable energy output and power exchange between the microgrid and the main grid. A longer cycle life for hybrid energy storage systems will minimize operating costs for renewable energy (or micro-grids).

Keywords: Battery, Super-capacitor, Micro-grid.

Corresponding author: [1]Vineetkumar31july@gmail.com, [2]hashfaq@jmi.ac.in, [3]rsingh@jmi.ac.in

DOI: 10.1201/9781003357346-80

1. Introduction

Concerns for the future of the current condition of renewable energy storage have an impact on the hybrid energy system's performance. An important technology for integration is the microgrid. Loads from distributed energy resources and energy storage technologies both are grid-connected and independent microgrids. Island mode has a different set of difficulties than grid-connected mode, namely how to link all dispersed devices while keeping the frequency and voltage constant. A multi-terminal control system solution is needed to maintain the island microgrid's stability.

Because of its clean, endless properties, solar energy has become the most popular kind of renewable energy, and PV power is commonly employed in microgrids.

Because PV power is intermittent, power balancing and voltage/frequency consistency in an isolated microgrid system needs the use of an energy storage system (ESS). As a result, the PV/storage micro-grid has gotten a lot of interest from academics, and power management solutions for microgrids with PV and batteries have been built on the islands.

Whenever the system is exposed to fast load fluctuations or source power, the battery merely maintains the instant power variations. Various ways of coordinated regulation between PV and batteries are used for power variations. Slower power changes may be absorbed utilizing a battery energy storage device due to its power characteristic limitation (BESS). Rapid power fluctuations cause transitory stress in the battery, reducing its lifespan.

To solve the aforementioned problem, a combination of several energy storage technologies is necessary. A battery and a management system are used for storing and processing the energy of the supercapacitor. To satisfy microgrid systems' high performance and energy density requirements, this combo type is often used. The battery as well as the SC's power-sharing is a serious issue. A filter is used to separate the maximum and minimum frequency elements of the energy (LPF). Moving average filtering separates the average versus fluctuation output of HESS; however, the LPF has a phase lag, which affects system stability.

In parallel-connected converters, virtual impedance has been employed to correctly disperse the output current. This method may be used on either an AC or DC microgrid (L. Di et al., 2020). Because of its communication features, decentralized virtual impedance control is much more desirable than centralized control.

When it comes to allocating load power, virtual impedance management is only useful in stable states. It has been claimed that increased droop control approaches, including such virtual output impedances, may result in decentralized power-sharing among battery and SC with different dynamic responses.

This study presents energy management as well as coordinated control approach for an isolated AC microgrid utilizing HESS. MPPT and LPT modes are used in PV systems to prevent the battery from overcharging, depending on the power connections between the supply (PV, battery) and the load demand.

2. Configuration of PV/HESS Micro-grid

All of the components of an information AC microgrid are included in a typical system design. When the microgrid is linked to the utility grid, a switch allows it to function as an islanded system or as part of the grid. This study only looks at the stranded version. A converter is used to increase the output of a signal, and PV is linked to the DC line (Punna and Manthati, 2020). The HESS is connected via bidirectional DC/DC converters, which comprises the battery and the SC, to the DC connector. The HESS in use is designed to transfer instantaneous current as from battery to the SC, Consequently, battery stress is reduced. In order to connect all of the system's components, a three-phase AC bus is used. A solid-state switch connects the loads, which include critical and non-critical loads. PV front-to-back converters, HESS front to back converters, as well as load shedding are all controlled by the energy management system (EMS).

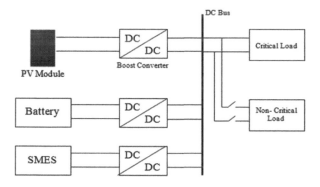

Fig. 80.1 PV-based microgrid aided by HESS

Accordingly, in Section 2 of the paper, the various ways of action laid forth in chronological order of DC microgrid are explained briefly, Section 3 is discussion about operation modes of DC microgrid, Section 4 displays results and discussion, and Section 5 concludes the work.

3. Operation Modes of DC Microgrid

Consistent, reliable, and safe operation of a freestanding photovoltaic DC mini grid is the primary goal of coordinating administration of a freestanding photovoltaic DC microgrid and to maximize the usage of solar energy and optimize the lithium-ion battery's operating condition. Microgrid operation control is achieved by analyzing the changing light intensity and combining the condition of each unit, making logical conclusions, and then creating and delivering control signals to every subsystem (Xu and Chen, 2011).

Microgrid operating mode should be determined based on light intensity, lithium-ion present battery charge state, and connection to the sun array's output power. According to this research, the light intensity is categorized into two groups. The first group's light is becoming brighter all the time. The light intensity gradually diminishes or disappears in the second group. For lithium-ion batteries, the working range of SOC is fixed at 25% to 90%, although it may be adjusted under specific situations to enhance the energy use of lithium-ion batteries.

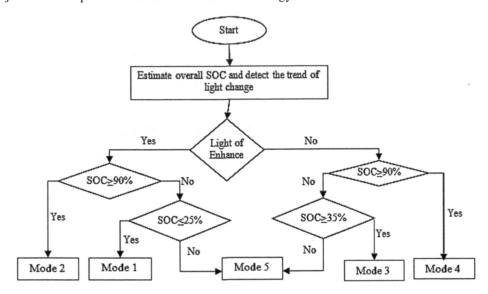

Fig. 80.2   Flow chart of operation modes for microgrid

Mode 1: Increases in light intensity are occurring, as well as decreases in the lithium-ion battery's state of charge (SOC), when non-critical loads are being carried out. In this case, the system senses an increase in light intensity and makes necessary adjustments (Y. Han et al., 2017). The power consumption rate of a lithium-ion battery, which switches off when the SOC drops below 25%, non-critical load, including future energy replenishment in a hurry owing to the output capabilities of a photovoltaic array. A photovoltaic array is a term that refers to a collection of solar panels.

Mode 2: Despite the fact that the lithium-ion battery's state of charge (SOC) declines to less than a quarter of its maximum, the device fails to stop the non-critical load from increasing in intensity. Here, the system can detect an increase in light intensity and adjust the situation. Due to the output capabilities of Photovoltaic arrays, the lithium-ion battery's power use rate, which switches off when the SOC decreases to 25%, non-critical load, and renewable power replenishing in a hurry owing to the output capabilities of a photovoltaic array. Here photovoltaic array is a phrase for a collection of solar panels (Elmouatamid et al., 2021).

Mode 3: With a SOC of less than 35%, the light is dimming or going off completely. It disconnects the non-critical load because the lithium-ion battery's SOC doesn't really fall below a certain threshold (25 percent). In this instance, extra electric energy is stored in a lithium-ion battery for later use. Night-time essential loads should be protected by a backup generator (Naik et al, 2021).

Mode 4: The light intensity progressively diminishes or vanishes, and the SOC reaches a maximum of 90%. To prevent the battery from being overtaxed and losing its ability to giving power at night, the SOC is adjusted to 95 percent in certain cases.

Mode 5: The superior and inferior SOC limits do not have to be changed since the SOC is within the operating range, which is 25 percent to 90 percent. The energy balance relationship will manage the charging and discharge of batteries in a microgrid. Before making a decision, the different ways of microgrid operation are evaluated in this study. The lithium-ion battery's SOC and the trend in light intensity change are used to define the mode of operation as well as appropriately control each unit, providing coordinated and stable system operation.

A. Pre-Judgment Coordination Control Strategy for DC Microgrid

Each component of the system's control approach must be developed to achieve comprehensive microgrid pre-judgment coordination control. The most significant components are light intensity sensing, DC/DC and bi-directional DC/DC conversion management, inverter control, and load hysteresis control (Hongpeng et al., 2019).

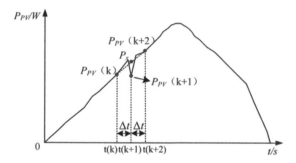

Fig. 80.3   Fluctuation point in output PV power

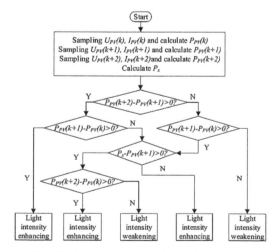

Fig. 80.4   Algorithm for intensity change

B. Detection of the Change of Light Intensity

Detecting the electrical output of a solar system and identifying the recent trends of light intensity is an essential aspect of applying the coordination control approach (Takeuchi et al., 2021).

By sensing the output power value, the sample interval t is set to 0.3s., , and of the three points of the PV system;

Considering the PV system's current,

$$P_{pv}(k) = U_{PV}(K) \times I_{pv}(k) \tag{1}$$

$$P_{pv}(k+1) = U_{PV}(K+1) \times I_{pv}(k+1) \tag{2}$$

$$P_{pv}(k+2) = U_{PV}(K+2) \times I_{pv}(k+2) \tag{3}$$

Clouds as well as other shelters might cause the system's operating mode to be repeatedly changed owing to a short-term drop in light intensity (Prakash and Veni, 2017). Due to the unpredictability of the sun. To counteract the effects of the short light drop, the following mechanism is required.

$$P_x = \frac{P_{pv}(k) + P_{pv}(k+2)}{2} \tag{4}$$

It is also possible to determine whether or not the current light change trend as well as judging approach is accurate by comparing PV system output power at various times.

C. Control Strategy of Converter

Photovoltaic Array Converter

A dual closed-loop controller controls the photovoltaic array converter (Z. Uli et al., 2017). This operator makes use of an outer loop with a load power consistently demonstrated (LPTC) and an inner loop with a current loop to monitor the maximum power point. MPPT makes use of both a constant-voltage and a disturbance monitoring technique in tandem. MPPT mode is activated whenever the lithium-ion battery SOC falls below 90%. Whenever the SOC is more than 90 percentage points and the PV array's output power exceeds the load's required power, the LPTC mode is activated on the PV array (R. M. Button, 1996).

Lithium-Ion Battery Converter

The rechargeable battery converter also makes use of a dual closed-loop controller (voltage outer loop and current inner loop). In contrast to the prior control strategy, the battery is charged and drained when the SOC approaches the higher and lower operational ranges. If PV output power surpasses load demand because there is extra power available for charging, SOC is within the operational range (Moghaddam and Bossche, 2019).

In the event that the load's power requirements are greater than the solar panel's ability to provide, the battery will release stored energy to meet those demands, and fulfil the load requirement and there is a surplus of power available.

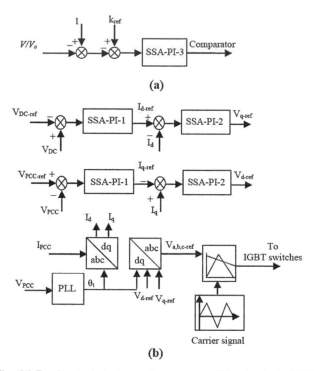

Fig. 80.5 Control strategy of converters (Elazab et al., 2020)

4. Result And Discussion

In order to simulate and verify the proposed coordination control strategy, two cases are shown in Fig. 80.6(a) and Fig. 80.6(b). In Fig. 80.6(a), step decrement of input solar irradiance is considered, the irradiance to the solar panel is stepped down from 500 to 1000 at 0.4 seconds, and the temperature is maintained at 25°C throughout the simulation, and their corresponding solar power output and load powers are visualized Fig. 80.7(a). The second case, shown in Fig. 80.6(b), considered the gradual change in solar irradiation. The irradiation gradually increases from 0 to 1000 in time 0.1 to 0.2 and remains at 1000 from time 0.2 to 0.4 and then gradually decreases from 1000 to 0 in time 0.4 to 0.5 and the temperature is maintained at 25°C throughout the simulation, and their corresponding solar power output and load powers are visualized along with the battery and super-capacitor powers in Fig. 80.7(b).

From Fig. 80.7(a), it is clear that our proposed strategy is able to stabilize the system operation as soon as possible when the disturbance occurs (step decrement of input irradiance). From Fig. 7(b), it can be observed that initially total load power is supplied from the battery as well as super-capacitor supports the transients and provides smooth operation to battery. When irradiation starts increasing and is able to supply load power, the battery slope becomes negative and battery starts charging itself. When irradiation starts to decrease, battery slope becomes positive and starts discharging.

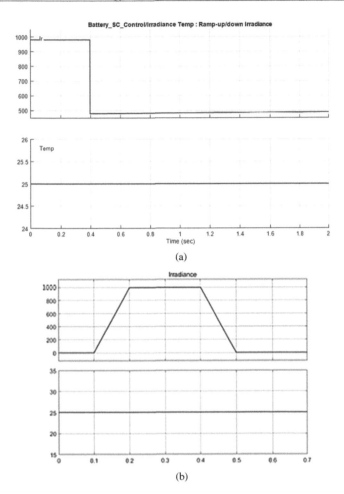

Fig. 80.6   Irradiance and temperature vs time

5. Conclusion

A microgrid with a hybrid energy system is the subject of this study, which presents an energy management as well as coordinated control approach. A variety of flow of energy relationships here between battery, PV, as well as load may be accommodated using the provided strategies. When the battery's charging falls within a certain range, the PV switches to MPPT mode, as well as the battery is charged and discharged in accordance with the energy equilibrium. If the battery's charging hits the upper limit, the PV system changes to LPT mode, and it supplies electricity according to the load demand. When the battery's SOC falls below a certain threshold, the system automatically eliminates non-critical loads in accordance with the load

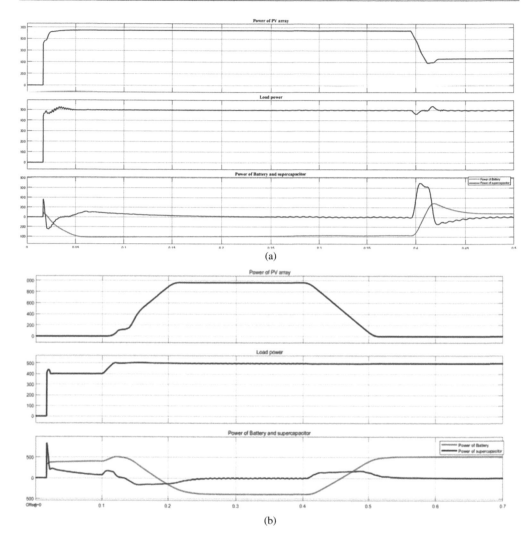

Fig. 80.7 Scope results of Power

management strategy, ensuring that the battery's SOC remains within the specified range. In a mixed energy storage device, a virtual resistor and a virtual capacitance are used to accomplish power sharing. The SC delivers transient power, whereas the battery provides stable power changes during quick changes in load as well as PV output, resulting in reduced transitional strain in the battery as well as a longer battery life. Under both normal and critical situations, the simulation results support the suggested energy management system. The results obtained are provided in the results section.

REFERENCES

1. L. Di, G. Liu, J. Zhang, G. Tian, and S. Wang, "Research on pre-judgment coordinated control strategy of stand-alone DC microgrid," *Proc. - 2020 5th Asia Conf. Power Electr. Eng. ACPEE 2020*, no. 201802030, pp. 638–644, 2020, doi: 10.1109/ACPEE48638.2020.9136571.

2. S. Punna and U. B. Manthati, "Optimum design and analysis of a dynamic energy management scheme for HESS in renewable power generation applications," *SN Appl. Sci.*, vol. 2, no. 3, pp. 1–13, 2020, doi: 10.1007/s42452-020-2313-3.

3. L. Xu and D. Chen, "Control and operation of a DC microgrid with variable generation and energy storage," *IEEE Trans. Power Deliv.*, vol. 26, no. 4, pp. 2513–2522, 2011, doi: 10.1109/TPWRD.2011.2158456.

4. Y. Han, X. Xie, H. Deng, and W. Ma, "Central energy management method for photovoltaic DC micro-grid system based on power tracking control," *IET Renew. Power Gener.*, vol. 11, no. 8, pp. 1138–1147, 2017, doi: 10.1049/iet-rpg.2016.0351.

5. A. Elmouatamid, R. Ouladsine, M. Bakhouya, N. El Kamoun, M. Khaidar, and K. Zine-Dine, "Review of control and energy management approaches in micro-grid systems," *Energies*, vol. 14, no. 1, 2021, doi: 10.3390/en14010168.

6. K. R. Naik, B. Rajpathak, A. Mitra, and M. L. Kolhe, "Assessment of energy management technique for achieving the sustainable voltage level during grid outage of hydro generator interfaced DC Micro-Grid," *Sustain. Energy Technol. Assessments*, vol. 46, no. April, p. 101231, 2021, doi: 10.1016/j.seta.2021.101231.

7. Z. Hongpeng, Z. Jiankun, L. Qun, and D. Shibo, "Distributed New Energy Micro-grid Absorption Pilot Project in Northeast China Power Grid," *4th Int. Hybrid Power Syst. Work.*, no. May, 2019.

8. Y. Takeuchi, M. Iwasaka, M. Matsusda, and A. Hamasaki, "Angle Distribution Measurement of Scattered Light Intensity from Needle-Shaped Crystals in a Magnetic Field for Gout Diagnosis," *IEEE Trans. Magn.*, vol. 57, no. 2, pp. 1–5, 2021, doi: 10.1109/TMAG.2020.3006556.

9. T. Prakash and V. Veni, "COORDINATED V-F AND P-Q CONTROL OF," no. June, 2015.

10. Z. Uli, W. Zedi, L. Yang, and D. Li, "Design of control strategy for hybrid energy storage system based on coordination and cooperation," *Proc. - 2017 Int. Conf. Smart Grid Electr. Autom. ICSGEA 2017*, vol. 2017-January, pp. 159–162, 2017, doi: 10.1109/ICSGEA.2017.42.

11. R. M. Button, "Advanced photovoltaic array regulator module," *Proc. Intersoc. Energy Convers. Eng. Conf.*, vol. 1, pp. 519–524, 1996, doi: 10.1109/iecec.1996.552937.

12. A. F. Moghaddam and A. Van Den Bossche, "A Battery Equalization Technique Based on Ćuk Converter Balancing for Lithium Ion Batteries," *2019 8th Int. Conf. Mod. Circuits Syst. Technol. MOCAST 2019*, pp. 1–4, 2019, doi: 10.1109/MOCAST.2019.8741779.

13. O. S. Elazab, M. Debouza, H. M. Hasanien, S. M. Muyeen, and A. Al-Durra, "Salp swarm algorithm-based optimal control scheme for LVRT capability improvement of grid-connected photovoltaic power plants: Design and experimental validation," *IET Renew. Power Gener.*, vol. 14, no. 4, pp. 591–599, 2020, doi: 10.1049/iet-rpg.2019.0726.

Intelligent Systems and Smart Infrastructure – Brijesh Mishra et al. (eds)
© 2023 Taylor & Francis Group, London, ISBN 978-1-032-41287-0

CHAPTER

81

IoT-Based Oxygen Supply Management System

Beauti Kumari[1], Km Gunjan[2], Raju Ranjan[3]

School of Computing Science & Engineering,
Galgotias University, India

Abstract

Oxygen is spread all around us in the air that we breathe; it is something without which we can't even live. In this duration of covid-19 pandemic, the requirement of having oxygen is increased, and in our country, though there is enough oxygen, the main problem is to transport it to the hospitals or to the needy one's on time, and this is just because of high communication gap between suppliers and hospitals, so we are planning to implement an idea that will work to reduce this gap using real-time tracking as we can monitor the movement of oxygen tankers, by gathering the requirements of oxygen at various locations, checking the availability of oxygen in inventory, and by delivering it to someone who is in emergency, as soon as possible. To implement it successfully, we are using a pressure sensor using Micro-electromechanical systems (MEMS), that publishes the value of remaining oxygen from its location to the supplier using Message Queuing Telemetry Transport protocol (MQTT) and a Wi-Fi module (ESP32) using wireless communication.

Keywords: Medical oxygen liquid gas, IoT, Firebase database, Cloud storage

Corresponding author: [1]xbbeautisingh123@gmail.com, [2]Gunjanpanchal476@gmail.com, [3]drraju.ranjan@galgotiasuniversity.edu.in

DOI: 10.1201/9781003357346-81

1. Introduction

Medical oxygen (O_2) gas is used by patients in healthcare provisions for the life assist and for medical treatment purpose. Ensuring that the medical oxygen gas supply system gives a reliable and safe supply of medical oxygen gas to patients and healthcare provisions is vital, as end-user. Consequences of system breakdown could be quite earnest, as indicated by past exposures. It is therefore essential that both healthcare provisions and oxygen gas suppliers' management get to know about the needs on the design and installation of liquid medical oxygen gas supply and pipeline diffusion system. This is common in Asia where the gas suppliers are accountable for the installation and design of medical supply but it may or may not be involved in the pipeline diffusion system. The most threatening problem is that, in spite of having enough oxygen here in our country, it's quite difficult to transport it on time to the needy one's or to the hospitals due to lack of proper communication between the hospitals and oxygen supplier plants. There are different tools, techniques, and methods, which are used here in our project, to monitor the movement of oxygen cylinders and supply it to those who are in need, indeed. So, our objective is to inform the quick and increased demand of oxygen that arises during this critical situation of pandemic, and to analyse the relationship of different stakeholders, for example, the oxygen suppliers, hospitals, and the government during the critical or unexpected situation.[1]

2. Literature Review

Only some traders in our country are granted to produce oxygen, which is an important drug. The call for medical oxygen gas usually acts in accordance with a periodic pattern, which most of the suppliers follow along with some indicators and market data for forecasting the call or demand and supply. It allows them to increase their stock during the measured months so that they'll be having enough for the engaged period. However, the pandemic has brought about a projection that has already made some long-established prediction models fundamentally unusable. The medical oxygen (O_2) gas suppliers have had to restore and review forecasting plans regularly, even on daily bases, for finding out the number of panzers they are having with them for storage purpose, the number of tanks still with customers, as well as other integral data such as the capacity of warehouse, production and the activities of diffusion. As many producers analyse and regulate their predictions, they should definitely have some crucial points in their mind as stated below [2]:

(A) Not to just forsake maintenance schemes and take some extra usage into account during this critical situation of pandemic, as these failures will facilitate secured supply.

(B) This is quite significant to have a management system, which is highly efficient for returned empty cylinders that require thorough and complete examination to assure that they will not fail and they are safe as well. But this is quite essential to ensure that the process does not cause a congestion. The suppliers are also required to have a proper approach to contemporary information on vacant cylinders available, expiry of certification, number of cylinders that needs repair and the number of cylinders which are ready to be filled. This overall process of producing gas, arranging cylinders, maintaining its availability and other related information has resulted in variation of cost and unreasonable supply of oxygen. Having a play of fair market, the Governments of various states should definitely look into the encouragement of investment in this particular sector. Few states which are quite populous such as Uttar Pradesh and Bihar, for example, can set up oxygen plants at different locations individually in Central, Eastern, and Western Uttar Pradesh or in the north and south Bihar. Few other states can also plan equitable availability at different places as per their requirement. The following should also be considered while decisions are being made:

(a) Proficiency in the gas production is distinct from proficiency in its supply. So, timely and safe supply of medical oxygen is equally important as its production, and it should be supported by the Government.

(b) Estimation of supply system: Most of the hospitals even at district-level and having large investment do not have piped medical gas system and still uses cylinder based supply system with single cylinder. They are required to evaluate for constant oxygen supply system, which is necessary at all district-level hospitals. Such type of evaluation should be done at least at an interval or every six/seven years as decided by the healthcare departments to marked call and technological advancements.

(c) Demonstration of the requirements of the oxygen gas review and accession system: Every district-based hospital should have a system of guessing O_2 gas needs. This type of assessment should be mentioned seasonally dissimilarity in the regular use of oxygen. A report should also include the O2 supplier's representation, and the district-based hospital should use this type of datasets for the accession of oxygen gas.

(d) Superintend system: Maximum number of the district hospitals are incapable to keep an eye on continuous supply of oxygen gas, resulting in unwanted victims at the pedantic care areas. So, there should be an effectual alarm monitoring system, based on the system for supervising the supply of O_2 gas in various hospitals, particularly in critical care areas, to remove the shortage of oxygen supply, that's why we are using Internet of Things (IoTs) here to tackle this particular problem.[3]

Although the long-established outflow detector systems of oxygen gas are having quite good accuracy, they fail to grant a very few of the factors in this area of growing the people about the outflow. Therefore, we have used the IoTs technology to make a gas outflow detector having Smart changeable approaches, including sending text message, calling and an e-mail to the responsible authorization and a potential to forecast threatening affairs to the people so that they could be made familiar about sensor readings, in advance, by executing data coherent on the same.

The supply chain activity is being revised and improved with the exercise of newer technologies based on IoTs. It has helped in improving the process of production, refilling, reloading and calculating the count of gas cylinders at various stages of utilization with the help of RFID based identification. At present, the administration of the oxygen gas cylinders is quite an easy task, where only few of the data about the medical oxygen gas cylinder is raised and automatically administered by the trader. In this task, it is recommended an administered application construction for, on the one side, automated oxygen gas cylinders specification and categorization, fulfilling both accountability and the needs of superintending during the manufacturing process, on the other side, specifying the authentication of level and acceptance of alerts and new administered services from the trader to the customer. In order to achieve the same, oxygen gas cylinders have been RFID-tagged, and a Wireless Personal gadget has been evolved, in order to provide 6LoWPAN connectivity and a spontaneous LCD touch screen for communicating with the RFID-HF reader [4].

3. Libraries

A. Data Flow Diagram

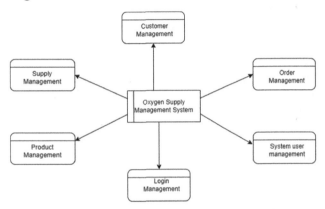

Fig. 81.1   Brief description of various modules

In this Data Flow Diagram, the entire process of analyzing and visualizing to develop an enhanced one application of oxygen supply management is described. As shown in this diagram, we are going to work on various modules, e.g., customer, supply, order, product, system user, and login management.

4. Working Mechanism

There are some steps defined here, which can lead to reduce the problems listed above in the introductory part:

- First step is to accumulate the needs of oxygen at different locations.
- Check the accessibility of oxygen in inventory.
- Delivery of oxygen to the targeted audience, who are in urgent situation and that would be prioritized on basis of oxygen left at particular locations.

To implement the particular flow, we will follow the following plan: The pressure sensor produces the value of oxygen remaining from location to the Message Queuing Telemetry Transport (MQTT) broker, which constantly captures the value published. However, if the value remaining becomes lesser than the threshold fixed by us, then an automated alert via mail will be sent to the nearest supplier plant. After the supplier confirms, the location of the truck will be tracked. All these things will be done using an application through any of the devices; i.e., the computer or mobile phone. For performing this task, we are required to have following components:

- The dual-mode Bluetooth and ESP32 Wi-Fi module, which is a sequence of cost-efficient and low-power system embedded with a chip microcontroller with combined Wi-Fi.
- MEMS pressure sensor that is used to enable the combination of high-frequent sensors, powerful processing, and wireless communication; for example, Bluetooth or Wi-Fi.
- The ad fruit IO focuses on ease of use and following simple data connections with little bit of programming required.
- For translating messages or information between various devices, servers, and applications, we are using the MQTT protocol, which is a lightweight messaging protocol that uses the pub/sub pattern.
- The IFTTT (If We Use This Then That) protocol defines a clear and concise protocol which your service's API will implement. Each trigger and action for our API will map one-to-one to an API endpoint on our service [5].

5. Results and Discussion

A. Login Page

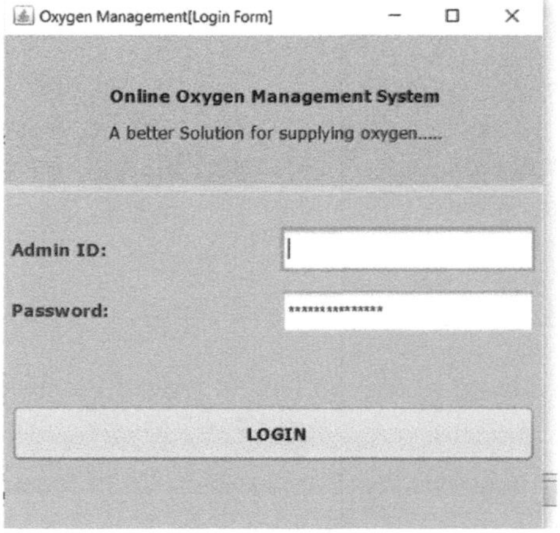

Fig. 81.2 Login Interface

This is the front page of the application, where users are required to login, at very first, to navigate through and to take the advantage of the project.

B. User Registration page

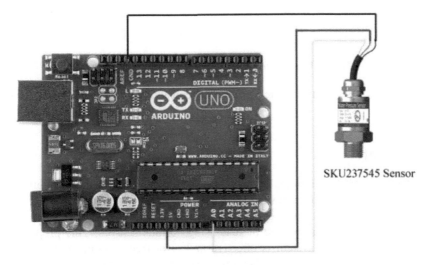

Fig. 81.3 Registration Interface for adding new users

This is the user registration page of the application, where users are required to register, if already not registered, to navigate through and to utilize the project.

C. Model

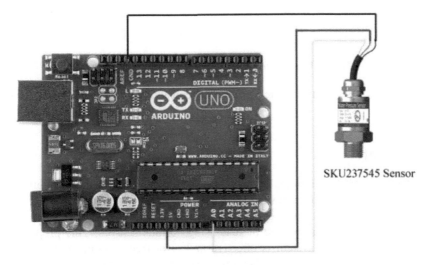

Fig. 81.4 Connection of sensor with Arduino board

In this proposed model, the sensor will be connected to the Arduino model like that the red wire will be connected to the 5Vsoucrce and the black wire will be connected to the ground and yellow one with the analogue pin A0, which, through serial monitor, we can view the value of the pressure sensor.

D. Statistics

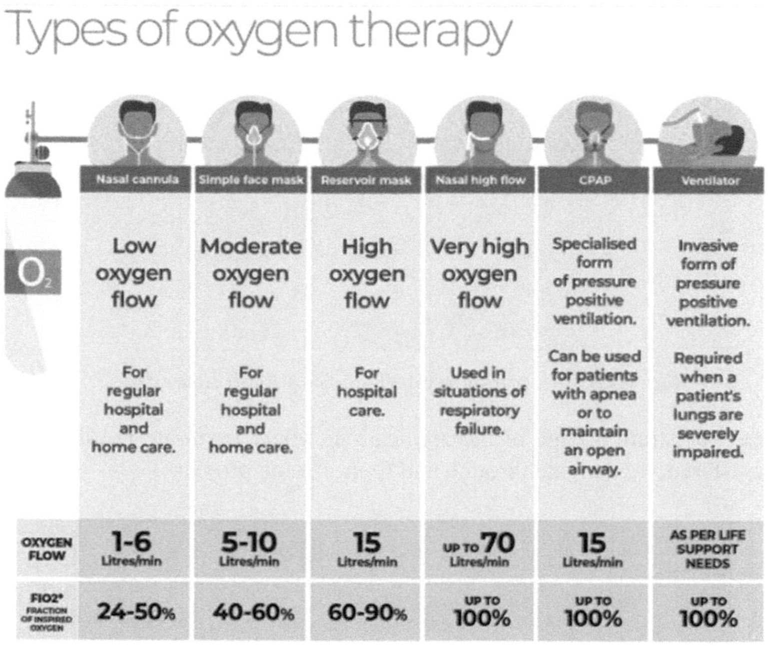

Types of oxygen therapy

	Nasal cannula	Simple face mask	Reservoir mask	Nasal high flow	CPAP	Ventilator
O₂	Low oxygen flow	Moderate oxygen flow	High oxygen flow	Very high oxygen flow	Specialised form of pressure positive ventilation.	Invasive form of pressure positive ventilation.
	For regular hospital and home care.	For regular hospital and home care.	For hospital care.	Used in situations of respiratory failure.	Can be used for patients with apnea or to maintain an open airway.	Required when a patient's lungs are severely impaired.
OXYGEN FLOW	1-6 Litres/min	5-10 Litres/min	15 Litres/min	UP TO 70 Litres/min	15 Litres/min	AS PER LIFE SUPPORT NEEDS
FIO2* FRACTION OF INSPIRED OXYGEN	24-50%	40-60%	60-90%	UP TO 100%	UP TO 100%	UP TO 100%

Fig. 81.5   Oxygen requirement at various stages

Figure 81.5 shows a few statistics related to flow of medical oxygen gas with respect to fraction of inspired oxygen.

6. Conclusion

Our main aim is to provide high-quality service by implementing a solution to eliminate the communication gap between oxygen supplier plants and hospitals or needy one's, so that it can lead to fulfil the requirement of oxygen on time without any delay, especially in this duration of pandemic, by taking the advantage of IoT, using various sensors. This proposed online oxygen supply management system is effective and error-free as compared to established solutions. The application will be very helpful in easy and comfortable supply management system of medical oxygen gas. Problems related to the heart have been increased to patients already suffering from heart diseases, during the pandemic, coupled with various factors. On the other side, people who are infected may healed by enough perquisites. In these provisions, medical oxygen gas is something which is a technology-dependent drug that needs a successful teamwork among technicians and healthcare workers, and upgrading medical oxygen gas systems is a feasible priority for hospitals in Low and Middle-income countries (LMICs). This paper intends to provide hands-on steps to assist remarkable and feasible upgrades in healthcare oxygen gas systems during the pandemic.

REFERENCES

1. Kelly, F. E., Bailey, C. R., Aldridge, P., Brennan, P. A., Hardy, R. P., Henrys, P., ... & Taft, D. (2021). Fire safety and emergency evacuation guidelines for intensive care units and operating theatres: for use in the event of fire, flood, power cut, oxygen supply failure, noxious gas, structural collapse or other critical incidents: Guidelines from the Association of Anaesthetists and the Intensive Care Society. *Anaesthesia*, *76*(10), 1377–1391.
2. Kabir, S., Azad, T., Walker, M., & Gheraibia, Y. (2015, December). Reliability analysis of automated pond oxygen management system. In *2015 18th International Conference on Computer and Information Technology (ICCIT)* (pp. 144–149). IEEE.
3. Weller, J., Merry, A., Warman, G., & Robinson, B. (2007). Anaesthetists' management of oxygen pipeline failure: room for improvement. *Anaesthesia*, *62*(2), 122–126.
4. Naik, V. N., Savoldelli, G. L., Joo, H. S., Lorraway, P. G., Chandra, D. B., & Chow, R. E. (2006). Management of simulated oxygen supply failure: is there a gap in curriculum?. *Simulation in Healthcare*, *1*(2), 95.
5. Shackell, E. The Oxygen Supply.
6. Kayton, A., Timoney, P., Vargo, L., Perez, J. A., Harris-Haman, P. A., & Zukowsky, K. (2018). A review of oxygen physiology and appropriate management of oxygen levels in premature neonates. *Advances in Neonatal Care*, *18*(2), 98.
7. Seyfang, A., Miksch, S., Horn, W., Urschitz, M. S., Popow, C., & Poets, C. F. (2001, July). Using time-oriented data abstraction methods to optimize oxygen supply for neonates. In *Conference on Artificial Intelligence in Medicine in Europe* (pp. 217–226). Springer, Berlin, Heidelberg.
8. Lorraway, P. G., Savoldelli, G. L., Joo, H. S., Chandra, D. B., Chow, R., & Naik, V. N. (2006). Management of simulated oxygen supply failure: is there a gap in the curriculum?. *Anesthesia & analgesia*, *102*(3), 865–867.
9. Manohar, P., Krithiga, S., Vijaya Lakshmi, D., Sahoo, P., & Kalambhe, A. C. (2021). Design And Development Of Healthcare Monitoring System Using Iot. *Ilkogretim Online*, *20*(1).

Intelligent Systems and Smart Infrastructure – Brijesh Mishra et al. (eds)
© 2023 Taylor & Francis Group, London, ISBN 978-1-032-41287-0

CHAPTER

82

Energy Audit of DIT University

Deepak Kumar Verma[1], Gagan Singh[2]

Department of Electrical Electronics & Communication Engineering,
DIT University, Dehradun, India

Abstract

Energy Savings is one of the challenging things in today's world. For energy savings, Energy Audit is extremely necessary. Energy Audit is used to identify various sectors where we can conserve energy. In this paper, we have conducted an energy audit of DIT University, before organizing energy audit of the university pre-audit phase is scheduled. In this phase, electricity bill of the university over a period is analyzed, and different power parameters like load factor, total power, energy consumption, and energy rebate are taken into consideration. In Auditing phase, different loads were identified according to its power rating and quality of power. In post-auditing phase, all the data collected is cross-verified with the electricity bill of the university. Following an energy audit, it was discovered that there are various areas where electrical energy may be consumed to increase the power quality and efficiency of the university's systems.

Keywords: Energy Audit and Management, Power Quality Improvement, Energy Saving, Energy Efficiency Measure, Energy Utilization.

1. Introduction

An energy audit, also known as an energy survey or an energy inventory, is a thorough evaluation of a property's entire energy use. The analysis is intended to give a quick and easy way to determine not only how much energy is spent, but also where and when it is utilized. Defects in operational methods and physical infrastructure will be identified during the energy

Corresponding author: [1]deepakkumar8290@gmail.com, [2]gagan.singh@dituniversity.edu.in

DOI: 10.1201/9781003357346-82

audit. Once these problems are identified, it will be evident where energy conservation efforts should be concentrated. The energy audit is the basis and starting point for a successful energy management program [1].

Human settlements are made up of a range of structures. The audit technique is essentially the same regardless of the structure in question. If there are many facilities involved, no two buildings are the same when it comes to energy use. Occupancy rates, building size and orientation, geographic location, kind of heating and cooling systems, quantity and types of equipment utilized, type of construction, insulation level, and other factors all have an impact on buildings. Because it is difficult to generalize about energy usage trends because each building is different, each one requires an energy audit. Most structures were probably conceived, erected, and outfitted when cheap energy was commonly available. The importance of energy efficiency was overlooked. As a result, there is a lot of space to save money on operational costs in older buildings.

The tools and calculations needed to understand all phases of energy utilization are explained in the following chapters. Both manual and computer-assisted energy audits can benefit from the information acquired. The emphasis is on do-it-yourself, which is reflected in the worksheets [2-3].

Electrical Energy Consumption in industrial homes is gaining attention, and hence the use of electric device will increase. Conducting an Energy Audit with the corporates is step one toward green electric power. To increase Energy Effificiency Measures (EEMs), power audit is needed to gather foundation or complete electric power information from the organisation real power consumption, that is, as compared with the same old power usage. It's crucial to build EEMs for businesses if you want to stay competitive, as operational costs are rising significantly due to projected increases in electric power prices and in power costs due to projected increases in gas prices and globally constrained gas resources. The Efficient Management of Electrical Energy Regulations 2022 had been delivered on February 28, 2022, to sell power performance in India [4-10]. These rules make it obligatory for the control of big industrial and commercial electric purchasers in India to broaden and enforce EEMs to lessen strength losses and enforce green electric strength usage of their organizations. Energy performance cognizance inside the organisation is essential to the achievement of EEMs. Only top management' determination and the involvement of the entire organization's workforce can ensure the success of a strength audit. There is a need of imposing electricity control, and for this reason, electricity audits ought to be emphasised through the employer's control. The lighting fixtures and air con structures eat the maximum electric electricity in a structure. Lighting structures account for nearly 1/4 of all electric electricity utilized in buildings. The biggest source of electricity waste at a facility is the attitude of employer people who are no longer turning off lighting fixtures while it is no longer need to be in running centres. The adoption of well-designed lights can also additionally lower electricity losses, and, at the same time, additionally enhancing gadget toughness and illumination. Air conditioning consumes the maximum electric electricity in buildings, accounting for around fifty six percentage of all electric electricity utilized in buildings. The most commonplace place waste is failing to turn off air conditioning equipment when it isn't in use, as well as operating air conditioning systems inefficiently due to inside and outdoor windows, both of which can reduce equipment losses and increase machine efficiency [11-16]. When developing an EEM, use an energy recorder to find out how much energy your lighting and air conditioning system is using. Electrical energy audits can be basic audits to establish low-cost actions, or full audits to establish medium- and high-priced EEMs, but they are costly and time-consuming. Similar to EEM, energy waste and leaks are found throughout the audit. Establishing IEEE standard practices for energy management by organizational managers is the topic of this energy audit study.

2. Literature Review

University Teknogi MARA has done their audit [17] on their university campus. As per their case study, 25% and 56% power consumption had occurred there but after completing their audit and energy management, they has saved up to 10% of their power consumption. So that paper has been shown the importance of energy savings in this time [18].

3. Objective

The goal of the energy audit is to promote energy conservation on the DIT university campus. The goal of the energy audit is to find, quantify, characterize, and prioritize cost-cutting measures in colleges, departments, and institution central facilities.

4. Phase of Auditing

(a) Pre-Audit Data Collection
(b) Post-Audit
(c) Audit Detailed Measurement
(d) Suggestion for Implementation and Energy Saving

5. Case Study of DIT University

Energy audits are a way to identify where a facility is consuming energy and identify the EEMs that can be used to reduce energy consumption. Energy auditing is an important energy management service focused on the use of energy to implement energy-efficiency measures for buildings. The DIT University campus building received an energy rating in 2022. The energy audit profiled the energy consumption of electrical equipment in terms of kwh consumption, including uptime and cost, as shown in the graph. According to audits, electrical energy is used in lighting, air conditioners, and other electrical equipment, with an average total energy consumption. Lighting and air conditioning consume most of the energy in a building, so there is plenty of room to save energy. As a result, EEMs for lighting and air conditioning systems were created to reduce energy consumption and costs within the facility [19-20].

Energy Audit is described by the Energy Conservation Act of 2001 as "The verification monitoring and analysis of energy usage, as well as the production of a technical report that includes cost-effective strategies for increasing energy efficiency a cost-benefit analysis and a strategy for reducing energy use consumption". The proposed steps can be implemented in a variety of ways. To assist consumers in achieving considerable energy savings levels of consumption in the building, an energy audit was carried out. In February 2022, the DIT University Campus will be completed. The audit of energy use created a profile of electrical energy use in terms of kilowatt-hour usage [21].

Electrical energy was discovered to be used for lights, air conditioning, PCs, and fans during the audit. To minimize energy consumption and costs in buildings, energy-efficient measures were created for lighting, air conditioning, computers, and fans [3].

Table 82.1 Electricity Bill for year 2019

S. No	Duration	Year 2019					
		Units Consumed	Rs/Unit	Bill Amount	Load Factor	Power Factor	Load
1	Jan–Feb	161,150	4.05	652,657.5	30.13	0.99	>25 kW
2	Feb–March	212,340	4.05	859,977	63.35	0.99	>25 kW
3	Mar–Apr	119,940	4.05	485,757	37.11	0.99	>25 kW
4	Apr–May	185,480	4.25	788,290	46.22	0.99	>25 kW
5	May–June	83,160	4.25	353,430	24.06	0.99	>25 kW
6	June–July	170,550	4.25	724,837.5	53.49	0.99	>25 kW
7	July–Aug	289,290	4.25	1,229,482.5	51.43	0.99	>25 kW
8	Aug–Sep	146,700	4.25	623,475	24.71	0.99	>25 kW
9	Sep–Oct	246,740	4.25	1,048,645	41.59	0.99	>25 kW
10	Oct–Nov	161,770	4.25	687,522.5	31.57	0.99	>25 kW
11	Nov–Dec	163,700	4.25	695,725	51.77	0.99	>25 kW
12	Dec	151,980	4.25	6,459,115	24.81	0.99	>25 kW

Table 82.1 shows the electricity bill analysis of the year 2019. From the above table, it is observed that in the summer season, the energy consumed is more as compared to other seasons of the year. Therefore, the possibility of energy wastage is mostly possible in this period. During this period of time, students are mostly available at their homes.

Table 82.2 shows the same analysis for the year 2020, for the duration when higher energy consumption is noticed. Therefore, it is recommended to purpose the energy conservation methods for this duration. As during the summer season, natural light is present for most of the time, it is recommended to make minimum use of artificial lights like CFL and LED. As the usage of ACs when the temperature desired is achieved.

Table 82.2 Electricity Bill for year 2021

Year 2021					
Units Consumed	**Rs/unit**	**Bill Amount**	**Load Factor**	**Power Factor**	**Load**
105,740	4.4	465,260	27.03	1	>25 kW
109,680	4.4	477,108	43.1	1	>25 kW
109,910	4.4	478,108.5	47.09	1	>25 kW
86,650	4.4	381,260	39.27	1	>25 kW
63,620	4.4	279,928	41.19	1	>25 kW
56,450	4.4	248,380	42.41	1	>25 kW
72,570	4.4	319,308	36.67	1	>25 kW
74,780	4.4	329,032	35.65	1	>25 kW
152,350	4.4	670,340	34.41	1	>25 kW
175,480	4.4	772,112	46.25	1	>25 kW
151,340	4.4	665,896	52.16	1	>25 kW

Figure 82.1 and Fig. 82.2 represent the load curve for the year 2019. The load curve varies every month, but it peaks during the summer season, which shows that the consumption of ACs and fans were more than the heater used in winter. In winter, the demand was almost constant.

Figure 82.3 represents the load curve of DIT university for the year 2021. From the above tables, it was identified that the energy consumption is more for July–August, i.e., during peak summer. The load curve shows how the energy demand/load increases or decreases over a period. From Fig. 82.3, it is observed that during October to December interval, the energy load increases.

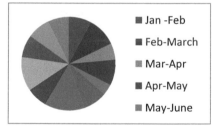

Fig. 82.1 Pie Chart of the load curve for the year 2019

The load increases from 70,000 kwh to 150,000 kwh. This shows that the load almost gets doubled during the starting of winter. Because during winter, heaters are heavily used, so the reason behind this high increase can be the web of heaters in the university. By comparing

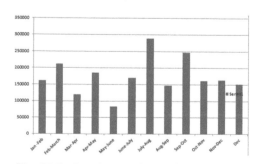

Fig. 82.2 Load curve Jan 2019–Dec 2019

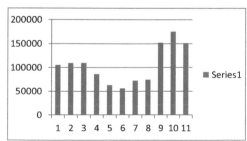

Fig. 82.3 Bar graph of load curve for Jan 2021–Dec 2021

the above load curves, it is observed that during the summer, the load demand increases as compared to the winter season. The highest demand was in the July–August interval in the year 2019.

In this section, the total percentage of LED, CFL, and Fan percentage currently in the University is shown.

Table 82.3 Equipment connected in DIT University

Building	Lights		Fans		Air Conditioner (AC)		Computer	Projector
	CFL	LED	Electrinoic Regulator	Without Regulator	Qty	Rating	PC	-
Grand Total Chankya	377	549	363	65	24		136	7
Total Girls Hostel	**469**	**62**	**501**	**10**				
Grand Total Vastu	326	127	143		15		8	10
Grand Total Vedanta	1439	511	402		74		740	33
Totl Visvesarya	1341	189	308		14		123	12
Grand Total Vishwekarma	428		159		1		39	18
Bhabha Boys Hostel	756		246					
Raman Hostel	277		185					
Sara Bhai	496		229	182				
Bose Hostel	213	114	135				2	
Pharmacy Grand Total	171		90				16	1
Total	6293	1552	2761	257	128		1064	81

Equipment Connected		Non-Modular/ Ordinary Switch & Sockets				Modular Switch & Sockets			
Type	Rating	6 Amp. Switch	16 Amp. Switch	6 Amp. Socket	16 Amp. Socket	6 Amp Switch	16 Amp. Switch	6 Amp. Socket	16 Amp. Socket
42		372	2	5	25	1319	181	362	181
38						2182	197	1301	220
14		27	2	3	3	1089	95	621	95
23Exst, 12000 watt Disco light						3157	114	1719	114
27						1063	211	327	211
33						769	100	174	100
70-Exst,1-Printer,1-TV		1518	8	521	8				
8-Exst		1278		409		81	9		9
29-Exst						1880	3	550	3

Equipment Connected		Non-Modular/ Ordinary Switch & Sockets				Modular Switch & Sockets			
Type	Rating	6 Amp. Switch	16 Amp. Switch	6 Amp. Socket	16 Amp. Socket	6 Amp Switch	16 Amp. Switch	6 Amp. Socket	16 Amp. Socket
8-Exst, 1-Geaser		899	33	284	33				
17		184	40	89	40	287	270	70	270
346-Exst		4278	85	1311	109	11827	1180	5124	1203

The total LEDs currently being used is about 20% of the overall artificial lights used in the university, whereas the usage of CFL is around 80%. If the CFLs used in the university are replaced by LEDs, the power consumed is shown below:

CFL and LED calculation:

Total No. of CFL = 6293

Total No. of LED = 1552

CFL and LED = 7845

$$\% \text{ of CFL in University} = \frac{\text{No. of CFl}}{\text{Total Lights}}$$

$$= \frac{6293}{7845} \times 100$$

% of CFL in University = 80.2167%

$$\% \text{ of LED in University} = \frac{1552}{7845}$$

% of LED in university= 19.783%

The total number of fans with regulator currently present in the university is about 90%, whereas rest of them are without a regulator.

Total fan with regulator = 2761

Total fan without regulator = 252

Total fan regulator and without regulator = 3018

$$\% \text{ of fan with regulator} = \frac{2761}{3018} \times 100$$

% of fan with regulator = 91.484%

$$\% \text{ of fan without regulator} = \frac{252}{3018} \times 100$$

% of fan without regulator =8.349%

The calculation shows that if the majority of CFLs are replaced with LEDs, CFL has the power rating of 18 watts and 36 watts, whereas LED has the power rating of 12 watts.

A total of 37,758 watts can be saved, and the cost of LEDs will be covered within a span of about 1 year. However, the initial cost of replacing CFLs with LEDs is very high but recovery time is also very less.

6. Energy Audit for Lightning System

Buildings typically use fluorescent lights, LEDs, fluorescent lights, and inefficient high-power magnetic ballasts rated at 18 watts, 12 watts, and 36 watts. Building lighting levels are verified using a light meter and comply with best practice guidance recommended by the Lighting Engineering Association (IES). We used an exposure meter or Luxmeter to verify that the chart readings corresponded to the actual lighting level of the structure.

Except for open areas where there is plenty of natural light other than standard lighting, there is no electricity in the lighting, and all public areas and toilets use conventional fluorescent lights, so they will not turn off. Hmmm inefficient magnetic ballast shows that most parts of the building meet lighting standards.

7. Fan Energy Audit

The majority of the fans in the building are 60 watt, and some of them include a resistance regulator. During the energy audit of the fan system, it was discovered that several of the fans lacked a regulator.

Table 82.4   Fan Details

Equipment	Fixture	Rated Power
Fan	3018	watt

8. Energy Audit of Lightning System

The lighting system in the building is primarily composed of 36W T12 fluorescent bulb tubes with inefficient high-power magnetic ballasts. (1) Most portions of the facility satisfied lighting requirements, with the exception of a few open spaces with sufficient natural lighting for better than average illumination, according to the energy audit of the lighting system. (2) A lack of energy conservation knowledge - energy waste from lights that was left on although the space was well lit and no one was in the public zone, and rooms with inefficient magnetic ballast. [4-6].

9. Energy Audit of Computer System

The overall number of computers utilized in the university's various departments is 1300. The computer system energy audit analyzes energy usage and cooling expenses, as well as noise reduction.

10. Energy Audit of Air Conditioning System

A 128 AC system for various rooms, a central air conditioning system for the theatre, and a 128 AC system for the laboratory make up the air conditioning system. Two compressors with a total installed capacity of 1 Ton and 0.5 Ton are installed.

The following energy inefficiencies and wastes in the split unit air conditioning system were discovered during the energy audit: (1) insufficient filter maintenance; (2) EER (energy efficiency ratio) is too low. Table 82.3 shows the split unit equipment. EER must be more than 3.2 for high-efficiency split units. [7]

EEMs for Split Unit Air Conditioning Systems: To improve the energy efficiency of split unit air conditioning systems, as an air conditioning refrigerant, R22 is being phased out in favor of "Cold*22." R22 refrigerant is an important ingredient that keeps the air coming out of your air conditioner cool. R22, often known as Freon* and referred to by the EPA as HCFC-22, is the refrigerant used in most air conditioners older than 10 years. The refrigerant R22, commonly known as HCFC-22, absorbs and removes heat from your air conditioner, heat pump, and automotive air conditioning system. The temperature of your air conditioner should be set according to how comfortable you want to be inside and how much you are willing to pay to achieve that comfort.

Replace a low-efficiency air conditioner with a higher efficiency one. Cold*22 is an environmentally friendly refrigerant that may save you up to 20% on your electricity cost. The cost savings from replacing low EER split units with high EER split units with an EER more than 3.2 can be considerable.

To improve the energy efficiency of the chiller plant air conditioning system, EEMs such as chiller evaporator and condenser pipework maintenance are advised. Replace a low-efficiency compressor with a more efficient one. Cleaning the dirty chiller and evaporator is the first step in achieving a decent SPC in KW/ton. If routine maintenance fails to achieve adequate chiller efficiency, retrofitting with a high-efficiency compressor, despite its high cost, should be able to achieve satisfactory chiller SPC.

11. Conclusion

As more and more high-power electrical equipment is installed, the consumption of electrical energy in buildings continues to increase and will continue to do so in the near future. Energy audits reduce power consumption by finding a variety of energy-saving measures that can be implemented within the enterprise, eliminating waste and improving efficiency. Energy audits help buildings achieve maximum energy efficiency by reducing energy waste and increasing the efficiency of lighting and air conditioning systems. Finally, EEMs are created through electrical energy audits, and managers can only benefit from energy and cost savings by implementing EEMs in their facilities.

REFERENCES

1. J. Randolph and G. M. Masters, "Energy for Sustainability: Technology, Planning, Policy," 165–212, 2008.
2. Energy Commission Malaysia, "Efficient Management of Electrical Energy Regulations," ST, 2008.
3. Harapajan Singh, Manjeevan Seera," Electrical Energy Audit in a Malaysian University- A Case Study" International Conference on Power and Energy, 2–5 December 2012. 2. S. R. Bhawarkar and S. Y. Kamdi, "Electrical Energy Audit of a Electroplating Unit – A case study," 2011 International Conference on Recent Advancement in Electrical, Electronics and Control Engineers, 25–29, 2011.
4. W. N. W. Muhamad, M. Y. M. Zain, N. Wahab, N. H. A. Aziz, and R. A. Kadir, "Energy Efficient Lighting Design for Building", 2010 International Conference on Intelligent Systems, Modeling and Simulation, 282–286, 2010.
5. Chen Zhongping The analysis of energy consumption and energy-saving potential in existing publich builidngs[J]. Journal of Wuhan polytechnic university 2008,27(3): 93–95.
6. Narendra b Soni "the transition to led illumination: a case study on energy conservation" Journal of Theoretical and Applied Information Technology, 2008.
7. Jayesh. R, Jagdish. V, Julian George "Energy Auditing in an Educational Institution with Special Focus on Reduction in Maximum Power Demand" ISSN 2248-9967 Volume 4, Number 3 (2014)
8. p. wang, j. y. huang, y. ding, p. loh, l. goel, " demand side load management of smart grids using intelligent trading/metering/billing system", power and energy society general meeting, ieee paper, 2010
9. A.Prudenzi, V.Caracciolo, A.Silvestri, "Electrical Load analysis in a hospital complex", Power Tech, 2009 IEEE Bucharest
10. H.E. Hua, L. Tian-yu, Z. Zhi-yong, and Z. Juan, "Energy Saving Potential of a Public Building in Jiangbei District of Chongqing," International Conference on Management and Service Science, 2009.
11. X. Wang, C. Huang, and W. Cao, "Energy Audit of Building: A case study of a Commercial Building in Shanghai," Power and Energy Engineering Conference (APPEEC), 1–4, 2010.
12. Lou Chengzhi, Yang Hongxing etc Energy audit of buildings: a case study of a commercial building in causeway bay of HongKang [J]. Joural of HV&AC ,2006(36) :44-50.
13. IEEE Standard 739–1995, IEEE Standard Practice for Energy Management in Industrial and Commercial Facilities, 1995.
14. Energy Commission Malaysia, "Efficient Management of Electrical Energy Regulations," ST, 2008.
15. S. R. Bhawarkar and S. Y. Kamdi, "Electrical Energy Audit of a Electroplating Unit – A case study," 2011 International Conference on Recent Advancement in Electrical, Electronics and Control Engineers, 25–29, 2011.
16. W. N. W. Muhamad, M. Y. M. Zain, N. Wahab, N. H. A. Aziz, and R. A. Kadir, "Energy Efficient Lighting Design for Building", 2010 International Conference on Intelligent Systems, Modelling and Simulation, 282–286, 2010.
17. H.E. Hua, L. Tian-yu, Z. Zhi-yong, and Z. Juan, "Energy Saving Potential of a Public Building in Jiangbei District of Chongqing," International Conference on Management and Service Science, 2009.
18. X. Wang, C. Huang, and W. Cao, "Energy Audit of Building:A case study of a Commercial Building in Shanghai," Power and Energy Engineering Conference (APPEEC), 1–4, 2010.
19. IEEE Standard 739-1995, IEEE Stanadard Practice for Energy Management in Industrial and Commercial Facilities, 1995.
20. A. Thumann and W. J. Younger, "Handbook of Energy Audits (7th Ed.)," Fairmont Press, 1–12, 2008.
21. J. Randolph and G. M. Masters, "Energy for Sustainability: Technology, Planning, Policy," 165–212, 2008.